# An Introduction to The Mathematics and Methods of ASTRODYNAMICS

**Richard H. Battin, Ph.D.**
The Charles Stark Draper Laboratory, Inc.
Adjunct Professor of Aeronautics and Astronautics
Massachusetts Institute of Technology

**AIAA EDUCATION SERIES**
J. S. Przemieniecki
Series Editor-in-Chief
Air Force Institute of Technology
Wright-Patterson Air Force Base, Ohio

Published by
American Institute of Aeronautics and Astronautics, Inc.
1633 Broadway, New York, N.Y. 10019

**Texts Published in the AIAA Education Series**

Re-Entry Vehicle Dynamics
    Frank J. Regan, 1984
Aerothermodynamics of Gas Turbine and Rocket Propulsion
    Gordon C. Oates, 1984
Aerothermodynamics of Aircraft Engine Components
    Gordon C. Oates, Editor, 1985
Aircraft Combat Survivability Analysis and Design
    Robert E. Ball, 1985
Intake Aerodynamics
    J. Seddon and E. L. Goldsmith, 1985
Composite Materials for Aircraft Structures
    Brian C. Hoskin and Alan A. Baker, Editors, 1986
Gasdynamics: Theory and Applications
    George Emanuel, 1986
Aircraft Engine Design
    Jack D. Mattingly, William H. Heiser, and
        Daniel H. Daley, 1987

American Institute of Aeronautics and Astronautics, Inc.
New York, New York

**Library of Congress Cataloging in Publication Data**

Battin, Richard H.
    An introduction to the mathematics and methods
of astrodynamics.

(AIAA education series)
Includes index.
1. Astrodynamics—Mathematics. I. Title. II. Series.
        629.4'11    87-1207

Second Printing

Copyright © 1987 by Richard H. Battin and the American Institute of Aeronautics and Astronautics, Inc. All rights reserved. Printed in the United States of America. No part of this publication may be reproduced, distributed, or transmitted, in any form or by any means, or stored in a data base or retrieval system, without prior written permission of the publisher.

To Martha and Les

Margery, Thomas, Pamela, and Jeffrey

# Foreword

The all-inclusive treatise on "Astrodynamics" by Richard H. Battin clearly must be counted among the great classics in scientific literature. This text documents the fundamental theoretical developments in astrodynamics and space navigation which eventually led to man's ventures into space. It includes all the essential elements of celestial mechanics, spacecraft trajectories, and space navigation as well as the history of the underlying mathematical developments over the past three centuries culminating finally with the $20^{\text{th}}$ century space exploration.

The first half of his text deals with the necessary mathematical preliminaries of hypergeometric functions, analytical dynamics, the two-body problem leading to the solution of two-body orbits, Kepler's equation, and Lambert's problem. The second half includes non-Keplerian motion, patched-conic orbits and perturbation methods, variation of parameters, two-body orbital transfer, numerical integration of the equations of motion in orbital mechanics, the celestial position fix for spacecraft, and space navigation. All mathematical concepts are fully explained in the text so that there is no need for any additional reference materials. The most abstruse mathematical derivations are made simple through clarity of style, logical exposition, and attention to details.

Dr. Battin has produced a textbook that will be used by the present and future generations of aerospace engineers as they venture beyond the Apollo program to conquer the "high frontier." This text is a great testimony of Dr. Battin's exceptional pioneering work as a scientist and engineer and his outstanding personal contributions to the US space program.

**J. S. Przemieniecki**
**Editor-in-Chief**
**AIAA Education Series**

# Prologue

*This article appeared in the* New York Times *on the eve of the Apollo 8 mission. It was reprinted in a special* Look *magazine issue titled:* Apollo 8—Voyage to the Moon.

The Apollo voyage to the moon represents a new and exciting plateau in the ancient art of navigation.

By applying principles as old as the planetary theories of Kepler and technologies as new as the high-speed electronic digital computer, an astronaut can determine the position and course of his craft in the vastness of outer space with an accuracy that Columbus or Prince Henry the Navigator would have deemed impossible in their time.

Ever since man first went to sea, the need to navigate accurately has been a constant challenge. For many centuries only the brave or foolhardy dared to venture out of sight of land except for short distances. Progress in navigation was extremely slow, and not until just 200 years ago could a ship's location at sea be determined with anything approaching precision.

The ability to determine latitude—the distance north or south of the Equator—by observing the angle that the North Star makes with the horizon was known in early times. The instrument used for this purpose, the astrolabe, was invented by the Greeks more than 2,000 years ago and well may be the oldest scientific instrument in the world.

The mariner's compass was introduced much later, in the $12^{th}$ or $13^{th}$ century. With it, a seaman could set a course and, by estimating the speed of the ship, obtain a crude approximation of his position. However, only the ship's latitude could be verified by direct observation.

It was not until the $18^{th}$ century that east-west distances, called longitude, could be measured accurately. In fact, it was quite recently in history before it was even recognized that the essential element required to obtain longitude was a reliable and transportable clock.

During the $16^{th}$ and $17^{th}$ centuries, the longitude problem assumed enormous proportions to each of the maritime powers. Fantastic rewards were offered for the solution as each nation vied to become the first to develop the important capability of accurate navigation at sea.

The world's leading scientists devoted their attention to the problem. Economic, political, and military considerations were at stake, and the struggle for supremacy may, in some sense, be likened to the modern day race for the moon.

The first successful seaborne clock, or chronometer, was finally invented by John Harrison, a carpenter from Yorkshire, England. It took 30 years to develop and it was first demonstrated in 1761. With this instrument the problem of navigation at sea was solved, and during succeeding years the science of navigation was perfected through the development of more accurate instruments.

Each of the navigation instruments carried aboard the Apollo 8 spacecraft has its counterpart in these earlier devices. The astrolabe has evolved into a space sextant with which the astronaut can sight simultaneously on stars and landmarks on the surface of the earth or moon.

The purpose of the instrument is to measure angles between the lines-of-sight to celestial objects. The data gathered from such measurements would aid the Apollo navigator in determining the position and speed of the spacecraft.

The mariner's compass with its north-seeking magnetic needle would find little utility in outer space. However, the function performed by the compass of providing a constant reference direction is as important in navigating a spacecraft as it is for a ship or an aircraft. Moreover, the problem of direction in space is three-dimensional, rather than two, and the accuracy requirements are more severe.

In Apollo a reference direction is maintained by means of a device called an inertial measuring unit. The instrument is basically a small platform supported and pivoted so that the spacecraft is free to rotate about it just as a compass needle is pivoted to indicate always a northerly direction independent of the orientation of the ship.

On this small platform are mounted three gyroscopes that sense and prevent any rotation of the platform from occurring. Thus, as the orientation of the Apollo spacecraft changes during flight, the direction in which it is pointing can always be measured with respect to this platform, which unerringly maintains a fixed direction in space.

The need for accurate timekeeping also is as essential for the Apollo navigator as for the ship at sea. Indeed, this point can be appreciated by noting that a spacecraft on the way to the moon travels at speeds as high as seven miles a second. The moon also is moving, at a rate of one-half mile a second, with respect to the earth. Thus, a small error in the clock can result directly in significant errors in position.

In the Apollo spacecraft, the clock is a part of the onboard digital computer. Just as the navigator at sea is required to perform mathematical calculations using charts and tables, so is the problem of navigating a spacecraft largely a mathematical one. A small but versatile digital computer is provided for this purpose and the precision-timing circuits in the computer serve as a clock.

# Prologue

Some of the similarities in instruments used to navigate in orbit and in a ship at sea have been noted. There exist, however, fundamental differences that are apparent when one contrasts the environment through which each vehicle moves, the speed of travel, and the selection of an appropriate route to the destination.

In navigating at sea or in the air, the effects of wind and water currents must be taken into account even though they cannot be directly measured. By using two successive position fixes and the time elapsed between them, the navigator can estimate the currents and compensate for them. Even if he is unable to do this very accurately, the resulting errors are usually not serious and can be corrected ultimately by whatever changes in course and speed are required.

On the other hand, the spacecraft navigator enjoys the advantage of not having his vehicle subject to substantial unknown and unpredictable forces such as air or sea currents. Motion in space is much more certain since the forces involved are well understood.

Despite this important advantage, the high speeds characteristic of space travel present serious problems not encountered on earth. When the Apollo spacecraft is hurtled toward the moon, it must travel many times faster than a rifle bullet if it is to coast to the moon without falling back to earth. Because of these tremendous speeds, coupled with the limited amount of fuel that can be transported, significant changes in direction and speed are limited.

The command pilot cannot freely order a "hard right rudder" as a sea captain might to correct for a mistake in course. With such severe restrictions on maneuverability, it is mandatory that each phase of the Apollo mission be carefully planned in advance.

Finally, there are the problems of charting a proper route to the destination. In planning a sea or air voyage, fuel considerations generally dictate that the shortest path be followed.

The selection of an appropriate trajectory to the moon also is influenced by the need for fuel economy, but in a much more esoteric way.

For example, the efficient use of propulsion requires that the moon's gravitational field be exploited so as to deflect the trajectory of the Apollo vehicle when it passes behind the moon to place it on proper course back to earth should the decision be made not to enter lunar orbit. To accomplish this task successfully demands highly accurate navigation so that the spacecraft will pass the moon with the correct speed, altitude, and direction of flight.

The moon appears in the sky to be a rather substantial target. Therefore, we might reasonably wonder if guiding a spacecraft to its vicinity is really very difficult. To answer this question we should consider briefly the effect on the Apollo lunar trajectory of errors incurred at the instant of departure from an earth parking orbit.

For simplicity, suppose that Apollo were, indeed, a projectile fired at the moon with the only requirement being to strike the lunar surface. Even for this relatively more elementary task, an error of only one-tenth of 1% in the speed of projection, or an error of only a small fraction of a degree in the direction of aim, would result in missing the moon.

The accuracy requirements of Apollo 8 are far more stringent. The object is not simply to hit a target 2,000 miles in diameter. On the contrary, the Apollo vehicle must miss the moon by a carefully controlled amount.

It is entirely unrealistic to suppose that such precision can be achieved without the need for at least some small changes in direction and speed as the spacecraft approaches the moon.

Granted that corrections will be required, they can be accurately made only when the position, speed, and direction of motion of the spacecraft are accurately known. Since trajectory errors mean wasted fuel, a precise knowledge of these quantities is of the utmost importance to a spacecraft navigator.

Having looked at some of the characteristics of space navigation and the basic instruments required, we should now examine in more detail how the Apollo navigation task is actually performed. There are, fundamentally, two phases of flight to consider—the coasting phase and the accelerated phase.

The periods of time of coasting flight—when the course of the vehicle is affected only by gravity—are measured in hours and days. On the other hand, accelerated flight times—when the main engine is firing—are of only a few minutes duration. As might be suspected, the techniques involved are quite different in the two cases.

Navigating the Apollo spacecraft during the long coast to the moon involves two processes. First, frequent navigation measurements are made to improve the estimate of the spacecraft's position and velocity. Second, a prediction is made periodically of the position and velocity of the spacecraft at the expected time of rendezvous with the moon.

If these predictions indicate that the spacecraft is not following the intended course, then small corrections to the speed and direction of motion can be applied using the rocket engine.

Predicting the course of Apollo during prolonged periods of coasting flight is the same as the astronomer's problems of predicting the position of the moon and planets. The motion of the spacecraft, as well as the planets, is caused by the interaction of the various gravitational fields of the bodies that make up the solar system.

The basic physical principles governing this motion were discovered by Sir Isaac Newton. He was the first to describe the solar system as consisting of many bodies each attracting all the others in accord with his law of gravitation. As a consequence of Newton's work, the possibility of accurate predictions of the positions of the planets by mathematical means was finally at hand.

There are several considerations influencing the ability to make long-range predictions. First of all are the mathematical techniques used for solving the equations formulated by Newton. These equations cannot be solved exactly, and the resulting errors will rapidly degrade the solution unless elaborate computational techniques are employed. Without the availability of modern high speed digital computers the required calculations could not be performed rapidly enough to keep pace with the Apollo voyage.

Second, the accuracy of predicting position and velocity also is subject to man's knowledge of the planets themselves, such as size, shape, density, and mass, all of which play an important role in the mathematics.

Finally, and most important of all, there are the problems that mathematicians refer to as "initial conditions"—the values of position and velocity at the time from which the prediction is made. Unless these initial values are accurately known, it is obvious that they cannot be accurately predicted.

In order to ensure accurate initial conditions, it is necessary periodically to correct the estimate of spacecraft position and velocity using data gathered from optical or radar measurements made either with the onboard space sextant or with the extensive earth-based worldwide tracking network.

Earth-based radar installations are capable of measuring the distance, the direction, and rate of change of the distance from the radar site to the spacecraft.

Use of the space sextant allows the astronaut, for example, to measure the apparent elevation of a star above the earth's horizon to a landmark on the moon. These measurements are utilized much as a ship's navigator uses compass bearings from lighthouses or radio beacons to correct his estimate of position.

At the time a measurement is made, the best estimate of the spacecraft's position and velocity is contained in the digital computer. Then, since the directions of the stars and the locations of landmarks and tracking stations are known, it is possible to calculate the expected value of the quantity to be measured—such as an angle or distance from a tracking station.

When the expected value of this measurement is compared with the value actually measured, the difference can be used to correct the estimate of the spacecraft's position and velocity. A sequence of such measurements separated in time, together with an accurate mathematical description of the solar system, will eventually produce estimates with sufficient precision to permit corrective maneuvers to be made with confidence.

The other major navigation phase to be discussed is the task of navigating and steering the Apollo vehicle when the main engine is firing. With the thrust provided by the service propulsion system, it is possible to make rather substantial changes in the speed and direction of motion of the spacecraft.

This capability was provided to the Apollo 8 flight for three possible maneuvers: (1) slowing the vehicle down as it passes the moon, which is necessary to achieve a lunar orbit; (2) acquiring the necessary speed while in lunar orbit to escape the moon and be on a proper course back to earth; and (3) to return to earth before reaching the moon should it be necessary to abort the mission.

As a specific example, consider the phase of the mission called trans-earth injection, of accelerating the spacecraft out of lunar orbit for the trip back to earth. At any location in lunar orbit, a velocity can be calculated that would be the correct velocity required by the vehicle to coast back to earth from that position.

The spacecraft, of course, does not have that velocity but is instead moving at a speed and in a direction appropriate for orbiting the moon. However, if it were possible to make a sudden and instantaneous change in its speed and direction of motion of the required amounts, the vehicle would immediately begin on its return voyage.

The difference between the velocity that the spacecraft actually has and the velocity it should have for return to earth is called the velocity-to-be-gained. If the velocity-to-be-gained were zero or could be made so, the desired objective would be accomplished and the long coast home would be under way.

Of course, the speed and direction of motion cannot be suddenly altered. In fact, it requires about $2\frac{1}{2}$ minutes of thrusting to change the velocity by the necessary amount. However, by pointing the spacecraft engine and thrusting in the direction in which the additional speed must be added, the velocity-to-be-gained will gradually decrease to zero. When this condition is achieved, the engine is turned off and coasting flight begins.

During an accelerated maneuver, the Apollo navigation system must steer the vehicle in the proper direction, measure the thrust acceleration imparted by the engine, repeatedly compute the velocity still to be gained, and provide an engine-off signal when the maneuver is completed.

The orientation of the spacecraft when the rockets are firing is measured with respect to the inertial platform, as described earlier. The direction is controlled both by firing clusters of small jets and swiveling the engine causing the vehicle to rotate in the proper direction to eliminate pointing errors. The thrust acceleration is measured by small instruments called accelerometers mounted on the inertial platform.

## Prologue

No instrument is capable of directly measuring the forces of gravity. However, since the gravitational forces depend only on the position of the spacecraft with respect to the earth, the sun, and the moon, they can be accurately computed mathematically. These gravity calculations are made in the Apollo computer and are combined with the thrust acceleration measurements for computing the additional velocity needed before engine cutoff is commanded.

The navigation task of Apollo during the return trip again consists of measuring, predicting, and correcting the trajectory of the spacecraft. However, the margin for error is much more critical than for the outbound flight. The vehicle is required to enter the earth's atmosphere along a path that must not deviate by more than 1 degree to either side of the planned entry direction.

If the path is too steep, the deceleration forces might be too great for the structure or crew to withstand. On the other hand, too shallow an entry could result in the spacecraft's skipping out of the atmosphere. Accurate midcourse navigation is, therefore, essential to the final success of the mission.

Will the moon prove to be the limit of man's ventures into space? To assume so would ignore one of his most basic drives—to explore, to understand, and to conquer his environment. On the contrary, man is now embarking on a new Age of Discovery, which, like the first, will provide new challenges for the science of navigation.

**Richard H. Battin**
**December 21, 1968**

# Contents

Foreword . . . . . . . . . . . . . . . . . . . . . . . . . . v
Prologue . . . . . . . . . . . . . . . . . . . . . . . . . vii
Contents . . . . . . . . . . . . . . . . . . . . . . . . . . xv
Preface . . . . . . . . . . . . . . . . . . . . . . . . . xxvii
Introduction . . . . . . . . . . . . . . . . . . . . . . . . 1

## PART I

### 1 Hypergeometric Functions and Elliptic Integrals — 33

1.1 Hypergeometric Functions . . . . . . . . . . . . . . . 34
    Examples of Hypergeometric Functions    34
    Gauss' Relations for Contiguous Functions    36
    Gauss' Differential Equation    38
    Bilinear Transformation Formulas    40
    Quadratic Transformation Formulas    42
    Confluent Hypergeometric Functions    43

1.2 Continued Fraction Expansions . . . . . . . . . . . . 44
    Gauss' Continued Fraction Expansion Theorem    47
    Continued Fractions Versus Power Series    51
    Continued Fraction Solutions of the Cubic Equation    53

1.3 Convergence of Continued Fractions . . . . . . . . . 54
    Recursive Properties of the Convergents    55
    Convergence of Class I Continued Fractions    57
    Convergence of Class II Continued Fractions    59
    Equivalent Continued Fractions    61

1.4 Evaluating Continued Fractions . . . . . . . . . . . 63
    Wallis' Method    63
    The Bottom-Up Method    64
    Euler's Transformation    64
    The Top-Down Method    67

1.5 Elliptic Integrals . . . . . . . . . . . . . . . . . . . . 68
    Elliptic Integral of the First Kind    70
    Landen's Transformation    71
    Gauss' Method of the Arithmetic-Geometric Mean    72
    Elliptic Integral of the Second Kind    73
    Evaluating Complete Elliptic Integrals    74
    Jacobi's Zeta Function    77

## 2 Some Basic Topics in Analytical Dynamics — 79

- 2.1 Transformation of Coordinates ........... 80
  - Euler's Theorem  81
  - The Rotation Matrix  81
  - Euler Angles  84
  - Elementary Rotation Matrices  85
- 2.2 Rotation of a Vector ................. 86
  - Kinematic Form of the Rotation Matrix  87
  - Euler Parameters  88
- 2.3 Multiple Rotations of a Vector ............ 91
  - Relations Among the Euler Parameters  91
  - Quaternions  93
- 2.4 The n-Body Problem ................. 95
  - Equations of Motion  95
  - Conservation of Total Linear Momentum  96
  - Conservation of Total Angular Momentum  97
  - Potential Functions  97
  - Conservation of Total Energy  98
- 2.5 Kinematics in Rotating Coordinates ......... 101

## 3 The Problem of Two Bodies — 107

- 3.1 Equation of Relative Motion ............. 108
- 3.2 Solution by Power Series .............. 110
  - Lagrange's Fundamental Invariants  111
  - Recursion Equations for the Coefficients  112
- 3.3 Integrals of the Two-Body Problem ......... 114
  - Angular Momentum Vector  115
  - Eccentricity Vector  115
  - The Parameter and Energy Integral  116
  - Equation of Orbit  117
  - Period and Mean Motion  119
  - Time of Pericenter Passage  120
- 3.4 Orbital Elements and Coordinate Systems ...... 123
- 3.5 The Hodograph Plane ................ 126
  - Two-Body Orbits in the Hodograph Plane  126
  - The Flight-Direction Angle  128
- 3.6 The Lagrangian Coefficients ............. 128
- 3.7 Preliminary Orbit Determination .......... 131
  - Orbit from Three Coplanar Positions  131
  - Orbit from Three Position Vectors  133
  - Approximate Orbit from Three Position Fixes  134
  - Approximate Orbit from Three Range Measurements  136
  - Approximate Orbit from Three Observations  138

## 4 Two-Body Orbits and the Initial-Value Problem — 141

4.1 Geometrical Properties . . . . . . . . . . . . . . . 142
    Focus-Directrix Property    144
    Focal-Radii Property    145
    Orbital Tangents    145
    Sections of a Cone    147

4.2 Parabolic Orbits and Barker's Equation . . . . . . . 149
    Trigonometric Solution    151
    Improved Algebraic Solution    151
    Graphical Solution    152
    Continued Fraction Solution    153
    Descartes' Method    154
    Lagrangian Coefficients    155
    Orbital Tangents    156

4.3 Elliptic Orbits and Kepler's Equation . . . . . . . . 158
    Analytic Derivation of Kepler's Equation    160
    Geometric Derivation of Kepler's Equation    161
    Lagrangian Coefficients    162

4.4 Hyperbolic Orbits and the Gudermannian . . . . . . 165
    The Gudermannian Transformation    167
    Geometrical Representation of $H$    168
    Lagrangian Coefficients    170
    Asymptotic Coordinates    171

4.5 Universal Formulas for Conic Orbits . . . . . . . . 174
    The Universal Functions $U_n(\chi; \alpha)$    175
    Linear Independence of $U_n(\chi; \alpha)$    177
    Lagrangian Coefficients and Other Orbital Quantities    178

4.6 Identities for the Universal Functions . . . . . . . . 183
    Identities Involving Compound Arguments    183
    Identities for $U_n^2(\chi; \alpha)$    184
    Identities for $U_{n+1}U_{n+1-m} - U_{n+2}U_{n-m}$    185
    Identities Involving the True Anomaly Difference    186

4.7 Continued Fractions for Universal Functions . . . . . 187
    Continued Fraction Determination of $U_3$ and $U_4$    188
    Continued Fraction Determination of $U_5$ and $U_6$    189

## 5 Solving Kepler's Equation — 191

5.1 Elementary Methods . . . . . . . . . . . . . . . . 192
    Graphical Methods    193
    Inverse Linear Interpolation (Regula Falsi)    193
    Successive Substitutions    196

## Astrodynamics

- 5.2 Lagrange's Expansion Theorem ... 199
  - Euler's Trigonometric Series 200
  - Generalized Expansion Theorem 202
  - Convergence of the Lagrange Series 204
- 5.3 Fourier-Bessel Series Expansion ... 206
  - Series Expansion of the Eccentric Anomaly 206
  - Bessel Functions 208
  - Series Expansion of the True Anomaly 210
- 5.4 Series Reversion and Newton's Method ... 212
  - Series Reversion Algorithm 213
  - Newton's Method 216
  - Power Series for the Generalized Anomaly $\chi$ 218
  - An Alternate Form of Kepler's Equation 219
- 5.5 Near-Parabolic Orbits ... 220
  - Method of Successive Approximations 221
  - Motivating Gauss' Method 222
  - Gauss' Method 224
- 5.6 Extending Gauss' Method ... 227
  - Transformation of Kepler's Equation 228
  - Solution of the Cubic Equation 230
  - Series Representations 231
  - Algorithm for the Kepler Problem 234

## 6  Two-Body Orbital Boundary-Value Problem  237

- 6.1 Terminal Velocity Vectors ... 238
  - Minimum-Energy Orbit 240
  - Locus of Velocity Vectors 241
  - Parameter in Terms of Velocity-Components Ratio 246
  - Parameter in Terms of Flight-Direction Angle 248
  - Relation Between Velocity and Eccentricity Vectors 249
- 6.2 Orbit Tangents and the Transfer-Angle Bisector ... 250
  - Ellipse and Hyperbola 251
  - Parabola 253
  - Parameter in Terms of Eccentric-Anomaly Difference 255
- 6.3 The Fundamental Ellipse ... 256
  - The Fundamental (Minimum-Eccentricity) Ellipse 258
  - Intersection of the Transfer-Angle Bisector and the Chord 261
  - Parameter in Terms of Eccentricity 262
  - Tangent Ellipses 263
- 6.4 A Mean Value Theorem ... 264
  - Geometry of the Mean-Point Locus 266
  - The Mean-Point Radius 267
  - Elegant Expressions for the Mean-Point Radii 269
  - Parameter in Terms of Mean-Point Radius 270

## Contents

6.5 Locus of the Vacant Focus . . . . . . . . . . . . . . 271
    Elliptic Orbits   272
    Hyperbolic Orbits   273
    Parameter in Terms of Semimajor Axis   274
6.6 Lambert's Theorem . . . . . . . . . . . . . . . . . 276
    Euler's Equation for Parabolic Orbits   276
    Lagrange's Equation for Elliptic Orbits   277
    The Orbital Parameter   279
6.7 Transforming the Boundary-Value Problem . . . . . 281
    Transforming to a Rectilinear Ellipse   283
    Transforming to a Rectilinear Hyperbola   285
6.8 Terminal Velocity Vector Diagrams . . . . . . . . . 287
    Elliptic Orbits   288
    Parabolic Orbit   292
    Hyperbolic Orbits   292
    Boundary Conditions at Infinity   294

## 7    Solving Lambert's Problem    295

7.1 Formulations of the Transfer-Time Equation . . . . . 297
    Lagrange's Equation   298
    Gauss' Equation   300
    Combined Equations   302
    Multiple-Revolution Transfer Orbits   305
    The Velocity Vector   306
7.2 The Q Function . . . . . . . . . . . . . . . . . . 307
    Improving the Convergence   308
    Continued Fraction Representation   310
    Derivative Formulas   312
7.3 Gauss' Method . . . . . . . . . . . . . . . . . . . 313
    The Classical Equations of Gauss   315
    Solving Gauss' Equations   316
    Solving Gauss' Cubic Equation   320
7.4 An Alternate Geometric Transformation . . . . . . . 322
    Transforming the Mean Point to an Apse   322
    Relating $h$ and $\sigma$ to the Original Orbit   324
7.5 Improving Gauss' Method . . . . . . . . . . . . . 325
    Removing the Singularity   325
    Computing $\ell$, $m$, and the Orbital Elements   329
    Improving the Convergence   332
    Transforming the Function $\xi(x)$   335
    Solving the Cubic   338
    Comparing the Two Methods   340
    Behaviour Near the Singularity   341

## Appendices

- A  Mathematical Progressions . . . . . . . . . . . . . . 343
    - A.1 Arithmetic Progression   343
    - A.2 Geometric Progression   343
    - A.3 Harmonic Progression   344
- B  Vector and Matrix Algebra . . . . . . . . . . . . 345
    - B.1 Vector Algebra   345
    - B.2 Matrix Algebra   347
- C  Power Series Manipulations . . . . . . . . . . . . 351
- D  Linear Algebraic Systems . . . . . . . . . . . . . 353
- E  Conic Sections . . . . . . . . . . . . . . . . . . . 355
- F  Tschebycheff Approximations . . . . . . . . . . 359
    - F.1 Tschebycheff Polynomials   359
    - F.2 Economization of Power Series   362
- G  Plane Trigonometry . . . . . . . . . . . . . . . . 363

## PART II

### 8  Non-Keplerian Motion   365

- 8.1 Lagrange's Solution of the Three-Body Problem . . . 366
    - Equilateral Triangle Solution   367
    - Straight Line Solutions   367
    - Conic Section Solutions   370
- 8.2 The Restricted Problem of Three Bodies . . . . . . . 371
    - Jacobi's Integral   372
    - Rectilinear Oscillation of an Infinitesimal Mass   373
    - Surfaces of Zero Relative Velocity   376
    - Lagrangian Points   379
- 8.3 Stability of the Lagrangian Points . . . . . . . . . . 382
    - The Equilateral Libration Points   383
    - The Colinear Libration Points   385
- 8.4 The Disturbing Function . . . . . . . . . . . . . . 387
    - Explicit Calculation of the Disturbing Acceleration   388
    - Expansion of the Disturbing Function   389
    - Jacobi's Expansion and Rodrigues' Formula   392
    - Legendre Polynomials   393
- 8.5 The Sphere of Influence . . . . . . . . . . . . . . . 395
- 8.6 The Canonical Coordinates of Jacobi . . . . . . . . 398
- 8.7 Potential of Distributed Mass . . . . . . . . . . . . 401
    - MacCullagh's Approximation   402
    - Expansion as a Series of Legendre Functions   404

8.8 Spacecraft Motion Under Continuous Thrust . . . . . 408
   Constant Radial Acceleration   409
   Transforming the Integral to Normal Form   410
   Constant Tangential Acceleration   415

# 9 Patched-Conic Orbits and Perturbation Methods   419

9.1 Approach Trajectories Near a Target Planet . . . . . 421
   Close Pass of a Target Planet   421
   Tisserand's Criterion   423
   Surface Impact at a Target Planet   425
9.2 Interplanetary Orbits . . . . . . . . . . . . . . . . 427
   Planetary Flyby Orbits   429
   Impulse Control of Flyby Altitude   430
   Examples of Free-Return, Flyby Orbits   431
9.3 Circumlunar Trajectories . . . . . . . . . . . . . 437
   Calculating the Conic Arcs   442
9.4 The Osculating Orbit and Encke's Method . . . . . . 447
9.5 Linearization and the State Transition Matrix . . . . 450
   Solution of the Forced Linear System   452
   Symplectic Property of the Transition Matrix   453
9.6 Fundamental Perturbation Matrices . . . . . . . . . 456
   Partitions of the Transition Matrix   457
   The Adjoint Matrices   461
9.7 Calculating the Perturbation Matrices . . . . . . . . 463
9.8 Precision Orbit Determination . . . . . . . . . . . 467
   Precision Orbits for Lambert's Problem   468
   Precision Free-Return Orbits   468

# 10 Variation of Parameters   471

10.1 Variational Methods for Linear Equations . . . . . . 473
10.2 Lagrange's Planetary Equations . . . . . . . . . . 476
   The Lagrange Matrix and Lagrangian Brackets   477
   Computing the Lagrangian Brackets   479
10.3 Gauss' Form of the Variational Equations . . . . . . 484
   Gauss' Equations in Polar Coordinates   485
   Eliminating the Secular Term   487
   Summary of Gauss' Equations   488
10.4 Nonsingular Elements . . . . . . . . . . . . . . . 490
10.5 The Poisson Matrix and Vector Variations . . . . . . 495
   The Poisson Matrix   495
   Variation of the Semimajor Axis   497
   Variation of the Angular Momentum Vector   497
   Variation of the Eccentricity Vector   499

Variation of the Inclination and Longitude of the Node   500
Variation of the Argument of Pericenter   500
Variations of the Anomalies   501
10.6 Applications of the Variational Method . . . . . . .   503
Effect of $J_2$ on Satellite Orbits   503
Effect of Atmospheric Drag on Satellite Orbits   505
Modified Bessel Functions   507
10.7 Variation of the Epoch State Vector . . . . . . . . .   508
Relation to the Perturbation Matrices   509
Avoiding Secular Terms   510
Variation of the True Anomaly Difference   511
Variational Equations of Motion   512

## 11  Two-Body Orbital Transfer   515

11.1 The Envelope of Accessibility . . . . . . . . . . . .   516
11.2 Optimum Single-Impulse Transfer . . . . . . . . .   518
Optimum Transfer from a Circular Orbit   521
Sufficient Condition for an Optimum Elliptic Transfer   523
11.3 Two-Impulse Transfer between Coplanar Orbits . . .   524
Cotangential Transfer Orbits   527
The Hohmann Transfer Orbit   529
11.4 Orbit Transfer in the Hodograph Plane . . . . . . .   531
Single Velocity Impulse   531
Transfer to a Specified Orbit   533
Transfer from a Circular to a Hyperbolic Orbit   533
11.5 Injection from Circular Orbits . . . . . . . . . . .   536
Optimum Injection   539
Tangential Injection from Perigee $(\beta + i_0 \geq 90°)$   540
Nontangential Injection from Perigee $(\beta + i_0 < 90°)$   542
11.6 Midcourse Orbit Corrections . . . . . . . . . . . .   543
Fixed-Time-of-Arrival Orbit Corrections   543
Variable-Time-of-Arrival Orbit Corrections   544
Pericenter Guidance   547
11.7 Powered Orbital Transfer Maneuvers . . . . . . . .   550
Constant Gravity Field Example   551
Cross-Product Steering   552
Estimation of Burn Time   553
Hyperbolic Injection Guidance   555
Circular-Orbit Insertion Guidance   556
11.8 Optimal Guidance Laws . . . . . . . . . . . . . .   558
Terminal State Vector Control   559
The Linear-Tangent Law   562

Contents     xxiii

## 12  Numerical Integration of Differential Equations    567

12.1 Fundamental Considerations . . . . . . . . . . . . .  569
12.2 Third-Order R-K-N Algorithms  . . . . . . . . . .  571
    Taylor's Expansion of $f^i(x^i + \delta^i)$    573
    Deriving the Condition Equations    574
    Solving the Condition Equations    574
12.3 Fourth-Order R-K-N Algorithms . . . . . . . . . .  577
    Vandermonde Matrices and Constraint Functions    579
    Solution of the Condition Equations    583
12.4 Fourth-Order R-K Algorithms . . . . . . . . . . .  584
    Solving the Condition Equations    586
    The Classical Runge-Kutta Algorithm    587
12.5 Fifth-Order R-K-N Algorithms . . . . . . . . . . .  590
    A Simple Solution of the Condition Equations    591
    The General Solution of the Condition Equations    593
12.6 Sixth-Order R-K-N Algorithms . . . . . . . . . . .  598
    Reformulation of the Condition Equations    599
    Solving the Condition Equations    602
12.7 Seventh-Order R-K-N Algorithms . . . . . . . . . .  603
    Solving the Condition Equations    606
12.8 Eighth-Order R-K-N Algorithms  . . . . . . . . .  608
    Eliminating Equations (**w**) and (**z**)    610
    Solution of the Condition Equations    611
12.9 Integration Step-Size Control . . . . . . . . . . . .  613
    Second-Order Algorithm with Third-Order **x**-Control    614
    Third-Order Algorithm with Fourth-Order **x**-Control    615
    Fourth-Order Algorithm with Fifth-Order **x**-Control    615
    Fifth-Order Algorithm with Sixth-Order **x**-Control    617
    Sixth-Order Algorithm with Seventh-Order **x**-Control    619
    Seventh-Order Algorithm with Eighth-Order **x**-Control    620
    Higher-Order **x**-Control Algorithms    622

## 13  The Celestial Position Fix    623

13.1 Geometry of the Navigation Fix . . . . . . . . . .  624
13.2 Navigation Measurements . . . . . . . . . . . . .  627
    Measuring the Angle between a Near Body and a Star    627
    Measuring the Apparent Angular Diameter of a Planet    628
    Star-Elevation Measurement    628
    Star-Occultation Measurement    629
    Measuring the Angle between Two Near Bodies    630
    Radar-Range, Azimuth, and Elevation Measurements    631

13.3 Error Analysis of the Navigation Fix . . . . . . . . 632
    Planet-Star, Planet-Star, Planet-Diameter Measurement 634
    Planet-Star, Planet-Star, Sun-Star Measurement 635
    Planet-Star, Planet-Star, Planet-Sun Measurement 638
13.4 A Method of Correcting Clock Errors . . . . . . . . 641
13.5 Processing Redundant Measurements . . . . . . . . 644
    The Pseudo-Inverse of a Matrix 644
    Gauss' Method of Least Squares 646
13.6 Recursive Formulations . . . . . . . . . . . . . . . 648
    The Matrix Inversion Lemma 648
    The Information Matrix and its Inverse 650
    Recursive Form of the Estimator 651
    The Characteristic Polynomial of the $\mathbf{P}$ Matrix 652
13.7 Square-Root Formulation of the Estimator . . . . . . 655
    Symmetric Square Roots of a Matrix 655
    Test for a Positive Definite Matrix 656
    Triangular Square Roots of a Matrix 658
    Recursion Formula for the Square Root of the $\mathbf{P}$ Matrix 659

# 14     Space Navigation     661

14.1 Review of Probability Theory . . . . . . . . . . . 663
14.2 Maximum-Likelihood Estimate . . . . . . . . . . . 665
    The Gauss-Markov Theorem 669
    Properties of the Maximum-Likelihood Estimate 670
14.3 Position and Velocity Estimation . . . . . . . . . . 672
    Range-Rate Measurement 673
    Recursive Estimation 674
    Partitioning and Propagating the Covariance Matrix 675
    The Minimum-Variance Estimator 676
    A Property of the Optimum Estimator 677
    Energy and Angular Momentum Pseudo-Measurements 678
    Square-Root Filtering with Plant Noise 679
14.4 Statistical Error Analysis . . . . . . . . . . . . . 681
    Error Propagation during Planetary Contact 685
    Variation in the Point of Impact 686
14.5 Optimum Selection of Measurements . . . . . . . . 687
14.6 Optimization of the Measurement Schedule . . . . . 690
14.7 Correlated Measurement Errors . . . . . . . . . . . 693
14.8 Effect of Parameter Errors . . . . . . . . . . . . . 696
    Effect of Incorrect Measurement Variance 696
    Effect of Incorrect Cross-Correlation Error Model
        Parameters 697

## Appendices

H  Probability Theory and Applications . . . . . . . . .  699
    H.1  Sampling and Probabilities   699
    H.2  Coin-tossing Experiment   702
    H.3  Combinatorial Analysis   705
    H.4  Random Variables   710
    H.5  Probability Distribution and Density Functions   711
    H.6  Expectation, Mean, and Variance   715
    H.7  Independence and Covariance of Random Variables   717
    H.8  Applications to Coin-tossing and Card-matching   720
    H.9  Characteristic Function of a Random Variable   724
    H.10 The Binomial Distribution   726
    H.11 The Poisson Distribution   729
    H.12 Example of the Central Limit Theorem   730
    H.13 The Gaussian Probability Density Function   732
    H.14 The Law of Large Numbers   735
    H.15 The Chi-square Distribution   737
    H.16 The Markov Chain   741

I  Miscellaneous Problems . . . . . . . . . . . . . . . .  745

   Epilogue . . . . . . . . . . . . . . . . . . . .  751
   Index . . . . . . . . . . . . . . . . . . . . .  785

# Preface

In the three centuries following Kepler and Newton, the world's greatest mathematicians brought celestial mechanics to such an elegant state of maturity that, for several decades preceding the USSR's Sputnik in 1957, it all but disappeared from the university curriculum. Of course, celestial mechanics to the classical astronomer was confined to the prediction of the paths followed by celestial bodies existing naturally in the solar system. Not until recently did the problem exist of designing orbits subject to elaborate constraints to accomplish sophisticated mission objectives at a target planet—except possibly in the fantasy of the boldest imaginations.

The feasibility of space flight by man-made vehicles became apparent in the early 1950's with the rapid development of rockets capable of intercontinental ranges, and gradually serious space-mission planning began. The term "Astrodynamics," attributed to the late Sam Herrick,† came into common usage at that time to categorize aspects of celestial mechanics relevant to a new breed—the aerospace engineer.

One class of imaginative proposals for space missions exploited the gravity fields of planets to achieve multiple planetary flybys. Apparently, the first such study was presented in 1956 at the Seventh International Astronautical Congress in Rome by the Italian General Gaetano Arturo Crocco. His subject—a "One Year Exploration Trip Earth-Mars-Venus-Earth." Although his results were based on a solar system modelled by coplanar, concentric circular planetary orbits and pieced-conic spacecraft trajectories, the germ of an important idea was born. The exotic mission planned for Project Galileo involving dozens or more close encounter flybys of the Jovian moons will be a dramatic highlight of both space exploration and the field of Astrodynamics.

Another Astrodynamics milestone had its origin in 1772 when Joseph-Louis Lagrange submitted his prize memoir "Essai sur le Problème des Trois Corps" to the Paris academy. In it he described particular solutions to the problem of three bodies today known as the "Lagrangian libration points." Lagrange showed that if two bodies of finite mass circularly orbit their common center of mass, then there will be (a) two points in space forming equilateral triangles with the two masses plus (b) three points

---

† Samuel Herrick (1911–1974) was educated at Williams College and the University of California at Berkeley. He served on the Faculty of UCLA from 1937 as a Professor of Astronomy until his untimely death on March 20, 1974.

on the straight line connecting the two masses, where, placing a third mass, will conserve the configuration with respect to the rotating frame of reference.

The equilateral points are known to be stable in many cases. As if in tribute to Lagrange's monumental work, it was early in the Twentieth Century that the so-called "Trojan asteroids" were discovered in the vicinity of the Jupiter-sun equilateral libration points. The colinear points, on the other hand, are unstable points of equilibrium, as was first demonstrated by the mathematician Joseph Liouville in 1845.

The earth-moon equilateral points have been the subject of much popular interest recently as potential sites for space colonies. In fact, one of the sun-earth colinear points was exploited (in 1978) by a spacecraft known as the International Sun-Earth Explorer.†

Libration points, we expect, will play an increasingly important role in spaceflight. In addition to possible scientific applications, these orbits are advantageous for lunar-farside communications, staging sites for lunar and interplanetary transportation systems, and locales for possible space colonies.

$$\Longleftrightarrow$$

The purpose of this book is to provide the engineer and scientist as well as the student with the background for understanding and contributing to the field of Astrodynamics. The material presented is the outgrowth of a course given by the author in the Department of Aeronautics and Astronautics at MIT which he has taught and developed over a period of 25 years. (Three of the astronauts‡ who walked on the moon were students in this course.) It should be considered as a major revision and extension of his first book on this subject titled "Astronautical Guidance" and published in 1964. The text was "typeset" by the author using the typesetting computer program called TEX which was designed by Professor Donald E. Knuth of Stanford University specifically for mathematically oriented texts.

Hypergeometric functions, continued fractions, elliptic integrals, and certain basic topics in analytical dynamics are dealt with in the first two chapters for logical reasons only. It is not required or expected that the

---

† More recently, Bob Farquhar of the Goddard Space Flight Center renamed that spacecraft the International Cometary Explorer and retargeted it, including a close pass of the moon on December 22, 1983 to attain sufficient energy, to pass through the tail of the comet Giacobini-Zinner in September of 1985. Along the way the spacecraft also explored, for the first time ever, the geomagnetic tail, a region downstream from the earth where the planet's magnetic field is swept into a long tail by the solar wind. According to Dr. Farquhar, "It's the most complicated thing that's ever been done, I think, in the way of orbital dynamics in moving a spacecraft around."

‡ Edwin E. "Buzz" Aldrin, Jr., 1961, Apollo 11; Edgar D. Mitchell, 1963, Apollo 14; and David R. Scott, 1962, Apollo 15.

reader or student begin at the beginning. Chapter 3 is a good place to start—indeed, Chapters 3 through 7, with references to Chapters 1 and 2 as needed, constitute most of the first term material in the author's course in Astrodynamics. The chapters in Part II are largely independent of each other and may be read or taught in any order. By picking and choosing, an undergraduate or graduate course may be organized to meet the needs of students having various levels of background and preparation. A textbook containing more subject matter than is covered in a course of instruction is, generally, of benefit to the student. The motivated ones are, thereby, tempted to stray from the beaten path of the classroom.

The Introduction to this book is not an "Introduction" in the generally accepted sense of the word. Instead, it is a reprinting of an AIAA History of Key Technologies paper presenting a personal history of the author's involvement with Astrodynamics since the early 1950's. The intent is that it motivate an interest in the subject matter to follow. Although it is not easy reading for the technically unsophisticated, every reader with any interest at all in the history of space guidance and navigation should find something worthwhile there.

The Prologue and Epilogue are a tribute to the flight of Apollo 8. This was the first manned spaceflight beyond the confines of an earth orbit and the first demonstration of the feasibility of onboard, self-contained space navigation. To many of us who were involved in the Apollo program it was the most exciting of all of the flights. The New York Times commissioned this author to write a popular article for its readers describing how we intended to navigate the Apollo spacecraft to the moon. That article was published on the eve of the Apollo 8 mission and appears here as the Prologue to this book.

The Epilogue begins with a detailed description of just how well the onboard navigation system actually did function during the flight of Apollo 8. The evidence presented is conclusive that the astronauts could have performed successfully on their own without ground contact. Then, in the spirit of the Prologue (which was, of course, originally written for the layman), a fairly complete technical description of the onboard guidance and navigation system of the command and lunar modules is given. The Epilogue also was originally for another purpose—a chapter in a book on the theory and application of Kalman filtering which was commissioned by the Guidance and Control Panel of AGARD-NATO early in 1969. Then, the Epilogue ends appropriately with a digest of an article by Sam Phillips, the Apollo Program Director at NASA Headquarters, on the flight of Apollo 8—what it meant to America and to the history of the world.

A wide variety of problems is a distinctive characteristic of this book. Many of the problems consist of statements or equations to be proved or derived even though no such instruction appears in the text. The student is expected to verify everything which is either stated or implied. Some are simple exercises intended to test the reader's knowledge of the more important concepts. However, many of the problems extend the scope of the text and provide the reader with ample opportunity to develop considerable facility with the subject. These problems are labeled with the "dangerous bend in the road" sign

used in Knuth's book on TEX and possibly originated by Nicolas Bourbaki —the mysterious *nom de plume* of the collective authors of the classical books "Éléments de Mathématique."

A few remarks relevant to notational conventions are appropriate. Vectors of various dimensions are dealt with generally. A column vector of any dimension is represented by a lowercase boldface letter. The corresponding italic letter usually denotes the magnitude of the vector. Matrices are represented by uppercase boldface and can be either square or rectangular arrays. The transpose of a vector or a matrix is denoted by the superscript $^T$. Thus, the scalar product of two vectors **a** and **b** may be written either as $\mathbf{a} \cdot \mathbf{b}$ or $\mathbf{a}^T \mathbf{b}$. In like manner, a quadratic form associated with a square matrix **A** is written $\mathbf{x}^T \mathbf{A} \mathbf{x}$. Further, the notation $\mathbf{M}^{-T}$ is used in place of the more awkward $(\mathbf{M}^{-1})^T$ which is, of course, equivalent to $(\mathbf{M}^T)^{-1}$.

Differentiation of a scalar with respect to a vector results, by definition, in a row vector. Thus, suppose $f(\mathbf{x})$ is a scalar function of a vector $\mathbf{x}$ which is itself a function of $t$. Then, we have

$$\frac{df}{dt} = \frac{\partial f}{\partial \mathbf{x}} \frac{d\mathbf{x}}{dt}$$

as a compact form of the chain rule—to be regarded as either the scalar product of two vectors or the matrix product of a row matrix by a column matrix. For example, if $\mathbf{x}(t)$ has three components $x_1(t)$, $x_2(t)$, and $x_3(t)$, then

$$\frac{df}{dt} = \frac{\partial f}{\partial x_1} \frac{dx_1}{dt} + \frac{\partial f}{\partial x_2} \frac{dx_2}{dt} + \frac{\partial f}{\partial x_3} \frac{dx_3}{dt}$$

Likewise, when a vector function of a vector $\mathbf{f}(\mathbf{x})$ is differentiated, we write

$$\frac{d\mathbf{f}}{dt} = \frac{\partial \mathbf{f}}{\partial \mathbf{x}} \frac{d\mathbf{x}}{dt}$$

The factor $\partial \mathbf{f}/\partial \mathbf{x}$ is a matrix whose rows are the row vectors resulting from the differentiation of each of the scalar components of $\mathbf{f}$ with respect to the vector $\mathbf{x}$. For example, if

$$\mathbf{f} = \begin{bmatrix} f_1 \\ f_2 \\ f_3 \end{bmatrix} \quad \text{and} \quad \mathbf{x} = \begin{bmatrix} x_1 \\ x_2 \\ x_3 \end{bmatrix}$$

then

$$\frac{\partial \mathbf{f}}{\partial \mathbf{x}} = \begin{bmatrix} \dfrac{\partial f_1}{\partial x_1} & \dfrac{\partial f_1}{\partial x_2} & \dfrac{\partial f_1}{\partial x_3} \\ \dfrac{\partial f_2}{\partial x_1} & \dfrac{\partial f_2}{\partial x_2} & \dfrac{\partial f_2}{\partial x_3} \\ \dfrac{\partial f_3}{\partial x_1} & \dfrac{\partial f_3}{\partial x_2} & \dfrac{\partial f_3}{\partial x_3} \end{bmatrix}$$

Specific references are included in the text where appropriate. However, certain books of general value to the author are listed here:

- Abramowitz, M. and Stegun, I. A., *Handbook of Mathematical Functions*, Dover Publications, New York, 1965.
- Baker, R. M. L., Jr. and Makemson, M.W., *An Introduction to Astrodynamics*, Academic Press, New York, 1960.
- Coolidge, J. L., *A History of the Conic Sections and Quadric Surfaces*, Oxford University Press, England, 1945.
- Cramér, H., *Mathematical Methods of Statistics*, Princeton University Press, Princeton, New Jersey, 1946.
- Danby, J. M. A., *Fundamentals of Celestial Mechanics*, The Macmillan Company, New York, 1962.
- Deprit, A., *Fundamentals of Astrodynamics*, (Part I), Mathematical Note No. 556, Mathematics Research Laboratory, Boeing Scientific Research Laboratories, April, 1968.
- Dubyago, A. D., *The Determination of Orbits*, The Macmillan Company, New York, 1961.
- El'yasberg, P. E., *Introduction to the Theory of Flight of Artificial Earth Satellites*, Israel Program for Scientific Translations, Jerusalem, 1967.
- Gauss, C. F., *Theory of the Motion of the Heavenly Bodies Moving about the Sun in Conic Sections*, Dover Publications, New York, 1963.
- Henrici, P., *Discrete Variable Methods in Ordinary Differential Equations*, Wiley, New York, 1962.
- Herget, P., *The Computation of Orbits*, published privately by the author, Ann Arbor, Mich., 1948.

- Kline, M., *Mathematical Thought from Ancient to Modern Times*, Oxford University Press, New York, 1972.
- MacMillan, W. D., *Statics and the Dynamics of a Particle*, McGraw-Hill, New York, 1927.
- MacRobert, T. M., *Spherical Harmonics*, Dover Publications, New York, 1948.
- Moulton, F. R., *An Introduction to Celestial Mechanics*, The Macmillan Company, New York, 1914.
- Plummer, H. C., *An Introductory Treatise on Dynamical Astronomy*, Cambridge University Press, England, 1918.
- Smart, W. M., *Text-Book on Spherical Astronomy*, Cambridge University Press, England, 1956.
- Smart, W. M., *Celestial Mechanics*, Longmans, Green & Co., London, 1953.
- Stumpff, K., *Himmelsmechanik*, Vol. I, Veb Deutscher Verlag der Wissenschaften, Berlin, 1959.
- Wall, H. S., *Analytic Theory of Continued Fractions*, D. Van Nostrand Co., New York, 1948.
- Whittaker, E. T., *A treatise on the Analytical Dynamics of Particles and Rigid Bodies*, Cambridge University Press, 1965.
- Whittaker, E. T. and Watson, G. N., *A Course of Modern Analysis*, Cambridge University Press, England, 1946.

**Richard H. Battin**
**May 29, 1987**

---

Richard H. Battin received an S.B. degree in Electrical Engineering in 1945 and a Ph.D. in Applied Mathematics in 1951—both from the Massachusetts Institute of Technology. Currently, he is Associate Department Head of The Charles Stark Draper Laboratory, Inc. and Adjunct Professor of Aeronautics and Astronautics at MIT. In 1956, in collaboration with Dr. J. Halcombe Laning, he co-authored *Random Processes in Automatic Control*—a book which has appeared in Russian, French, and Chinese editions. His 1964 book *Astronautical Guidance* was also published in a Russian edition. Dr. Battin is a Fellow of the American Institute of Aeronautics and Astronautics and the American Astronautical Society. He is a member of the International Academy of Astronautics and the National Academy of Engineering. In 1972, he and David G. Hoag were presented by the AIAA with the Louis W. Hill Space Transportation Award (now called the Goddard Astronautics Award) "for leadership in the hardware and software design of the Apollo spacecraft primary control, guidance, and navigation system which first demonstrated the feasibility of onboard space navigation during the historic flight of Apollo 8." He received the AIAA Mechanics and Control of Flight Award for 1978, the Institute of Navigation Superior Achievement Award for 1980, and the AIAA Pendray Aerospace Literature Award for 1987. "In recognition of outstanding teaching" the students of the MIT Department of Aeronautics and Astronautics honored him in 1981 with their first Teaching Award.

# Introduction

*Originally published as* "Space Guidance Evolution — A Personal Narrative" *by the present author in the* Journal of Guidance, Control, and Dynamics *for March-April, 1982. It was invited as a History of Key Technologies paper as part of the AIAA's Fiftieth Anniversary celebration.*

The prospect of preparing a comprehensive history of space guidance and navigation was, initially, a delight to contemplate. But, as the unproductive weeks went by, the original euphoria was gradually replaced by a sense of pragmatism. I reasoned that the historical papers which had the greatest appeal were written by "old timers" telling of their personal experiences. Since I had lived through the entire space age, and had the good fortune of being involved in many of the nation's important aerospace programs, I decided to narrow the scope to encompass only that of which I had personal knowledge. (It is, however, a sobering thought that you might qualify as an "old timer.")

The story begins in the early 1950's when the MIT Instrumentation Laboratory (later to become The Charles Stark Draper Laboratory, Inc.) was chosen by the Air Force Western Development Division to provide a self-contained guidance system backup to Convair in San Diego for the new Atlas intercontinental ballistic missile. The work was contracted through the Ramo-Wooldridge Corporation, and the technical monitor for the MIT task was a young engineer named Jim Fletcher who later served as the NASA Administrator.

The Atlas guidance system was to be a combination of an onboard autonomous system, and a ground-based tracking and command system. This was the beginning of a philosophic controversy, which, in some areas, remains unresolved. The self-contained system finally prevailed in ballistic missile applications for obvious reasons. In space exploration, a mixture of the two remains.

The electronic digital computer industry was in its infancy then, so that an onboard guidance system could be mechanized only with analog components. Likewise, the design and analysis tools were highly primitive by today's standards. It is difficult to appreciate the development problems without considering the available computational aids.

## Computing in the Fifties

When I joined the MIT Instrumentation Lab in 1951, digital computation was performed with electrically driven mechanical desk calculators by a battery of young female operators. For analog computation, an electronic analog computer marketed by the Reeves Instrument Company, called the REAC, was used. The big innovation, which signalled the demise of the desk computers, was the IBM Card Programmed Calculator (CPC) acquired in 1952. Floating point calculations could now be made at the fantastic rate of one hundred per minute. But read-write memory was at a premium, and consisted of 27 mechanical counters each holding a ten decimal digit number with sign and housed in bulky units known as "ice boxes."

Development of the all-electronic digital computer was well underway at MIT in the early 1950's. Project Whirlwind produced an enormous machine, completely filling a large building off-campus, which boasted 1024 sixteen-bit words electrostatically stored on cathode-ray tubes. We were fortunate to have access (albeit somewhat limited) to this marvel of the electronic age. (Today, of course, that same capability can be had on a single silicon chip.)

In the summer of 1952, following about six months experience as a user of Whirlwind, my boss, Dr. J. Halcombe Laning, Jr., became enamored of the idea that computers should be capable of accepting conventional mathematical language directly, without the time-consuming intermediate step of recasting engineering problems in an awkward, and all too error-prone, logic that was far removed from the engineer's daily experiences. Over the next few months he personally brought this idea to fruition with the successful development of the first algebraic compiler called, affectionately, "George"—from the old saw *Let George do it.*

Of some interest are the first compiler statements successfully executed by "George":

$$x = 1,$$
$$\text{Print } x.$$

Unfortunately, this is not as well-known as

"*Mr. Watson, come here, I want you.*"

since few programmers are aware of this bit of folklore.

The first nontrivial program executed by George was a set of six nonlinear differential equations describing the lead-pursuit dynamics of an air-to-air fire-control problem. The power of this grandfather of all compilers was aptly demonstrated—the equations were programmed in less than one hour, and successfully executed on the very first trial.

When "peripherals" were added to the Whirlwind computer, Hal Laning encouraged Neal Zierler to collaborate in extending, perfecting, and documenting[1] George. In June of 1954, almost two years after Hal had begun his work, John Backus and a team of programming researchers from IBM came to MIT for a demonstration of George. They were beginning work on a programming system for IBM's newly announced 704 calculator. As a result of this visit, algebraic expressions found their way into the Fortran language.[2]

For historical interest, a program I wrote in March 1954 using the George compiler to compute the Atlas missile trajectory is reproduced in Fig. 1. The notation was constrained by the symbol availability on a Flexowriter, a specially designed typewriter that produced a coded pattern of holes in a paper tape. Since only superscripts were available, subscripts were indicated with a vertical slash prefix. The upper case letter $D$ in the program denotes $d/dt$. The symbols $F^2$ and $F^3$ designate the sine and cosine functions.

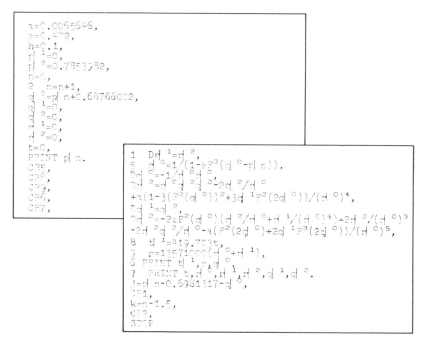

**Fig. 1:** Atlas trajectory program illustrating the "George" compiler.

The use of and interest in George began to wane when our laboratory acquired its own stored program digital computer—an IBM type 650 Magnetic Drum Data Processing Machine—in the fall of 1954. But three years later, when tapes were available, Hal, with the help of Phil Hankins and Charlie Werner, initiated work on MAC—an algebraic programming language for the IBM 650, which was completed by early spring of 1958.

Over the years MAC became the work-horse of the laboratory, and many versions were written to be hosted on the IBM 650, 704, 7090, and 360, as well as the Honeywell H800, H1800, and the CDC 3600.

MAC is an extremely readable language having a three-line format, vector-matrix notations, and mnemonic and indexed subscripts.[3] (I had left the laboratory for "greener pastures" during the period of MAC's creation, and will always regret not participating in its development. But I take some solace in having originated the three-line format, which permits exponents and subscripts to assume their proper position in an equation. The idea was offered to IBM to use in Fortran but was dismissed as being "too hard to keypunch.") Unfortunately, after all these years of yeoman service, MAC seems destined to share the fate of Sanskrit, Babylonic cuneiform and other ancient but dead languages.

The high-order language called HAL, developed by Intermetrics, Inc. and used to program the NASA space shuttle avionics computers, is a direct offshoot of MAC. Since the principal architect of HAL was Jim Miller, who co-authored with Hal Laning a report[3] on the MAC system, it is a reasonable speculation that the space shuttle language is named for Jim's old mentor, and not, as some have suggested, for the electronic superstar of the Arthur Clarke movie "2001—A Space Odyssey."

Since MAC was not then available on our IBM 650, some of the early analysis of the Atlas guidance system was made using a program, which Bob O'Keefe, Mary Petrick, and I developed, known as the MIT Instrumentation Laboratory Automatic Coding 650 Program or, simply, MITILAC.[4] We modeled the coding format to resemble that used for the CPC to minimize the transitional shock to those laboratory engineers who, though still uncomfortable with the digital computer, were beginning to wean themselves away from their more familiar analog devices.

MITILAC was soon superceded by BALITAC,[5] a mnemonic for Basic Literal Automatic Coding, because MITILAC programs were inefficient consumers of machine time. Besides, laboratory problems like Atlas guidance generally involved three-dimensional dynamics so that direct codes were provided (for the first time ever) to perform vector and matrix operations. (The coding format was alpha-numeric, which was no easy trick to implement without an "alphabetic device"—obtainable from IBM for an additional monthly rental of $350 but too expensive for our budget.)

**Delta Guidance**

Initially, Hal Laning and I were the only ones at the laboratory involved in the analytical work for the Atlas guidance system. With no vast literature to search on "standard" methods of guiding ballistic missiles, we "invented" one.

Suppose **r** and **v** are the position and velocity vectors of a vehicle, and $\mathbf{r}_T$ is the position vector of the target. Then along any free-fall, target-intersecting trajectory there is a functional relation among the vectors **r**, **v**, and $\mathbf{r}_T$. Call it

$$F(\mathbf{r}, \mathbf{v}, \mathbf{r}_T) = 0 \qquad (1)$$

At the end of the powered or thrusting portion of the flight, this relationship must be satisfied if the missile is to hit the target.

A reference powered-flight trajectory is chosen for which the "cut-off criterion" of Eq. (1) will be satisfied. Let the function $F$ be expanded in a Taylor series about the reference cut-off values $\mathbf{r}_0$, $\mathbf{v}_0$. Thus,

$$F(\mathbf{r}_0 + \Delta\mathbf{r}, \mathbf{v}_0 + \Delta\mathbf{v}, \mathbf{r}_T) = F(\mathbf{r}_0, \mathbf{v}_0, \mathbf{r}_T) + \left.\frac{\partial F}{\partial \mathbf{r}}\right|_0 \Delta\mathbf{r} + \left.\frac{\partial F}{\partial \mathbf{v}}\right|_0 \Delta\mathbf{v} + \cdots \qquad (2)$$

where the function and its derivatives on the right-hand side are all evaluated on the reference path. (For each value of **r** along the reference trajectory there is a value of **v** for which $F$ will vanish. Thus, each point is a potential cut-off point.)

In essence, then, the zero$^{\text{th}}$-order term on the right of Eq. (2) is zero by definition of the reference path, and the linear terms are driven to zero by an autopilot. Thus, the function $F$, with off-nominal arguments, will eventually vanish (assuming second- and higher-order terms are negligible). There are, of course, complications of detail, which shall be ignored in this discussion.

The particular function $F$ chosen for this purpose was

$$F(\mathbf{r}, \mathbf{v}, \mathbf{r}_T) = (\mathbf{v} \times \mathbf{r}) \cdot [\mathbf{v} \times (\mathbf{r}_T - \mathbf{r})] + \mu r_T \cdot \left(\frac{\mathbf{r}_T}{r_T} - \frac{\mathbf{r}}{r}\right) \qquad (3)$$

where $\mu$ is the earth's gravitation constant. It is not a difficult exercise to show that $F = 0$ is necessary and sufficient for a target intercept. However, I am unable to recall from whence the expression came. Since at that time neither Hal nor I were celestial mechanists (nor acquainted with any), the mystery is all the more puzzling.

Though simple in concept, the Delta guidance method (as it came to be called) is not easy to mechanize especially with analog hardware. First, considerable reference data must be stored; second, a complete navigation system is required; and third, time-of-flight errors are uncompensated, which will most certainly compromise system accuracy unless separately handled (with additional hardware, of course). Nevertheless, this is the system we were determined to make work, despite all of its deficiencies, until I made my first trip to Convair San Diego in the summer of 1955.

## The Convair Legacy

The key figures at Convair were Charlie Bossart, the Chief Engineer, and Walter Schweidetzky, head of the guidance group. Walter had worked with Wernher von Braun at Peenemuende during World War II, and had a most delightful Spanish wife who served as our interpreter during the inevitable evening adventure in Tijuana.

I returned to Cambridge spouting a new vocabulary: "correlated flight path" and "correlated velocity" — "velocity-to-be-gained" and "distance-to-be-gained." The correlated flight path was a predetermined, free-fall reference trajectory designed to intercept the target. The nominal missile flight path would intersect the correlated path at the nominal cut-off point. To each point in time on the missile trajectory corresponded a point on the reference trajectory so that the missile velocity vector $\mathbf{v}_m$ was related in a one-to-one manner to a corresponding reference velocity $\mathbf{v}_c$—the correlated velocity. The velocity-to-be-gained vector was the difference $\mathbf{v}_g = \mathbf{v}_c - \mathbf{v}_m$; distance-to-be-gained was the time integral of $\mathbf{v}_g$. A page from my old notebook illustrating these concepts is reproduced as Fig. 2.

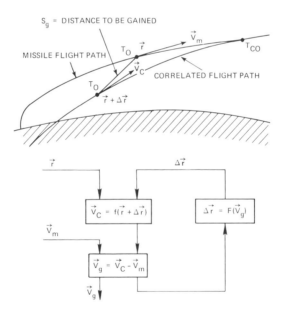

**Fig. 2:** Early concept of the correlated trajectory.

As nearly as I can recall, the Convair mechanization proposal went something like this: If $\mathbf{r}$ is the position vector of the missile, and $\mathbf{r} + \Delta \mathbf{r}$ is the position of the correlated point on the reference path, then the correlated velocity would be obtained by a polynomial approximation utilizing a family of constant-momentum (or constant-energy) trajectories all passing through the target. In addition, a functional relationship between $\mathbf{v}_g$ and $\Delta \mathbf{r}$ could be determined since $\mathbf{v}_g$ was well-approximated by the integral of the thrust acceleration $\mathbf{a}_T$. (Near the cut-off point, the difference between the gravity terms along the actual and reference paths rapidly approach zero.) In short, $\Delta \mathbf{r}$ could be represented as

$$\Delta \mathbf{r} = \frac{s_g}{v_g} \mathbf{v}_g \qquad (4)$$

where $s_g = \int v_g \, dt$. An iteration loop was implied since $\mathbf{v}_c$ is computed from a polynomial function of $\mathbf{r} + \Delta \mathbf{r}$ while $\mathbf{v}_g$ is determined as $\mathbf{v}_c - \mathbf{v}_m$.

The Convair engineers recognized that the velocity-to-be-gained vector eventually remains essentially parallel to a fixed direction in inertial space, and proposed a number of control schemes to drive the velocity-to-be-gained to zero.

The immediate outcome of my trip to San Diego, and subsequent debriefing by Hal Laning, was his total abandonment of the Delta system. From that moment, the Delta guidance development was my millstone to bear. Hal no longer appeared interested in guiding the Atlas missile, or in anything else for that matter. But after what seemed like an eternity (several weeks at least), Hal reappeared with an idea and needed a sympathetic ear. It had to do with a redefinition of the concept of correlated velocity, and a simple differential equation for velocity-to-be-gained. In a few days, Delta guidance would be an orphan.

## The Q-System

If $\mathbf{r}$ is the radius vector representing the position of the missile at an arbitrary time $t$ after launch, the correlated velocity vector $\mathbf{v}_c$ was now to be defined as the velocity required by the missile at the position $\mathbf{r}(t)$ in order that it might travel thereafter by free-fall in a vacuum to a desired terminal condition (here considered to be coincidence of the missile and a target on the earth's surface although other applications to be discussed later are possible). For the definition of $\mathbf{v}_c$ to be unique, a further condition must be stipulated, such as the time at which the missile and target shall coincide. (This requirement would alleviate one of the deficiencies of Delta guidance.)

The point $M$ in Fig. 3 represents the missile position at time $t$; the heavy line through $M$ is the powered-flight path terminating at the cut-off point (CO) in the elliptical free-fall trajectory shown as a dashed line to the target position $T$. Tangent to the correlated velocity vector $\mathbf{v}_c$ is a second ellipse, which would be followed by the missile in free-fall if it, indeed, possessed the velocity $\mathbf{v}_c$ at the point $M$.

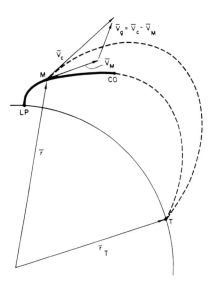

**Fig. 3**: Correlated trajectory and velocity-to-be-gained (from Ref. 6).

Suppose, now, that when the missile is at the point $M$ at time $t$, a "correlated missile" is simultaneously located at the same position. The correlated missile is assumed to experience only gravity acceleration $\mathbf{g}$ and moves with velocity $\mathbf{v}_c$. The actual missile has velocity $\mathbf{v}_m$ and is affected by both gravity $\mathbf{g}$ and engine thrust acceleration $\mathbf{a}_T$. During a time interval $\Delta t$, the two missiles are allowed to move "naturally" with the result that they will diverge in position by the amount

$$\Delta \mathbf{r} = (\mathbf{v}_m - \mathbf{v}_c)\, \Delta t \tag{5}$$

Each experiences a velocity change given by

$$\Delta \mathbf{v}_c = \mathbf{g}\, \Delta t \qquad \Delta \mathbf{v}_m = (\mathbf{g} + \mathbf{a}_T)\, \Delta t \tag{6}$$

At the end of this time interval, imagine that the correlated missile is brought back into coincidence with the actual missile. This change in position must be accompanied by a corresponding change in velocity if terminal conditions imposed on the correlated missile are to remain satisfied. The appropriate change may be expressed as

$$\Delta \mathbf{v}_c = \mathbf{Q}\, \Delta \mathbf{r} \tag{7}$$

where the elements of the matrix $\mathbf{Q}$ are the partial derivatives of the components of the velocity $\mathbf{v}_c$ with respect to the components of the position vector $\mathbf{r}$. It is understood, in carrying out the differentiations, that the target location $\mathbf{r}_T$ and the time-of-free flight $t_{ff}$ remaining (as well as $t$

itself) are held fixed in the process. Thus, we have

$$\mathbf{Q} = \left.\frac{\partial \mathbf{v}_c}{\partial \mathbf{r}}\right|_{\mathbf{r}_T, t_{ff}} \tag{8}$$

The total change in $\mathbf{v}_c$ as a result of these two steps is then

$$\Delta \mathbf{v}_c = \mathbf{g}\,\Delta t + \mathbf{Q}(\mathbf{v}_m - \mathbf{v}_c)\,\Delta t$$
$$= \mathbf{g}\,\Delta t - \mathbf{Q}\mathbf{v}_g\,\Delta t \tag{9}$$

Finally, the change in velocity-to-be-gained $\Delta \mathbf{v}_g$ is simply the difference between $\Delta \mathbf{v}_c$ and $\Delta \mathbf{v}_m$ so that

$$\Delta \mathbf{v}_g = -\mathbf{a}_T\,\Delta t - \mathbf{Q}\mathbf{v}_g\,\Delta t \tag{10}$$

Division by $\Delta t$, and letting $\Delta t$ approach zero, produces the fundamental differential equation for velocity-to-be-gained

$$\frac{d\mathbf{v}_g}{dt} = -\mathbf{a}_T - \mathbf{Q}\mathbf{v}_g \tag{11}$$

Behold the absence of the gravity vector! The necessity to compute earth's gravity, an implied feature of Delta guidance, had vanished. In effect, almost all of the difficulties of the guidance problem were now bound up in the matrix $\mathbf{Q}$. (Hal had a marvelous blackboard derivation of the fundamental equation utilizing block diagrams and an eraser. The audience never failed to be impressed when the block labeled $\mathbf{g}$ magically disappeared.)

To say that calculating the elements of the $Q$ matrix was not a simple exercise would be a gross understatement. In our final report[6] on the $Q$-system it took fourteen pages of an appendix just to describe the necessary equations. Of what possible use could the $\mathbf{v}_g$ differential equation be if the coefficient matrix was that complex? (Had Delta guidance been abandoned too cavalierly?) We were encouraged to proceed because the $Q$ matrix was so simple in the hypothetical case of a flat earth with constant gravity.

With the vector $\mathbf{g}$ a constant, it is not difficult to show that

$$\mathbf{r}_T = \mathbf{r} + \mathbf{v}_c t_{ff} + \tfrac{1}{2}\mathbf{g}\,t_{ff}^2 \tag{12}$$

is the appropriate relation for the problem variables. Therefore, it follows at once that

$$\mathbf{Q} = -\frac{1}{t_{ff}}\mathbf{I} \tag{13}$$

where $\mathbf{I}$ is the identity matrix, and the $\mathbf{v}_g$ equation is simply

$$\frac{d\mathbf{v}_g}{dt} = \frac{1}{t_{ff}}\mathbf{v}_g - \mathbf{a}_T \tag{14}$$

(This differential equation is technically unstable, so that errors in $\mathbf{v}_g$ will increase with time. But the "time constant" associated with this instability

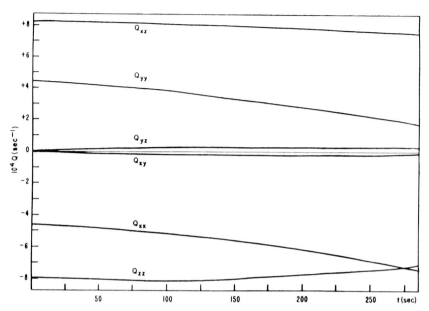

**Fig. 4:** $Q$ coefficients for 5500 mile ICBM trajectory (from Ref. 6).

is the missile time-of-free flight. Since no more than one fourth of the flight time is spent in the powered mode, the magnification of any errors will not exceed $\frac{4}{3}$ in any case.)

The general nature of the $Q$'s for ICBM applications is illustrated in Fig. 4. They correspond to a range of 5500 n.mi. and a coordinate system for which the $x, z$ plane is approximately directed toward the target with the $x$-axis elevated 20 deg above the local horizontal at the launch point. (The matrix **Q** is symmetrical—more about this later.) It is seen from the figure that the $Q$'s are slowly varying functions of time suggesting that they may be adequately approximated by simple polynomials.† Indeed, for IRBM (intermediate-range ballistic missile) applications, for which the range is 1500 miles or less, the $Q$'s could be taken as constants with acceptable accuracy (less than a nautical mile).

The computation of the velocity-to-be-gained vector is only one element of the $Q$-guidance scheme. Of equal importance is a method to control the missile in pitch and yaw, in order that the thrust acceleration will cause all three components of $\mathbf{v}_g$ to vanish simultaneously.

The elegant solution to this control problem came as a brilliant burst of insight. It was all so simple and obvious! If you want to drive a vector

---

† For a single reference trajectory, the $Q$'s may be regarded as functions only of time. However, for an actual missile trajectory with missile parameters different from nominal values, the mathematically correct $Q$'s depend also on missile position. It is only an engineering approximation to regard the $Q$'s as time-programmable.

to zero, it is sufficient to align the time rate of change of the vector with the vector itself. Therefore, the components of the vector cross product $\mathbf{v}_g \times d\mathbf{v}_g/dt$ could be used as the basic autopilot rate signals—a technique that became known as "cross-product steering."

With this control method, it was clear that the $\mathbf{v}_g$ vector would eventually vanish. However, the effect on fuel economy was not so obvious. Therefore, an optimization program was constructed utilizing the calculus of variations to study optimum fuel trajectories[7] (one of the earliest such applications made on a digital computer). The upshot was a confirmation of our suspicion that a good approximation to optimum fuel usage was, indeed, provided by cross-product steering.

Almost ten years later, Fred Martin reconsidered this problem in his MIT doctoral thesis.[8] One interesting little tidbit bears repeating here. Fred was able to show, using elementary methods only, that cross-product steering is optimum for the flat-earth hypothesis. His argument went as follows:

Form the scalar product of Eq. (14) and the vector $\mathbf{v}_g$ to obtain

$$\frac{d}{dt}(\mathbf{v}_g \cdot \mathbf{v}_g) = \frac{d}{dt}v_g^2 = \frac{2}{T-t}v_g^2 - 2\mathbf{a}_T \cdot \mathbf{v}_g$$

where $T$ is the time of impact at the target ($t_{ff} = T - t$). Then integrate from the present time $t$ to the time of engine cut-off $t_{co}$. After integrating by parts we have

$$\int_t^{t_{co}} [2(T-t)\mathbf{a}_T \cdot \mathbf{v}_g - v_g^2]\, dt = (T-t)v_g^2 \tag{15}$$

Now, for any particular time $t$, the right-hand side of Eq. (15) is determined. Therefore, to minimize the integration interval $t_{co} - t$, we should maximize $\mathbf{a}_T \cdot \mathbf{v}_g$—i.e., align the thrust direction with the $\mathbf{v}_g$ vector. In the special case of a flat earth, [check Eq. (14)] this requirement is equivalent to cross-product steering.

The vector block diagram of Fig. 5 shows the basic simplicity of an analog mechanization of the $Q$-system for an IRBM application. By use of the $\mathbf{Qv}_g$ signals as a torque feedback to the pendulous integrating gyro (PIG) units, the output of the latter can be made available as shaft rotations proportional to the components of $\mathbf{v}_g$. Voltage signals can, therefore, be obtained, which are proportional to the $\mathbf{v}_g$ components by exciting potentiometers on the $\mathbf{v}_g$ shafts. These signals can, in turn, be fed into constant gains at the torquing amplifier inputs to provide the necessary multiplications and summations that constitute the matrix-vector product $\mathbf{Qv}_g$. The thrust acceleration sensed by the PIG's varies from approximately 50 to 200 ft/sec$^2$, while the product $\mathbf{Qv}_g$ is of the general order of magnitude of 20 ft/sec$^2$ at launch, and decreases to zero at cut-off. Thus, the $\mathbf{Qv}_g$ product is of the nature of a correction term, which, although

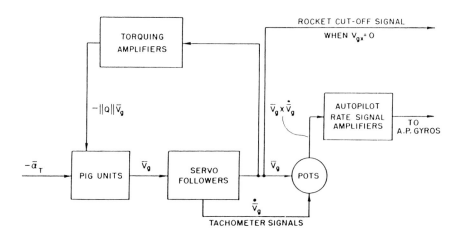

**Fig. 5:** Vector schematic of IRBM guidance computer (from Ref. 6).

far from negligible, does not have to be instrumented with the same precision as the integral of the thrust acceleration itself. As a result, accuracy requirements on each component in the computation of $\mathbf{Q}\mathbf{v}_g$ is about one-quarter of one percent for a one mile miss at the target—well within the range of analog technology available at the time.

Control signals for pitch and yaw are obtained simply using tachometers mounted on the follow-up servos, which produce signals proportional to the derivative of $\mathbf{v}_g$. These are used as excitation for potentiometers mounted on the $\mathbf{v}_g$ shafts. The resulting signals are combined to give the appropriate components of the vector cross product, which are then transmitted to the autopilot as appropriate command rates in pitch and yaw.

A report[6] on the $Q$-system was presented at the first Technical Symposium on Ballistic Missiles held at the Ramo-Wooldridge Corporation in Los Angeles on June 21 and 22, 1956. In the afternoon of the second day came the only session on Inertial Guidance, and all of the papers except ours dealt with inertial instruments—the $Q$-system had no competition! We could easily have returned to Boston by walking on the clouds.

The $Q$-system was first implemented on the Thor IRBM and then on the Polaris fleet ballistic missile, but not the Atlas for which it had been designed. What system was used for Atlas? Some form of Delta guidance, I've been told.

# Introduction

## Symmetry of the Q Matrix

In 1955, the program output from the IBM 650 was a stack of punched cards that had to be printed separately using a Type 418 accounting machine. Hal and I watched the 418 type bars rise and fall, with their characteristic noisy clank, as the first set of $Q$ matrix elements was being tabulated in a neat array format. At the pace of 150 lines per minute, plenty of time was available for a surprising, and totally unexpected, observation. The $Q$ matrix was symmetric! In fact, the off-diagonal elements were asymmetric only in the last decimal place.

Considering the enormous complexity of the program, the phenomenon could not be happenstance. Two conclusions were immediate: 1) the symmetry of the $Q$ matrix *must* be analytically demonstrable, and 2) our computer program *must* be correct. It was only much later that the mathematical proof was supplied. Meanwhile, an instant check was always available on the complicated numerical procedures required to produce the $Q$ matrix.

In an appendix to our report,[6] two different proofs of the symmetry property were given. The first utilized a special coordinate system for which it could be shown that four of the off-diagonal elements of the $Q$ matrix were identically zero. The two remaining corner elements were then shown to be equal by a rather messy, nonintuitive argument requiring five pages of uninspiring and tedious mathematics.

The second proof provided greater physical insight, and involved a hydrodynamical analogy. Correlated velocity was to be visualized as a vector-velocity field describing the motion of an inviscid, compressible fluid. The symmetry of the $Q$ matrix was then equivalent to the velocity field having a zero curl $\nabla \times \mathbf{v}_c = \mathbf{0}$. The equation of motion

$$\frac{d\mathbf{v}_c}{dt} = \mathbf{g}$$

is the same as that for an inviscid fluid moving under the action of conservative body forces throughout which the internal pressure gradient is zero. Together with the equation describing the variation in fluid density $\rho$

$$\frac{d\rho}{dt} + \rho \nabla \cdot \mathbf{v}_c = 0$$

it follows (with just a little exercise in ingenuity) that

$$\frac{1}{\rho} \nabla \times \mathbf{v}_c = \text{constant}$$

The demonstration concludes with an argument that the fluid is converging on the target point $\mathbf{r}_T$ so that the density in the vicinity of $\mathbf{r}_T$ is becoming infinite. Hence, the constant is zero, implying that the curl is everywhere zero.

The distribution of the $Q$-system report to those individuals having both a secret clearance and a "need to know" triggered an informal competition. Who would be first with the shortest and most elegant proof of $Q$ matrix symmetry? (Among the contestants, the only one I recall vividly supplied a carefully detailed but erroneous demonstration of nonsymmetry.)

In 1960, the original $Q$-system report[6] was reprinted in a shortened form[9] with a new appendix describing my own most recent proof. The key was to establish $\mathbf{Q}^{-1}$ as a solution of the matrix Ricatti equation

$$\frac{d}{dt}\mathbf{Q}^{-1} + \mathbf{Q}^{-1}\mathbf{G}\mathbf{Q}^{-1} = \mathbf{I} \tag{16}$$

where $\mathbf{G}$ is the gravity-gradient matrix

$$\mathbf{G} = \frac{\partial \mathbf{g}}{\partial \mathbf{r}} \tag{17}$$

The symmetry of $\mathbf{Q}^{-1}$ follows at once from the differential equation and the terminal condition for $\mathbf{Q}^{-1}$. The matrix $\mathbf{G}$ is necessarily symmetric since $\mathbf{g}$ is the gradient of a scalar potential function. Hence, Eq. (16) and its transpose are identical. Also $\mathbf{Q}^{-1} = \mathbf{O}$ at the terminal point is symmetric. Therefore, $\mathbf{Q}^{-1}$ (hence also $\mathbf{Q}$) is everywhere symmetric.

A by-product of this symmetry proof was an alternate computational procedure for determining $\mathbf{Q}$, which is independent of the assumption that the gravity field through which the missile moves is inverse square. Hence, a more precise modeling of earth gravity could be incorporated when computing the $Q$ matrix.

Five years later, Fred Martin published an explicit expression for the $Q$ matrix as an appendix to his doctoral thesis.[8] The symmetry was now obvious from inspection.

The latest bulletin on the subject is a recent recognition that no elaborate proof is really necessary! The property follows from the inherent nature of symplectic matrices.

It has been known for years that the $Q$ matrix can be formed algebraically from partitions of the state transition matrix for the linearized equations describing a missile in free fall. It has also been known (since 1962 when Jim Potter first called attention to the fact[10]) that the transition matrix is an example of a class of so-called "symplectic" matrices. The virtue of a symplectic matrix is that the inverse is easily obtained by a simple rearrangement of its elements.

One day last year while preparing a lecture for my class, I noticed that the product of the transition matrix, and its inverse, produced a number of symmetric matrices—one of which was $\mathbf{Q}$. The interested reader may wish to verify this for himself.

## October 4, 1957 and the Aftermath

Like so many other Americans, the first half of October 1957 found me standing in my yard in the cold but clear early morning hours watching and waiting for the Russian Sputnik to pass overhead. I had been away from MIT for just one year exploring new and different career opportunities in the alleged greener pastures of industry. A few months later during one of my infrequent telephone conversations with Hal Laning, I learned that he had a simulation of the solar system running on the IBM 650 and was "flying" round trips to Mars.

It didn't take long to wind up my affairs and head back to the Instrumentation Lab. My return practically coincided with the publication of a laboratory report[11] on the technical feasibility of an unmanned photographic reconnaissance flight to the planet Mars. It was asserted by the authors that a research and development program to that end could reasonably be expected to lead to the launching of such a vehicle within the next five to seven years. (It is interesting that the study and report had been sponsored by the Ballistic Missile Division of the U.S. Air Force.)

A small group was forming to flesh out the system proposal for the Mars mission. Hal and I were responsible for the trajectory determination, as well as the mathematical development of a suitable navigation and guidance technique. The project culminated a year or so later in a three volume report,[12] and a full-scale model of the spacecraft.

To my surprise, it quickly became evident that we did not really know how to compute trajectories for the simple two-body, two-point boundary-value problem! How could that be possible after all the work on ballistic missile trajectories only a few years earlier? As I reviewed those equations in the $Q$-system report, the difficulty (but not the solution) was apparent. We had, indeed, developed expressions involving the correlated velocity vector but they were all implicit—$\mathbf{v}_c$ never appeared explicitly. These equations were fine for calculating the $Q$ matrix by implicit differentiation but in no way did it seem possible to isolate the velocity vector. (Hal had been calculating round-trip Martian trajectories by "trial and error"—adjusting and readjusting the spacecraft initial conditions and determining the orbit by numerically solving the equations of motion. There had to be a better way!)

I found the clue in the classical treatise on dynamics by Whittaker[13]:

> "Lambert in 1761 shewed (sic) that in elliptic motion under the Newtonian law, the time occupied in describing any arc depends only on the major axis, the sum of the distances from the center of force to the initial and final points, and the length of the chord joining these points: so that if these three elements are given, the time is determinate, whatever the form of the ellipse."

The proof followed, and the section ended with a neat analytical expression for time of flight as an explicit function of the problem geometry and the semimajor axis $a$ of the orbit. Given the geometry and the time of flight, then $a$ could be determined—not directly but by iteration.

It was the footnote that gave me pause:

*"It will be noticed that owing to the presence of the radicals, Lambert's theorem is not free from ambiguity of sign. The reader will be able to determine without difficulty the interpretation of sign corresponding to any given position of the initial and final points."*

By no means was it obvious to me how to resolve the ambiguity or, more to the point, how to instruct a computer to choose unerringly from among the several alternatives. Whittaker's only reference was to Lagrange (Oeuvres de Lagrange, IV, p. 559) who also failed to address my concerns; but going to the original source did pay dividends. Instead of proceeding immediately to his proof of Lambert's theorem, Lagrange first chatted about the problem from different perspectives† —one of which led me to transform the problem to rectilinear motion. The ambiguity then ceased to exist.

A nontrivial problem remained—to obtain the initial velocity vector in terms of the semimajor axis $a$. An intense effort produced finally a delightfully elegant expression. We were now able to generate interplanetary trajectories with great aplomb. (My first trajectory program suffered from an annoying deficiency. Time of flight is a double-valued function of the semimajor axis $a$ with infinite slope for the minimum-energy trajectory— far from ideal for a Newton-Raphson iteration. The difficulty was resolved by a different choice of independent variable against which the time of flight is a monotonic function. This small, but necessary, wrinkle was first reported in an appendix to Ref. 9, and practically eliminated the audible vulgarisms that so frequently accompanied the use of the original program.)

With some trepidation, I presented this method[14] of trajectory determination in New York on January 28, 1959 at the annual meeting of the Institute of the Aeronautical Sciences. My scant background in celestial mechanics did little to inspire self-confidence in the novelty of the technique. But, as I later learned, Rollin Gillespie and Stan Ross were in the audience, and had carried a preprint back home to their associate John Breakwell at the Lockheed Missiles and Space Division. They, too, had been grappling with the trajectory problem and (according to Rollin) this was the "breakthrough" they also needed.

---

† Lagrange's paper would never appear in the *Journal of Guidance, Control, and Dynamics*, or in any other modern archival publication, without strong protestations from the editor—"Needs at least a 50% reduction!"

The method became the basis of the major orbit-determination programs of the Jet Propulsion Laboratory for its series of unmanned interplanetary probes, and of the Navy and Air Force for targeting ballistic missiles. Indeed, in the early sixties, JPL used this technique to generate an enormous set of volumes—similar to the Airline Guide—in which were tabulated daily launch conditions for Venus and Mars missions extending many years into the future.

To support the Mars reconnaissance study project, we confined our attention to trajectories whose flight times were of the order of three years, and which had launch dates in the years 1962–1963. These missions, for which the space vehicle makes two circuits about the sun while the earth makes three, seemed to provide the greatest flexibility in launch window and passing distance at Mars without placing unreasonable requirements on launch system capabilities. Later we investigated round-trip missions to Venus, which could be accomplished with flight times of only a year and a quarter.

One day, when plotting a few of these Venusian reconnaissance trajectories, I was impressed by the proximity of the spacecraft orbit and the Martian orbit resulting from the increased velocity induced during the Venusian flyby. The interesting possibility of a dual contact with both planets seemed feasible—a kind of celestial game of billiards. The infrequency of proper planetary configurations would, of course, severely limit the practicality of such a mission if, indeed, one existed at all.

Using trusty "cut and try" methods, I found that ideal circumstances did prevail on June 9, 1972. On that date, a vehicle in a parking orbit launched from Cape Canaveral on a 110° launch azimuth course could be injected into just such a trajectory at the geographical location of 5° W and 18° S and with an injection velocity relative to the earth of 15,000 ft/sec. The first planet encountered would be Venus after 0.4308 year. The vehicle would pass 4426 miles above the surface of the planet and would, thereby, receive from the Venusian gravity field alone a velocity impulse sending it in the direction of Mars. The second leg of the journey would consume 0.3949 year and the spacecraft would then contact Mars, passing 1538 miles above the surface. The trip from Mars back to earth would last 0.4348 year so that the vehicle would return on September 13, 1973. This truly remarkable orbit is illustrated in Fig. 6. (At the time, the launch date seemed incredibly far off—twelve whole years! But the day finally came and, sad though it may seem, passed without fanfare or even a comment.)

Although this was the first realistic multiple flyby mission ever designed, it was not the first ever conceived. That distinction goes to General Gaetano Arturo Crocco who was Director of Research of the Air Ministry and a Professor of Aeronautics at the University of Rome, Italy. His paper[15] described an earth to Mars to Venus to earth mission of one year duration. The orbits were all coplanar; the velocity requirements were

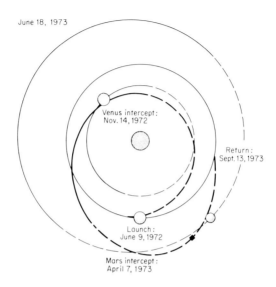

**Fig. 6:** Double-reconnaissance trajectory (from Ref. 10).

enormous; and the reversed itinerary prevented the best utilization of the gravity assist maneuvers. But it was published in 1956—one year before Sputnik. (AIAA members might appreciate knowing that General Crocco was a founding member of the Institute of the Aeronautical Sciences—one of our parent organizations.)

The Mars reconnaissance preliminary design was ready for customer review in the summer of 1959. The Air Force had been our sponsor, and it was there that we expected to turn for authorization to proceed. We were ready to go—"Mars or bust!"—with an enthusiasm that was exceeded only by our naiveté. While we had been busy nailing down the myriad of technical problems one by one, the political climate was changing. A new government agency called the "National Aeronautics and Space Administration," not the Air Force, would control the destiny of the Mars probe.

With view-graphs, reports, and a wooden spacecraft model, we headed for Washington instead of Los Angeles, and arrived there on the same day as Chairman Khrushchev. Although our presentation was well received, the high-level NASA audience we had expected (including Hugh Dryden, the Deputy Administrator) was attending to the necessary protocol mandated by the Russian visit. We were sent home with a pat on the head and the promise of some future study money. As our dreams of instant glory in interplanetary space began to fade, we secretly took perverted pleasure in having Nikita Khrushchev himself as a ready-made scapegoat. The Russians were formidable opponents indeed!

# Introduction

The NASA study contract allowed our small team to continue the work that had begun under Air Force auspices. For this we were most grateful; but the absence of a specific goal diminished much of the enthusiasm. Now we were simply doing "interplanetary navigation system studies." There certainly was no reason to expect that a new goal lay just over the horizon, which would challenge and excite us beyond our wildest imaginations.

### Prelude to Apollo

The general method of navigation that Hal and I had created for the Mars probe mission[16] was based on perturbation theory, so that only deviations in position and velocity from a reference path were used. Data was to be gathered by an optical angle-measuring device, and processed by a spacecraft digital computer. Periodic small changes in the spacecraft velocity were to be implemented by a propulsion system as directed by the computer.

The appropriate velocity changes were calculated from a pair of matrices obtained as solutions of the differential equations

$$\frac{d\mathbf{R}^\star}{dt} = \mathbf{V}^\star \qquad \frac{d\mathbf{V}^\star}{dt} = \mathbf{G}\mathbf{R}^\star \tag{18}$$

where $\mathbf{G}$ is the gravity-gradient matrix evaluated along the reference path. Boundary conditions were specified at the reference time of arrival $t_A$ at the target as

$$\mathbf{R}^\star(t_A) = \mathbf{O} \qquad \mathbf{V}^\star(t_A) = \mathbf{I} \tag{19}$$

Then if $\delta\mathbf{r}(t)$ is the position deviation from the reference path at time $t$, the required velocity deviation $\delta\mathbf{v}(t)$ was found to be

$$\delta\mathbf{v}(t) = \mathbf{V}^\star(t)\mathbf{R}^{\star-1}(t)\,\delta\mathbf{r}(t) \tag{20}$$

It is a trifle embarrassing to admit that we did not immediately recognize our old friend the $Q$ matrix in Eq. (20). When the dawn came, we were truly nonplused. Here we were now working in an unclassified area with every intent to freely publish the results—but the $Q$-system was still classified! That last point had been dramatically emphasized only a year or so earlier. An author, who shall remain nameless, wrote a book on guidance containing a section that made full disclosure of the $Q$-system. When the U.S. Navy was finally ready to act, the books were on the publisher's loading dock awaiting shipment. All copies—several thousand at least—were seized and burned.

Then and there the matrix product $\mathbf{V}^\star\mathbf{R}^{\star-1}$ was christened $\mathbf{C}^\star$. We reasoned that the $Q$ matrix by itself was just a mathematical collection of partial derivatives. The security classification derived from its use in the velocity-to-be-gained differential equation as applied to ballistic missile guidance. Since the letter "$Q$" signified absolutely nothing, it would have been pointless to persist in its use in an entirely different context.

The velocity correction, as calculated from Eq. (20), was perfectly adequate for interplanetary missions, except when the spacecraft was in proximity to the destination planet. With a relatively short time of flight remaining, the constraint imposed on the vehicle by the $\mathbf{C}^*$ matrix to reach the target at a predetermined time caused an inordinate expenditure of rocket fuel.

The deficiency was later corrected during our NASA studies with the invention of variable-time-of-arrival guidance.[17] Equation (20) could now be replaced by

$$\delta \mathbf{v}(t) = \mathbf{C}^*(t)\, \delta \mathbf{r}(t) + \mathbf{R}^{*-\mathrm{T}}(t)\mathbf{v}_r(t_A)\, \delta t \qquad (21)$$

where $\mathbf{v}_r(t_A)$ is the velocity of the spacecraft relative to the target planet at the nominal time of arrival $t_A$ and $\delta t$ is the change in arrival time. The increment $\delta t$ is chosen to minimize the magnitude of the required velocity correction.

To navigate the Mars probe, a sequence of measurements of angles between selected pairs of celestial bodies, together with the measurement of the angular diameter of a nearby planet, was to be made on board the spacecraft (automatically, of course, under computer control) to obtain a celestial fix. For specificity, the measured angles, illustrated in Fig. 7, were chosen as follows: 1) from the sun to the nearest visible planet $P$; 2) from Alpha Centauri to $P$; 3) from that one of Sirius or Arcturus to $P$ such that the plane of measurement is most nearly orthogonal to the plane of the angle measured in 2; 4) from the sun to the same star selected in 3; 5) from the sun to the second closest planet provided that more than one planet is "visible" (at least 15° away from the line-of-sight to the sun); and 6) the angular diameter of $P$, provided that it exceeds 1 mrad. This strategy for observations ensured that a minimum of four angles would be measured, provided at least one planet was visible. (The three particular stars were selected because they are among the brightest and form roughly an orthogonal triad.) The intended result of these observations was a determination of the coordinates of spaceship position together with a correction to the spaceship clock.

Although the terminology was not yet in vogue, we were in fact dealing with an estimation problem involving a four-dimensional "state vector." We linearized the measurements about a reference point, and developed a weighted least-squares procedure to obtain the celestial fix.[16]

So much for the Mars probe, which was now, at best, in a state of limbo. We began working for NASA, and the close technical collaboration that had existed between Hal and myself gradually subsided.

Hal Laning renewed his old love affair with the digital computer—however, it was basic computer architecture, and not software, that

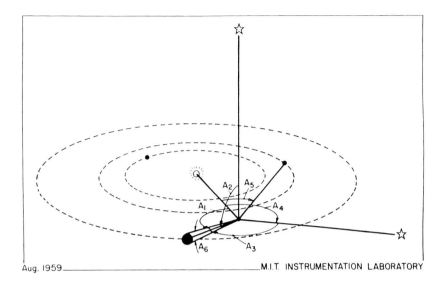

**Fig. 7:** Mars probe navigation fix (from Ref. 16).

attracted his interest this time. He joined forces with Ramon Alonzo to develop a design for a small control computer with some unique characteristics for space and airborne applications.[18] Some of those features were variable speed with power consumption proportional to speed, relatively few transistors, parallel word transfer, automatic incrementing of counters, and automatic interruption of normal computer processes upon receipt of inputs. The program and constants were stored in a wired-in form of memory called a "core rope," which permitted unusually high bit densities for that time. Such a computer would have been ideally suited for the Mars probe but, in fact, Hal and Ray were unknowingly designing the computer whose technical offspring would take man to the moon.

Meanwhile, I continued alone on the guidance and navigation analysis. There were some annoying problems with our interplanetary navigation algorithm having to do with numerical difficulties encountered in the required matrix inversion associated with the method of least squares.

In the notation used at that time,[19] the least-squares method resulted in the following expression†

$$\delta \overline{r}_4 = (\mathbf{U}_{4m} \mathbf{\Phi}_{mm}^{-1} \mathbf{U}_{m4})^{-1} \mathbf{U}_{4m} \mathbf{\Phi}_{mm}^{-1} \delta \overline{A}_m \qquad (B\text{--}10)$$

Here $m$ is the total number of measurements for any particular fix; $\mathbf{U}_{m4}$ is the $m \times 4$ measurement geometry matrix; $\mathbf{\Phi}_{mm}$ is the diagonal moment

---

† Equations with the prefix B are taken from, and numbered in accordance with, the appendix of Ref. 19.

matrix of measurement errors; $\delta \overline{A}_m$ is the vector of measurements; $\delta \overline{r}_4$ is the least-squares estimate of the four-dimensional state vector. Four measurements are sufficient to determine the spaceship position and the clock correction, so that $r = m - 4$ is the number of redundant measurements.

To cope with the numerical difficulties of Eq. (B–10), I was determined to obtain, if at all possible, an explicit inverse of the matrix product $\mathbf{U}_{4m} \mathbf{\Phi}_{mm}^{-1} \mathbf{U}_{m4}$. The task became an obsession and the derivation was so involved that only the final expression appeared[19] in an unclassified appendix to an otherwise classified† report. The result was recorded as follows:

"Now if we define the two square matrices

$$\mathbf{P}_{44} = \mathbf{U}_{44}^{-1} \mathbf{\Phi}_{44} \mathbf{U}_{44}^{T\,-1} \qquad (B\text{–}12)$$

$$\mathbf{Q}_{rr} = \mathbf{\Phi}_{rr} + \mathbf{U}_{r4} \mathbf{P}_{44} \mathbf{U}_{4r} \qquad (B\text{–}13)$$

it can be verified directly that

$$(\mathbf{U}_{4m} \mathbf{\Phi}_{mm}^{-1} \mathbf{U}_{m4})^{-1} = \mathbf{P}_{44} - \mathbf{P}_{44} \mathbf{U}_{4r} \mathbf{Q}_{rr}^{-1} \mathbf{U}_{r4} \mathbf{P}_{44} \qquad (B\text{–}14)$$

Then, by substituting (B–14) into Eq. (B–10), we obtain, after a little manipulation,

$$\delta \overline{r}_4 = \mathbf{U}_{44}^{-1} \{ \| \mathbf{I}_{44} \quad \mathbf{O}_{4r} \| + \mathbf{B}_{4r} \mathbf{Q}_{rr}^{-1} \| -\mathbf{A}_{r4} \quad \mathbf{I}_{rr} \| \} \delta \overline{A}_m \qquad (B\text{–}15)$$

where we have defined the two rectangular matrices

$$\mathbf{A}_{r4} = \mathbf{U}_{r4} \mathbf{U}_{44}^{-1} \qquad (B\text{–}16)$$

$$\mathbf{B}_{4r} = \mathbf{\Phi}_{44} \mathbf{A}_{4r} \qquad (B\text{–}17)$$

The matrix $\mathbf{Q}_{rr}$ may be expressed in terms of $\mathbf{A}_{r4}$ and $\mathbf{B}_{4r}$ by

$$\mathbf{Q}_{rr} = \mathbf{\Phi}_{rr} + \mathbf{A}_{r4} \mathbf{B}_{4r} \qquad (B\text{–}18)$$

Equation (B–15) displays explicitly the effect of adding redundant measurements."

Those familiar with the Kalman filter will recognize Eq. (B–14) at once as the covariance matrix update formula. Although the expression (B–15) for the state vector update is not in the customary form, it is evident that the first term on the right is the state estimate using four measurements. The second term may be rewritten as

$$-\mathbf{U}_{44}^{-1} \mathbf{B}_{4r} \mathbf{Q}_{rr}^{-1} \mathbf{A}_{r4} \delta \overline{A}_4 + \mathbf{U}_{44}^{-1} \mathbf{B}_{4r} \mathbf{Q}_{rr}^{-1} \delta \overline{A}_r$$

with $\delta \overline{A}_4$ and $\delta \overline{A}_r$ denoting the two partitions of the vector of measurements $\delta \overline{A}_m$. Then, substituting from Eqs. (B–12), (B–16), and (B–17),

---

† The report was classified because it quoted some confidential Centaur missile data.

and introducing the matrix
$$\mathbf{W}_{4r} = \mathbf{P}_{44}\mathbf{U}_{4r}\mathbf{Q}_{rr}^{-1}$$
called the "weighting matrix" in the current vernacular, the term in question becomes
$$\mathbf{W}_{4r}(\delta\overline{A}_r - \mathbf{U}_{r4}\mathbf{U}_{44}^{-1}\delta\overline{A}_4)$$
Since this is precisely the weighted difference between the actual redundant measurements and the predicted values of those measurements, Eq. (B-15) is then exactly equivalent to the now conventional state vector update formula.

Unbeknownst to me at the time, Rudolf Kalman was also addressing the estimation problem, albeit with greater generality and from a more esoteric standpoint. His now classical paper[20] was published almost simultaneously with Ref. 19. About a year later, I learned of Kalman's work from Stan Schmidt at the Ames Research Center.

## The Race to the Moon

After the publication of our studies for NASA in the three volume report R-273,[19] a hiatus in further navigation work resulted from an unexpected invitation by the Research and Advanced Development Division of Avco Corporation. They enlisted the support of the Instrumentation Laboratory to design a system for guiding a vehicle propelled electrically by a low-thrust arc-jet engine from an earth-satellite orbit into a lunar orbit. We were so eager to work on a real space program that we fairly leaped at this opportunity. A bright and amiable young engineer, Mike Yarymovych (who served as the AIAA president for the 1982–83 term), was our principal contact.

At that time a great deal of work had already been accomplished in optimizing low-thrust escape trajectories utilizing variational techniques, but virtually no attention had been directed to the guidance problem. We succeeded in designing a multiphased guidance scheme—one aspect of which relied heavily on the concepts of velocity-to-be-gained and steering developed for the $Q$-system. Preliminary results were first published[21] in January of 1961, and Jim Miller carried through the complete development as his doctoral dissertation,[22] which he presented on August 29, 1961 in Los Angeles at the Sixth Symposium on Ballistic Missile and Aerospace Technology.

Meanwhile, in February of that same year, NASA came through with another six-month contract—this time for a preliminary design study of a guidance and navigation system for Apollo to be sponsored by the Space Task Group of NASA. It was time to dust off and re-examine the navigation problem once again.

I learned from Gerald Smith that the Ames Research Center's Dynamic Analysis Branch was working on midcourse navigation and guidance for a circumlunar mission.[23] The Branch Chief, Stanley Schmidt, and his associates were most hospitable during my visit, and gave me a private blackboard lecture describing the filter equations taken from Kalman's year old paper[20] which they were using for their own navigation studies. I also received a copy of Kalman's paper, along with the admonition that it would not be easy reading.

The key idea gleaned from the meeting at Ames was the possibility of eliminating the notion of the navigation fix. I learned that the covariance matrix could be easily extrapolated using the state transition matrix. Navigation measurements could be spaced in time and the update equations could be applied recursively to a full six-dimensional state vector. Indeed, if only one scalar measurement was processed at any one time, the matrix $\mathbf{Q}_{rr}$ of Eq. (B–13) would be simply a positive scalar with no matrix inversion required at all!

Kalman's paper was to me so abstruse that it was not clear whether his equations were equivalent to those obtained using the maximum-likelihood method. (Indeed, Stan told me during our visit that this question had not yet been resolved to everyone's complete satisfaction.) To settle this in my own mind, I wrote down a linear state-vector update equation to process a single measurement and left the weighting vector to be determined so as to minimize the variance of the estimation error. The result agreed with Kalman's.

As a second check, I applied the equations (with the state transition matrix replaced by the identity matrix) to one of the Mars mission position fixes, and processed the measurements one at a time. Again the result was the same.

The recursive navigation algorithm was clearly the best formulation for an onboard computer. But a number of questions still remained. When a single measurement is to be made, which star and planet combination provides the "best" available observation? Does the best observation give a sufficient reduction in the predicted target error to warrant its being made at all? Is the uncertainty in the computer velocity correction a small enough percentage of the correction itself to justify an engine restart and propellant expenditure? Can a statistical simulation of a space-flight mission be made without resorting to Monte Carlo techniques? How would cross-correlation effects of random measurement errors affect the estimator? These questions were all addressed in a paper[24] presented in October of 1961, on Friday the thirteenth, at the American Rocket Society's Space Flight Report to the Nation held in the New York Coliseum.

During the preparation of that paper, political events provided a new urgency to our work. On May 25, 1961, President John F. Kennedy in his Special Message to Congress on Urgent National Needs said:

## Introduction

*"I believe that this nation should commit itself to achieve the goal, before this decade is out, of landing a man on the moon and returning him safely to earth."*

Less than three months later on August 10, NASA contracted with our Laboratory for the development of the Apollo guidance and navigation system—the first major Apollo contract awarded by the space agency.

The history of the Apollo onboard guidance, navigation, and control system was well told[25] by Dave Hoag at the International Space Hall of Fame Dedication Conference in Alamogordo, N.M., during October 1976. With that as background, a description of the Apollo system and its development will not be necessary here. However, a few items require emphasis to provide a proper perspective for the rest of this narrative.

Initially, the specifications were for a completely self-contained system—there would be absolutely no ground communications either verbally or by telemetry with the vehicle. This requirement, which was presumably to prevent an over enthusiastic competitor in the race to the moon from intentionally interfering with an Apollo flight, gradually eroded away—but not before computer algorithms had been designed and implemented which would permit completely autonomous missions.

Several fundamental characteristics of the Apollo guidance computer (AGC) made the implementation of self-contained algorithms a definite challenge: 1) a short word length, 16 bits, necessitating double precision for most calculations; 2) a modest memory size—36,864 read-only, and 2048 read-write registers; and 3) moderate speed—23.4 $\mu$sec add time. Small as that may seem, it was a major improvement in speed and capacity over that which was available in the fall of 1961. Then the AGC had 4096 words of fixed memory, 256 words of erasable memory, and twice the cycle time. (Over the years technology advances permitted the expansion in capacity while maintaining the original size of one cubic foot. The physical dimensions could change only at great cost—that was all the space provided for in the spacecraft.)

The first mission programming for the AGC was to implement the recursive navigation algorithm. That, at least, we knew how to do! Of course, the program changed many times during the ensuing years but not the concept. A complete description, including all the nitty-gritty, of the final implementation for Apollo is found in Ref. 26.† A diagram which we used countless times for customer briefings is reproduced as Fig. 8. Note from the figure that the reference trajectory has been replaced by the integrated vehicle state. The necessity for this important change was obvious when we first addressed the implementation problem. The modification is generally referred to as the "extended" Kalman filter.

---

† That reference comprises the Epilogue of this book.

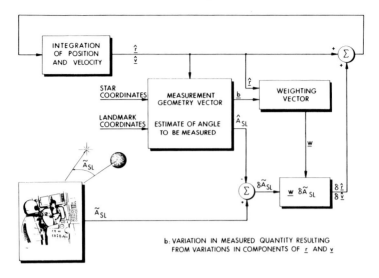

**Fig. 8:** Apollo coasting-flight navigation.

Guiding the Apollo vehicle during its many and varied powered maneuvers was another matter. The idea of using the original $Q$-system for these purposes was soon rejected. Its principal virtue was the ease of mechanization on board the vehicle. But this advantage had to be traded off against the burden placed on ground facilities. (Consider the significant staff and computers of the Dahlgren Naval Weapons Laboratory devoted solely to the task of supplying the necessary targeting data, and the curve-fitted elements of the $Q$ matrix for the fleet ballistic missiles of the U.S. Navy.) With the AGC we had at our disposal for the first time ever a powerful general purpose digital computer as the key ingredient of a vehicle-borne guidance system. Why not use it?

In fact, the $Q$ matrix could be avoided altogether in the $\mathbf{v}_g$ differential equation by simply differentiating the defining equation for velocity-to-be-gained $\mathbf{v}_g = \mathbf{v}_r - \mathbf{v}$ to obtain

$$\frac{d\mathbf{v}_g}{dt} = \frac{d\mathbf{v}_r}{dt} - \mathbf{g} - \mathbf{a}_T \tag{22}$$

(The terminology "correlated velocity" $\mathbf{v}_c$ was replaced by "required impulse velocity" $\mathbf{v}_r$; missile velocity $\mathbf{v}_m$ became vehicle velocity $\mathbf{v}$.) If the vector $\mathbf{v}_r$ could be expressed in analytical form, then the vector

$$\mathbf{p} = \frac{d\mathbf{v}_r}{dt} - \mathbf{g} \tag{23}$$

## Introduction

could be calculated so that the rate of change of $\mathbf{v}_g$ is determined from

$$\frac{d\mathbf{v}_g}{dt} = \mathbf{p} - \mathbf{a}_T \tag{24}$$

(We were no longer concerned about computing gravity—it posed no problem for the AGC). We had an expression for $\mathbf{v}_r$ when the target vector and the time of flight are specified. It remained to be seen how many of the major orbital transfer maneuvers could be accomplished conceptually by a single impulsive velocity change, and if simple formulas could be obtained for the corresponding required velocities.

One by one, we accumulated suitable required velocity expressions for a variety of possible Apollo maneuvers. For example, when the Apollo command module returned to earth, it had to impact the atmosphere at a specified flight-path angle—otherwise it might either skip out of the atmosphere or be destroyed by overheating. A simple formula for $\mathbf{v}_r$ was obtained (see problem 3.11 in Ref. 10).

Braking into a circular lunar orbit was another mission requirement. We used for $\mathbf{v}_r$ the velocity the vehicle should have in order to be in a circular orbit at its present position and in a specified plane. In this manner, we were able to control the shape and orientation of the final orbit but not its radius. However, it turned out that an empirical relation could be found between the final radius and the pericenter of the approach trajectory so that the desired radius could be established by an appropriate selection of the approach orbit. (This technique was based on an idea developed during our low-thrust guidance work for Avco.)

On the first unmanned guided Apollo flight in August 1966, the required velocity vector was defined so as to achieve an orbit of specified eccentricity and semimajor axis. The list goes on, but does not include, for example, the lunar landing since this maneuver cannot be performed conceptually with a single impulsive burn.

We experimented with a variety of guidance laws to drive $\mathbf{v}_g$ to zero: 1) align the thrust acceleration $\mathbf{a}_T$ with the $\mathbf{v}_g$ direction; 2) direct $\mathbf{a}_T$ to cause $\mathbf{v}_g$ to be aligned with its derivative—cross-product steering; and 3) a combination of both as illustrated in Fig. 9. The scalar mixing parameter $\gamma$ was chosen empirically to maximize fuel economy. A constant $\gamma$ was usually sufficient for a particular mission phase; however, $\gamma$ could be allowed to vary as a function of some convenient system variable. Fred Martin found that this third method gave a highly efficient steering law that compared favorably with calculus of variations optimum solutions.[27]

A functional diagram illustrating the computation of the error signal required for control purposes is shown in Fig. 10. Numerical differentiation of the required velocity was simpler than programming the analytically obtained derivative. Near the end of the maneuver, when $v_g$ is small, cross-product steering is terminated, the vehicle holds a constant attitude,

# Astrodynamics

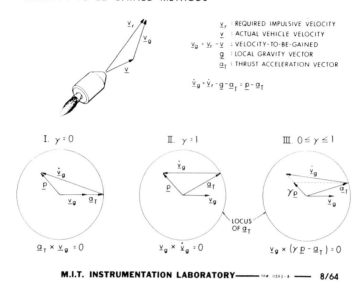

**Fig. 9:** Velocity-to-be-gained guidance laws (from Ref. 28).

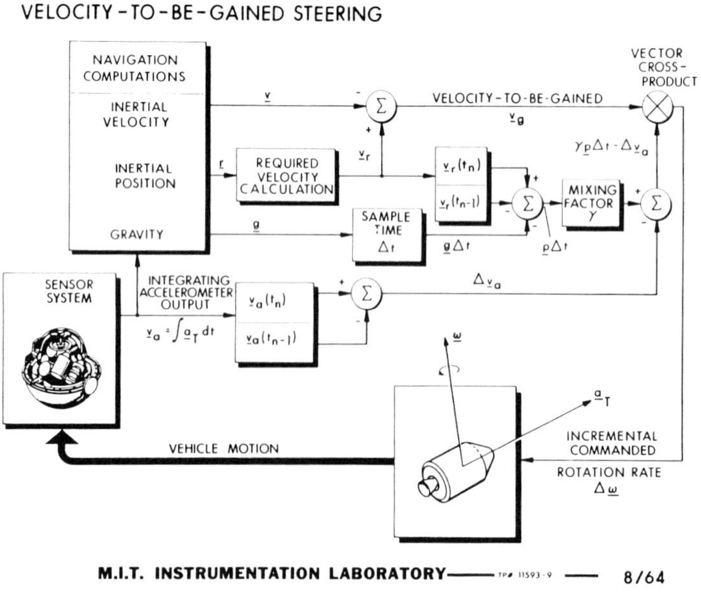

**Fig. 10:** Apollo cross-product steering (from Ref. 28).

and engine cut-off is made on the basis of the magnitude of the vector $\mathbf{v}_g$. A detailed description of just how this guidance scheme was mechanized in the AGC is provided in Ref. 28.

Steering to intercept a given target at a specified time came to be known as Lambert guidance after Johann Heinrich Lambert, the famous eighteenth century Alsatian scholar who discovered the theorem that bears his name. Since $\mathbf{v}_r$ had to be calculated cyclically in real time, Lambert guidance (which also required an iterative solution of Lambert's time equation) placed one of the heaviest burdens on the AGC. The task of completing all the necessary calculations in the time available became a programmer's nightmare. Ever since, the problem has fascinated me, and I am always on the lookout for new and better solutions of Lambert's equation.[29-32]

As the years went by, more and more of the guidance and navigation responsibilities were transferred from the onboard system to the Real Time Control Center (RTCC) in Houston. Much of the capability remained and was used in the AGC as a backup, but the RTCC was primary. Targeting calculations were made in the ground-based computers and Apollo performed most of its maneuvers in the so-called "external $\Delta v$" mode.

**The Circle Closes**

The intense pressure under which we all worked began to ease somewhat after the first landing on the moon. Timothy Brand, who played an active role in mechanizing the powered-flight AGC algorithms, had time now to reflect on Lambert guidance performance. Was it possible to avoid the frequent solutions of Lambert's problem, which were necessary to maintain an accurate value of the vector $\mathbf{v}_g$? What could be done about the small yet persistent error in cutoff when estimating the time to go until thrust termination? How can a more nearly constant attitude maneuver be attained that would avoid any relatively large turning rates?

Perhaps all of these difficulties had to do with the definition of the vector $\mathbf{v}_g$ itself. What if we defined a single coasting trajectory, which coincides with the powered trajectory at thrust termination, and used it as the basis of the velocity-to-be-gained computation? We would have then $\mathbf{v}_g = \mathbf{v}_r - \mathbf{v}$, but now $\mathbf{v}_r$ is the velocity along the single coasting path. If the corresponding position vector on the coasting trajectory is $\mathbf{r}'$, then

$$\frac{d\mathbf{v}_r}{dt} = \mathbf{g}(\mathbf{r}') \qquad (25)$$

would describe the rate of change of $\mathbf{v}_r$. The $\mathbf{v}_g$ equation would be

$$\frac{d\mathbf{v}_g}{dt} = \mathbf{g}(\mathbf{r}') - \mathbf{g}(\mathbf{r}) - \mathbf{a}_T = \Delta \mathbf{g} - \mathbf{a}_T \qquad (26)$$

with $\Delta \mathbf{g}$ replacing the term $-\mathbf{Q}\mathbf{v}_g$ in the older version.

The advantages of the new formulation became evident. An easy calculation showed that the contribution of the term $\Delta \mathbf{g}$ is generally much smaller than that of $\mathbf{Q}\mathbf{v}_g$. Furthermore, $\Delta \mathbf{g}$ approaches zero at a rate proportional to $v_g^2$, while the $\mathbf{Q}\mathbf{v}_g$ term, on the other hand, vanishes like $v_g$ to the first power. Simulations verified that $\Delta \mathbf{g}$ is so small for short maneuvers that a nearly constant attitude can be obtained by merely steering the vehicle so as to align the thrust vector along $v_g$. Velocity-to-be-gained, under these circumstances, is particularly easy to compute—the accelerometer-sensed velocity change is subtracted from the previous value of $\mathbf{v}_g$ on each computer guidance cycle.

Tim's technique[33] works well, even for long duration maneuvers, if we periodically create a new coasting-flight trajectory. A suitable approximation for $\Delta \mathbf{r} = \mathbf{r}' - \mathbf{r}$ is found to be

$$\Delta \mathbf{r} = -\frac{v_g}{2a_T}\mathbf{v}_g \qquad (27)$$

which, when added to current vehicle position, produces the position vector $\mathbf{r}'$. Knowing $\mathbf{r}'$ and the target $\mathbf{r}_T$, together with the time of flight, permits a new Lambert solution—hence a new $\mathbf{v}_r$ and a new coasting trajectory. Subtracting the current vehicle velocity provides an updated value of $\mathbf{v}_g$ with which to begin anew.

If none of these ideas seem familiar, you have forgotten the Convair legacy. What has just been described is essentially what the Convair engineers were advocating those many years ago. I must confess that I did not make the connection between Tim's new technique and the old Convair proposal until I began rummaging through my memorabilia in preparation for this paper. Obviously, Tim knew nothing of this—he was only about ten years old at the time.

Of course, the Tim Brand or the Convair scheme would have been impractical for an onboard implementation to guide the early ballistic missiles. It was feasible only after the small airborne digital computer replaced all those servos, amplifiers, potentiometers, and other analog devices of the good old days.

## References

[1] Laning, J. H., Jr. and Zierler, N., "A Program for Translation of Mathematical Equations for Whirlwind I," Engineering Memorandum E-364, MIT Instrumentation Laboratory, Cambridge, Mass., Jan. 1954.

[2] Knuth, D. E. and Pardo, L. T., "The Early Development of Programming Languages," STAN-CS-76-562, Stanford University, Stanford, Calif., Aug. 1976, pp. 55–60.

[3] Laning, J. H., Jr. and Miller, J. S., "The MAC Algebraic Language System" Report R-681, MIT Charles Stark Draper Laboratory, Cambridge, Mass., Nov. 1970.

⁴ Battin, R. H., O'Keefe, R. J., and Petrick, M. B., "The MIT Instrumentation Laboratory Automatic Coding 650 Program," *IBM Applied Science Division Technical Newsletter*, No. 10, Oct. 1955, pp. 63–79.

⁵ Battin, R. H., "Programming Manual for BALITAC 650 Routine," Report R-126, MIT Instrumentation Laboratory, Cambridge, Mass., Aug. 1956.

⁶ Laning, J. H., Jr. and Battin, R. H., "Theoretical Principle for a Class of Inertial Guidance Computers for Ballistic Missiles," Report R-125, MIT Instrumentation Laboratory, Cambridge, Mass., June 1956.

⁷ Laning, J. H., Jr. and Battin, R. H., "Optimum Fuel Trajectories for a $Q$-Type Guidance System," Engineering Memorandum E-520, MIT Instrumentation Laboratory, Cambridge, Mass., Feb. 1956.

⁸ Martin, F. H., "Closed-Loop Near-Optimum Steering for a Class of Space Missions," Doctor of Science Thesis, Massachusetts Institute of Technology, Cambridge, Mass., May 1965.

⁹ Laning, J. H., Jr. and Battin, R. H., "Theoretical Principle for a Class of Inertial Guidance Computers for Ballistic Missiles," Engineering Memorandum E-988, MIT Instrumentation Laboratory, Cambridge, Mass., 1960.

¹⁰ Battin, R. H., *Astronautical Guidance*, McGraw-Hill, New York, 1964.

¹¹ Laning, J. H., Jr., Frey, E. J., and Trageser, M. B., "Preliminary Considerations on the Instrumentation of a Photographic Reconnaissance of Mars," Report R-174, MIT Instrumentation Laboratory, Cambridge, Mass., April 1959.

¹² "A Recoverable Interplanetary Space Probe," Report R-235, MIT Instrumentation Laboratory, Cambridge, Mass., July 1959.

¹³ Whittaker, E. T., *A Treatise on the Analytical Dynamics of Particles and Rigid Bodies*, Cambridge University Press, Cambridge, England, 1937.

¹⁴ Battin, R. H., "The Determination of Round-Trip Planetary Reconnaissance Trajectories," *Journal of the Aerospace Sciences*, Vol. 26, September 1959, pp. 545–567.

¹⁵ Crocco, G. A., "One Year Exploration Trip Earth-Mars-Venus-Earth." *Proceedings of the Seventh International Astronautical Congress*, Rome, 1956, pp. 227–252.

¹⁶ Battin, R. H. and Laning, J. H., Jr., "A Navigation Theory for Round Trip Reconnaissance Missions to Venus and Mars," *Planetary and Space Science*, Vol. 7, Pergamon Press, London, 1961, pp. 40–56. (Originally published as MIT Instrumentation Laboratory Report R-240, Aug. 1959.)

¹⁷ Battin, R. H., "A Comparison of Fixed and Variable Time of Arrival Navigation for Interplanetary Flight," *Proceedings of the Fifth AFBMD/STL Aerospace Symposium on Ballistic Missile and Space Technology*, Academic Press, New York, 1960, pp. 3–31. (Originally published as MIT Instrumentation Laboratory Report R-283, May 1960.)

¹⁸ Alonso, R. and Laning, J. H., Jr., "Design Principles for a General Control Computer," Report R-276, MIT Instrumentation Laboratory, Cambridge, Mass., April 1960.

[19] Battin, R. H., "Computational Procedures for the Navigational Fix," Appendix B of "Interplanetary Navigation System Study," Report R-273, MIT Instrumentation Laboratory, Cambridge, Mass., April 1960.

[20] Kalman, R. E., "A New Approach to Linear Filtering and Prediction Problems," *Journal of Basic Engineering, Transactions of the American Society of Mechanical Engineers*, Vol. 82D, March 1960, pp. 35–45.

[21] Battin, R. H. and Miller, J. S., "Trajectory and Guidance Theory for a Lunar Reconnaissance Vehicle," Report E-993, MIT Instrumentation Laboratory, Cambridge, Mass., Jan. 1961.

[22] Miller, J. S., "Trajectory and Guidance Theory for a Low-Thrust Lunar Reconnaissance Vehicle," Report T-292, MIT Instrumentation Laboratory, Cambridge, Mass., Aug. 1961.

[23] Schmidt, S. F., "The Kalman Filter: Its Recognition and Development for Aerospace Applications," *Journal of Guidance and Control*, Vol. 4, Jan-Feb. 1981, pp. 4–7.

[24] Battin, R. H., "A Statistical Optimizing Navigation Procedure for Space Flight," *ARS Journal*, Vol. 32, Nov. 1962., pp. 1681–1692. Reprinted in *Kalman Filtering: Theory and Application*, edited by H.W. Sorenson—a volume in the IEEE PRESS Selected Reprint Series, 1985.

[25] Hoag, D. G., "The History of Apollo On-Board Guidance, Navigation and Control," *The Eagle Has Returned*, Science and Technology, Vol. 43, American Astronautical Society Publication, 1976. Reprinted in *Journal of Guidance, Control, and Dynamics*, Vol. 6, Jan.-Feb. 1983, pp. 4–13.

[26] Battin, R. H. and Levine, G. M., "Application of Kalman Filtering Techniques to the Apollo Program," *Theory and Applications of Kalman Filtering*, edited by C.T. Leondes, AGARD-ograph 139, 1970, Chap. 14.

[27] Martin, F. H., "Closed-Loop Near-Optimum Steering for a Class of Space Missions," *AIAA Journal*, Vol. 4, Nov. 1966, pp. 1920–1927.

[28] Martin, F. H. and Battin, R. H., "Computer-Controlled Steering of the Apollo Spacecraft," *Journal of Spacecraft and Rockets*, Vol. 5, April 1968, pp. 400–407.

[29] Battin, R. H., "A New Solution for Lambert's Problem," *Proceedings of the XIX International Astronautical Congress*, Vol. 2, Astrodynamics and Astrionics, Pergamon Press, New York, 1970, pp. 131–150.

[30] Battin, R. H., "Lambert's Problem Revisited," *AIAA Journal*, Vol. 15, May 1977, pp. 707–713.

[31] Battin, R. H., Fill, T. S., and Shepperd, S. W., "A New Transformation Invariant in the Orbital Boundary-Value Problem," *Journal of Guidance and Control*, Vol. 1, Jan.-Feb. 1978, pp. 50–55.

[32] Battin, R. H. and Vaughan, R. M., "An Elegant Lambert Algorithm," *Journal of Guidance, Control, and Dynamics*, Vol. 7, Nov.-Dec. 1984, pp. 662–670.

[33] Brand, T. J., "A New Approach to Lambert Guidance," Report R-694, MIT Charles Stark Draper Laboratory, Cambridge, Mass., June 1971.

PART I

Chapter 1

# Hypergeometric Functions and Elliptic Integrals

REGRETFULLY, HYPERGEOMETRIC FUNCTIONS, CONTINUED FRACTION expansions, and elliptic integrals have received minor, if any, attention in the education of the modern engineer. They do, however, play an important role in many aspects of Astrodynamics. As examples: Gauss' classical solution to the two-body, two-point, time-constrained boundary-value problem relies heavily on a particular continued fraction expansion of the ratio of two contiguous hypergeometric functions; and, the gravitational attraction of a solid homogeneous ellipsoid upon an exterior particle is represented in terms of elliptic integrals.

Continued fraction expansions are, also, not given the prominence they deserve in the university curricula despite the fact that they are, generally, far more efficient tools for evaluating the classical functions than the more familiar infinite power series. Their convergence is typically faster and more extensive than the series and, ironically, they were in use centuries before the invention of the power series.

We shall have a number of occasions throughout this book to utilize these mathematical entities. It seems, therefore, appropriate to devote this first chapter to their development and application. Both have a long and fascinating history with contributions made by many of the world's best mathematicians. Here, we develop some of the properties of hypergeometric functions, their transformations, and continued fraction expansions. Later, as the occasions arise, we provide appropriate continued fraction representations of the special functions needed in Astrodynamics.

In this chapter, we also derive several of the convergence tests for infinite continued fractions and consider, as well, several algorithms for their evaluation—one of which is quite recent in development.

The elementary problem in analytical mechanics—the motion of the simple pendulum—cannot be accurately described without resorting to elliptic integrals. The small amplitude assumption is introduced so subtly that the student easily forgets that the simple formula for the period is only approximately correct. Our treatment of elliptic integrals is far from complete and concentrates primarily on the truly delightful algorithms, devised by Gauss, by which they can be evaluated.

## 1.1 Hypergeometric Functions

The term *hypergeometric series* was first given by the famous English mathematician John Wallis† in 1655 to the series whose $n^{\text{th}}$ term is

$$a(a+b)(a+2b)\ldots[a+(n-1)b]$$

in an effort to generalize the familiar geometric series

$$1 + x + x^2 + x^3 + \cdots$$

The modern use of the term applies to the series

$$1 + \frac{\alpha\beta}{\gamma}\frac{x}{1!} + \frac{\alpha(\alpha+1)\beta(\beta+1)}{\gamma(\gamma+1)}\frac{x^2}{2!} + \frac{\alpha(\alpha+1)(\alpha+2)\beta(\beta+1)(\beta+2)}{\gamma(\gamma+1)(\gamma+2)}\frac{x^3}{3!} + \cdots$$

which is easily seen by the ratio test to be absolutely convergent for $|x| < 1$. Within this interval, the series defines a function denoted by $F(\alpha, \beta; \gamma; x)$ which was called the *hypergeometric function* by Johann Friedrich Pfaff (1765–1825), a friend and teacher of Carl Friedrich Gauss.

### Examples of Hypergeometric Functions

The hypergeometric function is of great importance because it is a generalization of many of the familiar (and not so familiar) mathematical functions. By comparing the Taylor series expansion of a function with the hypergeometric series, we can frequently identify specific values of $\alpha$, $\beta$, and $\gamma$ for which the two series will be identical. For example, the geometric series

$$\frac{1}{1-x} = 1 + x + x^2 + \cdots = F(1, \beta; \beta; x)$$

is so represented. More generally,

$$(1-x)^{-\alpha} = 1 + \alpha x + \alpha(\alpha+1)\frac{x^2}{2!} + \alpha(\alpha+1)(\alpha+2)\frac{x^3}{3!} + \cdots$$
$$= F(\alpha, \beta; \beta; x) \tag{1.1}$$

and

$$(1+x)^{-\alpha} = 1 - \alpha x + \alpha(\alpha+1)\frac{x^2}{2!} - \alpha(\alpha+1)(\alpha+2)\frac{x^3}{3!} + \cdots$$
$$= F(\alpha, \beta; \beta; -x) \tag{1.2}$$

---

† John Wallis (1616–1703), Savilian professor of geometry at Oxford for 50 years and a contemporary of Sir Isaac Newton, ranks next to Newton as the ablest British mathematician of the seventeenth century. One of his notable contributions is the remarkable expression for $\pi$ known as *Wallis' theorem*:

$$\frac{\pi}{2} = \frac{2 \cdot 2 \cdot 4 \cdot 4 \cdot 6 \cdot 6 \cdot 8 \cdot 8 \cdots}{1 \cdot 3 \cdot 3 \cdot 5 \cdot 5 \cdot 7 \cdot 7 \cdot 9 \cdots}$$

By adding and subtracting, we also obtain

$$\frac{1}{2}[(1-x)^{-\alpha} + (1+x)^{-\alpha}] = 1 + \alpha(\alpha+1)\frac{x^2}{2!}$$

$$+ \alpha(\alpha+1)(\alpha+2)(\alpha+3)\frac{x^4}{4!} + \cdots = F\left(\frac{\alpha}{2}, \frac{\alpha+1}{2}; \frac{1}{2}; x^2\right) \quad (1.3)$$

$$\frac{1}{2\alpha x}[(1-x)^{-\alpha} - (1+x)^{-\alpha}] = 1 + (\alpha+1)(\alpha+2)\frac{x^2}{3!}$$

$$+ (\alpha+1)(\alpha+2)(\alpha+3)(\alpha+4)\frac{x^4}{5!} + \cdots = F\left(\frac{\alpha+1}{2}, \frac{\alpha+2}{2}; \frac{3}{2}; x^2\right) \quad (1.4)$$

The logarithm and inverse trigonometric functions can be similarly represented. Specifically,

$$\frac{1}{x}\log(1+x) = 1 - \frac{x}{2} + \frac{x^2}{3} - \frac{x^3}{4} + \cdots$$
$$= F(1,1;2;-x) \quad (1.5)$$

$$\frac{1}{x}\arctan x = 1 - \frac{x^2}{3} + \frac{x^4}{5} - \frac{x^6}{7} + \cdots$$
$$= F(\tfrac{1}{2}, 1; \tfrac{3}{2}; -x^2) \quad (1.6)$$

$$\frac{1}{x}\arcsin x = 1 + \frac{1}{2}\frac{x^2}{3} + \frac{1\cdot 3}{2\cdot 4}\frac{x^4}{5} + \frac{1\cdot 3\cdot 5}{2\cdot 4\cdot 6}\frac{x^6}{7} + \cdots$$
$$= F(\tfrac{1}{2}, \tfrac{1}{2}; \tfrac{3}{2}; x^2) \quad (1.7)$$

The inverse hyperbolic functions are closely related to the inverse trigonometric functions. Indeed, we have

$$\frac{1}{x}\operatorname{arctanh} x = F(\tfrac{1}{2}, 1; \tfrac{3}{2}; x^2) \quad (1.8)$$

$$\frac{1}{x}\operatorname{arcsinh} x = F(\tfrac{1}{2}, \tfrac{1}{2}; \tfrac{3}{2}; -x^2) \quad (1.9)$$

◇ **Problem 1–1**

Define

$$x = \frac{e^{i\phi} - e^{-i\phi}}{e^{i\phi} + e^{-i\phi}} = i\tan\phi \quad \text{so that} \quad \frac{1-x}{1+x} = e^{-2i\phi}$$

Then use Eqs. (1.3) and (1.4) to obtain

$$\frac{\tan k\phi}{k\tan\phi} = \frac{F\left(\frac{1-k}{2}, \frac{2-k}{2}; \frac{3}{2}; -\tan^2\phi\right)}{F\left(\frac{1-k}{2}, -\frac{k}{2}; \frac{1}{2}; -\tan^2\phi\right)}$$

which we shall need later to develop a beautiful continued fraction found by Leonhard Euler.

## Gauss' Relations for Contiguous Functions

The six functions
$$F^{\alpha\pm} = F(\alpha \pm 1, \beta; \gamma; x)$$
$$F^{\beta\pm} = F(\alpha, \beta \pm 1; \gamma; x)$$
$$F^{\gamma\pm} = F(\alpha, \beta; \gamma \pm 1; x)$$
are called *contiguous* to $F = F(\alpha, \beta; \gamma; x)$. Gauss† discovered that a linear relationship exists between $F$ and any pair of contiguous functions. There are fifteen such linear relations whose coefficients are rational functions of $\alpha$, $\beta$, $\gamma$. They are as follows:

(1) $[2\alpha - \gamma + (\beta - \alpha)x]F = \alpha(1-x)F^{\alpha+} + (\alpha - \gamma)F^{\alpha-}$

(2) $(\beta - \alpha)F = \beta F^{\beta+} - \alpha F^{\alpha+}$

(3) $(\gamma - \alpha - \beta)F = (\gamma - \beta)F^{\beta-} - \alpha(1-x)F^{\alpha+}$

(4) $\gamma[\alpha - (\gamma - \beta)x]F = \alpha\gamma(1-x)F^{\alpha+} - (\gamma - \alpha)(\gamma - \beta)xF^{\gamma+}$

(5) $(\gamma - \alpha - 1)F = (\gamma - 1)F^{\gamma-} - \alpha F^{\alpha+}$

(6) $(\gamma - \alpha - \beta)F = (\gamma - \alpha)F^{\alpha-} - \beta(1-x)F^{\beta+}$

(7) $(\beta - \alpha)(1-x)F = (\gamma - \alpha)F^{\alpha-} - (\gamma - \beta)F^{\beta-}$

(8) $\gamma(1-x)F = \gamma F^{\alpha-} - (\gamma - \beta)xF^{\gamma+}$

(9) $[\alpha - 1 - (\gamma - \beta - 1)x]F = (\gamma - 1)(1-x)F^{\gamma-} - (\gamma - \alpha)F^{\alpha-}$

(10) $[\gamma - 2\beta + (\beta - \alpha)x]F = (\gamma - \beta)F^{\beta-} - \beta(1-x)F^{\beta+}$

(11) $\gamma[\beta - (\gamma - \alpha)x]F = \beta\gamma(1-x)F^{\beta+} - (\gamma - \alpha)(\gamma - \beta)xF^{\gamma+}$

(12) $(\gamma - \beta - 1)F = (\gamma - 1)F^{\gamma-} - \beta F^{\beta+}$

(13) $\gamma(1-x)F = \gamma F^{\beta-} - (\gamma - \alpha)xF^{\gamma+}$

(14) $[\beta - 1 - (\gamma - \alpha - 1)x]F = (\gamma - 1)(1-x)F^{\gamma-} - (\gamma - \beta)F^{\beta-}$

(15) $\gamma[\gamma - 1 - (2\gamma - \alpha - \beta - 1)x]F =$
$\gamma(\gamma - 1)(1-x)F^{\gamma-} - (\gamma - \alpha)(\gamma - \beta)xF^{\gamma+}$

To verify these identities, it is convenient to introduce the symbol $(\alpha)_k$, defined as

$$(\alpha)_0 = 1 \quad (\alpha)_k = \alpha(\alpha+1)(\alpha+2)\ldots(\alpha+k-1) \quad (1.10)$$

which permits the general term of the hypergeometric series to be written

---

† Gauss' paper *Disquisitiones Generales Circa Seriem Infinitam* on the hypergeometric function in 1812 was also the first important and strictly rigorous study of the convergence of infinite series. He is generally regarded as the first to recognize the need to restrict the use of series to their regions of convergence, and, for the hypergeometric series, showed that it converges for $|x| < 1$ and diverges for $|x| > 1$. At the endpoints $x = 1$ and $x = -1$ he found that the series converges if and only if $\alpha + \beta < \gamma$ and $\alpha + \beta < \gamma + 1$, respectively.

in the compact form

$$F(\alpha, \beta; \gamma; x) = \sum_{k=0}^{\infty} F_k \frac{x^k}{k!} \quad \text{where} \quad F_k = \frac{(\alpha)_k (\beta)_k}{(\gamma)_k} \quad (1.11)$$

With this notation, the contiguous functions may be written as

$$F^{\alpha+} \equiv F(\alpha+1, \beta; \gamma; x) = \sum_{k=0}^{\infty} F_k^{\alpha+} \frac{x^k}{k!}$$

for example. Also, since $(\alpha)_k$ has the readily verified properties

$$(\alpha)_{k+1} = \alpha(\alpha+1)_k = (\alpha+k)(\alpha)_k$$

we may write

$$F_k^{\alpha+} = \frac{\alpha+k}{\alpha} F_k \qquad F_k^{\alpha-} = \frac{\alpha-1}{\alpha-1+k} F_k$$

$$F_k^{\beta+} = \frac{\beta+k}{\beta} F_k \qquad F_k^{\beta-} = \frac{\beta-1}{\beta-1+k} F_k$$

$$F_k^{\gamma+} = \frac{\gamma}{\gamma+k} F_k \qquad F_k^{\gamma-} = \frac{\gamma-1+k}{\gamma-1} F_k$$

Further, it is easy to verify that

$$xF = \sum_{k=0}^{\infty} \frac{k(\gamma-1+k)}{(\alpha-1+k)(\beta-1+k)} F_k \frac{x^k}{k!}$$

and, for the contiguous functions,

$$xF^{\alpha+} = \sum_{k=0}^{\infty} \frac{k(\gamma-1+k)}{\alpha(\beta-1+k)} F_k \frac{x^k}{k!}$$

$$(1-x)F^{\alpha+} = \sum_{k=0}^{\infty} \left[1 + \frac{k(\beta-\gamma)}{\alpha(\beta-1+k)}\right] F_k \frac{x^k}{k!}$$

for example. Thus, to establish Eq. (3)

$$(\gamma - \alpha - \beta)F = (\gamma - \beta)F^{\beta-} - \alpha(1-x)F^{\alpha+}$$

as an illustration, we need only to verify the relation

$$\gamma - \alpha - \beta = (\gamma - \beta)\frac{\beta-1}{\beta-1+k} - \alpha\left[1 + \frac{k(\beta-\gamma)}{\alpha(\beta-1+k)}\right]$$

The other fourteen identities are similarly handled.

The importance of these relationships is that through repeated application, any function $F(\alpha+l, \beta+m; \gamma+n; x)$ for integral $l$, $m$, $n$ can be expressed as a linear combination of $F(\alpha, \beta; \gamma; x)$ and one of its contiguous functions.

◇ **Problem 1–2**

Later in this section, when we derive Gauss' differential equation for hypergeometric functions, we shall need certain relations which are the subject of this problem. First obtain

$$\frac{d}{dx}F(\alpha,\beta;\gamma;x) = \frac{\alpha\beta}{\gamma}F(\alpha+1,\beta+1;\gamma+1;x)$$

and then verify, either directly or using Gauss' relations, that

$$xF(\alpha+1,\beta+1;\gamma+1;x) = \frac{\gamma}{\alpha}(F^{\beta+} - F) = \frac{\gamma}{\beta}(F^{\alpha+} - F)$$

◇ **Problem 1–3**

The following two identities are fundamental in developing Gauss' continued fraction expansion theorem:

$$F(\alpha,\beta+1;\gamma+1;x) - F(\alpha,\beta;\gamma;x) = \frac{\alpha(\gamma-\beta)}{\gamma(\gamma+1)}xF(\alpha+1,\beta+1;\gamma+2;x)$$

and

$$F(\alpha+1,\beta+1;\gamma+2;x) - F(\alpha,\beta+1;\gamma+1;x) =$$
$$\frac{(\beta+1)(\gamma+1-\alpha)}{(\gamma+1)(\gamma+2)}xF(\alpha+1,\beta+2;\gamma+3;x)$$

### Gauss' Differential Equation

Using some of the identities established previously, we can derive the differential equation satisfied by the hypergeometric function. For this purpose, let

$$y = F(\alpha,\beta;\gamma;x)$$

and use the results of Prob. 1–2 to obtain

$$\frac{dy}{dx} = \frac{\alpha\beta}{\gamma}F(\alpha+1,\beta+1;\gamma+1;x)$$

$$x\frac{dy}{dx} = \beta F(\alpha,\beta+1;\gamma;x) - \beta y$$

Then by defining

$$z = x\frac{dy}{dx} + \beta y$$

we have

$$\frac{dz}{dx} = \frac{\alpha\beta(\beta+1)}{\gamma}F(\alpha+1,\beta+2;\gamma+1;x)$$

$$x\frac{dz}{dx} = \alpha\beta F(\alpha+1,\beta+1;\gamma;x) - \alpha z$$

Now, multiply this last equation by $1-x$ and use Gauss' identity (3) with $\beta$ replaced by $\beta+1$, to obtain

$$x(1-x)\frac{dz}{dx} = (\gamma - \beta - 1)(\beta y - z) + \alpha x z$$

Finally, replace $z$ by $x(dy/dx) + \beta y$. There results

$$x(1-x)\frac{d^2 y}{dx^2} + [\gamma - (\alpha + \beta + 1)x]\frac{dy}{dx} - \alpha\beta y = 0 \qquad (1.12)$$

as the desired differential equation. Although Euler treated this equation and its series solution much earlier, it is, nevertheless, named for Gauss.

Gauss' equation has *regular singular points* at 0, 1, and $\infty$. Since it is of second order, there are two linearly independent solutions of the form

$$y = x^\rho(c_0 + c_1 x + c_2 x^2 + \cdots)$$

Using the so-called *method of Frobenius*,† we substitute in Eq. (1.12) and require that the coefficients of all powers of $x$ must vanish. We find that the exponent $\rho$ must satisfy the so-called *indicial equation*

$$\rho(\rho - 1 + \gamma) = 0$$

whose roots are $\rho = 0$ and $\rho = 1 - \gamma$. The first corresponds to the hypergeometric series $F(\alpha, \beta; \gamma; x)$ from which the differential equation was derived. The second corresponds to the solution

$$y = x^{1-\gamma} F(\alpha - \gamma + 1, \beta - \gamma + 1; 2 - \gamma; x)$$

unless $\gamma$ is a positive integer. If $\gamma = 1$, the two solutions are identical. If $\gamma$ is a negative integer, then the first solution $F(\alpha, \beta; \gamma; x)$ does not exist.

As an example of the use of Gauss' equation, consider the equation

$$\frac{d^2 y}{dx^2} + n^2 y = 0$$

satisfied by both $\sin nx$ and $\cos nx$. We transform the equation by changing the independent variable $x$ to $q$ where

$$q = \sin^2 x$$

Then

$$\frac{dy}{dx} = \sin 2x \frac{dy}{dq}$$

$$\frac{d^2 y}{dx^2} = 2\cos 2x \frac{dy}{dq} + \sin^2 2x \frac{d^2 y}{dq^2}$$

---

† Georg Ferdinand Frobenius (1849–1917), professor at the University of Berlin, is noted, chiefly, for the modern concept of abstract structures in mathematics developed during his major achievements in Group Theory.

But
$$\sin^2 2x = 4q(1-q) \quad \text{and} \quad \cos 2x = 1 - 2q$$
so that $y$ as a function of $q$ satisfies
$$q(1-q)\frac{d^2y}{dq^2} + (\tfrac{1}{2} - q)\frac{dy}{dq} + \tfrac{1}{4}n^2 y = 0$$
—a special case of Gauss' equation with
$$\alpha = \tfrac{1}{2}n \qquad \beta = -\tfrac{1}{2}n \qquad \gamma = \tfrac{1}{2}$$
for which the general solution is
$$y = c_1 F\left(\frac{n}{2}, -\frac{n}{2}; \frac{1}{2}; \sin^2 x\right) + c_2 \sin x \, F\left(\frac{1+n}{2}, \frac{1-n}{2}; \frac{3}{2}; \sin^2 x\right)$$

Now, $y = \sin nx$ is a solution so that $c_1$ must be zero since $y = 0$ when $x = 0$. Hence
$$\frac{\sin nx}{\sin x} = c_2 F\left(\frac{1+n}{2}, \frac{1-n}{2}; \frac{3}{2}; \sin^2 x\right)$$
To determine $c_2$, let $x$ tend to zero and we find that $c_2 = n$.

Again $y = \cos nx$ is also a solution. Setting $x = 0$, we obtain $c_1 = 1$. By differentiating and setting $x = 0$, we find that $c_2 = 0$.

In summary, then, we have obtained
$$\sin nx = n \sin x \, F\left(\frac{1+n}{2}, \frac{1-n}{2}; \frac{3}{2}; \sin^2 x\right) \tag{1.13}$$
$$\cos nx = F\left(\frac{n}{2}, -\frac{n}{2}; \frac{1}{2}; \sin^2 x\right) \tag{1.14}$$
which are valid for $-\tfrac{1}{2}\pi \leq x \leq \tfrac{1}{2}\pi$, and will prove of great value in developing continued fraction solutions of those cubic equations which are important in orbital mechanics.

### Bilinear Transformation Formulas

In this and the next subsection, we shall develop transformations of the hypergeometric functions which replace the argument $x$ with bilinear and quadratic functions of $x$. Among the possible applications, this will allow an extension of the interval of convergence of the power series representation of these functions.

The simple bilinear transformation
$$q = \frac{x}{x-1} \quad \text{or} \quad x = \frac{q}{q-1}$$
has properties such as
$$1 - x = \frac{1}{1-q} \quad \text{and} \quad \frac{dx}{dq} = -\frac{1}{(1-q)^2} = -(1-x)^2$$

which motivate a transformation of Gauss' equation. Specifically, we ask for what values, if any, of the constant $c$ will the function

$$z = (1-x)^c y$$

be a hypergeometric function of $q$ if $y = F(\alpha, \beta; \gamma; x)$?

For this investigation, we calculate

$$y = (1-q)^c z$$

$$\frac{dy}{dx} = (1-q)^{c+1}\left[cz - (1-q)\frac{dz}{dq}\right]$$

$$\frac{d^2y}{dx^2} = (1-q)^{c+2}\left[c(c+1)z - 2(c+1)(1-q)\frac{dz}{dq} + (1-q)^2\frac{d^2z}{dq^2}\right]$$

and substitute into Gauss' equation (1.12) to obtain, thereby,

$$q(1-q)^2\frac{d^2z}{dq^2} + (1-q)\left[\gamma - (\gamma + 2c - \alpha - \beta + 1)q\right]\frac{dz}{dq} - Rz = 0$$

where

$$R = (c\gamma - \alpha\beta) - cq(c + \gamma - \alpha - \beta)$$

Now, the differential equation for $z$ as a function of $q$ will be of the form of Gauss' equation provided $R$, the coefficient of $z$, has $1-q$ as a factor. This occurs for exactly two values of $c$—namely,

$$c = \alpha: \qquad R = \alpha(\gamma - \beta)(1-q)$$
$$c = \beta: \qquad R = \beta(\gamma - \alpha)(1-q)$$

The corresponding solutions for $z(q)$ are then

$$c = \alpha: \qquad z = F(\alpha, \gamma - \beta; \gamma; q)$$
$$c = \beta: \qquad z = F(\gamma - \alpha, \beta; \gamma; q)$$

As a consequence, we have derived the following two transformations of the hypergeometric function:

$$F(\alpha, \beta; \gamma; x) = (1-x)^{-\alpha}F\left(\alpha, \gamma - \beta; \gamma; \frac{x}{x-1}\right) \qquad (1.15)$$

$$F(\alpha, \beta; \gamma; x) = (1-x)^{-\beta}F\left(\gamma - \alpha, \beta; \gamma; \frac{x}{x-1}\right) \qquad (1.16)$$

Finally, there is an important consequence of these last two results. The right-hand sides of Eqs. (1.15) and (1.16) are equal so that

$$(1-q)^\alpha F(\alpha, \gamma - \beta; \gamma; q) = (1-q)^\beta F(\gamma - \alpha, \beta; \gamma; q)$$

Therefore, if we replace $\beta$ by $\gamma - \beta$ and $q$ by $x$, we have

$$F(\alpha, \beta; \gamma; x) = (1-x)^{\gamma - \alpha - \beta}F(\gamma - \alpha, \gamma - \beta; \gamma; x) \qquad (1.17)$$

—a fundamental relation first discovered by Euler.

## Quadratic Transformation Formulas

The transformation

$$q = 4x(1-x) \quad \text{or} \quad 1-q = (1-2x)^2$$

applied to Gauss' equation leads to other useful identities for hypergeometric functions. Since

$$\frac{dy}{dx} = \frac{dq}{dx}\frac{dy}{dq} = 4\sqrt{1-q}\,\frac{dy}{dq}$$

then Gauss' equation becomes

$$q(1-q)\frac{d^2y}{dq^2} + [\gamma - \tfrac{1}{2}(\alpha+\beta+1)]\sqrt{1-q}\,\frac{dy}{dq}$$
$$+ [\tfrac{1}{2}(\alpha+\beta+1) - \tfrac{1}{2}(\alpha+\beta+2)q]\frac{dy}{dq} - \tfrac{1}{4}\alpha\beta y = 0$$

If the term containing $\sqrt{1-q}$ as a factor were missing, the differential equation for $y(q)$ would also be Gauss' equation with different parameters. Thus, by requiring

$$\gamma = \tfrac{1}{2}(\alpha+\beta+1)$$

the differential equation for $y$ as a function of $q$ becomes

$$q(1-q)\frac{d^2y}{dq^2} + [\gamma - (\tfrac{1}{2}\alpha + \tfrac{1}{2}\beta + 1)q]\frac{dy}{dq} - \tfrac{1}{2}\alpha\,\tfrac{1}{2}\beta\, y = 0$$

Therefore, we have

$$y(q) = F\left[\tfrac{1}{2}\alpha,\,\tfrac{1}{2}\beta;\,\tfrac{1}{2}(\alpha+\beta+1);\,q\right]$$

provided that

$$y(x) = F\left[\alpha,\beta;\,\tfrac{1}{2}(\alpha+\beta+1);\,x\right]$$

Finally, then, two equivalent identities are obtained—namely

$$F\left[\alpha,\beta;\,\tfrac{1}{2}(\alpha+\beta+1);\,x\right] = F\left[\tfrac{1}{2}\alpha,\,\tfrac{1}{2}\beta;\,\tfrac{1}{2}(\alpha+\beta+1);\,4x(1-x)\right] \quad (1.18)$$

$$F(\alpha,\beta;\,\alpha+\beta+\tfrac{1}{2};\,x) = F\left[2\alpha,2\beta;\,\alpha+\beta+\tfrac{1}{2};\,\tfrac{1}{2}(1-\sqrt{1-x})\right] \quad (1.19)$$

◇ **Problem 1-4**
Obtain the identity

$$F(\alpha,\beta;\,\alpha+\beta-\tfrac{1}{2};\,x) = (1-x)^{-\tfrac{1}{2}} F\left[2\alpha-1,\,2\beta-1;\,\alpha+\beta-\tfrac{1}{2};\,\tfrac{1}{2}(1-\sqrt{1-x})\right]$$

by first using Eq. (1.17) and then Eq. (1.19).

## Confluent Hypergeometric Functions

The function
$$M(\beta, \gamma, x) = \lim_{\alpha \to \infty} F(\alpha, \beta; \gamma; x/\alpha) \qquad (1.20)$$
is called a *confluent hypergeometric function* and the series representation
$$M(\beta, \gamma, x) = 1 + \frac{\beta}{\gamma}\frac{x}{1!} + \frac{\beta(\beta+1)}{\gamma(\gamma+1)}\frac{x^2}{2!} + \cdots \qquad (1.21)$$
is convergent for all values of $x$. The same limiting process applied to Gauss' equation results in
$$x\frac{d^2y}{dx^2} + (\gamma - x)\frac{dy}{dx} - \beta y = 0 \qquad (1.22)$$
as the differential equation for the confluent hypergeometric function. The equation has a *regular singular point* at $x = 0$. The singularity of Gauss' equation at $x = 1$ has come into confluence with the singularity at $x = \infty$. The word "confluence" implies a flowing together or a coming together of the two regular singular points at one and infinity. However, the singularity of Eq. (1.22) at $x = \infty$ is now an *irregular singular point*. Using the method of Frobenius, the general solution is found to be
$$y = c_1 M(\beta, \gamma, x) + c_2 x^{1-\gamma} M(\beta - \gamma + 1, 2 - \gamma, x)$$

Linear relations between $M(\beta, \gamma, x)$ and pairs of the four contiguous functions $M(\beta \pm 1, \gamma, x)$, $M(\beta, \gamma \pm 1, x)$ can be obtained by the limiting process applied to an appropriate subset of the fifteen corresponding identities of Gauss. There will, of course, be only six such relationships for the confluent hypergeometric functions.

◊ **Problem 1–5**
Verify that the function
$$N(\gamma, x) = \lim_{\beta \to \infty} M(\beta, \gamma, x/\beta)$$
$$= 1 + \frac{1}{\gamma}\frac{x}{1!} + \frac{1}{\gamma(\gamma+1)}\frac{x^2}{2!} + \frac{1}{\gamma(\gamma+1)(\gamma+2)}\frac{x^3}{3!} + \cdots$$
converges for all $x$ and satisfies the differential equation
$$x\frac{d^2y}{dx^2} + \gamma\frac{dy}{dx} - y = 0$$

## 1.2 Continued Fraction Expansions

The *Fibonacci series*

$$0, 1, 1, 2, 3, 5, 8, 13, 21, 34, 55, \ldots, F_n, F_{n+1}, \ldots$$

which was first obtained by Leonardo of Pisa† in the thirteenth century as the solution of a certain rabbit-breeding problem, provides an excellent introduction to the Gauss continued fraction expansion theorem. Each term in the series is the sum of the two previous terms—or, equivalently, each term is the difference between the two terms on either side. Thus,

$$F_{n+1} = F_n + F_{n-1} \quad \text{where} \quad F_0 = 0 \quad \text{and} \quad F_1 = 1 \qquad (1.23)$$

Equation (1.23) is a linear constant coefficient difference equation which can be solved by seeking solutions of the form $F_n = c\beta^n$. Substituting this into the difference equation, we obtain an algebraic equation for $\beta$

$$c\beta^{n+1} - c\beta^n - c\beta^{n-1} = 0$$

or

$$\beta^2 - \beta - 1 = 0$$

which has two solutions: $\beta = \frac{1}{2}(1 \pm \sqrt{5})$. The general solution of (1.23) is, consequently, of the form

$$F_n = c_1 \left(\frac{1+\sqrt{5}}{2}\right)^n + c_2 \left(\frac{1-\sqrt{5}}{2}\right)^n$$

The constants $c_1$ and $c_2$ are obtained from the conditions $F_0 = 0$ and $F_1 = 1$; hence, $c_1 = -c_2 = 1/\sqrt{5}$. Therefore, the general term in the Fibonacci series is

$$F_n = \frac{\sqrt{5}}{5}\left[\left(\frac{1+\sqrt{5}}{2}\right)^n - \left(\frac{1-\sqrt{5}}{2}\right)^n\right] \qquad (1.24)$$

Now let $G_n$ be the ratio of two successive terms in the series so that

$$G_n = \frac{F_{n+1}}{F_n} \quad \text{for} \quad n = 1, 2, 3, \ldots \qquad (1.25)$$

Then, from Eq. (1.23), we have

$$\frac{F_{n+1}}{F_n}\frac{F_n}{F_{n-1}} = 1 + \frac{F_n}{F_{n-1}} \quad \text{or} \quad G_n G_{n-1} = 1 + G_{n-1}$$

---

† Leonardo of Pisa, better known as Fibonacci (literally meaning "son of Bonaccio") was the greatest mathematician of the Middle Ages. His best known work, *Liber abaci* completed in Pisa in the year 1202, defended the merits of the Hindu-Arabic decimal system of numbers over the clumsy Roman system still in use in Italy at the time. Although he made many valuable contributions to mathematics, he is mainly remembered today because of the number sequence which bears his name.

Sect. 1.2]     Continued Fraction Expansions     45

which provides the recursive relation
$$G_n = 1 + \frac{1}{G_{n-1}} \quad \text{for} \quad n = 2, 3, 4, \ldots$$

Therefore,
$$G_1 = 1 \quad G_2 = 1 + \frac{1}{1} = 2 \quad G_3 = 1 + \frac{1}{1 + \frac{1}{1}} = \frac{3}{2}$$

$$G_4 = 1 + \frac{1}{1 + \frac{1}{1 + \frac{1}{1}}} = \frac{5}{3} \quad \text{etc.}$$

From the definition of $G_n$ and Eq. (1.24), we have
$$G_n = \frac{1+\sqrt{5}}{2}\left(\frac{1-x^{n+1}}{1-x^n}\right) \quad \text{where} \quad x = \frac{1-\sqrt{5}}{1+\sqrt{5}}$$

and, since $|x| < 1$,
$$\lim_{n \to \infty} G_n = \frac{1+\sqrt{5}}{2}$$

The number $\frac{1}{2}(1+\sqrt{5})$ is called the *Golden Section* and has a fascinating history† which, unfortunately, we cannot afford the space to develop here.

Suffice it to say that the Golden Section can be expressed as the simplest possible of all continued fractions:
$$\text{Golden Section} = \frac{1+\sqrt{5}}{2} = 1 + \cfrac{1}{1 + \cfrac{1}{1 + \cfrac{1}{1 + \cfrac{1}{1 + \cdots}}}} \quad (1.26)$$

---

† If a line segment is divided into two parts of lengths $x$ and $y$ such that the ratio of the whole to the greater part $x$ is the same as the ratio of the greater to the lesser part, i.e., $(x+y)/x = x/y$, then $x/y = \frac{1}{2}(1+\sqrt{5})$. This was called the "sacred ratio" in the Papyrus of Ahmes, which gives an account of the building of the Great Pyramid of Gizeh about 3070 B.C., and the Golden Section by the ancient Greeks. Most books on recreational mathematics (for example, Martin Gardner's book *Mathematical Circus* published in 1979 by Alfred A. Knopf) will provide the reader with many fascinating properties of Fibonacci numbers and the Golden Section.

## ◊ Problem 1-6

Although continued fractions were familiar to the Hindu mathematicians as early as the fifth century, it was not until the sixteenth century that they were used to approximate irrational numbers. Besides Eq. (1.26), other interesting continued fractions can be obtained. For example,

$$\sqrt{2} = 1 + (\sqrt{2} - 1) = 1 + \cfrac{1}{\sqrt{2} + 1} = 1 + \cfrac{1}{2 + (\sqrt{2} - 1)}$$

$$= 1 + \cfrac{1}{2 + \cfrac{1}{\sqrt{2}+1}} = 1 + \cfrac{1}{2 + \cfrac{1}{2 + \cfrac{1}{\sqrt{2}+1}}} = 1 + \cfrac{1}{2 + \cfrac{1}{2 + \cfrac{1}{2 + \cfrac{1}{2 + \cdots}}}}$$

In the same manner derive the following continued fractions:

$$\sqrt{3} = 1 + \cfrac{1}{1 + \cfrac{1}{2 + \cfrac{1}{1 + \cfrac{1}{2 + \cdots}}}} \qquad \sqrt{5} = 2 + \cfrac{1}{4 + \cfrac{1}{4 + \cfrac{1}{4 + \cfrac{1}{4 + \cdots}}}}$$

*Raphael Bombelli* 1572

## ◊ Problem 1-7

For any real number $x$, the system of equations

$$x = a_0 + \xi_0 \qquad (0 \le \xi_0 < 1)$$
$$\frac{1}{\xi_0} = a_1 + \xi_1 \qquad (0 \le \xi_1 < 1)$$
$$\frac{1}{\xi_1} = a_2 + \xi_2 \qquad (0 \le \xi_2 < 1)$$
$$\frac{1}{\xi_2} = a_3 + \xi_3 \qquad (0 \le \xi_3 < 1) \quad \text{etc.}$$

with $a_0, a_1, \ldots$ as integers, is known as the *continued fraction algorithm*. The algorithm continues so long as $\xi_n \ne 0$ and provides a continued fraction representation of $x$ of the form

$$x = a_0 + \cfrac{1}{a_1 + \cfrac{1}{a_2 + \cfrac{1}{a_3 + \xi_3}}}$$

Use this algorithm to obtain the rational approximation $\pi \approx \frac{355}{113}$ which gives $\pi$ correctly to six decimal places.

Sect. 1.2]  Continued Fraction Expansions  47

◊ **Problem 1-8**
The positive root of the quadratic equation
$$x^2 - bx - c = 0 \quad \text{where} \quad b = ac$$
can be obtained as
$$x = b + \cfrac{1}{a + \cfrac{1}{b + \cfrac{1}{a + \cfrac{1}{b + \cfrac{1}{a + \cdots}}}}}$$
assuming that $a$ and $b$ are both positive.

### Gauss' Continued Fraction Expansion Theorem

Consider the following sequence of hypergeometric functions defined for $n = 0, 1, 2, \ldots$
$$F_{2n} = F(\alpha + n, \beta + n; \gamma + 2n; x)$$
$$F_{2n+1} = F(\alpha + n, \beta + n + 1; \gamma + 2n + 1; x)$$
From the identities of Prob. 1–3, we have
$$F_{2n+1} - F_{2n} = \delta_{2n+1} x F_{2n+2}$$
$$F_{2n} - F_{2n-1} = \delta_{2n} x F_{2n+1}$$
(1.27)

where the odd- and even-labelled $\delta$'s are determined from
$$\delta_{2n+1} = \frac{(\alpha + n)(\gamma - \beta + n)}{(\gamma + 2n)(\gamma + 2n + 1)} \qquad \delta_{2n} = \frac{(\beta + n)(\gamma - \alpha + n)}{(\gamma + 2n - 1)(\gamma + 2n)}$$

Equations (1.27) are linear difference equations analogous to Eq. (1.23) for Fibonacci numbers; moreover, the development to follow exactly parallels the steps used there.

Divide the first identity by $F_{2n}$, the second by $F_{2n-1}$, and define
$$G_{2n} = \frac{F_{2n+1}}{F_{2n}} \qquad G_{2n-1} = \frac{F_{2n}}{F_{2n-1}}$$

We obtain, thereby,
$$G_{2n} - 1 = \delta_{2n+1} x G_{2n+1} G_{2n}$$
$$G_{2n-1} - 1 = \delta_{2n} x G_{2n} G_{2n-1}$$

or
$$G_{2n} = \frac{1}{1 - \delta_{2n+1} x G_{2n+1}} \qquad G_{2n-1} = \frac{1}{1 - \delta_{2n} x G_{2n}}$$

If we put successively $n = 0$, $n = 1$, etc., we derive a continued fraction expansion for $G_0 = F_1/F_0$. Thus:

$$\frac{F(\alpha, \beta+1; \gamma+1; x)}{F(\alpha, \beta; \gamma; x)} = \cfrac{1}{1 - \cfrac{\delta_1 x}{1 - \cfrac{\delta_2 x}{1 - \cfrac{\delta_3}{\ddots \\ 1 - \delta_{2n} x G_{2n}}}}} \qquad (1.28)$$

and letting $n$ become infinite results in an infinite continued fraction. The question of convergence of such an expansion will be addressed in the following section.

It is important to note that if $\beta = 0$, the denominator of Eq. (1.28) $F(\alpha, 0; \gamma; x) = 1$ so that the continued fraction then represents *not* the ratio of two associated hypergeometric functions but rather the function† $F(\alpha, 1; \gamma+1; x)$. Therefore, if we replace $\gamma+1$ by $\gamma$, we have

$$F(\alpha, 1; \gamma; x) = 1 + \frac{\alpha}{\gamma} x + \frac{\alpha(\alpha+1)}{\gamma(\gamma+1)} x^2 + \cdots$$

$$= \cfrac{1}{1 - \cfrac{\beta_1 x}{1 - \cfrac{\beta_2 x}{1 - \cdots}}} \qquad (1.29)$$

where

$$\beta_{2n+1} = \frac{(\alpha+n)(\gamma+n-1)}{(\gamma+2n-1)(\gamma+2n)} \qquad \beta_{2n} = \frac{n(\gamma-\alpha+n-1)}{(\gamma+2n-2)(\gamma+2n-1)}$$

The corresponding continued fractions for the confluent functions

$$\frac{M(\beta+1, \gamma+1, x)}{M(\beta, \gamma, x)} = \cfrac{1}{1 - \cfrac{\gamma_1 x}{1 - \cfrac{\gamma_2 x}{1 - \cdots}}} \qquad (1.30)$$

where

$$\gamma_{2n+1} = \frac{\gamma - \beta + n}{(\gamma+2n)(\gamma+2n+1)} \qquad \gamma_{2n} = -\frac{\beta+n}{(\gamma+2n-1)(\gamma+2n)}$$

are obtained from Gauss' expansion by replacing $x$ by $x/\alpha$ in Eq. (1.28) and letting $\alpha$ become infinite.

---

† Examples of such functions have already been encountered in Eqs. (1.5), (1.6), and (1.8) for the logarithm, the inverse tangent, and the inverse hyperbolic tangent.

Sect. 1.2]          Continued Fraction Expansions          49

In some of the examples which follow, the notion of *equivalent fractions* is used. Specifically, if $c_0, c_1, \ldots$ is a sequence of non-zero constants, then

$$\cfrac{1}{1 - \cfrac{\beta_1 x}{1 - \cfrac{\beta_2 x}{1 - \cdots}}} \equiv \cfrac{c_0}{c_0 - \cfrac{c_0 c_1 \beta_1 x}{c_1 - \cfrac{c_1 c_2 \beta_2 x}{c_2 - \cdots}}}$$

◇ **Problem 1–9**
Develop the expansion

$$\log(1+x) = \cfrac{x}{1 + \cfrac{1^2 x}{2 + \cfrac{1^2 x}{3 + \cfrac{2^2 x}{4 + \cfrac{2^2 x}{5 + \cfrac{3^2 x}{6 + \cfrac{3^2 x}{7 + \cdots}}}}}}}$$

◇ **Problem 1–10**
Derive and compare the expansions

$$\arctan x = \cfrac{x}{1 + \cfrac{x^2}{3 + \cfrac{(2x)^2}{5 + \cfrac{(3x)^2}{7 + \cdots}}}} \qquad \operatorname{arctanh} x = \cfrac{x}{1 - \cfrac{x^2}{3 - \cfrac{(2x)^2}{5 - \cfrac{(3x)^2}{7 - \cdots}}}}$$

◇ **Problem 1–11**
Use the results of Prob. 1–1 to obtain

$$\tan k\phi = \cfrac{k \tan \phi}{1 - \cfrac{(k^2 - 1) \tan^2 \phi}{3 - \cfrac{(k^2 - 4) \tan^2 \phi}{5 - \cfrac{(k^2 - 9) \tan^2 \phi}{7 - \cdots}}}}$$

which terminates if $k$ is an integer.

*Leonhard Euler 1744*

## Problem 1-12

Derive the continued fraction expansion of the confluent hypergeometric function

$$M(1,\gamma,x) = 1 + \frac{x}{\gamma} + \frac{x^2}{\gamma(\gamma+1)} + \cdots$$

$$= \cfrac{1}{1 - \cfrac{x}{\gamma + \cfrac{x}{\gamma + 1 - \cfrac{\gamma x}{\gamma + 2 + \cfrac{2x}{\gamma + 3 - \cfrac{(\gamma+1)x}{\gamma + 4 + \cfrac{3x}{\gamma + 5 - \cdots}}}}}}}$$

## Problem 1-13

Obtain the continued fraction expansion of the exponential function

$$e^x = \cfrac{1}{1 - \cfrac{x}{1 + \cfrac{x}{2 - \cfrac{x}{3 + \cfrac{x}{2 - \cfrac{x}{5 + \cfrac{x}{2 - \cdots}}}}}}}$$

## Problem 1-14

Obtain the continued fraction expansion

$$\frac{N(\gamma+1,x)}{N(\gamma,x)} = \cfrac{\gamma}{\gamma + \cfrac{x}{\gamma+1 + \cfrac{x}{\gamma+2 + \cfrac{x}{\gamma+3 + \cfrac{x}{\gamma+4 + \cfrac{x}{\gamma+5 + \cfrac{x}{\gamma+6 + \cdots}}}}}}}$$

where $N(\gamma,x)$ is defined in Prob. 1-5.

## Continued Fractions Versus Power Series

One of the advantages of continued fraction expansions over the power series is dramatically illustrated by the function $\tan x$ whose series expansion is given by

$$\tan x = x + \frac{1}{3}x^3 + \frac{2}{15}x^5 + \frac{17}{315}x^7 + \cdots$$
$$+ \frac{(-1)^{n-1}2^{2n}(2^{2n}-1)B_{2n}}{(2n)!}x^{2n-1} + \cdots$$

Here, $B_{2n}$ are *Bernoulli numbers*, named for James Bernoulli (1655–1705) who introduced them in *Ars Conjectandi* in connection with a problem in probability. The first few of the even order Bernoulli numbers are

$$B_2 = \tfrac{1}{6} \qquad B_4 = -\tfrac{1}{30} \qquad B_6 = \tfrac{1}{42} \qquad B_8 = -\tfrac{1}{30} \qquad B_{10} = \tfrac{5}{66}$$

and the reader is forgiven if he does not immediately see the pattern. Actually, Euler† found that $t(e^t - 1)^{-1}$ is the *generating function* for the Bernoulli numbers in the sense that

$$\frac{t}{e^t - 1} = \sum_{k=0}^{\infty} B_k \frac{t^k}{k!}$$

Not only are the series coefficients for $\tan x$ quite complicated, but the series converges only in the interval $-\tfrac{1}{2}\pi \leq x \leq \tfrac{1}{2}\pi$. Compare this with the simple continued fraction developed in the next problem which converges for all $x$ not equal to $\tfrac{1}{2}\pi \pm n\pi$. Convergence criteria are the subject of the next section of this chapter.

◇ **Problem 1–15**
For the functions $N(\gamma, x)$ of Prob. 1–5, verify that

$$N(\tfrac{1}{2}, -\tfrac{1}{4}x^2) = \cos x \qquad \text{and} \qquad N(\tfrac{3}{2}, -\tfrac{1}{4}x^2) = \frac{\sin x}{x}$$

(with similar expressions obtaining for the hyperbolic functions) and then use Prob. 1–14 to derive the expansions

$$\tan x = \cfrac{x}{1 - \cfrac{x^2}{3 - \cfrac{x^2}{5 - \cfrac{x^2}{7 - \cdots}}}} \qquad \tanh x = \cfrac{x}{1 + \cfrac{x^2}{3 + \cfrac{x^2}{5 + \cfrac{x^2}{7 + \cdots}}}}$$

---

† One of Euler's finest triumphs involved the Bernoulli numbers when he obtained the summation formula

$$\frac{1}{1^n} + \frac{1}{2^n} + \frac{1}{3^n} + \frac{1}{4^n} + \cdots = \frac{(2\pi)^n}{2n!}|B_n|$$

where $n$ is an even integer.

**52**     Hypergeometric Functions and Elliptic Integrals     [Chap. 1]

NOTE: Johann Heinrich Lambert (1728–1777) actually proved the convergence of the continued fraction expansion for $\tan x$. He also used such fractions to prove the irrationality of integral powers of $\pi$ and $e$ which he reported to the Berlin Academy of Sciences in 1761.

### ◇ Problem 1–16

We are unable to derive directly a continued fraction expansion of the inverse sine since $\alpha = \beta = \frac{1}{2}$ in Eq. (1.7). However, by using Euler's identity (1.17), we obtain

$$\arcsin x = x\sqrt{1-x^2}\, F(1,1;\tfrac{3}{2};x^2) \quad \text{and} \quad \operatorname{arcsinh} x = x\sqrt{1+x^2}\, F(1,1;\tfrac{3}{2};-x^2)$$

where

$$F(1,1;\tfrac{3}{2};x) = \cfrac{1}{1 - \cfrac{1 \cdot 2x}{3 - \cfrac{1 \cdot 2x}{5 - \cfrac{3 \cdot 4x}{7 - \cfrac{3 \cdot 4x}{9 - \cdots}}}}}$$

### ◇ Problem 1–17

Develop the expansion

$$\sin \tfrac{1}{3}x = \frac{\sin x}{3} \frac{F(\tfrac{2}{3}, \tfrac{4}{3}; \tfrac{3}{2}; \sin^2 x)}{F(\tfrac{2}{3}, \tfrac{1}{3}; \tfrac{1}{2}; \sin^2 x)}$$

$$= \cfrac{\sin x}{3 - \cfrac{4 \cdot 1 \sin^2 x}{9 - \cfrac{5 \cdot 8 \sin^2 x}{15 - \cfrac{10 \cdot 7 \sin^2 x}{21 - \cfrac{11 \cdot 14 \sin^2 x}{27 - \cfrac{16 \cdot 13 \sin^2 x}{33 - \cfrac{17 \cdot 20 \sin^2 x}{39 - \cdots}}}}}}}$$

as well as the corresponding one for hyperbolic functions by replacing $x$ by $ix$ and using the relation $\sin ix = i \sinh x$. These are needed for the representation of the solution of the algebraic cubic equation as a continued fraction.

HINT: Use Eq. (1.17) to transform Eq. (1.13) and set $n = \tfrac{2}{3}$. In a similar manner, transform Eq. (1.14) and set $n = -\tfrac{1}{3}$.

## Continued Fraction Solutions of the Cubic Equation

The idea of using trigonometric identities and tables to solve cubic equations originated with François Viète.† By including hyperbolic identities as well, we can obtain an elegant and useful expression for the positive root of a large class of cubic equations.

The general cubic equation

$$w^3 + Aw^2 + Bw = C$$

may be reduced, by the substitution $w = z - \frac{1}{3}A$, to the normal form

$$z^3 \pm pz = q \tag{1.31}$$

where $p$ is positive. A further substitution $z = \sqrt{\frac{1}{3}p}\, y$ results in the canonical form

$$y^3 \pm 3y = 2b \tag{1.32}$$

If we assume that $b$ is positive, then Eq. (1.32) will have only one positive real root according to Descartes' rule of signs.

Consider first the equation $y^3 + 3y = 2b$. By writing

$$y = 2\sinh \tfrac{1}{3}x \quad \text{and} \quad b = \sinh x$$

the cubic equation becomes

$$4\sinh^3 \tfrac{1}{3}x + 3\sinh \tfrac{1}{3}x = \sinh x$$

which we recognize as a standard identity for the hyperbolic sine. Therefore, using the results of Prob. 1–17, we can write the solution of the cubic equation as

$$y = \frac{2b}{3}\frac{F(\tfrac{2}{3},\tfrac{4}{3};\tfrac{3}{2};-b^2)}{F(\tfrac{2}{3},\tfrac{1}{3};\tfrac{1}{2};-b^2)} = \cfrac{2b}{3 + \cfrac{4\cdot 1 b^2}{9 + \cfrac{5\cdot 8 b^2}{15 + \cfrac{10\cdot 7 b^2}{21 + \cdots}}}} \tag{1.33}$$

---

† Franciscus Vieta (1540–1603) was a lawyer by profession but is recognized as the foremost mathematician of the sixteenth century and the father of modern algebraic notation. His *Canon Mathematicus seu ad Triangula* in 1579 was the first of his many works on plane and spherical trigonometry and contained for the first time many of the now familiar trigonometric identities. *De Aequationum Recognitione et Emendatione*, which was written in 1591 but not published until 1615, contains his method of solving the irreducible cubic by using a trigonometric identity. Like Wallis, he too had a remarkable infinite product representation of $\pi$:

$$\frac{2}{\pi} = \sqrt{\tfrac{1}{2}} \cdot \sqrt{\tfrac{1}{2} + \tfrac{1}{2}\sqrt{\tfrac{1}{2}}} \cdot \sqrt{\tfrac{1}{2} + \tfrac{1}{2}\sqrt{\tfrac{1}{2} + \tfrac{1}{2}\sqrt{\tfrac{1}{2}}}} \cdots$$

If the cubic equation has the form $y^3 - 3y = 2b$, we must initially address the cases $b \leq 1$ and $b \geq 1$ separately. For the former, we write

$$y = 2\cos \tfrac{2}{3}x = 2(1 - 2\sin^2 \tfrac{1}{3}x) \qquad \text{and} \qquad b = \cos 2x = 1 - 2\sin^2 x$$

and obtain

$$4\cos^3 \tfrac{2}{3}x - 3\cos \tfrac{2}{3}x = \cos 2x$$

which is an appropriate identity for the cosine function. Therefore,

$$\begin{aligned} y &= 2 - 4\sin^2 x \left(\frac{\sin \tfrac{1}{3}x}{\sin x}\right)^2 \\ &= 2 - \frac{2(1-b)}{9}\left(\frac{F[\tfrac{2}{3}, \tfrac{4}{3}; \tfrac{3}{2}; \tfrac{1}{2}(1-b)]}{F[\tfrac{2}{3}, \tfrac{1}{3}; \tfrac{1}{2}; \tfrac{1}{2}(1-b)]}\right)^2 \end{aligned} \qquad (1.34)$$

If $b \geq 1$, we write

$$y = 2\cosh \tfrac{2}{3}x = 2(1 + 2\sinh^2 \tfrac{1}{3}x) \qquad \text{and} \qquad b = \cosh 2x = 1 + 2\sinh^2 x$$

Then, the cubic is transformed to

$$4\cosh^3 \tfrac{2}{3}x - 3\cosh \tfrac{2}{3}x = \cosh 2x$$

Hence,

$$y = 2 + 4\sinh^2 x \left(\frac{\sinh \tfrac{1}{3}x}{\sinh x}\right)^2$$

But, when $y$ is expressed in terms of $b$, the result is the same as Eq. (1.34)—exactly! Therefore, (1.34) is the solution of (1.32) regardless of the size of $b$.

## 1.3 Convergence of Continued Fractions

The criteria for convergence of continued fractions are not nearly so complete as for power series. Of course, by convergence of the general continued fraction

$$\cfrac{a_0}{b_0 + \cfrac{a_1}{b_1 + \cfrac{a_2}{b_2 + \cfrac{a_3}{b_3 + \cdots}}}} \qquad (1.35)$$

we mean convergence of the infinite sequence of *partial convergents* $p_0/q_0$, $p_1/q_1$, ... defined as

$$\frac{p_0}{q_0} = \frac{a_0}{b_0}$$

$$\frac{p_1}{q_1} = \frac{a_0}{b_0 + \dfrac{a_1}{b_1}} = \frac{a_0 b_1}{b_0 b_1 + a_1}$$

$$\frac{p_2}{q_2} = \frac{a_0}{b_0 + \dfrac{a_1}{b_1 + \dfrac{a_2}{b_2}}} = \frac{a_0(b_1 b_2 + a_2)}{b_0 b_1 b_2 + b_0 a_2 + b_2 a_1} \quad \text{etc.}$$

(1.36)

Therefore, the *value* of the infinite continued fraction is the *limit* of the infinite sequence.

### Recursive Properties of the Convergents

John Wallis became interested in continued fractions when Lord William Brouncker (1620–1684), the first president of the Royal Society, transformed Wallis' infinite product representation of $\pi$ to the continued fraction

$$\frac{4}{\pi} = \frac{3 \cdot 3 \cdot 5 \cdot 5 \cdot 7 \cdots}{2 \cdot 4 \cdot 4 \cdot 6 \cdot 6 \cdots} = 1 + \cfrac{1^2}{2 + \cfrac{3^2}{2 + \cfrac{5^2}{2 + \cfrac{7^2}{2 + \cdots}}}}$$

In his *Opera Mathematica*, in which he also introduced the term "continued fraction", Wallis gave the general rule for calculating the convergents which we shall now prove. However, he gave no definitive results on the subject of convergence.

We observe from Eqs. (1.36) that

$$p_2 = b_2 p_1 + a_2 p_0$$
$$q_2 = b_2 q_1 + a_2 q_0$$

which suggests the possibility of a general recurrence relation of the form

$$p_n = b_n p_{n-1} + a_n p_{n-2}$$
$$q_n = b_n q_{n-1} + a_n q_{n-2}$$

(1.37)

with initial conditions

$$p_0 = a_0 \quad p_1 = a_0 b_1 \quad q_0 = b_0 \quad q_1 = b_0 b_1 + a_1$$

We can most easily prove Wallis' rule using *mathematical induction*. Equations (1.37) are certainly true for $n = 2$. Applying mathematical induction, we assume that they are true for all integers up to and including $n$ and attempt to show that they are true for $n + 1$. For this purpose, we first note that $p_{n+1}/q_{n+1}$ is generated from $p_n/q_n$ by replacing $b_n$ by $b_n + a_{n+1}/b_{n+1}$. Hence

$$\frac{p_{n+1}}{q_{n+1}} = \frac{(b_{n+1}b_n + a_{n+1})p_{n-1} + b_{n+1}a_n p_{n-2}}{(b_{n+1}b_n + a_{n+1})q_{n-1} + b_{n+1}a_n q_{n-2}}$$

But the proposition (1.37) is true for $n$ by hypothesis so that we have

$$\frac{p_{n+1}}{q_{n+1}} = \frac{b_{n+1}p_n + a_{n+1}p_{n-1}}{b_{n+1}q_n + a_{n+1}q_{n-1}}$$

Therefore, the proposition is true for all $n$.

There are two other important recurrence relations that we shall also require for the discussion of convergence. For the first, define

$$f_n = p_n q_{n-1} - p_{n-1} q_n$$

and use Eqs. (1.37) to write

$$p_n q_{n-1} = b_n p_{n-1} q_{n-1} + a_n p_{n-2} q_{n-1}$$
$$p_{n-1} q_n = b_n p_{n-1} q_{n-1} + a_n p_{n-1} q_{n-2}$$

so that

$$f_n = -a_n(p_{n-1}q_{n-2} - p_{n-2}q_{n-1}) = -a_n f_{n-1}$$

Hence

$$f_n = -a_n f_{n-1} = (-a_n)(-a_{n-1})f_{n-2} = (-a_n)(-a_{n-1})\cdots(-a_2)f_1$$

or, alternately,

$$f_n \equiv p_n q_{n-1} - p_{n-1} q_n = (-1)^n a_0 a_1 \ldots a_n \quad (1.38)$$

For the second, we define

$$g_n = p_n q_{n-2} - p_{n-2} q_n$$

and again use Eqs. (1.37) to write

$$p_n q_{n-2} = b_n p_{n-1} q_{n-2} + a_n p_{n-2} q_{n-2}$$
$$p_{n-2} q_n = b_n p_{n-2} q_{n-1} + a_n p_{n-2} q_{n-2}$$

Hence

$$g_n = b_n(p_{n-1}q_{n-2} - p_{n-2}q_{n-1}) = b_n f_{n-1}$$

and, using Eq. (1.37), we obtain

$$g_n \equiv p_n q_{n-2} - p_{n-2} q_n = (-1)^{n-1} a_0 a_1 \ldots a_{n-1} b_n \quad (1.39)$$

Sect. 1.3]    Convergence of Continued Fractions    57

With these results, we can derive sufficient conditions for the convergence of continued fractions of two different kinds or classes.

## Convergence of Class I Continued Fractions

The general continued fraction (1.35) is said to be of the first class if $a_n$ and $b_n$ are all positive. Now, from Eq. (1.38), we have

$$\frac{p_{2n+1}}{q_{2n+1}} - \frac{p_{2n}}{q_{2n}} = -\frac{a_0 a_1 \cdots a_{2n+1}}{q_{2n+1} q_{2n}}$$

$$\frac{p_{2n}}{q_{2n}} - \frac{p_{2n-1}}{q_{2n-1}} = \frac{a_0 a_1 \cdots a_{2n}}{q_{2n} q_{2n-1}}$$

which demonstrates that every even convergent is greater than every odd convergent. Furthermore, from Eq. (1.39), we have

$$\frac{p_{2n+1}}{q_{2n+1}} - \frac{p_{2n-1}}{q_{2n-1}} = \frac{a_0 a_1 \cdots a_{2n} b_{2n+1}}{q_{2n+1} q_{2n-1}}$$

$$\frac{p_{2n}}{q_{2n}} - \frac{p_{2n-2}}{q_{2n-2}} = -\frac{a_0 a_1 \cdots a_{2n-1} b_{2n}}{q_{2n} q_{2n-2}}$$

so that the odd convergents continually increase while the even convergents decrease. In summary,

$$\frac{p_1}{q_1} < \frac{p_2}{q_2} > \frac{p_3}{q_3} < \frac{p_4}{q_4} > \frac{p_5}{q_5} < \frac{p_6}{q_6} > \frac{p_7}{q_7} < \cdots$$

Clearly, the odd and even convergents could each have a separate limit. Hence, *the fraction will either converge or oscillate between two different values.*

There is a sufficient condition for convergence of Class I fractions which we can now derive. From the second of Eqs. (1.37) we have

$$q_n = b_n q_{n-1} + a_n q_{n-2}$$
$$= b_n (b_{n-1} q_{n-2} + a_{n-1} q_{n-3}) + a_n q_{n-2}$$
$$> (b_n b_{n-1} + a_n) q_{n-2}$$

so that

$$q_n q_{n-1} > (b_n b_{n-1} + a_n) q_{n-1} q_{n-2}$$
$$> (b_n b_{n-1} + a_n)(b_{n-1} b_{n-2} + a_{n-1}) q_{n-2} q_{n-3}$$
$$> (b_n b_{n-1} + a_n) \cdots (b_2 b_1 + a_2) q_1 q_0$$
$$> (b_n b_{n-1} + a_n) \cdots (b_1 b_0 + a_1) b_0$$

Therefore,

$$\frac{q_n q_{n-1}}{a_0 a_1 \cdots a_n} > \frac{b_0}{a_0}\left(1 + \frac{b_0 b_1}{a_1}\right)\left(1 + \frac{b_1 b_2}{a_2}\right) \cdots \left(1 + \frac{b_{n-1} b_n}{a_n}\right)$$

Clearly, the infinite product diverges if

$$\lim_{n\to\infty} \frac{b_{n-1}b_n}{a_n} > 0 \tag{1.40}$$

Hence

$$\frac{p_n}{q_n} - \frac{p_{n-1}}{q_{n-1}} = (-1)^n \frac{a_0 a_1 \cdots a_n}{q_n q_{n-1}}$$

will approach zero as $n$ becomes infinite provided that the inequality (1.40) is satisfied. Thus, (1.40) is a sufficient condition for the convergence of the infinite continued fraction.

◇ **Problem 1–18**

Show that

$$\cfrac{1}{2+\cfrac{2}{3+\cfrac{3}{4+\cfrac{4}{5+\cfrac{5}{6+\cdots}}}}}$$

is a convergent continued fraction.

◇ **Problem 1–19**

When $x$ is negative, the continued fraction representation of the ratio of two contiguous hypergeometric functions will eventually be of Class I—that is, after no more than a finite number of levels of the fraction, the $a$'s and $b$'s will become and remain positive. Prove that the fraction will always converge.

◇ **Problem 1–20**

The class I fraction

$$\cfrac{x}{a+\cfrac{x^2}{a+\cfrac{x^3}{a+\cfrac{x^4}{a+\cfrac{x^5}{a+\cdots}}}}}$$

has a fascinating property unlike what we might encounter with infinite series. If $x \leq 1$, then the fraction converges. On the other hand, if $x > 1$, the fraction will oscillate.

Verify these statements and illustrate the oscillation process with a numerical example.

### Convergence of Class II Continued Fractions

The continued fraction of the form

$$\cfrac{a_0}{b_0 - \cfrac{a_1}{b_1 - \cfrac{a_2}{b_2 - \cfrac{a_3}{b_3 - \cdots}}}} \qquad (1.41)$$

is said to be of the second class if $a_n$ and $b_n$ are all positive. The recursion formulas of Eqs. (1.37) take the form

$$\begin{aligned} p_n &= b_n p_{n-1} - a_n p_{n-2} \\ q_n &= b_n q_{n-1} - a_n q_{n-2} \end{aligned} \qquad (1.42)$$

with the initial conditions

$$p_0 = a_0 \qquad p_1 = a_0 b_1 \qquad q_0 = b_0 \qquad q_1 = b_0 b_1 - a_1$$

We can establish the relation

$$f_n \equiv p_n q_{n-1} - p_{n-1} q_n = a_0 a_1 \cdots a_n$$

in exactly the same manner that was used to derive Eq. (1.38). Hence,

$$\frac{p_n}{q_n} - \frac{p_{n-1}}{q_{n-1}} = \frac{a_0 a_1 \cdots a_n}{q_n q_{n-1}}$$

and we see immediately that the partial convergents form an increasing sequence. The infinite continued fraction will then *either converge or diverge to infinity*.

Convergence of a Class II fraction will be assured if the inequality

$$b_n \geq a_n + 1 \qquad (1.43)$$

becomes and remains true as $n$ increases. To validate this sufficient condition, there are two inequalities which we must establish. First, since

$$\begin{aligned} p_n - p_{n-1} &= (b_n - 1) p_{n-1} - a_n p_{n-2} \\ q_n - q_{n-1} &= (b_n - 1) q_{n-1} - a_n q_{n-2} \end{aligned}$$

then the inequality (1.43) will insure that

$$\begin{aligned} p_n - p_{n-1} &\geq a_n (p_{n-1} - p_{n-2}) \\ &\geq a_n a_{n-1} \cdots a_2 (p_1 - p_0) = a_n a_{n-1} \cdots a_2 a_0 (b_1 - 1) \\ &\geq a_0 a_1 \cdots a_n \end{aligned}$$

From this, and the *telescoping series*

$$p_n - p_0 = (p_n - p_{n-1}) + (p_{n-1} - p_{n-2}) + \cdots + (p_1 - p_0)$$

it follows that
$$p_n \geq a_0 + a_0 a_1 + a_0 a_1 a_2 + \cdots + a_0 a_1 \ldots a_n \quad (1.44)$$

Of course, $q_n$ satisfies the same inequality; however, an even stronger inequality for $q_n$ exists and will be needed.

For this purpose, we first observe that
$$\begin{aligned} q_n - p_n &= b_n(q_{n-1} - p_{n-1}) - a_n(q_{n-2} - p_{n-2}) \\ &\geq (q_{n-1} - p_{n-1}) + a_n[(q_{n-1} - p_{n-1}) - (q_{n-2} - p_{n-2})] \end{aligned} \quad (1.45)$$

provided $q_{n-1} - p_{n-1} \geq 0$. That this requirement is true follows from
$$\begin{aligned} (q_1 - p_1) - (q_0 - p_0) &= (b_0 b_1 - a_1 - a_0 b_1) - (b_0 - a_0) \\ &= (b_0 - a_0)(b_1 - 1) - a_1 \\ &\geq 1 \cdot a_1 - a_1 = 0 \end{aligned}$$

and
$$q_1 - p_1 = b_1(b_0 - a_0) - a_1 \geq b_1 - a_1 \geq 1$$

Hence
$$q_2 - p_2 \geq q_1 - p_1 \geq q_0 - p_0 = b_0 - a_0 \geq 1$$

Continuing this process recursively, Eq. (1.45) is established in general. We have, therefore,
$$q_n - p_n \geq q_{n-1} - p_{n-1} \geq \cdots \geq q_0 - p_0 \geq 1$$

from which we establish the second inequality
$$q_n \geq p_n + 1 \quad (1.46)$$

Now, since $p_n$ and $q_n$ are both positive and $q_n \geq 1$, it follows from (1.46) that
$$\frac{p_n}{q_n} \leq 1 - \frac{1}{q_n} \quad (1.47)$$

Therefore, the infinite continued fraction *converges to a finite limit not greater than one* and the sufficiency of the inequality (1.43) is verified.

The inequality $b_n \geq a_n + 1$ need not be satisfied for all $n$ but only for $n$ greater than some $N$. In this case, the infinite continued fraction converges but the limit is not necessarily less than or equal to one. For example, consider the expansion for the $\tan x$ obtained in Prob. 1–15 which is Class II. We have $a_0 = x$, $a_n = x^2$ for $n = 1, 2, \ldots$ and $b_n = 2n + 1$ for $n = 0, 1, \ldots$. Clearly, $b_n \geq a_n + 1$ holds for $n > N$ for any value of $x$ and a deterministic value of $N$. Therefore, the fraction converges. (Of course, $x$ must not coincide with a singularity of the function $\tan x$.)

Sect. 1.3]     Convergence of Continued Fractions          61

**Problem 1-21**
Another sufficient condition for the convergence of the Class II fraction (1.41) is

$$\frac{b_1}{a_0 a_1} + \frac{b_2}{a_1 a_2} + \cdots + \frac{b_{n+1}}{a_n a_{n+1}} \leq 1$$

for all values of $n$.

◇ **Problem 1-22**
For positive $x$ the fraction

$$\cfrac{x}{x+1 - \cfrac{x}{x+1 - \cfrac{x}{x+1 - \cdots}}}$$

is of Class II. Can you demonstrate the fascinating property that the value of this fraction is equal either to $x$ or 1 according as $x < 1$ or $x \geq 1$, respectively?
HINT: Recall the continued fraction development for the Golden Section.

### Equivalent Continued Fractions

The convergence tests given for Class I and Class II fractions cannot always be applied directly. The tests may fail and yet the fraction could still converge. This is because an equivalent form of the fraction may pass the test.

We have already introduced the subject of equivalent continued fractions in the previous section of this chapter. But there are two cases which are worthy of special consideration. In general, we have

$$\cfrac{a_0}{b_0 + \cfrac{a_1}{b_1 + \cfrac{a_2}{b_2 + \cdots}}} \equiv \cfrac{a_0 c_0}{b_0 c_0 + \cfrac{a_1 c_0 c_1}{b_1 c_1 + \cfrac{a_2 c_1 c_2}{b_2 c_2 + \cdots}}}$$

Now if we choose

$$c_0 = \frac{1}{b_0} \qquad c_1 = \frac{1}{b_1} \qquad c_2 = \frac{1}{b_2} \qquad \text{etc.}$$

we convert the fraction to the form

$$\cfrac{a_0}{b_0 + \cfrac{a_1}{b_1 + \cfrac{a_2}{b_2 + \cdots}}} \equiv \cfrac{a_0/b_0}{1 + \cfrac{a_1/b_0 b_1}{1 + \cfrac{a_2/b_1 b_2}{1 + \cdots}}} \qquad (1.48)$$

On the other hand, if we choose

$$c_0 = \frac{1}{a_0} \quad c_1 = \frac{1}{a_1 c_0} \quad c_2 = \frac{1}{a_2 c_1} \quad c_3 = \frac{1}{a_3 c_2} \quad \text{etc.}$$

then we have the equivalent fraction

$$\cfrac{a_0}{b_0 + \cfrac{a_1}{b_1 + \cfrac{a_2}{b_2 + \cdots}}} \equiv \cfrac{1}{\cfrac{b_0}{a_0} + \cfrac{1}{\cfrac{b_1 a_0}{a_1} + \cfrac{1}{\cfrac{b_2 a_1}{a_0 a_2} + \cdots}}} \qquad (1.49)$$

If these alternate or other equivalent forms of the fraction satisfy the convergence criteria, then the original fraction will, indeed, converge.

For example, when $x$ is positive, the continued fraction expansion of the hypergeometric function $F(3, 1; \frac{5}{2}; x)$, encountered in Chapter 7, is of Class II for positive $x$ after the first two stages since

$$F(3, 1; \tfrac{5}{2}; x) = \cfrac{3}{3 - \cfrac{18x}{5 + \cfrac{6x}{7 - \cfrac{40x}{9 - \cfrac{4x}{11 - \cfrac{70x}{13 - \cfrac{18x}{15 - \cfrac{108x}{17 - \cdots}}}}}}}}$$

Suppose that $0 < x \leq 1$ and consider the following fraction which is equivalent to the tail of the fraction in question with $x = 1$:

$$\cfrac{4c_1}{11c_1 - \cfrac{70c_1 c_2}{13c_2 - \cfrac{18c_2 c_3}{15c_3 - \cfrac{108c_3 c_4}{17c_4 - \cfrac{40c_4 c_5}{19c_5 - \cfrac{154c_5 c_6}{21c_6 - \cdots}}}}}}$$

By requiring that $b_n = a_n + 1$, the constants $c_i$ can be determined recursively. We have

$$11c_1 = 4c_1 + 1 \qquad\qquad c_1 = \tfrac{1}{7}$$
$$13c_2 = 70c_1c_2 + 1 \qquad\qquad c_2 = \tfrac{1}{3}$$
$$15c_3 = 18c_2c_3 + 1 \qquad\qquad c_3 = \tfrac{1}{9}$$
$$17c_4 = 108c_3c_4 + 1 \quad \text{which give} \quad c_4 = \tfrac{1}{5} \quad \text{etc.}$$
$$19c_5 = 40c_4c_5 + 1 \qquad\qquad c_5 = \tfrac{1}{11}$$
$$21c_6 = 154c_5c_6 + 1 \qquad\qquad c_6 = \tfrac{1}{7}$$

Therefore, the tail of the fraction is equivalent to

$$\cfrac{\tfrac{4}{7}}{\tfrac{11}{7} - \cfrac{\tfrac{10}{3}}{\tfrac{13}{3} - \cfrac{\tfrac{2}{3}}{\tfrac{5}{3} - \cfrac{\tfrac{12}{5}}{\tfrac{17}{5} - \cfrac{\tfrac{8}{11}}{\tfrac{19}{11} - \cdots}}}}}$$

which is convergent by the Class II sufficiency test.

## 1.4  Evaluating Continued Fractions

The continued fraction (assumed to be convergent)

$$\cfrac{a_0}{b_0 - \cfrac{a_1}{b_1 - \cfrac{a_2}{b_2 - \cdots}}}$$

may be evaluated by any of several methods. Three are considered here, the last of which has none of the disadvantages of the first two.

### Wallis' Method

The numerator and denominator of the partial convergents $p_n/q_n$ may be obtained recursively for $n = 2, 3, \ldots$ from Wallis' formulas (1.42) until convergence within the required tolerance is achieved. The principal disadvantage of this method is that $p_n$ and $q_n$ are likely to grow rapidly with $n$. When implemented on a computer, repeated scaling may be necessary.

## The Bottom-Up Method

The necessity for scaling can be avoided by calculating the finite fraction $p_n/q_n$ from the bottom to the top by successive division. Thus, if we set

$$f_k^{(n)} = \cfrac{a_k}{b_k - \cfrac{a_{k+1}}{b_{k+1} - \cdots}} \qquad \text{for} \qquad 0 \leq k \leq n$$

and generate these quantities recursively from

$$f_k^{(n)} = \frac{a_k}{b_k - f_{k+1}^{(n)}} \qquad \text{for} \qquad k = n-1, n-2, \ldots, 0$$

starting with $f_n^{(n)} = 0$; then $f_0^{(n)} = p_n/q_n$. To obtain the value of the continued fraction, the process must be repeated for increasing values of $n$ until $f_0^{(n)}$ converges to within the desired accuracy.

Although the method is simple and easily programmed, the obvious disadvantage is the required iteration on $n$ which can necessitate an inordinate number of arithmetic operations.

## Euler's Transformation

The foundation of a theory of continued fractions was laid by Leonhard Euler through a series of papers. In 1785 his memoir *De Transformatione Serierum in Fractiones Continuas* appeared showing how to convert infinite series into continued fractions which is the basis of an efficient method for evaluating continued fractions.

Assume that $b_n = a_n + 1$ in the continued fraction of the form (1.41). Then all of the inequalities derived in the subsection on Class II fractions in Sect. 1.3 become equalities. Indeed,

$$\frac{p_n}{q_n} = 1 - \frac{1}{q_n} \qquad \text{and} \qquad q_n = 1 + a_0 + a_0 a_1 + \cdots + a_0 a_1 \ldots a_n$$

so that

$$q_n = \cfrac{1}{1 - \cfrac{p_n}{q_n}}$$

Therefore, from Eq. (1.41), we have

$$q_n = \cfrac{1}{1 - \cfrac{a_0}{a_0 + 1 - \cfrac{a_1}{a_1 + 1 - \cfrac{a_2}{a_2 + 1 - \cfrac{a_3}{\ddots \; a_{n-1} + 1 - \cfrac{a_n}{a_n + 1}}}}}}$$

An equivalent representation of $q_n$ is clearly possible. We have

$$q_n = \cfrac{1}{1 - \cfrac{c_0 a_0}{c_0 a_0 + c_0 - \cfrac{c_0 c_1 a_1}{c_1 a_1 + c_1 - \cfrac{c_1 c_2 a_2}{\ddots \; c_{n-1} a_{n-1} + c_{n-1} - \cfrac{c_{n-1} c_n a_n}{c_n a_n + c_n}}}}}$$

where $c_0, c_1, \ldots$ can be arbitrarily selected. In particular, if we define

$$c_0 = 1 \quad c_1 = a_0 \quad c_2 = a_1 a_0 \quad c_3 = a_2 a_1 a_0 \quad \text{etc.}$$

and, further, write

$$u_0 = a_0 \quad u_1 = a_0 a_1 \quad u_2 = a_0 a_1 a_2 \quad u_3 = a_0 a_1 a_2 a_3 \quad \text{etc.}$$

we obtain

$$1 + u_0 + u_1 + \cdots + u_n$$

$$\equiv \cfrac{1}{1 - \cfrac{u_0}{1 + u_0 - \cfrac{u_1}{u_0 + u_1 - \cfrac{u_0 u_2}{u_1 + u_2 - \cfrac{u_1 u_3}{\ddots \; u_{n-2} + u_{n-1} - \cfrac{u_{n-2} u_n}{u_{n-1} + u_n}}}}}}$$

Finally, since

$$\cfrac{1}{1 - \cfrac{u_0}{1 + u_0 - P}} - 1 = \cfrac{u_0}{1 - P}$$

we have

$$u_0 + u_1 + \cdots + u_n \equiv \cfrac{u_0}{1 - \cfrac{u_1}{u_0 + u_1 - \cfrac{u_0 u_2}{u_1 + u_2 - \cfrac{u_1 u_3}{\ddots \quad u_{n-2} + u_{n-1} - \cfrac{u_{n-2} u_n}{u_{n-1} + u_n}}}}} \quad (1.50)$$

This is Euler's famous transformation of a series into a continued fraction which we shall use to derive a most convenient algorithm for the efficient evaluation of a continued fraction. It is important to realize that Eq. (1.50) is, in fact, merely an algebraic identity and should not be confused with the powerful expansion developed by Gauss in the previous section. Euler's continued fraction will converge or not under the exact same circumstances as the power series and at the same rate. On the other hand, Gauss' expansion generally broadens the range and increases the speed of convergence when compared to the corresponding series.

◊ **Problem 1–23**

By transforming the power series for $\arctan x$ into a continued fraction, derive Brouncker's "formula for the quadrature of the circle" given at the beginning of Sect. 1.3.

◊ **Problem 1–24**

Euler's continued fraction for $\sin x$ is

$$\sin x = \cfrac{x}{1 + \cfrac{x^2}{2 \cdot 3 - x^2 + \cfrac{2 \cdot 3 x^2}{4 \cdot 5 - x^2 + \cfrac{4 \cdot 5 x^2}{6 \cdot 7 - x^2 + \ddots}}}}$$

◊ **Problem 1–25**

Euler's continued fraction for $e^x$ is

$$e^x = \cfrac{1}{1 - \cfrac{x}{1 + x - \cfrac{x}{2 + x - \cfrac{2x}{3 + x - \cfrac{3x}{4 + x - \ddots}}}}}$$

### The Top-Down Method

Euler's transformation also permits a continued fraction to be converted into an equivalent series such that the $n^{\text{th}}$ convergents of both the fraction and the series are identical. If, in Eq. (1.50), we define

$$\Sigma_n = u_0 + u_1 + \cdots + u_n$$

and

$$u_0 = \rho_0 \qquad u_1 = \rho_0\rho_1 \qquad u_2 = \rho_0\rho_1\rho_2 \qquad \text{etc.}$$

we obtain the alternate form

$$\Sigma_n = \cfrac{\rho_0}{1 - \cfrac{\rho_1}{1 + \rho_1 - \cfrac{\rho_2}{1 + \rho_2 - \cfrac{\rho_3}{\ddots \cfrac{}{1 + \rho_{n-1} - \cfrac{\rho_n}{1 + \rho_n}}}}}}$$

Now let

$$1 + \rho_k = \delta_k \qquad \text{for} \qquad k = 1, 2, \ldots, n$$

so that

$$\Sigma_n = \cfrac{\rho_0}{1 - \cfrac{(\delta_1 - 1)/\delta_1}{1 - \cfrac{(\delta_2 - 1)/\delta_1\delta_2}{\ddots \cfrac{}{1 - (\delta_n - 1)/\delta_{n-1}\delta_n}}}}$$

Finally, by making the identifications,

$$\rho_0 = \frac{a_0}{b_0}, \qquad \frac{\delta_k - 1}{\delta_{k-1}\delta_k} = \frac{a_k}{b_{k-1}b_k} \qquad \text{for} \qquad k = 1, 2, \ldots, n$$

we obtain

$$\Sigma_n = \cfrac{a_0/b_0}{1 - \cfrac{a_1/b_0b_1}{1 - \cfrac{a_2/b_1b_2}{\ddots \cfrac{}{1 - a_n/b_{n-1}b_n}}}}$$

which is equivalent to the $n^{\text{th}}$ convergent $p_n/q_n$ of the original continued fraction.

These relations provide the basis for a recursive algorithm† to generate the convergents of the continued fraction. For $n = 1, 2, \ldots$, we have

$$\delta_n = \frac{1}{1 - \frac{a_n}{b_{n-1}b_n}\delta_{n-1}}$$

$$u_n = u_{n-1}(\delta_n - 1) \tag{1.51}$$

$$\frac{p_n}{q_n} \equiv \Sigma_n = \Sigma_{n-1} + u_n$$

where

$$\delta_0 = 1 \qquad u_0 = \Sigma_0 = \frac{a_0}{b_0}$$

The iteration continues until that value of $n$ is reached for which

$$u_n = \frac{p_n}{q_n} - \frac{p_{n-1}}{q_{n-1}}$$

is within a specified tolerance.

## 1.5 Elliptic Integrals

Although problems involving elliptic integrals had been pursued for almost a century by such notables as the Bernoullis, Leibnitz, Fagnano,‡ and Euler, the definitive work was done by Adrien-Marie Legendre (1752–1833) over a period spanning four decades. Legendre's chief result, recorded in *Traité des fonctions elliptiques* in 1825–26, may be stated as:

> If $P(x)$ is a polynomial of at most fourth degree with real coefficients and if $R$ is a rational function of two variables with real coefficients, while $x$ is restricted to a range in which $P(x)$ is positive, the integral
>
> $$\int_0^x R\bigl[x, \sqrt{P(x)}\,\bigr]\,dx$$
>
> can be expressed as a linear combination of terms, each of which is either an elementary function, or an elliptic integral of the first, second, or third kind.§

---

† This algorithm was published by W. Gautschi in a paper entitled "Computational Aspects of Three-Term Recurrence Relations" which appeared in *SIAM Review*, Vol. 9, Jan. 1967.

‡ Count Giulio Carlo de' Fagnano (1682–1766), an amateur mathematician, was led to the general elliptic integral of the first kind through his treatment of the difference of two lemniscate arcs.

§ It was not until the year of Legendre's death that Joseph Liouville (1809–1882) showed that elliptic integrals could not be expressed with a finite number of algebraic, circular, logarithmic, or exponential functions—the so-called *elementary functions*.

## Elliptic Integrals

The normal forms of the elliptic integrals $F$ and $E$ of the first and second kind, respectively, are

$$F(k,\phi) = \int_0^\phi \frac{d\phi}{\sqrt{1-k^2\sin^2\phi}} = \int_0^{\sin\phi} \frac{dx}{\sqrt{(1-x^2)(1-k^2x^2)}} \quad (1.52)$$

$$E(k,\phi) = \int_0^\phi \sqrt{1-k^2\sin^2\phi}\, d\phi = \int_0^{\sin\phi} \sqrt{\frac{1-k^2x^2}{1-x^2}}\, dx \quad (1.53)$$

which are functions of the *amplitude* $\phi$, where $0 < \phi \leq \frac{1}{2}\pi$, and of the *modulus* $k$, where $0 \leq k \leq 1$. (The symbol $m$, called the *parameter*, is also used in place of the square of the modulus, i.e., $m \equiv k^2$). The first form of these special functions, which were originally tabulated by Legendre, is called *Legendre's form*. The alternate, obtained from the first by setting $x = \sin\phi$, is called *Jacobi's form*.

When the amplitude $\phi = \frac{1}{2}\pi$, the integrals are then *complete elliptic integrals* and denoted by $K(k)$ and $E(k)$. Thus,

$$F(k, \tfrac{1}{2}\pi) \equiv K(k) \quad \text{and} \quad E(k, \tfrac{1}{2}\pi) \equiv E(k) \quad (1.54)$$

Otherwise, they are referred to as *incomplete elliptic integrals*.

The elliptic integral of the third kind is

$$\Pi(n,k,\phi) = \int_0^\phi \frac{d\phi}{(1-n\sin^2\phi)\sqrt{1-k^2\sin^2\phi}}$$

$$= \int_0^{\sin\phi} \frac{dx}{(1-nx^2)\sqrt{(1-x^2)(1-k^2x^2)}} \quad (1.55)$$

where $n$ is the *characteristic* and can range in value from $-\infty$ to $\infty$. The properties of the integral depend on the location of the characteristic in this interval. It is interesting to note that, for $\phi = \frac{1}{2}\pi$, the complete elliptic integrals of the third kind can be represented in terms of incomplete integrals of the first and second kind together with elementary functions.

◇ **Problem 1-26**

For fixed amplitude, we have

$$\phi \leq F(k,\phi) \leq \log(\tan\phi + \sec\phi)$$

and

$$\sin\phi \leq E(k,\phi) \leq \phi$$

◇ **Problem 1-27**

The identities

$$F(k,-\phi) = -F(k,\phi) \qquad E(k,-\phi) = -E(k,\phi)$$
$$F(k,n\pi+\phi) = 2nK(k) + F(k,\phi) \qquad E(k,n\pi+\phi) = 2nE(k) + E(k,\phi)$$

where $n$ is any integer, are useful in computations.

### Elliptic Integral of the First Kind

The oscillating pendulum provides a simple physical model, the accurate description of which involves the elliptic integral of the first kind. Let $\ell$ be the length of a light rod to which is attached a heavy bob and let $g$ be the constant gravitation acceleration. Then, if the angle $\theta$ measures the displacement of the bob from the vertical, the equation of motion is

$$\ell \frac{d^2\theta}{dt^2} + g \sin \theta = 0$$

A first integral can be obtained by writing the differential equation in the equivalent form

$$\frac{\ell}{2} \frac{d}{d\theta} \left(\frac{d\theta}{dt}\right)^2 + g \sin \theta = 0$$

Then, by integrating and determining the constant of integration so that $d\theta/dt = 0$ when $\theta = \theta_0$ we have

$$\left(\frac{d\theta}{dt}\right)^2 = \frac{2g}{\ell}(\cos\theta - \cos\theta_0) = \frac{4g}{\ell}(\sin^2 \tfrac{1}{2}\theta_0 - \sin^2 \tfrac{1}{2}\theta)$$

Introduce a new variable $\phi$ defined by

$$\sin \tfrac{1}{2}\theta = \sin \tfrac{1}{2}\theta_0 \sin \phi$$

so that, in terms of $\phi$, the equation of motion is

$$\left(\frac{d\phi}{dt}\right)^2 = \frac{g}{\ell}(1 - \sin^2 \tfrac{1}{2}\theta_0 \sin^2 \phi)$$

Hence, the period of the pendulum is

$$\text{Period} = 4\sqrt{\frac{\ell}{g}} \int_0^{\frac{1}{2}\pi} \frac{d\phi}{\sqrt{1 - \sin^2 \tfrac{1}{2}\theta_0 \sin^2 \phi}} = 4\sqrt{\frac{\ell}{g}} K(\sin \tfrac{1}{2}\theta_0) \quad (1.56)$$

The first mathematician to deal with elliptic *functions* as opposed to elliptic *integrals* was Gauss but the first results were published by Niels Henrik Abel (1802–1829) and Carl Gustav Jacob Jacobi (1804–1851). Jacobi regarded the inverse of the elliptic integral

$$u = \int_0^\phi \frac{d\phi}{\sqrt{1 - k^2 \sin^2 \phi}}$$

as fundamental and denoted the amplitude $\phi$ as $\operatorname{am} u$. Then he defined

$$\cos \phi = \cos \operatorname{am} u \equiv \operatorname{cn} u$$
$$\sin \phi = \sin \operatorname{am} u \equiv \operatorname{sn} u$$

and

$$\Delta \phi = \sqrt{1 - k^2 \sin^2 \phi} = \Delta \operatorname{am} u \equiv \operatorname{dn} u$$

called *Jacobian elliptic functions* about which there exists an extensive literature which is beyond the scope of this chapter. Suffice it to say that elliptic functions are periodic since

$$\operatorname{sn}(u + 4K) = \operatorname{sn} u \qquad \operatorname{cn}(u + 4K) = \operatorname{cn} u \qquad \operatorname{dn}(u + 2K) = \operatorname{dn} u$$

which can be demonstrated as some of the many important properties of these functions. Indeed, it is for this reason that the complete elliptic integral of the first kind $K$ is frequently referred to as the *quarter period*.

## Landen's Transformation

An interesting and important transformation of elliptic integals was discovered and reported in the *Philosophical Transactions of the Royal Society* of 1775 by the English mathematician John Landen (1719–1790). Landen's transformation can be based on trigonometric identities associated with the triangle one of whose angles is $\theta$ with opposite side unity, and the other, $2\phi - \theta$ with opposite side $\beta$ where $0 \leq \beta \leq 1$. The law of tangents and the law of sines for this triangle are

$$\tan(\theta - \phi) = \frac{1 - \beta}{1 + \beta} \tan \phi \tag{1.57}$$

and

$$\beta \sin \theta = \sin(2\phi - \theta) \tag{1.58}$$

The law of sines can also be written as

$$\tan \theta = \frac{\sin 2\phi}{\beta + \cos 2\phi} \tag{1.59}$$

from which

$$\sin^2 \theta = \frac{\tan^2 \theta}{1 + \tan^2 \theta} = \frac{\sin^2 2\phi}{1 + \beta^2 + 2\beta \cos 2\phi}$$

and

$$d(\tan \theta) = (1 + \tan^2 \theta) \, d\theta = \frac{2(1 + \beta \cos 2\phi)}{(\beta + \cos 2\phi)^2} \, d\phi$$

Therefore, we obtain

$$1 - \beta^2 \sin^2 \theta = \frac{(1 + \beta \cos 2\phi)^2}{1 + \beta^2 + 2\beta \cos 2\phi}$$

and

$$d\theta = \frac{2(1 + \beta \cos 2\phi)}{1 + \beta^2 + 2\beta \cos 2\phi} \, d\phi$$

Then, by writing

$$1 + \beta^2 + 2\beta \cos 2\phi = (1 + \beta)^2 (1 - k^2 \sin^2 \phi)$$

with $k^2$ defined as

$$k^2 = \frac{4\beta}{(1 + \beta)^2} \qquad \text{so that} \qquad \beta = \frac{1 - \sqrt{1 - k^2}}{1 + \sqrt{1 - k^2}} \tag{1.60}$$

we have
$$\frac{d\theta}{\sqrt{1-\beta^2 \sin^2 \theta}} = \frac{2\, d\phi}{(1+\beta)\sqrt{1-k^2 \sin^2 \phi}}$$
As a consequence, we obtain the following identity for elliptic integrals of the first kind:
$$F(k,\phi) = \tfrac{1}{2}(1+\beta)F(\beta,\theta) \tag{1.61}$$

The effect of the transformation from $(k,\phi)$ to $(\beta,\theta)$ is to *decrease the modulus and increase the amplitude*—the latter assertion being verified from Eq. (1.57). It is precisely for this reason that Landen's transformation leads to clever and efficient algorithms for the numerical evaluation of elliptic integrals.

### Gauss' Method of the Arithmetic-Geometric Mean

The identity (1.61) can, of course, be used recursively. For this purpose, we write
$$F(k_n,\phi_n) = \tfrac{1}{2}(1+k_{n+1})F(k_{n+1},\phi_{n+1})$$
$$k_{n+1} = \frac{1-\sqrt{1-k_n^2}}{1+\sqrt{1-k_n^2}} \tag{1.62}$$
$$\tan(\phi_{n+1}-\phi_n) = \frac{1-k_{n+1}}{1+k_{n+1}}\tan\phi_n$$

with $k_0 = k$ and $\phi_0 = \phi$. Since the modulus $k_n$ is steadily decreasing, let $N$ be the value of $n$ for which $k_N$ is essentially zero (to a specified tolerance). Then, since $F(0,\phi_N) = \phi_N$, we have
$$F(k,\phi) = \tfrac{1}{2}(1+k_1)\cdot\tfrac{1}{2}(1+k_2)\ldots\tfrac{1}{2}(1+k_{N-1})\cdot\tfrac{1}{2}\cdot 1\cdot \phi_N$$
as a method for evaluating $F(k,\phi)$.

Gauss converted this process into a beautifully simple algorithm which is based on the two sequences $a_0, a_1, \ldots$ and $b_0, b_1, \ldots$ generated as follows:
$$a_{n+1} = \tfrac{1}{2}(a_n+b_n) \qquad b_{n+1} = \sqrt{a_n b_n} \tag{1.63}$$
with $a_0 = 1$ and $b_0 = \sqrt{1-k_0^2}$. Thus, each new $a$ and $b$ is, respectively, the *arithmetic mean* and the *geometric mean* between the previous $a$ and $b$. (See Appendix A.) It is not difficult to verify that
$$k_{n+1} = \frac{a_n - b_n}{a_n + b_n}$$
$$\frac{1+k_{n+1}}{2} = \frac{a_n}{a_n+b_n} = \frac{a_n}{2a_{n+1}} \tag{1.64}$$
$$\frac{1-k_{n+1}}{1+k_{n+1}} = \frac{b_n}{a_n}$$

Sect. 1.5]  Elliptic Integrals  73

Thus, when $N$ is the value of $n$ for which $a_N - b_N$ is essentially zero, we have

$$F(k,\phi) \approx \frac{\phi_N}{2^N a_N} \tag{1.65}$$

It is important to remember that $\phi_{n+1} > \phi_n$ when using the recursion formula

$$\tan(\phi_{n+1} - \phi_n) = \frac{b_n}{a_n} \tan \phi_n \tag{1.66}$$

to generate the sequence $\phi_0, \phi_1, \ldots, \phi_N$.

◇ **Problem 1-28**
For the pendulum problem with an amplitude of $\theta_0 = \frac{1}{2}\pi$, calculate the time required for the bob to travel from the position for which $\theta = \frac{1}{2}\pi$ to $\theta = \frac{1}{4}\pi$.

### Elliptic Integral of the Second Kind

The simplest example of an elliptic integral of the second kind occurs in the calculation of the arc length of an ellipse—indeed, it is for this reason that the terminology "elliptic" has been used to describe these special integrals.

Write the equations of the ellipse in parametric form as

$$x = a\cos\theta \qquad y = b\sin\theta$$

Then the differential of arc length $ds$ is

$$ds^2 = dx^2 + dy^2 = (a^2 \sin^2\theta + b^2 \cos^2\theta)\,d\theta^2 = [a^2 - (a^2 - b^2)\cos^2\theta]\,d\theta^2$$

Since $a^2 - b^2 = a^2 e^2$, where $e$ is the eccentricity of the ellipse, then

$$ds = a\sqrt{1 - e^2 \cos^2\theta}\,d\theta$$

Therefore, if we define $\phi = \frac{1}{2}\pi - \theta$, in order to put the integral into the normal form of Eq. (1.53), the perimeter of the ellipse is

$$\text{Perimeter} = 4a \int_0^{\frac{1}{2}\pi} \sqrt{1 - e^2 \sin^2\phi}\,d\phi = 4aE(e) \tag{1.67}$$

Landen's transformation for elliptic integrals of the second kind is somewhat more involved than for the first kind. Proceeding as before, we can establish

$$\sqrt{1 - \beta^2 \sin^2\theta}\,d\theta = \frac{2(1 + \beta \cos 2\phi)^2}{(1+\beta)^3(1 - k^2 \sin^2\phi)^{\frac{3}{2}}}\,d\phi$$

which, after some modest algebra, yields

$$E(\beta,\theta) = (1-\beta)F(k,\phi) + \tfrac{1}{2}(1+\beta)\left[E(k,\phi) + (1-k^2)\Pi(k^2,k,\phi)\right]$$

Fortunately, the elliptic integral of the third kind in this equation is rather special (the characteristic $n$ is equal to the parameter $m = k^2$) and can be expressed as

$$(1 - k^2)\Pi(k^2, k, \phi) = E(k, \phi) - \frac{k^2 \sin 2\phi}{2\sqrt{1 - k^2 \sin^2 \phi}} \quad (1.68)$$

Now, since

$$\sin \theta = \frac{\sin 2\phi}{(1 + \beta)\sqrt{1 - k^2 \sin^2 \phi}}$$

we obtain

$$E(\beta, \theta) = (1 - \beta)F(k, \phi) + (1 + \beta)E(k, \phi) - \beta \sin \theta$$

By rearranging and using Eq. (1.61), we obtain

$$E(k, \phi) = \frac{1}{1 + \beta}E(\beta, \theta) - \frac{1 - \beta}{2}F(\beta, \theta) + \frac{\beta}{1 + \beta}\sin \theta \quad (1.69)$$

as the identity for elliptic integrals of the second kind corresponding to that for $F(k, \phi)$ in Eq. (1.61).

◊ **Problem 1–29**

Derive the expression for the special case of the elliptic integral of the third kind $\Pi(k^2, k, \phi)$ given in Eq. (1.68).

HINT: First show that

$$\frac{d}{d\phi}\sqrt{1 - k^2 \sin^2 \phi} = -\frac{k^2 \sin 2\phi}{2\sqrt{1 - k^2 \sin^2 \phi}}$$

$$\frac{d^2}{d\phi^2}\sqrt{1 - k^2 \sin^2 \phi} = \frac{1 - k^2}{(1 - k^2 \sin^2 \phi)^{\frac{3}{2}}} - \sqrt{1 - k^2 \sin^2 \phi}$$

### Evaluating Complete Elliptic Integrals

Recursion formulas for the complete elliptic integrals of the first and second kind can also be developed.

First, from the triangle defining Landen's transformation, we note that $\phi = \frac{1}{2}\pi$ corresponds to $\theta = \pi$. Then, using the identities derived in Prob. 1–27, we have $F(k, \pi) = 2K(k)$ so that Eq. (1.61) becomes

$$K(k_n) = (1 + k_{n+1})K(k_{n+1}) \quad (1.70)$$

or, in terms of the Gaussian sequences $a_0, a_1, \ldots$ and $b_0, b_1, \ldots,$

$$K(k_n) = \frac{a_n}{a_{n+1}}K(k_{n+1}) \quad (1.71)$$

This equation, applied recursively, gives

$$K(k) \approx \frac{\pi}{2a_N} \quad (1.72)$$

and should be compared to Eq. (1.65).

In a similar manner, since $E(k, \pi) = 2E(k)$, Eq. (1.69) becomes

$$E(k_n) = \frac{2}{1+k_{n+1}} E(k_{n+1}) - (1 - k_{n+1})K(k_{n+1}) \qquad (1.73)$$

or

$$E(k_n) = \frac{2a_{n+1}}{a_n} E(k_{n+1}) - \frac{b_n}{a_{n+1}} K(k_{n+1}) \qquad (1.74)$$

However, rather than work directly with $E(k)$, the function $Q(k)$, defined as

$$Q(k) = \frac{K(k) - E(k)}{K(k)} \qquad (1.75)$$

is more convenient because it has such a simple recursion formula. Using Landen's transformation, it is readily seen that

$$Q(k) = \frac{2}{(1+\beta)^2} [Q(\beta) - \beta] \qquad (1.76)$$

or, in terms of the Gaussian sequences,

$$Q(k_n) = \frac{2a_{n+1}}{a_n^2} [a_{n+1} Q(k_{n+1}) + c_{n+1}] \qquad (1.77)$$

where $c_{n+1}$ is determined from

$$c_{n+1} = \tfrac{1}{2}(a_n - b_n) \qquad (1.78)$$

and generates a third sequence to be appended to the other two.

By successive application of this recursion formula,† we obtain

$$Q(k) \approx \sum_{i=1}^{N} 2^i a_i c_i \qquad (1.79)$$

Finally, from the definition of $Q(k)$, we have

$$E(k) \approx \frac{\pi}{2a_N} \left(1 - \sum_{i=1}^{N} 2^i a_i c_i\right) \qquad (1.80)$$

as a convenient algorithm for calculating $E(k)$—the complete elliptic integral of the second kind.

---

† Note that $Q(0) = 0$.

## ◇ Problem 1–30

By differentiating under the integral sign with respect to the parameter $m \equiv k^2$, show that $K(m)$ and $E(m)$, the complete elliptic integrals of the first and second kind, satisfy the following first order differential equations:

$$2m(1-m)\frac{dK}{dm} = E - (1-m)K$$

$$2m\frac{dE}{dm} = E - K$$

Differentiate a second time and obtain

$$m(1-m)\frac{d^2K}{dm^2} + (1-2m)\frac{dK}{dm} - \tfrac{1}{4}K = 0$$

$$m(1-m)\frac{d^2E}{dm^2} + (1-m)\frac{dE}{dm} + \tfrac{1}{4}E = 0$$

each of which is a special case of Gauss' equation for hypergeometric functions. Therefore,

$$\frac{2}{\pi}K(m) = F(\tfrac{1}{2},\tfrac{1}{2};1;m) \quad \text{and} \quad \frac{2}{\pi}E(m) = F(-\tfrac{1}{2},\tfrac{1}{2};1;m)$$

HINT: Use the equations in the "hint" of Prob. 1–29.

## ◇ Problem 1–31

Use the results of Prob. 1–30 and the Gauss identity (8) from Sect. 1.1 to derive

$$(1-m)\frac{K(m)}{E(m)} = 1 - \frac{m}{2}\frac{F(\tfrac{1}{2},\tfrac{1}{2};2;m)}{F(\tfrac{1}{2},-\tfrac{1}{2};1;m)}$$

Then, obtain the continued fraction expansion

$$E(m) = \cfrac{(1-m)K(m)}{1 - \cfrac{m}{2 - \cfrac{3m}{4 - \cfrac{m}{2 - \cfrac{5m}{8 - \cfrac{3m}{2 - \cfrac{7m}{12 - \cfrac{5m}{2 - \ddots}}}}}}}}$$

NOTE: This continued fraction can be used to determine $E(m)$ after $K(m)$ has been calculated from Eq. (1.72).

## Sect. 1.5] Elliptic Integrals

◇ **Problem 1–32**
Calculate
(a) the period of the pendulum whose amplitude is $\theta_0 = \frac{1}{2}\pi$
and
(b) the perimeter of the ellipse for which $a = 1$ and $e = \frac{1}{2}$.
In the latter case, use the series approximation of Eq. (1.80) as well as the continued fraction of Prob. 1–31 and compare the two results.

ANSWER:

(a) $$\text{Period} = 7.4162987\sqrt{\frac{\ell}{g}}$$

(b) $$\text{Perimeter} = 5.8698488$$

◇ **Problem 1–33**
For the sine curve
$$y = a \sin x$$
the perimeter of one arch is given by
$$2\sqrt{1+a^2}\, E\!\left(a\big/\sqrt{1+a^2}\right)$$
Determine a numerical value for the perimeter when the amplitude $a$ is unity.

### Jacobi's Zeta Function

Finally, we shall obtain a formula for approximating the value of an incomplete elliptic integral of the second kind analogous to Eq. (1.65) for integrals of the first kind. For this purpose, we follow a similar pattern of argument, which proved to be useful for the complete elliptic integral of the second kind, by introducing

$$Z(k,\phi) = E(k,\phi) - \frac{E(k)}{K(k)}F(k,\phi) \qquad (1.81)$$

called *Jacobi's Zeta function*.

The Zeta function $Z(k,\phi)$ is appropriate because it has a simpler form, after application of Landen's transformation, than the elliptic integral $E(k,\phi)$. In this respect, it is similar to the function $Q(k)$ defined in the previous subsection. Indeed, we have

$$Z(k,\phi) = \frac{1}{1+\beta}Z(\beta,\theta) + \frac{\beta}{1+\beta}\sin\theta \qquad (1.82)$$

or, in recursive form,

$$Z(k_n,\phi_n) = \frac{1}{1+k_{n+1}}Z(k_{n+1},\phi_{n+1}) + \frac{k_{n+1}}{1+k_{n+1}}\sin\phi_{n+1} \qquad (1.83)$$

and should be compared to Eqs. (1.69) and (1.73).

In terms of the Gaussian sequences, Eqs. (1.63) and (1.78), we have

$$Z(k_n, \phi_n) = \frac{a_{n+1}}{a_n} Z(k_{n+1}, \phi_{n+1}) + \frac{c_{n+1}}{a_n} \sin \phi_{n+1} \qquad (1.84)$$

Therefore, since $Z(0, \phi) = 0$, then Jacobi's Zeta function is approximated by

$$Z(k, \phi) \approx \sum_{i=1}^{N} c_i \sin \phi_i \qquad (1.85)$$

Hence, using Eqs. (1.65), (1.79), and (1.85), we obtain

$$E(k, \phi) \approx \frac{\phi_N}{2^N a_N} \left(1 - \sum_{i=1}^{N} 2^i a_i c_i\right) + \sum_{i=1}^{N} c_i \sin \phi_i \qquad (1.86)$$

as an appropriate algorithm for calculating the incomplete elliptic integral of the second kind.

◇ **Problem 1-34**

For the ellipse with $a = 1$ and $e = \frac{1}{2}$, calculate the length of the arc from $(x = 1,\ y = 0)$ to $(x = \frac{1}{2}\sqrt{2},\ y = \frac{1}{4}\sqrt{6})$.

Chapter 2

# Some Basic Topics in Analytical Dynamics

INTERACTIONS OF MATERIAL BODIES AND THEIR RESULTING MOTIONS due to their mutual attractions is the subject of *Analytical Dynamics*. Leonhard Euler (1707–1783) was the first to use mathematical rather than geometrical methods for addressing problems in dynamics and, therefore, is considered the father of analytical dynamics. Indeed, Euler applied mathematics to study the entire realm of physics. For example, he calculated the perturbative effects of other celestial bodies on the motion of a planet as well as the motion of a projectile subject to atmospheric drag. He studied the bending of beams, the compression of columns, and the propagation of sound. The basic equation of motion of the flow of an ideal fluid is his and he even applied it to study the flow of blood through the human body. He wrote three volumes on optical instruments and made important contributions to the refraction and dispersion of light waves. He gave the first significant treatment of the calculus of variations—the list is almost endless. As a measure of the dauntless spirit of the man, several of his books and approximately four hundred of his papers where written during the last seventeen years of his life when he was totally blind.

Leonhard Euler ranks with Archimedes, Newton, and Gauss, and was the key mathematician and theoretical physicist of the eighteenth century. Although his father was a preacher and wanted him to study theology, he decided instead to pursue mathematics and completed his work at the Swiss University of Basel, near his birth place, at the age of fifteen. He studied there under John Bernoulli and, except for the period 1741–1766 when he was in Berlin at the request of Frederick the Great, his brilliant career of mathematical research was centered in Russia at the St. Petersburg Academy under the auspices of Catherine the Great.

Despite his enormous contribution to mathematics, Euler was a devoted family man with thirteen children and many grandchildren. They benefited from the many hours he spent instructing them, playing scientific games and reading to them from the Bible in the evenings. He also had a prodigious memory and could carry out difficult mathematical calculations in his head without having to resort to tables. Euler enjoyed universal respect from his colleagues and at the end of his life he could regard all of the European mathematicians as his pupils. His name is everywhere in mathematics: Euler's constants, Eulerian integrals, Euler's formulas, Euler's theorems, Euler angles, Euler parameters, Euler axes, ...

Joseph-Louis Lagrange (1736–1813) contributed much to mathematics in such diverse fields as the theory of numbers, algebraic equations, the calculus, differential equations, and the calculus of variations but his chief interest was celestial mechanics. He was born in Turin Italy of both French and Italian lineage. As a boy he had little interest in mathematics until he read an essay by Sir Edmond Halley extolling the merits of Newton's calculus. From this inspiration his career found direction and, when but nineteen years old, he became a professor of mathematics at the Royal Artillery School of Turin. All of his life, however, he envied Sir Isaac Newton since there was, indeed, only one universe and Newton had already discovered its mathematical laws.

Curiously, Lagrange's most famous work *Mécanique Analytique*, which was essentially complete in 1782, did not appear until 1788 for lack of a publisher. Indeed, a friend of his had to agree to purchase all copies remaining unsold after a certain date before the printer would risk the expense of publication.

Lagrange's name is often linked with Euler's since they shared many interests. For example, the Euler-Lagrange equation is fundamental in the calculus of variations. To some extent, Euler was a mentor of Lagrange being the first to give him encouragement at the tender age of nineteen. They remained friendly rivals for life.

In this chapter, we develop certain fundamentals of analytical dynamics—much of which is attributed to Euler—the transformation of coordinates and rotation of vectors. Because of its current interest in the attitude control of space vehicles, we also discuss Sir William Hamilton's *quaternions*. The last two sections treat the important relations of the $n$-body problem and various forms of certain kinematical relations. All of this is basic to the rest of our work.

Although vector-matrix notation is used throughout, the reader should remember that these tools came into being much later in history. Neither Euler nor Lagrange would recognize the notation but both would have been impressed with the elegance and convenience of the expressions.

## 2.1   Transformation of Coordinates

The vectors with which we are concerned, such as position and velocity vectors, are usually expressed as components in an orthogonal coordinate system. Occasions arise when skewed coordinate axes are more natural for special purposes, but these cases are exceptional. Changes in coordinate systems will be desired both in translation and rotation. Translation of coordinates, in which the origin is moved to a new location without changing the direction of the axes, is straight-forward and needs no elaboration here. In this section, we concentrate on the rotation of coordinate systems.

## Euler's Theorem

Consider two rectangular cartesian coordinate systems with a common origin and call them $S_1(x, y, z)$ and $S_2(\xi, \eta, \varsigma)$. Let the $S_1$ axes be fixed in space while the $S_2$ axes are free to turn in any manner about the origin. Now, if the $S_2$ axes are initially aligned with the $S_1$ axes, we will show that any other configuration can be attained by a simple rotation of the $S_2$ triad about some definite line through the common origin. This is Euler's fundamental theorem on rigid body rotation—that a rotation about a point is always equivalent to a rotation about a line through the point.

For the proof, consider an arbitrary configuration of the $S_2$ frame with respect to $S_1$. The $x, \xi$ axes determine a plane as do also the $y, \eta$ axes. Now construct two planes: one perpendicular to the $x, \xi$ plane, the other perpendicular to the $y, \eta$ plane—oriented so that each bisects, respectively, the angles between the axes $x, \xi$ and $y, \eta$. The line of intersection of these two planes makes equal angles with the $x$ and $\xi$ axes. It also makes equal angles with the $y$ and $\eta$ axes. Therefore, if this line of intersection were considered to be rigidly attached to the $S_2$ reference frame, its direction would be the same after the new configuration was established as it was initially. Thus, a rotation about this direction will bring the $S_2$ axes from an initial alignment with the $S_1$ axes to the specified configuration.

## The Rotation Matrix

Let $\mathbf{i}_x$, $\mathbf{i}_y$, $\mathbf{i}_z$ and $\mathbf{i}_\xi$, $\mathbf{i}_\eta$, $\mathbf{i}_\varsigma$ be two sets of orthogonal unit vectors parallel to their respective coordinate axes for some particular configuration of the $x, y, z$ system. Then we may write

$$\begin{aligned} \mathbf{i}_\xi &= l_1 \mathbf{i}_x + m_1 \mathbf{i}_y + n_1 \mathbf{i}_z \\ \mathbf{i}_\eta &= l_2 \mathbf{i}_x + m_2 \mathbf{i}_y + n_2 \mathbf{i}_z \\ \mathbf{i}_\varsigma &= l_3 \mathbf{i}_x + m_3 \mathbf{i}_y + n_3 \mathbf{i}_z \end{aligned} \quad (2.1)$$

where $l_1, m_1, \ldots, n_3$ are called *direction cosines*. They are cosines of the nine angles determined by the axes of one triad and the axes of the other.

Consider an arbitrary vector $\mathbf{r}$ expressed in terms of components along the $x, y, z$ axes as well as the $\xi, \eta, \varsigma$ axes. Thus we have two equivalent representations

$$\mathbf{r} = x \mathbf{i}_x + y \mathbf{i}_y + z \mathbf{i}_z = \xi \mathbf{i}_\xi + \eta \mathbf{i}_\eta + \varsigma \mathbf{i}_\varsigma$$

To obtain any coordinate in one system in terms of those of the other system, we simply take the scalar product of the above identity with the corresponding unit vector. In this manner, we may obtain two sets of three linear equations which may be written in vector-matrix form as

$$\begin{bmatrix} x \\ y \\ z \end{bmatrix} = \mathbf{R} \begin{bmatrix} \xi \\ \eta \\ \varsigma \end{bmatrix} \quad \text{or} \quad \begin{bmatrix} \xi \\ \eta \\ \varsigma \end{bmatrix} = \mathbf{R}^T \begin{bmatrix} x \\ y \\ z \end{bmatrix} \qquad (2.2)$$

The matrix $\mathbf{R}$, called the *rotation matrix*, is then

$$\mathbf{R} = \begin{bmatrix} \mathbf{i}_x \cdot \mathbf{i}_\xi & \mathbf{i}_x \cdot \mathbf{i}_\eta & \mathbf{i}_x \cdot \mathbf{i}_\varsigma \\ \mathbf{i}_y \cdot \mathbf{i}_\xi & \mathbf{i}_y \cdot \mathbf{i}_\eta & \mathbf{i}_y \cdot \mathbf{i}_\varsigma \\ \mathbf{i}_z \cdot \mathbf{i}_\xi & \mathbf{i}_z \cdot \mathbf{i}_\eta & \mathbf{i}_z \cdot \mathbf{i}_\varsigma \end{bmatrix} = \begin{bmatrix} l_1 & l_2 & l_3 \\ m_1 & m_2 & m_3 \\ n_1 & n_2 & n_3 \end{bmatrix} \qquad (2.3)$$

From Eq. (2.2) it is clear that

$$\mathbf{R}\mathbf{R}^T = \mathbf{R}^T\mathbf{R} = \mathbf{I}$$

where $\mathbf{I}$ is the three-dimensional identity matrix, so that

$$\mathbf{R}^T = \mathbf{R}^{-1} \qquad (2.4)$$

Thus, $\mathbf{R}$ is an *orthogonal matrix* and the two sets of vectors, formed from the rows and columns of $\mathbf{R}$, each constitute an orthonormal set. Specifically, then we have

$$\begin{aligned} l_1^2 + l_2^2 + l_3^2 &= 1 & l_1 m_1 + l_2 m_2 + l_3 m_3 &= 0 \\ m_1^2 + m_2^2 + m_3^2 &= 1 & l_1 n_1 + l_2 n_2 + l_3 n_3 &= 0 \\ n_1^2 + n_2^2 + n_3^2 &= 1 & m_1 n_1 + m_2 n_2 + m_3 n_3 &= 0 \\ l_1^2 + m_1^2 + n_1^2 &= 1 & l_1 l_2 + m_1 m_2 + n_1 n_2 &= 0 \\ l_2^2 + m_2^2 + n_2^2 &= 1 & l_1 l_3 + m_1 m_3 + n_1 n_3 &= 0 \\ l_3^2 + m_3^2 + n_3^2 &= 1 & l_2 l_3 + m_2 m_3 + n_2 n_3 &= 0 \end{aligned}$$

Furthermore, it is useful to note that each element of $\mathbf{R}$ is its own cofactor; that is,

$$\begin{aligned} l_1 &= m_2 n_3 - m_3 n_2 & l_2 &= m_3 n_1 - m_1 n_3 & l_3 &= m_1 n_2 - m_2 n_1 \\ m_1 &= l_3 n_2 - l_2 n_3 & m_2 &= l_1 n_3 - l_3 n_1 & m_3 &= l_2 n_1 - l_1 n_2 \\ n_1 &= l_2 m_3 - l_3 m_2 & n_2 &= l_3 m_1 - l_1 m_3 & n_3 &= l_1 m_2 - l_2 m_1 \end{aligned}$$

These expressions will prove useful in some of the problems which follow.

## ◇ Problem 2-1

If $\mathbf{i}_\xi - \mathbf{i}_x$ and $\mathbf{i}_\eta - \mathbf{i}_y$ are not parallel, the rotation axis of Euler's theorem has the direction of the vector

$$\begin{aligned}
\mathbf{w} &= (\mathbf{i}_\xi - \mathbf{i}_x) \times (\mathbf{i}_\eta - \mathbf{i}_y) \\
&= (n_1 + l_3)\mathbf{i}_x + (n_2 + m_3)\mathbf{i}_y + (n_3 - m_2 - l_1 + 1)\mathbf{i}_z \\
&= (n_1 + l_3)\mathbf{i}_\xi + (n_2 + m_3)\mathbf{i}_\eta + (n_3 - m_2 - l_1 + 1)\mathbf{i}_\varsigma
\end{aligned}$$

Also, $\mathbf{w}$ is the characteristic vector of the rotation matrix $\mathbf{R}$ corresponding to the characteristic value of unity. By direct calculation, verify that

$$\mathbf{R}\mathbf{w} = \mathbf{w}$$

In other words, the Euler axis $\mathbf{w}$ is unaltered by the rotation defined by the rotation matrix $\mathbf{R}$.

## ◇ Problem 2-2

For the *spherical coordinate system* illustrated in Fig. 2.1, the unit vectors, $\mathbf{i}_\phi$, $\mathbf{i}_\theta$, $\mathbf{i}_r$ are related to $\mathbf{i}_x$, $\mathbf{i}_y$, $\mathbf{i}_z$ by

$$\mathbf{i}_\phi = \mathbf{i}_\theta \times \mathbf{i}_r$$

where

$$\mathbf{i}_\theta = -\sin\theta\, \mathbf{i}_x + \cos\theta\, \mathbf{i}_y$$

and

$$\mathbf{i}_r = \sin\phi \cos\theta\, \mathbf{i}_x + \sin\phi \sin\theta\, \mathbf{i}_y + \cos\phi\, \mathbf{i}_z$$

The associated rotation matrix is

$$\mathbf{R} = \begin{bmatrix} \cos\phi\cos\theta & -\sin\theta & \sin\phi\cos\theta \\ \cos\phi\sin\theta & \cos\theta & \sin\phi\sin\theta \\ -\sin\phi & 0 & \cos\phi \end{bmatrix}$$

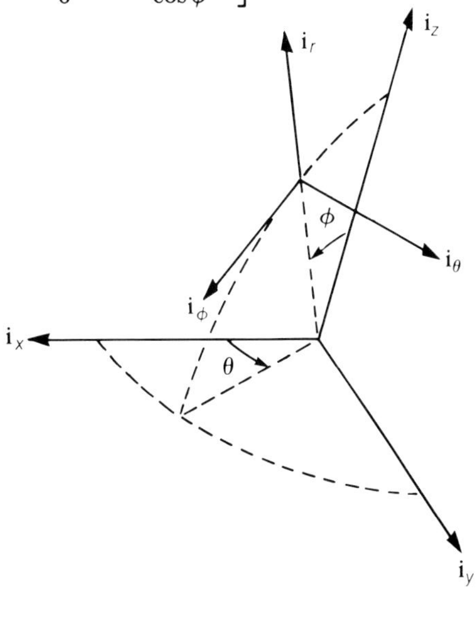

**Fig. 2.1:** Spherical coordinate system.

## Euler Angles

In celestial mechanics, as well as in the study of rigid body dynamics, Euler angles are frequently used to relate two coordinate systems. As illustrated in Fig. 2.2, the $\mathbf{i}_\xi$, $\mathbf{i}_\eta$, $\mathbf{i}_\zeta$ axes are the final orientation of an orthogonal triad, which originally coincided with $\mathbf{i}_x$, $\mathbf{i}_y$, $\mathbf{i}_z$, after undergoing three successive rotations described as follows:

1. A positive rotation about $\mathbf{i}_z$ through an angle $\Omega$ which establishes the direction of the unit vector $\mathbf{i}_n$ as

$$\mathbf{i}_n = \cos\Omega\, \mathbf{i}_x + \sin\Omega\, \mathbf{i}_y \tag{2.5}$$

2. A positive rotation about $\mathbf{i}_n$ through an angle $i$ which establishes the direction of $\mathbf{i}_\zeta$ as

$$\mathbf{i}_\zeta = \sin i\, \mathbf{i}_n \times \mathbf{i}_z + \cos i\, \mathbf{i}_z$$

or, using Eq. (2.5),

$$\mathbf{i}_\zeta = \sin\Omega \sin i\, \mathbf{i}_x - \cos\Omega \sin i\, \mathbf{i}_y + \cos i\, \mathbf{i}_z \tag{2.6}$$

3. A positive rotation about $\mathbf{i}_\zeta$ through an angle $\omega$ which establishes the direction of $\mathbf{i}_\xi$ as

$$\mathbf{i}_\xi = \cos\omega\, \mathbf{i}_n + \sin\omega\, \mathbf{i}_m \tag{2.7}$$

where $\mathbf{i}_m = \mathbf{i}_\zeta \times \mathbf{i}_n$. Then, from Eqs. (2.5) and (2.6), we obtain

$$\mathbf{i}_m = -\sin\Omega \cos i\, \mathbf{i}_x + \cos\Omega \cos i\, \mathbf{i}_y + \sin i\, \mathbf{i}_z \tag{2.8}$$

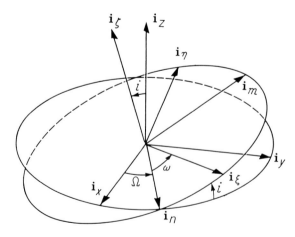

Fig. 2.2: Euler angles.

The components of the unit vectors $\mathbf{i}_\xi$ and $\mathbf{i}_\zeta$ are now determined in the $x$, $y$, $z$ reference frame. We have also $\mathbf{i}_\eta = \mathbf{i}_\zeta \times \mathbf{i}_\xi$ so that all of the direction cosine elements of the rotation matrix **R** in Eq. (2.3) are obtained in terms of the three Euler angles $\Omega$, $i$, $\omega$ as follows:

$$\begin{aligned}
l_1 &= \cos\Omega\cos\omega - \sin\Omega\sin\omega\cos i \\
l_2 &= -\cos\Omega\sin\omega - \sin\Omega\cos\omega\cos i \\
l_3 &= \sin\Omega\sin i \\
m_1 &= \sin\Omega\cos\omega + \cos\Omega\sin\omega\cos i \\
m_2 &= -\sin\Omega\sin\omega + \cos\Omega\cos\omega\cos i \\
m_3 &= -\cos\Omega\sin i \\
n_1 &= \sin\omega\sin i \\
n_2 &= \cos\omega\sin i \\
n_3 &= \cos i
\end{aligned} \quad (2.9)$$

◊ **Problem 2-3**

The direction cosines appearing as elements of the rotation matrix may be determined by using the cosine law of spherical trigonometry. For example, the angle between $\mathbf{i}_x$ and $\mathbf{i}_\xi$ forms one side of a spherical triangle whose other two sides are $\Omega$ and $\omega$. The included angle between these sides is $\pi - i$. Therefore, from the cosine law, derived in Appendix B on vector algebra, we have

$$l_1 = \mathbf{i}_x \cdot \mathbf{i}_\xi = \cos\Omega\cos\omega + \sin\Omega\sin\omega\cos(\pi - i)$$
$$= \cos\Omega\cos\omega - \sin\Omega\sin\omega\cos i$$

Derive the other direction cosines in a similar fashion.

## Elementary Rotation Matrices

If the Euler axis coincides with a coordinate axis, the associated rotation matrix is called an *elementary rotation matrix*. For example, if the Euler axis of rotation is aligned with $\mathbf{i}_x = \mathbf{i}_\xi$ and the rotation angle is $\theta$, then from Eq. (2.3) we have

$$\mathbf{R}_x(\theta) = \begin{bmatrix} 1 & 0 & 0 \\ 0 & \cos\theta & -\sin\theta \\ 0 & \sin\theta & \cos\theta \end{bmatrix}$$

Similarly, for a rotation about the $y$ axis: $\mathbf{i}_y = \mathbf{i}_\eta$, and about the $z$ axis: $\mathbf{i}_z = \mathbf{i}_\varsigma$, the elementary matrices are

$$\mathbf{R}_y(\theta) = \begin{bmatrix} \cos\theta & 0 & \sin\theta \\ 0 & 1 & 0 \\ -\sin\theta & 0 & \cos\theta \end{bmatrix} \quad \text{and} \quad \mathbf{R}_z(\theta) = \begin{bmatrix} \cos\theta & -\sin\theta & 0 \\ \sin\theta & \cos\theta & 0 \\ 0 & 0 & 1 \end{bmatrix}$$

Consider again the two rectangular coordinate systems $S_1(x, y, z)$ and $S_2(\xi, \eta, \varsigma)$ initially aligned. Suppose, for example, that the $S_2$ triad is first rotated about the $x$ axis through an angle $\theta$ to establish an intermediate coordinate system $S_i(u, v, w)$. Then it is rotated about the $v$ axis through an angle $\phi$. The equations analogous to (2.2) are

$$\begin{bmatrix} x \\ y \\ z \end{bmatrix} = \mathbf{R}_x(\theta) \begin{bmatrix} u \\ v \\ w \end{bmatrix} \quad \text{and} \quad \begin{bmatrix} u \\ v \\ w \end{bmatrix} = \mathbf{R}_y(\phi) \begin{bmatrix} \xi \\ \eta \\ \varsigma \end{bmatrix}$$

or

$$\begin{bmatrix} x \\ y \\ z \end{bmatrix} = \mathbf{R}_x(\theta) \mathbf{R}_y(\phi) \begin{bmatrix} \xi \\ \eta \\ \varsigma \end{bmatrix}$$

◇ **Problem 2–4**

As an example of elementary matrices, the rotation matrix for the spherical coordinate system of Prob. 2–2 is the product of two separate rotation matrices, i.e.,

$$\mathbf{R} = \begin{bmatrix} \cos\theta & -\sin\theta & 0 \\ \sin\theta & \cos\theta & 0 \\ 0 & 0 & 1 \end{bmatrix} \begin{bmatrix} \cos\phi & 0 & \sin\phi \\ 0 & 1 & 0 \\ -\sin\phi & 0 & \cos\phi \end{bmatrix}$$

Interpret this result geometrically.

Also, the rotation matrix, whose elements are given in Eq. (2.9), may be obtained as the product of three separate rotation matrices

$$\mathbf{R} = \begin{bmatrix} \cos\Omega & -\sin\Omega & 0 \\ \sin\Omega & \cos\Omega & 0 \\ 0 & 0 & 1 \end{bmatrix} \begin{bmatrix} 1 & 0 & 0 \\ 0 & \cos i & -\sin i \\ 0 & \sin i & \cos i \end{bmatrix} \begin{bmatrix} \cos\omega & -\sin\omega & 0 \\ \sin\omega & \cos\omega & 0 \\ 0 & 0 & 1 \end{bmatrix}$$

## 2.2 Rotation of a Vector

We have seen that a transformation of the form Eq. (2.2) can be regarded as a rotation of coordinate system axes in that the quantities $x$, $y$, $z$ and $\xi$, $\eta$, $\varsigma$ are the coordinates of the *same* vector $\mathbf{r}$ referred to different systems of coordinates. However, this equation can be interpreted from a different point of view. It may also be regarded as the rotation of a vector, whose coordinates are $\xi$, $\eta$, $\varsigma$, to produce a new vector, whose coordinates are $x$, $y$, $z$ both *referred to the same coordinate system*.

In order to substantiate this interpretation, consider a vector $\mathbf{r}$ to be rigidly attached to the triad $\mathbf{i}_\xi$, $\mathbf{i}_\eta$, $\mathbf{i}_\varsigma$ so that it will have constant projections $\xi$, $\eta$, $\varsigma$ on these axes

$$\mathbf{r} = \xi \mathbf{i}_\xi + \eta \mathbf{i}_\eta + \varsigma \mathbf{i}_\varsigma$$

despite the orientation of the moving frame. Initially, when the two coordinate frames are aligned, the components of $\mathbf{r}$ along $\mathbf{i}_x$, $\mathbf{i}_y$, $\mathbf{i}_z$ will also be $\xi$, $\eta$, $\varsigma$. Let this be the vector $\mathbf{r}_0$ with

$$\mathbf{r}_0 = \xi \mathbf{i}_x + \eta \mathbf{i}_y + \varsigma \mathbf{i}_z$$

After the rotation, the components of the vector **r** along $\mathbf{i}_x$, $\mathbf{i}_y$, $\mathbf{i}_z$ will be

$$\mathbf{r} = (\xi l_1 + \eta l_2 + \varsigma l_3)\,\mathbf{i}_x + (\xi m_1 + \eta m_2 + \varsigma m_3)\,\mathbf{i}_y + (\xi n_1 + \eta n_2 + \varsigma n_3)\,\mathbf{i}_z$$

which are obtained using Eqs. (2.1). This equation is equivalent to

$$\mathbf{r} = \mathbf{R}\mathbf{r}_0 \tag{2.10}$$

and expresses the relationship between two vectors $\mathbf{r}_0$ and $\mathbf{r}$—one obtained from the other by a rotation about a fixed axis.

### Kinematic Form of the Rotation Matrix

It is useful to explore the concept of the rotation of a vector from a different and, perhaps, more direct approach. Consider a vector **r** rotating with a constant vector angular velocity $\boldsymbol{\omega}$ so that

$$\frac{d\mathbf{r}}{dt} = \boldsymbol{\omega} \times \mathbf{r}$$

Let $l$, $m$, $n$ be the direction cosines of $\boldsymbol{\omega}$ and $d\psi/dt$, the constant angular speed. Then

$$\boldsymbol{\omega} = \frac{d\psi}{dt}(l\,\mathbf{i}_x + m\,\mathbf{i}_y + n\,\mathbf{i}_z) \equiv \frac{d\psi}{dt}\,\mathbf{i}_\omega$$

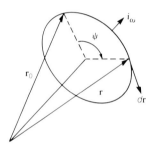

If we define a *skew-symmetric* matrix **S** as

$$\mathbf{S} = \begin{bmatrix} 0 & -n & m \\ n & 0 & -l \\ -m & l & 0 \end{bmatrix} \tag{2.11}$$

then the vector $\mathbf{i}_\omega \times \mathbf{r}$ may be replaced by the matrix-vector product $\mathbf{Sr}$. We have

$$\frac{d\mathbf{r}}{d\psi} = \mathbf{S}\mathbf{r}$$

as a linear vector differential equation for **r** with constant coefficients. The solution,† as a function of $\psi$ for which $\mathbf{r}(0) = \mathbf{r}_0$, may be written as

$$\mathbf{r}(\psi) = \left(\mathbf{I} + \psi\mathbf{S} + \frac{1}{2!}\psi^2\mathbf{S}^2 + \frac{1}{3!}\psi^3\mathbf{S}^3 + \cdots\right)\mathbf{r}_0$$

The matrix coefficient of $\mathbf{r}_0$ is the rotation matrix **R** expressed as an infinite matrix series. To obtain a more useful expression, we recall that

$$\mathbf{S}\mathbf{r}_0 = \mathbf{i}_\omega \times \mathbf{r}_0$$

Therefore,

---

† If the quantities involved in the differential equation were scalars only, we could, correctly, write the solution as $\mathbf{r} = e^{\psi \mathbf{S}}\mathbf{r}_0$. For the case at hand, and in general, if we define the exponential function of a matrix by the infinite series, the formal operation implied by $e^{\psi \mathbf{S}}$ will be valid.

$$\mathbf{S}^2\mathbf{r}_0 = \mathbf{i}_\omega \times (\mathbf{i}_\omega \times \mathbf{r}_0) = (\mathbf{i}_\omega \cdot \mathbf{r}_0)\mathbf{i}_\omega - \mathbf{r}_0$$

and

$$\mathbf{S}^3\mathbf{r}_0 = \mathbf{i}_\omega \times [\mathbf{i}_\omega \times (\mathbf{i}_\omega \times \mathbf{r}_0)] = -\mathbf{i}_\omega \times \mathbf{r}_0 = -\mathbf{S}\mathbf{r}_0$$

from which we conclude that

$$\mathbf{S}^3 = -\mathbf{S}$$

Thus, all powers of the matrix $\mathbf{S}$ are either $\pm\mathbf{S}$ or $\pm\mathbf{S}^2$ and we have

$$\mathbf{R} = \mathbf{I} + \left(\psi - \frac{1}{3!}\psi^3 + \cdots\right)\mathbf{S} + \left(\frac{1}{2!}\psi^2 - \frac{1}{4!}\psi^4 + \cdots\right)\mathbf{S}^2$$

or

$$\mathbf{R} = \mathbf{I} + \sin\psi\,\mathbf{S} + (1 - \cos\psi)\mathbf{S}^2 \tag{2.12}$$

This form of the rotation matrix† shows the explicit dependence on the kinematical quantities—the direction cosines $l$, $m$, $n$ of the rotation axis (as elements of $\mathbf{S}$) and the rotation angle $\psi$.

### Euler Parameters

Suppose that we express the trigonometric relations in Eq. (2.12) in terms of half angles, i.e.,

$$\sin\psi = 2\sin\tfrac{1}{2}\psi\cos\tfrac{1}{2}\psi \qquad 1 - \cos\psi = 2\sin^2\tfrac{1}{2}\psi$$

Then it is reasonable to combine the term $\sin\tfrac{1}{2}\psi$ with the direction cosines of the rotation axis $l$, $m$, $n$ by defining the matrix

$$\mathbf{E} = \sin\tfrac{1}{2}\psi\mathbf{S} = \begin{bmatrix} 0 & -n\sin\tfrac{1}{2}\psi & m\sin\tfrac{1}{2}\psi \\ n\sin\tfrac{1}{2}\psi & 0 & -l\sin\tfrac{1}{2}\psi \\ -m\sin\tfrac{1}{2}\psi & l\sin\tfrac{1}{2}\psi & 0 \end{bmatrix}$$

so that Eq. (2.12) may be written as

$$\mathbf{R} = \mathbf{I} + 2\cos\tfrac{1}{2}\psi\,\mathbf{E} + 2\mathbf{E}^2 \tag{2.13}$$

Now, for convenience, define

$$\alpha = l\sin\tfrac{1}{2}\psi \quad \beta = m\sin\tfrac{1}{2}\psi \quad \gamma = n\sin\tfrac{1}{2}\psi \quad \delta = \cos\tfrac{1}{2}\psi \tag{2.14}$$

so that

$$\mathbf{E} = \begin{bmatrix} 0 & -\gamma & \beta \\ \gamma & 0 & -\alpha \\ -\beta & \alpha & 0 \end{bmatrix} \tag{2.15}$$

These quantities are called the *Euler parameters* and clearly they satisfy the identity

$$\alpha^2 + \beta^2 + \gamma^2 + \delta^2 = 1 \tag{2.16}$$

---

† The reader will find it instructive to show that this form of $\mathbf{R}$ is an orthogonal matrix. Remember that $\mathbf{S}^T = -\mathbf{S}$ since $\mathbf{S}$ is skew-symmetric.

By expanding the new expression for $\mathbf{R}$ and using this identity, we can express the rotation matrix in terms of the Euler parameters as

$$\mathbf{R} = \begin{bmatrix} 1 - 2(\beta^2 + \gamma^2) & 2(\alpha\beta - \gamma\delta) & 2(\alpha\gamma + \beta\delta) \\ 2(\alpha\beta + \gamma\delta) & 1 - 2(\alpha^2 + \gamma^2) & 2(\beta\gamma - \alpha\delta) \\ 2(\alpha\gamma - \beta\delta) & 2(\beta\gamma + \alpha\delta) & 1 - 2(\alpha^2 + \beta^2) \end{bmatrix} \quad (2.17)$$

Finally, we note the ease with which the Euler parameters may be recovered from the rotation matrix; that is, if we are given an arbitrary rotation matrix, the Euler parameters are readily calculated from the matrix elements. The parameter $\delta$ can be determined from the *trace* of the matrix $\mathbf{R}$ as

$$\delta^2 = \tfrac{1}{4}(1 + \operatorname{tr} \mathbf{R})$$

which, from Eq. (2.3), gives

$$\delta^2 = \tfrac{1}{4}(1 + l_1 + m_2 + n_3) \quad (2.18)$$

Similarly, the quantities $4\gamma\delta$, $4\beta\delta$, $4\alpha\delta$ can be obtained by successively calculating the differences of the three pairs of off-diagonal elements of $\mathbf{R}$. Thus,

$$\alpha = \frac{n_2 - m_3}{4\delta} \qquad \beta = \frac{l_3 - n_1}{4\delta} \qquad \gamma = \frac{m_1 - l_2}{4\delta} \quad (2.19)$$

Since $\alpha$, $\beta$, $\gamma$ are proportional to the direction cosines of the rotation axis, we may write

$$\mathbf{i}_\omega = \frac{1}{\Lambda}[(n_2 - m_3)\mathbf{i}_x + (l_3 - n_1)\mathbf{i}_y + (m_1 - l_2)\mathbf{i}_z] \quad (2.20)$$

where

$$\Lambda = \sqrt{(1 + \operatorname{tr} \mathbf{R})(3 - \operatorname{tr} \mathbf{R})}$$

is the normalizing factor.

◇ **Problem 2–5**
We have now two different expressions for the Euler rotation axis: $\mathbf{i}_\omega$ obtained from Eq. (2.20) as well as the vector $\mathbf{w}$ determined in Prob. 2–1. Verify by direct calculation that these two vectors are, indeed, parallel.

◇ **Problem 2–6**
The rotation matrix $\mathbf{R}$ has the characteristic equation

$$|\mathbf{R} - \lambda \mathbf{I}| = -\lambda^3 + (\operatorname{tr} \mathbf{R})\lambda^2 - (\operatorname{tr} \mathbf{R})\lambda + 1 = 0$$

and the roots or characteristic values of $\mathbf{R}$ are

$$\lambda = 1, \; e^{i\psi}, \; e^{-i\psi}$$

where $i = \sqrt{-1}$. Furthermore, observe that if $\operatorname{tr} \mathbf{R} = 3$, the root $\lambda = 1$ is thrice repeated and if $\operatorname{tr} \mathbf{R} = -1$, there is a double root at $\lambda = -1$. [Note that the roots of the characteristic equation are also the zeros of the normalizing factor $\Lambda$ in Eq. (2.20).]

In calculating $\delta$ from Eq. (2.18), there are, of course, two possible solutions. This ambiguity is easily resolved since the choice of a positive value for $\delta$ corresponds to a rotation angle $\psi$ of less than $180°$ while a negative value is interpreted as a rotation of $360°$ minus $\psi$ in the opposite sense.

The above equations are not appropriate for either vanishingly small values of $\psi$ or values near $180°$, i.e., $\operatorname{tr}\mathbf{R}$ near 3 or $-1$. (See Prob. 2–6.) A singularity free computation is the subject of the next problem.

## Problem 2–7
Let the element in the $i^{\text{th}}$ row and $j^{\text{th}}$ column of the rotation matrix $\mathbf{R}$ be defined as $r_{ij}$ and introduce the notation

$$p_0 = 2\delta \qquad p_1 = 2\alpha \qquad p_2 = 2\beta \qquad p_3 = 2\gamma$$

$$r_{00} = \operatorname{tr}\mathbf{R} = r_{11} + r_{22} + r_{33}$$

(a) The $p$'s may be calculated from

$$p_j^2 = 1 + 2r_{jj} - \operatorname{tr}\mathbf{R} \qquad \text{for} \qquad j = 0, 1, 2, 3$$

so that the maximum $r_{jj}$ is seen to coincide with the maximum $p_j^2$.

(b) The largest value of the $p_j^2$'s lies in the closed interval $(1, 4)$ and the lower bound is attained only if all diagonal elements of $\mathbf{R}$ are zero.

(c) Derive the relations

$$p_0 p_1 = r_{32} - r_{23} \qquad p_2 p_3 = r_{32} + r_{23}$$
$$p_0 p_2 = r_{13} - r_{31} \qquad p_3 p_1 = r_{13} + r_{31}$$
$$p_0 p_3 = r_{21} - r_{12} \qquad p_1 p_2 = r_{21} + r_{12}$$

and observe that there always exists a subset of three equations for determining the other $p$'s once $p_j$ has been established. (Note that if $p_0$ happens to be negative, the signs of all $p$'s may be changed if so desired.)

*Stanley W. Shepperd*† 1978

## ◊ Problem 2–8
The Euler parameters may be expressed in terms of the spherical coordinate angles as

$$\alpha = -\sin \tfrac{1}{2}\phi \sin \tfrac{1}{2}\theta \qquad \gamma = \cos \tfrac{1}{2}\phi \sin \tfrac{1}{2}\theta$$
$$\beta = \sin \tfrac{1}{2}\phi \cos \tfrac{1}{2}\theta \qquad \delta = \cos \tfrac{1}{2}\phi \cos \tfrac{1}{2}\theta$$

and in terms of the Euler angles as

$$\alpha = \sin \tfrac{1}{2}i \cos \tfrac{1}{2}(\Omega - \omega) \qquad \gamma = \cos \tfrac{1}{2}i \sin \tfrac{1}{2}(\Omega + \omega)$$
$$\beta = \sin \tfrac{1}{2}i \sin \tfrac{1}{2}(\Omega - \omega) \qquad \delta = \cos \tfrac{1}{2}i \cos \tfrac{1}{2}(\Omega + \omega)$$

---

† "Quaternion from Rotation Matrix," *Journal of Guidance and Control*, Vol. 1, May–June, 1978, pp. 223–224.

## ◇ Problem 2-9

Instead of using the rotation matrix $\mathbf{R}$ to effect the rotation of a vector $\mathbf{r}_0$, the same result can be obtained with vector operations. If we define the vector

$$\mathbf{q} = \alpha\,\mathbf{i}_x + \beta\,\mathbf{i}_y + \gamma\,\mathbf{i}_z = [\alpha \quad \beta \quad \gamma]^T$$

then Eq. (2.10) may be written as

$$\mathbf{r} = \mathbf{r}_0 + 2\delta(\mathbf{q} \times \mathbf{r}_0) + 2\mathbf{q} \times (\mathbf{q} \times \mathbf{r}_0)$$

where $\alpha$, $\beta$, $\gamma$, $\delta$ are the Euler parameters.

## 2.3  Multiple Rotations of a Vector

Let $\mathbf{R}_3$ and $\alpha_3$, $\beta_3$, $\gamma_3$, $\delta_3$ be the rotation matrix and the associated Euler parameters corresponding to the resultant of two successive rotations—the first defined by the matrix $\mathbf{R}_1$ and the parameters $\alpha_1$, $\beta_1$, $\gamma_1$, $\delta_1$, and the second by $\mathbf{R}_2$ and $\alpha_2$, $\beta_2$, $\gamma_2$, $\delta_2$. Clearly, the rotation matrices are related by†

$$\mathbf{R}_3 = \mathbf{R}_2 \mathbf{R}_1 \qquad (2.21)$$

Not so obvious are the following relationships for the Euler parameters

$$\begin{aligned}
\alpha_3 &= \phantom{-}\delta_2\alpha_1 - \gamma_2\beta_1 + \beta_2\gamma_1 + \alpha_2\delta_1 \\
\beta_3 &= \phantom{-}\gamma_2\alpha_1 + \delta_2\beta_1 - \alpha_2\gamma_1 + \beta_2\delta_1 \\
\gamma_3 &= -\beta_2\alpha_1 + \alpha_2\beta_1 + \delta_2\gamma_1 + \gamma_2\delta_1 \\
\delta_3 &= -\alpha_2\alpha_1 - \beta_2\beta_1 - \gamma_2\gamma_1 + \delta_2\delta_1
\end{aligned} \qquad (2.22)$$

These remarkable equations were obtained independently and at different times by Gauss, Rodrigues, Hamilton, and Cayley. We shall present an elegant original derivation in the following subsection.‡

### Relations Among the Euler Parameters

As a first step in the derivation of Eq. (2.22), we develop an exceedingly beautiful factorization of the rotation matrix. For this purpose, note that

$$\mathbf{E}^2 = \begin{bmatrix} -\gamma^2 - \beta^2 & \beta\alpha & \gamma\alpha \\ \alpha\beta & -\gamma^2 - \alpha^2 & \gamma\beta \\ \alpha\gamma & \beta\gamma & -\beta^2 - \alpha^2 \end{bmatrix} = (\delta^2 - 1)\mathbf{I} + \mathbf{q}\mathbf{q}^T$$

from Eq. (2.15) and by using the identity (2.16). The vector $\mathbf{q}$ is defined in Prob. 2-9.

---

† Note that the order of the matrix product is the reverse of that used to create a rotation matrix from elementary matrices.

‡ From the author's paper "New Properties of the Rotation Matrix" presented at the Thirty-Eighth Congress of the International Astronautical Federation held in Brighton, England in 1987.

The so-called *dyadic product* $\mathbf{qq}^T$ is the product of a matrix having three rows and one column with a matrix having one row and three columns—resulting in a three-dimensional square matrix. Specifically,

$$\mathbf{qq}^T = \begin{bmatrix} \alpha \\ \beta \\ \gamma \end{bmatrix} \begin{bmatrix} \alpha & \beta & \gamma \end{bmatrix} = \begin{bmatrix} \alpha^2 & \alpha\beta & \alpha\gamma \\ \beta\alpha & \beta^2 & \beta\gamma \\ \gamma\alpha & \gamma\beta & \gamma^2 \end{bmatrix}$$

As a consequence, the expression for the rotation matrix given in Eq. (2.13) may be written as

$$\mathbf{R} = \mathbf{I} + 2\delta\mathbf{E} + 2\mathbf{E}^2 = \mathbf{I} + \mathbf{E}^2 + 2\delta\mathbf{E} + \mathbf{E}^2$$
$$= \mathbf{qq}^T + \delta^2\mathbf{I} + 2\delta\mathbf{E} + \mathbf{E}^2 = \mathbf{qq}^T + [\delta\mathbf{I} + \mathbf{E}][\delta\mathbf{I} + \mathbf{E}]$$

or, since $\mathbf{E}$ is skew-symmetric,

$$\mathbf{R} = \mathbf{qq}^T + [\delta\mathbf{I} + \mathbf{E}][\delta\mathbf{I} - \mathbf{E}]^T = \begin{bmatrix} \mathbf{q} & \delta\mathbf{I} + \mathbf{E} \end{bmatrix} \begin{bmatrix} \mathbf{q}^T \\ \delta\mathbf{I} + \mathbf{E} \end{bmatrix}$$

Hence

$$\mathbf{R} = \begin{bmatrix} \alpha & \delta & -\gamma & \beta \\ \beta & \gamma & \delta & -\alpha \\ \gamma & -\beta & \alpha & \delta \end{bmatrix} \begin{bmatrix} \alpha & \beta & \gamma \\ \delta & -\gamma & \beta \\ \gamma & \delta & -\alpha \\ -\beta & \alpha & \delta \end{bmatrix} \quad (2.23)$$

gives $\mathbf{R}$ in factored form.

Next we define two matrices $\mathbf{P}$ and $\mathbf{Q}$

$$\mathbf{P} = \begin{bmatrix} \delta & -\mathbf{q}^T \\ \mathbf{q} & \delta\mathbf{I} + \mathbf{E} \end{bmatrix} = \begin{bmatrix} \delta & -\alpha & -\beta & -\gamma \\ \alpha & \delta & -\gamma & \beta \\ \beta & \gamma & \delta & -\alpha \\ \gamma & -\beta & \alpha & \delta \end{bmatrix} \quad (2.24)$$

$$\mathbf{Q} = \begin{bmatrix} \delta & \mathbf{q}^T \\ -\mathbf{q} & \delta\mathbf{I} + \mathbf{E} \end{bmatrix} = \begin{bmatrix} \delta & \alpha & \beta & \gamma \\ -\alpha & \delta & -\gamma & \beta \\ -\beta & \gamma & \delta & -\alpha \\ -\gamma & -\beta & \alpha & \delta \end{bmatrix} \quad (2.25)$$

obtained by augmenting a row and column to the first and second matrix factors, respectively, of the rotation matrix $\mathbf{R}$. It is interesting and easy to verify that $\mathbf{P}$ and $\mathbf{Q}$ are both orthogonal matrices. But also, remarkably, they are commutative; that is $\mathbf{PQ} = \mathbf{QP}$. Indeed, it is readily seen that

$$\mathbf{PQ} = \mathbf{QP} = \begin{bmatrix} 1 & \mathbf{0}^T \\ \mathbf{0} & \mathbf{R} \end{bmatrix} \quad (2.26)$$

where $\mathbf{0}$ is the three-dimensional zero vector.

Sect. 2.3]   Multiple Rotations of a Vector   93

The matrix product $\mathbf{R}_3 = \mathbf{R}_2\mathbf{R}_1$, representing two successive rotations, can, therefore, be written as

$$\mathbf{P}_3\mathbf{Q}_3 = \begin{bmatrix} 1 & \mathbf{0}^T \\ \mathbf{0} & \mathbf{R}_3 \end{bmatrix} = \begin{bmatrix} 1 & \mathbf{0}^T \\ \mathbf{0} & \mathbf{R}_2 \end{bmatrix} \begin{bmatrix} 1 & \mathbf{0}^T \\ \mathbf{0} & \mathbf{R}_1 \end{bmatrix} = \mathbf{P}_2\mathbf{Q}_2\mathbf{P}_1\mathbf{Q}_1$$

Not only do the matrices $\mathbf{P}$ and $\mathbf{Q}$ commute, but it is also easy to verify that

$$\mathbf{Q}_2\mathbf{P}_1 = \mathbf{P}_1\mathbf{Q}_2 \tag{2.27}$$

Hence

$$\mathbf{P}_3\mathbf{Q}_3 = \mathbf{P}_2\mathbf{Q}_2\mathbf{P}_1\mathbf{Q}_1 = \mathbf{P}_2\mathbf{P}_1\mathbf{Q}_2\mathbf{Q}_1$$

which leads to the following identifications

$$\mathbf{P}_3 = \mathbf{P}_2\mathbf{P}_1 \qquad \mathbf{Q}_3 = \mathbf{Q}_2\mathbf{Q}_1 \tag{2.28}$$

Finally, by defining the four-dimensional vector $\mathbf{q}^*$ as

$$\mathbf{q}^* = \begin{bmatrix} \delta \\ \mathbf{q} \end{bmatrix} = \begin{bmatrix} \delta \\ \alpha \\ \beta \\ \gamma \end{bmatrix}$$

it is readily shown that the expressions for $\mathbf{P}_3$ and $\mathbf{Q}_3$ are equivalent to

$$\mathbf{q}_3^* = \mathbf{P}_2\mathbf{q}_1^* \qquad \mathbf{q}_3^* = \mathbf{Q}_1^T\mathbf{q}_2^* \tag{2.29}$$

When written out in component form, these equations are seen to be identical with Eqs. (2.22). The derivation is now complete.

Quaternions

The Euler parameters $\alpha$, $\beta$, $\gamma$, $\delta$ may be regarded as the components of a *quaternion*

$$\tilde{\mathbf{q}} = \delta + \mathbf{q} = \delta + \alpha\mathbf{i} + \beta\mathbf{j} + \gamma\mathbf{k} \tag{2.30}$$

where $\mathbf{i}$, $\mathbf{j}$, $\mathbf{k}$ have some of the properties of a right-handed, orthogonal triad of unit vectors. The parameter $\delta$ is called the *scalar part* and $\mathbf{q}$ is called the *vector part*. When multiplied pairwise, we define

$$\mathbf{i}\mathbf{j} = -\mathbf{j}\mathbf{i} = \mathbf{k} \qquad \mathbf{j}\mathbf{k} = -\mathbf{k}\mathbf{j} = \mathbf{i} \qquad \mathbf{k}\mathbf{i} = -\mathbf{i}\mathbf{k} = \mathbf{j} \tag{2.31}$$

but when each is multiplied by itself, we define

$$\mathbf{i}\mathbf{i} = \mathbf{j}\mathbf{j} = \mathbf{k}\mathbf{k} = -1 \tag{2.32}$$

By means of the symbols $\mathbf{i}$, $\mathbf{j}$, $\mathbf{k}$, invented by Sir William Hamilton,[†] the result of two successive rotations defined by the quaternions $\tilde{\mathbf{q}}_1$ and

---

† Sir William Rowan Hamilton (1805–1865), one of the greatest of the English mathematicians and physicists, announced his invention of quaternions and the unusual properties of his $I$, $J$, $K$ creations in 1843 at a meeting of the Royal Irish Academy. He spent the remainder of his life developing the subject and mistakenly regarded it as his greatest contribution—comparable in importance to Newton's calculus.

$\widetilde{\mathbf{q}}_2$ is equivalent to the single rotation defined by $\widetilde{\mathbf{q}}_3$ where

$$\widetilde{\mathbf{q}}_3 = \widetilde{\mathbf{q}}_2 \widetilde{\mathbf{q}}_1 \tag{2.33}$$

It is readily seen that this equation is equivalent to Eqs. (2.22). Actually, the algebra of quaternions is exactly the same as the algebra of vectors with the exception of multiplication. But the product of two quaternions can be represented vectorially as

$$\widetilde{\mathbf{q}}_2 \widetilde{\mathbf{q}}_1 = \delta_2 \delta_1 - \mathbf{q}_2 \cdot \mathbf{q}_1 + \delta_1 \mathbf{q}_2 + \delta_2 \mathbf{q}_1 + \mathbf{q}_2 \times \mathbf{q}_1 \tag{2.34}$$

the proof of which is a good exercise for the reader.

The *conjugate* or *inverse* of a quaternion is defined as

$$\widetilde{\mathbf{q}}^{-1} = \delta - \alpha \mathbf{i} - \beta \mathbf{j} - \gamma \mathbf{k} = \delta - \mathbf{q} \tag{2.35}$$

and it is not difficult to verify that

$$\widetilde{\mathbf{q}} \widetilde{\mathbf{q}}^{-1} = \widetilde{\mathbf{q}}^{-1} \widetilde{\mathbf{q}} = 1 \quad \text{and} \quad (\widetilde{\mathbf{q}}_1 \widetilde{\mathbf{q}}_2)^{-1} = \widetilde{\mathbf{q}}_2^{-1} \widetilde{\mathbf{q}}_1^{-1} \tag{2.36}$$

In the same manner in which we defined elementary rotation matrices, we may also define elementary quaternions. The quaternion representing a rotation about the $x$ axis through an angle $\psi$ is

$$\widetilde{\mathbf{q}}_x(\psi) = \cos \tfrac{1}{2}\psi + \sin \tfrac{1}{2}\psi \, \mathbf{i}$$

Similarly, for rotations about the $y$ and $z$ axes, we have

$$\widetilde{\mathbf{q}}_y(\psi) = \cos \tfrac{1}{2}\psi + \sin \tfrac{1}{2}\psi \, \mathbf{j} \quad \text{and} \quad \widetilde{\mathbf{q}}_z(\psi) = \cos \tfrac{1}{2}\psi + \sin \tfrac{1}{2}\psi \, \mathbf{k}$$

In this way, for example, a rigid body rotation about the $x$ axis through an angle $\theta$ followed by a rotation about the new $y$ axis through an angle $\phi$ is equivalent to the quaternion

$$\widetilde{\mathbf{q}} = \widetilde{\mathbf{q}}_x(\theta) \widetilde{\mathbf{q}}_y(\phi)$$

[Again, notice the reverse order of the factors from that of Eq. (2.33).]

Some examples of quaternions are to be found as problems in this and the following section.

◊ **Problem 2–10**

By regarding the position vector $\mathbf{r}_0$ as the vector part of a quaternion whose scalar part is zero, show that the rotation of $\mathbf{r}_0$ can be obtained by

$$\widetilde{\mathbf{r}} = \widetilde{\mathbf{q}} \widetilde{\mathbf{r}}_0 \widetilde{\mathbf{q}}^{-1}$$

where

$$\widetilde{\mathbf{r}}_0 = 0 + \mathbf{r}_0 \quad \text{and} \quad \widetilde{\mathbf{r}} = 0 + \mathbf{r}$$

◇ **Problem 2–11**

The quaternion equivalent of the spherical coordinate system rotation matrix of Prob. 2–2 is

$$\widetilde{\mathbf{q}} = \widetilde{\mathbf{q}}_z(\theta)\widetilde{\mathbf{q}}_y(\phi) = (\cos\tfrac{1}{2}\theta + \sin\tfrac{1}{2}\theta\,\mathbf{k})(\cos\tfrac{1}{2}\phi + \sin\tfrac{1}{2}\phi\,\mathbf{j})$$
$$= \cos\tfrac{1}{2}\theta\cos\tfrac{1}{2}\phi - \sin\tfrac{1}{2}\theta\sin\tfrac{1}{2}\phi\,\mathbf{i} + \cos\tfrac{1}{2}\theta\sin\tfrac{1}{2}\phi\,\mathbf{j} + \sin\tfrac{1}{2}\theta\cos\tfrac{1}{2}\phi\,\mathbf{k}$$

## 2.4 The n-Body Problem

According to Newton's law of gravitation, two particles attract each other with a force, acting along the line joining them, which is proportional to the product of their masses and inversely proportional to the square of the distance between them.

### Equations of Motion

For the purpose of providing an analytical description of the interactions and resulting motion of a system of $n$ particles whose masses are $m_1, m_2, \ldots, m_n$, let

$$\mathbf{r}_i = x_i\,\mathbf{i}_x + y_i\,\mathbf{i}_y + z_i\,\mathbf{i}_z$$
$$\mathbf{v}_i = \frac{d\mathbf{r}_i}{dt} = \frac{dx_i}{dt}\mathbf{i}_x + \frac{dy_i}{dt}\mathbf{i}_y + \frac{dz_i}{dt}\mathbf{i}_z$$

be the position and velocity vectors of the $i^{\text{th}}$ particle expressed with respect to unaccelerated coordinate axes. The coordinate system is right-handed and orthogonal with $\mathbf{i}_x$, $\mathbf{i}_y$, $\mathbf{i}_z$ as unit vectors parallel to the reference axes. Alternately, using matrix notation, we may write $\mathbf{r}$ and $\mathbf{v}$ as column vectors

$$\mathbf{r}_i = \begin{bmatrix} x_i \\ y_i \\ z_i \end{bmatrix} \qquad \mathbf{v}_i = \begin{bmatrix} dx_i/dt \\ dy_i/dt \\ dz_i/dt \end{bmatrix}$$

Furthermore, let

$$r_{ij} = |\mathbf{r}_j - \mathbf{r}_i| = \sqrt{(\mathbf{r}_j - \mathbf{r}_i)\cdot(\mathbf{r}_j - \mathbf{r}_i)}$$

denote the distance between $m_i$ and $m_j$ so that the magnitude of the force of attraction between the $i^{\text{th}}$ and $j^{\text{th}}$ particles is $Gm_im_j/r_{ij}^2$ where the proportionality factor $G$ is called the *universal gravitation constant*.

The directions of the forces are conveniently expressed in terms of unit vectors. Thus, the force acting on $m_i$ due to $m_j$ has the direction of $(\mathbf{r}_j - \mathbf{r}_i)/r_{ij}$ while the force on $m_j$ due to $m_i$ is oppositely directed.

Hence, the total force $\mathbf{f}_i$ affecting $m_i$, due to the presence of the other $n-1$ masses, is

$$\mathbf{f}_i = G \sum_{j=1}^{n}{}' \frac{m_i m_j}{r_{ij}^3}(\mathbf{r}_j - \mathbf{r}_i) \tag{2.37}$$

where the prime on the summation symbol indicates that the term for which $i = j$ is to be omitted.

In accordance with Newton's second law of motion,

$$\mathbf{f}_i = m_i \frac{d^2 \mathbf{r}_i}{dt^2} \equiv m_i \frac{d\mathbf{v}_i}{dt} \tag{2.38}$$

so that the $n$ vector differential equations

$$\frac{d^2 \mathbf{r}_i}{dt^2} = G \sum_{j=1}^{n}{}' \frac{m_j}{r_{ij}^3}(\mathbf{r}_j - \mathbf{r}_i) \tag{2.39}$$

together with appropriate initial conditions, constitute a complete mathematical description of the motion of the system of $n$ mass particles.

For a complete solution of the $n$-body problem, a total of $6n$ integrals is required. Although only 10 are obtainable in general, the known integrals have important physical interpretations. We shall now derive those 10 integrals and, as a consequence, show that, when no external forces are acting on the system, the total linear and angular momenta as well as the total energy are conserved.

### Conservation of Total Linear Momentum

It is readily seen from Eq. (2.37) that the sum of the force vectors $\mathbf{f}_1$, $\mathbf{f}_2$, ..., $\mathbf{f}_n$ has a zero resultant. Thus, we have

$$\frac{d^2}{dt^2}(m_1 \mathbf{r}_1 + m_2 \mathbf{r}_2 + \cdots + m_n \mathbf{r}_n) = \mathbf{0}$$

which demonstrates that the *center of mass* of the $n$-body system

$$\mathbf{r}_{cm} = \frac{m_1 \mathbf{r}_1 + m_2 \mathbf{r}_2 + \cdots + m_n \mathbf{r}_n}{m_1 + m_2 + \cdots + m_n} \tag{2.40}$$

is unaccelerated. Therefore, the linear momentum of the system is conserved and

$$\mathbf{r}_{cm} = \mathbf{c}_1 t + \mathbf{c}_2 \tag{2.41}$$

where $\mathbf{c}_1$ and $\mathbf{c}_2$ are the vector constants of integration.

## Conservation of Total Angular Momentum

Again from Eq. (2.37), we may verify that the sum of all the vector moments $\mathbf{r}_i \times \mathbf{f}_i$ for $i = 1, 2, \ldots, n$ also has a zero resultant so that

$$\frac{d}{dt}\left(m_1 \mathbf{r}_1 \times \frac{d\mathbf{r}_1}{dt} + m_2 \mathbf{r}_2 \times \frac{d\mathbf{r}_2}{dt} + \cdots + m_n \mathbf{r}_n \times \frac{d\mathbf{r}_n}{dt}\right) = \mathbf{0}$$

By performing the integration

$$m_1 \mathbf{r}_1 \times \mathbf{v}_1 + m_2 \mathbf{r}_2 \times \mathbf{v}_2 + \cdots + m_n \mathbf{r}_n \times \mathbf{v}_n = \mathbf{c}_3 \qquad (2.42)$$

we see that the total angular momentum vector is constant in magnitude and direction.

The *invariable plane* of the system is the terminology frequently applied to that plane which contains the center of mass $\mathbf{r}_{cm}$ and whose normal is parallel to the total angular momentum vector $\mathbf{c}_3$.

## Potential Functions

The *gravitational potential* $V_i$ at the point $(x_i, y_i, z_i)$ is defined as

$$V_i = G \sum_{j=1}^{n}{}' \frac{m_j}{r_{ij}} \qquad (2.43)$$

Since the potential function depends only on the distances to the other particles, it is, consequently, independent of the choice of coordinate axes. The importance of the gravitational potential derives from the property that the gradient of $V_i$ gives the force of attraction on a particle of unit mass at the point $(x_i, y_i, z_i)$. Thus, we have†

$$\mathbf{f}_i^T = m_i \frac{\partial V_i}{\partial \mathbf{r}_i} \qquad (2.44)$$

where

$$\frac{\partial V_i}{\partial \mathbf{r}_i} = \left[\frac{\partial V_i}{\partial x_i} \quad \frac{\partial V_i}{\partial y_i} \quad \frac{\partial V_i}{\partial z_i}\right]$$

is defined to be a row vector. The superscript $^T$ indicates the matrix transpose and is required so that $\mathbf{f}_i^T$ will be a row vector.

For some purposes it is convenient to introduce the function $U$, defined by

$$U = \tfrac{1}{2} \sum_{i=1}^{n} m_i V_i \qquad (2.45)$$

---

† The idea that a force can be derived from a potential function, and even the term "potential function," were used by Daniel Bernoulli in *Hydrodynamica* (1738). In vector analysis, $\partial V_i/\partial \mathbf{r}_i$ is called the gradient of $V_i$ with respect to the vector $\mathbf{r}_i$ and is written $\nabla V_i$. For our purposes, the alternate notation is preferred since we can then apply the *chain rule* of partial differentiation using vectors as well as scalars. This is illustrated later in this section when we demonstrate the property of conservation of energy.

and called the *force function*, which is equal to the total work done by the gravitational forces in assembling the system of $n$ point masses from a state of infinite dispersion to a given configuration. The *potential energy* of the system is then $-U$. In terms of $U$, the force vector $\mathbf{f}_i$ is simply†

$$\mathbf{f}_i^T = \frac{\partial U}{\partial \mathbf{r}_i} \tag{2.46}$$

For many purposes, expressing the force vector as the gradient of the force function rather than the gravitational potential will be more convenient since $U$ is independent of the coordinates of any particular point.

## Conservation of Total Energy

The force function $U$ is a function of the components $x_1$, $y_1$, $z_1$, $x_2$, ..., $y_n$, $z_n$ of the position vectors $\mathbf{r}_1, \mathbf{r}_2, \ldots, \mathbf{r}_n$. Since each component is, in turn, a function of t, the total derivative of $U$ is simply

$$\frac{dU}{dt} = \frac{\partial U}{\partial x_1}\frac{dx_1}{dt} + \frac{\partial U}{\partial y_1}\frac{dy_1}{dt} + \cdots + \frac{\partial U}{\partial z_n}\frac{dz_n}{dt}$$

Now $\partial U/\partial \mathbf{r}_i$ is a row vector and $d\mathbf{r}_i/dt$ is a column vector, so that

$$\frac{\partial U}{\partial \mathbf{r}_i}\frac{d\mathbf{r}_i}{dt} = \frac{\partial U}{\partial x_i}\frac{dx_i}{dt} + \frac{\partial U}{\partial y_i}\frac{dy_i}{dt} + \frac{\partial U}{\partial z_i}\frac{dz_i}{dt}$$

Hence, we may write the total derivative of $U$ as

$$\frac{dU}{dt} = \frac{\partial U}{\partial \mathbf{r}_1}\frac{d\mathbf{r}_1}{dt} + \frac{\partial U}{\partial \mathbf{r}_2}\frac{d\mathbf{r}_2}{dt} + \cdots + \frac{\partial U}{\partial \mathbf{r}_n}\frac{d\mathbf{r}_n}{dt}$$

Then, using Eq. (2.46), we have

$$\frac{dU}{dt} = \mathbf{f}_1^T \mathbf{v}_1 + \mathbf{f}_2^T \mathbf{v}_2 + \cdots + \mathbf{f}_n^T \mathbf{v}_n$$

or, in vector notation,

$$\frac{dU}{dt} = \mathbf{f}_1 \cdot \mathbf{v}_1 + \mathbf{f}_2 \cdot \mathbf{v}_2 + \cdots + \mathbf{f}_n \cdot \mathbf{v}_n$$

Finally, from Eq. (2.38), we may write

$$\frac{dU}{dt} = m_1 \frac{d\mathbf{v}_1}{dt} \cdot \mathbf{v}_1 + m_2 \frac{d\mathbf{v}_2}{dt} \cdot \mathbf{v}_2 + \cdots + m_n \frac{d\mathbf{v}_n}{dt} \cdot \mathbf{v}_n$$
$$= \frac{dT}{dt}$$

where

$$T = \tfrac{1}{2}(m_1 v_1^2 + m_2 v_2^2 + \cdots + m_n v_n^2) \tag{2.47}$$

---

† It is suggested that the reader write out the terms in these various equations for, say, $n = 3$ and, in particular, see why the $\tfrac{1}{2}$ is needed in Eq. (2.45).

is the *kinetic energy* of the system. Thus

$$T - U = c \tag{2.48}$$

verifying that the sum of the kinetic and potential energies is a constant.

It is known that no further integrals are obtainable in general for the $n$-body problem. The 10 constants of integration consist of the components of the three vectors $\mathbf{c}_1$, $\mathbf{c}_2$, $\mathbf{c}_3$ together with the scalar constant $c$.

### Problem 2–12

Let $q_1, q_2, \ldots, q_{3n}$ be independent geometrical quantities specifying the configuration of the $n$ masses. They are frequently referred to as the *generalized coordinates* of the system and can be regarded as the components of a $3n$-dimensional vector $\mathbf{q}$. Thus, in general, we have

$$\mathbf{r}_i = \mathbf{r}_i[t, \mathbf{q}(t)]$$

and

$$\mathbf{v}_i = \frac{d\mathbf{r}_i}{dt} = \frac{\partial \mathbf{r}_i}{\partial t} + \frac{\partial \mathbf{r}_i}{\partial \mathbf{q}} \dot{\mathbf{q}} = \mathbf{v}_i(t, \mathbf{q}, \dot{\mathbf{q}})$$

where $\dot{\mathbf{q}} = d\mathbf{q}/dt$ and the symbol $\partial \mathbf{r}_i/\partial \mathbf{q}$ denotes the 3 by $3n$-dimensional matrix

$$\frac{\partial \mathbf{r}_i}{\partial \mathbf{q}} = \begin{bmatrix} \frac{\partial x_i}{\partial q_1} & \frac{\partial x_i}{\partial q_2} & \cdots & \frac{\partial x_i}{\partial q_{3n}} \\ \frac{\partial y_i}{\partial q_1} & \frac{\partial y_i}{\partial q_2} & \cdots & \frac{\partial y_i}{\partial q_{3n}} \\ \frac{\partial z_i}{\partial q_1} & \frac{\partial z_i}{\partial q_2} & \cdots & \frac{\partial z_i}{\partial q_{3n}} \end{bmatrix}$$

(a) Verify that

$$\frac{\partial \mathbf{v}_i}{\partial \dot{\mathbf{q}}} = \frac{\partial \mathbf{r}_i}{\partial \mathbf{q}}$$

and obtain

$$\frac{\partial U}{\partial \mathbf{q}} = \sum_{i=1}^{n} m_i \frac{d\mathbf{v}_i^T}{dt} \frac{\partial \mathbf{r}_i}{\partial \mathbf{q}}$$

(b) Further, verify

$$\frac{d\mathbf{v}_i^T}{dt} \frac{\partial \mathbf{r}_i}{\partial \mathbf{q}} = \frac{d}{dt}\left[\mathbf{v}_i^T \frac{\partial \mathbf{v}_i}{\partial \dot{\mathbf{q}}}\right] - \mathbf{v}_i^T \frac{\partial \mathbf{v}_i}{\partial \mathbf{q}} = \frac{d}{dt} \frac{\partial}{\partial \dot{\mathbf{q}}}(\tfrac{1}{2}\mathbf{v}_i^T \mathbf{v}_i) - \frac{\partial}{\partial \mathbf{q}}(\tfrac{1}{2}\mathbf{v}_i^T \mathbf{v}_i)$$

and derive, therefrom, *Lagrange's form of the equations of motion* of $n$ bodies

$$\frac{d}{dt}\frac{\partial L}{\partial \dot{\mathbf{q}}} = \frac{\partial L}{\partial \mathbf{q}}$$

where the *Lagrangian function* $L$, is defined as

$$L = T + U$$

and sometimes called the *kinetic potential*.

*Joseph-Louis Lagrange* 1788

## Problem 2-13

Let $p_1, p_2, \ldots, p_{3n}$ be the components of a $3n$-dimensional vector $\mathbf{p}$ defined by

$$\mathbf{p}^T = \frac{\partial T}{\partial \dot{\mathbf{q}}}$$

and referred to as the *generalized momenta* of the system.

(a) The kinetic energy $T(t, \mathbf{q}, \dot{\mathbf{q}})$ is the sum of

(1) $T_0$ (a function of $t$ and $\mathbf{q}$ only)
(2) $T_1 = \mathbf{w}^T \dot{\mathbf{q}}$ (a *linear form* in $\dot{\mathbf{q}}$)
(3) $T_2 = \dot{\mathbf{q}}^T \mathbf{W} \dot{\mathbf{q}}$ (a *quadratic form* in $\dot{\mathbf{q}}$)

where

$$T_0 = \tfrac{1}{2} \sum_{i=1}^{n} m_i \frac{\partial \mathbf{r}_i^T}{\partial t} \frac{\partial \mathbf{r}_i}{\partial t}$$

and the row vector $\mathbf{w}^T$ and the *symmetric matrix* $\mathbf{W}$ are the following functions of $t$ and $\mathbf{q}$:

$$\mathbf{w}^T = \sum_{i=1}^{n} m_i \frac{\partial \mathbf{r}_i^T}{\partial t} \frac{\partial \mathbf{r}_i}{\partial \mathbf{q}} \qquad \mathbf{W} = \tfrac{1}{2} \sum_{i=1}^{n} m_i \left[\frac{\partial \mathbf{r}_i}{\partial \mathbf{q}}\right]^T \frac{\partial \mathbf{r}_i}{\partial \mathbf{q}}$$

Hence, obtain

$$\mathbf{p} = 2\mathbf{W}\dot{\mathbf{q}} + \mathbf{w}$$

so that

$$\dot{\mathbf{q}} = \tfrac{1}{2} \mathbf{W}^{-1}(\mathbf{p} - \mathbf{w}) = \dot{\mathbf{q}}(t, \mathbf{q}, \mathbf{p})$$

provided that the transformation $\mathbf{r}_i = \mathbf{r}_i(t, \mathbf{q})$ is nonsingular.

(b) Consider the equation

$$2T_2 + T_1 = \mathbf{p}^T \dot{\mathbf{q}}$$

and let the vectors $\mathbf{q}$ and $\dot{\mathbf{q}}$ receive small independent variations $\delta \mathbf{q}$ and $\delta \dot{\mathbf{q}}$ so that

$$2\delta T_2 + \delta T_1 = \mathbf{p}^T \delta \dot{\mathbf{q}} + \dot{\mathbf{q}}^T \delta \mathbf{p}$$

and

$$\delta T = \frac{\partial T}{\partial \dot{\mathbf{q}}} \delta \dot{\mathbf{q}} + \frac{\partial T}{\partial \mathbf{q}} \delta \mathbf{q}$$

Hence, obtain

$$\delta(T_2 - T_0) = \dot{\mathbf{q}}^T \delta \mathbf{p} - \frac{\partial T}{\partial \mathbf{q}} \delta \mathbf{q}$$

(c) By regarding $T_2 - T_0$ as a function of $t$, $\mathbf{q}$, and $\mathbf{p}$, i.e.,

$$T_2 - T_0 = T^*(t, \mathbf{q}, \mathbf{p})$$

obtain

$$\delta(T_2 - T_0) = \frac{\partial T^*}{\partial \mathbf{q}} \delta \mathbf{q} + \frac{\partial T^*}{\partial \mathbf{p}} \delta \mathbf{p}$$

Therefore, show that

$$\dot{\mathbf{q}}^T = \frac{\partial T^*}{\partial \mathbf{p}} \qquad \frac{\partial T}{\partial \mathbf{q}} = -\frac{\partial T^*}{\partial \mathbf{q}}$$

and derive *Hamilton's canonical form of the equations of motion*

$$\frac{d\mathbf{q}^T}{dt} = \frac{\partial H}{\partial \mathbf{p}} \qquad \frac{d\mathbf{p}^T}{dt} = -\frac{\partial H}{\partial \mathbf{q}}$$

where the *Hamiltonian function* $H(t, \mathbf{q}, \mathbf{p})$ is defined as

$$H(t, \mathbf{q}, \mathbf{p}) = T^* - U$$

(d) If the transformation $\mathbf{r}_i = \mathbf{r}_i(\mathbf{q})$ does *not* involve the time $t$, show that

$$H(\mathbf{q}, \mathbf{p}) = T - U = \text{constant}$$

by calculating $dH/dt$ and using the canonic equations.

NOTE: Such a system is called *scleronomic*. The more general case, for which the transformation $\mathbf{r}_i = \mathbf{r}_i(t, \mathbf{q})$ is an explicit function of time, is called *rheonomic*.

*Sir William Rowan Hamilton* 1834

## 2.5 Kinematics in Rotating Coordinates

In the previous section we considered the motion of particles for which the reference coordinate axes were regarded as fixed. The components of the velocity and acceleration vectors could then be computed as the time derivatives of the components of the position vector.

For some problems, however, it is more convenient to express the motion of bodies relative to a rotating coordinate frame. The components of velocity and acceleration, under such circumstances, will include several additional terms arising solely from the motion of the coordinate system in addition to the time derivatives of the position vector components.

In order to obtain the appropriate forms for the velocity and acceleration vectors, we introduce the time dependent rotation matrix $\mathbf{R}$ which will effect an orthogonal transformation of vector components from the moving axes to the reference axes. Furthermore, to avoid any possible confusion, we will use an asterisk to distinguish a vector resolved along fixed axes from the same vector resolved along the rotating axes. Thus, we write

$$\mathbf{r}^* = \mathbf{R}\mathbf{r} \qquad \mathbf{v}^* = \mathbf{R}\mathbf{v} \qquad \mathbf{a}^* = \mathbf{R}\mathbf{a} \qquad (2.49)$$

where $\mathbf{r}$, $\mathbf{v}$, $\mathbf{a}$ are the position, velocity, and acceleration vectors whose components are understood to be projections along the moving axes.

To obtain the required expressions for the velocity $\mathbf{v}$, we calculate

$$\mathbf{v}^* = \frac{d\mathbf{r}^*}{dt} = \mathbf{R}\left[\frac{d\mathbf{r}}{dt} + \mathbf{\Omega}\mathbf{r}\right] = \mathbf{R}\mathbf{v} \qquad (2.50)$$

where the matrix $\mathbf{\Omega}$ is defined as

$$\mathbf{\Omega} = \mathbf{R}^T \frac{d\mathbf{R}}{dt} \qquad (2.51)$$

By differentiating the identity $\mathbf{R}^T\mathbf{R} = \mathbf{I}$, it is readily seen that $\mathbf{\Omega}$, called the *angular velocity matrix*, is a skew-symmetric matrix, $\mathbf{\Omega}^T = -\mathbf{\Omega}$, and as such may be written in the form

$$\mathbf{\Omega} = \begin{bmatrix} 0 & -\omega_\zeta & \omega_\eta \\ \omega_\zeta & 0 & -\omega_\xi \\ -\omega_\eta & \omega_\xi & 0 \end{bmatrix}$$

Therefore, we can define a vector $\boldsymbol{\omega}$, whose components along the moving axes are $\omega_\xi$, $\omega_\eta$, $\omega_\zeta$, so that the relationship between the velocity vector components in the two frames of reference may be alternately expressed as

$$\mathbf{v}^* = \mathbf{R}\left[\frac{d\mathbf{r}}{dt} + \boldsymbol{\omega} \times \mathbf{r}\right] = \mathbf{R}\mathbf{v} \qquad (2.52)$$

The *angular velocity vector* $\boldsymbol{\omega}$ is identified as the angular velocity of the moving coordinate system with respect to the fixed system.

For the acceleration vector, we differentiate Eq. (2.50) to obtain

$$\mathbf{a}^* = \frac{d^2\mathbf{r}^*}{dt^2} = \mathbf{R}\left[\frac{d^2\mathbf{r}}{dt^2} + 2\mathbf{\Omega}\frac{d\mathbf{r}}{dt} + \frac{d\mathbf{\Omega}}{dt}\mathbf{r} + \mathbf{\Omega}\mathbf{\Omega}\mathbf{r}\right] = \mathbf{R}\mathbf{a} \qquad (2.53)$$

or, in terms of the angular velocity vector $\boldsymbol{\omega}$,

$$\mathbf{a}^* = \mathbf{R}\left[\frac{d^2\mathbf{r}}{dt^2} + 2\boldsymbol{\omega} \times \frac{d\mathbf{r}}{dt} + \frac{d\boldsymbol{\omega}}{dt} \times \mathbf{r} + \boldsymbol{\omega} \times (\boldsymbol{\omega} \times \mathbf{r})\right] = \mathbf{R}\mathbf{a} \qquad (2.54)$$

The four terms which comprise the acceleration referred to rotating axes are called the *observed*, the *Coriolis*,† the *Euler*, and the *centripetal* accelerations, respectively. (The observed velocity and acceleration vectors $d\mathbf{r}/dt$ and $d^2\mathbf{r}/dt^2$ will sometimes be denoted by $\mathbf{v}_{rel}$ and $\mathbf{a}_{rel}$ since they are quantities measured *relative* to the rotating axes. The symbols $\mathbf{v}$ and $\mathbf{a}$ will be reserved for the total velocity and acceleration vectors which include the effects of the moving axes relative to the fixed axes.)

◊ **Problem 2-14**

For motion referred to a rotating spherical coordinate system, as defined in Prob. 2-2, the angular velocity vector of the moving system with respect to the fixed axes is

$$\boldsymbol{\omega} = -\sin\phi\frac{d\theta}{dt}\mathbf{i}_\phi + \frac{d\phi}{dt}\mathbf{i}_\theta + \cos\phi\frac{d\theta}{dt}\mathbf{i}_r$$

or

$$\boldsymbol{\omega}^* = -\sin\theta\frac{d\phi}{dt}\mathbf{i}_x + \cos\theta\frac{d\phi}{dt}\mathbf{i}_y + \frac{d\theta}{dt}\mathbf{i}_z$$

---

† Besides his contributions to relative motion, Gaspard Gustave de Coriolis (1792–1843) gave the first modern definitions of "work" and "kinetic energy" in mechanics.

## Sect. 2.5]   Kinematics in Rotating Coordinates   103

The position, velocity, and acceleration vectors referred to the moving axes are

$$\mathbf{r} = r\,\mathbf{i}_r$$

$$\mathbf{v} = r\frac{d\phi}{dt}\mathbf{i}_\phi + r\sin\phi\frac{d\theta}{dt}\mathbf{i}_\theta + \frac{dr}{dt}\mathbf{i}_r$$

$$\mathbf{a} = \left[r\frac{d^2\phi}{dt^2} + 2\frac{dr}{dt}\frac{d\phi}{dt} - r\sin\phi\cos\phi\left(\frac{d\theta}{dt}\right)^2\right]\mathbf{i}_\phi$$

$$+ \left[\frac{1}{r\sin\phi}\frac{d}{dt}\left(r^2\sin^2\phi\frac{d\theta}{dt}\right)\right]\mathbf{i}_\theta + \left[\frac{d^2r}{dt^2} - r\left(\frac{d\phi}{dt}\right)^2 - r\sin^2\phi\left(\frac{d\theta}{dt}\right)^2\right]\mathbf{i}_r$$

◇ **Problem 2–15**
If the motion is confined to the $x, y$ plane ($\phi = 90°$), the position, velocity, and acceleration vectors, referred to rotating *polar coordinates*, are given by

$$\mathbf{r} = r\,\mathbf{i}_r$$

$$\mathbf{v} = \frac{dr}{dt}\mathbf{i}_r + r\frac{d\theta}{dt}\mathbf{i}_\theta$$

$$\mathbf{a} = \left[\frac{d^2r}{dt^2} - r\left(\frac{d\theta}{dt}\right)^2\right]\mathbf{i}_r + \left[\frac{1}{r}\frac{d}{dt}\left(r^2\frac{d\theta}{dt}\right)\right]\mathbf{i}_\theta$$

where the angular velocity vector of the moving system with respect to the fixed axes is

$$\boldsymbol{\omega}^* = \boldsymbol{\omega} = -\frac{d\theta}{dt}\mathbf{i}_\phi = \frac{d\theta}{dt}\mathbf{i}_z$$

◇ **Problem 2–16**
For motion confined to the $x, y$ plane in which the acceleration vector is directed along the radius vector, the rate at which the radius vector sweeps out area is a constant. That is, the so-called *areal velocity* is

$$\frac{dA}{dt} = \frac{1}{2}r^2\frac{d\theta}{dt} = \frac{1}{2}\left(x\frac{dy}{dt} - y\frac{dx}{dt}\right)$$

and is constant. Such motion is said to obey the *law of areas*.

**Problem 2–17**
The matrices $\Omega$ and $\Omega^{*\mathrm{T}}$ are *similar*, i.e.,

$$R\Omega R^{\mathrm{T}} = \Omega^{*\mathrm{T}}$$

where the elements of $\Omega$ and $\Omega^*$ are the appropriate components of $\boldsymbol{\omega}$ and $\boldsymbol{\omega}^*$—the angular velocity vector resolved along rotating axes and fixed axes, respectively.

NOTE: It is important to remember that $\boldsymbol{\omega} = -\boldsymbol{\omega}^*$.

### Problem 2-18

The characteristic equation of the matrix $\mathbf{\Omega}$, the angular velocity matrix, is

$$|\mathbf{\Omega} - \lambda \mathbf{I}| = \lambda^3 + \omega^2 \lambda = 0$$

where $\omega$ is the magnitude of the angular velocity vector $\boldsymbol{\omega}$. Furthermore, $\boldsymbol{\omega}$ is the characteristic vector of the matrix $d\mathbf{R}/dt$ corresponding to the zero characteristic value.

### Problem 2-19

For motion confined to a plane, let $\mathbf{i}_t$ and $\mathbf{i}_n$ be orthogonal unit vectors directed along the velocity vector and normal to it, respectively, as shown in Fig. 2.3. The velocity vector $\mathbf{v}$ makes an angle $\phi$ with the reference $x$ axis and the *radius of curvature* $\rho$ of the path is defined as

$$\rho = \frac{ds}{d\phi}$$

where $s$ denotes the arc length of the path described in time $t$.

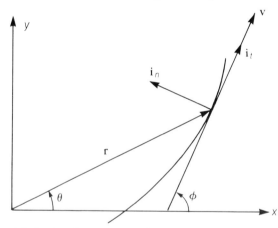

**Fig. 2.3:** Tangential and normal coordinates.

(a) Derive the equations for velocity and acceleration in tangential and normal components in the form

$$\mathbf{v} = v\,\mathbf{i}_t = \frac{ds}{dt}\mathbf{i}_t$$

$$\mathbf{a} = v\frac{dv}{ds}\mathbf{i}_t + \frac{v^2}{\rho}\mathbf{i}_n = \frac{dv}{dt}\mathbf{i}_t + \frac{v^2}{\rho}\mathbf{i}_n$$

(b) Derive the following expressions for the *curvature* $1/\rho$:

(1) $\qquad \dfrac{1}{\rho} = \dfrac{1}{v^2}\sqrt{a^2 - \left(\dfrac{dv}{dt}\right)^2}$

(2) $\qquad \dfrac{1}{\rho} = \dfrac{d^2y}{dx^2}\left[1 + \left(\dfrac{dy}{dx}\right)^2\right]^{-\frac{3}{2}}$

if the equation of the path is $y = y(x)$;

(3) $$\frac{1}{\rho} = \left(\frac{dx}{dt}\frac{d^2y}{dt^2} - \frac{dy}{dt}\frac{d^2x}{dt^2}\right)\left[\left(\frac{dx}{dt}\right)^2 + \left(\frac{dy}{dt}\right)^2\right]^{-\frac{3}{2}}$$

if the equation of the path is $x = x(t)$, $y = y(t)$;

(4) $$\frac{1}{\rho} = \left[r^2 + 2\left(\frac{dr}{d\theta}\right)^2 - r\frac{d^2r}{d\theta^2}\right]\left[r^2 + \left(\frac{dr}{d\theta}\right)^2\right]^{-\frac{3}{2}}$$

if the equation of the path is expressed in polar coordinates; and

(5) $$\frac{1}{\rho} = \frac{1}{r}\left[1 - \left(\frac{dr}{ds}\right)^2 - r\frac{d^2r}{ds^2}\right]\left[1 - \left(\frac{dr}{ds}\right)^2\right]^{-\frac{1}{2}}$$

if the equation of the path is expressed in terms of the arc length as $r(s)$.

NOTE: In 1691 both James and John Bernoulli gave the formula for the radius of curvature of a plane curve. James, who also gave the result in polar coordinates, called it his "golden theorem."

### Problem 2–20

The kinematic expressions for rotating coordinates can also be obtained using quaternions. As before, let $\mathbf{r}^*$, $\mathbf{v}^*$ and $\mathbf{r}$, $\mathbf{v}$ be the position and velocity vectors whose components are resolved along fixed and rotating axes, respectively. Let $\tilde{\mathbf{q}}$ be the time dependent quaternion which effects the transformation from the moving axes to the reference axes. Then we have

$$\tilde{\mathbf{r}}^* = \tilde{\mathbf{q}}\,\tilde{\mathbf{r}}\,\tilde{\mathbf{q}}^{-1} \quad \text{and} \quad \tilde{\mathbf{v}}^* = \tilde{\mathbf{q}}\,\tilde{\mathbf{v}}\,\tilde{\mathbf{q}}^{-1}$$

where the quaternions

$$\tilde{\mathbf{r}}^* = 0 + \mathbf{r}^* \qquad \tilde{\mathbf{r}} = 0 + \mathbf{r}$$
$$\tilde{\mathbf{v}}^* = 0 + \mathbf{v}^* \qquad \tilde{\mathbf{v}} = 0 + \mathbf{v}$$

all have zero scalar parts.

(a) The first step in the derivation is

$$\tilde{\mathbf{v}} = \frac{d\tilde{\mathbf{r}}}{dt} + \tilde{\mathbf{q}}^{-1}\frac{d\tilde{\mathbf{q}}}{dt}\tilde{\mathbf{r}} + \tilde{\mathbf{r}}\frac{d\tilde{\mathbf{q}}^{-1}}{dt}\tilde{\mathbf{q}} = \frac{d\tilde{\mathbf{r}}}{dt} + \tilde{\mathbf{q}}^{-1}\frac{d\tilde{\mathbf{q}}}{dt}\tilde{\mathbf{r}} - \left(\tilde{\mathbf{q}}^{-1}\frac{d\tilde{\mathbf{q}}}{dt}\tilde{\mathbf{r}}\right)^{-1}$$

(b) Then, verify that

$$\tilde{\mathbf{q}}^{-1}\frac{d\tilde{\mathbf{q}}}{dt} = \delta\frac{d\mathbf{q}}{dt} - \frac{d\delta}{dt}\mathbf{q} - \mathbf{q}\times\frac{d\mathbf{q}}{dt}$$

which is, therefore, a quaternion whose scalar part is zero.

(c) Define

$$\tilde{\boldsymbol{\omega}} = 0 + 2\tilde{\mathbf{q}}^{-1}\frac{d\tilde{\mathbf{q}}}{dt}$$

and show that

$$\tilde{\mathbf{v}} = \frac{d\tilde{\mathbf{r}}}{dt} + \tilde{\boldsymbol{\omega}}\,\tilde{\mathbf{r}}$$

or, alternately,

$$\mathbf{v} = \frac{d\mathbf{r}}{dt} + \boldsymbol{\omega}\times\mathbf{r}$$

where
$$\omega = 2\tilde{q}^{-1}\frac{d\tilde{q}}{dt}$$
is the angular velocity vector of the moving coordinate system with respect to the fixed system whose components are along the moving axes.

(d) Finally, derive the components of $\omega$ in the form
$$\omega = 2\left(\delta\frac{d\alpha}{dt} + \gamma\frac{d\beta}{dt} - \beta\frac{d\gamma}{dt} - \alpha\frac{d\delta}{dt}\right)\mathbf{i}_\xi$$
$$+ 2\left(-\gamma\frac{d\alpha}{dt} + \delta\frac{d\beta}{dt} + \alpha\frac{d\gamma}{dt} - \beta\frac{d\delta}{dt}\right)\mathbf{i}_\eta + 2\left(\beta\frac{d\alpha}{dt} - \alpha\frac{d\beta}{dt} + \delta\frac{d\gamma}{dt} - \gamma\frac{d\delta}{dt}\right)\mathbf{i}_\zeta$$

Also obtain this result directly using Eqs. (2.51) and (2.17).

(e) Using the Euler parameters for the spherical coordinate system (Prob. 2-8), derive the angular velocity vector $\omega$ obtained in Prob. 2-14.

Chapter 3

# The Problem of Two Bodies

CURIOUSLY, THE ANALYTIC SOLUTION OF THE TWO-BODY PROBLEM for spheres of finite size was not accomplished until many years after Newton's geometrical solution (given in his *Principia*, Book I, Section 11) which he obtained about 1685. Although the methods of the calculus were enthusiastically developed in continental Europe at the beginning of the eighteenth century, Newton's system of mechanics did not find immediate acceptance. Indeed, the French preferred the vortex theory of René Descartes (1596–1650) until Voltaire, after his London visit in 1727, vigorously supported the Newtonian theory. This, coupled with the fact that the English continued to employ the geometrical methods of the *Principia*, delayed the analytical solution of the problem. It was probably first given by Daniel Bernoulli in the memoir for which he received the prize from the French Academy in 1734. It was certainly solved in detail by Euler in 1744 in his *Theoria motuum planetarum et cometarum*.

Sir Isaac Newton (1642–1727) was educated at local schools of low educational standards near the hamlet of Woolsthrope, England where he was born. He was not a particularly distinguished student and entered the Trinity College of Cambridge University in 1661 with a deficiency, from the entrance examinations, in Euclidean geometry. Just after graduation, the university was closed because the plague was then rampant in the London area. He left school for the family home where he began his work in mechanics, mathematics, and optics. Others had advanced the concept of the inverse square law of gravitation—including Kepler—but Newton recognized it as the key to celestial mechanics. He also developed general methods for treating problems of the calculus and discovered that white light is really composed of all colors.

> "All this was in the two plague years of 1665 and 1666, for in those days I was in the prime of my age for invention, and minded mathematics and philosophy [science] more than at any other time since."

Having an abnormal fear of criticism, he neither published nor even discussed his discoveries. They only came to light after Isaac Barrow (1630–1677), Newton's friend, teacher, and predecessor in the Lucasian chair of mathematics at Cambridge, and later Edmond Halley (1656–1742), the astronomer for whom Halley's comet is named, recognized his greatness and encouraged him. The first edition of the *Philosophiae Naturalis Principia Mathematica* appeared in 1687. He purposely made it difficult to understand "*to avoid being bated by little smatterers in mathematics.*"

In this chapter, we develop the vector equation of two-body motion and solve it first by power series, then by vector methods of analysis. In the former, we encounter Lagrange's *fundamental invariants* and, in the latter, we are led to the basic integrals called *orbital elements*. We also consider several methods of two-body orbit determination—both exact and approximate—for illustrative purposes only. It is not within the scope of this book to examine the practical details of concern to the astronomer.

## 3.1 Equation of Relative Motion

The equations of motion of two mass particles governed solely by their mutual gravitational attraction are obtained immediately from Sect. 2.4 by setting $n = 2$. Thus, the motion of two bodies is fully described by the following pair of nonlinear vector differential equations

$$m_1 \frac{d^2 \mathbf{r}_1}{dt^2} = \frac{Gm_1 m_2}{r_{12}^3}(\mathbf{r}_2 - \mathbf{r}_1)$$
$$m_2 \frac{d^2 \mathbf{r}_2}{dt^2} = \frac{Gm_2 m_1}{r_{21}^3}(\mathbf{r}_1 - \mathbf{r}_2)$$
(3.1)

together with a set of initial conditions such as the position vectors $\mathbf{r}_1(t)$, $\mathbf{r}_2(t)$ and the velocity vectors $\mathbf{v}_1(t)$, $\mathbf{v}_2(t)$ specified at some particular instant of time. Finding the positions and velocities at future times is the famous two-body problem which was solved by Newton.

In most instances, we are concerned with either the motion of one mass relative to the other or, alternatively, the motion of each with respect to their common center of mass. Seldom are we interested in the absolute motion referred to an arbitrary fixed reference system.

The equation describing the motion of $m_2$ relative to $m_1$ is readily obtained by differencing Eqs. (3.1) after first cancelling the common mass factors. Thus, we have

$$\frac{d^2 \mathbf{r}}{dt^2} + \frac{\mu}{r^3}\mathbf{r} = \mathbf{0} \tag{3.2}$$

where

$$\mathbf{r} = \mathbf{r}_2 - \mathbf{r}_1 \tag{3.3}$$

is the vector position of $m_2$ relative to $m_1$ and

$$\mu = G(m_1 + m_2) \tag{3.4}$$

This is the fundamental differential equation of the two-body problem.

It is worth emphasizing that Eq. (3.2) is actually the vector form of three simultaneous second-order, nonlinear, scalar differential equations in the components of the vector

$$\mathbf{r} = x\,\mathbf{i}_x + y\,\mathbf{i}_y + z\,\mathbf{i}_z$$

Specifically,

$$\frac{d^2x}{dt^2} + \mu\frac{x}{r^3} = 0 \qquad \frac{d^2y}{dt^2} + \mu\frac{y}{r^3} = 0 \qquad \frac{d^2z}{dt^2} + \mu\frac{z}{r^3} = 0$$

where

$$r = \sqrt{x^2 + y^2 + z^2}$$

A somewhat more intuitive appreciation for the equation of relative motion can be had from the following heuristic argument. Let $\mathbf{a}_1$ and $\mathbf{a}_2$ be the acceleration vectors associated with $m_1$ and $m_2$, respectively. Then, since the forces acting on the two masses are equal in magnitude and oppositely directed, we may write

$$m_1 a_1 = m_2 a_2 = \frac{G m_1 m_2}{r^2} \qquad \text{and} \qquad m_1 \mathbf{a}_1 + m_2 \mathbf{a}_2 = \mathbf{0}$$

Let the same acceleration $-\mathbf{a}_1$ be applied to each mass. Their relative motion will be unaltered and, in addition, the mass $m_1$ will be unaccelerated. The acceleration of $m_2$ is then

$$\mathbf{a}_2 - \mathbf{a}_1 = \mathbf{a}_2 + \frac{m_2}{m_1}\mathbf{a}_2 = \frac{m_1 + m_2}{m_1}\mathbf{a}_2$$

and the net force acting on $m_2$ is now

$$m_2(\mathbf{a}_2 - \mathbf{a}_1) = \frac{m_1 + m_2}{m_1} \times \frac{G m_1 m_2}{r^2}\left(-\frac{\mathbf{r}}{r}\right)$$

with the unit vector $-\mathbf{r}/r$ specifying the direction of the force from $m_2$ toward $m_1$. Finally, since

$$m_2 \frac{d^2 \mathbf{r}}{dt^2} = m_2(\mathbf{a}_2 - \mathbf{a}_1)$$

the equation of relative motion (3.2) is again obtained.

One final remark is worthwhile. Remember that the body whose mass is $m_1$ is *not* fixed in space.† It would be an inappropriate application of Newton's law of gravitation to assume that it was and to do so would lead the unwary to a different and erroneous result.

---

† Equation (3.2) may be regarded as describing the motion of a body of mass $m$ about a *fixed* center of attraction for which the force magnitude is $Gm(m_1 + m_2)/r^2$.

## Problem 3-1

Derive the differential equations

$$\frac{d^2r}{dt^2} - r\left(\frac{d\theta}{dt}\right)^2 = -\frac{\mu}{r^2}$$

$$\frac{d}{dt}\left(r^2\frac{d\theta}{dt}\right) = 0$$

which describe the relative motion of two bodies in polar coordinates.

## Problem 3-2

Derive the equation of motion of a body of mass $m_1$ with respect to the center of mass of $m_1$ and $m_2$

$$\mathbf{r}_{cm} = \frac{m_1\mathbf{r}_1 + m_2\mathbf{r}_2}{m_1 + m_2}$$

in the form

$$\frac{d^2\mathbf{r}}{dt^2} + \frac{Gm_2^3}{(m_1+m_2)^2 r^3}\mathbf{r} = 0$$

where $\mathbf{r} = \mathbf{r}_1 - \mathbf{r}_{cm}$. A similar equation for the motion of $m_2$ obtains by reversing the subscripts.

NOTE: The equation of motion is the same as Eq. (3.2) with a different value for the constant $\mu$.

## Problem 3-3

The Lagrangian function for the motion of $m_1$ with respect to the centroid of $m_1$ and $m_2$ is

$$L = \frac{1}{2}\frac{m_1}{m_2}(m_1+m_2)\mathbf{v}^T\mathbf{v} + \frac{Gm_1 m_2^2}{(m_1+m_2)r}$$

Use Lagrange's form of the equations of motion [see Prob. 2-12] to provide an alternate solution to Prob. 3-2.

## 3.2 Solution by Power Series

The basic equations governing the relative motion of two bodies are nonlinear so that, a priori, we should not expect closed form expressions for the position and velocity vectors **r** and **v** to exist as time dependent quantities. Under any circumstances, though, power series developments may be obtained. Indeed, the coefficients in a Taylor series expansion†

$$\mathbf{r}(t) = \mathbf{r}_0 + (t-t_0)\left.\frac{d\mathbf{r}}{dt}\right|_0 + \frac{(t-t_0)^2}{2!}\left.\frac{d^2\mathbf{r}}{dt^2}\right|_0 + \frac{(t-t_0)^3}{3!}\left.\frac{d^3\mathbf{r}}{dt^3}\right|_0 + \cdots$$

can be found from the equation of motion (3.2) and its higher derivatives.

---

† Brook Taylor (1685–1731) was the secretary of the Royal Society from 1714–1718. During this period, his *Methodus Incrementorum Directa et Inversa* was published in which he derived the theorem that still bears his name and which he had stated in 1712. John Bernoulli had published practically the same result in the *Acta Eruditorum* of 1694. Taylor knew the result but did not mention it since his own "proof" was different.

## Lagrange's Fundamental Invariants

Successive differentiation of Eq. (3.2) involves higher derivatives of the quantity $\mu/r^3$, a calculation that, fortunately, can be expedited in a convenient and quite interesting manner. For, if we define

$$\epsilon = \frac{\mu}{r^3}$$

then

$$\frac{d\epsilon}{dt} = -3\frac{\mu}{r^4}\frac{dr}{dt} = -3\epsilon\frac{1}{r}\frac{dr}{dt}$$

Now, define

$$\lambda = \frac{1}{r}\frac{dr}{dt} = \frac{\mathbf{r}\cdot\mathbf{v}}{r^2}$$

(using Prob. 2–15 to obtain the alternate form) so that

$$\frac{d\lambda}{dt} = \frac{1}{r^2}\left(\mathbf{v}\cdot\mathbf{v} + \mathbf{r}\cdot\frac{d\mathbf{v}}{dt}\right) - 2(\mathbf{r}\cdot\mathbf{v})\frac{1}{r^3}\frac{dr}{dt} = \frac{v^2}{r^2} - \epsilon - 2\lambda^2$$

Finally, we define

$$\psi = \frac{v^2}{r^2} = \frac{\mathbf{v}\cdot\mathbf{v}}{r^2}$$

and calculate

$$\frac{d\psi}{dt} = \frac{2}{r^2}\mathbf{v}\cdot\frac{d\mathbf{v}}{dt} - 2v^2\frac{1}{r^3}\frac{dr}{dt} = -2\lambda(\epsilon + \psi)$$

The term *fundamental invariants* has been used for $\epsilon$, $\lambda$, $\psi$—they are "invariant" because they are independent of the selected coordinate system and "fundamental" because they form a closed set under the operation of time differentiation. Thus, to calculate the various derivatives of the position vector $\mathbf{r}$, we successively differentiate

$$\frac{d\mathbf{r}}{dt} = \mathbf{v} \qquad \frac{d\mathbf{v}}{dt} = -\epsilon\mathbf{r} \qquad (3.5)$$

using the relations

$$\frac{d\epsilon}{dt} = -3\epsilon\lambda \qquad \frac{d\lambda}{dt} = \psi - \epsilon - 2\lambda^2 \qquad \frac{d\psi}{dt} = -2\lambda(\epsilon + \psi) \qquad (3.6)$$

where the quantities $\epsilon$, $\lambda$, $\psi$ are defined as

$$\epsilon = \frac{\mu}{r^3} \qquad \lambda = \frac{\mathbf{r}\cdot\mathbf{v}}{r^2} \qquad \psi = \frac{\mathbf{v}\cdot\mathbf{v}}{r^2} \qquad (3.7)$$

In this manner, we obtain

$$\frac{d\mathbf{r}}{dt} = \mathbf{v}$$

$$\frac{d^2\mathbf{r}}{dt^2} = -\epsilon\mathbf{r}$$

$$\frac{d^3\mathbf{r}}{dt^3} = 3\epsilon\lambda\mathbf{r} - \epsilon\mathbf{v}$$

$$\frac{d^4\mathbf{r}}{dt^4} = (-15\epsilon\lambda^2 + 3\epsilon\psi - 2\epsilon^2)\mathbf{r} + 6\epsilon\lambda\mathbf{v}$$

$$\frac{d^5\mathbf{r}}{dt^5} = (105\epsilon\lambda^3 - 45\epsilon\psi\lambda + 30\epsilon^2\lambda)\mathbf{r} + (-45\epsilon\lambda^2 + 9\epsilon\psi - 8\epsilon^2)\mathbf{v} \quad \text{etc.}$$

indicating that the position vector $\mathbf{r}$ at any time $t$ can be represented in terms of the position and velocity vectors $\mathbf{r}_0$ and $\mathbf{v}_0$ at time $t_0$ in the form†

$$\mathbf{r}(t) = F(t)\mathbf{r}_0 + G(t)\mathbf{v}_0$$

where $F$ and $G$ have series representations in powers of $t - t_0$.

**Problem 3-4**
Define two functions $\gamma$ and $\delta$ as

$$\gamma = 2\epsilon - \psi \qquad \delta = \psi - \lambda^2$$

where $\epsilon$, $\lambda$, $\psi$ are the fundamental invariants of the two-body problem. Show that

$$\frac{1}{\gamma}\frac{d\gamma}{dt} = -\frac{2}{r}\frac{dr}{dt} \qquad \frac{1}{\delta}\frac{d\delta}{dt} = -\frac{4}{r}\frac{dr}{dt}$$

and, by integration, prove that the energy and angular momentum are constant.

Karl Stumpff 1959

### Recursion Equations for the Coefficients

The calculation rapidly becomes complex and tedious so that it is desirable to have a more orderly and formal procedure. To this end, let us first note that the functions $F$ and $G$ each satisfy the differential equation

$$\frac{d^2Q}{dt^2} + \epsilon Q = 0 \tag{3.8}$$

and, for $Q = F$, the initial conditions are

$$Q(t_0) = F(t_0) = 1 \quad \text{and} \quad \left.\frac{dQ}{dt}\right|_{t_0} = \left.\frac{dF}{dt}\right|_{t_0} = 0$$

---

† The functions $F$ and $G$ are the so-called *Lagrangian coefficients* or the *Lagrange F and G functions* to be discussed further in Sect. 3.6.

Sect. 3.2]   Solution by Power Series   113

while, for $Q = G$, we have

$$Q(t_0) = G(t_0) = 0 \quad \text{and} \quad \left.\frac{dQ}{dt}\right|_{t_0} = \left.\frac{dG}{dt}\right|_{t_0} = 1$$

We may utilize the standard technique of determining the coefficients of a power series solution of Eq. (3.8), coupled with Eqs. (3.6), in the form

$$Q = \sum_{n=0}^{\infty} Q_n (t - t_0)^n \qquad (3.9)$$

where we also write

$$\epsilon = \sum_{n=0}^{\infty} \epsilon_n (t - t_0)^n \qquad \lambda = \sum_{n=0}^{\infty} \lambda_n (t - t_0)^n \qquad \psi = \sum_{n=0}^{\infty} \psi_n (t - t_0)^n$$

By substituting these power series into the relevant differential equations and requiring coefficients of like powers of $t - t_0$ to be the same, we obtain recursive expressions for the coefficients which can be easily solved.

In developing these recursion equations, it is essential to deal properly with the operation of multiplying two power series. (Refer to Appendix C.) It is easy to show that, for example,

$$\epsilon Q = \sum_{n=0}^{\infty} \epsilon_n (t - t_0)^n \sum_{m=0}^{\infty} Q_m (t - t_0)^m$$

$$= \sum_{n=0}^{\infty} \sum_{m=0}^{\infty} \epsilon_n Q_m (t - t_0)^{n+m} = \sum_{n=0}^{\infty} \left( \sum_{j=0}^{n} \epsilon_j Q_{n-j} \right) (t - t_0)^n$$

with similar relations obtaining for the products in Eqs. (3.6).

The coefficients in the power series are then obtained successively by evaluating the following recursion equations for $n = 0, 1, 2, \ldots$:

$$(n+1)(n+2)Q_{n+2} = -(\epsilon_0 Q_n + \epsilon_1 Q_{n-1} + \cdots + \epsilon_n Q_0)$$
$$(n+1)\epsilon_{n+1} = -3(\epsilon_0 \lambda_n + \epsilon_1 \lambda_{n-1} + \cdots + \epsilon_n \lambda_0)$$
$$(n+1)\lambda_{n+1} = \psi_n - \epsilon_n - 2(\lambda_0 \lambda_n + \lambda_1 \lambda_{n-1} + \cdots + \lambda_n \lambda_0) \quad (3.10)$$
$$(n+1)\psi_{n+1} = -2[\lambda_0(\epsilon_n + \psi_n) + \lambda_1(\epsilon_{n-1} + \psi_{n-1})$$
$$+ \cdots + \lambda_n(\epsilon_0 + \psi_0)]$$

In this way, the series coefficients for the Lagrange $F$ and $G$ functions are found to be

$$F_0 = 1 \qquad F_3 = \tfrac{1}{2}\epsilon_0 \lambda_0$$
$$F_1 = 0 \qquad F_4 = -\tfrac{5}{8}\epsilon_0 \lambda_0^2 + \tfrac{1}{8}\epsilon_0 \psi_0 - \tfrac{1}{12}\epsilon_0^2$$
$$F_2 = -\tfrac{1}{2}\epsilon_0 \qquad F_5 = \tfrac{7}{8}\epsilon_0 \lambda_0^3 - \tfrac{3}{8}\epsilon_0 \psi_0 \lambda_0 + \tfrac{1}{4}\epsilon_0^2 \lambda_0 \qquad \text{etc.}$$

and

$$G_0 = 0 \qquad G_3 = -\tfrac{1}{6}\epsilon_0$$
$$G_1 = 1 \qquad G_4 = \tfrac{1}{4}\epsilon_0 \lambda_0$$
$$G_2 = 0 \qquad G_5 = -\tfrac{3}{8}\epsilon_0 \lambda_0^2 + \tfrac{3}{40}\epsilon_0 \psi_0 - \tfrac{1}{15}\epsilon_0^2 \qquad \text{etc.}$$

The higher-order coefficients are considerably more complex. Clearly, for numerical work, it is easier to use the recursion formulas (3.10) directly rather than, first, to develop literal expressions for these coefficients.

◇ **Problem 3-5**

The position and velocity vectors of a spacecraft in interplanetary space at time $t = 0.010576712$ year are:

$$\mathbf{r} = \begin{bmatrix} 0.159321004 \\ 0.579266185 \\ 0.052359607 \end{bmatrix} \text{ a.u.} \qquad \mathbf{v} = \begin{bmatrix} -9.303603251 \\ 3.018641330 \\ 1.536362143 \end{bmatrix} \text{ a.u./year}$$

Determine the position vector at time $t = 0.021370777$ year. Compare this with the exact value

$$\mathbf{r}(0.021370777 \text{ year}) = \begin{bmatrix} 0.057594337 \\ 0.605750797 \\ 0.068345246 \end{bmatrix} \text{ a.u.}$$

NOTE: The unit of length used here is the *astronomical unit* which is abbreviated as "a.u." and defined in the next section. Also, for the gravitation constant, use $\mu = 4\pi^2$, the justification for which is given, also, in the following section.

## 3.3 Integrals of the Two-Body Problem

Even though the second-order vector differential equation governing the relative motion of two bodies is nonlinear, the equation is capable of a completely general analytical solution. This is expedited by some ad hoc vector operations applied to the equation of motion written in the form

$$\frac{d\mathbf{v}}{dt} = -\frac{\mu}{r^3}\mathbf{r} \qquad (3.11)$$

In each case, the vector manipulations result in transformed versions of Eq. (3.11) which are perfect differentials and, hence, immediately integrable. The constants of integration, called *integrals*[†] of the motion, are of profound importance in conveying the properties of the solution.

---

[†] The "constants of integration" or "integrals" are also called *orbital elements*—a term introduced in the next section.

## Angular Momentum Vector

By taking the vector product of Eq. (3.11) with the position vector $\mathbf{r}$, we have

$$\mathbf{r} \times \frac{d\mathbf{v}}{dt} = \frac{d}{dt}(\mathbf{r} \times \mathbf{v}) = \mathbf{0}$$

and, by integrating, obtain

$$\mathbf{h} = \mathbf{r} \times \mathbf{v} \qquad (3.12)$$

where $\mathbf{h}$ is the integration constant. The vector $\mathbf{h}$ is interpreted as a *massless angular momentum*. Hence, the angular momentum is constant and the motion takes place in the plane $\mathbf{h} \cdot \mathbf{r} = 0$.

Using the polar coordinate expression for $\mathbf{r}$ and $\mathbf{v}$ given in Prob. 2–15, we find that

$$\mathbf{h} = r^2 \frac{d\theta}{dt} \mathbf{i}_z$$

and since $\frac{1}{2} r^2 d\theta/dt$ is the rate at which the radius vector sweeps out area, we have a verification of *Kepler's second law of planetary motion*. Thus, $h$ is twice the areal velocity and we may write

$$r^2 \frac{d\theta}{dt} = h \qquad (3.13)$$

If $h = 0$, the position and velocity vectors are parallel, and the resulting motion is said to be *rectilinear*.

## Eccentricity Vector

The vector product of Eq. (3.11) with the angular momentum vector $\mathbf{h}$ yields

$$\frac{d\mathbf{v}}{dt} \times \mathbf{h} = -\frac{\mu}{r^3} \mathbf{r} \times \mathbf{h} = -\frac{\mu}{r^3} \mathbf{r} \times (\mathbf{r} \times \mathbf{v}) = \frac{\mu}{r^3}[r^2 \mathbf{v} - (\mathbf{r} \cdot \mathbf{v})\mathbf{r}]$$

But, from Prob. 2–15,

$$\mathbf{r} \cdot \mathbf{v} = r \frac{dr}{dt}$$

and also

$$\frac{d\mathbf{v}}{dt} \times \mathbf{h} = \frac{d}{dt}(\mathbf{v} \times \mathbf{h})$$

since $\mathbf{h}$ is a constant vector. Therefore,

$$\frac{d}{dt}(\mathbf{v} \times \mathbf{h}) = \frac{\mu}{r^2}\left(r\mathbf{v} - \mathbf{r}\frac{dr}{dt}\right) = \mu \frac{d}{dt}\left(\frac{\mathbf{r}}{r}\right)$$

which may be integrated to obtain

$$\mu \mathbf{e} = \mathbf{v} \times \mathbf{h} - \frac{\mu}{r}\mathbf{r} \qquad (3.14)$$

The vector constant of integration $\mu\mathbf{e}$ is sometimes called the *Laplace vector*. We shall, instead, use the terminology *eccentricity vector* for the constant vector $\mathbf{e}$ since its magnitude $e$ is the *eccentricity* of the orbit.

### The Parameter and Energy Integral

An important relationship is revealed by calculating the magnitude of the eccentricity vector from Eq. (3.14). There results

$$e^2 = \mathbf{e} \cdot \mathbf{e} = \frac{1}{\mu^2}(\mathbf{v} \times \mathbf{h}) \cdot (\mathbf{v} \times \mathbf{h}) - \frac{2}{\mu r}\mathbf{r} \cdot \mathbf{v} \times \mathbf{h} + 1$$

But

$$(\mathbf{v} \times \mathbf{h}) \cdot (\mathbf{v} \times \mathbf{h}) = \mathbf{v} \cdot \mathbf{h} \times (\mathbf{v} \times \mathbf{h}) = h^2 v^2$$

since $\mathbf{h}$ and $\mathbf{v}$ are orthogonal and

$$\mathbf{r} \cdot \mathbf{v} \times \mathbf{h} = \mathbf{r} \times \mathbf{v} \cdot \mathbf{h} = h^2$$

Hence,

$$1 - e^2 = \frac{h^2}{\mu}\left(\frac{2}{r} - \frac{v^2}{\mu}\right)$$

The first factor

$$p = \frac{h^2}{\mu} \quad (3.15)$$

has the dimension of length and is known as the *parameter*. The second factor must be a constant of the motion. Thus, we define

$$a = \left(\frac{2}{r} - \frac{v^2}{\mu}\right)^{-1} \quad (3.16)$$

which has also the dimension of length. When expressed in the form

$$\frac{1}{2}v^2 - \frac{\mu}{r} = \text{constant} = -\frac{\mu}{2a}$$

we can identify $\frac{1}{2}v^2$ as the *kinetic energy* and $-\mu/r$ as the *potential energy*. It follows that the total energy is constant which was demonstrated for the general case in Sect. 2.4. The quantity $-\mu/2a$ is called the *total energy constant*.

When Eq. (3.16) is expressed in the equivalent form

$$v^2 = \mu\left(\frac{2}{r} - \frac{1}{a}\right) \quad (3.17)$$

the resulting relation is the *energy integral*, sometimes called the *vis-viva integral*.†

---

† Historically, in the field of mechanics, two types of forces were recognized called *vis viva* and *vis mortua*, living force and dead force. In general, the forces resulting in equilibrium were dead forces while those causing motion were living forces. Hence, the distinct branches of mechanics—statics and dynamics.

### Integrals of the Two-Body Problem

Clearly, the quantities $p$, $a$, and $e$ are related by

$$p = a(1 - e^2) \tag{3.18}$$

Since $p$ is never negative, we see that $e$ must be less than or greater than one according as $a$ is positive or negative. Furthermore, the eccentricity $e$ will be unity either for rectilinear motion ($h = 0$) or for zero total energy ($v^2 = 2\mu/r$).

### Equation of Orbit

By calculating the scalar product of Eq. (3.14) and the position vector $\mathbf{r}$, we obtain

$$p = r + \mathbf{e} \cdot \mathbf{r} \tag{3.19}$$

Now let $f$, called the *true anomaly*, be the angle between the vectors $\mathbf{r}$ and $\mathbf{e}$ so that we have

$$r = \frac{p}{1 + e \cos f} \tag{3.20}$$

as the *equation of orbit* in polar coordinates. Clearly, the orbit is symmetrical about the axis defined by the eccentricity vector $\mathbf{e}$. Furthermore, the orbit is bounded if $e < 1$ and unbounded if $e \geq 1$.

To convert the equation of orbit to rectangular cartesian coordinates, let the $x, y$ plane be the plane of motion with the $x$ axis directed along the eccentricity vector. Then if $x$, $y$ are the coordinates of a point on the orbit, we have $x = r \cos f$ and $r = p - ex$ from Eq. (3.19). Therefore, for the case $e \neq 1$, we may use Eq. (3.18) to write

$$y^2 = r^2 - x^2 = (p - ex)^2 - x^2 = (1 - e^2)[a^2 - (x + ea)^2]$$

or

$$\frac{(x + ea)^2}{a^2} + \frac{y^2}{a^2(1 - e^2)} = 1 \tag{3.21}$$

On the other hand, for the case $e = 1$, we have

$$y^2 = r^2 - x^2 = (p - x)^2 - x^2$$

or simply

$$y^2 = 2p(\tfrac{1}{2}p - x) \tag{3.22}$$

Equation (3.21) represents a *circle*, *ellipse*, or *hyperbola* according as the eccentricity is zero, less than one, or greater than one, while Eq. (3.22) is that of a *parabola*. In each case the *focus* $F$ is at the origin. For the circle, ellipse, and hyperbola, the *center* $C$ has the coordinates $(-ea, 0)$ while the *vertex* $A$ for the parabola is at $(\tfrac{1}{2}p, 0)$. The various cases are illustrated in Figs. 3.1, 3.2, and 3.3.

The point $A$ in the figure, corresponding to $f = 0$, at which $r$ is a minimum is called *pericenter* or *periapse*—an *apse* being that point in an

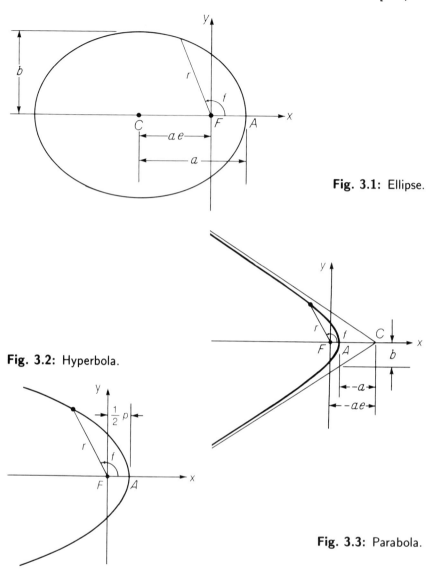

Fig. 3.1: Ellipse.

Fig. 3.2: Hyperbola.

Fig. 3.3: Parabola.

orbit where the motion is at right angles to the radius vector. The ellipse has a second apse called *apocenter* or *apoapse* where $r$ has its maximum value. (Of course, all points on a circle are apses.)

Because of its geometrical significance, the energy constant $a$ is termed the *semimajor axis* and is positive for ellipses, negative for hyperbolas, and infinite for parabolas. Historically, the semimajor axis of an ellipse is called the *mean distance*, although it is *not* the average length of the radius vector with respect to time. In astronomy, the semimajor axis of the earth's orbit is frequently chosen as the unit of length called the *astronomical unit*.

The *semiminor axis* $b$ of an orbit is defined as the positive square root of
$$b^2 = |a^2(1-e^2)| \tag{3.23}$$
The circle is, of course, the special case of an ellipse for which $a$ and $b$ are equal. The corresponding case for a hyperbola is called an *equilateral* or *rectangular* hyperbola.

Finally, we remark that the chord through the focus and perpendicular to the major axis is called the *latus rectum* and has length $2p$ so that $p$ is, sometimes, referred to as the *semilatus rectum*.

## Period and Mean Motion

The *period* of elliptic motion may be obtained from a simple application of Kepler's second law since it is the time required for the radius vector to sweep over the entire enclosed area. Denoting the period by $P$ and recalling that the area of an ellipse is $\pi ab$, we have
$$\frac{2\pi ab}{P} = h$$
Then, using Eqs. (3.15), (3.18), and (3.23), we readily obtain
$$P = 2\pi \sqrt{\frac{a^3}{\mu}} \tag{3.24}$$
If the masses of the planets are considered negligible when compared with the mass of the sun, then Eq. (3.24) is a verification of Kepler's third law of planetary motion.

The term *mean angular motion* or simply *mean motion* is frequently given to the quantity $n$ defined by
$$n = \frac{2\pi}{P} = \sqrt{\frac{\mu}{a^3}} \tag{3.25}$$
Thus, *Kepler's third law of motion* may be stated simply as
$$\mu = n^2 a^3 \tag{3.26}$$
For approximate numerical calculations, it is sometimes convenient to use the semimajor axis (or mean distance) of the earth's orbit for the unit of length and the earth's period as the unit of time. In this case, from Eq. (3.24), we conclude that $\mu$ must then be taken as $4\pi^2$.

### Time of Pericenter Passage

The vectors **h** and **e** together determine the size, shape, and orientation of the orbit with respect to the frame of reference. Their components provide six scalar constants of integration of the two-body equation of motion, but these constants are not independent since we always have $\mathbf{h} \cdot \mathbf{e} = 0$. Therefore, an additional integration constant will be required to complete the solution. What is missing, of course, is the location of the body in orbit at some particular instant of time.

By combining Kepler's second law, as expressed by Eq. (3.13), and the equation of orbit Eq. (3.20), we may write

$$\sqrt{\frac{\mu}{p^3}}\, dt = \frac{df}{(1 + e \cos f)^2} \qquad (3.27)$$

Integration of this equation provides the necessary relationship between the true anomaly $f$ and the time $t$, and yields as well the remaining integration constant. The classical choice for this constant of integration is the time $\tau$ at which the bodies are at their closest point of approach, i.e., pericenter. Thus, the constant $\tau$ is known as the *time of pericenter passage*.

The integrated form of Eq. (3.27) for elliptic orbits is the famous transcendental equation of Kepler which has played a major role in the development of many branches of mathematics. Indeed, solving Kepler's equation has occupied the attention of many of the world's foremost mathematicians and is the subject of Chapter 5.

◊ **Problem 3-6**

Kepler's second law, Eq. (3.13), provides a transformation of independent variable from $t$ to $\theta$ as given by

$$\frac{d}{dt} = \frac{h}{r^2} \frac{d}{d\theta}$$

Use the polar coordinate form of the equation of motion (Prob. 3-1) to derive

$$\frac{1}{r^2}\frac{d^2 r}{d\theta^2} - \frac{2}{r^3}\left(\frac{dr}{d\theta}\right)^2 - \frac{1}{r} = -\frac{1}{p}$$

or, equivalently,

$$\frac{d^2}{d\theta^2}\left(\frac{1}{r}\right) + \frac{1}{r} = \frac{1}{p}$$

as a linear, constant-coefficient, second-order differential equation for $1/r$. This will provide an independent derivation of the equation of orbit.

◇ **Problem 3–7**
Derive the following differential equations

(1) $$\frac{d^2r}{dt^2} = \frac{\mu}{r^3}(p-r)$$

(2) $$\left(\frac{dr}{dt}\right)^2 = \mu\left(\frac{2}{r} - \frac{p}{r^2} - \frac{1}{a}\right)$$

(3) $$\frac{d^2}{dt^2}(r^2) = 2\mu\left(\frac{1}{r} - \frac{1}{a}\right)$$

which are vital to a class of orbit-determination problems to be considered later in this chapter.

◇ **Problem 3–8**
From the first part of the previous problem, it is clear that

$$Q = r - p = -\mathbf{r} \cdot \mathbf{e}$$

satisfies the differential equation (3.8). Use the technique of Sect. 3.2 to develop a Taylor series expansion for $r$.

◇ **Problem 3–9**
Develop power series solutions for the set of differential equations

$$\frac{d^2Q}{dt^2} + \epsilon Q = 0 \qquad \frac{d^2r}{dt^2} + \epsilon(r-p) = 0 \qquad r\frac{d\epsilon}{dt} + 3\epsilon\frac{dr}{dt} = 0$$

and obtain, thereby, the following set of recursion equations, as an alternate to those developed in Sect. 3.2, for obtaining the coefficients $Q_2, Q_3, \ldots$:

$$(n+1)(n+2)Q_{n+2} = -(\epsilon_0 Q_n + \epsilon_1 Q_{n-1} + \cdots + \epsilon_n Q_0)$$
$$(n+1)(n+2)d_{n+2} = -(\epsilon_0 d_n + \epsilon_1 d_{n-1} + \cdots + \epsilon_n d_0)$$
$$r_0(n+1)\epsilon_{n+1} = -[(3n+3)\epsilon_0 d_{n+1} + (3n+1)\epsilon_1 d_n + \cdots + (n+3)\epsilon_n d_1]$$

where we have used $d$ for $r - p$. The initial conditions are

$$\epsilon_0 = \frac{\mu}{r_0^3} \qquad d_0 = r_0 - p \qquad d_1 = \frac{\mathbf{r}_0 \cdot \mathbf{v}_0}{r_0}$$

NOTE: This algorithm is independent of the Lagrange invariants and, as such, illustrates the utility of the method for nonlinear (but *not* singular) differential equations in general.

*Victor Bond*† 1966

◇ **Problem 3–10**
Consider two bodies of masses $m_1$ and $m_2$ in orbit about their common center of mass. If each is moving in an elliptical orbit, then the semimajor axes of the two orbits are in inverse ratio to their masses and their eccentricities are the same.

---

† "A Recursive Formulation for Computing the Coefficients of the Time-Dependent $f$ and $g$ Series Solution to the Two-Body Problem," in *The Astronomical Journal*, Vol. 71, February 1966, pp. 8–9.

◊ **Problem 3–11**

The magnitude of the velocity vector **v** for the motion of $m_2$ with respect to $m_1$ is inversely proportional to the length of the perpendicular from $m_1$ to the orbital tangent.

◊ **Problem 3–12**

Derive the relation

$$v^2 = \frac{\mu}{p}(1 + 2e\cos f + e^2)$$

◊ **Problem 3–13**

Let $r_p$ and $r_a$ be the pericenter and apocenter radii, respectively. Then,

$$a = \tfrac{1}{2}(r_p + r_a) \qquad b = \sqrt{r_p r_a} \qquad p = \frac{2 r_p r_a}{r_p + r_a}$$

That is, the semimajor axis of an ellipse is the arithmetic mean between the pericenter and apocenter radii, while the semiminor axis and the parameter are the corresponding geometric and harmonic means, respectively.

NOTE: For a discussion of the various properties of mathematical means refer to Appendix A.

◊ **Problem 3–14**

Consider the hypothetical problem for which the force of attraction is proportional to the distance separating $m_1$ and $m_2$ rather than inversely proportional to the square of the distance. Develop the properties of the relative motion of two bodies and, in particular, show that the orbit is a conic. Is one of the bodies at the focus of the conic? If not, where is it? Is hyperbolic motion possible? Is rectilinear motion possible?

◊ **Problem 3–15**

Let $v_p$ and $v_a$ be the velocity magnitudes of a vehicle at pericenter and apocenter, respectively. Then

$$(1-e)v_p = (1+e)v_a$$

◊ **Problem 3–16**

The period $P$ of an elliptic orbit with velocity $v$ at a given point can be expressed in the form

$$P = P_c \left[ 2 - \left(\frac{v}{v_c}\right)^2 \right]^{-\frac{3}{2}}$$

where $P_c$ and $v_c$ are, respectively, the period and velocity associated with a circular orbit through the same point.

## Problem 3-17
Neglecting the mass of the first satellite of Jupiter, calculate the mass of this planet in terms of the earth from the following data:
    Period of first satellite: 1 day, 18 hours, 28 minutes
    Mean distance of first satellite from Jupiter's center: 267,000 miles
    Radius of earth: 3,960 miles
    Acceleration of gravity at earth's surface: 32.2 fps$^2$

## Problem 3-18
If $q$ is the pericenter distance, the equation of orbit may be written as

$$r = q \frac{1 + \tan^2 \frac{1}{2} f}{1 + \lambda \tan^2 \frac{1}{2} f} \qquad \text{where} \qquad \lambda = \frac{1-e}{1+e}$$

a form which will be particularly useful for studying near-parabolic orbits.

## Problem 3-19
A vehicle is in a two-body orbit with position and velocity vectors $\mathbf{r}$ and $\mathbf{v}$. If the vehicle is to intercept a target position $\mathbf{r}_T$, the following relation

$$(\mathbf{r} \times \mathbf{v}) \cdot [(\mathbf{r}_T - \mathbf{r}) \times \mathbf{v}] + \mu r_T \cdot \left( \frac{\mathbf{r}_T}{r_T} - \frac{\mathbf{r}}{r} \right) = 0$$

among the three vectors must hold true.

NOTE: This is the expression first used by the author for the so-called "delta guidance" algorithm discussed in the Introduction to this book.

## 3.4 Orbital Elements and Coordinate Systems

In celestial mechanics the six integration constants of the two-body orbit, or various functions thereof, are referred to as the *elements* of the orbit. For example, $p$, $e$, $\tau$ are three possible orbital elements. They define the conic irrespective of its relation to the frame of reference. Three other quantities are required for the spatial orientation of the orbit. The classical choices for the remaining three elements are the Euler angles defined in Sect. 2.1.

Typically the coordinates for bodies in the solar system are either *heliocentric* (sun-centered) or *geocentric* (earth-centered), although occasionally the origin may be taken at the center of a planet or the moon. In the latter two cases the phraseology is *planetocentric* and *selenocentric*.

The two fundamental coordinate systems are the so-called *ecliptic system* and *equatorial system*. The fundamental plane in the ecliptic system is the plane of the earth's orbit; in the equatorial system it is the plane of the earth's equator. The inclination of the ecliptic to the equator is referred to as the *obliquity of the ecliptic*. In both systems the reference direction is toward the *vernal equinox*, which is the point of intersection of the two fundamental planes where the sun crosses the equator from south to north in its apparent annual motion along the ecliptic. The spherical coordinates

$\theta$ and $\frac{1}{2}\pi - \phi$ of Fig. 2.1 are called *longitude* and *latitude* in the ecliptic system and *right ascension* and *declination* in the equatorial system. In a rectangular coordinate reference system, the $x$ axis is the direction of the vernal equinox, the $z$ axis is normal to the fundamental plane and positive toward the north, and the $y$ axis then completes a right-handed system. Unit vectors in these directions will be denoted by $\mathbf{i}_x$, $\mathbf{i}_y$, $\mathbf{i}_z$.

Consider now a body moving under solar gravitation. The line of intersection of the plane in which it moves and the plane of the ecliptic is called the *line of nodes*. The *ascending node* is the point at which the body crosses the ecliptic with a positive component of velocity in the $z$ direction. The *longitude of the ascending node*, as measured from the vernal equinox, is denoted by $\Omega$. The *angle of inclination* of the orbital plane of the body to the ecliptic is symbolized by $i$.

To specify the location of the body, a different set of heliocentric coordinate axes will be used. The unit vectors $\mathbf{i}_e$ and $\mathbf{i}_p$ are selected in the body's own orbital plane with $\mathbf{i}_e$ in the direction of *perihelion*, the point of closest approach of the body to the sun. The line from the origin through perihelion is frequently referred to as the *line of apsides* or the *apsidal line*. The unit vectors $\mathbf{i}_p$ and $\mathbf{i}_h$ are then chosen as shown in Fig. 3.4 to make the coordinate system right-handed. The apsidal line makes an angle $\omega$, called the *argument of perihelion*, with the direction of the ascending node. The three angles $\Omega$, $i$, $\omega$ are the Euler angles.

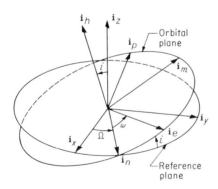

**Fig. 3.4:** Coordinate system geometry.

It is customary to denote the sum $\Omega + \omega$ by $\varpi$, called the *longitude of perihelion*. It should be noted, however, that this is *not* a longitude in the ordinary sense because it is measured in two different planes.

Obviously, similar quantities can be defined for the equatorial reference system. In this case, the point of closest approach to the earth is called *perigee*. For an arbitrary coordinate system the terminology is *pericenter*.

Sect. 3.4]     Orbital Elements and Coordinate Systems     125

Also, in the case of an elliptic orbit, the point of greatest separation is called, correspondingly, *aphelion*, *apogee*, and *apocenter*.

When specifying the position of a body in orbit the following terminology has common usage: the *argument of latitude* $\theta = \omega + f$ and the *true longitude* $L = \varpi + f$. We shall also have occasion to use the *eccentric longitude* $K = \varpi + E$, where $E$ is the eccentric anomaly defined in the next chapter.

A transformation of coordinates between the orbital plane and either the ecliptic or equatorial system is affected by means of the rotation matrix developed in Sect. 2.1.

◇ **Problem 3–20**

Let the origin of coordinates be the center of the earth. Denote by $\lambda$ and $\beta$ the longitude and latitude of a point in the ecliptic system of coordinates, and by $\alpha$ and $\delta$, the right ascension and declination of the same point in the equatorial system. Then

$$\cos\delta\cos\alpha = \cos\beta\cos\lambda$$
$$\cos\delta\sin\alpha = \cos\beta\sin\lambda\cos\epsilon - \sin\beta\sin\epsilon$$
$$\sin\delta = \cos\beta\sin\lambda\sin\epsilon + \sin\beta\cos\epsilon$$

where $\epsilon$ is the obliquity of the ecliptic.

◇ **Problem 3–21**

The position and velocity vectors, expressed as components along the reference axes $x$, $y$, $z$, are

$$\mathbf{r} = r(\cos\Omega\cos\theta - \sin\Omega\sin\theta\cos i)\mathbf{i}_x$$
$$+ r(\sin\Omega\cos\theta + \cos\Omega\sin\theta\cos i)\mathbf{i}_y$$
$$+ r\sin\theta\sin i\,\mathbf{i}_z$$

and

$$\mathbf{v} = -\frac{\mu}{h}[\cos\Omega(\sin\theta + e\sin\omega) + \sin\Omega(\cos\theta + e\cos\omega)\cos i]\mathbf{i}_x$$
$$- \frac{\mu}{h}[\sin\Omega(\sin\theta + e\sin\omega) - \cos\Omega(\cos\theta + e\cos\omega)\cos i]\mathbf{i}_y$$
$$+ \frac{\mu}{h}(\cos\theta + e\cos\omega)\sin i\,\mathbf{i}_z$$

where

$$\theta = \omega + f$$

◇ **Problem 3–22**

The position and velocity vectors of a spacecraft in interplanetary space are the same as those given in Prob. 3–5. Determine the orbital elements $a$, $e$, $p$, $\Omega$, $i$, $\omega$, and the true anomaly $f$. Express all angles in degrees.
ANSWER: $a = 1.2$, $e = 0.5$, $p = 0.9$, $\Omega = 45°$, $i = 10°$, $\omega = 20°$, $f = 10°$.

## 3.5 The Hodograph Plane

Consider a body in orbit and imagine a vector with a fixed point as origin to represent its velocity. As the body moves in its orbit, the velocity vector changes its length and direction so that the terminus describes a curve which is called the *hodograph*.

By taking the vector product of the angular momentum vector **h** and the expression (3.14) for the eccentricity vector **e**, we can derive an equation for the velocity vector suitable for hodograph representation. Thus,

$$\mu \mathbf{h} \times \mathbf{e} = \mathbf{h} \times (\mathbf{v} \times \mathbf{h}) - \frac{\mu}{r}\mathbf{h} \times \mathbf{r}$$

But

$$\mathbf{h} \times (\mathbf{v} \times \mathbf{h}) = h^2 \mathbf{v} - (\mathbf{h} \cdot \mathbf{v})\mathbf{h} = h^2 \mathbf{v}$$

since **h** and **v** are orthogonal. Therefore, we have

$$\mathbf{v} = \frac{\mu}{h^2}\mathbf{h} \times \left(\mathbf{e} + \frac{\mathbf{r}}{r}\right) \tag{3.28}$$

Equation (3.28) provides an elegant equation for the velocity vector **v** in terms of the radius vector **r** for any orbit with known angular momentum and eccentricity vectors **h** and **e**. Indeed, since the radius appears as **r**/r, only the direction of the radius vector is required to determine the velocity vector.

### Two-Body Orbits in the Hodograph Plane

The hodograph representation for two-body motion is based on a graphical interpretation of Eq. (3.28) which we may write as

$$\frac{h\mathbf{v}}{\mu} = \mathbf{i}_h \times (e\,\mathbf{i}_e + \mathbf{i}_r)$$
$$= e\,\mathbf{i}_h \times \mathbf{i}_e + \mathbf{i}_h \times \mathbf{i}_r$$
$$= e\,\mathbf{i}_p + \mathbf{i}_\theta$$

Then, since

$$\mathbf{i}_p = \sin f \,\mathbf{i}_r + \cos f \,\mathbf{i}_\theta$$

we may express the normalized velocity vector $h\mathbf{v}/\mu$ in polar coordinate form

$$\frac{h\mathbf{v}}{\mu} = e \sin f \,\mathbf{i}_r + (1 + e \cos f)\,\mathbf{i}_\theta \tag{3.29}$$

The dimensionless variables of Eq. (3.29) will be convenient to describe the components of the velocity vector in the hodograph plane. Thus, if we plot

$$\frac{h v_r}{\mu} = e \sin f \quad \text{and} \quad \frac{h v_\theta}{\mu} = 1 + e \cos f$$

Sect. 3.5]  The Hodograph Plane  127

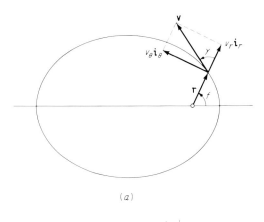

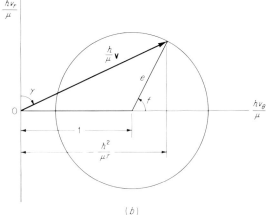

**Fig. 3.5:** Hodograph geometry. (a) Physical plane; (b) hodograph plane.

along the ordinate and abscissa, respectively, as shown in Fig. 3.5, we see that the hodograph is a circle of radius $e$ which is centered at $(1,0)$. The vector of length $hv/\mu$ from the origin of coordinates terminates on the circumference of the circle. The terminus of this scaled-velocity vector moves along the circle in the hodograph plane in direct correspondence with the motion of the position vector $\mathbf{r}$ along the orbit in the physical plane.

For a circular orbit, the hodograph is simply the point $(1,0)$. For elliptic orbits, the circle is confined to the right-half plane. For a parabola, the circle is tangent to the $hv_r/\mu$ axis, and for hyperbolas, the circle intersects this axis. Of course, the hodograph is then that part of the circle in the right-half plane.

Several applications to orbital transfer problems, using the geometrical interpretations made possible through the use of the hodograph plane, are discussed in Chapter 11.

## The Flight-Direction Angle

The angle $\gamma$ between the position vector $\mathbf{r}$ and the velocity vector $\mathbf{v}$ will be referred to as the *flight-direction angle*. This name distinguishes it from the more traditional *flight-path angle* which is the complement of $\gamma$. Clearly, from the figure we have

$$\sin \gamma = \frac{\mu}{hv}(1 + e \cos f)$$
$$\cos \gamma = \frac{\mu}{hv} e \sin f \qquad (3.30)$$

which relate the flight-direction angle and the true anomaly.

◊ **Problem 3-23**
Derive the expressions

$$\sigma \equiv \frac{\mathbf{r} \cdot \mathbf{v}}{\sqrt{\mu}} = \frac{\sqrt{\mu}\, re \sin f}{h} = \frac{rv \cos \gamma}{\sqrt{\mu}} = \sqrt{p} \cot \gamma$$

$$h = rv \sin \gamma$$

◊ **Problem 3-24**
From the results of Prob. 3-23 and the vis-viva integral, derive the following expressions for the parameter $p$ and the velocity vector $\mathbf{v}$ in terms of the flight-direction angle $\gamma$:

$$p = r\left(2 - \frac{r}{a}\right)\sin^2 \gamma$$

$$\mathbf{v} = \frac{h}{r}(\cot \gamma\, \mathbf{i}_r + \mathbf{i}_\theta) = \frac{\mu e \sin f}{h} \mathbf{i}_r + \frac{h}{r}\mathbf{i}_\theta$$

◊ **Problem 3-25**
The quantity $Q \equiv \sigma = \sqrt{p} \cot \gamma$ is a solution of Eq. (3.8). Use the method of Sect. 3.2 to expand $\cot \gamma$ in a Taylor series.

## 3.6 The Lagrangian Coefficients

The components of the position and velocity vectors $\mathbf{r}_0$ and $\mathbf{v}_0$ at a given instant of time $t_0$ serve to describe completely the motion of one body relative to another. In fact, these components can be used as orbital elements and, indeed, for some applications may be the most natural choice. When such is the case, we will require equations for $\mathbf{r}(t)$ and $\mathbf{v}(t)$ in terms of $\mathbf{r}_0$ and $\mathbf{v}_0$. For this purpose, we note that the position and velocity vectors may be expressed in terms of orbital plane coordinates as

$$\mathbf{r} = r \cos f\, \mathbf{i}_e + r \sin f\, \mathbf{i}_p$$
$$\mathbf{v} = -\frac{\mu}{h}\sin f\, \mathbf{i}_e + \frac{\mu}{h}(e + \cos f)\mathbf{i}_p \qquad (3.31)$$

(The equation for $\mathbf{v}$ follows at once from Eq. (3.28) with $\mathbf{e} = e\,\mathbf{i}_e$ and $\mathbf{h} = h\,\mathbf{i}_h$.) These equations, of course, are valid at the initial point for which the position and velocity are $\mathbf{r}_0$ and $\mathbf{v}_0$. When they are inverted, the coordinate unit vectors are obtained in terms of these initial vectors.

The inversion is readily accomplished by first observing that the determinant of the two-dimensional matrix of coefficients in Eqs. (3.31) is simply $h$. Hence,

$$\begin{aligned} \mathbf{i}_e &= \frac{\mu}{h^2}(e + \cos f_0)\,\mathbf{r}_0 - \frac{r_0}{h}\sin f_0\,\mathbf{v}_0 \\ \mathbf{i}_p &= \frac{\mu}{h^2}\sin f_0\,\mathbf{r}_0 + \frac{r_0}{h}\cos f_0\,\mathbf{v}_0 \end{aligned} \quad (3.32)$$

and substitution into Eq. (3.31) gives the desired result in the form

$$\begin{aligned} \mathbf{r} &= F\mathbf{r}_0 + G\mathbf{v}_0 \\ \mathbf{v} &= F_t\mathbf{r}_0 + G_t\mathbf{v}_0 \end{aligned} \quad (3.33)$$

The two-dimensional matrix of coefficients

$$\mathbf{\Phi} = \begin{bmatrix} F & G \\ F_t & G_t \end{bmatrix} \quad (3.34)$$

acts as a *transition matrix* and the matrix elements are the *Lagrangian coefficients*. Clearly, the coefficients $F_t$ and $G_t$ are simply the respective time derivatives of $F$ and $G$.

Two basic properties of $\mathbf{\Phi}$ are readily established:

1. The value of the determinant

$$|\mathbf{\Phi}| = FG_t - GF_t = 1 \quad (3.35)$$

follows from the conservation of angular momentum

$$\mathbf{r} \times \mathbf{v} = (FG_t - GF_t)\mathbf{r}_0 \times \mathbf{v}_0 = \mathbf{r}_0 \times \mathbf{v}_0$$

The inverse of $\mathbf{\Phi}$ is simply

$$\mathbf{\Phi}^{-1} = \begin{bmatrix} G_t & -G \\ -F_t & F \end{bmatrix} \quad (3.36)$$

so that $\mathbf{\Phi}$ is a symplectic matrix.†

2. For any three points on an orbit $\mathbf{r}_0, \mathbf{r}_1, \mathbf{r}_2$

$$\mathbf{\Phi}_{2,0} = \mathbf{\Phi}_{2,1}\mathbf{\Phi}_{1,0} \quad (3.37)$$

which is proved by successive applications of Eqs. (3.33).

Closed-form equations for the elements of $\mathbf{\Phi}$ do not generally exist in terms of the times $t, t_0$. However, they are readily obtained as functions

---

† The symplectic matrix is defined in Chapter 9. It is shown there that any two-dimensional matrix, whose determinant is equal to one, is symplectic.

of the true anomalies $f$, $f_0$ by multiplying the two matrices composed of the coefficients of Eqs. (3.31) and (3.32).

It is more convenient to express the Lagrangian coefficients in terms of the true anomaly difference

$$\theta = f - f_0 \tag{3.38}$$

For this purpose, write

$$\cos f = \cos(\theta + f_0) = \cos\theta \cos f_0 - \sin\theta \sin f_0$$

Then, obtain $\cos f_0$ from the equation of orbit and $\sin f_0$ by calculating the scalar product of the two equations in (3.31). Thus,

$$e \cos f_0 = \frac{p}{r_0} - 1 \quad \text{and} \quad e \sin f_0 = \frac{\sqrt{p}\,\sigma_0}{r_0} \tag{3.39}$$

where $\sigma_0$, which occurs frequently in other contexts, is defined by

$$\sigma_0 \equiv \frac{\mathbf{r}_0 \cdot \mathbf{v}_0}{\sqrt{\mu}} \tag{3.40}$$

Then the polar form of the equation of orbit, Eq. (3.20), may be written as

$$r = \frac{p r_0}{r_0 + (p - r_0)\cos\theta - \sqrt{p}\,\sigma_0 \sin\theta} \tag{3.41}$$

and the Lagrangian coefficients as

$$F = 1 - \frac{r}{p}(1 - \cos\theta) \qquad G = \frac{r r_0}{\sqrt{\mu p}} \sin\theta$$

$$F_t = \frac{\sqrt{\mu}}{r_0 p}[\sigma_0(1 - \cos\theta) - \sqrt{p}\sin\theta] \qquad G_t = 1 - \frac{r_0}{p}(1 - \cos\theta) \tag{3.42}$$

Equations (3.41) and (3.42) are of major importance in our later work.

◇ **Problem 3–26**
Let the skew-symmetric matrix $\mathbf{H}$ be defined as

$$\mathbf{H} = \mathbf{v}\mathbf{r}^T - \mathbf{r}\mathbf{v}^T$$

where the individual matrices $\mathbf{v}\mathbf{r}^T$ and $\mathbf{r}\mathbf{v}^T$ are *dyadic products*. Then, the product $\mathbf{H}\mathbf{w}$ is the vector product of the angular momentum $\mathbf{h}$ and a vector $\mathbf{w}$.

◇ **Problem 3–27**
Derive the following expressions for the reference unit vectors of the orbital plane:

$$\mathbf{i}_e = \frac{1}{e}\left[\left(\frac{v^2}{\mu} - \frac{1}{r}\right)\mathbf{r} - \frac{\sigma}{\sqrt{\mu}}\mathbf{v}\right]$$

$$\mathbf{i}_p = \frac{1}{e\sqrt{p}}\left[\frac{\sigma}{r}\mathbf{r} + \frac{p-r}{\sqrt{\mu}}\mathbf{v}\right]$$

$$\mathbf{i}_h = \frac{1}{\sqrt{\mu p}}\mathbf{r} \times \mathbf{v}$$

## ◇ Problem 3-28
Derive the following expressions for the eccentricity:

(1) $$e^2 = 1 + \frac{p}{r^2}(p - 2r + \sigma^2)$$

(2) $$e^2 = 1 + \frac{rv^2}{\mu}\left(\frac{rv^2}{\mu} - 2\right)\sin^2\gamma$$

## ◇ Problem 3-29
If $v_r$ and $v_\theta$ are the radial and circumferential components of the velocity at a distance $r$ from the center of attraction, then the eccentricity $e$ may be computed from

$$e^2 = \left[\left(\frac{v_\theta}{v_c}\right)^2 - 1\right]^2 + \left(\frac{v_\theta}{v_c}\right)^2\left(\frac{v_r}{v_c}\right)^2$$

where $v_c$ is the velocity of a body in a circular orbit at distance $r$.

## 3.7 Preliminary Orbit Determination

The problem of orbit determination from the point of view of the astronomer is one of calculating the orbital elements of a celestial body from a given set of observations made from the surface of the earth. Several closely related problems will be considered in this section. For each we show how to derive the conventional set of orbital elements from certain postulated geometric and dynamic conditions—one example, Laplace's method, uses exclusively observational data. The techniques described are illustrative only in that they contain the essential ingredients but are devoid of many of the practical details.

### Orbit from Three Coplanar Positions

Assume that we are given three measurements: $r_1, \theta_1$, $r_2, \theta_2$, and $r_3, \theta_3$ specified in polar coordinates, and we wish to determine the parameter $p$, the eccentricity $e$, and the argument of perihelion $\omega$ of an orbit which passes through these three points. Using the equation of orbit in the form

$$r = \frac{p}{1 + e\cos(\theta - \omega)} \quad \text{where} \quad \theta = \omega + f$$

we have directly three equations in the three unknowns which are, of course, nonlinear.

They can be linearized with a simple change of variable. By defining

$$P = e\cos\omega \quad \text{and} \quad Q = e\sin\omega \qquad (3.43)$$

we then have the following three equations in the three unknowns $p$, $P$, and $Q$:

$$\frac{p}{r_1} - P\cos\theta_1 - Q\sin\theta_1 = 1$$
$$\frac{p}{r_2} - P\cos\theta_2 - Q\sin\theta_2 = 1 \quad (3.44)$$
$$\frac{p}{r_3} - P\cos\theta_3 - Q\sin\theta_3 = 1$$

Hence, according to *Cramer's rule*,†

$$p = \frac{\begin{vmatrix} 1 & -\cos\theta_1 & -\sin\theta_1 \\ 1 & -\cos\theta_2 & -\sin\theta_2 \\ 1 & -\cos\theta_3 & -\sin\theta_3 \end{vmatrix}}{\begin{vmatrix} 1/r_1 & -\cos\theta_1 & -\sin\theta_1 \\ 1/r_2 & -\cos\theta_2 & -\sin\theta_2 \\ 1/r_3 & -\cos\theta_3 & -\sin\theta_3 \end{vmatrix}}$$

By evaluating the determinants and using obvious trigonometric identities, we obtain

$$p = \frac{r_1 r_2 r_3 [\sin(\theta_3 - \theta_2) + \sin(\theta_1 - \theta_3) + \sin(\theta_2 - \theta_1)]}{r_2 r_3 \sin(\theta_3 - \theta_2) + r_1 r_3 \sin(\theta_1 - \theta_3) + r_1 r_2 \sin(\theta_2 - \theta_1)} \quad (3.45)$$

Similarly,

$$P = \frac{r_1(r_2 - r_3)\sin\theta_1 + r_2(r_3 - r_1)\sin\theta_2 + r_3(r_1 - r_2)\sin\theta_3}{r_2 r_3 \sin(\theta_3 - \theta_2) + r_1 r_3 \sin(\theta_1 - \theta_3) + r_1 r_2 \sin(\theta_2 - \theta_1)}$$

$$Q = \frac{r_1(r_3 - r_2)\cos\theta_1 + r_2(r_1 - r_3)\cos\theta_2 + r_3(r_2 - r_1)\cos\theta_3}{r_2 r_3 \sin(\theta_3 - \theta_2) + r_1 r_3 \sin(\theta_1 - \theta_3) + r_1 r_2 \sin(\theta_2 - \theta_1)}$$

from which we can calculate $e$ and $\omega$ using the equations

$$e = \sqrt{P^2 + Q^2} \quad \text{and} \quad \tan\omega = \frac{Q}{P} \quad (3.46)$$

This problem is solved in the next subsection using vector algebra as was done by Willard Gibbs‡ who, it is interesting to note, was the first

---

† Gabriel Cramer (1704–1752) published the rule in his *Introduction à l'analyse des lignes courbes algébriques* in 1750 in connection with the problem of determining the coefficients of the general conic, $A + By + Cx + Dy^2 + Exy + x^2 = 0$, passing through five given points. However, for simultaneous linear equations in two, three, and four unknowns, the solution by the method of determinants was created by Colin Maclaurin (1698–1746), probably in 1729, and published in his posthumous *Treatise of Algebra* in 1748.

‡ Josiah Willard Gibbs (1839–1903), professor of mathematical physics at Yale College, made the formal break with quaternions and inaugurated the new subject—three-dimensional vector analysis. He had printed for private distribution to his students a small pamphlet on the *Elements of Vector Analysis* in 1881 and 1884. Gibbs' pamphlet became widely known and was finally incorporated in the book *Vector Analysis*, authored by J. W. Gibbs and E. B. Wilson, and published in 1901.

Sect. 3.7]     Preliminary Orbit Determination     133

recipient of our country's engineering degree and the first American scholar to contribute to the field of celestial mechanics.

## Orbit from Three Position Vectors

Given three successive position vectors $\mathbf{r}_1$, $\mathbf{r}_2$, $\mathbf{r}_3$ assumed to be coplanar, that is, $\mathbf{r}_1 \times \mathbf{r}_2 \cdot \mathbf{r}_3 = 0$, the elements of the two-body orbit which includes these positions can be determined using vector algebra. Gibbs' method for this purpose is to apply vector operations to an equation, which expresses one of the position vectors as a linear combination of the other two

$$\mathbf{r}_2 = \alpha \mathbf{r}_1 + \beta \mathbf{r}_3 \tag{3.47}$$

The constants $\alpha$ and $\beta$ are found by calculating (1) the vector product of Eq. (3.47) with $\mathbf{r}_3$ and $\mathbf{r}_1$, respectively, and (2) the scalar product of the resulting equations with $\mathbf{r}_1 \times \mathbf{r}_3$. Thus, we have

$$\alpha = \frac{\mathbf{r}_2 \times \mathbf{r}_3 \cdot \mathbf{n}}{n^2} \qquad \beta = \frac{\mathbf{r}_1 \times \mathbf{r}_2 \cdot \mathbf{n}}{n^2} \tag{3.48}$$

where

$$\mathbf{n} = \mathbf{r}_1 \times \mathbf{r}_3 \tag{3.49}$$

To obtain the parameter, take the scalar product of Eq. (3.47) with the eccentricity vector $\mathbf{e}$. Since $\mathbf{e} \cdot \mathbf{r}_i = p - r_i$ for $i = 1, 2, 3$, we have a scalar equation involving $p$ as the only unknown. Hence

$$p = \frac{\alpha r_1 - r_2 + \beta r_3}{\alpha - 1 + \beta} \tag{3.50}$$

The vector $\mathbf{e}$ is determined by first observing that

$$\mathbf{n} \times \mathbf{e} = (\mathbf{r}_1 \times \mathbf{r}_3) \times \mathbf{e} = (\mathbf{e} \cdot \mathbf{r}_1)\mathbf{r}_3 - (\mathbf{e} \cdot \mathbf{r}_3)\mathbf{r}_1 = (p - r_1)\mathbf{r}_3 - (p - r_3)\mathbf{r}_1$$

Then, since $\mathbf{e}$ is normal to $\mathbf{n}$, we have

$$(\mathbf{n} \times \mathbf{e}) \times \mathbf{n} = n^2 \mathbf{e}$$

so that

$$\mathbf{e} = \frac{1}{n^2}[(p - r_1)\mathbf{r}_3 \times \mathbf{n} - (p - r_3)\mathbf{r}_1 \times \mathbf{n}] \tag{3.51}$$

The size, shape, and orientation of the orbit are now completely determined. The velocity vector at any of the three positions may, of course, be calculated using Eq. (3.28).

◇ **Problem 3–30**
Derive the following alternate expression for the eccentricity vector:

$$\mathbf{e} = \frac{1}{r_1 \sin^2 \theta}\left[\left(\frac{p}{r_1} - 1\right) - \left(\frac{p}{r_2} - 1\right)\cos\theta\right]\mathbf{r}_1$$
$$+ \frac{1}{r_2 \sin^2 \theta}\left[\left(\frac{p}{r_2} - 1\right) - \left(\frac{p}{r_1} - 1\right)\cos\theta\right]\mathbf{r}_2$$

where $\theta$ is the angle between the position vectors $\mathbf{r}_1$ and $\mathbf{r}_2$.

**134**    The Problem of Two Bodies    [Chap. 3

Approximate Orbit from Three Position Fixes

Although the orbit-determination equations, as developed above for three specified position vectors, are exact, it is clear that numerical accuracy is impaired if the angles between the vectors are small. As an alternative, an approximate orbit can be obtained if the times associated with the vector positions are known. We may then use the dynamic properties of the orbit to determine the orbital elements rather than relying solely on geometry.

Specifically, let $\mathbf{r}_1$, $\mathbf{r}_2$, $\mathbf{r}_3$ be the position vectors of a body in orbit at the times $t_1$, $t_2$, $t_3$, respectively. These data comprise three so-called *position fixes* and can be related by means of appropriate power series expansions. Thus, if we neglect powers of $t$ beyond the fifth, we may write

$$\mathbf{r} = \mathbf{a}_0 + t\,\mathbf{a}_1 + t^2\mathbf{a}_2 + t^3\mathbf{a}_3 + t^4\mathbf{a}_4 + t^5\mathbf{a}_5$$
$$\mathbf{v} = \mathbf{a}_1 + 2t\,\mathbf{a}_2 + 3t^2\mathbf{a}_3 + 4t^3\mathbf{a}_4 + 5t^4\mathbf{a}_5$$
$$-\epsilon\,\mathbf{r} = 2\mathbf{a}_2 + 6t\,\mathbf{a}_3 + 12t^2\mathbf{a}_4 + 20t^3\mathbf{a}_5$$

In the last series we have used the equation of motion to replace $d^2\mathbf{r}/dt^2$ by $-\epsilon\,\mathbf{r}$ where, as before, we have defined

$$\epsilon = \frac{\mu}{r^3} \tag{3.52}$$

Next, define a sequence of time intervals†

$$\tau_1 = t_3 - t_2 \qquad \tau_2 = t_3 - t_1 \qquad \tau_3 = t_2 - t_1 \tag{3.53}$$

and replace $t$ in the power series successively by $-\tau_3$, $0$, $\tau_1$. Thus, we have

$$\mathbf{r}_1 = \mathbf{a}_0 - \tau_3\mathbf{a}_1 + \tau_3^2\mathbf{a}_2 - \tau_3^3\mathbf{a}_3 + \tau_3^4\mathbf{a}_4 - \tau_3^5\mathbf{a}_5$$
$$\mathbf{r}_2 = \mathbf{a}_0$$
$$\mathbf{r}_3 = \mathbf{a}_0 + \tau_1\mathbf{a}_1 + \tau_1^2\mathbf{a}_2 + \tau_1^3\mathbf{a}_3 + \tau_1^4\mathbf{a}_4 + \tau_1^5\mathbf{a}_5$$
$$\mathbf{v}_2 = \mathbf{a}_1 \tag{3.54}$$
$$-\epsilon_1\mathbf{r}_1 = 2\mathbf{a}_2 - 6\tau_3\mathbf{a}_3 + 12\tau_3^2\mathbf{a}_4 - 20\tau_3^3\mathbf{a}_5$$
$$-\epsilon_2\mathbf{r}_2 = 2\mathbf{a}_2$$
$$-\epsilon_3\mathbf{r}_3 = 2\mathbf{a}_2 + 6\tau_1\mathbf{a}_3 + 12\tau_1^2\mathbf{a}_4 + 20\tau_1^3\mathbf{a}_5$$

as seven equations for the seven unknowns $\mathbf{a}_0, \ldots, \mathbf{a}_5, \mathbf{v}_2$. But $\mathbf{a}_0$, $\mathbf{a}_1$, and $\mathbf{a}_2$ are eliminated immediately so that Eqs. (3.54) reduce to a system of just four equations for the four vectors $\mathbf{a}_3$, $\mathbf{a}_4$, $\mathbf{a}_5$, $\mathbf{v}_2$.

Using Cramer's rule to solve for $\mathbf{v}_2$, we obtain the following formula for $\mathbf{v}_2$ as a function of the three position fixes $\mathbf{r}_1$, $\mathbf{r}_2$, $\mathbf{r}_3$:

$$A_0\mathbf{v}_2 = A_1\mathbf{r}_1 + A_2\mathbf{r}_2 + A_3\mathbf{r}_3 \tag{3.55}$$

---

† Note that $\tau_2 = \tau_1 + \tau_3$.

where

$$A_0 = 12\tau_1\tau_2\tau_3(\tau_3 - \tau_1)(2\tau_1^2 + 5\tau_1\tau_3 + 2\tau_3^2)$$
$$A_1 = \tau_1^4[12(2\tau_2 + 3\tau_3) + \epsilon_1\tau_3^2(\tau_1 + 3\tau_2)]$$
$$A_2 = \epsilon_2\tau_1^2\tau_2\tau_3^2(8\tau_2^2 + 3\tau_1\tau_3) - 12\tau_2[2\tau_2^4 - 5\tau_1\tau_3(\tau_2^2 + \tau_1\tau_3)]$$
$$A_3 = \tau_3^4[12(3\tau_1 + 2\tau_2) + \epsilon_3\tau_1^2(3\tau_2 + \tau_3)]$$

Unfortunately, when $\tau_1 = \tau_3 = \tau$, Eq. (3.55) reduces to

$$(12 + \epsilon_1\tau^2)\mathbf{r}_1 + (10\epsilon_2\tau^2 - 24)\mathbf{r}_2 + (12 + \epsilon_3\tau^2)\mathbf{r}_3 = \mathbf{0} \tag{3.56}$$

which does not involve the velocity vector but represents instead a condition imposed on the three position vectors. Thus, the value of Eq. (3.55) for numerical calculation is somewhat questionable even when the time intervals of the position fixes are not equal.

An alternative expression for $\mathbf{v}_2$, which is free of this difficulty, was developed by Samuel Herrick. When the power series are truncated beyond the fourth order, we then have only five coefficients to eliminate from seven equations, and this degree of freedom can be effectively exploited.

To this end, we first eliminate $\mathbf{a}_2$ between the first and third equations of (3.54) by multiplying them by $\tau_1^2$ and $\tau_3^2$, respectively, and subtracting. The resulting equation

$$\tau_1^2\mathbf{r}_1 - \tau_3^2\mathbf{r}_3 = (\tau_1 - \tau_3)\tau_2\mathbf{a}_0 - \tau_1\tau_2\tau_3\mathbf{a}_1 - \tau_1^2\tau_2\tau_3^2\mathbf{a}_3 + \tau_1^2\tau_2\tau_3^2(\tau_3 - \tau_1)\mathbf{a}_4$$

is then used in place of the two equations from which it was derived. We now have six equations for $\mathbf{a}_0$, $\mathbf{a}_1$, $\mathbf{a}_2$, $\mathbf{a}_3$, $\mathbf{a}_4$, $\mathbf{v}_2$. Solving for $\mathbf{v}_2$, we obtain

$$\mathbf{v}_2 = -\tau_1\left(\frac{1}{\tau_2\tau_3} + \frac{\epsilon_1}{12}\right)\mathbf{r}_1$$
$$- (\tau_3 - \tau_1)\left(\frac{1}{\tau_1\tau_3} + \frac{\epsilon_2}{12}\right)\mathbf{r}_2 + \tau_3\left(\frac{1}{\tau_1\tau_2} + \frac{\epsilon_3}{12}\right)\mathbf{r}_3 \tag{3.57}$$

which is valid to fourth order in the time intervals.

With $\mathbf{r}_2$ and $\mathbf{v}_2$ determined, we are able to calculate the more conventional orbital elements in the usual manner.

◇ **Problem 3–31**

From two position fixes $\mathbf{r}_1(t_1)$ and $\mathbf{r}_2(t_2)$, develop the approximate formula for the velocity vector $\mathbf{v}_1$ at time $t_1$

$$\mathbf{v}_1 = -\left(\frac{1}{\tau} - \frac{\epsilon_1}{3}\tau\right)\mathbf{r}_1 + \left(\frac{1}{\tau} + \frac{\epsilon_2}{6}\tau\right)\mathbf{r}_2$$

valid to third order in the time interval $\tau = t_2 - t_1$.

◊ **Problem 3–32**

Two position fixes of a spacecraft in interplanetary space at times

$$t_1 = 0.010576712 \text{ year} \qquad t_2 = 0.021370777 \text{ year}$$

are found to be

$$\mathbf{r}(t_1) = \begin{bmatrix} 0.159321004 \\ 0.579266185 \\ 0.052359607 \end{bmatrix} \text{ a.u.} \qquad \mathbf{r}(t_2) = \begin{bmatrix} 0.057594337 \\ 0.605750797 \\ 0.068345246 \end{bmatrix} \text{ a.u.}$$

Determine the velocity vector at the location corresponding to the earlier time.

If the third position fix at time $t_3 = 0.005274926$ year is included in the data, where

$$\mathbf{r}(t_3) = \begin{bmatrix} 0.208200171 \\ 0.561804188 \\ 0.044088057 \end{bmatrix} \text{ a.u.}$$

calculate the velocity vector at the same location as determined above.
Compare the two results with the exact value given by

$$\mathbf{v}(0.010576712 \text{ year}) = \begin{bmatrix} -9.303603251 \\ 3.018641330 \\ 1.536362143 \end{bmatrix} \text{ a.u./year}$$

## Approximate Orbit from Three Range Measurements

If $r_1$, $r_2$, $r_3$ are the distances of a body in orbit from the center of attraction at the times $t_1$, $t_2$, $t_3$, respectively, we can determine an approximate orbit using a method also attributed to Willard Gibbs. As before we neglect powers of $t$ beyond the fourth and write

$$r = a_0 + a_1 t + a_2 t^2 + a_3 t^3 + a_4 t^4$$

$$\frac{d^2 r}{dt^2} = \epsilon(p - r) = 2a_2 + 6a_3 t + 12 a_4 t^2$$

In the second series we have used an equation developed in Prob. 3–7 to replace $d^2r/dt^2$ by $\epsilon(p-r)$ where $\epsilon$ is defined in Eq. (3.52). Again, we define the sequence of time intervals as in Eq. (3.53) and generate three pairs of equations by replacing $t$ by $-\tau_3$, $0$, and $\tau_1$. These provide six equations in the six unknowns $a_0$, $a_1$, $a_2$, $a_3$, $a_4$, $p$.

Rather than use Cramer's rule, it is convenient to regard this system of equations as six linear equations in five unknowns with the parameter $p$ to be determined to make the system consistent. Then, according to Prob. D–1 in Appendix D on linear algebraic systems, the determinantal equation

$$\begin{vmatrix} 1 & -\tau_3 & \tau_3^2 & -\tau_3^3 & \tau_3^4 & r_1 \\ 1 & 0 & 0 & 0 & 0 & r_2 \\ 1 & \tau_1 & \tau_1^2 & \tau_1^3 & \tau_1^4 & r_3 \\ 0 & 0 & 2 & -6\tau_3 & 12\tau_3^2 & \epsilon_1(p-r_1) \\ 0 & 0 & 2 & 0 & 0 & \epsilon_2(p-r_2) \\ 0 & 0 & 2 & 6\tau_1 & 12\tau_1^2 & \epsilon_3(p-r_3) \end{vmatrix} = 0 \qquad (3.58)$$

which expresses the necessary and sufficient condition for the consistency of our set of linear equations, can be used directly to calculate $p$.

The determinant may be evaluated most readily using the result of Prob. B–20, part b of Appendix B. We obtain

$$144\tau_1\tau_2\tau_3[\tau_1 A_1 \epsilon_1 (p-r_1) + \tau_2 A_2 \epsilon_2 (p-r_2) \\ + \tau_3 A_3 \epsilon_3 (p-r_3) - \tau_1 r_1 + \tau_2 r_2 - \tau_3 r_3] = 0$$

where, for convenience, we have defined

$$12 A_1 = \tau_2 \tau_3 - \tau_1^2 \qquad 12 A_2 = \tau_1 \tau_3 + \tau_2^2 \qquad 12 A_3 = \tau_1 \tau_2 - \tau_3^2 \qquad (3.59)$$

Thus, we have the parameter $p$ of the orbit obtained from

$$p = \frac{r_1 \tau_1 (1 + \epsilon_1 A_1) - r_2 \tau_2 (1 - \epsilon_2 A_2) + r_3 \tau_3 (1 + \epsilon_3 A_3)}{\tau_1 \epsilon_1 A_1 + \tau_2 \epsilon_2 A_2 + \tau_3 \epsilon_3 A_3} \qquad (3.60)$$

which is valid provided that we may neglect terms of fifth order in the time intervals.

The semimajor axis $a$ of the orbit is similarly obtained using another equation developed in Prob. 3–7. From the series expansions

$$r^2 = b_0 + b_1 t + b_2 t^2 + b_3 t^3 + b_4 t^4$$

$$\frac{d^2}{dt^2}(r^2) = 2\mu \left( \frac{1}{r} - \frac{1}{a} \right) = 2b_2 + 6b_3 t + 12 b_4 t^2$$

and following the same procedure as used for the parameter, we find

$$\frac{\mu}{a} = -\frac{r_1^2}{\tau_2 \tau_3}(1 - 2\epsilon_1 A_1) + \frac{r_2^2}{\tau_1 \tau_3}(1 + 2\epsilon_2 A_2) - \frac{r_3^2}{\tau_1 \tau_2}(1 - 2\epsilon_3 A_3) \qquad (3.61)$$

◇ **Problem 3–33**

Three range measurements are available for a spacecraft in interplanetary space as follows:

$r_1 = 0.600762027$ a.u.  $t_1 = 0.005274926$ year
$r_2 = 0.603053915$ a.u.  $t_2 = 0.010576712$ year
$r_3 = 0.612308916$ a.u.  $t_3 = 0.021370777$ year

Determine the semimajor axis $a$, the eccentricity $e$, and the parameter $p$ of the orbit as well as the true anomalies at the three measurement points.

### ◊ Problem 3–34

Use an equation developed in Prob. 3–7 to invent an orbit-determination method using two distinct pairs of range and range-rate measurements made at two different times.

NOTE: Knowledge of the measurement times is not required to calculate the orbital elements.

### ◊ Problem 3–35

A pair of range and range-rate measurements is available for a spacecraft in interplanetary space as follows:

$$r_1 = 0.600762027 \text{ a.u.} \qquad r_2 = 0.603053915 \text{ a.u.}$$

$$\frac{dr_1}{dt} = 0.288618834 \text{ a.u./year} \qquad \frac{dr_2}{dt} = 0.575041077 \text{ a.u./year}$$

Determine the elements $a$, $e$, and $p$ of the orbit.

### Approximate Orbit from Three Observations

Let $\mathbf{i}_{\rho_1}$, $\mathbf{i}_{\rho_2}$, $\mathbf{i}_{\rho_3}$ be unit vectors corresponding to three observed line-of-sight directions between two bodies in orbit—the observations occurring at times $t_1$, $t_2$, $t_3$. We desire an approximate orbit for the observed body, assuming that its mass is negligible in comparison to the masses of the other two. The method we describe for the determination of a preliminary orbit of a celestial object in solar orbit as observed from the earth is named for Laplace.[†] It was developed in 1780—the year before the discovery of the planet Uranus.

We define three position vectors $\mathbf{r}$, $\boldsymbol{\rho}$, $\mathbf{d}$ with

$$\mathbf{r} = \boldsymbol{\rho} + \mathbf{d} \tag{3.62}$$

where $\mathbf{r}$ and $\mathbf{d}$ are, respectively, the positions, relative to the center of force, of the observed body and the body from which the observations are made. The vector $\mathbf{d}$ is known, $\mathbf{r}$ is to be determined, and $\boldsymbol{\rho}$ has known directions at the times of the three observations. Furthermore, if $m_1$ and $m_2$ are masses of the finite bodies, we have

$$\frac{d^2\mathbf{r}}{dt^2} + \frac{\mu_1}{r^3}\mathbf{r} = 0 \qquad \frac{d^2\mathbf{d}}{dt^2} + \frac{\mu_2}{d^3}\mathbf{d} = 0$$

---

[†] Pierre-Simon de Laplace (1749–1827), mathematician and French politician during the Napoleonic era, made many important discoveries in mathematical physics and chemistry but most of his life was devoted to celestial mechanics. His *Mécanique céleste*, consisting of five volumes published between 1799 and 1825, was so complete that his immediate successors found little to add. Unfortunately, his vanity kept him from sufficiently crediting the works of those whom he considered rivals. Laplace was the originator of that most troublesome phrase, which continues to plague students of mathematics: "*It is easy to see that* ..." when, in fact, the missing details are anything but obvious.

where
$$\mu_1 = Gm_1 \qquad \mu_2 = G(m_1 + m_2)$$

Therefore, if we write
$$\boldsymbol{\rho} = \rho \mathbf{i}_\rho \qquad (3.63)$$

and calculate the second derivative of Eq. (3.63), we obtain

$$\left(\frac{d^2\rho}{dt^2} + \frac{\mu_1}{r^3}\rho\right)\mathbf{i}_\rho + 2\frac{d\rho}{dt}\frac{d\mathbf{i}_\rho}{dt} + \rho\frac{d^2\mathbf{i}_\rho}{dt^2} = \left(\frac{\mu_2}{d^3} - \frac{\mu_1}{r^3}\right)\mathbf{d}$$

Next, we calculate the scalar product of this last vector equation, first with $\mathbf{i}_\rho \times d\mathbf{i}_\rho/dt$ and second with $\mathbf{i}_\rho \times d^2\mathbf{i}_\rho/dt^2$. There results

$$\rho\left(\mathbf{i}_\rho \times \frac{d\mathbf{i}_\rho}{dt} \cdot \frac{d^2\mathbf{i}_\rho}{dt^2}\right) = \left(\frac{\mu_2}{d^3} - \frac{\mu_1}{r^3}\right)\left(\mathbf{i}_\rho \times \frac{d\mathbf{i}_\rho}{dt} \cdot \mathbf{d}\right) \qquad (3.64)$$

$$2\frac{d\rho}{dt}\left(\mathbf{i}_\rho \times \frac{d\mathbf{i}_\rho}{dt} \cdot \frac{d^2\mathbf{i}_\rho}{dt^2}\right) = \left(\frac{\mu_2}{d^3} - \frac{\mu_1}{r^3}\right)\left(\mathbf{i}_\rho \times \mathbf{d} \cdot \frac{d^2\mathbf{i}_\rho}{dt^2}\right) \qquad (3.65)$$

together with
$$r^2 = \rho^2 + d^2 + 2\rho(\mathbf{i}_\rho \cdot \mathbf{d}) \qquad (3.66)$$

obtained by squaring Eq. (3.62).

As shown in the first problem at the end of this section, each of the triple-scalar products in Eqs. (3.64) and (3.65) can be calculated approximately at the intermediate time $t_2$ from the observational data. Therefore, the last three equations serve to determine $r$, $\rho$, and $d\rho/dt$ at time $t_2$.

First, we may solve Eqs. (3.64) and (3.66) for $r_2$ and $\rho_2$ using some appropriate iteration technique. On the other hand, however, the convenient formulation of Eq. (3.64) as a successive substitution algorithm is an attractive alternative.† Either way, with $r_2$ obtained, $d\rho/dt$ at time $t_2$ is computed from Eq. (3.65).

Finally, the position and velocity vectors at time $t_2$ may be calculated from

$$\mathbf{r}_2 = \rho_2 \mathbf{i}_{\rho 2} + \mathbf{d}_2 \qquad (3.67)$$

$$\mathbf{v}_2 = \frac{d\rho_2}{dt}\mathbf{i}_{\rho 2} + \rho_2 \frac{d\mathbf{i}_{\rho 2}}{dt} + \frac{d\mathbf{d}_2}{dt} \qquad (3.68)$$

which can then be used to obtain the other orbital elements.

---

† This method was developed by Gauss and the basic idea is the subject of a problem at the end of this section.

◇ **Problem 3-36**

Using appropriate Taylor series expansions, show that

$$\left.\frac{d\mathbf{i}_\rho}{dt}\right|_{t_2} = -\frac{\tau_1}{\tau_2\tau_3}\mathbf{i}_{\rho 1} + \frac{\tau_1 - \tau_3}{\tau_1\tau_3}\mathbf{i}_{\rho 2} + \frac{\tau_3}{\tau_1\tau_2}\mathbf{i}_{\rho 3}$$

$$\left.\frac{d^2\mathbf{i}_\rho}{dt^2}\right|_{t_2} = \frac{2}{\tau_2\tau_3}\mathbf{i}_{\rho 1} - \frac{2}{\tau_1\tau_3}\mathbf{i}_{\rho 2} + \frac{2}{\tau_1\tau_2}\mathbf{i}_{\rho 3}$$

are valid to second order in the time intervals where the $\tau$'s are defined in Eq. (3.53). More accurate values for these derivatives can be obtained if more than three sets of observational data are available.

NOTE: The determination of the derivatives of the observational data is the greatest weakness in Laplace's method of orbit determination. In fact, it is necessary to use additional observations to obtain any reasonable accuracy at all.

*Joseph-Louis Lagrange* 1778

◇ **Problem 3-37**

Let $D_1$ and $D_2$ denote the triple-scalar products appearing, respectively, on the left and right sides of Eq. (3.64). Further, let $\pi - \psi$ be the angle between $\mathbf{d}$ and $\boldsymbol{\rho}$ ($\psi$ is known from the observations), and let $\phi$ be the angle between $\mathbf{r}$ and $\boldsymbol{\rho}$.

Equations, from which $\rho$ and $r$ can be determined, are given by

$$\rho = d\,\frac{\sin(\psi + \phi)}{\sin\phi} \quad \text{and} \quad r = d\,\frac{\sin\psi}{\sin\phi}$$

where $\phi$ is the solution of

$$\sin(\phi - \beta) = m\sin^4\phi$$

The quantities $m$ and $\beta$ are determined from

$$m_0 \sin\beta = d\sin\psi$$

$$m_0 \cos\beta = -d\cos\psi + \frac{D_2\mu_2}{D_1 d^3}$$

$$mm_0 = \frac{D_2\mu_1}{D_1 d^3 \sin^3\psi}$$

The sign of $m_0$, which is at our disposal, can be chosen so that $m$ will be positive.

NOTE: The equation for the determination of $\phi$ is transcendental and may be solved using the methods developed in Chapter 5. However, in general, there will be multiple roots which must be reconciled.

*Carl Friedrich Gauss* 1809

Chapter 4

# Two-Body Orbits and the Initial-Value Problem

H AD IT NOT BEEN FOR KEPLER, THE DESTINY OF THE HELIOCENTRIC theory of Nicholas Copernicus (1473–1543) would have been considerably in doubt. Johannes Kepler (1571-1630) was fascinated by the beauty and harmony of the Copernican system of the world and devoted his life to uncover whatsoever additional geometric harmonies might also exist. It is truly amazing that he managed to pursue his scientific work with such extraordinary enthusiasm and diligence considering the vicissitudes of his life. His hands were crippled and his eyesight impaired from smallpox as a boy. He suffered from religious persecution for his protestant beliefs. He lost his first wife and several children. Often in desperate financial difficulties, he endured a bare subsistence livelihood. He even had to defend his mother who was accused of witchcraft.

Kepler's initial efforts to use the five regular geometric solids to account for the placement of the planets were both extensive and unproductive. Indeed, it was not until he was expelled from his professorship in Graz, Austria, when that city fell under Catholic dominance, and he became an assistant to the astronomer Tycho Brahe in the observatory at Prague, that his research finally bore fruit. Tycho Brahe's observations of the planet Mars, whose orbital eccentricity, fortunately, was pronounced, provided Kepler with the means of testing his theories of planetary motion. They were of the proper order of accuracy for this purpose, being sufficiently accurate to discriminate between true and false hypotheses and yet not so refined as to involve the problem in a maze of unmanageable detail.

Kepler's most important work, *Astronomia Nova De Motibus Stellae Martis*, was published in 1609 and contained the first valid approximations to the kinematical relations of the solar system. His first result was that the heliocentric motions of the planets take place in fixed planes passing through the actual position of the sun. Before this, reference had always been made to the "mean position" of the sun so that no astronomer prior to Kepler had been able to represent the latitudes of the planets with even tolerable success. Next, he discovered the *law of areas*—that the area of the sector traced by the radius vector from the sun, between any two positions of a planet in its orbit, is proportional to the time occupied in passing from one position to the other—known today as *Kepler's second law*.

His third major result was surprisingly difficult to obtain since it required that he abandon the concept of a circular orbit, or any combination of moving circles called epicycles, as well as various oval-like curves, to describe a planetary orbit. (Curiously, the hypothesis of an eccentric circular orbit plays an important role in the theory of ellipses as we shall see.) Almost in desperation did he try to fit the data with the little known "ellipse" which had been recorded in the Alexandrian Library by the ancient greek mathematician Apollonius of Perga. It was a perfect match with the observations! The form of a planetary orbit is an ellipse with the sun at one focus—familiar now to almost everyone as *Kepler's first law*.

It required ten more years of extraordinary effort before he announced *Kepler's third law*: that the square of the periodic time is proportional to the cube of the mean distance (or the semimajor axis). This was published in 1619 in *Harmonices Mundi*, "The Harmony of the World", and succeeded in uniting the planets in a way that Kepler could scarcely have anticipated when he first began his work.

A direct consequence of the law of areas is *Kepler's equation* which relates position in orbit with time. Although it is easily derived using calculus, we include, in this chapter, a geometrical derivation along those lines used by Kepler himself. Since Kepler's equation is transcendental, the determination of position for a given time cannot be expressed in a finite number of terms.

This chapter treats the initial-value problem of two-body mechanics—given appropriate initial conditions, such as position and velocity at a specific instant of time, to determine those quantities at a later time. After a general discussion of many of the fascinating properties of conic sections, the various kinds of orbits are dealt with separately. As we will see, each has a different form of Kepler's equation. Finally, we derive all of the relevant two-body equations using *universal functions* so that the initial-value problem will be free of the troubles which arise with the classical equations when transition from one kind of orbit to another occurs.

## 4.1 Geometrical Properties

Two-body orbits are plane curves shown in Chapter 3 to be the locus of points whose position vectors **r** satisfy

$$r = p - \mathbf{e} \cdot \mathbf{r} \qquad (4.1)$$

The constant eccentricity vector **e** of magnitude $e \geq 0$ lies in the orbital plane and the parameter $p$ is a nonnegative constant. The parameter and the eccentricity can be conveniently related via the semimajor axis $a$ or the pericenter distance $q$. Thus

$$p = a(1 - e^2) = q(1 + e) \tag{4.2}$$

where $a$ may have any real value between plus or minus infinity and $q$ may be positive or zero.

The vectors **r** and **e** originate at the focus $F$, and if this point is also chosen as the origin of an $x$, $y$ cartesian coordinate system with the positive $x$ axis in the direction of **e**, then Eq. (4.1) is simply

$$\text{Origin at focus:} \quad r = p - ex \tag{4.3}$$

As noted in the discussion following Eq. (3.21), the center of the orbit is the point $x = -ea$, $y = 0$ provided $e \neq 1$. Now translate the origin of coordinates from the focus $F$ to the center $C$, i.e., replace $x$ by $x - ea$, and we have

$$r = p - e(x - ea) = a(1 - e^2) - ex + e^2 a$$

so that

$$\text{Origin at center:} \quad r = a - ex \tag{4.4}$$

On the other hand, if the origin is translated from the focus $F$ to the pericenter $A(x = q, \ y = 0)$, then

$$r = p - e(x + q) = q(1 + e) - ex - eq$$

Hence

$$\text{Origin at pericenter:} \quad r = q - ex \tag{4.5}$$

(We emphasize, in each of the three cases, that the radial distance $r$ is always measured from $F$ while the abscissa $x$ is referenced to the origin of coordinates $F$, $C$, or $A$.)

We can exploit to advantage the various representations. For example, with the origin at pericenter, the cartesian equation of orbit is found from

$$r^2 = (q + x)^2 + y^2 = (q - ex)^2$$

Hence

$$y^2 = -(1 + e)[2qx + (1 - e)x^2] \tag{4.6}$$

provides an equation, due to Euler, valid for all orbits. It is universal in that no difficulty is encountered for a transition from ellipse to parabola to hyperbola by holding $q$ constant and allowing $e$ to increase continuously from $e < 1$ to $e > 1$. This is not the case when the center of the orbit is the origin of coordinates. [See Eqs. (3.21) and (3.22).]

The other two forms of the equation of orbit can be used to demonstrate three basic and well-known geometric properties. In the usual systematic development of analytic geometry, one of these properties is generally chosen as the definition of this special class of plane curves.

## Focus-Directrix Property

If the eccentricity $e = 0$, the focus $F$ and the center $C$ coincide and the orbit is the circle $r = p$. Otherwise, for $e \neq 0$, Eq. (4.3) may be written as

$$\frac{p}{e} = x + \frac{r}{e}$$

Refer now to Fig. 4.1 which shows the orbit drawn in the $x, y$ plane with the origin at $F$. The straight line parallel to and at a distance $p/e$ from the $y$ axis is called the *directrix*. Now if $P$ is any point in the orbit and the line $PN$ is perpendicular to the directrix, we observe that

$$\frac{PF}{PN} = e \qquad (4.7)$$

Hence, *the orbit is the locus of points whose distances from a fixed point and a fixed straight line are in constant ratio.*

Although Apollonius made no mention of the directrix in his writings, there is some evidence that Euclid was familiar with the focus-directrix-eccentricity theorem.

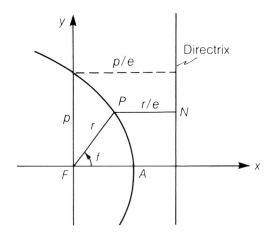

**Fig. 4.1:** Illustration of the focus-directrix property.

## Focal-Radii Property

In orbital mechanics, the focus $F$ is frequently called the *occupied focus* referring, of course, to the body located at $F$ when the relative motion of two bodies is under consideration. With the origin of coordinates at the center of the orbit, the coordinates of $F$ are $(ea, 0)$. The point $F^*(-ea, 0)$, symmetrically placed on the $x$ axis, is called the unoccupied or *vacant focus*. The distances to a point $P$ on the orbit from the two foci are called the *focal-radii*.

Clearly, the squares of the focal-radii are
$$PF^2 = (x-ea)^2 + y^2 \quad \text{and} \quad PF^{*\,2} = (x+ea)^2 + y^2$$
so that
$$PF^{*\,2} = PF^2 + 4aex$$
But, from Eq. (4.4),
$$PF = r = a - ex$$
Hence,
$$PF^{*\,2} = (a-ex)^2 + 4aex = (a+ex)^2$$
Remembering that $PF^*$ is positive, we have
$$PF^* = \begin{cases} a + ex & \text{ellipse} & a > 0 \\ -(a+ex) & \text{hyperbola} & a < 0, \ x < 0 \end{cases}$$
Thus,
$$\begin{aligned} PF^* + PF &= 2a & \text{ellipse} \\ PF^* - PF &= -2a & \text{hyperbola} \end{aligned} \quad (4.8)$$
and we have the property that *an ellipse (hyperbola) is the locus of points for which the sum (difference) of the focal-radii is constant.*†

### Orbital Tangents

Another important and useful property of orbits is that *the focal-radii to a point on an orbit make equal angles with the orbital tangent at that point.* An elegant demonstration is possible which includes, as an extra bonus, a simple geometric construction of the orbital tangent together with an alternate definition of an ellipse and a hyperbola.

Consider first an ellipse of semimajor axis $a$. Construct a circle of radius $2a$ centered at the focus $F$. Through an arbitrary point $P$ on the ellipse draw the radius of this circle and label $Q$ the point where this radius meets the circle as shown in Fig. 4.2. Therefore, $PF + PQ = 2a$. But $PF + PF^* = 2a$ by the focal-radii property already established so

---

† Philippe de La Hire (1640–1718), a painter turned mathematician and astronomer, first gave this property. Indeed, his greatest work *Sectiones Conicae* published in 1685 contained all of the now familiar properties of conic sections.

The usual development, found in most books on analytic geometry, is to introduce two radicals for $PF$ and $PF^*$. Then the focal-radii property is used to derive the equation of the conic. The ingenious proof given here is by Guillaume Antoine François L'Hospital (1661–1704), a pupil of John Bernoulli, in his *Traité des sections coniques* published in Paris in 1707.

Despite the fact that he was active in introducing the methods of the calculus and published, in 1696, an influential book on the subject—*Analyse des infiniment petits*, the Marquis de l'Hospital made no application of the calculus to the conics. All students are familiar with "L'Hospital's rule" but not too many know that it was really John Bernoulli who produced that famous theorem for obtaining the limit approached by a fraction whose numerator and denominator approach zero.

that $PF^* = PQ$. It follows that a circle centered at $P$ of radius $PF^*$ will be tangent to the circle centered at $F$ of radius $2a$. In this way we are led to the following property or alternate definition: *An ellipse is the locus of centers of circles which are tangent to a fixed circle and pass through a fixed point.*

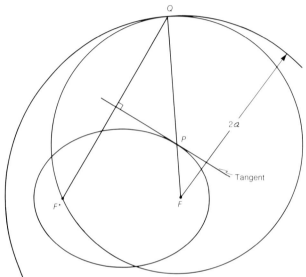

**Fig. 4.2:** Construction of the orbital tangent of an ellipse.

Since $QPF^*$ is an isosceles triangle, the perpendicular bisector of $QF^*$ passes through the point $P$ and, indeed, coincides with the tangent to the ellipse at point $P$. For the proof, assume that it was not tangent at $P$. Then it would intersect the ellipse at a second point $P'$ to which would correspond a second point $Q'$ on the large circle. Now, $P'F^* = P'Q'$ since $P'$ is a point on the ellipse. Also $P'F^* = P'Q$ since $P'$ lies on the perpendicular bisector of $F^*Q$. Thus, the triangle $P'Q'Q$ is isosceles; but the triangle $FQ'Q$ is also isosceles with the equal sides of length $2a$. Therefore, these two isosceles triangles have a common base $Q'Q$ and a common base angle $\angle P'Q'Q$ which is clearly impossible unless $P$ and $P'$ (also $Q$ and $Q'$) coincide.

Finally, the original assertion that $PF$ and $PF^*$ make equal angles with the orbital tangent at $P$ is readily verified from the figure. The entire argument applies equally well to the hyperbola by reversing the roles of the two foci. A circle of radius $-2a$ centered at $F^*$ is constructed. Then if $P$ is an arbitrary point on the hyperbola, a circle centered at $P$ of radius $PF$ will be tangent to the first circle. The construction of the orbital tangent, together with its properties, is the same as for the ellipse.

◇ **Problem 4–1**
If $P_1(x_1, y_1)$ is a point on the ellipse (hyperbola)

$$\frac{x^2}{a^2} \pm \frac{y^2}{b^2} = 1$$

then the equation of the tangent line at $P_1$ is

$$\frac{xx_1}{a^2} \pm \frac{yy_1}{b^2} = 1$$

If $P_1$ is not on the ellipse (hyperbola), this line connects the two points of contact of tangents to the curve which pass through $P_1$.


Sections of a Cone

The circle, ellipse, parabola, and hyperbola are often called *conic sections* because they can all be obtained as sections cut from a right circular cone by a plane. The type of conic depends on the dihedral angle between the cutting plane and the base of the cone. Thus, if the plane section is parallel to the base, the conic is a circle. If the plane is inclined to the base, but at an angle less than that between the generators of the cone and its base, the section is an ellipse. If the cutting plane is parallel to one of the generators, the section is a parabola. Finally, if the plane is inclined to the base at a still greater angle, the plane will also cut the cone formed by the extension of the generators. The section consisting of these two parts is a hyperbola. These cases are illustrated in Fig. 4.3.

It is interesting to note that the very words "ellipse," "hyperbola," and "parabola" have their origin in connection with the property of being sections of a cone. Ellipse is from the greek word "elleipsis" meaning to fall short or to leave out. Thus, our word ellipsis is the omission of words in a sentence—replacing them by a series of dots or asterisks. Extreme economy of speech or writing is called elliptic. Therefore, when the cutting plane is inclined to the base of the cone at an angle which is less than or "falls short of" the angle between the generators and the base, the section is called an ellipse.

Similarly, the greek word "hyperbole" means excess. Our word hyperbole means excessive or extravagant exaggeration. Thus, when the angle of the cutting plane is "excessive" or in excess of the angle between the generators and the base, the section is a hyperbola.

Finally, the greek word "parabole" literally means comparison and is the origin of our words parable and parallel. Therefore, when the cutting plane is parallel to the generator, the section is a parabola. From a different point of view, to distinguish the ellipse and the hyperbola we "compare" the cutting plane angles with the angle that the generator makes with the base.

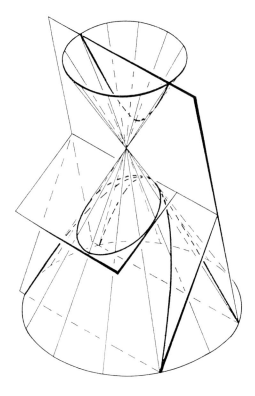

**Fig. 4.3:** Sections of a right circular cone.

Although the conic sections were known to the ancient greeks, the simplest proof which relates the geometrical property to the focal definition of the ellipse was supplied in 1822 by the Belgian mathematician Germinal P. Dandelin (1794–1847). To show that the ellipse is a section of a cone, introduce two spheres within the cone—one above and one below the cutting plane. The spheres each touch the cone in parallel circles and each is tangent to the cutting plane at one of two points, which we label $F$ and $F^*$, as seen in Fig. 4.4. If $P$ is any point on the curve, found by the intersection of the cone and the cutting plane, and if $Q$, $Q^*$ are the points at which the generator through $P$ intersects the two parallel circles, then the lines $PF$, $PQ$, and $PF^*$, $PQ^*$ are tangent, respectively, to the two spheres. Since all tangents from a given point to a given sphere are of equal length, then

$$PF = PQ \qquad PF^* = PQ^*$$

Hence,

$$PF + PF^* = PQ + PQ^* = QQ^*$$

so that the sum of the focal-radii is equal to the distance along the generator

Sect. 4.2]    Parabolic Orbits and Barker's Equation    149

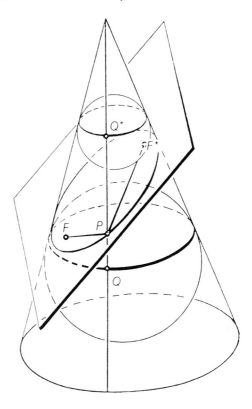

**Fig. 4.4:** Ellipse as a section of a cone.

between the two circles. But this distance is the same for all points $P$ on the section. Therefore, the curve is an ellipse.

### Problem 4–2
If the cutting plane has the same inclination to the base of the cone as the generators, there is a single sphere tangent to the cone along a circle and tangent to the plane at a point. Show that this point is the focus of the parabola and that the directrix is the line in which the plane of the circle cuts the plane of the parabola.

Furthermore, show that when the plane cuts both portions of the cone, the curve of intersection is a hyperbola. Note that one sphere is in each portion of the cone.

## 4.2  Parabolic Orbits and Barker's Equation

Except for the circle, for which the true anomaly is proportional to the time, the position of a body in orbit at a given time is simplest for the parabola. The polar equation of a parabola is

$$r = \frac{p}{1 + \cos f} = \frac{p}{2}(1 + \tan^2 \tfrac{1}{2} f) \tag{4.9}$$

so that from the law of areas

$$r^2 \frac{df}{dt} = h = \sqrt{\mu p}$$

it follows that

$$4\sqrt{\frac{\mu}{p^3}}\, dt = \sec^4 \tfrac{1}{2} f\, df$$

Performing the integration, we obtain

$$\tan^3 \tfrac{1}{2} f + 3 \tan \tfrac{1}{2} f = 2B \quad \text{where} \quad B = 3\sqrt{\frac{\mu}{p^3}}(t - \tau) \qquad (4.10)$$

and $\tau$ is the time of pericenter passage.

This relation between the true anomaly $f$ and the time $t$ is called *Barker's equation*.† The solution for $f$ when $t$ is given requires the root of a cubic equation in $\tan \tfrac{1}{2} f$ and it is easy to show that one and only one real root exists. To obtain it, we substitute

$$\tan \tfrac{1}{2} f = z - \frac{1}{z} \qquad (4.11)$$

and derive, thereby, a quadratic equation in $z^3$

$$z^6 - 2Bz^3 - 1 = 0$$

for which

$$z = \left(B \pm \sqrt{B^2 + 1}\right)^{\frac{1}{3}}$$

Either sign produces the same solution for $\tan \tfrac{1}{2} f$. Therefore,

$$\tan \tfrac{1}{2} f = \left(B + \sqrt{1 + B^2}\right)^{\frac{1}{3}} - \left(B + \sqrt{1 + B^2}\right)^{-\frac{1}{3}} \qquad (4.12)$$

which is in accord with the classic formula of Jerome Cardan.‡

Many variations of the solution of Barker's equation are considered in the following subsections which are both interesting and useful.

---

† The parabolic form of Kepler's equation is called Barker's equation after the man who prepared extensive tables for its solution in the eighteenth century. It contained values of the expression $75 \tan \tfrac{1}{2} f + 25 \tan^3 \tfrac{1}{2} f$ for the true anomalies at intervals of five minutes of arc from $0°$ to $180°$.

‡ Gerolamo Cardano (1501–1576) published the method for solving cubic equations which he obtained from Niccolò Fontana of Brescia (1499?–1557). Fontana is better known as Tartaglia, which means "Stammerer"—an unfortunate name he acquired because of a speech defect. Tartaglia had a method for solving the cubic which he revealed to Cardan in 1539 after a pledge from Cardan to keep it secret. Despite the pledge, Cardan published his version of the method in his *Ars Magna* in 1545.

## Trigonometric Solution

If we write $z = \cot\beta$ in Eq. (4.11), then

$$\tan\tfrac{1}{2}f = \cot\beta - \tan\beta = 2\cot 2\beta \tag{4.13}$$

But, from Eqs. (4.10) and (4.11), we have

$$\left(z - \frac{1}{z}\right)^3 + 3\left(z - \frac{1}{z}\right) = z^3 - \frac{1}{z^3} = \cot^3\beta - \tan^3\beta = 2B$$

Now, if we define

$$\tan\beta = \tan^{\frac{1}{3}}\alpha \tag{4.14}$$

then

$$B = \tfrac{1}{2}(\cot\alpha - \tan\alpha) = \cot 2\alpha \tag{4.15}$$

Therefore, the solution of Barker's equation can be had by first, computing $\alpha$ and $\beta$ from Eqs. (4.15) and (4.14); then, $\tan\tfrac{1}{2}f$ from Eq. (4.13).

## Improved Algebraic Solution

If the time interval $t - \tau$ is small, neither of the solutions so far considered is suitable. This is readily seen from Eq. (4.12) which would require calculating the difference of two almost equal quantities. However, we may write

$$\tan\tfrac{1}{2}f = \tan\beta(\cot^2\beta - 1) = \tan\beta\frac{\cot^6\beta - 1}{1 + \cot^2\beta + \cot^4\beta}$$

and since

$$\begin{aligned}\tan\beta(\cot^6\beta - 1) &= \cot^2\beta(\cot^3\beta - \tan^3\beta) \\ &= \cot^2\beta(\cot\alpha - \tan\alpha) \\ &= 2\cot^2\beta\cot 2\alpha\end{aligned}$$

we may define $A = \cot^2\beta$ or, equivalently, from Eqs. (4.12) and (4.13),

$$A = \left(B + \sqrt{1 + B^2}\right)^{\frac{2}{3}} \tag{4.16}$$

to obtain

$$\tan\tfrac{1}{2}f = \frac{2AB}{1 + A + A^2} \tag{4.17}$$

Equations (4.16) and (4.17) are a variation of the form of solution suggested by Karl Stumpff.

## Graphical Solution

An interesting interpretation of the quantities $\alpha$ and $\beta$ is possible which leads to a simple graphical solution of Barker's equation. For this purpose, we first observe that

$$2\alpha + 2\beta + f = \pi \qquad (4.18)$$

since

$$\tan(\alpha + \tfrac{1}{2}f) = \frac{\tan\alpha + \tan\tfrac{1}{2}f}{1 - \tan\alpha \tan\tfrac{1}{2}f}$$

$$= \frac{\tan^4\beta - \tan^2\beta + 1}{\tan^5\beta - \tan^3\beta + \tan\beta}$$

$$= \cot\beta = \tan(\tfrac{1}{2}\pi - \beta)$$

Then, from Eqs. (4.18) and (4.13),

$$\sin 2\alpha = \sin(f + 2\beta)$$
$$= \sin 2\beta(\cos f + \cot 2\beta \sin f)$$
$$= \sin 2\beta \cos^2 \tfrac{1}{2}f$$
$$= \frac{\tfrac{1}{2}p}{r}\sin 2\beta$$

so that

$$\frac{\sin 2\alpha}{\sin 2\beta} = \frac{\tfrac{1}{2}p}{r} \qquad (4.19)$$

Equations (4.18) and (4.19) permit the geometrical representation of the quantities $2\alpha$ and $2\beta$ shown in Fig. 4.5. The angles $f$, $2\alpha$, $2\beta$ are the interior angles of the triangle $FPA$ whose sides are $FP = r$ and $FA = \tfrac{1}{2}p$. The center of the circumscribed circle of this triangle lies on the perpendicular bisector of the line $FA$ and at a distance $\tfrac{1}{4}pB$ from $FA$ since $\cot 2\alpha = B$.

The graphical solution of Barker's equation follows at once from the figure. Given the parabola and a time interval $t - \tau$, the center of the circle is located. The position of the body in orbit at time $t$ is then found as the intersection of the circle and the parabola.

From a different point of view, we have just shown that there is a one-to-one correspondence between the position of a body moving in a parabolic orbit and the position of a fictitious body moving along a straight line with constant linear velocity $\tfrac{3}{4}\sqrt[3]{\mu/p}$.†

---

† This observation was made by one of the author's students Adel A. M. Saleh in 1966.

## Sect. 4.2] Parabolic Orbits and Barker's Equation

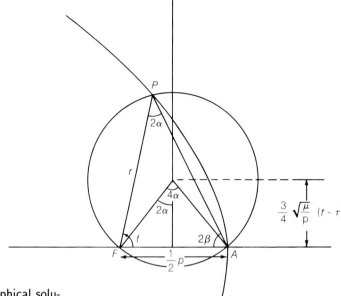

**Fig. 4.5:** Graphical solution of Barker's equation.

### Continued Fraction Solution

The solution of Barker's equation may also be expressed as the following infinite continued fraction:

$$\tan \tfrac{1}{2} f = \cfrac{\tfrac{2}{3} B}{1 + \cfrac{\tfrac{4 \cdot 1}{3 \cdot 9} B^2}{1 + \cfrac{\tfrac{5 \cdot 8}{9 \cdot 15} B^2}{1 + \cfrac{\tfrac{10 \cdot 7}{15 \cdot 21} B^2}{1 + \cfrac{\tfrac{11 \cdot 14}{21 \cdot 27} B^2}{1 + \cfrac{\tfrac{16 \cdot 13}{27 \cdot 33} B^2}{1 + \cfrac{\tfrac{17 \cdot 20}{33 \cdot 39} B^2}{1 + \cdots}}}}}}} \qquad (4.20)$$

which is obtained using the method described at the end of Sect. 1.2. Indeed, Barker's equation is exactly of the form of Eq. (1.32) whose continued fraction solution is given by Eq. (1.33). Here we have written the continued fraction in its "equivalent form" (4.20) so that the convenient top-down evaluation algorithm of Eqs. (1.51) can be utilized directly.†

---

† Be careful of signs! The algorithm (1.51) applies to the continued fraction of the form given at the beginning of Sect. 1.4 in which all of the signs on the diagonal are negative.

## Descartes' Method

A geometrical method of solving cubic and quartic equations, which is applicable to Barker's equation, was devised by René Descartes.†

Consider the reduced quartic equation (with the $x^3$ term missing)

$$x^4 + px^2 + qx + r = 0$$

which can be obtained from the general quartic by a simple linear change of variable. The same equation form results by eliminating $y$ from the following equations of a circle and a parabola:

$$(x - x_0)^2 + (y - y_0)^2 = R^2$$
$$y = x^2$$

Specifically, we have

$$x^4 + (1 - 2y_0)x^2 - 2x_0 x + x_0^2 + y_0^2 - R^2 = 0$$

which can be identified with the original quartic by defining the coordinates of the circle's center and its radius as

$$x_0 = -\tfrac{1}{2}q \qquad y_0 = \tfrac{1}{2}(1 - p)$$
$$R = \tfrac{1}{2}\sqrt{(1 - p)^2 + q^2 - 4r}$$

Thus, the real solutions of the reduced quartic are obtained as the intersections of the circle and the parabola.

For the reduced cubic equation

$$x^3 + px + q = 0$$

we need only multiply by $x$ to obtain the original quartic with $r = 0$ and the result is the same. In this case, the circle will always pass through the origin.

By applying this idea to Barker's equation, we find that $\tan \tfrac{1}{2} f$ is obtained as the intersection of the parabola $y = x^2$ and a circle centered at $(B, -1)$ of radius $R = \sqrt{1 + B^2}$ as shown in Fig. 4.6.

---

† René Descartes (1596–1650), the famous French philosopher whose scientific ideas came to dominate the seventeenth century, made basic contributions in philosophy, biology, physics, and mathematics. *La Géométrie*, published in 1637, was his only book on mathematics and contained his ideas on Analytic Geometry—the concept of coordinates and the concept of representing any algebraic equation with two unknowns in the form of a curve in the plane. He investigated the kinds of curves represented by the second order equation and showed that such equations describe an ellipse, hyperbola, or parabola. He also investigated the equations of other geometric loci, the transformations of algebraic equations, and gave without proof his famous *law of signs* for the number of positive roots of an algebraic equation. It has been said that Descartes did not simply revise geometry—he created it! Descartes understood the significance of what he had accomplished and boasted that he had so far surpassed all geometry before him as Cicero's rhetoric surpasses the ABC's.

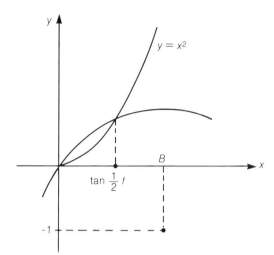

**Fig. 4.6:** Descartes solution of Barker's equation.

◇ **Problem 4–3**

If the pericenter distance of a parabolic orbit is $q$, then the time interval $\Delta t$ spent within a sphere of radius $r_S$ centered at the focus is

$$\sqrt{\mu}\,\Delta t = \tfrac{2}{3}\sqrt{2(r_S - q)}\,(r_S + 2q)$$

where $\mu$ is the gravitational constant.

## Lagrangian Coefficients

The position and velocity vectors in orbital-plane coordinates are given in Eqs. (3.31) with $e = 1$. However, since $\tan\tfrac{1}{2}f$ is obtained directly as the solution of Barker's equation, it is more convenient to express all trigonometric functions in terms of this function of the true anomaly. There results

$$\begin{aligned}\mathbf{r} &= \frac{p}{2}(1 - \tan^2 \tfrac{1}{2}f)\,\mathbf{i}_e + p\tan\tfrac{1}{2}f\,\mathbf{i}_p \\ \mathbf{v} &= -\frac{\sqrt{\mu p}}{r}\tan\tfrac{1}{2}f\,\mathbf{i}_e + \frac{\sqrt{\mu p}}{r}\,\mathbf{i}_p\end{aligned} \quad (4.21)$$

with $r$ determined from Eq. (4.9).

Now, let $\mathbf{r}_0$, $\mathbf{v}_0$ be the position and velocity vectors at time $t_0$. The Lagrangian coefficients for parabolic orbits are most conveniently expressed in terms of the variable

$$\chi = \sigma - \sigma_0 \quad (4.22)$$

The quantity $\sigma$ is determined from Eqs. (4.21) as

$$\sigma = \frac{\mathbf{r}\cdot\mathbf{v}}{\sqrt{\mu}} = \sqrt{p}\tan\tfrac{1}{2}f \quad (4.23)$$

and $\sigma_0$ is, of course, the value of $\sigma$ at time $t_0$.

From Eq. (4.9) we can easily establish
$$p = 2r - \sigma^2 = 2r_0 - \sigma_0^2$$
so that Barker's equation and the equation of orbit may be written as
$$6\sqrt{\mu}(t - t_0) = 6r_0\chi + 3\sigma_0\chi^2 + \chi^3 \qquad (4.24)$$
$$r = r_0 + \sigma_0\chi + \tfrac{1}{2}\chi^2 \qquad (4.25)$$

The Lagrangian coefficients are computed as described in Sect. 3.6 and we obtain
$$F = 1 - \frac{\chi^2}{2r_0} \qquad G = \frac{\chi}{2\sqrt{\mu}}(2r_0 + \sigma_0\chi)$$
$$F_t = -\frac{\sqrt{\mu}\,\chi}{rr_0} \qquad G_t = 1 - \frac{\chi^2}{2r} \qquad (4.26)$$

◇ **Problem 4–4**
The solution of the generalized form of Barker's equation (4.24) is
$$\chi = \sqrt{p}\,z - \sigma_0$$
where $z$ is the solution of
$$z^3 + 3z = 2B$$
and
$$B = \frac{1}{p^{\frac{3}{2}}}[\sigma_0(r_0 + p) + 3\sqrt{\mu}\,(t - t_0)]$$

NOTE: The equation for $z$ is exactly Eq. (4.10) so that all solution methods developed for Barker's equation are applicable without modification provided that $B > 0$.

## Orbital Tangents

We can use Eq. (4.23) to demonstrate a simple relationship between the true anomaly $f$ and the flight direction angle $\gamma$ between the position and velocity vectors for a parabolic orbit. By comparing Eq. (4.23) with
$$\sigma = \sqrt{p}\cot\gamma$$
obtained in Prob. 3–23, we see that $\tfrac{1}{2}f$ and $\gamma$ are complementary angles.

A simple construction of the parabola and its tangents then follows at once. Let $N$ be a point on the directrix of a parabola whose focus is at $F$. The perpendicular bisector of the line $FN$ meets the line through $N$ parallel to the axis at the point $P$ as shown in Fig. 4.7. Clearly, $P$ is a point on the parabola and the perpendicular bisector is the tangent to the parabola at $P$.

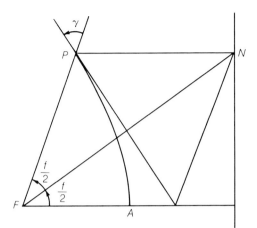

**Fig. 4.7:** Construction of the parabola and its tangents.

◊ **Problem 4–5**

A series of circles centered on the $x$ axis and passing through the point $x = 4q$, $y = 0$ with radii equal to and greater than $2q$ is shown in Fig. 4.8. The $x$ and $y$ intercepts of these circles are the coordinates of points on the parabola

$$y^2 = -4qx$$

*Gregory St. Vincent*† 1647

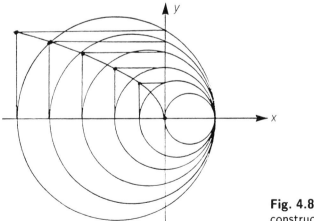

**Fig. 4.8:** St. Vincent's construction of a parabola.

---

† One of those who awakened interest in conic sections, after a period of twelve centuries of no progress at all, was Gregorius a San Vincento (1584–1667). His great work *Opus quadraturae circuli et sectionum coni* was published in 1647.

## 4.3 Elliptic Orbits and Kepler's Equation

A direct integration of Eq. (3.27), which expresses the law of areas relating time and angular position in orbit, does not result in a useful expression except for a circle or parabola. A new angle to replace the true anomaly $f$ is customary for elliptic motion, and the following construction provides the geometrical significance of this auxiliary variable.

Let $C$ be the center and $F$ the focus of an ellipse as shown in Fig. 4.9. Construct a circle of radius $a$ and center $C$. Let $P$ be the position of a body on the ellipse, and let $Q$ be the point where the perpendicular to the major axis cuts this *auxiliary circle*. The angle $\angle QCA$ was called the *eccentric anomaly* by Kepler and is denoted by $E$, while the angle $\angle PFA$ is, of course, the true anomaly $f$ of the point $P$.

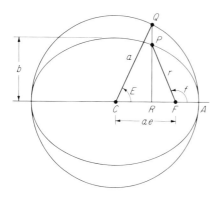

**Fig. 4.9:** Orbital anomalies for elliptic motion.

In terms of $E$, the equation of the ellipse can be expressed in parametric form as
$$x = a\cos E \qquad y = b\sin E \qquad (4.27)$$
in an $x,y$ cartesian coordinate system centered at $C$. Indeed, if $E$ is eliminated between these two equations, the standard form of the ellipse results. The radial position of the point $P$ is easily expressed in terms of $E$ using Eq. (4.4). Thus
$$r = a(1 - e\cos E) \qquad (4.28)$$
and if this is compared with the polar equation of the ellipse
$$r = \frac{a(1-e^2)}{1+e\cos f}$$
we obtain the identities
$$\cos f = \frac{\cos E - e}{1 - e\cos E} \qquad \cos E = \frac{e + \cos f}{1 + e\cos f} \qquad (4.29)$$

Further, since
$$y = b \sin E = a\sqrt{1-e^2} \sin E = r \sin f$$
we also obtain
$$\sin f = \frac{\sqrt{1-e^2} \sin E}{1 - e \cos E} \qquad \sin E = \frac{\sqrt{1-e^2} \sin f}{1 + e \cos f} \qquad (4.30)$$
Finally, from the first of Eqs. (4.29),
$$\sin^2 \tfrac{1}{2} f = \frac{a(1+e)}{r} \sin^2 \tfrac{1}{2} E \qquad \cos^2 \tfrac{1}{2} f = \frac{a(1-e)}{r} \cos^2 \tfrac{1}{2} E \qquad (4.31)$$
By dividing the two equations and taking the square root, it follows that
$$\tan \tfrac{1}{2} f = \sqrt{\frac{1+e}{1-e}} \tan \tfrac{1}{2} E \qquad (4.32)$$
This last identity is a most useful relation between $f$ and $E$, since $\tfrac{1}{2} f$ and $\tfrac{1}{2} E$ are always in the same quadrant.

◇ **Problem 4–6**

Greater elegance is attainable in many of the classical formulas for the ellipse by introducing in place of $e$ the angle $\phi$ where $e = \sin \phi$. Illustrate $\phi$ geometrically and verify that

$$\sqrt{1-e^2} = \cos \phi \qquad \sqrt{1+e} = \sqrt{2} \cos(\tfrac{1}{4}\pi - \tfrac{1}{2}\phi) \qquad \sqrt{1-e} = \sqrt{2} \cos(\tfrac{1}{4}\pi + \tfrac{1}{2}\phi)$$
$$\sqrt{1+e} + \sqrt{1-e} = 2 \cos \tfrac{1}{2}\phi \qquad \sqrt{1+e} - \sqrt{1-e} = 2 \sin \tfrac{1}{2}\phi$$
$$\sqrt{\frac{1-e}{1+e}} = \tan(\tfrac{1}{4}\pi - \tfrac{1}{2}\phi)$$

*Carl Friedrich Gauss* 1809

◇ **Problem 4–7**

Derive
$$\tan \tfrac{1}{2}(f - E) = \frac{\beta \sin f}{1 + \beta \cos f} = \frac{\beta \sin E}{1 - \beta \cos E}$$
as an alternate identity to Eq. (4.32), where
$$\beta = \frac{1 - \sqrt{1-e^2}}{e} = \frac{e}{1 + \sqrt{1-e^2}} = \frac{\sin \phi}{1 + \cos \phi} = \tan \tfrac{1}{2}\phi$$
and $\phi$ is defined in Prob. 4–6. Further, demonstrate that $\tfrac{1}{2}(f - E)$ is always less than $90°$ for all elliptic orbits which makes this equation particularly convenient to use. [See Sect. 5.3 for further properties of the parameter $\beta$.]

*Roger A. Broucke* and *Paul J. Cefola*† 1973

---

† "A Note on the Relations Between True and Eccentric Anomalies in the Two-Body Problem," *Celestial Mechanics*, Vol. 7, April 1973, pp. 388–389.

◇ **Problem 4–8**

The polar equation of an ellipse with the origin of coordinates at the center of the ellipse is

$$r = \frac{b}{\sqrt{1 - e^2 \cos^2 \theta}}$$

where $r$ is the radius from the center to a point on the ellipse.

HINT: $PF = a - ex = a - er \cos \theta$, $CF = ae$, and $CP = r$.

### Analytic Derivation of Kepler's Equation

Returning now to the problem of determining position in orbit as a function of time, we calculate the differential of Eq. (4.32)

$$\sec^2 \tfrac{1}{2} f \, df = \sqrt{\frac{1+e}{1-e}} \sec^2 \tfrac{1}{2} E \, dE$$

and use the second of the identities (4.31) to obtain

$$r \, df = b \, dE \tag{4.33}$$

Hence, from the law of areas, we have

$$r^2 df = h \, dt = br \, dE = ab(1 - e \cos E) \, dE$$

The integration is now trivial and the result is traditionally expressed as

$$M = E - e \sin E \tag{4.34}$$

The quantity $M$, called the *mean anomaly* by Kepler, includes the constant of integration. Thus,

$$M = \sqrt{\frac{\mu}{a^3}}(t - \tau) = n(t - \tau) \tag{4.35}$$

where $n$ is the *mean motion* and $\tau$ is, of course, the time of passage through pericenter.

One may interpret $M$ as the angular position of a body moving with constant angular velocity along the auxiliary circle. The relation between the mean anomaly and the eccentric anomaly, as expressed by Eq. (4.34), is called *Kepler's equation*.[†]

As noted in Sect. 3.4, the integration constant $\tau$ may be regarded as one of the six orbital elements. However, another quantity related to $\tau$ is frequently used instead. In the same section of Chapter 3 the *true longitude* of a body in orbit was defined as

$$L = \varpi + f \tag{4.36}$$

---

† During the period from 1618 to 1621 Kepler published a seven volume work entitled *Epitome Astronomiae Copernicanae*. It is in Book V that Kepler's equation appears for the first time.

Correspondingly, the *mean longitude* is defined by

$$l = \varpi + M \tag{4.37}$$

At time $t = 0$, referred to as the *epoch*, the mean longitude is

$$\epsilon = \varpi - n\tau \tag{4.38}$$

The quantity $\epsilon$, termed the *mean longitude at the epoch*, is then an orbital element which may be used instead of $\tau$. With this choice, the mean anomaly is determined from

$$M = nt + \epsilon - \varpi \tag{4.39}$$

### Geometric Derivation of Kepler's Equation

Kepler's original derivation of the equation which bears his name was geometrical and follows at once from his second law in its geometrical form as the law of areas. The area $PFA$ in Fig. 4.9, bounded by the arc of the ellipse and the focal-radii $FP$ and $FA$, is proportional to the time interval $t - \tau$. Thus,

$$\text{Area } PFA = \tfrac{1}{2} h(t - \tau) = \tfrac{1}{2} nab(t - \tau)$$

and it remains to determine this area in terms of the eccentric anomaly.

First, we note that the area of the circular sector $QCA$ is simply $\tfrac{1}{2} a^2 E$ and the area of the triangle $QCR$ is $\tfrac{1}{2} a \cos E$ multiplied by $a \sin E$. Therefore, the area bounded by the circular arc and the lines $RQ$ and $RA$ is

$$\text{Area } QRA = \text{Area } QCA - \text{Area } QCR$$
$$= \tfrac{1}{2} a^2 (E - \sin E \cos E)$$

Next, we invoke a property, easily verified from the parametric equations (4.27), that an ellipse is obtained from a circle by deforming the latter in such a way that the distances of its points from a fixed diameter are all changed in the same ratio—specifically $b/a$. Archimedes may have been the first to show that the areas of the circle and ellipse are also in this ratio. Applying this to the case at hand, we have

$$\text{Area } PRA = \frac{b}{a} \text{Area } QRA$$

The desired area $PFA$ is then found by subtracting the area of the triangle $PRF$

$$\text{Area } PRF = \tfrac{1}{2}(ae - x)y = \tfrac{1}{2} ab(e - \cos E) \sin E$$

from the area $PRA$. Hence

$$\text{Area } PFA = \tfrac{1}{2} ab(E - e \sin E)$$

which is then equated to $\tfrac{1}{2} nab(t - \tau)$ and the derivation is complete.

## ◇ Problem 4–9

In Fig. 4.10, the points $H$, $K$, $L$ are determined so that $FHKL$ is a rectangle. Show that the angle $\angle ACH$ is an approximation of the mean anomaly, i.e.,

$$M \simeq \angle ACH = E - \arcsin(e \sin E)$$

Robert G. Stern 1963

## ◇ Problem 4–10

A circle of radius $a$ centered at $F$ is shown in Fig. 4.11. Let $P$ be an arbitrary point on the ellipse. The line joining $P$ and $F$ meets the circle at $R$ and the angle $\angle CSR$ is a right angle. The magnitude $v$ of the velocity vector and the flight-direction angle $\gamma$ at $P$ are

$$v = \frac{h}{ap} CR \quad \text{and} \quad \gamma = \angle SCR$$

Show that

$$\mathbf{v} = \frac{h}{ap}(CS\,\mathbf{i}_r + SR\,\mathbf{i}_\theta) = \frac{h}{ap}(-CD\,\mathbf{i}_e + CB\,\mathbf{i}_p)$$

Robert G. Stern† 1963

## Lagrangian Coefficients

The position and velocity vectors in orbital-plane coordinates in Eqs. (3.31) may be written directly in terms of the eccentric anomaly by using the identities established at the beginning of this section. Thus,

$$\begin{aligned} \mathbf{r} &= a(\cos E - e)\,\mathbf{i}_e + \sqrt{ap}\sin E\,\mathbf{i}_p \\ \mathbf{v} &= -\frac{\sqrt{\mu a}}{r}\sin E\,\mathbf{i}_e + \frac{\sqrt{\mu p}}{r}\cos E\,\mathbf{i}_p \end{aligned} \quad (4.40)$$

Let $E_0$ be the eccentric anomaly associated with the position vector $r_0$. Then, following the development in Sect. 3.6, we may express the Lagrangian coefficients in terms of the eccentric anomaly difference as

$$\begin{aligned} F &= 1 - \frac{a}{r_0}[1 - \cos(E - E_0)] \\ G &= \frac{a\sigma_0}{\sqrt{\mu}}[1 - \cos(E - E_0)] + r_0\sqrt{\frac{a}{\mu}}\sin(E - E_0) \\ F_t &= -\frac{\sqrt{\mu a}}{rr_0}\sin(E - E_0) \\ G_t &= 1 - \frac{a}{r}[1 - \cos(E - E_0)] \end{aligned} \quad (4.41)$$

---

† Problems 4–9 and 4–10, attributed to Robert Stern, are from his MIT Ph.D. Thesis "Interplanetary Midcourse Guidance Analysis."

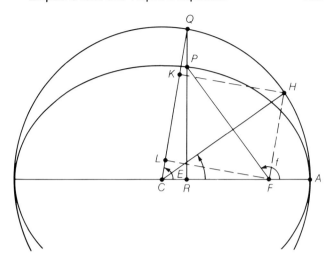

**Fig. 4.10:** Approximation of the mean anomaly.

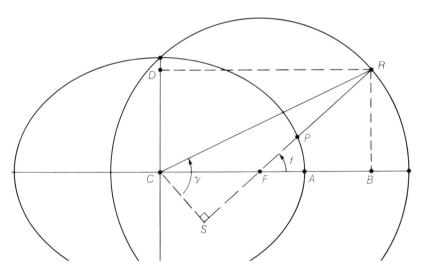

**Fig. 4.11:** Construction of velocity-vector components.

where
$$r = a + (r_0 - a)\cos(E - E_0) + \sigma_0\sqrt{a}\sin(E - E_0) \qquad (4.42)$$

In deriving these expressions, use has been made of the readily established relations

$$e\cos E_0 = 1 - \frac{r_0}{a} \qquad e\sin E_0 = \frac{\sigma_0}{\sqrt{a}}$$

Finally, Kepler's equation may be rewritten to relate the eccentric anomaly difference $E - E_0$ to the time interval $t - t_0$ where $t_0$ is the time corresponding to the position vector $\mathbf{r}_0$. We have

$$M - M_0 = E - E_0 - e(\sin E - \sin E_0)$$

which has the equivalent form

$$M - M_0 = E - E_0 + \frac{\sigma_0}{\sqrt{a}}[1 - \cos(E - E_0)] - \left(1 - \frac{r_0}{a}\right)\sin(E - E_0) \quad (4.43)$$

To find $\mathbf{r}$, $\mathbf{v}$ at time $t$ from $\mathbf{r}_0$, $\mathbf{v}_0$ at time $t_0$, we must first solve the generalized form of Kepler's equation for $E - E_0$. Then, the elements of the transition matrix are obtained from Eqs. (4.41) and the desired result follows from Eqs. (3.33).

◇ **Problem 4–11**
By equating Eqs. (3.42) and (4.41), derive the relation

$$\tan \tfrac{1}{2}(f - f_0) = \frac{\sqrt{ap}\tan\tfrac{1}{2}(E - E_0)}{r_0 + \sigma_0\sqrt{a}\tan\tfrac{1}{2}(E - E_0)}$$

and compare with Eq. (4.32).

NOTE: For a generalization of this identity see the last subsection of Sect. 4.6.

◇ **Problem 4–12**
The average value of the radius $r$ in elliptic motion is

$$a(1 + \tfrac{1}{2}e^2) \quad \text{with respect to time}$$

and

$$a\sqrt{1 - e^2} \quad \text{with respect to the true anomaly.}$$

NOTE: We call $a$ the "mean distance," but it is *not* the average value of $r$.

◇ **Problem 4–13**
The difference between the eccentric and mean anomalies satisfies the differential equation

$$\frac{d^2}{dt^2}(E - M) + \frac{\mu}{r^3}(E - M) = 0$$

Use Eqs. (3.10) to develop a solution of Kepler's equation in powers of $M$.

NOTE: See also Probs. 3–8 and 3–25.

## 4.4 Hyperbolic Orbits and the Gudermannian

An analogous procedure for hyperbolic orbits can be formulated which parallels the discussion presented for elliptic orbits. We begin with the equation of the hyperbola expressed in parametric form as

$$x = a \sec\varsigma \qquad y = b \tan\varsigma \tag{4.44}$$

in an $x, y$ cartesian coordinate system with origin at the center. Clearly, if $\varsigma$ is eliminated, the standard form of the hyperbola results.

To express the radius to a point $P$ in terms of the parameter $\varsigma$, we again use Eq. (4.4) to obtain†

$$r = a(1 - e\sec\varsigma) \tag{4.45}$$

Now, if in the equation of orbit

$$r + re\cos f = p = a(1 - e^2)$$

we express the first $r$ in terms of $\varsigma$ using Eq. (4.45), it follows that

$$-a\sec\varsigma + r\cos f = -ae$$

Thus, the angle $\varsigma$ and the true anomaly $f$ are related as shown in Fig. 4.12. Therefore, when $\varsigma$ is used in the analytical description of hyperbolic orbits it has a direct geometric analogy with the eccentric anomaly of the ellipse. In both cases auxiliary circles, whose centers are at the center of the orbit and whose radii are the semimajor axes of the orbits, play similar roles in the analysis.

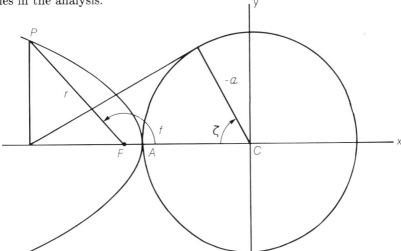

**Fig. 4.12:** Geometrical representation of the Gudermannian $\varsigma$.

---

† It is important to remember that $a$ is a negative number and $e$ is greater than one.

The identities relating $\varsigma$ and the true anomaly follow as before:

$$\cos f = \frac{e - \sec \varsigma}{e \sec \varsigma - 1} \qquad \sec \varsigma = \frac{e + \cos f}{1 + e \cos f}$$

$$\sin f = \frac{\sqrt{e^2 - 1} \tan \varsigma}{e \sec \varsigma - 1} \qquad \tan \varsigma = \frac{\sqrt{e^2 - 1} \sin f}{1 + e \cos f} \qquad (4.46)$$

Then since

$$\sin^2 \tfrac{1}{2} f = -\frac{a(e+1)}{r \cos \varsigma} \sin^2 \tfrac{1}{2} \varsigma \qquad \cos^2 \tfrac{1}{2} f = -\frac{a(e-1)}{r \cos \varsigma} \cos^2 \tfrac{1}{2} \varsigma \qquad (4.47)$$

we have

$$\tan \tfrac{1}{2} f = \sqrt{\frac{e+1}{e-1}} \tan \tfrac{1}{2} \varsigma \qquad (4.48)$$

To derive the analog of Kepler's equation for hyperbolic motion, we calculate the differential of Eq. (4.48) to obtain

$$r \, df = b \sec \varsigma \, d\varsigma \qquad (4.49)$$

Hence, in the same manner as for the ellipse, we have

$$N = e \tan \varsigma - \log \tan(\tfrac{1}{2} \varsigma + \tfrac{1}{4} \pi) \qquad (4.50)$$

where the quantity $N$ is analogous to the mean anomaly of elliptic motion and is defined as

$$N = \sqrt{\frac{\mu}{(-a)^3}} (t - \tau) \qquad (4.51)$$

◇ **Problem 4–14**
The two straight lines

$$y = \pm \frac{b}{a} x = \pm (\tan \psi) x$$

through the center $C$ are the *asymptotes* of the hyperbola where $\psi$ is related to the eccentricity as

$$\tan \psi = \sqrt{e^2 - 1} \qquad \text{or} \qquad \sec \psi = e$$

The equation of orbit can then be written as

$$r = \frac{p}{1 + e \cos f} = \frac{p \cos \psi}{2 \cos \tfrac{1}{2}(f + \psi) \cos \tfrac{1}{2}(f - \psi)}$$

which clearly displays the behaviour of the hyperbola in the vicinity of the asymptotes. Indeed, this equation defines the asymptotes.

*Carl Friedrich Gauss* 1809

### ◇ Problem 4–15
Define the quantity $u$ as
$$u = \tan(\tfrac{1}{2}\varsigma + \tfrac{1}{4}\pi)$$
Equation (4.50) can then be written as
$$N = \frac{e}{2}\left(u - \frac{1}{u}\right) - \log u$$
with the radius and true anomaly expressible in terms of $u$ as
$$r = a - \frac{ae}{2}\left(u + \frac{1}{u}\right) \quad \text{and} \quad \tan \tfrac{1}{2}f = \sqrt{\frac{e+1}{e-1}\frac{u-1}{u+1}}$$

### Problem 4–16
With the angle $\psi$ defined in Prob. 4-14, the quantity $u$, defined in Prob. 4-15, can be expressed in terms of the angles $f$ and $\psi$ as
$$u = \frac{\cos\tfrac{1}{2}(f-\psi)}{\cos\tfrac{1}{2}(f+\psi)} \quad \text{where} \quad 1 \leq u < \infty \quad \text{for } f > 0$$

*Carl Friedrich Gauss* 1809

### The Gudermannian Transformation

The analysis for hyperbolic orbits may be accomplished in terms of hyperbolic, rather than trigonometric, functions. Because of the familiar identity
$$\cosh^2 H - \sinh^2 H = 1$$
the parametric equations of the hyperbola can be written as
$$x = a\cosh H \qquad y = b\sinh H \qquad (4.52)$$
and the radius vector magnitude becomes
$$r = a(1 - e\cosh H) \qquad (4.53)$$
The identities between $H$ and the true anomaly are found simply by substituting
$$\tan\varsigma = \sinh H \qquad \sec\varsigma = \cosh H \qquad (4.54)$$
in Eqs. (4.46). We can also show that
$$\tan\tfrac{1}{2}\varsigma = \tanh\tfrac{1}{2}H \qquad (4.55)$$
so that Eq. (4.48) becomes
$$\tan\tfrac{1}{2}f = \sqrt{\frac{e+1}{e-1}}\tanh\tfrac{1}{2}H \qquad (4.56)$$

Applying the definition of the hyperbolic functions in terms of the exponential function, it follows from Eqs. (4.54) that

$$H = \log(\tan \varsigma + \sec \varsigma) = \log \tan(\tfrac{1}{2}\varsigma + \tfrac{1}{4}\pi) \quad (4.57)$$

Hence, the relation between time and the quantity $H$ is obtained from Eq. (4.50) as

$$N = e \sinh H - H \quad (4.58)$$

The inverse function, expressing $\varsigma$ in terms of $H$ and written symbolically as $\varsigma = \text{gd}\, H$, is called the *Gudermannian* of $H$. Explicitly,

$$\varsigma = \text{gd}\, H = 2 \arctan(e^H) - \tfrac{1}{2}\pi \quad (4.59)$$

This name was given by Arthur Cayley† in honor of the German mathematician Christof Gudermann (1798–1852) who was largely responsible for the introduction of the hyperbolic functions into modern analysis.

◊ **Problem 4–17**

The hyperbolic form of Kepler's equation can be obtained formally from Kepler's equation by writing

$$E = -iH \quad \text{and} \quad M = iN$$

where $i = \sqrt{-1}$.

### Geometrical Representation of $H$

If $A$ is the area swept out by the radius vector, then, from Prob. 2–16,

$$dA = \tfrac{1}{2}(x\,dy - y\,dx)$$

Hence, for the unit circle

$$x^2 + y^2 = 1 \quad \text{or} \quad x = \cos E, \quad y = \sin E$$

and for the unit equilateral hyperbola

$$x^2 - y^2 = 1 \quad \text{or} \quad x = \cosh H, \quad y = \sinh H$$

we have

$$dA = \tfrac{1}{2}dE \quad \text{(unit circle)}$$
$$dA = \tfrac{1}{2}dH \quad \text{(unit equilateral hyperbola)}$$

Furthermore, as shown in Fig. 4.13, with $AQ$ an arc of the circle and the shaded area equal to $\tfrac{1}{2}E$, there obtains

$$CR = \cos E \quad RQ = \sin E \quad AD = \tan E$$

---

† Although Sir Arthur Cayley (1821–1895) contributed much to mathematics, he is is generally remembered as the creator of the theory of matrices. Logically, the idea of a *matrix* should precede that of a *determinant* but historically the order was the reverse. Cayley was the first to recognize the matrix as an entity in its own right and the first to publish a series of papers on the subject.

Sect. 4.4]  Hyperbolic Orbits and the Gudermannian  169

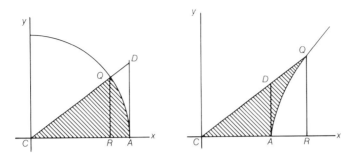

**Fig. 4.13:** Geometrical significance of $E$ and $H$.

Similarly, with $AQ$ an arc of the hyperbola and the shaded area equal to $\frac{1}{2}H$, then

$$CR = \cosh H \qquad RQ = \sinh H \qquad AD = \tanh H$$

Trigonometric functions are frequently called *circular functions* and this analogy between circular and hyperbolic functions is the reason for the designation of the latter as hyperbolic.

From this discussion, it is clear that the analog of the auxiliary circle, used in the analysis of the ellipse, should be the equilateral hyperbola having the same major axis as the hyperbolic orbit under consideration.

Refer to Fig. 4.14 where the points $C$ and $F$ are the center and focus of the hyperbola. The point $A$ is the vertex or pericenter position. The axis through $F$ and $A$ is called the *transverse axis*. The other axis through the center, called the *conjugate axis*, does not intersect the curve. Let $P$ be the position of a body on the hyperbola and let $Q$ be the point where the perpendicular to the transverse axis through $P$ cuts the auxiliary equilateral hyperbola. Then the area $CAQ$, bounded by the two straight lines $CA$, $CQ$, and the arc $AQ$, is

$$\text{Area } CAQ = \tfrac{1}{2}a^2 H \tag{4.60}$$

### Problem 4–18
Derive the hyperbolic form of Kepler's equation geometrically, using the same pattern of argument as for elliptic orbits. Further, show that if a fictitious body starts from $C$ when the real body is at $A$ and moves along the asymptote of the equilateral hyperbola with a constant speed equal to the ultimate speed of the real body, then

$$N = \frac{2}{a^2} \text{Area } F_0 C P'$$

where $F_0 C P'$ is a triangle whose vertices are $F_0$, the focus of the equilateral hyperbola, $C$, the center, and $P'$, the position of the fictitious body.

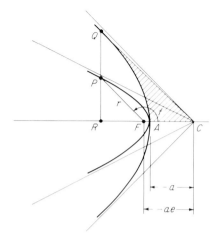

**Fig. 4.14:** Orbital relations for hyperbolic motion.

### Lagrangian Coefficients

The position and velocity vectors in orbital-plane coordinates are readily obtained as

$$\mathbf{r} = a(\cosh H - e)\,\mathbf{i}_e + \sqrt{-ap}\sinh H\,\mathbf{i}_p$$
$$\mathbf{v} = -\frac{\sqrt{-\mu a}}{r}\sinh H\,\mathbf{i}_e + \frac{\sqrt{\mu p}}{r}\cosh H\,\mathbf{i}_p \quad (4.61)$$

For the Lagrangian coefficients, we first establish

$$e\cosh H_0 = 1 - \frac{r_0}{a} \qquad e\sinh H_0 = \frac{\sigma_0}{\sqrt{-a}}$$

and then determine

$$F = 1 - \frac{a}{r_0}[1 - \cosh(H - H_0)]$$

$$G = \frac{a\sigma_0}{\sqrt{\mu}}[1 - \cosh(H - H_0)] + r_0\sqrt{\frac{-a}{\mu}}\sinh(H - H_0)$$

$$F_t = -\frac{\sqrt{-\mu a}}{rr_0}\sinh(H - H_0) \quad (4.62)$$

$$G_t = 1 - \frac{a}{r}[1 - \cosh(H - H_0)]$$

where

$$r = -a + (r_0 + a)\cosh(H - H_0) + \sigma_0\sqrt{-a}\sinh(H - H_0) \quad (4.63)$$

with the quantity $H - H_0$ obtained as the solution of

$$N - N_0 = -(H - H_0) + \frac{\sigma_0}{\sqrt{-a}}[\cosh(H - H_0) - 1]$$
$$+ \left(1 - \frac{r_0}{a}\right)\sinh(H - H_0) \quad (4.64)$$

## Asymptotic Coordinates

For some purposes it is convenient to represent the hyperbola in a coordinate system whose axes coincide with the hyperbolic asymptotes. This, of course, is not a cartesian system since the coordinate axes will be skewed in all cases except for the equilateral hyperbola.

Refer to Fig. 4.15 where we have labeled the asymptotic coordinate axes as $X$, $Y$. The coordinates of a point $P(X,Y)$ in this system are obtained as follows. The value of $X$ is the distance of $P$ from the $Y$ axis measured parallel to the $X$ axis. Similarly, $Y$ is the distance of $P$ from the $X$ axis measured parallel to the $Y$ axis. As seen from the figure

$$x = (Y + X)\cos\psi \qquad y = (Y - X)\sin\psi \qquad (4.65)$$

Now since $b^2 = a^2 \tan^2\psi$, the cartesian equation of the hyperbola with center at the origin is

$$x^2 - y^2 \cot^2\psi = a^2$$

Substituting from Eqs. (4.65), we have

$$XY = \tfrac{1}{4}a^2 e^2 = \tfrac{1}{4}(a^2 + b^2) \qquad (4.66)$$

as the desired result first obtained by Euler.

Suppose now that we rotate the $x,y$ coordinate axes clockwise bringing the $x$ and $X$ axes into coincidence and label the new rotated axes $\xi$, $\eta$. Then the $x,y$ and $\xi, \eta$ coordinates are related by

$$x = \xi \cos\psi + \eta \sin\psi \qquad y = -\xi \sin\psi + \eta \cos\psi \qquad (4.67)$$

and the equation of the hyperbola takes the form

$$\xi\eta \sin 2\psi - \eta^2 \cos 2\psi = a^2 \sin^2\psi \qquad (4.68)$$

The slope of the hyperbola is

$$\frac{d\eta}{d\xi} = \frac{\eta \sin 2\psi}{2\eta \cos 2\psi - \xi \sin 2\psi}$$

or, since

$$\xi = X + Y \cos 2\psi \qquad \eta = Y \sin 2\psi$$

we have

$$\frac{d\eta}{d\xi} = \frac{Y \sin 2\psi}{Y \cos 2\psi - X} \qquad (4.69)$$

With this last expression we can demonstrate a fascinating property of the hyperbola. Let $P$ be a point on the hyperbola, and let $Q$ and $R$ be the two points on the asymptotes obtained by projecting $P$ on the asymptotes as shown in Fig. 4.16. Then the straight line connecting $Q$ and $R$ is parallel to the slope of the hyperbola at point $P$. We have, thereby a simple and convenient method for constructing the tangent of a hyperbola.

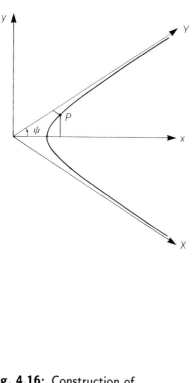

**Fig. 4.15:** Hyperbola in asymptotic coordinates.

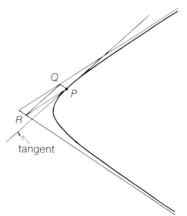

**Fig. 4.16:** Construction of the tangent to a hyperbola.

### Problem 4-19

Let $P$ be a point on a hyperbola, and let $A$ and $B$ be the lengths of the perpendiculars from $P$ to each of the asymptotes as shown in Fig. 4.17. Then,

$$A = x \sin \psi + y \cos \psi \qquad B = x \sin \psi - y \cos \psi$$

and the product of $A$ and $B$ is a constant, i.e.,

$$AB = \frac{b^2}{e^2}$$

Furthermore, $A$ and $B$ are related to the $\xi, \eta$ coordinates, defined by Eqs. (4.67), according to

$$\xi = A \cot 2\psi + B \csc 2\psi \qquad \eta = A$$

so that

$$\frac{d\eta}{d\xi} = \frac{A \sin 2\psi}{A \cos 2\psi - B}$$

Conclude, therefrom, that the lengths $A$ and $B$ can be used in place of the coordinates $X$ and $Y$ to construct the tangent to a hyperbola.

*Leonhard Euler* 1748

Sect. 4.4]     Hyperbolic Orbits and the Gudermannian     173

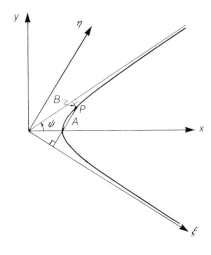

**Fig. 4.17:** Euler's method for the tangent to a hyperbola.

### Problem 4–20

An interesting construction of a hyperbola is possible using asymptotic coordinates. If $a$ and $e$ are specified, the asymptotes of the hyperbola and its vertex $A$ may be located. The line through $A$ perpendicular to the $x$ axis intersects the $X$ axis at $B$. Points $Q$ and $R$ are selected on the $X$ axis between $C$ and $B$ such that $QR = \frac{1}{2}CB$ but are otherwise arbitrary. Denote by $P$ the point of intersection of a line through $Q$ parallel to the $Y$ axis and a line connecting $R$ and $A$, as shown in Fig. 4.18. Show that $P$ lies on the hyperbola.

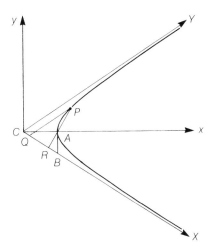

**Fig. 4.18:** Geometric construction of a hyperbola.

## 4.5 Universal Formulas for Conic Orbits

Thus far we have been obliged to use different formulations to describe the motion of a body in each of the various possible orbits. However, a generalization of the problem is possible using a new family of transcendental functions. With these functions, universally applicable formulas can be developed which are simultaneously valid for the parabola, the ellipse, and the hyperbola.

To motivate the development, the key differential relationships, derived in the previous three sections, can be summarized as

$$df = \frac{1}{r} \begin{cases} p\,d(\tan \tfrac{1}{2} f) \\ b\,dE \\ b\,dH \end{cases} = \frac{h}{r^2}\,dt$$

Since, for the three kinds of orbits, we have, respectively,

$$\frac{h}{p} = \sqrt{\frac{\mu}{p}} \qquad \frac{h}{b} = \sqrt{\frac{\mu}{a}} \qquad \frac{h}{b} = \sqrt{\frac{\mu}{-a}}$$

then we may write

$$\sqrt{\mu}\,dt = r \begin{cases} d(\sqrt{p}\tan \tfrac{1}{2} f) \\ d(\sqrt{a}\,E) \\ d(\sqrt{-a}\,H) \end{cases} = r\,d\chi \qquad (4.70)$$

where $\chi$ is to be regarded as a new independent variable—a kind of generalized anomaly. It is remarkable that when $\chi$ is used as the independent variable instead of the time $t$, then the nonlinear equations of motion can be converted into linear constant-coefficient differential equations.

The transformation defined by

$$\sqrt{\mu}\,\frac{dt}{d\chi} = r \qquad (4.71)$$

is called a *Sundman transformation*† and we shall now demonstrate that $\mathbf{r}$, $r$, $\sigma$, and $t$ can all be obtained as solutions of simple differential equations.

To begin, we differentiate the identity

$$r^2 = \mathbf{r} \cdot \mathbf{r}$$

and obtain

$$r\frac{dr}{d\chi} = \mathbf{r} \cdot \frac{d\mathbf{r}}{d\chi} = \frac{dt}{d\chi}\mathbf{r} \cdot \frac{d\mathbf{r}}{dt} = \frac{r}{\sqrt{\mu}}\mathbf{r} \cdot \mathbf{v} = r\sigma$$

---

† Karl Frithiof Sundman (1873–1949), professor of astronomy at the University of Helsinki and director of the Helsinki Observatory, introduced this transformation in his paper "Mémoire sur le Problème des Trois Corps" published in *Acta Mathematika*, Vol. 36, 1912.

## Sect. 4.5]  Universal Formulas for Conic Orbits

Cancelling the factor $r$ and differentiating a second time, we have

$$\frac{d^2r}{d\chi^2} = \frac{d\sigma}{d\chi} = \frac{r}{\mu}\frac{d}{dt}(\mathbf{r}\cdot\mathbf{v}) = \frac{r}{\mu}\left(v^2 + \mathbf{r}\cdot\frac{d\mathbf{v}}{dt}\right) = \frac{r}{\mu}\left(\frac{2\mu}{r} - \frac{\mu}{a} - \frac{\mu}{r}\right) = 1 - \frac{r}{a}$$

It is convenient here and in the sequel to write $\alpha$ for the reciprocal of $a$ so that $\alpha$ is defined as

$$\alpha \equiv \frac{1}{a} = \frac{2}{r} - \frac{v^2}{\mu} \tag{4.72}$$

and may be positive, negative, or zero.

In summary, then

$$\frac{dr}{d\chi} = \sigma = \sqrt{\mu}\,\frac{d^2t}{d\chi^2}$$

$$\frac{d^2r}{d\chi^2} = \frac{d\sigma}{d\chi} = \sqrt{\mu}\,\frac{d^3t}{d\chi^3} = 1 - \alpha r$$

$$\frac{d^3r}{d\chi^3} = \frac{d^2\sigma}{d\chi^2} = \sqrt{\mu}\,\frac{d^4t}{d\chi^4} = -\alpha\frac{dr}{d\chi} = -\alpha\sigma = -\alpha\sqrt{\mu}\,\frac{d^2t}{d\chi^2}$$

so that $\sigma$, $r$, and $t$ are solutions of the equations

$$\frac{d^2\sigma}{d\chi^2} + \alpha\sigma = 0 \qquad \frac{d^3r}{d\chi^3} + \alpha\frac{dr}{d\chi} = 0 \qquad \frac{d^4t}{d\chi^4} + \alpha\frac{d^2t}{d\chi^2} = 0 \tag{4.73}$$

The derivatives of the position vector $\mathbf{r}$

$$\frac{d\mathbf{r}}{d\chi} = \frac{r}{\sqrt{\mu}}\mathbf{v} \qquad \frac{d^2\mathbf{r}}{d\chi^2} = \frac{\sigma}{\sqrt{\mu}}\mathbf{v} - \frac{1}{r}\mathbf{r}$$

lead to

$$\frac{d^3\mathbf{r}}{d\chi^3} + \alpha\frac{d\mathbf{r}}{d\chi} = 0 \tag{4.74}$$

in a similar manner.

Linear differential equations with constant coefficients present no particular difficulty in their solution. Nevertheless, it is advantageous in this case to develop the solutions in a form utilizing a family of special functions defined solely for this purpose.

### The Universal Functions $U_n(\chi; \alpha)$

To construct the family of special functions, we begin by determining the power series solution of

$$\frac{d^2\sigma}{d\chi^2} + \alpha\sigma = 0$$

by substituting

$$\sigma = \sum_{k=0}^{\infty} a_k \chi^k$$

and equating coefficients of like powers of $\chi$. We are led to
$$a_{k+2} = -\frac{\alpha}{(k+1)(k+2)}a_k \quad \text{for} \quad k = 0, 1, \ldots$$
as a recursion formula for the coefficients. Hence
$$\sigma = a_0\left[1 - \frac{\alpha\chi^2}{2!} + \frac{(\alpha\chi^2)^2}{4!} - \cdots\right] + a_1\chi\left[1 - \frac{\alpha\chi^2}{3!} + \frac{(\alpha\chi^2)^2}{5!} - \cdots\right]$$
where $a_0$ and $a_1$ are two arbitrary constants. We shall designate the two series expansions by $U_0(\chi;\alpha)$ and $U_1(\chi;\alpha)$ so that
$$\sigma = a_0 U_0(\chi;\alpha) + a_1 U_1(\chi;\alpha)$$
The function $U_1$ is simply the integral of $U_0$ so that we are motivated to define a sequence of functions
$$U_1 = \int_0^\chi U_0 \, d\chi \quad U_2 = \int_0^\chi U_1 \, d\chi \quad U_3 = \int_0^\chi U_2 \, d\chi \quad \text{etc.}$$
The $n^{\text{th}}$ function of such a sequence is easily seen to be
$$U_n(\chi;\alpha) = \chi^n\left[\frac{1}{n!} - \frac{\alpha\chi^2}{(n+2)!} + \frac{(\alpha\chi^2)^2}{(n+4)!} - \cdots\right] \tag{4.75}$$

A basic identity for the $U$ functions is at once apparent from the series definition of $U_n(\chi;\alpha)$. Since Eq. (4.75) may be written as
$$U_n(\chi;\alpha) = \frac{\chi^n}{n!} - \alpha\chi^{n+2}\left[\frac{1}{(n+2)!} - \frac{\alpha\chi^2}{(n+4)!} + \frac{(\alpha\chi^2)^2}{(n+6)!} - \cdots\right]$$
we have
$$U_n(\chi;\alpha) + \alpha U_{n+2}(\chi;\alpha) = \frac{\chi^n}{n!} \tag{4.76}$$
It is clear, from the manner in which the family of functions was constructed, that
$$\frac{dU_n}{d\chi} = U_{n-1} \quad \text{for} \quad n = 1, 2, \ldots \tag{4.77}$$
and, by differentiating the series for $U_0$, we can easily show that
$$\frac{dU_0}{d\chi} = -\alpha U_1 \tag{4.78}$$
Now, if we differentiate the identity (4.76) $m+1$ times, where $m > n$, and use Eq. (4.77), we obtain
$$\frac{d^{m+1}U_n}{d\chi^{m+1}} + \alpha\frac{d^{m-1}U_n}{d\chi^{m-1}} = 0 \quad \text{for} \quad n = 0, 1, \ldots, m \tag{4.79}$$
It follows that $U_0$ and $U_1$ are each solutions of the second-order differential equation satisfied by $\sigma$, and we recall that $\sigma$ was, indeed, found to be a linear combination of $U_0$ and $U_1$.

Sect. 4.5]  Universal Formulas for Conic Orbits  177

Finally, by applying the identity (4.79) to the other two differential equations in (4.73), we conclude that $r$ is a linear combination of $U_0$, $U_1$. $U_2$ while $t$ is a linear combination of $U_0$, $U_1$, $U_2$, $U_3$. These will be the general solutions provided, of course, that the $U$ functions are linearly independent.

### Linear Independence of $U_n(\chi;\alpha)$

The functions $U_0$, $U_1$, ..., $U_n$ will be *linearly independent* if no one of the functions can be expressed as a linear combination of the others, or, equivalently, if no linear combination of the functions is identically zero over any interval of $\chi$ under consideration.

It is known that the functions will be linearly independent if the associated *Wronskian determinant*† is not identically zero. The elements of the first row of this determinant are the functions $U_0$, $U_1$, ..., $U_n$. The second row consists of the first derivatives of these functions, the third row, the second derivatives, and so forth with the last or $(n+1)^{\text{th}}$ row containing the $n^{\text{th}}$ derivatives.

For example, if $n = 3$, the Wronskian is

$$W = \begin{vmatrix} U_0 & U_1 & U_2 & U_3 \\ -\alpha U_1 & U_0 & U_1 & U_2 \\ -\alpha U_0 & -\alpha U_1 & U_0 & U_1 \\ \alpha^2 U_1 & -\alpha U_0 & -\alpha U_1 & U_0 \end{vmatrix}$$

where we have used the identities (4.77) and (4.78) to replace the derivatives by the appropriate $U$ functions.

To evaluate the determinant, we multiply the first row by $\alpha$ and add to the third row. Then, the second row is multiplied by $\alpha$ and added to the fourth row. Where appropriate, we utilize the identity (4.76) and obtain

$$W = \begin{vmatrix} U_0 & U_1 & U_2 & U_3 \\ -\alpha U_1 & U_0 & U_1 & U_2 \\ 0 & 0 & 1 & \chi \\ 0 & 0 & 0 & 1 \end{vmatrix}$$

Hence, the value of $W$ is simply $U_0^2 + \alpha U_1^2$. Indeed, it is easy to see that $W$ will have this value for any $n > 0$. Therefore, the question of linear independence will be resolved when we show that

$$U_0^2 + \alpha U_1^2 = 1 \qquad (4.80)$$

for all values of $\chi$.

---

† The name was given by Thomas Muir in 1882 to honor the Polish mathematician and philosopher Józef Maria Höené-Wroński (1776–1853) who first used this determinant in his studies of differential equations.

To this end, we multiply the identity [Eq. (4.76) with $n = 0$]
$$U_0 + \alpha U_2 = 1$$
by $U_1$ and integrate with respect to $\chi$. We have
$$U_1^2 + \alpha U_2^2 = 2U_2$$
or
$$U_1^2 + U_2(1 - U_0) = 2U_2$$
Hence
$$U_2 = U_1^2 - U_0 U_2$$
Substituting this, for $U_2$ in the equation $U_0 + \alpha U_2 = 1$, yields
$$\begin{aligned} U_0 + \alpha U_2 = 1 &= U_0 + \alpha(U_1^2 - U_0 U_2) \\ &= U_0 + \alpha U_1^2 - U_0(1 - U_0) \\ &= U_0^2 + \alpha U_1^2 \end{aligned}$$
and the identity (4.80) is established.

## Lagrangian Coefficients and Other Orbital Quantities

Since the $U$ functions are linearly independent, the general solution of the differential equation for $t$ may be written as
$$\sqrt{\mu}(t - t_0) = a_0 U_0 + a_1 U_1 + a_2 U_2 + a_3 U_3$$
If we require $t = t_0$ when $\chi = 0$, then we find that $a_0$ must be zero. The derivative of this expression, according to Eq. (4.71), yields
$$r = a_1 U_0 + a_2 U_1 + a_3 U_2$$
Setting $\chi = 0$, gives $a_1 = r_0$. Differentiating again produces
$$\sigma = -\alpha r_0 U_1 + a_2 U_0 + a_3 U_1$$
so that $a_2 = \sigma_0$. Finally, calculating one more derivative, we have
$$1 - \alpha r = -\alpha r_0 U_0 - \alpha \sigma_0 U_1 + a_3 U_0$$
from which $a_3 = 1$.

In this manner, we obtain the generalized form of Kepler's equation
$$\sqrt{\mu}(t - t_0) = r_0 U_1(\chi; \alpha) + \sigma_0 U_2(\chi; \alpha) + U_3(\chi; \alpha) \tag{4.81}$$
together with
$$r = r_0 U_0(\chi; \alpha) + \sigma_0 U_1(\chi; \alpha) + U_2(\chi; \alpha) \tag{4.82}$$
$$\sigma = \sigma_0 U_0(\chi; \alpha) + (1 - \alpha r_0) U_1(\chi; \alpha) \tag{4.83}$$

## Universal Formulas for Conic Orbits

In a similar fashion, we write

$$\mathbf{r} = U_0 \mathbf{a}_0 + U_1 \mathbf{a}_1 + U_2 \mathbf{a}_2$$

$$\frac{r}{\sqrt{\mu}} \mathbf{v} = -\alpha U_1 \mathbf{a}_0 + U_0 \mathbf{a}_1 + U_1 \mathbf{a}_2$$

$$\frac{\sigma}{\sqrt{\mu}} \mathbf{v} - \frac{1}{r} \mathbf{r} = -\alpha U_0 \mathbf{a}_0 - \alpha U_1 \mathbf{a}_1 + U_0 \mathbf{a}_2$$

and determine the vectors $\mathbf{a}_0$, $\mathbf{a}_1$, $\mathbf{a}_2$ by setting $\chi = 0$. Thus, we obtain the following expressions for the Lagrangian coefficients

$$\begin{aligned} F &= 1 - \frac{1}{r_0} U_2(\chi; \alpha) & G &= \frac{r_0}{\sqrt{\mu}} U_1(\chi; \alpha) + \frac{\sigma_0}{\sqrt{\mu}} U_2(\chi; \alpha) \\ F_t &= -\frac{\sqrt{\mu}}{r r_0} U_1(\chi; \alpha) & G_t &= 1 - \frac{1}{r} U_2(\chi; \alpha) \end{aligned} \qquad (4.84)$$

These equations are "universal" in the sense that they are valid for all conic orbits† and are void of singularities. For this reason the $U$ functions are referred to as *universal functions*. As we indicated at the beginning of this section, $\chi$ is a generalized anomaly and is related to the classical ones by

$$\chi = \begin{cases} \sqrt{p} \left( \tan \tfrac{1}{2} f - \tan \tfrac{1}{2} f_0 \right) = \sigma - \sigma_0 \\ \sqrt{a} \, (E - E_0) \\ \sqrt{-a} \, (H - H_0) \end{cases} \qquad (4.85)$$

Finally, an important relation for $\chi$ can be derived. If we multiply Eq. (4.81) by $\alpha$ and add Eq. (4.83), we have

$$\alpha \sqrt{\mu}(t - t_0) + \sigma = U_1 + \alpha U_3 + \sigma_0 (U_0 + \alpha U_2)$$

Hence, using Eq. (4.76),

$$\chi = \alpha \sqrt{\mu}(t - t_0) + \sigma - \sigma_0 \qquad (4.86)$$

is obtained as an explicit expression for $\chi$ which does not involve any of the $U$ functions.‡

---

† The case of the parabola was considered separately in Sect. 4.2.

‡ Equation (4.86) was discovered in August of 1967 by Charles M. Newman—a staff member of the MIT Instrumentation Laboratory during the era of Apollo. His derivation was more involved than the one presented here.

## ◊ Problem 4–21

The $U$ functions are, of course, related to the elementary functions but the particular relations depend on whether the orbit is a parabola $\alpha = 0$, an ellipse $\alpha > 0$, or a hyperbola $\alpha < 0$. The first four of the $U$ functions are given by

$$U_0(\chi;\alpha) = \begin{cases} 1 \\ \cos(\sqrt{\alpha}\,\chi) \\ \cosh(\sqrt{-\alpha}\,\chi) \end{cases} \qquad U_1(\chi;\alpha) = \begin{cases} \chi \\ \dfrac{\sin(\sqrt{\alpha}\,\chi)}{\sqrt{\alpha}} \\ \dfrac{\sinh(\sqrt{-\alpha}\,\chi)}{\sqrt{-\alpha}} \end{cases}$$

$$U_2(\chi;\alpha) = \begin{cases} \dfrac{\chi^2}{2} \\ \dfrac{1-\cos(\sqrt{\alpha}\,\chi)}{\alpha} \\ \dfrac{\cosh(\sqrt{-\alpha}\,\chi)-1}{-\alpha} \end{cases} \qquad U_3(\chi;\alpha) = \begin{cases} \dfrac{\chi^3}{6} \\ \dfrac{\sqrt{\alpha}\,\chi - \sin(\sqrt{\alpha}\,\chi)}{\alpha\sqrt{\alpha}} \\ \dfrac{\sinh(\sqrt{-\alpha}\,\chi) - \sqrt{-\alpha}\,\chi}{-\alpha\sqrt{-\alpha}} \end{cases}$$

## Problem 4–22

If we define a new universal anomaly $\psi$ as

$$\psi = \frac{r_0\chi}{\sqrt{\mu}(t-t_0)}$$

then the universal form of Kepler's equation may be written either as

$$1 = U_1(\psi;\varsigma) + \xi U_2(\psi;\varsigma) + \eta U_3(\psi;\varsigma)$$

where

$$\xi = \frac{\sqrt{\mu}(t-t_0)\sigma_0}{r_0^2} \qquad \eta = \frac{\mu(t-t_0)^2}{r_0^3} \qquad \varsigma = \frac{\alpha\mu(t-t_0)^2}{r_0^2}$$

or as

$$1 = \psi + \xi U_2(\psi;\varsigma) + \lambda U_3(\psi;\varsigma)$$

where

$$\lambda = \eta - \varsigma$$

Observe that for parabolic orbits the second form of Kepler's equation becomes

$$1 = \psi + \tfrac{1}{2}\xi\psi^2 + \tfrac{1}{6}\lambda\psi^3$$

the solution of which provides a good initial approximation for the near parabolic case.

Also, for circular orbits, the solution is simply $\psi = 1$, providing a good approximation for near circular orbits.

*Karl Stumpff* 1958

## Problem 4–23
Introduce the quantity
$$\psi = \alpha \chi^2$$
so that a family of functions $c_n(\psi)$ can be defined in terms of the $U$ functions by
$$\chi^n c_n(\psi) = U_n(\chi; \alpha)$$
Indeed, the entire subject of universal functions can be developed in terms of these alternate functions $c_n(\psi)$.

(a) Derive the series representation
$$c_n(\psi) = \frac{1}{n!} - \frac{\psi}{(n+2)!} + \frac{\psi^2}{(n+4)!} - \cdots$$
together with the recursion formula
$$c_n + \psi c_{n+2} = \frac{1}{n!}$$
and the identity
$$c_0^2 + \psi c_1^2 = 1$$

(b) Derive the following derivative formulas
$$\frac{dc_0}{d\psi} = -\frac{c_1}{2}$$
$$\frac{dc_n}{d\psi} = \frac{1}{2\psi}(c_{n-1} - nc_n) \quad \text{for} \quad n = 1, 2, \ldots$$
$$= \tfrac{1}{2}(nc_{n+2} - c_{n+1}) \quad \text{for} \quad n = 0, 1, \ldots$$

(c) The first four $c$ functions† are related to the elementary functions as follows

$$c_0(\psi) = \begin{cases} 1 \\ \cos\sqrt{\psi} \\ \cosh\sqrt{-\psi} \end{cases} \qquad c_1(\psi) = \begin{cases} 1 \\ \dfrac{\sin\sqrt{\psi}}{\sqrt{\psi}} \\ \dfrac{\sinh\sqrt{-\psi}}{\sqrt{-\psi}} \end{cases}$$

$$c_2(\psi) = \begin{cases} \dfrac{1}{2} \\ \dfrac{1-\cos\sqrt{\psi}}{\psi} \\ \dfrac{\cosh\sqrt{-\psi}-1}{-\psi} \end{cases} \qquad c_3(\psi) = \begin{cases} \dfrac{1}{6} \\ \dfrac{\sqrt{\psi}-\sin\sqrt{\psi}}{\psi\sqrt{\psi}} \\ \dfrac{\sinh\sqrt{-\psi}-\sqrt{-\psi}}{-\psi\sqrt{-\psi}} \end{cases}$$

where the alternate representations depend upon the sign of $\psi$.

---

† The functions $c_2(\psi)$ and $c_3(\psi)$ are identical with the functions $C(x)$ and $S(x)$ originally defined by the author in his book *Astronautical Guidance*. Their use in the Apollo program is documented in the Epilogue of this book.

◇ **Problem 4-24**
Consider another form of the Sundman transformation

$$\frac{dt}{d\chi} = r$$

(a) If $\sqrt{\mu}\sigma$ is replaced by $\sigma$ (that is, $\sigma$ is defined as $\sigma = \mathbf{r}\cdot\mathbf{v}$), then $t$, $r$, and $\sigma$ are given by

$$t - t_0 = r_0 U_1(\chi;\mu\alpha) + \sigma_0 U_2(\chi;\mu\alpha) + \mu U_3(\chi;\mu\alpha)$$
$$r = r_0 U_0(\chi;\mu\alpha) + \sigma_0 U_1(\chi;\mu\alpha) + \mu U_2(\chi;\mu\alpha)$$
$$\sigma = \sigma_0 U_0(\chi;\mu\alpha) + \mu(1 - \alpha r_0) U_1(\chi;\mu\alpha)$$

(b) Further obtain the Lagrangian coefficients

$$F = 1 - \frac{\mu}{r_0} U_2(\chi;\mu\alpha) \qquad G = r_0 U_1(\chi;\mu\alpha) + \sigma_0 U_2(\chi;\mu\alpha)$$
$$F_t = -\frac{\mu}{r r_0} U_1(\chi;\mu\alpha) \qquad G_t = 1 - \frac{\mu}{r} U_2(\chi;\mu\alpha)$$

NOTE: In this form, the solutions of the two-body equations of motion do not require that $\mu$ be positive so that they are equally valid for repulsive as well as attractive forces.

*William H. Goodyear†  1965*

 **Problem 4-25**
Parabolic coordinates $\xi, \eta$ are defined by the transformation

$$x = \xi^2 - \eta^2 \qquad y = 2\xi\eta$$

which provides a mapping of the $\xi, \eta$ plane onto the $x, y$ plane. The inverse transformation is most conveniently expressed in terms of polar coordinates $r, \theta$ in the $x, y$ plane.
(a) Show that

$$\xi = \sqrt{r}\cos\tfrac{1}{2}\theta \qquad \eta = \sqrt{r}\sin\tfrac{1}{2}\theta$$

is the appropriate mapping of the $x, y$ plane onto the $\xi, \eta$ plane.
(b) The two-body equations of motion in the $x, y$ plane are transformed into

$$\frac{d^2\xi}{d\chi^2} + \frac{\alpha}{4}\xi = 0 \qquad \frac{d^2\eta}{d\chi^2} + \frac{\alpha}{4}\eta = 0$$

in the $\xi, \eta$ plane, where $\alpha = 1/a$ is the reciprocal of the semimajor axis and $\chi$ is defined by the Sundman or *regularization transformation*

$$\sqrt{\mu}\frac{dt}{d\chi} = r$$

Thus, we see that the two-body motion in parabolic coordinates consists of two independent harmonic oscillators of the same frequency.

*André Deprit  1968*

---

† "Completely General Closed-Form Solution for Coordinates and Partial Derivatives of the Two-Body Problem," *The Astronomical Journal*, Vol. 70, April 1965, pp. 189–192.

## 4.6 Identities for the Universal Functions

There are a variety of identities involving the functions $U_n(\chi;\alpha)$, many of which will be required in further applications. These will be developed and collected in this section to serve as a ready reference when needed.

Because of the direct relationship between $U_0$, $U_1$ and the circular and hyperbolic functions, as seen in Prob. 4–21, we can immediately recognize

$$U_0^2 + \alpha U_1^2 = 1 \tag{4.87}$$

as the best known identity between sines and cosines or hyperbolic sines and cosines. Similarly, we can write

$$\begin{aligned} U_0(\chi \pm \psi) &= U_0(\chi)U_0(\psi) \mp \alpha U_1(\chi)U_1(\psi) \\ U_1(\chi \pm \psi) &= U_1(\chi)U_0(\psi) \pm U_0(\chi)U_1(\psi) \end{aligned} \tag{4.88}$$

and

$$\begin{aligned} U_0(2\chi) &= U_0^2(\chi) - \alpha U_1^2(\chi) = 2U_0^2(\chi) - 1 = 1 - 2\alpha U_1^2(\chi) \\ U_1(2\chi) &= 2U_0(\chi)U_1(\chi) \end{aligned} \tag{4.89}$$

as counterparts of other familiar identities. Just as Eq. (4.87) was derived earlier, without resort to its relation with the elementary functions, so also could these and all identities involving just $U_0$ and $U_1$.

For the higher order $U$ functions, the analogy with the elementary functions is not convenient to exploit and other techniques will have to be employed.

### Identities Involving Compound Arguments

The basic equation, from which all the identities will evolve, is

$$U_n + \alpha U_{n+2} = \frac{\chi^n}{n!} \tag{4.90}$$

For $n = 0$, we have

$$\alpha U_2(\chi \pm \psi) = 1 - U_0(\chi \pm \psi) = 1 - U_0(\chi)U_0(\psi) \pm \alpha U_1(\chi)U_1(\psi)$$

but this equation is not useful to calculate $U_2(\chi \pm \psi)$ since division by $\alpha$ would be required. (It will be a cardinal rule that we must *never* divide by $\alpha$ in any calculation involving universal functions.)

To obtain a proper identity, we write

$$\alpha U_2(\chi \pm \psi) = 1 - [1 - \alpha U_2(\chi)][1 - \alpha U_2(\psi)] \pm \alpha U_1(\chi)U_1(\psi)$$

so that $\alpha$ may be cancelled as a common factor. There results

$$U_2(\chi \pm \psi) = U_2(\chi)[1 - \alpha U_2(\psi)] + U_2(\psi) \pm U_1(\chi)U_1(\psi)$$

Hence, finally,

$$U_2(\chi \pm \psi) = U_2(\chi)U_0(\psi) + U_2(\psi) \pm U_1(\chi)U_1(\psi) \tag{4.91}$$

### ◊ Problem 4–26

A generalization of the well-known Euler identity for trigonometric functions is

$$e^{i\sqrt{\alpha}\,\chi} = U_0(\chi;\alpha) + i\sqrt{\alpha}\,U_1(\chi;\alpha)$$

where $i = \sqrt{-1}$. Use this relation to derive Eqs. (4.88).

### ◊ Problem 4–27

Derive the following identities for the universal functions of the sum and difference of two arguments:

$$U_3(\chi \pm \psi) = U_3(\chi) \pm U_3(\psi) + U_1(\chi)U_2(\psi) \pm U_2(\chi)U_1(\psi)$$
$$U_4(\chi \pm \psi) = U_2(\chi)U_2(\psi) + U_4(\chi) + U_4(\psi) \pm \psi U_3(\chi) \pm U_1(\chi)U_3(\psi)$$

### Identities for $U_n^2(\chi;\alpha)$

The method used to establish the identity

$$U_1^2 = U_2(1 + U_0)$$

which was derived in the previous section as a part of the calculation of the Wronskian of the $U$ functions, can be generalized to produce a sequence of identities. For this purpose, multiply Eq. (4.90) by $U_{n+1}$ and rewrite as

$$\frac{d}{d\chi}(U_{n+1}^2 + \alpha U_{n+2}^2) = 2\frac{\chi^n}{n!}U_{n+1}$$

Hence

$$U_{n+1}^2 + \alpha U_{n+2}^2 = 2\left[\frac{\chi^n}{n!}U_{n+2} - \frac{\chi^{n-1}}{(n-1)!}U_{n+3} + \cdots \pm U_{2n+2}\right]$$

is obtained by integrating the right-hand side by parts. Then, using Eq. (4.90) again, we have

$$U_{n+1}^2 = U_{n+2}\left(\frac{\chi^n}{n!} + U_n\right) - 2\left[\frac{\chi^{n-1}}{(n-1)!}U_{n+3} - \frac{\chi^{n-2}}{(n-2)!}U_{n+4} + \cdots\right] \quad (4.92)$$

Therefore, by setting $n = 0, 1, 2, \ldots$, we may establish successively

$$U_1^2 = U_2(1 + U_0)$$
$$U_2^2 = U_3(\chi + U_1) - 2U_4 \quad (4.93)$$
$$U_3^2 = U_4(\tfrac{1}{2}\chi^2 + U_2) - 2(\chi U_5 - U_6) \quad \text{etc.}$$

These equations are particularly useful to calculate $U_4$, $U_6$, $U_8$, ... in terms of the $U$ functions with lower subscripts. Similar explicit relations for the odd-ordered functions do not seem to exist. Of course, Eq. (4.90) permits a simple solution to the reverse problem, i.e., calculating lower-order functions from higher-order ones.

### Sect. 4.6]         Identities for the Universal Functions                    185

**Identities for** $U_{n+1}U_{n+1-m} - U_{n+2}U_{n-m}$

For any integer $m \leq n$, the identity (4.90) may be written as

$$U_{n-m} + \alpha U_{n+2-m} = \frac{\chi^{n-m}}{(n-m)!}$$

Now multiply this by $U_{n+1}$ and multiply Eq. (4.90) by $U_{n+1-m}$. Adding the resulting two equations gives

$$\frac{d}{d\chi}(U_{n+1}U_{n+1-m} + \alpha U_{n+2}U_{n+2-m}) = \frac{\chi^{n-m}}{(n-m)!}U_{n+1} + \frac{\chi^n}{n!}U_{n+1-m}$$

Hence

$$U_{n+1}U_{n+1-m} + \alpha U_{n+2}U_{n+2-m} =$$
$$\frac{\chi^{n-m}}{(n-m)!}U_{n+2} - \frac{\chi^{n-m-1}}{(n-m-1)!}U_{n+3} + \cdots \pm U_{2n+2-m}$$
$$+ \frac{\chi^n}{n!}U_{n+2-m} - \frac{\chi^{n-1}}{(n-1)!}U_{n+3-m} + \cdots \pm U_{2n+2-m}$$

or

$$U_{n+1}U_{n+1-m} - U_{n+2}U_{n-m} =$$
$$\frac{\chi^n}{n!}U_{n+2-m} - \frac{\chi^{n-1}}{(n-1)!}U_{n+3-m} + \cdots \pm U_{2n+2-m}$$
$$- \frac{\chi^{n-m-1}}{(n-m-1)!}U_{n+3} + \frac{\chi^{n-m-2}}{(n-m-2)!}U_{n+4} - \cdots \mp U_{2n+2-m} \quad (4.94)$$

which agrees with Eq. (4.92) for $m = 0$. The following identities result for $(m,n) = (1,1), (1,2), (2,2)$:

$$\begin{aligned}
U_2 U_1 - U_3 U_0 &= \chi U_2 - U_3 \\
U_3 U_2 - U_4 U_1 &= \tfrac{1}{2}\chi^2 U_3 - \chi U_4 \\
U_3 U_1 - U_4 U_0 &= \tfrac{1}{2}\chi^2 U_2 - \chi U_3 + U_4
\end{aligned} \quad (4.95)$$

◇ **Problem 4–28**
   Derive the identity
$$U_n(m\chi) + \alpha U_{n+2}(m\chi) = m^n[U_n(\chi) + \alpha U_{n+2}(\chi)]$$
where $m$ is an integer.

◇ **Problem 4–29**
   Show that
$$U_n(k\chi; \alpha) = k^n U_n(\chi; k^2\alpha)$$
obtains for any value of the parameter $k$.

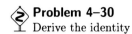

**Problem 4-30**
Derive the identity

$$U_3^2 = \tfrac{1}{6} x^3 U_3 + U_5(U_1 - x)$$

**Problem 4-31**
Derive the following double argument identities for the universal functions:

$$U_2(2x) = 2U_1^2(x) \qquad U_4(2x) = 2U_2^2(x) + 4U_4(x)$$
$$U_3(2x) = 2U_3(x) + 2U_1(x)U_2(x) \qquad \quad = 2U_3(x)[x + U_1(x)]$$
$$= 2U_0(x)U_3(x) + 2xU_2(x) \quad U_5(2x) = 2U_1(x)U_4(x) + x^2 U_3(x) + 2U_5(x)$$

### Identities Involving the True Anomaly Difference

Important relationships between the functions $U_n(x;\alpha)$ and trigonometric functions of the true anomaly difference $\theta = f - f_0$ can be obtained by comparing Eqs. (3.42) and (4.84). Thus,

$$U_1(x;\alpha) = \frac{r}{p}[\sqrt{p}\sin\theta - \sigma_0(1 - \cos\theta)]$$
$$U_2(x;\alpha) = \frac{rr_0}{p}(1 - \cos\theta) \tag{4.96}$$

Also, by using the identities

$$U_1(x) = 2U_0(\tfrac{1}{2}x)U_1(\tfrac{1}{2}x) \quad \text{and} \quad U_2(x) = 2U_1^2(\tfrac{1}{2}x)$$

we find that

$$U_0(\tfrac{1}{2}x;\alpha) = \sqrt{\frac{r}{r_0 p}}(\sqrt{p}\cos\tfrac{1}{2}\theta - \sigma_0 \sin\tfrac{1}{2}\theta)$$
$$U_1(\tfrac{1}{2}x;\alpha) = \sqrt{\frac{rr_0}{p}}\sin\tfrac{1}{2}\theta \tag{4.97}$$

which may be written alternately as

$$\sin\tfrac{1}{2}\theta = \sqrt{\frac{p}{rr_0}} U_1(\tfrac{1}{2}x;\alpha)$$
$$\cos\tfrac{1}{2}\theta = \frac{1}{\sqrt{rr_0}}[r_0 U_0(\tfrac{1}{2}x;\alpha) + \sigma_0 U_1(\tfrac{1}{2}x;\alpha)] \tag{4.98}$$

In particular, we obtain

$$\tan\tfrac{1}{2}\theta = \frac{\sqrt{p}\, U_1(\tfrac{1}{2}x;\alpha)}{r_0 U_0(\tfrac{1}{2}x;\alpha) + \sigma_0 U_1(\tfrac{1}{2}x;\alpha)} \tag{4.99}$$

as a convenient formula for determining $\theta$ from $x$.

Sect. 4.7]     Continued Fractions for Universal Functions     187

◊ **Problem 4-32**
Using Eq. (4.82), derive the identity

$$U_2(\chi;\alpha) = r + r_0 - 2\sqrt{rr_0}\cos\tfrac{1}{2}\theta\, U_0(\tfrac{1}{2}\chi;\alpha)$$

from which the parameter $p$ may be expressed in the form

$$p = \frac{rr_0 \sin^2 \tfrac{1}{2}\theta}{U_1^2(\tfrac{1}{2}\chi;\alpha)} = \frac{2rr_0 \sin^2 \tfrac{1}{2}\theta}{r + r_0 - 2\sqrt{rr_0}\cos\tfrac{1}{2}\theta\, U_0(\tfrac{1}{2}\chi;\alpha)}$$

**Problem 4-33**
For the family of functions $c_n(\psi)$ defined in Prob. 4-23, derive the following identities involving quadruple arguments

$$c_0(4\psi) = c_0^2(\psi) - \psi c_1^2(\psi) \qquad c_2(4\psi) = \tfrac{1}{2}c_1^2(\psi) = \tfrac{1}{2}[1 - \psi c_3(\psi)]^2$$
$$\quad = 2c_0^2(\psi) - 1 \qquad\qquad c_3(4\psi) = \tfrac{1}{4}[c_3(\psi) + c_1(\psi)c_2(\psi)]$$
$$c_1(4\psi) = c_0(\psi)c_1(\psi) \qquad\qquad\quad = \tfrac{1}{4}[c_2(\psi) + c_3(\psi) - \psi c_2(\psi)c_3(\psi)]$$

together with the identity set

$$c_1^2 - c_0 c_2 = c_2 \qquad\qquad c_3^2 - c_2 c_4 = \tfrac{1}{2}c_4 - 2(c_5 - c_6)$$
$$c_2^2 - c_1 c_3 = c_3 - 2c_4 \qquad c_1 c_2 - c_0 c_3 = c_2 - c_3$$

## 4.7 Continued Fractions for Universal Functions

Continued fraction representations of the sine and cosine functions are not possible using the Gauss expansion theorem of Chapter 1. Therefore, we should not expect such expansions for the functions $U_0(\chi;\alpha)$ and $U_1(\chi;\alpha)$. The Euler transformation of a series to a continued fraction is always possible, but no computational advantage will ensue since both the series and the fraction have identical convergence properties.

However, the tangent function does have a Gaussian expansion as we have seen in Prob. 1-15. By recalling their relationship to the elementary functions, we can obtain a continued fraction for the ratio of $U_1$ and $U_0$. Denoting this ratio by $u$, we have

$$u = \frac{U_1(\tfrac{1}{2}\chi;\alpha)}{U_0(\tfrac{1}{2}\chi;\alpha)} = \cfrac{\tfrac{1}{2}\chi}{1 - \cfrac{\alpha(\tfrac{1}{2}\chi)^2}{3 - \cfrac{\alpha(\tfrac{1}{2}\chi)^2}{5 - \cfrac{\alpha(\tfrac{1}{2}\chi)^2}{7 - \cdots}}}} \qquad (4.100)$$

**188**  Two-Body Orbits and the Initial-Value Problem  [Chap. 4]

Now, using some of the basic identities for the universal functions, we can express $U_0$, $U_1$, and $U_2$ as

$$U_0(\chi) = 2U_0^2(\tfrac{1}{2}\chi) - 1$$
$$U_1(\chi) = 2U_0(\tfrac{1}{2}\chi)U_1(\tfrac{1}{2}\chi) \quad (4.101)$$
$$U_2(\chi) = 2U_1^2(\tfrac{1}{2}\chi)$$

Hence, $U_0(\chi)$, $U_1(\chi)$, and $U_2(\chi)$ are determined from

$$U_0(\chi;\alpha) = \frac{1-\alpha u^2}{1+\alpha u^2} \quad U_1(\chi;\alpha) = \frac{2u}{1+\alpha u^2} \quad U_2(\chi;\alpha) = \frac{2u^2}{1+\alpha u^2} \quad (4.102)$$

As a consequence, values of the first three $U$-functions can be calculated using only a single continued-fraction evaluation.

### Continued Fraction Determination of $U_3$ and $U_4$

As we shall see in Chapter 7, it is fundamental to Gauss' method of solving the two-body, two-point, boundary-value problem that

$$\frac{2\psi - \sin 2\psi}{\sin^3 \psi} \quad \text{is a hypergeometric function of} \quad \sin^2 \tfrac{1}{2}\psi$$

We generalize this result to apply to the universal functions by obtaining a differential equation for

$$Q = \frac{U_3(2\chi;\alpha)}{U_1^3(\chi;\alpha)} \quad (4.103)$$

regarded as a function of

$$q = \alpha U_1^2(\tfrac{1}{2}\chi;\alpha) = \frac{\alpha u^2}{1+\alpha u^2} \quad (4.104)$$

By differentiating $q$ with respect to $\chi$, we obtain

$$\frac{dq}{d\chi} = \alpha U_1(\tfrac{1}{2}\chi)U_0(\tfrac{1}{2}\chi) = \tfrac{1}{2}\alpha U_1(\chi)$$

Then, differentiating $U_1^3(\chi)Q = U_3(2\chi)$ with respect to $q$, gives

$$U_1^3 \frac{dQ}{dq} + 3U_1^2(\chi)U_0(\chi)Q\frac{d\chi}{dq} = 2U_2(2\chi)\frac{d\chi}{dq}$$

which reduces to

$$\alpha U_1^2(\chi)\frac{dQ}{dq} + 6U_0(\chi)Q = 8$$

Now,

$$\alpha U_1^2(\chi) = 4\alpha U_0^2(\tfrac{1}{2}\chi)U_1^2(\tfrac{1}{2}\chi) = 4q(1-q)$$

and

$$U_0(\chi) = 1 - \alpha U_2(\chi) = 1 - 2\alpha U_1^2(\tfrac{1}{2}\chi) = 1 - 2q$$

so that an appropriate form for the differential equation is

$$q(1-q)\frac{dQ}{dq} + \tfrac{3}{2}(1-2q)Q = 2$$

Differentiating a second time produces Gauss' equation (1.12)

$$q(1-q)\frac{d^2Q}{dq^2} + (\tfrac{5}{2} - 5q)\frac{dQ}{dq} - 3Q = 0 \qquad (4.105)$$

with $\alpha = 3$, $\beta = 1$, and $\gamma = \tfrac{5}{2}$. The function $Q$ is, therefore, a hypergeometric function of $q$.

Indeed, since $\lim_{\chi \to 0} Q = \tfrac{4}{3}$, we have

$$Q = \tfrac{4}{3} F(3, 1; \tfrac{5}{2}; q) \qquad (4.106)$$

which, according to Eq. (1.29), admits of a continued fraction expansion. Therefore, $U_3(2\chi)$ may be obtained from†

$$U_3(2\chi) = \cfrac{\tfrac{4}{3} U_1^3(\chi)}{1 - \cfrac{\tfrac{6}{5} q}{1 + \cfrac{\tfrac{2}{5\cdot 7} q}{1 - \cdots}}} \qquad (4.107)$$

The final result follows from the identities of Prob. 4–31 and Eqs. (4.93). We see that the universal functions $U_3$ and $U_4$ can be calculated from

$$\begin{aligned}U_3(\chi) &= \tfrac{1}{2} U_3(2\chi) - U_1(\chi) U_2(\chi) \\ U_4(\chi) &= \tfrac{1}{2} U_3(\chi)[\chi + U_1(\chi)] - \tfrac{1}{2} U_2^2(\chi) \\ &= U_1(\chi) U_3(\chi) - \tfrac{1}{2}[U_2^2(\chi) - \alpha U_3^2(\chi)]\end{aligned} \qquad (4.108)$$

with $U_3(2\chi)$ evaluated using the continued fraction (4.107).


### Continued Fraction Determination of $U_5$ and $U_6$

Recently, Stanley W. Shepperd‡ was able to extend this technique to permit the function $U_5(\chi)$ to be evaluated by a continued fraction. For this purpose, define

$$R = \frac{U_2(2\chi) U_3(2\chi) + 2\chi U_4(2\chi) - 3 U_5(2\chi)}{U_1^5(\chi)} \qquad (4.109)$$

---

† In Sect. 7.2 the general expression for this continued fraction is given and methods of improving its convergence are explored.

‡ "Universal Keplerian State Transition Matrix," in *Celestial Mechanics*, Vol. 35, Feb. 1985, pp. 129–144.

In the same manner as before, we can obtain Gauss' differential equation in the form

$$q(1-q)\frac{d^2R}{dq^2} + (\tfrac{7}{2} - 7q)\frac{dR}{dq} - 5R = 0 \qquad (4.110)$$

for which $\alpha = 5$, $\beta = 1$, and $\gamma = \tfrac{7}{2}$. Thus, we have

$$R = \tfrac{16}{5} F(5, 1; \tfrac{7}{2}; q) \qquad (4.111)$$

In the course of the derivation, which we leave to the reader, the following identity is required:

$$U_2^2(2\chi) + 2U_4(2\chi) = 2\chi U_3(2\chi) + U_1(2\chi)U_3(2\chi)$$

which is obtained from the equations of Prob. 4-31.
Therefore,

$$U_5(2\chi) = \tfrac{4}{3}[U_1^2(\chi)U_3(\chi) + U_1^3(\chi)U_2(\chi) + \chi U_2^2(\chi)] + \tfrac{8}{3}\chi U_4(\chi)$$

$$-\cfrac{\tfrac{16}{15} U_1^5(\chi)}{1 - \cfrac{\tfrac{10\cdot 5}{5\cdot 7} q}{1 + \cfrac{\tfrac{2\cdot 3}{7\cdot 9} q}{1 - \cfrac{\tfrac{12\cdot 7}{9\cdot 11} q}{1 + \cfrac{\tfrac{4\cdot 1}{11\cdot 13} q}{1 - \cfrac{\tfrac{14\cdot 9}{13\cdot 15} q}{1 - \cfrac{\tfrac{6\cdot 1}{15\cdot 17} q}{1 - \cfrac{\tfrac{16\cdot 11}{17\cdot 19} q}{1 - \cdots}}}}}}}} \qquad (4.112)$$

and the final result

$$\begin{aligned} U_5(\chi) &= \tfrac{1}{2}U_5(2\chi) - U_1(\chi)U_4(\chi) - \tfrac{1}{2}\chi^2 U_3(\chi) \\ U_6(\chi) &= \tfrac{1}{2}[U_3^2(\chi) - U_2(\chi)U_4(\chi)] - \tfrac{1}{4}\chi^2 U_4(\chi) + \chi U_5(\chi) \end{aligned} \qquad (4.113)$$

follows as before with $U_5(2\chi)$ calculated from Eq. (4.112).

◇ **Problem 4-34**

Use the Gauss identities for contiguous hypergeometric functions from Sect. 1.1 to derive the identity

$$F(3, 1; \tfrac{5}{2}; q) = 1 - 2q + \tfrac{16}{5} q(1-q)F(5, 1; \tfrac{7}{2}; q)$$

Therefore, the continued fraction of Eq. (4.107) can be obtained from the continued fraction of Eq. (4.112). As a consequence, if the function $U_5(\chi)$ is to be calculated, no extra continued-fraction evaluation is necessary to obtain $U_3(\chi)$.

Chapter 5

# Solving Kepler's Equation

ALGORITHMS FOR THE SOLUTION OF KEPLER'S EQUATION ABOUND. The first such was, of course, by Kepler himself. The next solution was Newton's in his *Principia*. A very large number of analytical and graphical solutions have been discovered—nearly every prominent mathematician from Newton until the middle of the last century having given the subject more or less attention. Since the advent of the modern era of spaceflight, interest has revived and new algorithms are being published quite regularly. Indeed, this chapter contains a recent one by the present author.

Kepler's equation, the most famous of all transcendental equations, both spawned and motivated a number of significant developments in mathematics. Lagrange's expansion theorem, Bessel functions, Fourier series, some aspects of complex function theory, and various techniques of numerical analysis are but some of these which we will consider in this chapter.

One of the more interesting is the discovery of Bessel functions. In 1770, Lagrange developed his expansion theorem to produce a solution of Kepler's equation as a power series in the eccentricity $e$ with coefficients which turned out to be linear combinations of trigonometric functions of integral multiples of the mean anomaly. Later Laplace demonstrated that Lagrange's series would diverge if $e$ exceeded some critical value. The proof required analysis in the complex plane and, perhaps, provided added impetus for development of the then new field—functions of a complex variable.

Lagrange rearranged the terms in the series to obtain a form which we would now call a Fourier sine series. The Fourier coefficients were infinite power series in $e$ which converged for all elliptic orbits. Indeed, the effect of altering the order of the terms changed the convergence properties of Lagrange's expansion—which must have excited great interest.

In 1824, Friedrich Wilhelm Bessel (1784–1846), a mathematician and director of the astronomical observatory in Königsburg, attempted the direct solution of Kepler's equation as a Fourier series and obtained the coefficients in an integral form. The power series expansions of these integrals produced, of course, the same collection of series that Lagrange had obtained almost fifty years earlier—but, because Bessel made such an extensive study of these functions for many years, they bear his name and not that of Lagrange.

The question of priority in mathematics is always a difficult one. Special cases of Bessel functions occurred as early as 1703 in a letter by James Bernoulli to Gottfried Wilhelm Leibnitz. Then in 1733, Daniel Bernoulli wrote a paper on the modes of oscillation of a heavy chain in which the solution, for the case in which the chain was uniform, is $J_0$—the zero$^{\text{th}}$ order Bessel function of the first kind. For the nonuniform chain, the solution involved first-order Bessel functions of the second kind. Then, in 1736, Euler followed up on Bernoulli with his paper "On the Oscillations of a Flexible Thread Loaded with Arbitrarily Many Weights" in which he encountered the functions $I_n(x)$ known today as modified Bessel functions. We, too, will have need for these special functions in Chapter 10 when we study the effects of atmospheric drag on satellite orbits.

## 5.1 Elementary Methods

Kepler's equation
$$M = E - e\sin E$$
is transcendental in $E$, so that the solution for this quantity when $M$ is given cannot be expressed in a finite number of terms. However, there is one and only one solution as can be seen from the following argument.

Define the function
$$F_e(E) = E - e\sin E - M \tag{5.1}$$
and suppose that $k\pi \leq M < (k+1)\pi$, where $k$ is an integer. Then since
$$F_e(k\pi) = k\pi - M \leq 0$$
$$F_e[(k+1)\pi] = (k+1)\pi - M > 0$$
it follows that $F_e(E)$ vanishes at least once in the stated interval. However, the derivative $F_e'(E)$ is always positive, so that $F_e(E)$ is zero for only one value of $E$. Furthermore, since any value of the mean anomaly can be written as $2k\pi \pm M$, where $k$ is an integer (positive, negative, or zero), with a corresponding value of the eccentric anomaly given by $2k\pi \pm E$, there is no loss in generality by assuming that $E$ and $M$ are restricted to the interval $(0, \pi)$.

The hyperbolic form of Kepler's equation
$$N = e\sinh H - H$$
where $N$ corresponds in form to the mean anomaly, has also one and only one solution. For if we define
$$F_h(H) = e\sinh H - H - N \tag{5.2}$$

then we readily observe that
$$F_h(0) = -N \le 0$$
$$F_h\left(\frac{N}{e-1}\right) = e\left[\frac{1}{3!}\left(\frac{N}{e-1}\right)^3 + \frac{1}{5!}\left(\frac{N}{e-1}\right)^5 + \cdots\right] \ge 0$$
(We are assuming that $N$ and $H$ are both positive which is clearly justifiable by symmetry.) Since $F_h'(H)$ is always positive, it again follows that $F_h(H)$ vanishes just once in the interval $0 \le H \le N/(e-1)$.

## Graphical Methods

Simple graphical solutions of Kepler's equation are possible. In a rectangular system of coordinates, we can construct a sine curve and a straight line whose equations are
$$y = \sin E \quad \text{and} \quad y = \frac{1}{e}(E - M)$$
and observe that their point of intersection determines the value of $E$ for which $F_e(E) = 0$. Since the sine curve can be constructed once for all, only the slope and $y$ intercept of the straight line are problem dependent.

Obviously, a similar technique applies to the hyperbolic form by finding the intersection of the hyperbolic sine curve and the straight line
$$y = \sinh H \quad \text{and} \quad y = \frac{1}{e}(H + N)$$

◊ **Problem 5–1**

Consider a circle of unit radius rolling without slipping along a straight line. Consider a point $P$ fixed on the radius of the circle at a distance $e$ from the center and invent a graphical method of solving Kepler's equation.

NOTE: The locus of $P$ as the circle rolls is called a *trochoid* or a *curtate cycloid*.

Sir Isaac Newton 1687

## Inverse Linear Interpolation (Regula Falsi)

An extremely simple iteration technique, whose convergence is guaranteed, is the so-called *regula falsi* method which has obvious general applicability. Regula falsi, or the method of false position, is equivalent to inverse linear interpolation as is easily understood from its development.

Assume we are given a function $y = F(x)$ and seek a value $\xi$ such that $F(\xi) = 0$. If we choose $x_0$ and $x_1$ so that
$$F_0 = F(x_0) \quad \text{and} \quad F_1 = F(x_1)$$

have opposite signs, then the straight line†

$$\begin{vmatrix} x & y & 1 \\ x_0 & F_0 & 1 \\ x_1 & F_1 & 1 \end{vmatrix} = 0$$

connecting the two points $x_0$, $F_0$ and $x_1$, $F_1$ in the $x, y$ plane, intersects the $x$ axis to provide an approximate value $x_2$ for the root $\xi$. We have

$$x_2 = \frac{F_1 x_0 - F_0 x_1}{F_1 - F_0} \tag{5.3}$$

and can repeat the process with $x_2$, and either $x_0$ or $x_1$—the choice depending on whether $F(x_0)$ or $F(x_1)$ is of opposite sign to $F(x_2)$. Clearly, each step in the iteration process gives a value of $x$ which is an ever closer approximation to $\xi$.

As an example, consider the function $F_e(E)$. Since

$$F_e(M) = -e \sin M \leq 0 \quad \text{for} \quad 0 \leq M \leq \pi$$

and

$$F_e(M + e) = e[1 - \sin(M + e)] \geq 0$$

then we can choose $x_0 = M$ and $x_1 = M + e$. Hence,

$$e \begin{vmatrix} E & 0 & 1 \\ M & -\sin M & 1 \\ M+e & 1 - \sin(M+e) & 1 \end{vmatrix} = 0$$

and, by expanding the determinant, we obtain

$$E = M + \frac{e \sin M}{1 - \sin(M + e) + \sin M} \tag{5.4}$$

which provides an approximate root of Kepler's equation.

Similarly, with $x_0 = 0$ and $x_1 = N/(e-1)$, we have

$$\frac{1}{e-1} \begin{vmatrix} H & 0 & 1 \\ 0 & -N & 1 \\ N & e(e-1)\sinh[N/(e-1)] - eN & e-1 \end{vmatrix} = 0$$

or, in expanded form,

$$H = \frac{N^2}{e(e-1)\sinh \frac{N}{e-1} - N} \tag{5.5}$$

as an approximate root of the hyperbolic form of Kepler's equation.

---

† See Prob. D-3 in Appendix D.

## Problem 5-2
If $M$ is such that $0 \leq M \leq \frac{1}{2}\pi - e$, then $E$ lies in the range

$$\frac{\frac{1}{2}\pi}{\frac{1}{2}\pi - e} M \leq E \leq M + e$$

Also, if $M$ satisfies the inequalities $\frac{1}{2}\pi - e \leq M \leq \pi$, then $E$ is in the range

$$\frac{\pi}{\frac{1}{2}\pi + e}(\frac{1}{2}M + e) \leq E \leq M + e$$

HINT: Use *Jordan's inequality*†

$$\frac{2}{\pi} < \frac{\sin x}{x} \qquad \text{for} \qquad -\frac{\pi}{2} < x < \frac{\pi}{2}$$

$$\cos x \leq \frac{\sin x}{x} \leq 1 \qquad \text{for} \qquad 0 \leq x \leq \pi$$

*Gary R. Smith*‡ 1979

## Problem 5-3
For positive $N$, the hyperbolic anomaly $H$ is bounded according to

$$0 \leq H \leq \left(\frac{6N}{e}\right)^{\frac{1}{3}} \qquad \text{or} \qquad 0 \leq H \leq \frac{N}{e-1}$$

## Problem 5-4
If terms of order $e^2$ are ignored, then

$$E = M + e \sin M + O(e^2)$$

is a solution of Kepler's equation.

## Problem 5-5
In Prob. 4-9 there was developed a graphical approximation to the mean anomaly. This approximation is a solution of Kepler's equation if terms of order $e^3$ are omitted, i.e.,

$$M = E - \arcsin(e \sin E) + O(e^3)$$

Use this relation to develop the solution

$$\tan E = \frac{1}{\sqrt{1-e^2}} \tan 2\lambda \qquad \text{where} \qquad \tan \lambda = \sqrt{\frac{1+e}{1-e}} \tan \tfrac{1}{2} M$$

which gives $E$ to second order in the eccentricity.

---

† Published in 1894 in *Cours d'analyse de l'École Polytechnique*. Camille Jordan (1838–1921) had widespread influence and set new standards for what was to constitute a *rigorous* proof in mathematics.

‡ "A Simple, Efficient Starting Value for the Iterative Solution of Kepler's Equation," *Celestial Mechanics*, Vol. 19, February 1979, pp. 163–166.

◊ **Problem 5–6**

By defining $z = E - M$, Kepler's equation can be written as

$$z = e \sin(M + z)$$

Expand the right-hand side as a power series in $z$. Then, reverse the series, using the results of Prob. C–5 in Appendix C, to obtain

$$E = M + \frac{e \sin M}{1 - e \cos M} - \frac{1}{2}\left(\frac{e \sin M}{1 - e \cos M}\right)^3 + O(e^4)$$

as a solution of Kepler's equation to within third order in the eccentricity.

### Successive Substitutions

Another simple iterative solution to Kepler's equation is also possible using the method of successive substitutions. The infinite sequence

$$\begin{aligned} E_0 &= 0 \\ E_1 &= M + e \sin E_0 \\ &\cdots \\ E_{k+1} &= M + e \sin E_k \quad \text{etc.} \end{aligned} \qquad (5.6)$$

will have a limiting value, according to *Cauchy's theorem*,† provided that for any given positive number $\epsilon$, however small, it is possible to find a number $k$ such that

$$|E_{k+n} - E_k| < \epsilon \qquad (5.7)$$

for all positive integral values of $n$.

To show that this condition is satisfied for the sequence $E_0, E_1, \ldots,$ we first observe that

$$E_{k+n} - E_k = \sum_{m=k}^{k+n-1} (E_{m+1} - E_m)$$

---

† The eighteenth century mathematicians used infinite series and sequences in a cavalier manner. (At one period, even Gauss thought that a series would converge if the size of the terms tended to zero.) By the end of the century, there were so many plainly absurd results that some began to question the validity of operations with infinite processes. *Cours d'analyse algébrique*, written in 1821 by Augustine-Louis Cauchy (1789–1857), was the first extensive and significant treatment of the subject of convergence. Cauchy defined convergence and divergence and then stated the *Cauchy convergence criterion*—a sequence $S_k$ converges to a limit $S$ if and only if $S_{k+n} - S_k$ can be made less in absolute value than any assignable quantity for all $n$ and sufficiently large $k$. He proved the condition to be necessary but, strangely, ignored the sufficiency proof.

# Elementary Methods

Then, since

$$\begin{aligned}
E_{m+1} - E_m &= e(\sin E_m - \sin E_{m-1}) \\
&= 2e \sin \tfrac{1}{2}(E_m - E_{m-1}) \cos \tfrac{1}{2}(E_m + E_{m-1}) \\
&= e(E_m - E_{m-1}) \frac{\sin \tfrac{1}{2}(E_m - E_{m-1})}{\tfrac{1}{2}(E_m - E_{m-1})} \cos \tfrac{1}{2}(E_m + E_{m-1})
\end{aligned}$$

we have

$$|E_{m+1} - E_m| \le e|E_m - E_{m-1}| \le \cdots \le e^m |E_1 - E_0| = M e^m$$

Therefore

$$|E_{k+n} - E_k| \le M \sum_{m=k}^{k+n-1} e^m = M e^k \frac{1 - e^n}{1 - e}$$

so that if $e$ is numerically less than one,

$$|E_{k+n} - E_k| \le M \frac{e^k}{1 - e}$$

Thus, the condition of Cauchy's theorem is met by choosing the integer $k$ such that

$$k > \frac{\log \frac{\epsilon(1-e)}{M}}{\log e} \tag{5.8}$$

An alternate proof of convergence, which also can be adapted to the hyperbolic form of Kepler's equation, uses the *mean value theorem of differential calculus*. We write

$$E_{m+1} = M + e \sin E_m = f(E_m)$$

so that

$$\begin{aligned}
E_{m+1} - E_m &= f(E_m) - f(E_{m-1}) \\
&= (E_m - E_{m-1}) f'[E_{m-1} + \beta_m (E_m - E_{m-1})]
\end{aligned}$$

where $\beta_m$ is a constant in the interval $(0, 1)$. Hence

$$E_{m+1} - E_m = M \prod_{j=1}^{m} f'[E_{j-1} + \beta_j (E_j - E_{j-1})]$$

Then since $f'(E) = e \cos E$, we have

$$|E_{m+1} - E_m| \le M e^m$$

and the remainder of the proof is the same as before.

## ◊ Problem 5-7
By writing

$$x = 2\sqrt{\frac{\mu}{p^3}}(t-\tau) \quad \text{and} \quad y = \tan \tfrac{1}{2}f$$

Barker's equation takes the form $y = x - \tfrac{1}{3}y^3$. This can be reformatted as a successive substitutions algorithm

$$y_{k+1} = x - \tfrac{1}{3}y_k^3$$

Prove analytically that the algorithm does converge provided that $x$ does not exceed unity and $k$ is chosen so that

$$2k + 1 > \frac{\log \epsilon(1-x^2)}{\log x}$$

HINT: If $x$ and $y$ are both positive, then, clearly, $y$ is less than $x$. Also, the factorization

$$y_m^3 - y_{m-1}^3 = (y_m - y_{m-1})(y_m^2 + y_m y_{m-1} + y_{m-1}^2)$$

will be useful for the proof.

## Problem 5-8
The successive substitutions technique for the hyperbolic form of Kepler's equation cannot be applied directly. (The reader should verify this by a numerical example.) However, the sequence

$$H_0 = 0$$
$$H_1 = \operatorname{arcsinh} \frac{N + H_0}{e}$$
$$\ldots$$
$$H_{k+1} = \operatorname{arcsinh} \frac{N + H_k}{e} \quad \text{etc.}$$

does converge to the solution of the hyperbolic form of Kepler's equation if $e > 1$. Further, to achieve a given accuracy tolerance $\epsilon$, it is sufficient that

$$k > \frac{\log \dfrac{Ne}{\epsilon(e-1)^2}}{\log e}$$

NOTE: It may be convenient for computation to recall that

$$\operatorname{arcsinh} x = \log(x + \sqrt{x^2 + 1})$$

## 5.2 Lagrange's Expansion Theorem

Lagrange's approach to solving Kepler's equation in 1770 led to a generally useful expansion theorem. Consider the functional equation (for which Kepler's equation is, clearly, a special case)

$$y = x + \alpha\phi(y) \tag{5.9}$$

where $\alpha$ is to be considered a small parameter—originally identified with a planetary eccentricity. Then $y$, as a function of $x$ and $\alpha$, may be expanded in a Taylor series about $\alpha = 0$:

$$y(x,\alpha) = y(x,0) + \alpha \left.\frac{\partial y}{\partial \alpha}\right|_0 + \frac{\alpha^2}{2!}\left.\frac{\partial^2 y}{\partial \alpha^2}\right|_0 + \cdots$$

Lagrange's contribution was to obtain an elegant expression for the coefficients of the powers of $\alpha$.

Clearly,
$$y(x,0) = x$$

and
$$\frac{\partial}{\partial \alpha} y(x,\alpha) = \phi(y) + \alpha \frac{d\phi}{dy}\frac{\partial y}{\partial \alpha}$$

so that
$$\left.\frac{\partial y}{\partial \alpha}\right|_{\alpha=0} = \phi[x + 0 \cdot \phi(y)] = \phi(x)$$

Calculating the second derivative, we obtain

$$\frac{\partial^2 y}{\partial \alpha^2} = \frac{d\phi}{dy}\frac{\partial y}{\partial \alpha} + \frac{d\alpha}{d\alpha}\frac{d\phi}{dy}\frac{\partial y}{\partial \alpha} + \alpha\left[\frac{d^2\phi}{dy^2}\left(\frac{\partial y}{\partial \alpha}\right)^2 + \frac{d\phi}{dy}\frac{\partial^2 y}{\partial \alpha^2}\right]$$

Hence
$$\left.\frac{\partial^2 y}{\partial \alpha^2}\right|_{\alpha=0} = 2\phi(y)\left.\frac{d\phi}{dy}\right|_{y=x} = 2\phi(x)\frac{d\phi}{dx} = \frac{d}{dx}\phi(x)^2$$

Again, for the third derivative

$$\frac{\partial^3 y}{\partial \alpha^3} = 3\frac{d\phi}{dy}\frac{\partial^2 y}{\partial \alpha^2} + 3\frac{d^2\phi}{dy^2}\left(\frac{\partial y}{\partial \alpha}\right)^2 + \alpha[\cdots]$$

and setting $\alpha = 0$, gives

$$\left.\frac{\partial^3 y}{\partial \alpha^3}\right|_{\alpha=0} = 3\frac{d\phi}{dx}\frac{d}{dx}\phi^2 + 3\frac{d^2\phi}{dx^2}\phi^2 = \frac{d^2}{dx^2}\phi(x)^3$$

The calculation of higher derivatives becomes more and more complex; but, as will be proved later in this section, the $n^{\text{th}}$ derivative is simply

$$\left.\frac{\partial^n y}{\partial \alpha^n}\right|_0 = \frac{d^{n-1}}{dx^{n-1}}\phi(x)^n \tag{5.10}$$

so that
$$y = x + \sum_{n=1}^{\infty} \frac{a^n}{n!} \frac{d^{n-1}}{dx^{n-1}} \phi(x)^n \tag{5.11}$$
is the power series expansion formula of Lagrange.

Applying this result to Kepler's equation, we immediately obtain
$$E = M + e\sin M + \frac{e^2}{2!}\frac{d}{dM}\sin^2 M + \frac{e^3}{3!}\frac{d^2}{dM^2}\sin^3 M + \cdots \tag{5.12}$$
The question of convergence will be addressed at the end of this section. For now, it is sufficient to say that when $e$ is small the series is rapidly convergent.

◊ **Problem 5–9**

Using the notation of Prob. 5–7, the solution of Barker's equation can be expressed as the power series
$$y = x + \sum_{n=1}^{\infty} (-1)^n \frac{(3n)!}{3^n n!\,(2n+1)!} x^{2n+1}$$
Determine the range of values of $x = 2\sqrt{\mu/p^3}(t - \tau)$ for which the series will converge as well as the values of the true anomaly for which the series is valid.

**Problem 5–10**

By writing the hyperbolic form of Kepler's equation as
$$\sinh H = \frac{N + H}{e}$$
derive an expansion for $\sinh H$ in inverse powers of $e$. Then show that
$$H = A + \frac{A}{B} + \frac{A}{2B^2}\left(2 - \frac{AN}{B}\right) + \frac{A}{6B^3}\left(6 - A^2 - \frac{9AN}{B} + \frac{3A^2 N^2}{B^2}\right) + \cdots$$
where
$$A = \operatorname{arcsinh}\frac{N}{e} = \log\frac{N+B}{e} \quad \text{and} \quad B = \sqrt{N^2 + e^2}$$

### Euler's Trigonometric Series

In 1760 Leonhard Euler published some important trigonometric series which we can put to effective use. Using Euler's identity
$$x = e^{i\phi} = \cos\phi + i\sin\phi \quad \text{where} \quad i = \sqrt{-1}$$
it is apparent that
$$\cos\phi = \frac{1}{2}\left(x + \frac{1}{x}\right) \qquad \sin\phi = \frac{1}{2i}\left(x - \frac{1}{x}\right)$$
$$\cos n\phi = \frac{1}{2}\left(x^n + \frac{1}{x^n}\right) \qquad \sin n\phi = \frac{1}{2i}\left(x^n - \frac{1}{x^n}\right)$$

Then, we can use the binomial theorem

$$(a+b)^n = \sum_{k=0}^{n} \binom{n}{k} a^{n-k} b^k \quad \text{where} \quad \binom{n}{k} = \frac{n!}{k!(n-k)!}$$

to expand, for $m = 0, 1, 2, \ldots$, the function

$$\sin^{2m+2}\phi = \frac{(-1)^{m+1}}{2^{2m+2}} \left(x - \frac{1}{x}\right)^{2m+2}$$

Indeed, with very little difficulty, we obtain

$$\sin^{2m+2}\phi = \frac{(-1)^{m+1}}{2^{2m+1}} \sum_{k=0}^{m} (-1)^k \binom{2m+2}{k} \cos(2m+2-2k)\phi$$

$$+ \frac{1}{2^{2m+2}} \binom{2m+2}{m+1} \quad (5.13)$$

In a similar manner, we can produce the corresponding expression for the odd powers of $\sin\phi$. We have

$$\sin^{2m+1}\phi = \frac{(-1)^m}{2^{2m}} \sum_{k=0}^{m} (-1)^k \binom{2m+1}{k} \sin(2m+1-2k)\phi \quad (5.14)$$

Next, we differentiate Eq. (5.13) $2m+1$ times and Eq. (5.14) $2m$ times with the result that

$$\frac{d^{2m+1}}{d\phi^{2m+1}} \sin^{2m+2}\phi =$$

$$\sum_{k=0}^{m} (-1)^k \left(\frac{2m+2-2k}{2}\right)^{2m+1} \binom{2m+2}{k} \sin(2m+2-2k)\phi$$

$$\frac{d^{2m}}{d\phi^{2m}} \sin^{2m+1}\phi =$$

$$\sum_{k=0}^{m} (-1)^k \left(\frac{2m+1-2k}{2}\right)^{2m} \binom{2m+1}{k} \sin(2m+1-2k)\phi$$

If, in the first equation, we replace $2m+1$ by $n$ and, in the second equation, replace $2m$ by $n$, then the two equations are identical. The combined result may be expressed as

$$\frac{d^n}{d\phi^n} \sin^{n+1}\phi = \sum_{k=0}^{[\frac{1}{2}n]} (-1)^k \frac{n+1-2k}{2}^n \binom{n+1}{k} \sin(n+1-2k)\phi \quad (5.15)$$

where the notation $[m]$ indicates the greatest integer contained in $m$. Thus

$$[\tfrac{1}{2}n] = \begin{cases} \frac{1}{2}(n-1) & n \text{ odd} \\ \frac{1}{2}n & n \text{ even} \end{cases}$$

By substituting the expression (5.15) in Eq. (5.12), we obtain

$$E = M + \sum_{n=1}^{\infty} e^n \sum_{k=0}^{[\frac{1}{2}n-\frac{1}{2}]} (-1)^k \frac{(\frac{1}{2}n-k)^{n-1}}{k!(n-k)!} \sin(n-2k)M$$

Then, by regrouping the terms, the Lagrange series takes the form

$$E = M + \sum_{m=1}^{\infty} \frac{2}{m} \sin mM \sum_{k=0}^{\infty} \frac{(-1)^k(\frac{1}{2}me)^{m+2k}}{k!(m+k)!} \qquad (5.16)$$

In particular, if terms of order $e^7$ and higher are neglected, we have

$$E = M + (e - \tfrac{1}{8}e^3 + \tfrac{1}{192}e^5)\sin M + (\tfrac{1}{2}e^2 - \tfrac{1}{6}e^4 + \tfrac{1}{48}e^6)\sin 2M$$
$$+ (\tfrac{3}{8}e^3 - \tfrac{27}{128}e^5)\sin 3M + (\tfrac{1}{3}e^4 - \tfrac{4}{15}e^6)\sin 4M$$
$$+ \tfrac{125}{384}e^5 \sin 5M + \tfrac{27}{80}e^6 \sin 6M + O(e^7) \quad (5.17)$$

◇ **Problem 5–11**
Obtain the following expansions for the powers of $\cos\phi$:

$$\cos^{2m+2}\phi = \frac{1}{2^{2m+1}} \sum_{k=0}^{m} \binom{2m+2}{k} \cos(2m+2-2k)\phi + \frac{1}{2^{2m+2}} \binom{2m+2}{m+1}$$

$$\cos^{2m+1}\phi = \frac{1}{2^{2m}} \sum_{k=0}^{m} \binom{2m+1}{k} \cos(2m+1-2k)\phi$$

with $m = 0, 1, 2, \ldots$

### Generalized Expansion Theorem

Lagrange generalized his expansion theorem so that any function $F(y)$, rather than simply $y$, can be expanded as a power series in $\alpha$. With $y$ defined in Eq. (5.9), then

$$F(y) = F(x) + \sum_{n=1}^{\infty} \frac{\alpha^n}{n!} \frac{d^{n-1}}{dx^{n-1}} \left[\phi(x)^n \frac{dF(x)}{dx}\right] \qquad (5.18)$$

will follow immediately if we can prove that

$$\frac{\partial^n F}{\partial \alpha^n} = \frac{\partial^{n-1}}{\partial x^{n-1}}\left[\phi(x)^n \frac{\partial F(x)}{\partial x}\right] \qquad (5.19)$$

Our strategy will be to establish Eq. (5.19) by mathematical induction. Thus, for $n = 1$, we must prove that

$$\frac{\partial F}{\partial \alpha} = \phi \frac{\partial F}{\partial x} \qquad (5.20)$$

For this purpose, we first calculate

$$\frac{\partial y(x,\alpha)}{\partial \alpha} = \phi(y) + \alpha \frac{d\phi}{dy}\frac{\partial y}{\partial \alpha}$$

$$\frac{\partial y(x,\alpha)}{\partial x} = 1 + \alpha \frac{d\phi}{dy}\frac{\partial y}{\partial x}$$

and then, from the first equation subtract $\phi(y)$ times the second. Thus, we obtain

$$\left(1 - \alpha\frac{d\phi}{dy}\right)\left(\frac{\partial y}{\partial \alpha} - \phi\frac{\partial y}{\partial x}\right) = 0$$

and since the first factor can vanish only if $\alpha\phi = y + $ constant, then we must have

$$\frac{\partial y}{\partial \alpha} = \phi\frac{\partial y}{\partial x}$$

Equation (5.20) follows immediately from this result since

$$\frac{\partial F}{\partial \alpha} = \frac{dF}{dy}\frac{\partial y}{\partial \alpha} \quad \text{and} \quad \frac{\partial F}{\partial x} = \frac{dF}{dy}\frac{\partial y}{\partial x}$$

To complete the proof, we assume that Eq. (5.19) is true for $n$ and show that is also true for $n+1$. Differentiating with respect to $\alpha$, we have

$$\frac{\partial^{n+1} F}{\partial \alpha^{n+1}} = \frac{\partial^{n-1}}{\partial x^{n-1}}\left[\frac{\partial}{\partial \alpha}\left(\phi^n \frac{\partial F}{\partial x}\right)\right] = \frac{\partial^{n-1}}{\partial x^{n-1}}\left[\frac{\partial \phi^n}{\partial \alpha}\frac{\partial F}{\partial x} + \phi^n \frac{\partial}{\partial \alpha}\left(\frac{\partial F}{\partial x}\right)\right]$$

But Eq. (5.20) is true for all functions $F(y)$; in particular, then it is true for $\phi^n(y)$. Hence

$$\frac{\partial^{n+1} F}{\partial \alpha^{n+1}} = \frac{\partial^{n-1}}{\partial x^{n-1}}\left[\phi\frac{\partial \phi^n}{\partial x}\frac{\partial F}{\partial x} + \phi^n \frac{\partial}{\partial x}\left(\phi\frac{\partial F}{\partial x}\right)\right] = \frac{\partial^n}{\partial x^n}\left[\phi^{n+1}\frac{\partial F}{\partial x}\right]$$

so that Eq. (5.19) is, indeed, true for $n+1$. Therefore, by the principle of mathematical induction, the assertion is true for all $n$.

◊ **Problem 5–12**

Use the generalized expansion theorem to obtain

$$\frac{r}{a} = 1 + \frac{1}{2}e^2 - \left(e - \frac{3}{8}e^3 + \frac{5}{192}e^5\right)\cos M - \left(\frac{1}{2}e^2 - \frac{1}{3}e^4 + \frac{1}{16}e^6\right)\cos 2M$$
$$- \left(\frac{3}{8}e^3 - \frac{45}{128}e^5\right)\cos 3M - \left(\frac{1}{3}e^4 - \frac{2}{5}e^6\right)\cos 4M$$
$$- \frac{125}{384}e^5 \cos 5M - \frac{27}{80}e^6 \cos 6M + O(e^7)$$

## Convergence of the Lagrange Series

Lagrange's series is, of course, the Taylor series representation of the root of the functional equation

$$y - x - \alpha \phi(y) = 0$$

which, until now, we have assumed to be unique. Sufficient conditions for a unique root are obtained by a direct application of Rouché's theorem for analytic functions of a complex variable.†

The French mathematician Eugène Rouché (1832–1910) proved in 1862 that if $f_1(z)$ and $f_2(z)$ are analytic functions of the complex variable $z$ throughout a singly-connected bounded region of the complex $z$-plane and on its boundary $C$, and if

$$|f_2(z)| < |f_1(z)| \neq 0$$

on $C$, then $f_1(z)$ and $f_1(z) + f_2(z)$ have the same number of zeros inside the contour.

In our case, consider $y$ to be a complex variable, $x$ a point in the complex plane within the boundary $C$, and $\phi(y)$ an analytic function. The two functions of Rouché's theorem are selected as

$$f_1(y) = y - x \quad \text{and} \quad f_2(y) = -\alpha \phi(y)$$

Clearly, $f_1(y)$ has only one zero, namely, $y = x$. Hence,

$$f_1(y) + f_2(y) = y - x - \alpha \phi(y)$$

will have only one root provided that

$$|\alpha \phi(y)| < |y - x| \tag{5.21}$$

is satisfied at all points $y$ on the contour $C$. Within this contour the Taylor series representation of the root is known to converge.

We can readily apply the criterion to the expansion of $E$ as the power series in $e$ expressed in Eq. (5.12). Consider $E$ to be a complex variable and let $C$ be a circular contour of radius $\rho$ centered at the point $M$ on the real axis. Along the perimeter of the contour the values of $E$ are given by

$$E = M + \rho \cos \theta + i \rho \sin \theta$$

---

† Augustine-Louis Cauchy, whom we have already encountered in connection with convergence criteria, was a man of universal interests. In mathematics, he wrote over 700 papers—second only to Euler in number. For twenty-five years, from 1824, Cauchy developed complex function theory almost singlehandedly and many of the significant results in this field bear his name. The fundamental theorem of the subject, that the integral of an analytic function around a closed contour in the complex plane is zero, is called *Cauchy's integral theorem* even though Gauss first stated that proposition, albeit without proof, in a letter to Bessel in 1811.

and the problem is to determine the largest value of $e$ for which

$$|e \sin E| < |E - M| = \rho$$

will hold true at each point of $C$.

Now, since

$$\sin E = \sin(M + \rho \cos \theta) \cosh(\rho \sin \theta) + i \cos(M + \rho \cos \theta) \sinh(\rho \sin \theta)$$

we have

$$|\sin E|^2 = \cosh^2(\rho \sin \theta) - \cos^2(M + \rho \cos \theta)$$

The maximum value of the magnitude of $\sin E$ occurs when

$$\cos(M + \rho \cos \theta) = 0 \quad \text{and} \quad \sin \theta = \pm 1$$

or, equivalently, when

$$\theta = \pm \tfrac{1}{2}\pi \quad \text{and} \quad M = \pm \tfrac{1}{2}\pi$$

Thus, the largest value of $|\sin E|$ is $\cosh \rho$ so that the convergence criterion is

$$e < \frac{\rho}{\cosh \rho}$$

As yet the radius of the contour $C$ has been unspecified and is at our disposal. Clearly, the best choice of $\rho$ is that which maximizes $\rho/\cosh \rho$ and it is easy to see that this desired value is the solution of

$$\cosh \rho = \rho \sinh \rho$$

Hence, solving numerically for $\rho$ (using, for example, Newton's method described in Sect. 5.4), we find that

$$e < 0.6627434194\ldots$$

is the requirement that the Lagrange series for $E$ represents the unique root of Kepler's equation for all values of the mean anomaly $M$. Laplace was the first to show that if $e$ exceeds this critical value, then the series will diverge for some values of $M$.

## 5.3 Fourier-Bessel Series Expansion

The Lagrange expansion of the solution of Kepler's equation gives the eccentric anomaly as a power series in the eccentricity. As such the series was found to converge for all values of the mean anomaly only if $e$ was less than approximately $\frac{2}{3}$. In the previous section, we saw that the series can be reordered as a Fourier† sine series with coefficients which are infinite power series in the eccentricity. A remarkable fact emerges in this analysis. The coefficients are absolutely convergent for all values of $e$ and the periodic series in $M$ satisfies the necessary conditions to be a convergent Fourier expansion. Thus, *the convergence properties of the Lagrange expansion are altered when the series elements are reordered.*

The coefficients of the Fourier sine series are Bessel functions. In this section we shall obtain directly the Fourier expansion of the solution of Kepler's equation. The Fourier coefficients are the integral form of the Bessel functions—the form first obtained by Bessel.

### Series Expansion of the Eccentric Anomaly

To obtain the Fourier expansion we observe that

$$\frac{dE}{dM} = \frac{1}{1 - e \cos E}$$

is an even periodic function of $M$ with period $2\pi$ so that we may write

$$\frac{1}{1 - e \cos E} = A_0 + \sum_{k=1}^{\infty} A_k \cos kM$$

where $A_0, A_1, A_2, \ldots$ are the *Fourier coefficients*. Thus,

$$A_0 = \frac{1}{2\pi} \int_{-\pi}^{\pi} \frac{dM}{1 - e \cos E} = \frac{1}{2\pi} \int_{-\pi}^{\pi} dE = 1$$

$$A_k = \frac{1}{\pi} \int_{-\pi}^{\pi} \frac{\cos kM}{1 - e \cos E} dM = \frac{1}{\pi} \int_{-\pi}^{\pi} \cos kM \, dE$$

---

† Joseph Fourier (1768–1830) was chiefly concerned with heat flow problems and his book *Théorie analytique de la chaleur*, published in 1822, is one of the classics of mathematics. By solving the partial differential equation of heat flow using the method of separation of variables, he was led to the development of the *Fourier series*. Leonhard Euler and Alexis-Claude Clairaut (1713–1765) had already expanded some functions in such series and had obtained the general integral representations of the coefficients, but Fourier made the remarkable observation that *every* function could be so represented even if it were not periodic or even continuous. This possibility had been rejected by the eighteenth-century masters with the single exception of Daniel Bernoulli, and Fourier's work was not well received by the Academy of Sciences in Paris. In spite of their objections, Fourier persisted and was responsible for initiating a broadening of the concept of "function." No longer would mathematics be restricted to analytic functions or functions with Taylor series representations.

## Sect. 5.3]  Fourier-Bessel Series Expansion

To evaluate the last integral, we write $\cos kM$ as the real part of $\exp(ikM)$. Then

$$\exp(ikM) = \exp[ik(E - e\sin E)] = \exp(ikE)\exp\left[\frac{\nu}{2}\left(x - \frac{1}{x}\right)\right]$$

where

$$\nu = -ke \quad \text{and} \quad x = \exp(iE)$$

The second exponential factor has an infinite Laurent expansion† in $x$ whose coefficients are power series in $\nu$. Thus,

$$\exp\left[\frac{\nu}{2}\left(x - \frac{1}{x}\right)\right] = \exp\left(\frac{\nu x}{2}\right)\exp\left(-\frac{\nu}{2x}\right) = \sum_{\ell=0}^{\infty}\sum_{j=0}^{\infty}(-1)^j \frac{(\frac{1}{2}\nu)^{\ell+j}}{\ell!\,j!}x^{\ell-j}$$

$$= \sum_{n=-\infty}^{\infty} J_n(\nu)x^n$$

where the coefficients

$$J_n(\nu) = \sum_{j=0}^{\infty}(-1)^j \frac{(\frac{1}{2}\nu)^{n+2j}}{j!\,(n+j)!} \tag{5.22}$$

are *Bessel functions of the first kind of order $n$*.

It follows that

$$\exp(ikM) = \sum_{n=-\infty}^{\infty} J_n(\nu)\exp[i(n+k)E]$$

so that the Fourier coefficient $A_k$ is

$$A_k = \frac{1}{\pi}\sum_{n=-\infty}^{\infty} J_n(-ke)\int_{-\pi}^{\pi}\cos(n+k)E\,dE$$

The integrals vanish for all values of $n$ except for $n = -k$. Hence, we have

$$A_k = 2J_{-k}(-ke)$$

or, alternately,

$$A_k = 2J_k(ke)$$

according to one of the identities involving Bessel functions considered in the next subsection.

---

† For two decades Cauchy worked singlehandedly developing complex function theory. Then, in 1843, Pierre-Alphonse Laurent (1813–1854) published a major result in the *Journal de l'Ecole Polytechnique* which was an extension of the Taylor series expansion. He showed that if a function has a singularity at a point, then the function may be expanded in a series about that point which includes decreasing as well as increasing powers of the variable.

Finally, therefore, by integrating the differential form of Kepler's equation with the right-hand side expressed as a cosine series, we have

$$E = M + 2\sum_{k=1}^{\infty} \frac{1}{k} J_k(ke) \sin kM \qquad (5.23)$$

as the desired result. This expansion is identical with the one obtained in Eq. (5.16) using the Lagrange expansion theorem and reordering the terms.

◇ **Problem 5–13**
Derive the Fourier cosine expansion

$$\frac{a}{r} = 1 + 2\sum_{k=1}^{\infty} J_k(ke) \cos kM$$

## Bessel Functions

The function $J(x, \nu)$, where

$$J(x, \nu) = \exp\left[\frac{\nu}{2}\left(x - \frac{1}{x}\right)\right] = \sum_{n=-\infty}^{\infty} J_n(\nu) x^n \qquad (5.24)$$

is known as the *generating function* for the Bessel functions $J_n(\nu)$ and can be used to develop many of the properties of these functions. For example, by writing $-1/x$ for $x$ in Eq. (5.24), we can show that

$$J_{-n}(\nu) = (-1)^n J_n(\nu) \qquad (5.25)$$

In a similar manner, we also obtain

$$J_{-n}(-\nu) = J_n(\nu) \qquad (5.26)$$

Differentiating the generating function with respect to $x$ yields

$$\frac{\partial J}{\partial x} = \frac{\nu}{2}\left(1 + \frac{1}{x^2}\right) J$$

Then, by substituting the series for $J$ in this equation and equating coefficients of like powers of $x$, we can derive the recurrence formula

$$n J_n(\nu) = \tfrac{1}{2}\nu[J_{n+1}(\nu) + J_{n-1}(\nu)] \qquad (5.27)$$

Likewise, the derivative of the generating function with respect to $\nu$ gives

$$\frac{\partial J}{\partial \nu} = \frac{1}{2}\left(x - \frac{1}{x}\right) J$$

from which we obtain

$$J'_n(\nu) = \tfrac{1}{2}[J_{n-1}(\nu) - J_{n+1}(\nu)] \qquad (5.28)$$

as a formula for calculating the derivative of the Bessel functions.

## Sect. 5.3] Fourier-Bessel Series Expansion

We have already seen earlier in this section that the integral form of the Bessel function can be written as

$$J_k(\nu) = \frac{1}{\pi} \int_0^\pi \cos(kx - \nu \sin x)\, dx \tag{5.29}$$

from which follows the inequality

$$|J_n(\nu)| \leq 1 \tag{5.30}$$

Other properties and applications of Bessel functions are developed in the problems which follow.

◇ **Problem 5-14**
Bessel's differential equation

$$\nu^2 \frac{d^2 y}{d\nu^2} + \nu \frac{dy}{d\nu} + (\nu^2 - n^2) y = 0$$

is satisfied by $y = J_n(\nu)$.

◇ **Problem 5-15**
Derive the expression

$$\frac{d}{d\nu}[\nu^n J_n(\nu)] = \nu^n J_{n-1}(\nu)$$

◇ **Problem 5-16**
Derive the following two identities for the derivative of the Bessel function

$$\nu J'_n(\nu) = n J_n(\nu) - \nu J_{n+1}(\nu)$$
$$= -n J_n(\nu) + \nu J_{n-1}(\nu)$$

which are analogous to Eq. (5.28).

◇ **Problem 5-17**
The Bessel function $J_n(x)$ can be expressed as

$$J_n(x) = \frac{(\frac{1}{2}x)^2}{n!} N(n+1, -\tfrac{1}{4} x^2)$$

where the function $N(\gamma, x)$ is defined in Prob. 1-5. Derive the continued fraction expansion

$$\frac{J_n(x)}{J_{n-1}(x)} = \cfrac{\frac{1}{2}x}{n - \cfrac{(\frac{1}{2}x)^2}{n+1 - \cfrac{(\frac{1}{2}x)^2}{n+2 - \cfrac{(\frac{1}{2}x)^2}{n+3 - \cfrac{(\frac{1}{2}x)^2}{n+4 - \cdots}}}}}$$

◇ **Problem 5–18**
Derive the expansions

$$\cos E = -\frac{e}{2} + \sum_{k=1}^{\infty} \frac{1}{k}[J_{k-1}(ke) - J_{k+1}(ke)]\cos kM$$

$$= -\frac{e}{2} + \sum_{k=1}^{\infty} \frac{2}{k^2}\frac{dJ_k(ke)}{de}\cos kM$$

$$\sin E = \sum_{k=1}^{\infty} \frac{1}{k}[J_{k+1}(ke) + J_{k-1}(ke)]\sin kM$$

$$= \frac{2}{e}\sum_{k=1}^{\infty} \frac{1}{k}J_k(ke)\sin kM$$

$$\frac{r}{a} = 1 + \frac{e^2}{2} - 2e\sum_{k=1}^{\infty} \frac{1}{k^2}\frac{dJ_k(ke)}{de}\cos kM$$

NOTE: Compare the expansion of $r/a$ with the result of Prob. 5–12.

**Problem 5–19**
Obtain the Fourier-Bessel expansions of $\cos f$ and $\sin f$ in terms of the mean anomaly

$$\cos f = -e + \frac{2(1-e^2)}{e}\sum_{k=1}^{\infty} J_k(ke)\cos kM$$

$$\sin f = 2\sqrt{1-e^2}\sum_{k=1}^{\infty} \frac{1}{k}\frac{dJ_k(ke)}{de}\sin kM$$

directly from the expansions for $a/r$ and $r/a$ given in Probs. 5–13 and 5–18.

## Series Expansion of the True Anomaly

The true anomaly $f$, as a Fourier expansion in terms of the mean anomaly $M$, is similar to Eq. (5.23) but the coefficients are considerably more complex. The derivation parallels the previous development for the eccentric anomaly.

It is readily seen that

$$\frac{df}{dM} = \frac{\sqrt{1-e^2}}{(1-e\cos E)^2}$$

and is an even periodic function of $M$, so that

$$\frac{1}{(1-e\cos E)^2} = B_0 + \sum_{k=1}^{\infty} B_k \cos kM$$

## Sect. 5.3]  Fourier-Bessel Series Expansion

where
$$B_0 = \frac{1}{2\pi} \int_{-\pi}^{\pi} \frac{dM}{(1 - e\cos E)^2} = \frac{1}{2\pi} \int_{-\pi}^{\pi} \frac{dE}{1 - e\cos E}$$

$$= \frac{1}{2\pi} \int_{-\pi}^{\pi} \frac{df}{\sqrt{1 - e^2}} = \frac{1}{\sqrt{1 - e^2}}$$

$$B_k = \frac{1}{\pi} \int_{-\pi}^{\pi} \frac{\cos kM}{(1 - e\cos E)^2} dM = \frac{1}{\pi} \int_{-\pi}^{\pi} \frac{\cos kM}{1 - e\cos E} dE$$

$$= \frac{1}{\pi} \sum_{n=-\infty}^{\infty} J_n(-ke) \int_{-\pi}^{\pi} \frac{\cos(n+k)E}{1 - e\cos E} dE$$

To evaluate the integral portion of $B_k$ we make a complex change of variable
$$z = e^{iE} \qquad \cos E = \frac{1}{2}\left(z + \frac{1}{z}\right) \qquad dE = -\frac{i}{z} dz$$

so that
$$\int_{-\pi}^{\pi} \frac{\exp[i(n+k)E]}{1 - e\cos E} dE = \frac{2i}{e} \oint_C \frac{z^{n+k}}{(z-\alpha)(z-\beta)} dz$$

$$= \frac{2i}{e(\alpha - \beta)} \oint_C \left(\frac{1}{z - \alpha} - \frac{1}{z - \beta}\right) z^{n+k} dz$$

where
$$\alpha = \frac{1 + \sqrt{1 - e^2}}{e} \qquad \text{and} \qquad \beta = \frac{1 - \sqrt{1 - e^2}}{e}$$

The contour $C$ is the unit circle in the complex $z$-plane.

The integrals are evaluated using the *Cauchy residue theorem*. Since $\alpha\beta = 1$, then $\alpha > 1$ and $\beta < 1$. Therefore,

$$\oint_C \frac{z^{n+k}}{z - \alpha} dz = -\oint_C z^{n+k} \left(\frac{1}{\alpha} + \frac{z}{\alpha^2} + \cdots\right) dz = \begin{cases} -2\pi i \alpha^{n+k} & n+k < 0 \\ 0 & n+k \geq 0 \end{cases}$$

$$\oint_C \frac{z^{n+k}}{z - \beta} dz = \oint_C z^{n+k} \left(\frac{1}{z} + \frac{\beta}{z^2} + \cdots\right) dz = \begin{cases} 0 & n+k < 0 \\ 2\pi i \beta^{n+k} & n+k \geq 0 \end{cases}$$

Now, since $\alpha = 1/\beta$, we have
$$\int_{-\pi}^{\pi} \frac{\cos(k+n)E}{1 - e\cos E} dE = \frac{2\pi}{\sqrt{1 - e^2}} \beta^{|k+n|}$$

so that
$$B_k = \frac{2}{\sqrt{1 - e^2}} \sum_{n=-\infty}^{\infty} J_n(-ke) \beta^{|k+n|}$$

The desired expression for $f$

$$f = M + 2\sum_{k=1}^{\infty} \frac{1}{k}\left[\sum_{n=-\infty}^{\infty} J_n(-ke)\beta^{|k+n|}\right]\sin kM \qquad (5.31)$$

where

$$\beta = \frac{1-\sqrt{1-e^2}}{e} = \tan\tfrac{1}{2}\phi \qquad (5.32)$$

is obtained by integrating the original cosine series expansion. (The second equation for $\beta$ is from Prob. 4–7.)

To the astronomer $f - M$ is the *equation of the center* and plays a fundamental role in astronomical time computations.

◇ **Problem 5–20**

The quantity $\beta$, defined in Eq. (5.32), satisfies

$$\beta = \tfrac{1}{2}e + \tfrac{1}{2}e\beta^2$$

Use Lagrange's generalized expansion theorem to obtain

$$\beta^m = \left(\frac{e}{4}\right)^m\left[1 + \frac{m}{4}e^2 + \frac{m(m+3)}{4^2 2!}e^4 + \frac{m(m+4)(m+5)}{4^3 3!}e^6 + \cdots\right]$$

Since the functional equation for $\beta$ is a quadratic, there are two different roots. Which of the two roots does this expansion represent and for what values of $e$ does the series converge?

◇ **Problem 5–21**

As far as terms of order $e^6$, the equation of the center is

$$f - M = (2e - \tfrac{1}{4}e^3 + \tfrac{5}{96}e^5)\sin M + (\tfrac{5}{4}e^2 - \tfrac{11}{24}e^4 + \tfrac{17}{192}e^6)\sin 2M$$
$$+ (\tfrac{13}{12}e^3 - \tfrac{43}{64}e^5)\sin 3M + (\tfrac{103}{96}e^4 - \tfrac{451}{480}e^6)\sin 4M$$
$$+ \tfrac{1097}{960}e^5\sin 5M + \tfrac{1223}{960}e^6\sin 6M + O(e^7)$$

## 5.4 Series Reversion and Newton's Method

Formulas for reversing a power series are to be found in Appendix C. Expressions for the coefficients in that appendix are obtained by the formal process of substituting one series into another and equating terms of like powers of the variable. In this section, we take a different approach using the Lagrange expansion theorem developed earlier in this chapter.

Let $y(x)$ be a function which can be expanded in a Taylor series in the neighborhood of $x = x_0$. Thus

$$y(x) = y_0 + b_1(x - x_0) + \frac{b_2}{2!}(x - x_0)^2 + \frac{b_3}{3!}(x - x_0)^3 + \cdots \qquad (5.33)$$

where
$$b_n = \left.\frac{d^n y(x)}{dx^n}\right|_{x=x_0}$$

In the following, we assume that $b_1$ is different from zero and write Eq. (5.33) in the form
$$x = x_0 + (y - y_0)\phi(x) \tag{5.34}$$
where $\phi(x)$ is defined by
$$\phi(x) = \frac{1}{b_1 + \frac{b_2}{2!}(x - x_0) + \frac{b_3}{3!}(x - x_0)^2 + \cdots} \tag{5.35}$$

Equation (5.34) is precisely the same form as Eq. (5.9). Therefore, we may use the Lagrange expansion theorem to express $x$ as a power series in $\alpha = y - y_0$. Thus,
$$x(y) = x_0 + c_1(y - y_0) + \frac{c_2}{2!}(y - y_0)^2 + \frac{c_3}{3!}(y - y_0)^3 + \cdots \tag{5.36}$$
with the coefficients determined from
$$c_n = \left.\frac{d^{n-1}}{dx^{n-1}}\phi(x)^n\right|_{x=x_0} \tag{5.37}$$
where $\phi(x)$ is defined in Eq. (5.35). The series for $x(y)$ is said to be the *reverse* of the series for $y(x)$ and is sometimes called *Bürmann's series* after Heinrich Bürmann who published the result in 1799.

### Series Reversion Algorithm

Here we develop an original algorithm which will permit us to express the coefficients $c_1, c_2, \ldots$ of the reversed series in terms of the coefficients $b_1, b_2, \ldots$ of the original series. For this purpose, it is expedient to utilize *Leibnitz's rule* for differentiating products†
$$\frac{d^n(uv)}{dx^n} = \sum_{k=0}^{n} \binom{n}{k} \frac{d^k u}{dx^k} \frac{d^{n-k} v}{dx^{n-k}}$$

---

† Though his contributions were quite different, Gottfried Wilhelm Leibnitz (1646–1716) ranks with Newton as a builder of the calculus. Leibnitz was a philosopher, lawyer, diplomat, historian, philologist, and geologist. His interests included logic, mechanics, optics, mathematics, hydrostatics, pneumatics, nautical science, and calculating machines. He first gave the rules for differentiating sums, products, and quotients; indeed, most the manipulations taught in freshman calculus are his. In a manuscript, dated Nov. 1, 1675, he had difficulty with $d(uv)$ and $d(u/v)$ and, at first, thought that $d(uv) = du\, dv$—not unlike some students today. He spent considerable effort in devising an appropriate notation and it is interesting to remember that $dx$, $dy$, and $dy/dx$ are his original symbols.

First, we define
$$D_k^n = \frac{d^k}{dx^k}\phi(x)^n \quad \text{for} \quad k = 0, 1, \ldots, n-1 \quad (5.38)$$
and note that
$$D_k^n = \frac{d^j}{dx^j} D_{k-j}^n$$
Then, since $D_0^n = \phi^n$, we have
$$D_1^n = n\phi^{n-1}\frac{d\phi}{dx} = nD_0^{n-1}\frac{d\phi}{dx}$$
from which follows
$$D_k^n = \frac{d^{k-1}}{dx^{k-1}} D_1^n = n\frac{d^{k-1}}{dx^{k-1}}\left(D_0^{n-1}\frac{d\phi}{dx}\right)$$
Applying Leibnitz's rule, we obtain
$$D_k^n = n\sum_{j=0}^{k-1} \binom{k-1}{j} D_j^{n-1} \phi_0^{(k-j)} \quad \text{for} \quad k = 1, 2, \ldots, n-1 \quad (5.39)$$
where, for convenience, we adopt the notation
$$\phi^{(k)}(x) = \frac{d^k \phi(x)}{dx^k}$$
The coefficients in the reversed series are obtained from
$$c_n = D_{n-1}^n\big|_{x=x_0} \quad (5.40)$$
which can be generated recursively from Eq. (5.39) in terms of the quantities $\phi^{(k)}(x_0) \equiv \phi_0^{(k)}$.

There remains the problem of determining $\phi_0^{(k)}$ in terms of the coefficients $b_1, b_2, \ldots$ To this end, we apply Leibnitz's rule to Eq. (5.35) written in the form $\phi(x)[\cdots] = 1$, noting that the $k^{\text{th}}$ derivative of the bracketed quantity evaluated at $x = x_0$ is simply $b_{k+1}/(k+1)$. Therefore,
$$\sum_{i=0}^{k} \frac{1}{i+1} \binom{k}{i} b_{i+1} \phi_0^{(k-i)} = 0$$
Solving for $\phi_0^{(k)}$ gives
$$\phi_0^{(k)} = -\frac{1}{b_1} \sum_{i=1}^{k} \frac{1}{i+1} \binom{k}{i} b_{i+1} \phi_0^{(k-i)} \quad (5.41)$$
as a formula for generating $\phi_0^{(k)}$ recursively. Repeated evaluations of the equations (5.39) and (5.41), starting with the initial values
$$D_0^1\big|_{x=x_0} = \phi_0^{(0)} = \frac{1}{b_1}$$

**Sect. 5.4]**  Series Reversion and Newton's Method

will produce as many of the coefficients $c_1$, $c_2$, ... as may be desired.
Finally, it is useful to remark that since

$$b_n = \left.\frac{d^n y(x)}{dx^n}\right|_{x=x_0} \quad \text{and} \quad c_n = \left.\frac{d^n x(y)}{dy^n}\right|_{y=y_0}$$

as seen from Eqs. (5.33) and (5.36), the algorithm also provides a convenient method for determining the derivatives of $x$ with respect to $y$ explicitly in terms of the derivatives of $y$ with respect to $x$.

### ◊ Problem 5-22

Use the algorithm, consisting of Eqs. (5.39) and (5.41), to obtain

$$c_1 = \frac{1}{b_1} \qquad c_2 = -\frac{b_2}{b_1^3}$$

$$c_3 = -\frac{b_3}{b_1^4} + 3\frac{b_2^2}{b_1^5} \qquad c_4 = -\frac{b_4}{b_1^5} + 10\frac{b_2 b_3}{b_1^6} - 15\frac{b_2^3}{b_1^7}$$

and compare with the formulas given in Appendix C.

NOTE: This algorithm was mechanized using *MACSYMA*, a symbol manipulating digital computer program, which produced formulas for the following additional coefficients:

$$c_5 = -\frac{b_5}{b_1^6} + \frac{15 b_2 b_4 + 10 b_3^2}{b_1^7} - 105\frac{b_2^2 b_3}{b_1^8} + 105\frac{b_2^4}{b_1^9}$$

$$c_6 = -\frac{b_6}{b_1^7} + \frac{21 b_2 b_5 + 35 b_3 b_4}{b_1^8} - \frac{210 b_2^2 b_4 + 280 b_2 b_3^2}{b_1^9} + 1{,}260\frac{b_2^3 b_3}{b_1^{10}} - 945\frac{b_2^5}{b_1^{11}}$$

$$c_7 = -\frac{b_7}{b_1^8} + \frac{28 b_2 b_6 + 56 b_3 b_5 + 35 b_4^2}{b_1^9} - \frac{378 b_2^2 b_5 + 1{,}260 b_2 b_3 b_4 + 280 b_3^3}{b_1^{10}}$$
$$+ \frac{3{,}150 b_2^3 b_4 + 6{,}300 b_2^2 b_3^2}{b_1^{11}} - 17{,}325\frac{b_2^4 b_3}{b_1^{12}} + 10{,}395\frac{b_2^6}{b_1^{13}}$$

$$c_8 = -\frac{b_8}{b_1^9} + \frac{36 b_2 b_7 + 84 b_3 b_6 + 126 b_4 b_5}{b_1^{10}} - \frac{630 b_2^2 b_6 + 2{,}520 b_2 b_3 b_5}{b_1^{11}}$$
$$- \frac{1{,}575 b_2 b_4^2 + 2{,}100 b_3^2 b_4}{b_1^{11}} + \frac{6{,}930 b_2^3 b_5 + 34{,}650 b_2^2 b_3 b_4 + 15{,}400 b_2 b_3^3}{b_1^{12}}$$
$$- \frac{51{,}975 b_2^4 b_4 + 138{,}600 b_2^3 b_3^2}{b_1^{13}} + 270{,}270\frac{b_2^5 b_3}{b_1^{14}} - 135{,}135\frac{b_2^7}{b_1^{15}}$$

Many more coefficients were obtained but these are all that will be recorded.

*John R. Spofford* 1983

◊ **Problem 5–23**

Kepler's equation, referenced to an arbitrary epoch, can be written as

$$M = M_0 + E - E_0 - e\cos E_0 \sin(E - E_0) - e\sin E_0[\cos(E - E_0) - 1]$$

By series reversion, develop the expansion

$$E = E_0 + \xi - \tfrac{1}{2}\eta\xi^2 + \tfrac{1}{6}(3\eta^2 - \varsigma)\xi^3 + \tfrac{1}{24}(1 + 10\varsigma - 15\eta^2)\eta\xi^4 + \cdots$$

where

$$\xi = \frac{M - (E_0 - e\sin E_0)}{1 - e\cos E_0} \qquad \eta = \frac{e\sin E_0}{1 - e\cos E_0} \qquad \varsigma = \frac{e\cos E_0}{1 - e\cos E_0}$$

and compare with Prob. 5–6.

### Newton's Method

Series reversion can be used to obtain an approximate root of the equation

$$y(x) = 0 \tag{5.42}$$

If $\xi$ is such a root, then Eq. (5.36), with $x = \xi$ and $y = 0$, gives

$$\xi = x_0 - c_1 y_0 + \frac{c_2}{2!}y_0^2 - \frac{c_3}{3!}y_0^3 + \cdots \tag{5.43}$$

Therefore, if $y_0$ is a relatively small quantity (or, equivalently, $x_0$ is reasonably close to $\xi$) and the coefficients $c_1, c_2, c_3, \ldots$ are "well behaved," then the first several terms of Eq. (5.43) will provide an approximation to the true value of the root $\xi$.

Of course, Eq. (5.43) can be used recursively in the form

$$x_{k+1} = x_k - c_1 y_k + \frac{c_2}{2!}y_k^2 - \frac{c_3}{3!}y_k^3 + \cdots \qquad \text{for} \qquad k = 0, 1, 2, \ldots \tag{5.44}$$

Whether or not the sequence $x_0, x_1, x_2, \ldots$ will converge to the root $\xi$ of $y(x) = 0$ depends both on the specific function $y(x)$ and the initial approximation $x_0$.

The first two terms of Eq. (5.44) written as

$$x_{k+1} = x_k - \frac{y_k}{y'_k} \tag{5.45}$$

is recognized as the familiar root-finding algorithm of Sir Isaac Newton.[†] In celestial mechanics, it is frequently called the *method of differential corrections*.

---

[†] In his *De Analysi* and *Method of Fluxions*, Newton gave a general method for approximating the roots of $y(x) = 0$ which was published in John Wallis's *Algebra* in 1685. Joseph Raphson (1648–1715) made improvements in the method which he applied only to polynomial equations. His method was published in *Analysis Aequationum Universalis* in 1690—it is this which we call Newton's method or the Newton-Raphson method.

## Sect. 5.4]    Series Reversion and Newton's Method

To assess the behavior of the error $x_k - \xi$ at the $k^{\text{th}}$ step in the iteration cycle, we use the *Taylor series expansion with remainder* in the form

$$0 = y(\xi) = y_k + y'_k(\xi - x_k) + \tfrac{1}{2} y''(a)(\xi - x_k)^2$$

where

$$y''(a) = \left.\frac{d^2 y(x)}{dx^2}\right|_{x=a} \quad \text{and} \quad \xi \le a \le x_k$$

Hence

$$\frac{y_k}{y'_k} = x_k - \xi - \frac{y''(a)}{2 y'_k}(x_k - \xi)^2$$

so that, from Eq. (5.45),

$$x_{k+1} - \xi = \frac{y''(a)}{2 y'_k}(x_k - \xi)^2 \tag{5.46}$$

Thus, if $x_k - \xi$ is small, then $a \approx x_k$ and Eq. (5.46) demonstrates that the error in the $(k+1)^{\text{th}}$ approximation is proportional to the square of the error in the $k^{\text{th}}$ approximation. The process is said to have *quadratic convergence* properties.

◊ **Problem 5-24**

Newton's iteration for solving Kepler's equation can be written as

$$E_{k+1} = E_k + \frac{M - M_k}{1 - e \cos E_k}$$

where

$$M_k = E_k - e \sin E_k$$

[An excellent choice for $E_0$ is provided by Eq. (5.4)].

Further, derive the inequality

$$|E_{k+1} - E| \le \frac{e}{2(1-e)} |E_k - E|^2$$

Also, derive the corresponding algorithm for the hyperbolic form of Kepler's equation as well as the inequality

$$|H_{k+1} - H| \le \frac{\sinh\left(\dfrac{N/e}{1 - 1/e}\right)}{2(1 - 1/e)} |H_k - H|^2$$

with $H_0$ obtained, for example, from Eq. (5.5).

NOTE: This method for solving Kepler's equation was recommended by Johann Franz Encke in *Berliner Astronomisches Jahrbuch* in 1848.

◇ **Problem 5–25**
Derive the algorithm

$$x_{k+1} = x_k - \frac{y_k}{y'_k}\left(1 + \frac{1}{2}\frac{y_k y''_k}{(y'_k)^2}\right)$$

Furthermore, obtain the expression

$$x_{k+1} - \xi = \frac{1}{2}\left[\left(\frac{y''_k}{y'_k}\right)^2 - \frac{y'''(a)}{3y'_k}\right](x_k - \xi)^3 + O(x_k - \xi)^4$$

which displays the *cubic convergence* properties of the algorithm.

Specialize this higher order method to provide the following iterative solution of Kepler's equation:

$$E_{k+1} = E_k + \frac{M - M_k}{1 - e\cos E_k} - \frac{e\sin E_k}{2(1 - e\cos E_k)}\left(\frac{M - M_k}{1 - e\cos E_k}\right)^2$$

NOTE: Compare this result with that of Prob. 5–23.

## Power Series for the Generalized Anomaly $\chi$

The universal form of Kepler's equation

$$\sqrt{\mu}(t - t_0) = r_0 U_1(\chi; \alpha) + \sigma_0 U_2(\chi; \alpha) + U_3(\chi; \alpha)$$

developed in Sect. 4.5, is simply a closed form representation of the power series

$$\sqrt{\mu}(t - t_0) = r_0\chi + \frac{\sigma_0}{2!}\chi^2 + \frac{1 - \alpha r_0}{3!}\chi^3 - \frac{\alpha\sigma_0}{4!}\chi^4$$
$$- \frac{\alpha(1 - \alpha r_0)}{5!}\chi^5 + \frac{\alpha^2\sigma_0}{6!}\chi^6 + \cdots \quad (5.47)$$

To reverse the series we can, of course, utilize Eq. (5.36) with the expressions for the coefficients given in Prob. 5–22.

On the other hand, we can make use of the explicit expression for $\chi$, given in Eq. (4.86),

$$\chi = \alpha\sqrt{\mu}(t - t_0) + \sigma - \sigma_0$$

and recall from Prob. 3–25 that the algorithm developed in Sect. 3.2 can be used to expand $\sigma - \sigma_0$ as a power series in $t - t_0$.

In either case, we obtain

$$\psi = T - \tfrac{1}{2}\varphi_0 T^2 - \tfrac{1}{6}(1 - \gamma_0 - 3\varphi_0^2)T^3 + \tfrac{1}{24}\varphi_0(10 - 9\gamma_0 - 15\varphi_0^2)T^4 + \cdots \quad (5.48)$$

where we have defined

$$T = \sqrt{\frac{\mu}{r_0^3}}(t - t_0) \qquad \rho = \frac{r}{r_0} \qquad \varphi = \frac{\sigma}{\sqrt{r_0}}$$

$$\psi = \frac{\chi}{\sqrt{r_0}} \qquad \gamma_0 = \alpha r_0 = \frac{r_0}{a}$$

### Sect. 5.4] Series Reversion and Newton's Method

To affect a practical solution to the universal form of Kepler's equation, we can use the series expansion of $\psi$ as an initial value $\psi_0$ for a simple Newton iteration

$$\psi_{k+1} = \psi_k + \frac{T - T_k}{\rho_k}$$

or the higher-order algorithm of Prob. 5-25

$$\psi_{k+1} = \psi_k + \frac{T - T_k}{\rho_k} - \frac{\varphi_k}{\rho_k^3}(T - T_k)^2$$

where

$$T_k = U_1(\psi_k; \gamma_0) + \varphi_0 U_2(\psi_k; \gamma_0) + U_3(\psi_k; \gamma_0)$$
$$\rho_k = U_0(\psi_k; \gamma_0) + \varphi_0 U_1(\psi_k; \gamma_0) + U_2(\psi_k; \gamma_0)$$
$$\varphi_k = \varphi_0 U_0(\psi_k; \gamma_0) + (1 - \gamma_0) U_1(\psi_k; \gamma_0)$$

The continued fraction algorithms developed in Sect. 4.7 can be used to evaluate the universal functions.

### An Alternate Form of Kepler's Equation

Using the notation of the previous subsection, the universal form of Kepler's equation can be expressed in terms of the single variable†

$$w = \frac{U_1(\tfrac{1}{4}\psi; \gamma_0)}{U_0(\tfrac{1}{4}\psi; \gamma_0)} \tag{5.49}$$

and then solved by a Newton iteration. If we also define

$$z = \gamma_0 U_1^2(\tfrac{1}{4}\psi; \gamma_0) = \frac{\gamma_0 w^2}{1 + \gamma_0 w^2} \tag{5.50}$$

then, since

$$U_0(\tfrac{1}{2}\psi) = 1 - 2\gamma_0 U_1^2(\tfrac{1}{4}\psi) = 1 - 2z$$

and

$$U_1(\tfrac{1}{2}\psi) = 2U_0(\tfrac{1}{4}\psi)U_1(\tfrac{1}{4}\psi) = 2U_0^2(\tfrac{1}{4}\psi)\frac{U_1(\tfrac{1}{4}\psi)}{U_0(\tfrac{1}{4}\psi)} = 2(1-z)w$$

---

† This algorithm originated with the author using

$$w = U_0(\tfrac{1}{4}\psi; \alpha) \quad \text{and} \quad z = 2U_1(\tfrac{1}{4}\chi; \alpha)$$

as the variables with $w$ and $z$ related by

$$w^2 + \tfrac{1}{4}\alpha z^2 = 1$$

Recently, Stanley W. Shepperd redeveloped the method with, essentially, the variables we are using here. His choice of $w$ is preferable because an initial value of $w$ can be found as a continued fraction of $\tfrac{1}{4}\psi$ with $\psi$ determined from Eq. (5.48), although Shepperd did not use this fact himself.

we have

$$U_0(\psi) = 2U_0^2(\tfrac{1}{2}\psi) - 1 = 1 - 8z(1-z)$$
$$U_1(\psi) = 2U_0(\tfrac{1}{2}\psi)U_1(\tfrac{1}{2}\psi) = 4(1-2z)(1-z)w$$
$$U_2(\psi) = 2U_1^2(\tfrac{1}{2}\psi) = 8(1-z)^2 w^2$$
$$U_3(\psi) = \tfrac{4}{3}U_1^3(\tfrac{1}{2}\psi)F(3,1;\tfrac{5}{2};z) = \tfrac{32}{3}(1-z)^3 w^3 F(3,1;\tfrac{5}{2};z)$$

[The last equation is obtained from Eqs. (4.103), (4.104), and (4.106).]

Therefore, the universal form of Kepler's equation can be written in terms of $w$ as

$$T = \frac{4w}{(1+\gamma_0 w^2)^2}\left[(1-\gamma_0 w^2) + 2\varphi_0 w \right.$$
$$\left. + \frac{8}{3} \times \frac{w^2}{1+\gamma_0 w^2} F\left(3,1;\frac{5}{2};\frac{\gamma_0 w^2}{1+\gamma_0 w^2}\right)\right] \quad (5.51)$$

The hypergeometric function $F(3,1;\tfrac{5}{2};z)$ is best calculated, with $q$ replaced by $z$, from the continued fraction of Eq. (4.107).

A Newton iteration requires also the derivative of $T$ with respect to $w$. For this purpose, we calculate

$$\frac{dT}{dw} = \frac{dT}{d\psi}\frac{d\psi}{dw} = \rho\frac{d\psi}{dw}$$

and

$$\frac{dw}{d\psi} = \frac{\tfrac{1}{4}U_0^2(\tfrac{1}{4}\psi) + \tfrac{1}{4}\gamma_0 U_1^2(\tfrac{1}{4}\psi)}{U_0^2(\tfrac{1}{4}\psi)} = \frac{1}{4U_1^2(\tfrac{1}{4}\psi)}\frac{U_1^2(\tfrac{1}{4}\psi)}{U_0^2(\tfrac{1}{4}\psi)} = \frac{1+\gamma_0 w^2}{4}$$

Hence,

$$\frac{dT}{dw} = \frac{4\rho}{1+\gamma_0 w^2} \quad (5.52)$$

An initial, or starting value, for $w$ can be obtained using Eq. (5.48) to determine an approximate value of $\psi$. Then, $w$ is calculated from the continued fraction of Eq. (4.100) with an appropriate value for the variable.

## 5.5 Near-Parabolic Orbits

Both the elliptic and hyperbolic forms of Kepler's equation tend to become indeterminate when the orbit is nearly parabolic. As the semimajor axis increases, both the mean anomaly and the eccentric anomaly become vanishingly small.

In this section we develop two of the classical methods for dealing with near-parabolic orbits—the first by series expansion using a small parameter, and the second by an elegant and practical method of successive substitutions devised by Gauss.

Sect. 5.5]  Near-Parabolic Orbits

## Method of Successive Approximations

The efficacy of this first method depends on expressing Kepler's equation in terms of a parameter, which is small for orbits that are nearly parabolic, and representing the solution as a power series in that parameter with each term providing a higher order of approximation to the exact solution.

An appropriate parameter for this purpose is

$$\lambda = \frac{1-e}{1+e} \tag{5.53}$$

which is zero for $e = 1$ and also appears naturally in the relationship between the true and eccentric anomalies. Furthermore, since $\tan \frac{1}{2} f$ is the key variable in Barker's equation, we define

$$w = \tan \tfrac{1}{2} f \tag{5.54}$$

and seek a solution of Kepler's equation as a power series in $\lambda$

$$w = a_0 + a_1 \lambda + a_2 \lambda^2 + a_3 \lambda^3 + \cdots \tag{5.55}$$

The desired form of Kepler's equation is obtained using the equation of orbit, derived in Prob. 3–18,

$$r = q \frac{1+w^2}{1+\lambda w^2} \tag{5.56}$$

so that, from the law of areas, we have

$$\frac{\sqrt{\mu p}}{2q^2} dt = \frac{1+w^2}{(1+\lambda w^2)^2} dw$$

Then, expanding the right-hand side by polynomial division to produce a power series in $\lambda$ and integrating term by term, yields

$$\frac{\sqrt{\mu p}}{2q^2}(t - \tau) = w + \frac{w^3}{3} - 2\lambda \left( \frac{w^3}{3} + \frac{w^5}{5} \right)$$
$$+ 3\lambda^2 \left( \frac{w^5}{5} + \frac{w^7}{7} \right) - 4\lambda^3 \left( \frac{w^7}{7} + \frac{w^9}{9} \right) + \cdots$$

which, for $\lambda = 0$ and $p = 2q$, is recognized at once as Barker's equation.

Finally, we substitute† for $w$ from Eq. (5.55) and equate coefficients of corresponding powers of $\lambda$. The zero$^{\text{th}}$-order term $a_0$ is the one and only real root of

$$a_0 + \frac{1}{3} a_0^3 = \frac{\sqrt{\mu p}}{2q^2}(t - \tau)$$

---

† One might suspect that direct series reversion could be used here. It can, of course, be so applied but does not yield a useful result.

The first-order term is then computed from

$$a_1 = \frac{2a_0^3}{1+a_0^2}\left(\frac{1}{3} + \frac{a_0^2}{5}\right)$$

and the higher-order terms, successively, from

$$a_2 = -\frac{3a_0^5}{1+a_0^2}\left(\frac{1}{5} + \frac{a_0^2}{7}\right) + a_0 a_1\left(2a_0 - \frac{a_1}{1+a_0^2}\right)$$

$$a_3 = \frac{4a_0^7}{1+a_0^2}\left(\frac{1}{7} + \frac{a_0^2}{9}\right) + 2a_0 a_1^2\left(1 + \frac{2a_0^2}{1+a_0^2}\right)$$

$$- 3a_0^4 a_1 - \frac{1}{3}\left(\frac{a_1^3}{1+a_0^2}\right) + 2a_0^2 a_2 \quad \text{etc.}$$

It is hardly necessary to remark that these equations apply equally well to both elliptic and hyperbolic orbits with $\lambda$ positive in the former case and negative in the latter.

### Motivating Gauss' Method

In his *Theoria Motus* Gauss gave an extremely efficient technique for solving Kepler's equation for near-parabolic orbits. He was apparently quite impressed with the ingenuity of his method—so much so that he laboriously prepared extensive tables, which are required to implement the algorithm, and included them in an appendix to his book.

The essence of Gauss' method is a clever transformation of Kepler's equation to a form resembling an algebraic expression of third order. The transformed equation is exactly a cubic for parabolic motion and nearly so for elliptic and hyperbolic orbits whose eccentricities are close to unity. The solution is obtained by successive substitutions. At each stage (1) the equation is solved as though it were a cubic and (2) the tables are consulted to revise an algebraic coefficient. Convergence is remarkably rapid, with typically two iteration steps sufficing for seven decimal places of accuracy.

To provide some motivation for the derivation of Gauss' method, we note that, although Kepler's equation bears little resemblance to Barker's equation, they must coincide in the limit as the elliptic orbit approaches the parabolic orbit with the pericenter radius $q$ held fixed.

Recall that Barker's equation

$$\sqrt{\frac{\mu}{2q^3}}(t-\tau) = \tan\tfrac{1}{2}f\left(1 + \frac{1}{3}\tan^2\tfrac{1}{2}f\right)$$

is valid for the parabola while, for the ellipse, we have

$$\sqrt{\frac{\mu}{a^3}}(t-\tau) = E - e\sin E = (1-e)\sin E + E - \sin E$$

### Sect. 5.5] Near-Parabolic Orbits

with
$$\tan \tfrac{1}{2} f = \sqrt{\frac{1+e}{1-e}} \tan \tfrac{1}{2} E$$

relating the true and eccentric anomalies. Now, when the orbit is nearly parabolic, we have $e \simeq 1$ and $E \simeq 0$ so that

$$\tan \tfrac{1}{2} f \simeq \frac{E}{\sqrt{2(1-e)}}$$

If we substitute this for $\tan \tfrac{1}{2} f$ in Barker's equation, we are provided with a possible clue to the form that Kepler's equation must take when the orbit is nearly a parabola. Barker's equation becomes, approximately,

$$\sqrt{\frac{\mu(1-e)^3}{q^3}}(t-\tau) \approx E[(1-e) + \tfrac{1}{6}E^2] \tag{5.57}$$

and since

$$\sin E = E - \frac{1}{6}E^3 + \cdots$$

$$\frac{E - \sin E}{\sin E} = \frac{1}{6}E^2 + \frac{7}{360}E^4 + \cdots$$

we are led to write Kepler's equation in the form

$$\sqrt{\frac{\mu(1-e)^3}{q^3}}(t-\tau) = \sin E \left[(1-e) + \frac{E - \sin E}{\sin E}\right] \tag{5.58}$$

where the semimajor axis $a$ has been replaced by $q/(1-e)$.

When we compare Eqs. (5.57) and (5.58), it is clear that we must have

$$\sin E \approx \sqrt{\frac{6(E - \sin E)}{\sin E}}$$

Therefore, we replace $\sin E$ by

$$\sin E = \sqrt{\frac{\sin^3 E}{6(E - \sin E)}} \sqrt{\frac{6(E - \sin E)}{\sin E}}$$

If $B$ is used to denote the first of these two factors, then

$$B^2 = \frac{\sin^3 E}{6(E - \sin E)} = \frac{E^3 - \tfrac{1}{2}E^5 + \cdots}{E^3 - \tfrac{1}{20}E^5 + \cdots} = 1 - \frac{9}{20}E^2 + \frac{237}{2800}E^4 + \cdots$$

is a quantity nearly equal to one for small $E$.

With this substitution, Kepler's equation takes the form

$$\sqrt{\frac{\mu(1-e)^3}{2q^3}}(t-\tau) = B\sqrt{\frac{3(E - \sin E)}{\sin E}} \left[(1-e) + \frac{E - \sin E}{\sin E}\right] \tag{5.59}$$

Then, if we define a quantity $A$ as

$$A = \frac{3(E - \sin E)}{2 \sin E} = \frac{1}{4}E^2 + \frac{7}{240}E^4 + \frac{31}{10,080}E^6 + \cdots$$

Eq. (5.59) may be written as

$$\sqrt{\frac{\mu(1-e)^3}{2q^3}}(t - \tau) = B\sqrt{2A}\left[(1 - e) + \frac{2A}{3}\right] \qquad (5.60)$$

The factor $1 - e$ (which, of course, does not appear in Barker's equation) may be eliminated by replacing $A$ by a quantity $w$ defined as

$$A = \tfrac{1}{2}(1 - e)w^2 \qquad (5.61)$$

and the final form of Kepler's equation is

$$\frac{1}{B}\sqrt{\frac{\mu}{2q^3}}(t - \tau) = w + \frac{1}{3}w^3 \qquad (5.62)$$

The transformed equation is very nearly Barker's equation and is exactly so for $E = 0$. (For the parabola $w \equiv \tan \tfrac{1}{2}f$ and $B \equiv 1$.) The difference is the presence of the factor $B$ which is a function of the eccentric anomaly. Since $A$ is also a function of $E$, we may reverse the series for $A$ to obtain

$$\tfrac{1}{4}E^2 = A - \tfrac{7}{15}A^2 + \tfrac{376}{1575}A^3 + \cdots$$

and substitute in the series for $B^2$. Taking the square root, we have

$$B = 1 - \tfrac{9}{10}A + \tfrac{969}{1400}A^2 + \cdots$$

so that $B$ may be calculated directly as a power series in $A$.

The algorithm is simple to implement. First, assume $B = 1$ and solve the cubic equation for $w$. Next, calculate $A$ from $w$ and then a new value of $B$ from the power series in $A$. Repeat the process until $A$ ceases to change to within a specified tolerance.

### Gauss' Method

Unfortunately, the algorithm just derived is useful only for very small values of $E$. The convergence properties of the process will be adversely effected if $B$ changes substantially during the iteration. To render the algorithm practical, Gauss invented a technique to make $B$ as insensitive as possible to changes in $E$.

To develop the Gauss algorithm, we begin the derivation anew by writing Kepler's equation as

$$\sqrt{\frac{\mu(1-e)^3}{q^3}}(t - \tau) = Q\left[(1 - e) + [\beta + (1 - \beta)e]\frac{P}{Q}\right]$$

Sect. 5.5]  Near-Parabolic Orbits  225

where we have defined
$$P = E - \sin E \quad \text{and} \quad Q = E - \beta(E - \sin E)$$
and $\beta$ is a constant at our disposal. (Observe that $\beta = 1$ gives Kepler's equation in its usual form.) The manipulations are the same as before. We write
$$Q = \sqrt{\frac{Q^3}{6P} \times \frac{6P}{Q}} = 2B\sqrt{A} \quad \text{where} \quad A = \frac{3P}{2Q}$$
and
$$B^2 = \frac{Q^3}{6P} = \frac{E^3 - \frac{1}{2}\beta E^5 + \cdots}{E^3 - \frac{1}{20}E^5 + \cdots} = 1 + \left(\frac{1}{20} - \frac{1}{2}\beta\right)E^2 + \cdots$$

Therefore, choosing $\beta = \frac{1}{10}$ will eliminate the quadratic dependence of $B$ on the eccentric anomaly.

With this choice of $\beta$, the transformed equation of Kepler is
$$\sqrt{\frac{\mu(1-e)^3}{2q^3}}(t - \tau) = B\sqrt{2A}\left[(1-e) + \left(\frac{1}{10} + \frac{9}{10}e\right)\frac{2A}{3}\right]$$
where
$$A = \frac{15(E - \sin E)}{9E + \sin E} \quad \text{and} \quad B = \frac{9E + \sin E}{20\sqrt{A}}$$

Finally, we replace $A$ by $w$ where
$$A = \frac{5(1-e)}{1+9e}w^2 \tag{5.63}$$

and we have
$$\frac{1}{B}\sqrt{\frac{\mu}{q^3}\frac{1+9e}{20}}(t - \tau) = w + \frac{1}{3}w^3 \tag{5.64}$$

as can be readily verified following the steps outlined previously.

The following expansions of $A$ and $B$ in powers of $E$ were found using the symbol manipulating computer program *MACSYMA*:
$$A = \tfrac{1}{4}E^2 - \tfrac{1}{120}E^4 - \tfrac{1}{20,160}E^6 + \tfrac{1}{144,000}E^8 - \cdots$$
$$B = 1 + \quad\quad + \tfrac{3}{2800}E^4 - \tfrac{1}{84,000}E^6 + \tfrac{71}{258,720,000}E^8 - \cdots$$

(Observe that even when $E$ is as large as 60 degrees, the quantity $B$ differs from unity by only one part in a thousand.)

Gauss tabulated $B$ as a function of $A$ for $0 \leq A \leq 0.3$. However, we can, instead, reverse the series for $A(E)$ and substitute into the series for $B$ to obtain
$$B = 1 + \tfrac{3}{175}A^2 + \tfrac{2}{525}A^3 + \tfrac{471}{336,875}A^4$$
$$+ \tfrac{21,808}{43,793,750}A^5 + \tfrac{434,741}{2,299,171,875}A^6 + \cdots \tag{5.65}$$

The mechanics of the algorithm are simple as before. First, assume $B = 1$ and solve Eq. (5.64) for $w$. Next, calculate $A$ from Eq. (5.63) and then a new value of $B$ from the power series in $A$. Finally, repeat the process until $A$ ceases to change to within a specified tolerance.

To obtain the true anomaly and radius vector, Gauss again displayed his ingenuity. First, we observe that

$$\tfrac{1}{2}\sin E = \frac{\tan \tfrac{1}{2}E}{1 + \tan^2 \tfrac{1}{2}E} = \tan \tfrac{1}{2}E\,(1 - \tau + \tau^2 - \tau^3 + \cdots)$$

where, for brevity, we have written† $\tau$ for $\tan^2 \tfrac{1}{2}E$. Also, using the series for the arctangent,

$$\tfrac{1}{2}E = \tan \tfrac{1}{2}E\,(1 - \tfrac{1}{3}\tau + \tfrac{1}{5}\tau^2 - \tfrac{1}{7}\tau^3 + \cdots)$$

Then, we have

$$A = \frac{15(E - \sin E)}{9E + \sin E} = \frac{\tau - \tfrac{6}{5}\tau^2 + \tfrac{9}{7}\tau^3 - \tfrac{12}{9}\tau^4 + \tfrac{15}{11}\tau^5 - \cdots}{1 - \tfrac{6}{15}\tau + \tfrac{7}{25}\tau^2 - \tfrac{8}{35}\tau^3 + \tfrac{9}{45}\tau^4 - \cdots}$$

in which the law of progression for the coefficients is obvious. Now, dividing the numerator by the denominator

$$A = \tau - \tfrac{4}{5}\tau^2 + \tfrac{24}{35}\tau^3 - \tfrac{1{,}592}{2{,}625}\tau^4 + \tfrac{78{,}856}{144{,}375}\tau^5 - \tfrac{10{,}899{,}688}{21{,}896{,}875}\tau^6 + \cdots$$

and reversing the series, gives

$$\tau = A + \tfrac{4}{5}A^2 + \tfrac{104}{175}A^3 + \tfrac{1{,}112}{2{,}625}A^4 + \tfrac{297{,}032}{1{,}010{,}625}A^5 + \tfrac{875{,}944}{4{,}379{,}375}A^6 + \cdots$$

Finally, inverting the last series produces

$$\frac{A}{\tau} = 1 - \tfrac{4}{5}A + C \tag{5.66}$$

where

$$C = \tfrac{8}{175}A^2 + \tfrac{8}{525}A^3 + \tfrac{1{,}896}{336{,}875}A^4 + \tfrac{28{,}744}{13{,}138{,}125}A^5 + \tfrac{26{,}248}{29{,}859{,}375}A^6 + \cdots \tag{5.67}$$

is a quantity of fourth order in $E$ which Gauss also tabulated.‡

Now, to compute the true anomaly, we write

$$\tan^2 \tfrac{1}{2}f = \frac{1+e}{1-e}\tan^2 \tfrac{1}{2}E = \frac{1+e}{1-e} \times \tau = \frac{1+e}{1-e} \times \frac{A}{1 - \tfrac{4}{5}A + C}$$

and, substituting from Eq. (5.63) for the $A$ in the numerator, obtain

$$\tan^2 \tfrac{1}{2}f = \frac{5 + 5e}{1 + 9e} \times \frac{w^2}{1 - \tfrac{4}{5}A + C} \tag{5.68}$$

---

† Not to be confused with the time of pericenter passage.

‡ In his *Theoria Motus* Gauss gave the series for $C$ in powers of $A$ up to and including $A^5$—a truly impressive numerical feat!

In this way, we also gain a very convenient computation of the radius. It becomes, in fact, from Prob. 3–18,

$$r = q\frac{1+\tan^2\tfrac{1}{2}f}{1+\tan^2\tfrac{1}{2}E} = q\frac{1-\tfrac{4}{5}A+C}{1+\tfrac{1}{5}A+C}(1+\tan^2\tfrac{1}{2}f) \tag{5.69}$$

Gauss' method, as presented here in its original form, is efficient for those values of $A$ which correspond to values of $E$ not exceeding about 60 degrees. For large values of $A$, it may not converge at all.

### Problem 5–26
Show that Gauss' equations are equally valid for hyperbolic orbits—the difference being only that $A$ is negative.

### Problem 5–27
The perihelion of Halley's comet, which moves in a retrograde orbit of the sun, last occurred on February 9.43867, 1986. If

$$q = 0.587099 \text{ a.u.} \quad \text{and} \quad e = 0.96727$$

find the true anomaly and radius on February 1.0, 1986.

NOTE: $\sqrt{\mu} = 0.01720209895$ if distance is measured in astronomical units, time in mean solar days, and the unit of mass is that of the sun. This is the *Gaussian gravitational constant*.

ANSWER: $f = 334°.88527$ and $r = 0.615720$ a.u.

D. K. Yeomans 1985

## 5.6 Extending Gauss' Method

The purpose of this section is to explore the possibility of extending Gauss' method[†] to the general case for which the time interval is not reckoned from pericenter and the orbital eccentricity is arbitrary. The resulting algorithm is necessarily more complex than the original both because of the arbitrary epoch and the annoying fact that the more general quasicubic equation can have multiple real roots which must be reconciled. Nevertheless, the overall program logic is simple and convergence is relatively fast.

---

† This section is based on the paper "Extension of Gauss' Method for the Solution of Kepler's Equation" by R. H. Battin and T. J. Fill which was published in the AIAA *Journal of Guidance and Control*, Vol. 2, No. 3, May–June 1979, pp. 190–195. Some changes have been made here which, hopefully, improve the analysis.

## Transformation of Kepler's Equation

Let $\mathbf{r}_0$ and $\mathbf{v}_0$ be the position and velocity vectors corresponding to an epoch time $t_0$ for a Keplerian orbit. Then Kepler's equation (4.43) for an arbitrary epoch may be expressed as

$$M - M_0 = z - \sin z + \frac{r_0}{a}\sin z + \frac{\sigma_0}{\sqrt{a}}(1 - \cos z) \tag{5.70}$$

where

$$M - M_0 = \sqrt{\frac{\mu}{a^3}}(t - t_0) \qquad \sigma_0 = \frac{\mathbf{r}_0 \cdot \mathbf{v}_0}{\sqrt{\mu}} \qquad z = E - E_0$$

First, multiply by $\sqrt{a^3/r_0^3}$ and introduce the notation

$$T = \sqrt{\frac{\mu}{r_0^3}}(t - t_0) \quad \text{and} \quad \varphi_0 = \frac{\sigma_0}{\sqrt{r_0}} = \frac{\mathbf{r}_0 \cdot \mathbf{v}_0}{\sqrt{\mu r_0}} \tag{5.71}$$

to obtain

$$T = \sqrt{\frac{a}{r_0}}\left[\frac{a}{r_0}(z - \sin z) + \sin z\right] + \frac{a}{r_0}\varphi_0(1 - \cos z)$$

Then, paralleling the arguments of the previous section, we write

$$T = \sqrt{\frac{a}{r_0}}\left\{\left[\beta - \left(1 - \frac{a}{r_0}\right)\right]P + Q\right\} + \frac{a}{r_0}\varphi_0 R \tag{5.72}$$

with the constant $\beta$ introduced for the same purpose as before and

$$P = z - \sin z \qquad Q = z - \beta P \qquad R = 1 - \cos z \tag{5.73}$$

With the motivation provided by the earlier analysis, the next step in the transformation is to define

$$\gamma_0 = \frac{r_0}{a} = 2 - \frac{r_0 v_0^2}{\mu} \qquad \chi_0 = \tfrac{1}{2}[1 - (1 - \beta)\gamma_0] \tag{5.74}$$

and write

$$T = \frac{1}{\sqrt{\gamma_0}}\left(\frac{2\chi_0}{\gamma_0}P + Q\right) + \frac{\varphi_0}{\gamma_0}R$$

$$= \sqrt{\frac{Q^3}{6P}}\left[\frac{8\chi_0}{3}\left(\frac{3P}{2\gamma_0 Q}\right)^{\frac{3}{2}} + 2\left(\frac{3P}{2\gamma_0 Q}\right)^{\frac{1}{2}}\right] + \frac{\varphi_0}{\gamma_0}R$$

$$= B\left[\frac{8\chi_0}{3}\left(\frac{A}{\gamma_0}\right)^{\frac{3}{2}} + 2\left(\frac{A}{\gamma_0}\right)^{\frac{1}{2}} + 2\varphi_0 D\left(\frac{A}{\gamma_0}\right)\right]$$

Sect. 5.6]    Extending Gauss' Method    229

where, in the last step, we have used the notation†

$$A = \frac{3P}{2Q} \qquad B = \sqrt{\frac{Q^3}{6P}} \qquad D = \frac{R}{2AB} = \frac{\sqrt{2}\,R}{\sqrt{3PQ}} \qquad (5.75)$$

Finally, to eliminate $\gamma_0$ from the denominator, we introduce $w$ as before by defining

$$A = \tfrac{1}{4}\gamma_0 w^2 \qquad (5.76)$$

so that the original form of Kepler's equation (5.70) becomes

$$T = B(\tfrac{1}{3}\chi_0 w^3 + \tfrac{1}{2}\varphi_0 D w^2 + w) \qquad (5.77)$$

The quantities $B$ and $D$, defined in Eqs. (5.75), may be expressed as power series expansions

$$B = 1 + \tfrac{1}{4}(\tfrac{1}{10} - \beta)z^2 + \cdots \qquad (5.78)$$

$$D = 1 + \tfrac{1}{12}(\beta - \tfrac{7}{10})z^2 + \cdots \qquad (5.79)$$

In his development, Gauss chose $\beta$ to be $\tfrac{1}{10}$ so that the factor $B$ in Eq. (5.77) would be as insensitive as possible to changes in the anomaly. If we, too, make this choice, (justification for which is provided later) then

$$B = 1 \qquad\qquad + \tfrac{3}{2800}z^4 - \tfrac{1}{84{,}000}z^6 + \cdots$$

$$D = 1 - \tfrac{1}{20}z^2 + \tfrac{1}{4200}z^4 + \tfrac{11}{504{,}000}z^6 - \cdots$$

The quantity $A$, defined in Eq. (5.76), may also be expanded as a power series in $z$. Thus, we have, for $\beta = \tfrac{1}{10}$,

$$A = \tfrac{1}{4}z^2(1 - \tfrac{1}{30}z^2 - \tfrac{1}{5040}z^4 + \tfrac{1}{36{,}000}z^6 + \cdots)$$

Again, by series reversion and substitution, we obtain $B$ and $D$ as series representations in powers of $A$:

$$B = 1 \qquad\qquad + \tfrac{3}{175}A^2 + \tfrac{2}{525}A^3 + \cdots \qquad (5.80)$$

$$D = 1 - \tfrac{1}{5}A - \tfrac{4}{175}A^2 - \tfrac{2}{375}A^3 - \cdots \qquad (5.81)$$

Equations (5.77), (5.80), and (5.81) are the essence of the extended method of Gauss. From an initial approximation for $A$, values of $B$ and $D$ are calculated from the series expansions. Equation (5.77) is then solved as an algebraic cubic for $w$ resulting in a new, improved value for $A$ determined from Eq. (5.76).

As for the original method of Gauss, the fundamental relations involved are universal—one set of equations is valid for all orbits including even the rectilinear. The sign of $\gamma_0$ determines the sign of $A$: i.e., positive

---

† The first two quantities are from the original method of Gauss. However, since Gauss considered only the elementary version of Kepler's equation, the pericenter distance was $r_0$ and the last term was missing ($\varphi_0 = 0$). Hence, there was no need for the $D$ function.

for the ellipse and negative for the hyperbola. The parabola corresponds to the case $A = 0$ for which also $\gamma_0 = 0$ and $\chi_0 = \frac{1}{2}$. It is easy to demonstrate for this case that Eq. (5.77) is, indeed, the generalized form of Barker's equation (4.24) with

$$w = \sqrt{\frac{p}{r_0}}(\tan \tfrac{1}{2} f - \tan \tfrac{1}{2} f_0)$$

### Solution of the Cubic Equation

The solution of the transformed elementary version of Kepler's equation considered by Gauss presents no problem except for rectilinear orbits with which he was not concerned. In his case, with $r_0$ corresponding to pericenter, the factor $\chi_0$, appearing as the coefficient of $w^3$ in Eq. (5.77), is easily shown to be

$$\chi_0 = \tfrac{1}{20} + \tfrac{9}{20} e$$

where $e$ is the orbital eccentricity. Clearly, $\chi_0$ is positive nonzero for all orbits. Furthermore, the resulting cubic equation for $w$ has one and only one positive real root.

In the more general case, however, two difficulties arise which must be addressed. First, we observe from Eqs. (5.74) that $\chi_0$ will vanish at any point in an orbit for which

$$\frac{r_0 v_0^2}{\mu} = \frac{8}{9} \quad \text{or, equivalently,} \quad r_0 = \frac{11}{9} a$$

Second, due to the presence of $\varphi_0$ in the coefficient of $w^2$, which may be either positive or negative, the cubic equation can possess three real roots.

The first of these problems may be successfully countered by a simple change of variable which, at the same time, will also convert the cubic equation to its normal form. For this purpose, we substitute $x$ for $w$ in Eq. (5.77), where

$$w = \frac{3T}{B(1 + x/d)} \tag{5.82}$$

and $d$ is an arbitrary constant at our disposal. The resulting cubic equation for $x$ is simply

$$x^3 - 3\epsilon x - 2b = 0 \tag{5.83}$$

where

$$\epsilon = d^2 \left(1 + \frac{3\varphi_0 DT}{2B}\right) \tag{5.84}$$

$$2b = d^3 \left[2 + 9T\left(\frac{\varphi_0 D}{2B} + \frac{\chi_0 T}{B^2}\right)\right] \tag{5.85}$$

By choosing $d$ as either plus or minus one, we can assure that $b$ will never be negative. Then, although $\epsilon$ can still have either sign, Eq. (5.83) will have exactly one nonnegative real root.

To show that this nonnegative real root is the proper one under all circumstances, we first observe that, for $T = 0$, Eq. (5.83) has three real roots—one at $x = 2$ and a double root at $x = -1$. Only the positive root is appropriate for calculating $w$ from Eq. (5.82). Now, by allowing $T$ to increase or decrease, a simple continuity argument using Descartes method of representing the solution of a cubic, which is described in Sect. 4.2, will suffice to prove that the nonnegative real root of Eq. (5.83) is always the correct choice.

Finally, the possibility of a singularity arising in Eq. (5.82) must be considered. Indeed, $x = 1$ is the root of Eq. (5.83) if and only if both $\chi_0 = 0$ and $\varphi_0 T$ is negative with a magnitude sufficiently large to require $d$ to be $-1$. This singularity is a direct result of the use of $d$ as an artifice to force the desired root to be always positive. Since the problem does not arise if $T$ is small enough, we can always reduce the size of $T$ if we find that $x$ is too close to unity when $d = -1$. This minor bit of awkwardness is a small price to pay for the assurance of only one positive real root as the required solution of the cubic.

Equation (5.83) is of the same form as Eq. (1.31) of Sect. 1.2 so that the solution can be had with continued fractions. However, a simple Newton iteration seems more suitable for an efficient computer mechanization, especially since the accuracy with which $x$ must be obtained is considerably less than that required for $A$.

◊ **Problem 5-28**

A sequence of approximations $x_0$, $x_1$, $x_2$, ... to the appropriate root of Eq. (5.83) is generated recursively using the Newton iteration

$$x_{n+1} = \frac{2}{3}\left(\frac{x_n^3 + b}{x_n^2 - \epsilon}\right) \quad \text{where} \quad x_0 = 1 + |\epsilon|$$

With the initial value for $x = x_0$ as given, demonstrate that convergence of the Newton algorithm to the appropriate root is inevitable.

### Series Representations

Gauss' original method depended, for its efficiency, on the relative insensitivity of the function $B$ to changes in the anomaly. His choice of $\beta = \frac{1}{10}$ insured that $B$ would differ from unity by a quantity of fourth order in the anomaly. In the more general case, with two functions, $B$ and $D$, of the anomaly with which to contend, the choice of $\beta$ is not so obvious.

Referring to Eqs. (5.78) and (5.79), we observe several possibilities:

1. Choose $\beta = \frac{1}{10}$ to render $B$ as nearly constant as possible;
2. Choose $\beta = \frac{7}{10}$ so that $D$ will have this characteristic; or
3. Choose $\beta = \frac{1}{4}$ so that the behavior of $B$ and $D$ will be identical for second order variations in the anomaly.

Tom Fill† exercised the algorithm for each of these cases, and substantiated that the value originally assigned by Gauss to enhance convergence is also a good choice for the general problem.

Fortunately, it is possible to determine both $B$ and $D$ from a *single* series expansion as we now demonstrate. For this purpose, Eqs. (5.73) and (5.75), with $\beta = \frac{1}{10}$, are first used to establish

$$\sin z = Q - \tfrac{9}{10} P = Q(1 - \tfrac{3}{5} A) = 2B\sqrt{A}\,(1 - \tfrac{3}{5} A)$$

With this result, together with the expression for $D$ in Eqs. (5.75), we have

$$D^2 = \left(\frac{1 - \cos z}{2AB}\right)^2 = \frac{\sin^2 z \tan^2 \tfrac{1}{2} z}{4A^2 B^2} = \frac{\tau}{A}\left(1 - \tfrac{3}{5} A\right)^2 \tag{5.86}$$

where, for brevity, we have written $\tau$ for $\tan^2 \tfrac{1}{2} z$.

Again, from Eqs. (5.73) and (5.75), we obtain

$$\frac{1}{BD} = \frac{2A}{R} = \frac{2A}{1 - \cos z} = \frac{A \sec^2 \tfrac{1}{2} z}{\tan^2 \tfrac{1}{2} z} = \frac{A}{\tau} + A \tag{5.87}$$

Hence, both $B$ and $D$ can be calculated in terms of the quantities $A$ and $A/\tau$. The final task is to obtain the expansion of the latter as a power series in $A$.

To this end, we write the first of Eqs. (5.73) as

$$P = \sin z \left(\frac{z}{\sin z} - 1\right) = \sin z \sec^2 \tfrac{1}{2} z \left(\frac{\tfrac{1}{2} z}{\tan \tfrac{1}{2} z} - \frac{1}{1 + \tau}\right)$$

and expand the terms in parenthesis as a power series in $\tau$. Thus,

$$(1 - \tfrac{1}{3}\tau + \tfrac{1}{5}\tau^2 - \cdots) - (1 - \tau + \tau^2 - \cdots)$$
$$= \tfrac{2}{3}(\tau - \tfrac{6}{5}\tau^2 + \tfrac{9}{7}\tau^3 - \tfrac{12}{9}\tau^4 + \cdots)$$

Similarly, for $Q$, using Eqs. (5.73), we have

$$Q = \sin z \left(\frac{9P}{10 \sin z} + 1\right) = \sin z \sec^2 \tfrac{1}{2} z \left(\frac{9P}{10 \sin z \sec^2 \tfrac{1}{2} z} + \frac{1}{1 + \tau}\right)$$
$$= \sin z \sec^2 \tfrac{1}{2} z \left(1 - \tfrac{6}{15}\tau + \tfrac{7}{25}\tau^2 - \tfrac{8}{35}\tau^3 + \cdots\right)$$

---

† "Extension of Gauss' Method for the Solution of Kepler's Equation," MIT M.S. Thesis, May 1976.

Sect. 5.6]  Extending Gauss' Method  233

Then, since $A = 3P/2Q$, we have $A$ represented as the ratio of two power series in $\tau$ which is identical to the corresponding expression in Gauss' original method. Therefore,

$$\frac{A}{\tau} = 1 - \tfrac{4}{5}A + C \tag{5.88}$$

where the expansion of $C$ as a power series in $A$ is given in Eq. (5.67).

As a consequence, we may write

$$\frac{D}{B} = \frac{D^2}{BD} = \frac{1 + \tfrac{1}{5}A + C}{1 - \tfrac{4}{5}A + C}(1 - \tfrac{3}{5}A)^2 \tag{5.89}$$

and

$$\frac{1}{B^2} = \frac{D^2}{(BD)^2} = (1 + \tfrac{1}{5}A + C)\frac{D}{B} \tag{5.90}$$

which are the forms required for the coefficients in Eq. (5.83).

In retrospect, it appears that Gauss undoubtedly knew of the relation

$$B = \frac{\sqrt{1 - \tfrac{4}{5}A + C}}{(1 - \tfrac{3}{5}A)(1 + \tfrac{1}{5}A + C)} \tag{5.91}$$

by which $B$ could be computed from $C$. In the development of his original method, he gave explicitly only two terms in the expansion of $B$ and six in the expansion of $A/\tau$; yet both functions to the same accuracy are to be found in his tables.

◊ **Problem 5–29**

The Lagrangian coefficients, needed to extrapolate $\mathbf{r}$, $\mathbf{v}$ from $\mathbf{r}_0$, $\mathbf{v}_0$, are calculated from

$$F = 1 - \tfrac{1}{2}w^2 BD \qquad \sqrt{\tfrac{\mu}{r_0^3}}\, G = wB(1 - \tfrac{3}{5}A) + \tfrac{1}{2}\varphi_0 w^2 BD$$

$$F_t = -\sqrt{\tfrac{\mu}{r_0^3}}\, wB(1 - \tfrac{3}{5}A)\frac{r_0}{r} \qquad G_t = 1 - \tfrac{1}{2}w^2 BD \frac{r_0}{r}$$

where

$$\frac{r}{r_0} = 1 + \tfrac{1}{2}w^2(1 - \gamma_0)BD + \varphi_0 wB(1 - \tfrac{3}{5}A)$$

**234**  Solving Kepler's Equation  [Chap. 5]

## Algorithm for the Kepler Problem

The complete algorithm for the solution of Kepler's problem is elegant in its simplicity. The relevant equations are summarized below in a form designed to minimize the arithmetic operations required in a computer mechanization. Certain auxiliary quantities are introduced for this purpose solely to effect a more efficient and compact algorithm.

Given the vectors $\mathbf{r}_0$, $\mathbf{v}_0$ at time $t_0$, we desire the corresponding vectors $\mathbf{r}$, $\mathbf{v}$ at some other time $t$. We then begin the solution by calculating the preliminary quantities

$$\beta_0 \equiv \sqrt{\frac{\mu}{r_0^3}} \qquad T' \equiv 3T = 3\beta_0(t - t_0)$$

$$\delta_0 \equiv \frac{r_0 v_0^2}{\mu} \qquad \gamma_0' \equiv \tfrac{1}{4}\gamma_0 = \tfrac{1}{2} - \tfrac{1}{4}\delta_0 \qquad \chi_0' \equiv \tfrac{1}{2}\chi_0 \qquad (5.92)$$

$$\gamma_0'' \equiv 1 - \gamma_0 = \delta_0 - 1 \qquad\qquad = \tfrac{9}{40}\delta_0 - \tfrac{1}{5}$$

$$\varphi_0 \equiv \frac{\mathbf{r}_0 \cdot \mathbf{v}_0}{\sqrt{\mu r_0}} \qquad \varphi_0' \equiv \tfrac{1}{2}\varphi_0 \qquad\qquad \varphi_0'' \equiv \tfrac{1}{2}\varphi_0'$$

The iteration to determine $A$ starts with the initial value $A = 0$, corresponding to a parabolic orbit. Hence, the initial values of

$$\alpha \equiv 1 + \tfrac{1}{5}A + C \quad\text{and}\quad \alpha' \equiv 1 - \tfrac{3}{5}A \qquad (5.93)$$

are both unity while†

$$C = \tfrac{8}{175}A^2 + \tfrac{8}{525}A^3 + \tfrac{1{,}896}{336{,}875}A^4 + \tfrac{28{,}744}{13{,}138{,}125}A^5 + \tfrac{26{,}248}{29{,}859{,}375}A^6$$

$$+ \tfrac{134{,}584}{372{,}246{,}875}A^7 + \tfrac{129{,}802{,}986{,}344}{857{,}740{,}555{,}546{,}875}A^8 + \tfrac{55{,}082{,}676{,}856}{857{,}740{,}555{,}546{,}875}A^9$$

$$+ \tfrac{687{,}061{,}097{,}149{,}992}{24{,}934{,}041{,}427{,}216{,}796{,}875}A^{10} + \tfrac{8{,}033{,}038{,}585{,}237{,}352}{673{,}219{,}118{,}534{,}853{,}515{,}625}A^{11}$$

$$+ \tfrac{2{,}892{,}031{,}498{,}456{,}202{,}296}{555{,}405{,}772{,}791{,}254{,}150{,}390{,}625}A^{12} + \tfrac{4{,}093{,}458{,}329{,}892{,}811{,}912}{1{,}789{,}640{,}823{,}438{,}485{,}595{,}703{,}125}A^{13}$$

$$+ \tfrac{10{,}507{,}274{,}811{,}793{,}509{,}548{,}037{,}032}{10{,}398{,}126{,}371{,}321{,}703{,}046{,}014{,}404{,}296{,}875}A^{14}$$

$$+ \tfrac{25{,}259{,}289{,}756{,}593{,}760{,}965{,}336}{56{,}299{,}726{,}292{,}037{,}542{,}523{,}193{,}359{,}375}A^{15} + \cdots \qquad (5.94)$$

is initially zero. Next, we calculate

$$\psi \equiv \alpha T' \quad\text{and}\quad \eta \equiv \frac{\alpha'\alpha'\psi}{\alpha - A} \qquad (5.95)$$

---

† Sufficient accuracy is obtained for most applications using the truncated power series for $C$ given here even though more terms were actually determined using the MACSYMA program. Economization of this series is possible using Tschebycheff polynomials as developed in Appendix F.

from which the coefficients in the cubic equation (5.83) for $x$ are

$$\epsilon = 1 + \eta\varphi_0' \quad \text{and} \quad b = |\epsilon + \eta(\varphi_0'' + \chi_0'\psi)| \tag{5.96}$$

The solution of the cubic is obtained either using the Newton iteration derived in Prob. 5–28, continued fractions, or otherwise. After convergence, a new value of $A$ is computed from

$$\xi = 1 \pm x \quad \text{and} \quad \theta = \frac{\psi\eta}{\xi^2} \quad \text{then} \quad A = \gamma_0'\theta \tag{5.97}$$

The choice of sign in the equation for $\xi$ depends on the sign of the quantity within the absolute value symbols in the equation for $b$, i.e., the plus sign is chosen if that quantity is nonnegative and the minus sign otherwise. Also, if $\xi$ is dangerously small, reduce the size of $T$ and begin again.

Repeat the computation, beginning with Eqs. (5.93), until $A$ ceases to change by a preassigned amount. (During the second and subsequent cycles through these equations it is, of course, more efficient to select for $x_0$ the last value of $x$ determined from the previous Newton iteration.)

The elements of the transition matrix $\boldsymbol{\Phi}$, needed to extrapolate $\mathbf{r}$, $\mathbf{v}$ from $\mathbf{r}_0$, $\mathbf{v}_0$, are most conveniently obtained in terms of the auxiliary quantities

$$\lambda = \frac{\theta}{2\alpha} \quad \text{and} \quad \kappa = \frac{\alpha'T'}{\xi} \tag{5.98}$$

Then, from the results of Prob. 5–29, the magnitude of the new position vector $\mathbf{r}$ is obtained from

$$\rho \equiv \frac{r}{r_0} = 1 + \gamma_0''\lambda + \varphi_0\kappa \tag{5.99}$$

and the transition matrix is

$$\boldsymbol{\Phi} = \begin{bmatrix} 1 - \lambda & \dfrac{\kappa + \varphi_0\lambda}{\beta_0} \\ -\dfrac{\beta_0\kappa}{\rho} & 1 - \dfrac{\lambda}{\rho} \end{bmatrix} \tag{5.100}$$

The description of the basic algorithm is now complete. We note that no square or cube roots and only one polynomial are involved in the calculations. However, we must recall that the efficiency and practicality of the method require that values of $A$ (hence, values of $t - t_0$) be kept within reasonable bounds. For example, following Gauss, we might require that $|A| \leq \frac{3}{10}$.

In order to deal with the general problem, for which the time interval is unrestricted, let us define $A_m$ (and, correspondingly, $T_m$) as the maximum permissible values of $A$ and $T$. Then, the algorithm is easily modified, as will now be described.

Each time a new value of $A$ is computed in Eqs. (5.97), a test is made to determine if $|A| \leq A_m$. If the test fails, we set the magnitude of $A$ to the value of $A_m$, leaving the sign unchanged, and then compute

$$\theta_m = \frac{A}{\gamma_0'} \qquad \lambda_m = \frac{\theta_m}{2\alpha} \qquad B = \frac{\alpha\alpha'}{\sqrt{\alpha - A}}$$
$$T_m' = B\sqrt{\theta_m}\,(3 + \chi_0\theta_m) + 3\varphi_0\lambda_m \qquad (5.101)$$

where $\alpha$ and $\alpha'$ are calculated† using the extreme value of $A$. This is followed by Eqs. (5.98), (5.99), and (5.100). (Division by $\gamma_0'$ does not present a problem. If $\gamma_0'$ were zero, as it is for parabolic orbits, the test in question would not have failed.) Corresponding values of the other quantities are obtained from

$$r_m = \rho_m r_0 \qquad \varphi_m = \frac{\varphi_0(1 - \gamma_0\lambda_m) + \gamma_0''\kappa_m}{\sqrt{\rho_m}}$$
$$\beta_m = \frac{\beta_0}{\rho_m\sqrt{\rho_m}} \qquad \delta_m = 2 - \rho_m\gamma_0 \qquad (5.102)$$

Then, by replacing $T'$ by $T' - T_m'$ and $r_0$, $\beta_0$, $\varphi_0$, $\delta_0$ by $r_m$, $\beta_m$, $\varphi_m$, $\delta_m$, we are prepared to restart the algorithm anew, beginning with the appropriate ones of Eqs. (5.92).

If the time difference $t - t_0$ is sufficiently large, we may have to repeat this process of decrementing $T$ several times. The transition matrices thus sequentially generated are, of course, multiplied together to produce the final desired matrix.

The algorithm described here has been exercised for a variety of representative orbits. Some of the results are documented in the paper by Battin and Fill to which reference has already been made. When compared with a fairly standard Newton-Raphson iteration on a form of Kepler's equation utilizing universal variables, the computer memory utilization was almost exactly the same. Further, if the time difference $t - t_0$ is small enough so that no decrement in $T$ is necessary, the new algorithm uses only 40% to 85% of the time required by the more standard method. The lowest percentages are typical of the near parabolic and near rectilinear orbits while the largest percentages are characteristic of the hyperbolic cases. If several time increments are required, this computational advantage tends to diminish.

◇ **Problem 5–30**

From the initial condition data of Prob. 3–5, extrapolate the position and velocity vectors through the given time interval using the extension of Gauss' method for the solution of Kepler's equation.

---

† Of course, values of $B$ and $\alpha$ for $A = \pm A_m$ may be precomputed and stored for use at this point in the algorithm.

Chapter 6

# Two-Body Orbital Boundary-Value Problem

R EMARKABLE AND ELEGANT PROPERTIES ARE CHARACTERISTIC OF the two-body orbital boundary-value problem. Carl Friedrich Gauss must have been aware of the potential of this subject but, for some reason, chose not to develop it. In the *Theoria Motus* he remarks that

> "The discussion of the relations of two or more places of a heavenly body in its orbit as well as in space, furnishes an abundance of elegant propositions, such as might easily fill an entire volume. But our plan does not extend so far as to exhaust this fruitful subject, ... "

The author has been fascinated by this subject for many years and has collected (almost as a hobby as others would collect stamps) a number of delightful and often useful properties of the two-body, two-point, boundary-value problem. They constitute the subject matter of the current chapter. Some are original; but, of course, we do not know whether or not Gauss anticipated any of the "new" results.

Many of the properties of the boundary-value problem have been documented only recently.† Where memory serves, their source is given. Some came as sudden inspirations in the course of a lecture, others from concentrated effort to understand a particular phenomena. Some are from published papers and some resulted from conversations with students.

Perhaps many of these geometric properties remained so long undiscovered because of the tendency to adhere strictly to mathematical analysis and to avoid geometric arguments. For example, Lagrange took pride in the fact that his great work *Mécanique Analytique* contained not a single diagram. *Theoria Motus* is also devoid of any figures and, indeed, most advanced mathematical treatises today have this unfortunate characteristic. Although the great geometer Newton had extensively employed geometric

---

† The most recent publication is the author's paper "Elegant Propositions of the Boundary-Value Problem" presented at the Thirty-Seventh Congress of the International Astronautical Federation held in Innsbruck, Austria in 1986. It is scheduled to appear in *Astronautica Acta*.

proofs, it was felt by some that progress in analysis did not really begin until his methods were abandoned in favor of more analytic ones.

Johann Heinrich Lambert (1728–1779), often referred to as the great Alsatian scholar, deduced the theorem which bears his name in a geometric fashion. His ideas were ingenious but he was restricted, very much as Kepler was, to purely geometrical methods. It took men like Lagrange and Gauss to supply the missing proofs and turn his ideas into effective tools.

Transforming the boundary-value problem, as we do in this chapter, was the method Lambert used to obtain a series expansion of the transfer time as a function of the transformation invariants. That certain quantities were considered invariant, Lambert deduced from the forms of the energy integral and the transfer-time formula for the parabola which he and Euler had each independently derived using totally different arguments.

Lambert would certainly have appreciated the invariant property of the *mean point* which is introduced in Sect. 6.4 and is original with the author. As we shall see in Chapter 7, it is of great importance in extending the region of validity and improving the convergence of Gauss' classical method of orbit determination. Had Gauss or his followers been aware of this property of orbits, they, most assuredly, would have used it for the same purpose.

Lambert was highly respected in his lifetime and was ranked with Jean-Jacques Rousseau and Voltaire as one of the most famous philosophers of his century. Although forced to leave school at the age of twelve to assist his father in the tailoring business, he continued his studies with an excellent teacher and became extraordinarily successful in the sciences of philosophy, mathematics, astronomy, and physics. In Chapter 1 we have already mentioned his contributions to continued fractions and number theory. In astronomy, he contributed to the foundations of photometry and to the orbit determination of planets and comets which culminated in "Lambert's Theorem."

## 6.1 Terminal Velocity Vectors

Consider two position vectors $\mathbf{r}_1$ and $\mathbf{r}_2$ which locate the points $P_1$ and $P_2$ relative to a center of force fixed at a point $F$ as shown in Fig. 6.1. The angle $\theta$ between the position vectors is called the *transfer angle* and the line $P_1 P_2$, called the *chord*, is of length $c = |\mathbf{r}_2 - \mathbf{r}_1|$. Let $\mathbf{v}_1$ at $P_1$ and $\mathbf{v}_2$ at $P_2$ be the velocity vectors for an orbit connecting $P_1$ and $P_2$ with a focus at $F$.

We shall utilize the terminal velocity vectors expressed in polar coordinates as

$$\mathbf{v}_1 = v_{r_1} \mathbf{i}_{r_1} + v_{\theta_1} \mathbf{i}_{\theta_1} \qquad \mathbf{v}_2 = v_{r_2} \mathbf{i}_{r_2} + v_{\theta_2} \mathbf{i}_{\theta_2} \qquad (6.1)$$

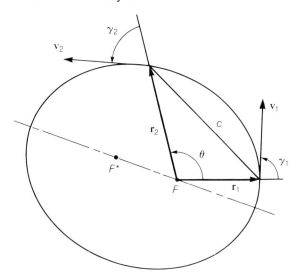

**Fig. 6.1:** Geometry of the boundary-value problem.

The radial component $v_r$ of the velocity vector $\mathbf{v}$ can be written in a variety of forms. For example, from the hodograph plane analysis of Sect. 3.5, we have

$$v_r = \frac{dr}{dt} = \frac{\mathbf{r}\cdot\mathbf{v}}{r} = \frac{\sqrt{\mu}\,\sigma}{r} = \frac{h}{r}\cot\gamma = \frac{he}{p}\sin f \qquad (6.2)$$

Also, for the circumferential component $v_\theta$,

$$v_\theta = r\frac{df}{dt} = \frac{h}{r} = \frac{\sqrt{\mu p}}{r} \qquad (6.3)$$

A relationship between the terminal circumferential components of the velocity vectors is an immediate consequence of Kepler's second law. Thus,

$$r_1 v_{\theta_1} = r_2 v_{\theta_2} \qquad (6.4)$$

A corresponding relation for the radial components is determined from the following argument.

Let $f_1$ and $f_2 = f_1 + \theta$ be the true anomalies of the points $P_1$ and $P_2$, so that

$$v_{r_1} + v_{r_2} = \frac{he}{p}[\sin f_1 + \sin(f_1 + \theta)]$$

In the special case for which the transfer angle $\theta$ is 180 degrees, we see at once that

$$v_{r_1} = -v_{r_2} \qquad \text{for} \qquad \theta = \pi \qquad (6.5)$$

Otherwise, we may write

$$v_{r_1} + v_{r_2} = \frac{2he}{p} \sin(f_1 + \tfrac{1}{2}\theta) \sin \tfrac{1}{2}\theta \cot \tfrac{1}{2}\theta$$

$$= \frac{h}{p}[e\cos f_1 - e\cos(f_1 + \theta)] \cot \tfrac{1}{2}\theta$$

Then, using the equation of orbit in the form

$$e \cos f = \frac{p}{r} - 1$$

we have the desired result

$$v_{r_1} + v_{r_2} = (v_{\theta_1} - v_{\theta_2}) \cot \tfrac{1}{2}\theta \tag{6.6}$$

Two corollaries of this last relationship follow at once. Using Eqs. (6.2) and (6.3), we may rewrite Eq. (6.6) either as

$$r_2 \sigma_1 + r_1 \sigma_2 = \sqrt{p}(r_2 - r_1) \cot \tfrac{1}{2}\theta \tag{6.7}$$

or as

$$r_2 \cot \gamma_1 + r_1 \cot \gamma_2 = (r_2 - r_1) \cot \tfrac{1}{2}\theta \tag{6.8}$$

The second form is particularly interesting since it is entirely geometric and is independent of the orbital elements. (It is not difficult to show that this last result reduces to $\gamma_1 - \gamma_2 = \tfrac{1}{2}\theta$ when the orbit is a parabola.)

### Minimum-Energy Orbit

As seen in Sect. 3.3, the total energy of a two-body orbit is $-\mu/2a$ so that the *minimum-energy orbit* would be an ellipse ($a > 0$) corresponding to the smallest possible value of the semimajor axis $a$. If $P_1$ and $P_2$ both lie on an orbit whose major axis is $2a$, then

$$P_1 F + P_1 F^* = 2a \qquad P_2 F + P_2 F^* = 2a$$

where $F^*$ is the vacant focus of the orbit. Since

$$P_1 F = r_1 \qquad P_2 F = r_2 \qquad P_1 P_2 = c$$

then

$$4a = P_1 F^* + P_2 F^* + r_1 + r_2$$

Clearly, the minimum value for $P_1 F^* + P_2 F^*$ is the chord length $c$. Therefore, the vacant focus $F_m^*$ lies on the chord $P_1 P_2$ and the corresponding value of $a$ is

$$2a_m = \tfrac{1}{2}(r_1 + r_2 + c) \tag{6.9}$$

or one half the perimeter of the triangle $\triangle F P_1 P_2$. For convenience, we introduce the notation

$$2s = r_1 + r_2 + c \tag{6.10}$$

so that $s$ is the semiperimeter of the triangle and $a_m = \frac{1}{2}s$. The point $F_m^*$ divides the chord $P_1 P_2$ in such a way that

$$P_1 F_m^* = s - r_1 \qquad P_2 F_m^* = s - r_2 \qquad (6.11)$$

A useful set of trigonometric formulas involving the semiperimeter of the triangle is given in Appendix G.

### Locus of Velocity Vectors

Consider first the case for which the transfer angle $\theta$ is 180 degrees or, equivalently, $r_1 + r_2 = c$. It is surprising indeed to find that the orbital parameter is the *same* for all orbits. This result is obtained directly from the equation of orbit—for, if between

$$p = r_1(1 + e\cos f_1) \qquad p = r_2[1 + e\cos(f_1 + \pi)]$$
$$= r_2(1 - e\cos f_1)$$

$e \cos f_1$ is eliminated, there obtains

$$p = \frac{2r_1 r_2}{r_1 + r_2} = \frac{2r_1 r_2}{c} \qquad \text{for } \theta = \pi \qquad (6.12)$$

Thus, *the parameter of all orbits connecting two terminals separated by 180 degrees is the harmonic mean† between the terminal radii.*

Using this result in Eq. (6.3), we have, from the vis-viva integral,

$$v_{\theta_1}^2 = \frac{2\mu r_2}{r_1 c} \quad v_{\theta_2}^2 = \frac{2\mu r_1}{r_2 c} \quad v_{r_1}^2 = v_{r_2}^2 = \mu\left(\frac{2}{c} - \frac{1}{a}\right) \quad \text{for } \theta = \pi \quad (6.13)$$

Hence, the circumferential components of the velocity vectors at each terminal are the *same* for all orbits and, in this case, are also equal to the minimum-energy velocities. Indeed, these velocities may be written as

$$v_{\theta_1}^2 = v_{m_1}^2 = \mu\left(\frac{2}{r_1} - \frac{1}{\frac{1}{2}c}\right) \quad v_{\theta_2}^2 = v_{m_2}^2 = \mu\left(\frac{2}{r_2} - \frac{1}{\frac{1}{2}c}\right) \quad \text{for } \theta = \pi$$

The locus of velocity vectors at $P_1$ (and at $P_2$) for all possible orbits connecting $P_1$ and $P_2$ with a transfer angle of 180 degrees is, clearly, a straight line as shown in Fig. 6.2. Velocity vectors occur in pairs $\mathbf{v}_1$ and $\tilde{\mathbf{v}}_1$ for the same value of $a$ and make equal angles with the vector $\mathbf{v}_{m_1}$.

There are two parabolic orbits with velocities $\mathbf{v}_{p_1}$ and $\tilde{\mathbf{v}}_{p_1}$ but $\tilde{\mathbf{v}}_{p_1}$ corresponds to motion from $P_2$ to $P_1$ in the counterclockwise direction. Likewise, the hyperbolic velocity vectors $\tilde{\mathbf{v}}_{h_1}$ correspond to orbits traversed from $P_2$ and $P_1$.

---

† We have already seen in Sect. 3.3 that, for the ellipse, the parameter is the harmonic mean between the pericenter and apocenter radii. Equation (6.12) is a generalization of this property.

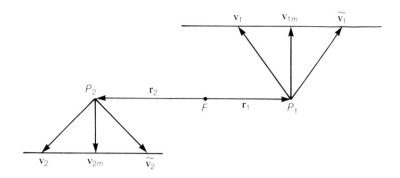

**Fig. 6.2:** Velocity vector locus for 180° transfer.

When the transfer angle is different from 180 degrees, the end result is similar but the analysis is quite different. Expressions for the terminal velocity vectors are obtained from Sect. 3.6 in which were introduced the Lagrangian coefficients. With our present notation, we have

$$\mathbf{r}_2 = F\mathbf{r}_1 + G\mathbf{v}_1 \qquad \mathbf{r}_1 = G_t\mathbf{r}_2 - G\mathbf{v}_2$$

where

$$F = 1 - \frac{r_2}{p}(1 - \cos\theta) \qquad G = \frac{r_1 r_2}{\sqrt{\mu p}}\sin\theta$$

$$G_t = 1 - \frac{r_1}{p}(1 - \cos\theta)$$

Therefore,

$$\mathbf{v}_1 = \frac{\sqrt{\mu p}}{r_1 r_2 \sin\theta}\left[(\mathbf{r}_2 - \mathbf{r}_1) + \frac{r_2}{p}(1 - \cos\theta)\mathbf{r}_1\right]$$

$$\mathbf{v}_2 = \frac{\sqrt{\mu p}}{r_1 r_2 \sin\theta}\left[(\mathbf{r}_2 - \mathbf{r}_1) - \frac{r_1}{p}(1 - \cos\theta)\mathbf{r}_2\right]$$

provided, of course, that $\theta$ is not equal to 180 degrees.

From the form of these expressions, we are led to replace the usual radial and circumferential components of orbital velocity by components along the skewed axes

$$\mathbf{i}_{r_1} = \frac{\mathbf{r}_1}{r_1} \qquad \mathbf{i}_{r_2} = \frac{\mathbf{r}_2}{r_2} \qquad \mathbf{i}_c = \frac{\mathbf{r}_2 - \mathbf{r}_1}{c}$$

Hence,

$$\mathbf{v}_1 = v_c \mathbf{i}_c + v_\rho \mathbf{i}_{r_1} \qquad \mathbf{v}_2 = v_c \mathbf{i}_c - v_\rho \mathbf{i}_{r_2} \qquad (6.14)$$

where

$$v_c = \frac{c\sqrt{\mu p}}{r_1 r_2 \sin\theta} \quad \text{and} \quad v_\rho = \sqrt{\frac{\mu}{p}}\frac{1 - \cos\theta}{\sin\theta} \qquad (6.15)$$

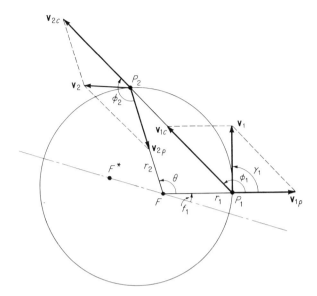

**Fig. 6.3:** Chordal and radial components of terminal velocity vectors.

Thus, when the terminal velocity vectors $\mathbf{v}_1$ and $\mathbf{v}_2$ are resolved into components parallel to the chord connecting the end points and parallel to their respective radius vectors, as shown in Fig. 6.3, we have the surprising result that *the magnitudes of components of the velocity vectors along these skewed axes at the two terminals are, respectively, equal.*[†]

Perhaps even more astonishing is the property displayed by multiplying $v_c$ and $v_\rho$. From Eqs. (6.15), we have the important result that *the product*

$$v_c v_\rho = \frac{\mu c}{2 r_1 r_2} \sec^2 \tfrac{1}{2} \theta \qquad (6.16)$$

*depends solely on the geometry of the boundary-value problem not on the orbit connecting the terminals.*

Equation (6.16) is recognized as a hyperbola in asymptotic coordinates of the type considered in Sect. 4.4. At $P_1$ the asymptotes are the chord $P_1 P_2$ and the radius $FP_1$ extended.

The angle $\phi_1$ between the asymptotes is the exterior angle at $P_1$ of the triangle $\Delta F P_1 P_2$. Similarly, at $P_2$ they are the radius $FP_2$ and the chord $P_1 P_2$ extended. Again the angle between the asymptotes $\phi_2$ is the exterior angle at $P_2$ of the triangle $\Delta F P_1 P_2$.

---

† The reader should remember that $v_\rho$ is not the same as the usual radial component $v_r$ of velocity. Velocity components along skewed axes were introduced by Thore Godal of the Norwegian Naval Academy in a paper entitled "Conditions of Compatibility of Terminal Positions and Velocities" and published in the *Proceedings of the Eleventh International Astronautical Congress* which was held in Stockholm, Sweden in 1960.

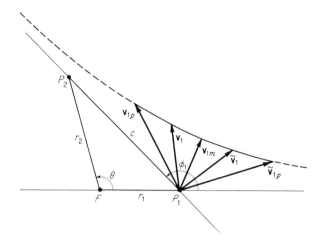

**Fig. 6.4:** Locus of velocity vectors for orbital transfer.

Clearly, at both terminals, *the minimum-energy velocity vectors bisect the angle $\phi$ between the asymptotes and have equal velocity components along the skewed axes.* By simple geometry,

$$v_{m_c} = v_{m_\rho} = \tfrac{1}{2} v_m \sec \tfrac{1}{2} \phi$$

Therefore, Eq. (6.16) may be written as either

$$v_c v_\rho = \tfrac{1}{4} v_{m_1}^2 \sec^2 \tfrac{1}{2} \phi_1 \tag{6.17}$$

which is the locus of velocity vectors at $P_1$ shown in Fig. 6.4, or

$$v_c v_\rho = \tfrac{1}{4} v_{m_2}^2 \sec^2 \tfrac{1}{2} \phi_2 \tag{6.18}$$

which is the corresponding locus at $P_2$.

The other branch of the hyperbolic locus described in Fig. 6.4 (but not shown in the illustration) corresponds to orbits from $P_1$ to $P_2$ traversed in the clockwise direction. In this case the transfer angle is $2\pi - \theta$ rather than $\theta$.

As seen in the diagram, velocity vectors occur in pairs, $\mathbf{v}$ and $\tilde{\mathbf{v}}$, corresponding to two different orbits having the same value of the semimajor axis. Such pairs are called *conjugate orbits*.

### Problem 6–1

From the results of Sect. 4.4, specifically, Eq. (4.66), the hyperbolic locus of velocity vectors is

$$v_c v_\rho = \tfrac{1}{4} a_h^2 e_h^2$$

Calculate directly, using Eq. (6.16), the eccentricity $e_h$ and semimajor axis $a_h$ of the hyperbola.

◇ **Problem 6–2**
Use the results of Prob. 4–19 to derive the equation of the hyperbolic locus of velocity vectors at $P_1$ in the form

$$r_1 v_1^2 \sin \gamma_1 \sin(\phi_1 - \gamma_1) = \mu \sin \phi_1 \tan \tfrac{1}{2}\theta$$

and at $P_2$

$$r_2 v_2^2 \sin \gamma_2 \sin(\phi_2 + \gamma_2) = -\mu \sin \phi_2 \tan \tfrac{1}{2}\theta$$

◇ **Problem 6–3**
Show that

$$r_2 = \frac{r_1}{\cos\theta - \cot\phi_1 \sin\theta}$$

Then, by eliminating $\phi_1$ between this and the result of Prob. 6–2, obtain the relation

$$r = \frac{r_1^2 v_1^2 \sin^2 \gamma_1}{\mu(1 - \cos\theta) - r_1 v_1^2 \sin\gamma_1 \sin(\theta - \gamma_1)}$$

where the subscript on $r_2$ has been dropped. Thus, $r$ as a function of the true anomaly difference $\theta$, with the initial conditions at $P_1$ corresponding to $\theta = 0$, is explicitly displayed.

In a similar manner, derive the corresponding relations

$$r_1 = \frac{r_2}{\cos\theta - \cot\phi_2 \sin\theta}$$

and

$$r = \frac{r_2^2 v_2^2 \sin^2 \gamma_2}{\mu(1 - \cos\theta) + r_2 v_2^2 \sin\gamma_2 \sin(\theta + \gamma_2)}$$

◇ **Problem 6–4**
The velocity components along skewed axes are related to ordinary polar coordinate components as

$$v_c = v_{\theta_1} \csc\phi_1 = \frac{c}{r_2} v_{\theta_1} \csc\theta$$

$$v_\rho = v_{r_1} - v_{\theta_1} \cot\phi_1 = v_{r_1} - \frac{r_2 \cos\theta - r_1}{r_2 \sin\theta} v_{\theta_1}$$

Derive the equation of the hyperbolic locus of velocity vectors in the form

$$v_{r_1} v_{\theta_1} \sin\theta - v_{\theta_1}^2 \left(\cos\theta - \frac{r_1}{r_2}\right) = \frac{\mu}{r_1}(1 - \cos\theta)$$

This equation exhibits no difficulties as $\theta$ approaches 180 degrees. Indeed, in the limit, show that the first of Eqs. (6.13) obtains.

### Parameter in Terms of Velocity-Components Ratio

By forming the ratio of the two equations in (6.15), we obtain

$$p = \frac{r_1 r_2}{c}(1 - \cos\theta)\frac{v_c}{v_\rho}$$

Hence, the parameter is a linear function of the ratio $v_c/v_\rho$ for a fixed geometry of the triangle $\triangle F P_1 P_2$. As we trace the locus of velocity vectors in Fig. 6.4 from right to left, this ratio (and, hence, the orbital parameter) increases monotonically from zero to infinity.

Clearly, the velocity component ratio is unity for the minimum-energy orbit. Therefore, the parameter $p_m$ of this special orbit is seen to be

$$p_m = \frac{r_1 r_2}{c}(1 - \cos\theta) \tag{6.19}$$

so that the general expression† for the parameter is simply

$$\frac{p}{p_m} = \frac{v_c}{v_\rho} \tag{6.20}$$

Conjugate orbits are characterized as having the magnitudes of their velocity components along skewed axes reversed. Therefore, if $p$ and $\widetilde{p}$ are the parameters of a pair of conjugate orbits, then

$$p_m = \sqrt{p\widetilde{p}} \tag{6.21}$$

Thus, *the parameter of the minimum-energy orbit is the geometric mean between the parameters of any pair of conjugate orbits.*

Furthermore, the flight-direction angles of conjugate orbits are also simply related. From Fig. 6.4 we see that

$$\gamma_1 + \widetilde{\gamma}_1 = \phi_1 \quad \text{and, similarly,} \quad \gamma_2 + \widetilde{\gamma}_2 = 2\pi - \phi_2 \tag{6.22}$$

For $\theta = \pi$, Eq. (6.19) gives a value for $p_m$ which agrees with Eq. (6.12). On the other hand, when $\theta$ is different from $\pi$, Eq. (6.19) may be written in the alternate form

$$p_m = d \tan \tfrac{1}{2}\theta \tag{6.23}$$

where $d$ is the perpendicular distance from the focus $F$ to the chord. This last expression permits the simple geometric construction of $p_m$ as illustrated in Fig. 6.5.

There is yet another interpretation of the parameter of the minimum-energy orbit. Since Eq. (6.19) can be written as

$$p_m = \frac{2r_1 r_2}{c}\sin^2 \tfrac{1}{2}\theta = \frac{2}{c}(s - r_1)(s - r_2) \tag{6.24}$$

---

† Although Thore Godal introduced and made extensive use of the concept of velocity vectors resolved along skewed axes, he, apparently, overlooked the elegant formula for the orbital parameter in terms of the ratio of these velocity components.

## Terminal Velocity Vectors

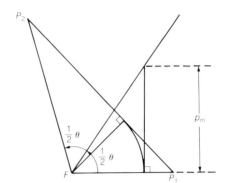

**Fig. 6.5:** Construction of the parameter for the minimum-energy orbit.

it follows from Eq. (6.11) that

$$\frac{1}{p_m} = \frac{1}{2}\left(\frac{1}{P_1 F_m^*} + \frac{1}{P_2 F_m^*}\right) \tag{6.25}$$

Hence, *the parameter of the minimum-energy orbit is the harmonic mean between the segments of the chord separated by the vacant focus $F_m^*$ of the minimum-energy orbit.*

◇ **Problem 6–5**

The hyperbolic locus of velocity vectors can be written as

$$v_c v_\rho = \frac{\mu}{d}\tan\tfrac{1}{2}\theta \qquad \text{where} \qquad d = \frac{r_1 r_2}{c}\sin\theta$$

and the chordal and radial components can be expressed in terms of the angular momentum $h$ as

$$v_c = \frac{h}{d} \qquad \text{and} \qquad v_\rho = \frac{\mu}{h}\tan\tfrac{1}{2}\theta$$

◇ **Problem 6–6**

The position and velocity vectors $\mathbf{r}$ and $\mathbf{v}$ for a conic may be expressed in terms of the initial values $\mathbf{r}_1$, $\mathbf{v}_1$, and the central angle $\theta$ by

$$\mathbf{r} = \mathbf{r}_1 + \frac{r_1^2}{h\cot\theta - \mathbf{r}_1\cdot\mathbf{v}_c}\mathbf{v}_c$$

$$\mathbf{v} = \mathbf{v}_c - v_\rho\,\mathbf{i}_r$$

where

$$h = \sqrt{(\mathbf{r}_1\times\mathbf{v}_1)\cdot(\mathbf{r}_1\times\mathbf{v}_1)}$$

$$v_\rho = \frac{\mu}{h}\tan\tfrac{1}{2}\theta \qquad \text{and} \qquad \mathbf{v}_c = \mathbf{v}_1 - v_\rho\,\mathbf{i}_{r_1}$$

## Problem 6-7

The hyperbolic locus of velocity vectors can be described in terms of the Gudermannian $\varsigma$ defined in Sect. 4.3. Specifically,

$$v^2 = v_m^2(1 + e_h^2 \tan^2 \varsigma)$$

where $e_h$ is the eccentricity of the hyperbolic locus and $v$ is the magnitude of the velocity vector corresponding to the value $\varsigma$. Also, the value of $\varsigma$ for the parabolic orbit, is obtained from

$$\tan^2 \varsigma_p = \frac{s(s - r_2)}{c(s - r_1)}$$

### Parameter in Terms of Flight-Direction Angle

According to the last subsection of Sect. 4.4, the vector $\mathbf{v}_c - \mathbf{v}_\rho$ is parallel to the tangent of the hyperbolic locus of velocity vectors. Also, at the terminal $P_1$ we have $A = v_1 \sin \gamma_1$ and $B = v_1 \sin(\phi_1 - \gamma_1)$, where $A$ and $B$ are defined in Prob. 4–19. From the result of that problem and the use of similar triangles,

$$\frac{v_c}{v_\rho} = \frac{\sin \gamma_1}{\sin(\phi_1 - \gamma_1)}$$

Therefore, the parameter of the orbit can be expressed in terms of the flight-direction angle $\gamma_1$ as

$$\frac{p}{p_m} = \frac{\sin \gamma_1}{\sin(\phi_1 - \gamma_1)} \tag{6.26}$$

Similarly, $A = v_2 \sin(\pi - \gamma_2)$ and $B = v_2 \sin(\phi_2 + \gamma_2 - \pi)$ at the terminal $P_2$, so that the parameter can also be written as

$$\frac{p}{p_m} = -\frac{\sin \gamma_2}{\sin(\phi_2 + \gamma_2)} \tag{6.27}$$

Finally, replace $\phi_1$ and $\phi_2$ with $\theta$, using the law of sines for the triangle $\Delta P_1 F P_2$, and obtain the alternate equations

$$\frac{p}{p_m} = \frac{c \sin \gamma_1}{r_1 \sin \gamma_1 + r_2 \sin(\theta - \gamma_1)} \tag{6.28}$$

$$\frac{p}{p_m} = \frac{c \sin \gamma_2}{r_2 \sin \gamma_2 - r_1 \sin(\theta + \gamma_2)} \tag{6.29}$$

Note that for $\theta = \pi$, Eqs. (6.28) and (6.29) both reduce to $p = p_m$ which is consistent with our earlier results. Furthermore, by equating these two expressions for $p$, we have an alternate derivation of Eq. (6.8).

As a practical application of these results, Eq. (6.29) provides an explicit solution to the problem of a spacecraft arriving at a target planet or returning to earth. Control of the atmospheric-entry angle is essential.

Sect. 6.1]  Terminal Velocity Vectors  249

If the vehicle is at position $\mathbf{r}_1$ and the desire is to enter the atmosphere at position $\mathbf{r}_2$ with a specified value of $\gamma_2$, Eq. (6.29) determines the parameter of the required orbit.

◇ **Problem 6–8**
Use the form of the hyperbolic locus of velocity vectors from Prob. 6–4 to derive the expression

$$\tan \gamma_1 = \frac{p r_2 \sin \theta}{r_1 r_2 (1 - \cos \theta) + p(r_2 \cos \theta - r_1)}$$

for the flight-direction angle and verify that this is equivalent to Eq. (6.28).

Relation Between Velocity and Eccentricity Vectors

Equation (3.28) for the velocity vector, obtained in Sect. 3.5, can be used for an interesting geometric determination of the velocity vector from the eccentricity vector. Since

$$\mathbf{v}_1 = \frac{1}{p} \mathbf{h} \times (\mathbf{e} + \mathbf{i}_{r_1})$$

the vector $\mathbf{v}_1$ must be perpendicular to the vector sum of the eccentricity vector $\mathbf{e}$ and the unit vector $\mathbf{i}_{r_1} = \mathbf{r}_1/r_1$.

Refer to Fig. 6.6 in which are shown the vectors $-\mathbf{i}_{r_1}$ and $\mathbf{e}$. A perpendicular from the point $P_1$ to the line connecting the termini of these two vectors is constructed. The direction of this perpendicular coincides with the direction of the velocity corresponding to the particular choice for $\mathbf{e}$. The extension of the perpendicular to the intersection with the hyperbolic locus determines both the magnitude and direction of $\mathbf{v}_1$.

A similar construction applies at the point $P_2$.

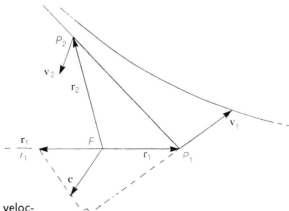

**Fig. 6.6:** Geometry of velocity and eccentricity vectors.

## 6.2 Orbit Tangents and the Transfer-Angle Bisector

A fundamental relationship exists between the tangents of an orbit at the terminals $P_1$, $P_2$ and the bisector of the transfer angle $\theta$. As shown in Fig. 6.7, *the line connecting the focus and the point of intersection of the orbital tangents at the terminals bisects the transfer angle.* This statement is certainly self-evident when $r_1 = r_2$.

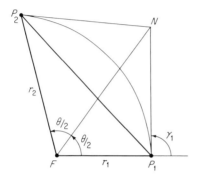

**Fig. 6.7:** Tangent-bisector property.

Before addressing the general proposition, it is instructive first to consider the 180 degree transfer for which $P_1$, $P_2$, and $F$ are colinear. From Eqs. (6.2) and (6.5) we have

$$\frac{\cot \gamma_1}{r_1} + \frac{\cot \gamma_2}{r_2} = 0 \quad \text{for} \quad \theta = \pi \tag{6.30}$$

or, equivalently,

$$r_1 \tan(\pi - \gamma_1) = r_2 \tan \gamma_2 = FN$$

where $N$ is the point of intersection of the orbital tangents at $P_1$ and $P_2$. Here $FN$ is perpendicular to the line connecting $P_1$ and $P_2$ so that $FN$ does, indeed, bisect the transfer angle $\theta = \pi$ as illustrated in Fig. 6.8.

Furthermore, from Sect. 4.2, we know that $\frac{1}{2}f + \gamma = \frac{1}{2}\pi$ obtains for the parabola. Therefore,

$$\gamma_1 - \gamma_2 = \tfrac{1}{2}(f_2 - f_1) = \tfrac{1}{2}\theta \quad \text{parabola} \tag{6.31}$$

or, in the special case being considered, $\gamma_1 = \gamma_2 + \frac{1}{2}\pi$. Thus, the angle $\angle P_1 N_p P_2 = \frac{1}{2}\pi$ for the parabolic orbit connecting $P_1$ and $P_2$ as shown in Fig. 6.8. It follows that $FN_p$ is the geometric mean between the radii $r_1$ and $r_2$, i.e.,

$$FN = \sqrt{r_1 r_2} \quad \text{parabola} \tag{6.32}$$

We shall see that this result holds also for arbitrary values of $\theta$.

Sect. 6.2]    Orbit Tangents and the Transfer-Angle Bisector    251

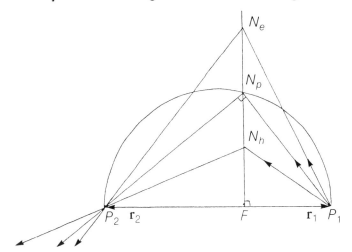

**Fig. 6.8:** Geometry for 180° transfer.

### Ellipse and Hyperbola

The demonstration of the tangent-bisector property is easily made for the ellipse using the geometry of orbital tangents developed in Sect. 4.1 and illustrated in Fig. 4.2.

As before, let $P_1$ and $P_2$ be points on the ellipse whose focus is at $F$ as shown in Fig. 6.9. The radii $FP_1$ and $FP_2$ are extended and intersect the circle (centered at $F$ and of radius equal to the major axis $2a$ of the ellipse) at the two points $Q_1$ and $Q_2$. Now, the perpendicular bisectors of $F^*Q_1$ and $F^*Q_2$ are the orbital tangents at $P_1$ and $P_2$, respectively, and intersect at $N$. Therefore, the point $N$ is the center of the circle through the three points $F^*$, $Q_1$, $Q_2$.

We have now two circles, one centered at $F$ and the other at $N$, each of which pass through $Q_1$ and $Q_2$. Hence, both $F$ and $N$ must lie on the perpendicular bisector of the line $Q_1Q_2$ which, of course, is also the bisector of the transfer angle $\angle Q_1FQ_2 \equiv \angle P_1FP_2$.

A similar argument applies to the hyperbola using the construction described at the end of the subsection on orbital tangents in Sect. 4.1.

A relationship involving the length of the line segment $FN$, analogous to Eq. (6.32), exists also for the ellipse and hyperbola. Apply the law of sines to the triangle $\Delta FP_1N$

$$\frac{FN}{\sin(\pi - \gamma_1)} = \frac{r_1}{\sin(\gamma_1 - \tfrac{1}{2}\theta)}$$

so that

$$FN = \frac{r_1}{\cos \tfrac{1}{2}\theta - \cot \gamma_1 \sin \tfrac{1}{2}\theta} \tag{6.33}$$

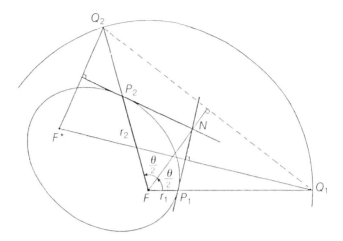

**Fig. 6.9:** Tangent-bisector property for elliptic orbits.

Next, adapt the analysis of the last subsection of Sect. 4.6 to the current situation so that Eqs. (4.98) may be written as

$$\sqrt{r_1 r_2} \cos \tfrac{1}{2}\theta = r_1 \cos \tfrac{1}{2}(E_2 - E_1) + \sqrt{ap}\cot \gamma_1 \sin \tfrac{1}{2}(E_2 - E_1)$$
$$\sqrt{r_1 r_2} \sin \tfrac{1}{2}\theta = \sqrt{ap} \sin \tfrac{1}{2}(E_2 - E_1) \quad (6.34)$$

When these are used in Eq. (6.33), we obtain

$$FN \cos \tfrac{1}{2}(E_2 - E_1) = \sqrt{r_1 r_2} \quad (6.35)$$

as the desired result to be compared with Eq. (6.32) for the parabola.

In a similar manner, using hyperbolic functions,

$$FN \cosh \tfrac{1}{2}(H_2 - H_1) = \sqrt{r_1 r_2} \quad (6.36)$$

is the corresponding relation for the hyperbola.

◇ **Problem 6-9**

To establish analytically the concurrency of the orbital tangents and the bisector of the transfer angle, assume that the tangents at $P_1$ and $P_2$ intersect the bisector at $N_1$ and $N_2$, respectively. Then derive

$$FN_1 = \frac{r_1 \csc \tfrac{1}{2}\theta}{\cot \tfrac{1}{2}\theta - \cot \gamma_1} \qquad FN_2 = \frac{r_2 \csc \tfrac{1}{2}\theta}{\cot \tfrac{1}{2}\theta + \cot \gamma_2}$$

and use Eq. (6.8) to demonstrate that

$$FN_1 = FN_2 = FN = \frac{(r_2 - r_1)\csc \tfrac{1}{2}\theta}{\cot \gamma_1 + \cot \gamma_2}$$

## Parabola

It is not without some benefit to establish the result of this section geometrically for the parabola. The geometric properties of the orbital tangents of a parabola are discussed in Sect. 4.2 and illustrated in Fig. 4.7. For the boundary-value problem geometry, refer to Fig. 6.10. Let $N_1$ and $N_2$ be points on the directrix of the parabola corresponding to points on the orbit $P_1$ and $P_2$. The orbital tangents at $P_1$ and $P_2$ intersect the axis of the parabola at $Q_1$ and $Q_2$. They also intersect the line, through the pericenter $A$ and normal to the axis, at $A_1$ and $A_2$. The point of intersection of the two orbital tangents is $N$. Finally, two other points of intersection are denoted by $S$ and $T$: (1) $S$ is the intersection of the line $FN_2$ and the tangent at $P_1$, while (2) $T$ is the intersection of the line $P_1 N_1$ and the tangent at $P_2$.

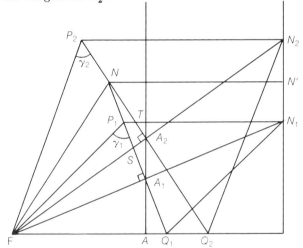

**Fig. 6.10:** Tangent-bisector property for parabolic orbits.

To prove the assertion that the line $FN$ bisects the angle $\angle P_1 F P_2 = \theta$, we first define two angles

$$\alpha = \angle P_1 N F \qquad \pi - \beta = \angle P_2 N F$$

Then, by summing the interior angles of the quadrilateral $FP_1NP_2$ and using Eq. (6.31), we establish that

$$\beta - \alpha = \gamma_1 - \gamma_2$$

Next, we apply the law of sines to the two triangles $\triangle FP_1N$ and $\triangle NP_2F$ to obtain

$$FN = \frac{r_1 \sin \gamma_1}{\sin \alpha} = \frac{r_2 \sin \gamma_2}{\sin \beta}$$

Then, from the equation of orbit for a parabola

$$r = \tfrac{1}{2} p \sec^2 \tfrac{1}{2} f = \tfrac{1}{2} p \csc^2 \gamma$$

we find that $\alpha$ and $\beta$ are further related by

$$\sin \gamma_1 \sin \alpha = \sin \gamma_2 \sin \beta$$

Solving for $\alpha$ and $\beta$ produces

$$\begin{aligned} \tan \alpha &= \tan \gamma_2 \\ \tan \beta &= \tan \gamma_1 \end{aligned} \quad \text{or, simply,} \quad \begin{aligned} \alpha &= \gamma_2 \\ \beta &= \gamma_1 \end{aligned}$$

Finally,

$$\angle P_1 FN = \gamma_1 - \gamma_2 = \angle P_2 FN = \tfrac{1}{2}\theta$$

which completes the proof.

The fact that the length of the line segment $FN$ is the geometric mean between the terminal radii follows from the similarity of the triangles $\triangle FP_1 N$ and $\triangle NP_2 F$. Thus,

$$\frac{FP_2}{FN} = \frac{FN}{FP_1}$$

and the generality of Eq. (6.32) is established.

One final result is of interest. Consider the triangle $\triangle NQ_1 Q_2$ and note that the base angles are

$$\angle Q_1 Q_2 P_2 = \gamma_2 \quad \text{and} \quad \angle Q_2 Q_1 P_1 = \pi - \gamma_1$$

Hence,

$$\angle Q_1 N Q_2 = \gamma_1 - \gamma_2 = \tfrac{1}{2}\theta$$

It follows that all three of the triangles $\triangle NP_1 T$, $\triangle FP_2 N$, $\triangle FP_1 N$ are similar. Therefore,

$$\frac{NT}{NP_1} = \frac{FN}{FP_1} = \frac{FN}{r_1} \quad \text{and} \quad \frac{NP_1}{FN} = \frac{NP_2}{FP_2} = \frac{NP_2}{r_2}$$

so that

$$NT = \sqrt{\frac{r_2}{r_1}}\, NP_1 = NP_2$$

If $N'$ is the intersection of the directrix and a line through $N$ parallel to the axis of the parabola, then

$$N_1 N' = N_2 N'$$

Hence, the line $NN'$ is equidistant from and parallel to the two lines $P_1 N_1$ and $P_2 N_2$.

Sect. 6.2]   Orbit Tangents and the Transfer-Angle Bisector   255

**Problem 6–10**

The equations of the orbital tangents at $P_1$ and $P_2$ for a parabola are

$$x + y \tan \tfrac{1}{2} f_1 = r_1 \quad \text{and} \quad x + y \tan \tfrac{1}{2} f_2 = r_2$$

where $x$, $y$ are cartesian coordinates centered at the focus $F$.

Consider a third straight line

$$y = mx$$

and show that the three straight lines have a common point of intersection if

$$m = \tan \tfrac{1}{2} (f_1 + f_2)$$

This provides an alternate proof that the orbital tangents and the bisector of the transfer angle intersect at a common point.

Finally, give an analytic demonstration of Eq. (6.32).

HINT: It will be convenient to use the determinantal equation of a straight line given in Prob. D–3 of Appendix D.

### Parameter in Terms of Eccentric-Anomaly Difference

Equation (6.28), for the parameter as a function of the flight-direction angle, may be written

$$\frac{p_m c}{p} = r_1 + r_2 (\sin \theta \cot \gamma_1 - \cos \theta)$$

$$= r_1 + r_2 - 2 r_2 \cos \tfrac{1}{2} \theta (\cos \tfrac{1}{2} \theta - \cot \gamma_1 \sin \tfrac{1}{2} \theta)$$

But, from Prob. 6–9 and Eq. (6.35)

$$FN = \frac{r_1}{\cos \tfrac{1}{2} \theta - \cot \gamma_1 \sin \tfrac{1}{2} \theta} = \frac{\sqrt{r_1 r_2}}{\cos \tfrac{1}{2} (E_2 - E_1)}$$

so that

$$\cos \tfrac{1}{2} \theta - \cot \gamma_1 \sin \tfrac{1}{2} \theta = \sqrt{\frac{r_1}{r_2}} \cos \tfrac{1}{2} (E_2 - E_1)$$

can be eliminated from the equation for $p$. We obtain

$$\frac{p}{p_m} = \frac{c}{r_1 + r_2 - 2\sqrt{r_1 r_2} \cos \tfrac{1}{2} \theta \cos \tfrac{1}{2} (E_2 - E_1)} \tag{6.37}$$

which expresses the parameter of an ellipse in terms of the difference between the eccentric anomalies of the two terminals.

Had we used Eq. (6.36) instead of (6.35), the equivalent expression

$$\frac{p}{p_m} = \frac{c}{r_1 + r_2 - 2\sqrt{r_1 r_2} \cos \tfrac{1}{2} \theta \cosh \tfrac{1}{2} (H_2 - H_1)} \tag{6.38}$$

for the hyperbola would have resulted.†

---

† These equations for the parameter are special cases of the expression given in Prob. 4–32 using universal functions. Also, the derivation here is quite different from that used in Chapter 4.

For the parabola, using Eq. (6.32), we have

$$\frac{p_p}{p_m} = \frac{c}{r_1 + r_2 - 2\sqrt{r_1 r_2}\cos\frac{1}{2}\theta}$$

and, by rationalizing the denominator,

$$\frac{p_p}{p_m} = \frac{1}{c}(r_1 + r_2 + 2\sqrt{r_1 r_2}\cos\frac{1}{2}\theta) \tag{6.39}$$

An interesting alternative is also possible. Since

$$r_1 + r_2 = 2s - c \quad \text{and} \quad r_1 r_2 \cos^2\tfrac{1}{2}\theta = s(s-c)$$

where $s$ is the semiperimeter of the triangle $\triangle P_1 F P_2$, then

$$\frac{p_p}{p_m} = \frac{c}{\left(\sqrt{s} - \sqrt{s-c}\right)^2} = \frac{1}{c}\left(\sqrt{s} + \sqrt{s-c}\right)^2 \tag{6.40}$$

## 6.3 The Fundamental Ellipse

Each orbit connecting $P_1$ and $P_2$ with focus at $F$ can be characterized by its associated eccentricity vector $\mathbf{e}$ which is directed from $F$ toward pericenter and whose length is the eccentricity of the orbit. The set of eccentricity vectors for the family of orbits connecting the two terminals possesses an extraordinary property that is easily demonstrated.

From the equation of orbit at $P_1$ and $P_2$ we have

$$\mathbf{e} \cdot \mathbf{r}_1 = p - r_1 \quad \text{and} \quad \mathbf{e} \cdot \mathbf{r}_2 = p - r_2$$

and by subtracting these two equations, obtain

$$\mathbf{e} \cdot (\mathbf{r}_2 - \mathbf{r}_1) = r_1 - r_2$$

In terms of the unit vector $\mathbf{i}_c$, directed along the chord from $P_1$ to $P_2$ and introduced in Sect. 6.1, this may be written as

$$-\mathbf{e} \cdot \mathbf{i}_c = \frac{r_2 - r_1}{c} \tag{6.41}$$

Thus, the set of eccentricity vectors has a constant projection on the chord $P_1 P_2$. In other words, *the locus of the termini of the eccentricity vectors is a straight line perpendicular to the chord*, as illustrated in Fig. 6.11. The distance from the focus $F$ to that line is $e_F$, where

$$e_F = \frac{|r_2 - r_1|}{c} \tag{6.42}$$

The orbit with eccentricity $e_F$ has, of course, the minimum eccentricity of all possible orbits connecting $P_1$ and $P_2$. Its eccentricity vector is

$$\mathbf{e}_F = \frac{r_1 - r_2}{c} \mathbf{i}_c \tag{6.43}$$

# Sect. 6.3]   The Fundamental Ellipse

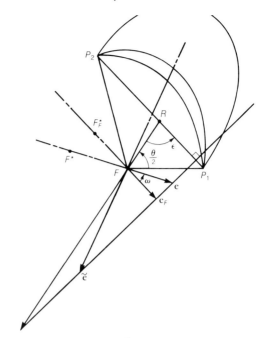

**Fig. 6.11:** Locus of the eccentricity vectors.

For any other orbit with eccentricity vector **e**, we have

$$\mathbf{e} = \mathbf{e}_F + \tan\omega\, \mathbf{i}_h \times \mathbf{e}_F \tag{6.44}$$

where $\omega$ is the angle between **e** and $\mathbf{e}_F$, and $\mathbf{i}_h$ is the unitized angular momentum vector. When $\theta$ is different from 180 degrees,

$$\mathbf{i}_h = \frac{\mathbf{r}_1 \times \mathbf{r}_2}{|\mathbf{r}_1 \times \mathbf{r}_2|} \tag{6.45}$$

(Of course, if $\theta = \pi$, the orbital plane is not determined by $\mathbf{r}_1$ and $\mathbf{r}_2$ so that the direction of $\mathbf{i}_h$ must be made using other considerations.)

◇ **Problem 6–11**

At the end of Sect. 6.1, an interesting scheme was developed for determining the velocity vector from the eccentricity vector. Solve the reverse problem—that is, given a terminal velocity vector at either $P_1$ or $P_2$, show how the orbital eccentricity vector can be obtained geometrically.

◇ **Problem 6–12**

For small transfer angles, the minimum possible eccentricity can be accurately computed as

$$e_F^2 = \frac{\epsilon^2}{\epsilon^2 + 4\dfrac{r_2}{r_1}\sin^2\tfrac{1}{2}\theta}$$

where $\epsilon$ is defined from

$$r_2 = r_1(1+\epsilon)$$

## The Fundamental (Minimum-Eccentricity) Ellipse

The orbit of minimum eccentricity will be called the *fundamental ellipse*. Since $P_1$ bears the same relation to the occupied focus $F$ as $P_2$ does to the vacant focus $F^*$, there is a certain symmetry which is apparent from the illustration of Fig. 6.12.

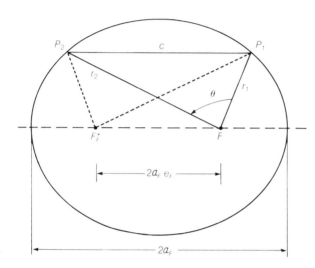

**Fig. 6.12:** Fundamental ellipse.

From the figure note that $P_1 P_2 F_F^* F$ is an isosceles trapezoid and, using the focal-radii property of the ellipse, it is clear that the semimajor axis $a_F$ is simply the arithmetic mean between the radii. That is,

$$a_F = \tfrac{1}{2}(r_1 + r_2) \tag{6.46}$$

To obtain the parameter it is convenient to use the notation

$$e_F = \sin \phi_F \tag{6.47}$$

recommended in Prob. 4-6. Then, since

$$p_F = a_F(1 - e_F^2) = a_F \cos^2 \phi_F$$

we calculate

$$\cos^2 \phi_F = 1 - e_F^2 = \frac{c^2 - (r_2 - r_1)^2}{c^2} = \frac{1}{c^2}(c + r_2 - r_1)(c + r_1 - r_2)$$

$$= \frac{4}{c^2}(s - r_1)(s - r_2) = \frac{4 r_1 r_2}{c^2} \sin^2 \tfrac{1}{2}\theta = \frac{2 p_m}{c} \tag{6.48}$$

to obtain

$$\frac{p_F}{p_m} = \frac{r_1 + r_2}{c} \tag{6.49}$$

where $p_m$ is the parameter of the minimum-energy orbit given in Eq. (6.19).

Sect. 6.3]   The Fundamental Ellipse   259

The two parabolas connecting $P_1$ and $P_2$ are conjugate orbits with eccentricity vectors $\mathbf{e}_p$ and $\tilde{\mathbf{e}}_p$ which bear a fascinating relationship to the fundamental ellipse. If $\omega_p$ and $-\omega_p$ are the corresponding angles that these vectors make with the chord, then it follows from the definition of $\phi_F$ that $\omega_p + \phi_F = \frac{1}{2}\pi$. Therefore, *the axes of the conjugate parabolic orbits coincide with the lines through the focus $F$ and the extremities of the minor axis of the fundamental ellipse* as shown in Fig. 6.13.

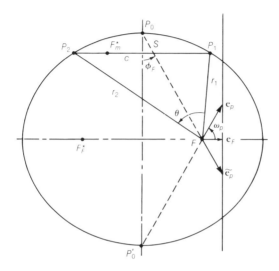

**Fig. 6.13:** Parabolic eccentricity vectors and the fundamental ellipse.

A geometric construction of the important elements of the fundamental ellipse directly from the triangle of the boundary-value problem is possible. To obtain the axes of the conjugate parabolas, refer to Fig. 6.14 and use the focus-directrix property of the parabola. The directrices $D_1 D_1'$ and $D_2 D_2'$ are the common tangents to the two circles centered at $P_1$ and $P_2$ of radii $r_1$ and $r_2$. The axes $A_1 A_1'$ and $A_2 A_2'$ of the two parabolas are the normals to these directrices through the focus $F$. The vertices $V_1$ and $V_2$ are the midpoints of the axes included between the focus $F$ and the directrices.

From this construction, the center and, therefore, the vertices of the fundamental ellipse are readily obtained. First, the extremities of the minor axis are located along the axes of the parabolas at a distance from $F$ equal to the arithmetic mean between the radii $r_1$ and $r_2$. The straight line connecting these points is the minor axis and intersects the major axis (the line through $F$ and parallel to the chord) at the center of the fundamental ellipse.

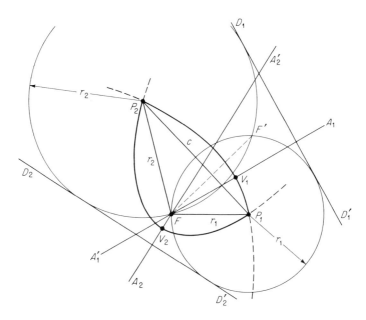

**Fig. 6.14:** Construction of the axes of the conjugate parabolas.

### Problem 6–13

The *conjugate* of the fundamental ellipse is that ellipse whose vacant focus $\widetilde{F}_F^*$ is the mirror image of $F_F^*$ with respect to the chord $P_1 P_2$. The quadrilateral $FP_1\widetilde{F}_F^*P_2$ is then a parallelogram and, of course, the semimajor axes of the fundamental ellipse and its conjugate are the same. Also, the orbital parameter and the semiminor axis are

$$\frac{\widetilde{p}_F}{p_m} = \frac{c}{r_1 + r_2} \quad \text{and} \quad \widetilde{b}_F = \sqrt{r_1 r_2} \sin \tfrac{1}{2}\theta = \sqrt{(s-r_1)(s-r_2)}$$

By simple geometry show that this orbit is characterized by the property that *the orbital tangents at the terminals are parallel to each other and to the bisector of the transfer angle $\theta$*. Finally, discuss the case for $\theta = 180°$.

### Problem 6–14

The vacant focus $F_F^*$ divides the major axis of the fundamental ellipse in the same proportion as the vacant focus $F_m^*$ of the minimum-energy ellipse divides the chord $P_1 P_2$.

### ◊ Problem 6–15

If $P_1$ and $P_2$ are any two points on an ellipse such that the straight line connecting them is parallel to the major axis, then the ratio of the focal-radii difference and the distance separating $P_1$ and $P_2$ is the eccentricity.

Sect. 6.3]     The Fundamental Ellipse     261

◊ **Problem 6–16**
The flight-direction angle of the fundamental ellipse at the terminal point $P_1$ is
$$\cot \gamma_1 = \frac{e_F \sin \phi_1}{1 - e_F \cos \phi_1}$$

HINT: Argue geometrically from Fig. 6.6.

## Intersection of the Transfer-Angle Bisector and the Chord

Let $R$ be the point of intersection of the chord and the bisector of the transfer angle $\theta$ with $\epsilon$ used to denote the angle $\angle FRP_1$. Then the distance from $R$ to the focus is either

$$FR = P_1 R \cos \epsilon + r_1 \cos \tfrac{1}{2}\theta \quad \text{or} \quad \begin{aligned} FR &= P_2 R \cos(\pi - \epsilon) + r_2 \cos \tfrac{1}{2}\theta \\ &= -P_2 R \cos \epsilon + r_2 \cos \tfrac{1}{2}\theta \end{aligned}$$

Hence, by subtraction,

$$\cos \epsilon = \frac{r_2 - r_1}{c} \cos \tfrac{1}{2}\theta = \sin \phi_F \cos \tfrac{1}{2}\theta \tag{6.50}$$

On the other hand, using the law of tangents for the triangle $\Delta P_1 F P_2$,

$$\frac{r_2 + r_1}{r_2 - r_1} = \frac{\tan \tfrac{1}{2}[\pi - (\epsilon + \tfrac{1}{2}\theta) + (\epsilon - \tfrac{1}{2}\theta)]}{\tan \tfrac{1}{2}[\pi - (\epsilon + \tfrac{1}{2}\theta) - (\epsilon - \tfrac{1}{2}\theta)]} = \frac{\tan \tfrac{1}{2}(\pi - \theta)}{\tan \tfrac{1}{2}(\pi - 2\epsilon)} = \frac{\cot \tfrac{1}{2}\theta}{\cot \epsilon}$$

Therefore,

$$\tan \epsilon = \frac{r_2 + r_1}{r_2 - r_1} \tan \tfrac{1}{2}\theta \tag{6.51}$$

and, combining with Eq. (6.50), gives

$$\sin \epsilon = \frac{r_1 + r_2}{c} \sin \tfrac{1}{2}\theta = \frac{p_F}{p_m} \sin \tfrac{1}{2}\theta \tag{6.52}$$

Finally, using the law of sines,

$$P_1 R = \frac{\sin \tfrac{1}{2}\theta}{\sin \epsilon} r_1 = \frac{r_1 c}{r_1 + r_2} \qquad P_2 R = \frac{\sin \tfrac{1}{2}\theta}{\sin(\pi - \epsilon)} r_2 = \frac{r_2 c}{r_1 + r_2} \tag{6.53}$$

so that

$$FR = \frac{r_1 c}{r_1 + r_2} \cos \epsilon + r_1 \cos \tfrac{1}{2}\theta = \left[\frac{r_1(r_2 - r_1)}{r_1 + r_2} + r_1\right] \cos \tfrac{1}{2}\theta$$

Hence, *the distance from $R$ to the focus is the product of the harmonic mean between the terminal radii and the cosine of half the transfer angle*, i.e.,

$$FR = \frac{2 r_1 r_2}{r_1 + r_2} \cos \tfrac{1}{2}\theta \tag{6.54}$$

There are also two interesting corollaries of this analysis. First, since

$$\frac{P_1 R}{P_2 R} = \frac{r_1}{r_2} \tag{6.55}$$

the transfer-angle bisector divides the chord in the same ratio as that of the terminal radii. Second, from

$$\frac{p_F}{p_m} = \frac{r_1}{P_1 R} = \frac{r_2}{P_2 R} \tag{6.56}$$

we see that *the ratio of the parameters of the fundamental and minimum-energy ellipses is the same as the ratio of either terminal radius to its corresponding fraction of the chord as cut by the transfer-angle bisector.*

## Parameter in Terms of Eccentricity

Consider now the problem of obtaining a functional relationship between the parameter and the eccentricity of an orbit. As a first step, note that the eccentricity vector may be expressed as a linear combination of $\mathbf{r}_1$ and $\mathbf{r}_2$ provided that $\mathbf{r}_1 \times \mathbf{r}_2 \neq \mathbf{0}$. Thus, we write

$$\mathbf{e} = A\,\mathbf{i}_{r_1} + B\,\mathbf{i}_{r_2} \tag{6.57}$$

and determine the coefficients $A$ and $B$ from

$$\mathbf{e} \cdot \mathbf{i}_{r_1} = \frac{p}{r_1} - 1 = A + B\cos\theta$$

$$\mathbf{e} \cdot \mathbf{i}_{r_2} = \frac{p}{r_2} - 1 = A\cos\theta + B$$

Hence,

$$\begin{aligned} A\sin^2\theta &= \left(\frac{p}{r_1} - 1\right) - \left(\frac{p}{r_2} - 1\right)\cos\theta \\ B\sin^2\theta &= \left(\frac{p}{r_2} - 1\right) - \left(\frac{p}{r_1} - 1\right)\cos\theta \end{aligned} \tag{6.58}$$

Next, calculate the magnitude of the vector product of $\mathbf{e}$ and $\mathbf{i}_c$

$$e\sin\omega = e_F \tan\omega = A\sin\phi_1 + B\sin\phi_2$$

where $\phi_1$ and $\phi_2$ are exterior angles of the triangle $\triangle FP_1P_2$ as defined in Sect. 6.1. Then, using the law of sines to replace $\phi_1$ and $\phi_2$ by $\theta$, we have

$$ce\sin\omega = ce_F \tan\omega = (Ar_2 + Br_1)\sin\theta$$

Finally, substitute for $A$ and $B$ from Eqs. (6.58). When solving for $p$, note that the coefficient of $p$ (according to the law of cosines) is simply $c^2$. Thus,

$$p = p_F + \frac{r_1 r_2}{c} e_F \tan\omega \sin\theta \tag{6.59}$$

Sect. 6.3]  The Fundamental Ellipse  263

is an expression for the parameter. The factor $e \sin \omega = e_F \tan \omega$ is simply the projection of the eccentricity vector $\mathbf{e}$ on the straight-line locus so that the parameter is seen to be a *linear function* of this projection.

An alternate form of Eq. (6.59) is also useful. Substituting for $p_F$ and $e_F$ from Eqs. (6.42) and (6.49) yields

$$\frac{p}{p_m} = \frac{r_2 + r_1}{c} + \frac{r_2 - r_1}{c} \tan \omega \cot \tfrac{1}{2}\theta$$

or, using Eq. (6.51),

$$\frac{p}{p_m} = e_F (\tan \epsilon + \tan \omega) \cot \tfrac{1}{2}\theta \qquad (6.60)$$

Therefore, we see that the parameter is zero when the eccentricity vector is colinear with the bisector of the transfer angle $\theta$. This vector corresponds to an orbit whose eccentricity is $e = \sec \tfrac{1}{2}\theta$ and whose semimajor axis is zero. The orbit is the singular hyperbola consisting of the two straight lines $P_2 F$ and $F P_1$ and, therefore, the orbit coincides with its asymptotes. Furthermore, since the parameter of an orbit cannot be negative, a lower limit on the straight-line locus of eccentricity vectors is established below which no such vector can extend.

◊ **Problem 6–17**
The eccentricity vector of the minimum-energy orbit is

$$\mathbf{e}_m = \frac{r_2 - s}{sc} \mathbf{r}_1 + \frac{r_1 - s}{sc} \mathbf{r}_2$$

HINT: Use the results of Prob. B–2 in Appendix B.

## Tangent Ellipses

Let the point $P_1$ be the pericenter of the ellipse whose apsidal line coincides with the line $FP_1$. Since this orbit is tangent at $P_1$ to a circle of radius $r_1$ and centered at $F$, we use the term *tangent ellipse* when referring to this orbit.

To obtain the eccentricity $e_{t_1}$, note that

$$e_{t_1} = e_F \sec \angle FP_1 P_2$$

with the angle $\angle FP_1 P_2$ determined from

$$r_1 = c \cos \angle FP_1 P_2 - r_2 \cos(\pi - \theta)$$

Hence,

$$e_{t_1} = \frac{r_2 - r_1}{r_1 - r_2 \cos \theta} = e_F \sec(\pi - \phi_1) \qquad (6.61)$$

The semimajor axis is found from

$$a_{t_1}(1 - e_{t_1}) = r_1$$

so that the parameter is

$$\frac{p_{t_1}}{p_m} = \frac{c}{r_1 - r_2 \cos\theta} = \cos(\pi - \phi_1) \qquad (6.62)$$

Of course, the parameter can also be obtained from Eq. (6.28) with the value $\gamma_1 = \tfrac{1}{2}\pi$.

In a similar manner,

$$e_{t_2} = \frac{r_2 - r_1}{r_2 - r_1 \cos\theta} = e_F \sec(\pi - \phi_2)$$

$$\frac{p_{t_2}}{p_m} = \frac{c}{r_2 - r_1 \cos\theta} = \cos(\pi - \phi_2)$$

(6.63)

obtain for the tangent ellipse at $P_2$.

## 6.4 A Mean Value Theorem

The mean value theorem of the differential calculus states that on any smooth arc of a curve joining the points $P_1$ and $P_2$, there is at least one intermediate point $P_0$ such that the tangent to the curve at $P_0$ is parallel to the chord joining $P_1$ and $P_2$. This property is trivially evident for the orbital boundary-value problem but here there exist several rather dramatic consequences not readily anticipated.

We are concerned with determining that point in an orbit connecting $P_1$ and $P_2$ for which the orbital tangent, i.e., the velocity vector, is parallel to the chord $P_1 P_2$. For this purpose, calculate the scalar product of the eccentricity vector and the vector $\mathbf{r}_2 - \mathbf{r}_1$ in the form

$$\mu \mathbf{e} \cdot (\mathbf{r}_2 - \mathbf{r}_1) = (\mathbf{r}_2 - \mathbf{r}_1) \times \mathbf{v} \cdot \mathbf{h} - \frac{\mu}{r}\mathbf{r} \cdot (\mathbf{r}_2 - \mathbf{r}_1)$$

where we have permuted the factors in the triple scalar product. The particular value of $\mathbf{r}$ desired (call it $\mathbf{r}_0$) corresponds to a value $\mathbf{v} = \mathbf{v}_0$ for which

$$(\mathbf{r}_2 - \mathbf{r}_1) \times \mathbf{v}_0 = 0$$

We also know that the eccentricity vector has the property

$$\mathbf{e} \cdot (\mathbf{r}_2 - \mathbf{r}_1) = r_1 - r_2 \qquad (6.64)$$

as demonstrated in Sect. 6.3. Therefore,

$$\mathbf{e} \cdot (\mathbf{r}_2 - \mathbf{r}_1) = -\frac{1}{r_0}\mathbf{r}_0 \cdot (\mathbf{r}_2 - \mathbf{r}_1) = r_1 - r_2$$

or, simply,

$$\mathbf{r}_0 \cdot (\mathbf{r}_2 - \mathbf{r}_1) = r_0(r_2 - r_1) \qquad (6.65)$$

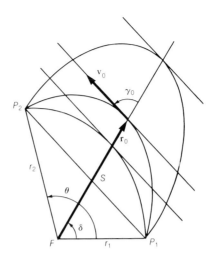

**Fig. 6.15:** Locus of the mean points.

—a relationship which is *independent* of the particular orbit in question†  and is, consequently, valid for *all* orbits.

From this equation we see that the direction of $\mathbf{r}_0$ is constant and depends only on the geometry of the boundary-value problem. Call the point $P_0$, where the parallelism occurs, the *mean point* of the orbit so that $\mathbf{r}_0$ is the position vector of the mean point as shown in Fig. 6.15. The fact that *all mean-point position vectors are colinear* is remarkable.

What is even more remarkable is seen by formally replacing $\mathbf{r}_0$ by either $-\tilde{\mathbf{e}}_p$ or $-\mathbf{e}_p$ in Eq. (6.65). The result is identical with Eq. (6.64) if $\mathbf{e}$ is the eccentricity vector of the parabolic orbits. Therefore, *the straight-line locus of mean points for all orbits coincides with the line connecting the focus and the extremities of the minor axis of the fundamental ellipse.*

The flight-direction angle $\gamma_0$ at the mean point is the angle between the orbital tangent and the radius to the mean point. Clearly, $\gamma_0$ is the same for all orbits of the boundary-value problem and, indeed, is identical to the angle $\omega_p$. Thus,

$$\gamma_0 = \omega_p = \tfrac{1}{2}\pi - \phi_F \tag{6.66}$$

The trigonometric relations of $\phi_F$ are recorded as Eqs. (6.47) and (6.48).

---

† This was discovered by Gerald M. Levine in connection with an optical-sighting problem for orbital navigation. He showed that the true anomaly of the point in an orbit where the velocity vector is parallel to the line of sight from the initial point to the terminal point is independent of the orbit. Levine's paper, "A Method of Orbital Navigation Using Optical Sightings to Unknown Landmarks," appeared in the *AIAA Journal*, Vol. 4, Nov. 1966, pp. 1928–1931. For many years the result had only academic interest.

## Geometry of the Mean-Point Locus

Again, refer to Fig. 6.13 and let $\delta$ be the angle between $\mathbf{r}_1$ and the axis of the parabola whose eccentricity vector is $\tilde{\mathbf{e}}_p$. Further, let $S$ denote the point of intersection of this axis and the chord $P_1 P_2$. Our objective will be to determine $\delta$ and $FS$ in terms of the geometry of the triangle $\triangle F P_1 P_2$.

Equation (6.65) can be written as

$$r_2 \cos(\theta - \delta) - r_1 \cos \delta = r_2 - r_1 \qquad (6.67)$$

or, equivalently,

$$r_1 \sin^2 \tfrac{1}{2}\delta = r_2 \sin^2 \tfrac{1}{2}(\theta - \delta) \qquad (6.68)$$

From this last expression, we derive

$$\tan \tfrac{1}{2}\delta = \frac{\sin \tfrac{1}{2}\theta}{\sqrt{\frac{r_1}{r_2}} + \cos \tfrac{1}{2}\theta} \quad \text{or} \quad \tan \tfrac{1}{2}(\theta - \delta) = \frac{\sin \tfrac{1}{2}\theta}{\sqrt{\frac{r_2}{r_1}} + \cos \tfrac{1}{2}\theta} \qquad (6.69)$$

as more convenient formulas for computing the angles $\delta$ or $\theta - \delta$.

Next, apply the law of sines to the triangle $\triangle F P_1 S$

$$FS = \frac{\sin(\gamma_0 + \delta)}{\sin \gamma_0} r_1$$

Then, from the triangle $\triangle F P_1 P_2$,

$$\sin(\gamma_0 + \delta) = \frac{r_2}{c} \sin \theta$$

so that

$$FS = \sqrt{r_1 r_2} \cos \tfrac{1}{2}\theta \qquad (6.70)$$

The length $FS$, which is the mean-point radius of the singular hyperbola consisting of the straight line connecting the terminals $P_1$ and $P_2$, is determined as *the product of the geometric mean between the terminal radii and the cosine of half the transfer angle*. It is an important quantity which will be frequently encountered.

The two line segments $FS$ and $FR$, introduced in the previous section, are beautifully related. They are the distances from the focus to the chord measured, respectively, along the mean-point locus and along the transfer-angle bisector. The first involves the geometric mean between the terminal radii and the second, the harmonic mean. Furthermore,

$$\frac{FR}{FS} = \frac{\sqrt{r_1 r_2}}{\tfrac{1}{2}(r_1 + r_2)} < 1 \qquad (6.71)$$

so that *the ratio of $FR$ and $FS$ is the same as the ratio of the geometric and arithmetic means between $r_1$ and $r_2$*.

## A Mean Value Theorem

Finally, consider the ratio of the segments of the chord divided by the locus of mean points. Since,

$$P_1 S = \frac{r_1 \sin \delta}{\sin \gamma_0} \quad \text{and} \quad P_2 S = \frac{r_2 \sin(\theta - \delta)}{\sin(\pi - \gamma_0)}$$

we have

$$\frac{P_1 S}{P_2 S} = \frac{r_1 \sin \delta}{r_2 \sin(\theta - \delta)}$$

But, from Eq. (6.68),

$$\frac{r_1}{r_2} = \frac{\sin^2 \frac{1}{2}(\theta - \delta)}{\sin^2 \frac{1}{2}\delta}$$

so that

$$\frac{P_1 S}{P_2 S} = \frac{\sin \frac{1}{2}(\theta - \delta) \cos \frac{1}{2}\delta}{\sin \frac{1}{2}\delta \cos \frac{1}{2}(\theta - \delta)} = \frac{\tan \frac{1}{2}(\theta - \delta)}{\tan \frac{1}{2}\delta}$$

Then, since the expressions for the tangents of $\delta$ and $\theta - \delta$ in Eqs. (6.69) may also be written as

$$\tan \tfrac{1}{2}\delta = \frac{\sqrt{r_1 r_2} \sin \tfrac{1}{2}\theta}{r_1 + FS} \quad \text{and} \quad \tan \tfrac{1}{2}(\theta - \delta) = \frac{\sqrt{r_1 r_2} \sin \tfrac{1}{2}\theta}{r_2 + FS}$$

the ratio is simply

$$\frac{P_1 S}{P_2 S} = \frac{r_1 + FS}{r_2 + FS} \tag{6.72}$$

### The Mean-Point Radius

For an arbitrary ellipse connecting $P_1$ and $P_2$ we can determine the eccentric anomaly of the mean point $P_0$. To this end, refer to Fig. 6.16, where we have constructed the locus of mean points together with the two angle bisectors –one bisecting the angle $\delta$ and the other bisecting $\theta - \delta$. Also shown in the figure is an arbitrary elliptic orbit as well as the parabolic orbit connecting $P_1$ and $P_2$. Tangents to each of these orbits are constructed at $P_1$ and $P_2$ as well as those parallel to $P_1 P_2$. The various points of intersection of these lines are labeled in the figure.

Let $E_1$ and $E_2$ be the eccentric anomalies of the ellipse corresponding to the points $P_1$ and $P_2$. Further, let $r_0$ and $E_0$ denote, respectively, the radius and eccentric anomaly of the mean point on the elliptic orbit. The corresponding radius for the parabola is $r_{0_p}$. Then, by suitably adapting Eqs. (6.32) and (6.35) to the purpose, we may write for the ellipse

$$FN_1 \cos \tfrac{1}{2}(E_0 - E_1) = \sqrt{r_1 r_0} \qquad FN_2 \cos \tfrac{1}{2}(E_2 - E_0) = \sqrt{r_0 r_2}$$

and for the parabola

$$FN_{1p} = \sqrt{r_1 r_{0_p}} \qquad FN_{2p} = \sqrt{r_{0_p} r_2}$$

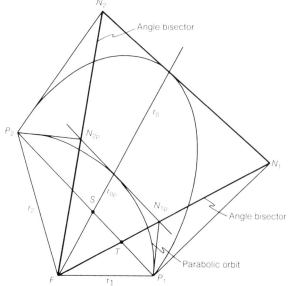

**Fig. 6.16:** Geometry for eccentric anomaly of the mean point.

Since triangle $\Delta FN_{1p}N_{2p}$ is similar to triangle $\Delta FN_1N_2$, then

$$\frac{FN_{1p}}{FN_{2p}} = \frac{FN_1}{FN_2}$$

or, simply,

$$\cos \tfrac{1}{2}(E_0 - E_1) = \cos \tfrac{1}{2}(E_2 - E_0)$$

Hence,

$$E_0 = \tfrac{1}{2}(E_1 + E_2) \tag{6.73}$$

so that

$$r_0 = a[1 - e \cos \tfrac{1}{2}(E_1 + E_2)] \tag{6.74}$$

where $a$ and $e$ are the semimajor axis and eccentricity of the ellipse. Thus, we obtain the fundamental result† that *the eccentric anomaly of the mean point of an orbit connecting two termini is the arithmetic mean between the eccentric anomalies of those termini.*

For elliptic orbits there are, of course, two mean points—the second for that portion of the orbit traversed from $P_2$ to $P_1$ as seen in the figure.

---

† This is an important corollary to *Lambert's Theorem* (to be developed in Sect. 6.6) which first appeared in a paper entitled "A New Transformation Invariant in the Orbital Boundary-Value Problem" by Richard H. Battin, Thomas J. Fill, & Stanley W. Shepperd. It was published in the first issue of the *Journal of Guidance and Control*, Vol. 1, Jan.-Feb., 1978, pp. 50–55.

It is not difficult to see that the eccentric anomaly of the second mean point is just $E_0 + \pi$.

Clearly, an analogous result holds for hyperbolic orbits. In a manner similar to that used for the ellipse, we can derive

$$H_0 = \tfrac{1}{2}(H_1 + H_2) \tag{6.75}$$
$$r_0 = a[1 - e \cosh \tfrac{1}{2}(H_1 + H_2)] \tag{6.76}$$

where $H_0$, $H_1$, and $H_2$ have the obvious meanings.

### Elegant Expressions for the Mean-Point Radii

The true anomaly of the mean point of the parabola connecting $P_1$ and $P_2$ is $2\phi_F$. Therefore, the mean point radius of the parabola is simply

$$r_{0p} = \frac{p_p}{1 + \cos 2\phi_F} = \frac{p_p}{2\cos^2 \phi_F} = \frac{p_p c}{4p_m}$$

—the last step following from Eq. (6.48). Finally, substituting from Eq. (6.39), we obtain

$$r_{0p} = \tfrac{1}{4}(r_1 + r_2 + 2\sqrt{r_1 r_2} \cos \tfrac{1}{2}\theta) = \tfrac{1}{2}(a_F + FS)$$

as the mean-point radius of the parabola whose eccentricity vector is $\mathbf{e}_p$. In other words, *the mean-point radius of the parabola is the arithmetic mean between the semimajor axis of the fundamental ellipse and the line segment* $FS$.

It turns out that the mean-point radius for the parabolic orbit characterized by $\tilde{\mathbf{e}}_p$ is the same equation with the plus sign preceding the last term replaced by a minus sign. This is shown most easily using Eq. (6.21). Thus, we have

$$r_{0p} = \tfrac{1}{4}(r_1 + r_2 \pm 2\sqrt{r_1 r_2} \cos \tfrac{1}{2}\theta) = \tfrac{1}{4}\left(\sqrt{s} \pm \sqrt{s-c}\right)^2 \tag{6.77}$$

It may also be of interest to record the parameter of the parabolic orbits in the form

$$p_p = \frac{4 p_m r_{0p}}{c} = 2 r_{0p} \cos^2 \phi_F \tag{6.78}$$

which is a direct consequence of the equation of orbit used at the beginning of this subsection.

A remarkable relation exists for the mean-point radius of any orbit in terms of the mean-point radius of the parabola. From the derivation of the eccentric anomaly of the mean point, we have

$$FN_2 \cos \tfrac{1}{2}(E_2 - E_0) = FN_2 \cos \tfrac{1}{4}(E_2 - E_1) = \sqrt{r_0 r_2}$$
$$FN_{2p} = \sqrt{r_{0p} r_2}$$

so that

$$\frac{FN_2}{FN_{2p}} \cos \tfrac{1}{4}(E_2 - E_1) = \sqrt{\frac{r_0}{r_{0p}}}$$

But, from similar triangles,

$$\frac{FN_2}{FN_{2p}} = \frac{r_0}{r_{0p}}$$

Therefore, we have the truly elegant expression

$$r_0 = r_{0p} \sec^2 \tfrac{1}{4}(E_2 - E_1) \tag{6.79}$$

and, as we might expect,

$$r_0 = r_{0p} \operatorname{sech}^2 \tfrac{1}{4}(H_2 - H_1) \tag{6.80}$$

obtains also for hyperbolic orbits. Equations (6.79) and (6.80) will prove essential in an important orbit determination scheme to be considered in the next chapter.

## Parameter in Terms of Mean-Point Radius

We can determine a simple functional relation between the parameter of an orbit and the radius to the associated mean point. By adapting Eq. (6.29) to our purpose, we have

$$p = \frac{2r_0 r_1 \sin \gamma_0 \sin^2 \tfrac{1}{2}\delta}{r_0 \sin \gamma_0 - r_1 \sin(\gamma_0 + \delta)} \tag{6.81}$$

But, in an earlier subsection of this section, we found that

$$\frac{\sin(\gamma_0 + \delta)}{\sin \gamma_0} r_1 = FS = \sqrt{r_1 r_2} \cos \tfrac{1}{2}\theta$$

Therefore, Eq. (6.81) can be written as

$$p = \frac{2r_0 r_1 \sin^2 \tfrac{1}{2}\delta}{r_0 - \sqrt{r_1 r_2} \cos \tfrac{1}{2}\theta} \tag{6.82}$$

which expresses the parameter $p$ as a *bilinear function* of $r_0$. (Note that for an elliptic orbit, the value of $r_0$ to be used in this equation corresponds only to the mean point on that portion of the orbit traversed from $P_1$ to $P_2$ in the counterclockwise direction.)

Using Eq. (6.82), it is easy to see that $dp/dr_0$ is always negative. Further, $p$ is infinite for the singular hyperbola whose mean-point radius is $FS = \sqrt{r_1 r_2} \cos \tfrac{1}{2}\theta$.

## Locus of the Vacant Focus

An alternate and somewhat simpler expression for the parameter is possible. From the first of Eqs. (6.69),

$$\sin^2 \tfrac{1}{2}\delta = \frac{\tan^2 \tfrac{1}{2}\delta}{1+\tan^2 \tfrac{1}{2}\delta} = \frac{r_2 \sin^2 \tfrac{1}{2}\theta}{r_1+r_2+2\sqrt{r_1 r_2}\cos \tfrac{1}{2}\theta} = \frac{r_2}{4 r_{Op}} \sin^2 \tfrac{1}{2}\theta$$

so that Eq. (6.82) may be written as

$$p = \frac{\widetilde{p}_p r_0}{r_0 - \sqrt{r_1 r_2}\cos \tfrac{1}{2}\theta} = \frac{\widetilde{p}_p r_0}{r_0 - FS} = \frac{\widetilde{p}_p}{1 - \dfrac{FS}{FP_0}} \qquad (6.83)$$

since

$$\frac{\widetilde{p}_p}{p_m} = \frac{c}{4 r_{Op}}$$

—a relation which follows from the expression for the parameter of the parabola and Eq. (6.21).

◇ **Problem 6–18**
The limiting value of $p$, as $r_0$ becomes infinite, is the parameter of the parabolic orbit transcribed in the counterclockwise direction from $P_2$ to $P_1$. Explain this result geometrically.

◇ **Problem 6–19**
For any orbit the parameter is

$$p = \frac{\widetilde{p}_F}{1 - \dfrac{FR}{FN}}$$

where $\widetilde{p}_F$ is the parameter of the conjugate of the fundamental ellipse. Also, $FR$ and $FN$ are distances, measured along the transfer-angle bisector, from the focus to the chord and to the intersection of the orbital tangents at the terminal points, respectively.

NOTE: The two expressions for the parameter—the one here and the similar one of Eq. (6.83)—are truly delightful and worthy of a few moments contemplation. Gauss was right—there are "an abundance of elegant propositions."

## 6.5 Locus of the Vacant Focus

We explore now a different characterization of the orbital boundary-value problem based on the fact that an orbit is uniquely determined when the second focus, the vacant focus, $F^*$ is specified. Since $F^*$ cannot be placed arbitrarily, it will be of interest to find the locus of the vacant foci for all orbits that satisfy the conditions of the problem.

### Elliptic Orbits

The points $P_1$ and $P_2$ both lie on the ellipse so that the point $F^*$ must be selected such that

$$P_1 F + P_1 F^* = P_2 F + P_2 F^* = 2a$$

or, equivalently,

$$P_1 F^* = 2a - r_1 \qquad P_2 F^* = 2a - r_2 \qquad (6.84)$$

Thus, for an ellipse of major axis $2a$, the point $F^*$ is determined as the point of intersection of two circles centered at $P_1$ and $P_2$ with respective radii $2a - r_1$ and $2a - r_2$. A number of such circles have been constructed in Fig. 6.17 for different values of the major axis $2a$.

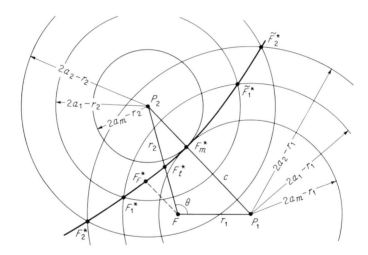

**Fig. 6.17:** Locus of the vacant foci for elliptic orbits.

If the selected value of $2a$ is too small, the circles will not intersect. The circles will be tangent for the minimum-energy ellipse, as defined in Sect. 6.1, with

$$a = a_m = \tfrac{1}{2} s$$

and the point of tangency $F_m^*$ lying on the chord $P_1 P_2$.

When $a > a_m$, the circle pairs intersect in two points $F^*$ and $\widetilde{F}^*$. Thus, there are in general a pair of conjugate elliptic orbits connecting $P_1$ and $P_2$—orbits with the same major axis but with vacant foci equidistant from and on opposite sides of the chord.†

---

† An example, already considered in Sect. 6.3, is the fundamental ellipse and its conjugate for which the vacant foci are $F_F^*$, $\widetilde{F}_F^*$ and $a_F = \tfrac{1}{2}(r_1 + r_2)$.

By subtracting the equations in (6.84), we have

$$P_2 F^* - P_1 F^* = -(r_2 - r_1) \tag{6.85}$$

indicating that the locus of $F^*$ is a hyperbola whose major axis is $-(r_2-r_1)$ and whose foci are $P_1$ and $P_2$. The distance between the foci is the chord length $c$ so that

$$2a^* = -(r_2 - r_1) \qquad 2a^* e^* = -c$$

where $a^*$ and $e^*$ are the the semimajor axis and eccentricity of the hyperbolic locus. It follows that

$$e^* = \frac{c}{r_2 - r_1} = \frac{1}{e_F} \tag{6.86}$$

Thus, *the eccentricity of the hyperbolic locus of vacant foci is the reciprocal of the eccentricity of the fundamental ellipse.*

The slope of the asymptotes of the hyperbolic locus is

$$\frac{b^*}{-a^*} = \sqrt{e^{*2} - 1} = \frac{\sqrt{1 - e_F^2}}{e_F} = \cot \phi_F = \tan \gamma_0 = \tan \omega_p$$

which shows that these *asymptotes are parallel to the axes of the two parabolic orbits connecting $P_1$ and $P_2$*. They may be constructed following the argument in Sect. 6.3 centered around Fig. 6.14. We see that the asymptotes of the hyperbolic locus of vacant foci are the normals to the directrices which pass through the midpoint of the chord $P_1 P_2$.

### Hyperbolic Orbits

As before, we argue that $P_1$ and $P_2$ must both lie on the hyperbolic orbit so that

$$P_1 F - P_1 F^* = P_2 F - P_2 F^* = 2a$$

or, equivalently,

$$P_1 F^* = r_1 - 2a \qquad P_2 F^* = r_2 - 2a \tag{6.87}$$

Again, $F^*$ is determined as the intersection of two circles centered at $P_1$ and $P_2$ of respective radii $r_1 - 2a$ and $r_2 - 2a$ shown in Fig. 6.18.

The semimajor axis $a$ of the hyperbolic orbit is, of course, negative with $-\infty < a \leq 0$. All points of intersection of the circle pairs fall outside the circle centered at $P_1$ and of radius $r_1$. One may easily verify that, for hyperbolic paths from $P_1$ to $P_2$ that are convex with respect to the focus $F$, the vacant foci all lie within this circle.

By subtracting the equations in (6.87), we have

$$P_1 F^* - P_2 F^* = -(r_2 - r_1) \tag{6.88}$$

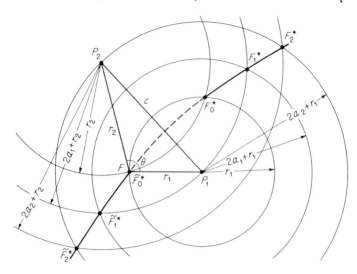

**Fig. 6.18:** Locus of the vacant foci for hyperbolic orbits.

which shows that the locus of these foci is the conjugate branch of the hyperbolic locus of the foci for the elliptic orbits.

The foci $F_0^*$ and $\widetilde{F}_0^*$, corresponding to $a = 0$, are extreme cases in that infinite velocities are required to describe the associated paths. The orbit whose vacant focus is at $F_0^*$ is the straight line connecting $P_1$ and $P_2$, i.e., the hyperbola for which $a = 0$ and $e = \infty$. Corresponding to the focus $\widetilde{F}_0^*$, the orbit is composed of the two straight-line segments from $P_2$ to $F$ and from $F$ to $P_1$. This is the hyperbola for which $a = 0$ and $e = \sec \frac{1}{2}\theta$.

### Parameter in Terms of Semimajor Axis

We can determine an equation for the parameter as a function of the semimajor axis† $a$ by starting with the expression for the eccentricity vector given in Eq. (6.57). We calculate

$$\mathbf{e} \cdot \mathbf{e} = e^2 = 1 - \frac{p}{a} = A^2 + 2AB \cos \theta + B^2$$

---

† Specifying $a$ is, of course, equivalent to specifying the velocity magnitudes at the end points. This form of the equation for the orbital parameter was originally presented by the author at the Nineteenth International Astronautical Congress held in New York City in October, 1968 as a part of a paper entitled "A New Solution for Lambert's Problem." It appeared in Volume 2 of *Astrodynamics and Astrionics* published by Pergamon Press in 1970. The more conventional form of the parameter, as a function of the semimajor axis, is treated at the end of the next section. Some of the material covered in the next chapter is also based on the contents of that paper.

and, using the expressions for $A$ and $B$ in Eqs. (6.58), obtain

$$\left(\frac{p}{r_1}-1\right)^2 - 2\left(\frac{p}{r_1}-1\right)\left(\frac{p}{r_2}-1\right)\cos\theta + \left(\frac{p}{r_2}-1\right)^2 = \left(1-\frac{p}{a}\right)\sin^2\theta$$

If we now collect terms in powers of $p$ (noting that the coefficient of $p^2$ is simply $ac^2$) and make further obvious simplifications, we find

$$p^2 - 2Dp_m p + p_m^2 = 0$$

as a quadratic equation for the parameter $p$ where

$$acD = a(r_1+r_2) - r_1 r_2 \cos^2\theta = a(2s-c) - s(s-c)$$

The coefficient $D$ can be written in the form

$$D = \frac{s-c}{c}\left(1-\frac{s}{2a}\right) + \frac{s}{c}\left(1-\frac{s-c}{2a}\right)$$

where each of the two terms is positive and the second is just one plus the first. Since these terms may have any value from zero to infinity, it is convenient to write

$$D = \cot^2\nu + \csc^2\nu \quad \text{for} \quad 0 \le \nu < \tfrac{1}{2}\pi$$

Then, the solution of the quadratic is

$$\frac{p}{p_m} = (\csc\nu \pm \cot\nu)^2 \tag{6.89}$$

where

$$\cot^2\nu = \frac{s-c}{c}\left(1-\frac{s}{2a}\right) \qquad \csc^2\nu = \frac{s}{c}\left(1-\frac{s-c}{2a}\right) \tag{6.90}$$

Finally, if the definition of $\nu$ is extended to the second as well as the first quadrant, both solutions for $p$ are contained in the single equation

$$\frac{p}{p_m} = \tan^2\tfrac{1}{2}\nu \quad \text{for} \quad 0 \le \nu \le \pi \tag{6.91}$$

Unlike the semimajor axis, the parameter is unique for each orbit of the boundary-value problem and increases monotonically with $\nu$. The values $\nu = 0$, $\tfrac{1}{2}\pi$, and $\pi$ correspond, respectively, to the orbits whose vacant foci are at $\widetilde{F}_0^*$, $F_m^*$, and $F_0^*$. The foci marked with a tilde are associated with values of $\nu$ in the first quadrant while those without correspond to $\nu$ in the second quadrant. For $\cot\nu = \pm\sqrt{(s-c)/c}$ the vacant foci are at infinity and the orbits are parabolic.

## Problem 6-20
The chordal and the extended radial components of the velocity vectors at $P_1$ and $P_2$ can be expressed in terms of the angle $\nu$ as follows:

$$v_c = \tfrac{1}{2}v_{m_1}\sec\tfrac{1}{2}\phi_1 \tan\tfrac{1}{2}\nu = \tfrac{1}{2}v_{m_2}\sec\tfrac{1}{2}\phi_2 \tan\tfrac{1}{2}\nu$$

$$v_\rho = \tfrac{1}{2}v_{m_1}\sec\tfrac{1}{2}\phi_1 \cot\tfrac{1}{2}\nu = \tfrac{1}{2}v_{m_2}\sec\tfrac{1}{2}\phi_2 \cot\tfrac{1}{2}\nu$$

## 6.6 Lambert's Theorem

Among the many surprising properties of conics is the lack of dependence on eccentricity for certain of the key orbital quantities. For example, one would scarcely anticipate that the period of elliptic motion depends only on the semimajor axis of the ellipse. The fact that a body moving in a conic orbit has a velocity magnitude or speed that is a function only of the distance from the center of force and the semimajor axis is another example.

Perhaps the most remarkable theorem in this connection is the one discovered by Lambert having to do with the time to traverse an elliptic arc.† Lambert conjectured that *the orbital transfer time depends only upon the semimajor axis, the sum of the distances of the initial and final points of the arc from the center of force, and the length of the chord joining these points.* If $t_2 - t_1$ is the time to describe the arc from $P_1$ to $P_2$, then Lambert's theorem states that

$$\sqrt{\mu}(t_2 - t_1) = F(a, r_1 + r_2, c) \tag{6.92}$$

and again the eccentricity is not involved.

### Euler's Equation for Parabolic Orbits

Euler developed the special case of Lambert's theorem for the parabola but subsequently neglected it and did not extend it to the ellipse and hyperbola. Therefore, Lambert deserves full credit for independently obtaining the result when it was otherwise "*buried in oblivion*" (as Gauss expressed it) and extending it to the remaining conic sections.

As a preliminary to the general case, we begin by deriving Euler's equation which Euler published in 1743 in an article on the determination of comet orbits.

Barker's equation (4.10) is written in the form

$$6\sqrt{\mu}(t - \tau) = 3p\sigma + \sigma^3$$

where

$$\sigma = \sqrt{p} \tan \tfrac{1}{2} f$$

With subscripts distinguishing quantities at $P_1$ and $P_2$, we derive

$$6\sqrt{\mu}(t_2 - t_1) = (\sigma_2 - \sigma_1)[3(p + \sigma_1 \sigma_2) + (\sigma_2 - \sigma_1)^2] \tag{6.93}$$

either directly or from Eq. (4.24) using the equation of orbit

$$r = \tfrac{1}{2} p \sec^2 \tfrac{1}{2} f = \tfrac{1}{2}(p + \sigma^2)$$

---

† Actually, the theorem is true for a general conic. It is important that the transfer-time relation depend only on a single orbital element since the solution of the boundary-value problem will, necessarily, require an iterative process.

Next, from the definition of $\sigma$ and the equation of orbit,
$$p + \sigma_1 \sigma_2 = p \sec \tfrac{1}{2} f_1 \sec \tfrac{1}{2} f_2 \cos \tfrac{1}{2}(f_2 - f_1) = 2\sqrt{r_1 r_2} \cos \tfrac{1}{2}\theta$$
$$= \pm 2\sqrt{s(s-c)} \qquad (6.94)$$

so that one of the factors in the time equation is expressed in terms of $r_1 + r_2$ and $c$. The choice of the upper or lower sign depends, respectively, on whether the transfer angle $\theta$ is less than or greater than 180 degrees.

To obtain the second factor $\sigma_2 - \sigma_1$, we use the equation of orbit to write
$$2(r_1 + r_2) = 2p + \sigma_1^2 + \sigma_2^2 = (\sigma_2 - \sigma_1)^2 + 2(p + \sigma_1 \sigma_2)$$
so that
$$(\sigma_2 - \sigma_1)^2 = 2[(r_1 + r_2) - (p + \sigma_1 \sigma_2)]$$
$$= 4(\sqrt{s} \mp \sqrt{s-c})^2$$
Then, since $\sigma_2$ is never smaller than $\sigma_1$, we have
$$\sigma_2 - \sigma_1 = 2(\sqrt{s} \mp \sqrt{s-c}) \qquad (6.95)$$
Thus, $\sigma_2 - \sigma_1$ also depends only on $r_1 + r_2$ and $c$.

The special case of Lambert's theorem for the parabola
$$\sqrt{\mu}(t_2 - t_1) = F(r_1 + r_2, c) \qquad (6.96)$$
is the immediate consequence of Eqs. (6.93), (6.94), and (6.95).

The explicit form of Eq. (6.96) is Euler's equation. It is readily obtained by substituting Eqs. (6.94) and (6.95) in Eq. (6.93). Therefore,
$$6\sqrt{\mu}(t_2 - t_1) = (r_1 + r_2 + c)^{\frac{3}{2}} \mp (r_1 + r_2 - c)^{\frac{3}{2}} \qquad (6.97)$$
with the upper or lower sign taking effect according as $\theta$ is less or more than 180 degrees.

It is interesting that Lambert not only rediscovered this result on his own but he also skillfully applied it to the problem of comet-orbit determination which Euler failed to do.

### Lagrange's Equation for Elliptic Orbits

Lagrange was first to supply the analytic proof of Lambert's theorem for elliptic orbits just one year before Lambert died. Lagrange wrote three memoirs on the theory of orbits, two in 1778 and one in 1783. His derivation, which was subsequently repeated by Gauss in his Theoria Motus but without acknowledgement, proceeded essentially along the following lines.

Let $E_1$ and $E_2$ be the respective eccentric anomalies associated with the end points $P_1$ and $P_2$. Then, from Kepler's equation, the time $t_2 - t_1$ to traverse the arc from $P_1$ to $P_2$ may be written as
$$\sqrt{\mu}(t_2 - t_1) = 2a^{\frac{3}{2}}[\tfrac{1}{2}(E_2 - E_1) - e \sin \tfrac{1}{2}(E_2 - E_1) \cos \tfrac{1}{2}(E_1 + E_2)]$$

Since the eccentric anomalies appear only in combination as the sum or difference, it is convenient to define two quantities $\psi$ and $\phi$ by

$$\psi = \tfrac{1}{2}(E_2 - E_1) \qquad \cos\phi = e\cos\tfrac{1}{2}(E_1 + E_2) \qquad (6.98)$$

so that the time equation becomes

$$\sqrt{\mu}(t_2 - t_1) = 2a^{\tfrac{3}{2}}(\psi - \sin\psi\cos\phi) \qquad (6.99)$$

The other parameters of Lambert's theorem, $r_1 + r_2$ and $c$, can also be expressed simply in terms of $\psi$ and $\phi$. Thus, from the equation of orbit,

$$\begin{aligned} r_1 + r_2 &= a(1 - e\cos E_1) + a(1 - e\cos E_2) \\ &= 2a(1 - \cos\psi\cos\phi) \end{aligned} \qquad (6.100)$$

and, using the relations

$$\sqrt{r}\cos\tfrac{1}{2}f = \sqrt{a(1-e)}\cos\tfrac{1}{2}E \qquad \sqrt{r}\sin\tfrac{1}{2}f = \sqrt{a(1+e)}\sin\tfrac{1}{2}E$$

from Eqs. (4.31), we obtain

$$\sqrt{r_1 r_2}\cos\tfrac{1}{2}\theta = a(\cos\psi - \cos\phi) \qquad (6.101)$$

which we can use to derive

$$c^2 = (r_1 + r_2)^2 - 4r_1 r_2 \cos^2\tfrac{1}{2}\theta = 4a^2 \sin^2\psi \sin^2\phi$$

Hence,

$$c = 2a\sin\psi\sin\phi \qquad (6.102)$$

The proof of Lambert's theorem follows at once since $\psi$ and $\phi$ can be determined in terms of $a$, $r_1 + r_2$, and $c$ from Eqs. (6.100) and (6.102) and the results substituted into Eq. (6.99). A similar argument shows the theorem to be true for the hyperbola as well.

Lagrange's form of the time equation is obtained using two quantities $\alpha$ and $\beta$ defined as

$$\alpha = \phi + \psi \qquad \beta = \phi - \psi \qquad (6.103)$$

so that

$$\psi = \tfrac{1}{2}(\alpha - \beta) \qquad \phi = \tfrac{1}{2}(\alpha + \beta) \qquad (6.104)$$

Thus, we have

$$\begin{aligned} 2s &= r_1 + r_2 + c = 2a[1 - \cos(\phi + \psi)] = 2a(1 - \cos\alpha) \\ 2(s - c) &= r_1 + r_2 - c = 2a[1 - \cos(\phi - \psi)] = 2a(1 - \cos\beta) \end{aligned}$$

or, simply,

$$\sin^2\tfrac{1}{2}\alpha = \frac{s}{2a} \qquad \sin^2\tfrac{1}{2}\beta = \frac{s-c}{2a} \qquad (6.105)$$

The time equation is expressed in terms of $\alpha$ and $\beta$ by substituting Eqs. (6.104) into (6.99). Therefore,

$$\sqrt{\mu}(t_2 - t_1) = a^{\tfrac{3}{2}}[(\alpha - \sin\alpha) - (\beta - \sin\beta)] \qquad (6.106)$$

Sect. 6.6] Lambert's Theorem 279

exactly as obtained by Lagrange in 1778. There is, of course, an ambiguity in quadrant for $\alpha$ and $\beta$; but this will be resolved in an elegant manner in the following section.

◊ **Problem 6–21**
For hyperbolic orbits, $\psi$ and $\phi$ are defined as

$$\psi = \tfrac{1}{2}(H_2 - H_1) \qquad \cosh\phi = e\cosh\tfrac{1}{2}(H_1 + H_2)$$

and the basic equations are

$$\sqrt{\mu}(t_2 - t_1) = 2(-a)^{\frac{3}{2}}(\sinh\psi\cosh\phi - \psi)$$
$$r_1 + r_2 = 2a(1 - \cosh\psi\cosh\phi)$$
$$c = -2a\sinh\psi\sinh\phi$$
$$\sqrt{r_1 r_2}\cos\tfrac{1}{2}\theta = a(\cosh\psi - \cosh\phi)$$

The Lagrange parameters are defined, as for the ellipse, by Eqs. (6.103). Then

$$\sqrt{\mu}(t_2 - t_1) = (-a)^{\frac{3}{2}}[(\sinh\alpha - \alpha) - (\sinh\beta - \beta)]$$

is the time equation for the hyperbola analogous to the one for the ellipse given in Eq. (6.106) where

$$\sinh^2\tfrac{1}{2}\alpha = -\frac{s}{2a} \qquad \sinh^2\tfrac{1}{2}\beta = -\frac{s-c}{2a}$$

**The Orbital Parameter**

In Sect. 6.2 we developed the following equation for the parameter of an elliptic orbit

$$p = \frac{2r_1 r_2 \sin^2\tfrac{1}{2}\theta}{r_1 + r_2 - 2\sqrt{r_1 r_2}\cos\tfrac{1}{2}\theta\cos\psi} \qquad (6.107)$$

which was labeled (6.37) with $\psi$ written out as one-half the difference of the eccentric anomalies at $P_1$ and $P_2$. Two different and, for some purposes, more convenient equations result when Eqs. (6.100) and (6.101) are substituted in Eq. (6.107). They are

$$p = \frac{r_1 r_2 \sin^2\tfrac{1}{2}\theta}{a\sin^2\psi} \qquad \text{and} \qquad p = \frac{4ar_1 r_2}{c^2}\sin^2\tfrac{1}{2}\theta\sin^2\phi$$

which can also be written

$$\frac{p}{p_m} = \frac{c}{2a\sin^2\psi} = \frac{2a}{c}\sin^2\phi \qquad (6.108)$$

Equations for the hyperbola are obtained when the trigonometric functions of $\psi$ and $\phi$ are replaced by hyperbolic functions. Specifically,

$$\frac{p}{p_m} = -\frac{c}{2a\sinh^2\psi} = -\frac{2a}{c}\sinh^2\phi \qquad (6.109)$$

These last equations are basic to the topic developed in Sect. 6.8. In the present context they provide the following convenient expressions for the orbital parameter in terms of the Lagrange parameters $\alpha$ and $\beta$:

$$\frac{p}{p_m} = \begin{cases} \dfrac{2a}{c}\sin^2 \tfrac{1}{2}(\alpha+\beta) = \dfrac{\sin\tfrac{1}{2}(\alpha+\beta)}{\sin\tfrac{1}{2}(\alpha-\beta)} \\ -\dfrac{2a}{c}\sinh^2 \tfrac{1}{2}(\alpha+\beta) = \dfrac{\sinh\tfrac{1}{2}(\alpha+\beta)}{\sinh\tfrac{1}{2}(\alpha-\beta)} \end{cases} \quad (6.110)$$

with the alternate form obtained using Eq. (6.102).

◊ **Problem 6-22**
Euler's time equation for the parabola can be obtained from Lagrange's equation for the ellipse by allowing the semimajor axis $a$ to become infinite. The same limiting process applied to Eq. (6.110) will produce the parameter of the parabola given in Eq. (6.39).

**Problem 6-23**
If the Gudermannian $\varsigma$ is used to represent the hyperbolic locus of vacant foci, then

$$\sec \varsigma = \frac{4a - (r_1 + r_2)}{c}$$

Derive the expressions

$$\tan^2 \tfrac{1}{2}\varsigma = \begin{cases} \dfrac{\cos^2 \tfrac{1}{2}\alpha}{\cos^2 \tfrac{1}{2}\beta} = -\cos\nu\,\dfrac{\sin\alpha}{\sin\beta} \\ \dfrac{\cosh^2 \tfrac{1}{2}\alpha}{\cosh^2 \tfrac{1}{2}\beta} = -\cos\nu\,\dfrac{\sinh\alpha}{\sinh\beta} \end{cases}$$

◊ **Problem 6-24**
The formula

$$FN = \sqrt{\frac{r_1 r_2}{1 - \dfrac{r_1 + r_2}{2a}}}$$

where $N$ is the point of intersection of the two orbital tangents at $P_1$ and $P_2$, is valid for all orbits when the transfer angle $\theta$ is 180 degrees.
HINT: Use Eqs. (6.108) and (6.12).
NOTE: Consider $a = a_F$.

◊ **Problem 6-25**
The eccentricity of the family of elliptic orbits connecting $P_1$ and $P_2$ can be written as

$$e^2 = 1 - \cos^2 \phi_F \sin^2 \tfrac{1}{2}(\alpha+\beta) = \sin^2 \phi_F \sin^2 \tfrac{1}{2}(\alpha+\beta) + \cos^2 \tfrac{1}{2}(\alpha+\beta)$$

Use this equation to calculate the eccentricities of the fundamental ellipse and its conjugate.

### Problem 6–26
The distance between the vacant foci of any pair of conjugate orbits is

$$4a \sin \gamma_0 \cos \tfrac{1}{2}\alpha \cos \tfrac{1}{2}\beta$$

where $\gamma_0$ is the flight-direction angle at the orbital mean point. Use this expression to calculate the distance between the vacant foci $F_F^*$ of the fundamental ellipse and $\widetilde{F}_F^*$ of its conjugate.

### Problem 6–27
The radius, $r_0$, of the mean point and the velocity vector, $\mathbf{v}_0$, at the mean point may be expressed in terms of the Lagrange parameters $\alpha$ and $\beta$ as

$$r_0 = \begin{cases} a \sin^2 \tfrac{1}{4}(\alpha + \beta) = a[1 - \cos \tfrac{1}{2}(\alpha + \beta)] \\ -a \sinh^2 \tfrac{1}{4}(\alpha + \beta) = a[1 - \cosh \tfrac{1}{2}(\alpha + \beta)] \end{cases}$$

$$\mathbf{v}_0 = \begin{cases} \sqrt{\dfrac{\mu}{a}} \cot \tfrac{1}{4}(\alpha + \beta) \, \mathbf{i}_c \\ \sqrt{\dfrac{\mu}{-a}} \coth \tfrac{1}{4}(\alpha + \beta) \, \mathbf{i}_c \end{cases}$$

Furthermore,

$$FS = \sqrt{r_1 r_2} \cos \tfrac{1}{2}\theta = \begin{cases} 2a \sin \tfrac{1}{2}\alpha \sin \tfrac{1}{2}\beta \\ -2a \sinh \tfrac{1}{2}\alpha \sinh \tfrac{1}{2}\beta \end{cases}$$

where $FS$ is the distance from the focus to the point of intersection of the chord and the locus of mean points.

## 6.7 Transforming the Boundary-Value Problem

Lambert's theorem permits interesting and important geometric transformations of the boundary-value problem. For example, consider an elliptic arc from $P_1$ and $P_2$. Then, according to Lambert's theorem, if $P_1$ and $P_2$ are held fixed, the shape of the ellipse may be altered by moving the foci $F$ and $F^*$ without altering the transfer time, provided, of course, that $r_1 + r_2$ and $a$ are unchanged. The locus of permissible locations for the focus $F$ is an ellipse with foci at $P_1$ and $P_2$ whose major axis is $r_1 + r_2$. For the minimum-energy orbit, $a_m = \tfrac{1}{2}s$ so that the geometric constraints (namely $r_1 + r_2$ and $c$ being unchanged) automatically constrain the semimajor axis. Thus, as $F$ moves along its elliptic locus with major axis $r_1 + r_2$, the locus of the vacant focus $F_m^*$ is the rectilinear ellipse with major axis $c$. Clearly then, the transfer orbit is the minimum-energy ellipse for all intermediate orbits encountered during the transformation.

A similar situation prevails for the fundamental ellipse. Indeed, since $a_F = \tfrac{1}{2}(r_1 + r_2)$, the semimajor axis is also implicitly constrained by the conditions imposed on the transformation. Furthermore, $F$ and $F_F^*$ move along the same elliptic locus with major axis $r_1 + r_2$, and all intermediate orbits encountered are also the corresponding fundamental ellipses.

For an arbitrary ellipse, the locus of the focus $F^*$ is an ellipse with major axis $4a - (r_1 + r_2)$ and confocal with the elliptic locus of $F$. Thus, referring to the left-hand part of Fig. 6.19, the focus $F$ may be moved to an intermediate point $F_i$ and the focus $F^*$ to $F_i^*$—the time to traverse the new arc from $P_1$ and $P_2$ will be unchanged.

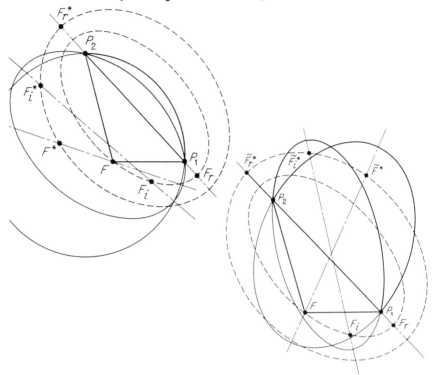

**Fig. 6.19:** Transformation of a pair of conjugate ellipses.

For the hyperbolic arc connecting $P_1$ and $P_2$, the locus of $F^*$ is again an ellipse confocal with the elliptic locus of $F$. However, the major axis of the locus is $r_1 + r_2 - 4a$. In the special case of the straight-line hyperbola, whose vacant focus is at $F_0^*$, the foci $F$ and $F_0^*$ share the same locus.

Certain important quantities besides the transfer time will be unchanged by the transformations just described. The Lagrange parameters $\alpha$ and $\beta$ are explicit functions of $s$, $c$, and $a$ so they too are invariants. The quantities $r_0$, $\mathbf{v}_0$, $FS$ considered in Prob. 6–27, which are functions of $\alpha$, $\beta$, and $a$, are invariant. The anomaly differences $E_2 - E_1$ and $H_2 - H_1$ are invariant under the transformations even though the individual anomalies are not.

Sect. 6.7]     Transforming the Boundary-Value Problem     283

### Transforming to a Rectilinear Ellipse

Consider an elliptic arc connecting $P_1$ and $P_2$—more specifically, one whose vacant focus lies along the lower branch of the hyperbolic locus shown in Fig. 6.17. By moving $F$ counterclockwise and $F^*$ clockwise on their respective loci illustrated in the left-half of Fig. 6.19, the ellipse becomes very flat. Ultimately, the limiting case is obtained, with the foci at $F_r$ and $F_r^*$, and the entire curve flattens out to coincide with the major axis. The orbit is then a rectilinear ellipse ($e = 1$), the arc in question coincides with the chord $c$, and the time interval to traverse the straight-line path from $P_1$ and $P_2$ is the same as the original value $t_2 - t_1$.

The end points of the rectilinear path are so located that

$$P_1 F_r + P_2 F_r = r_1 + r_2 \qquad P_2 F_r - P_1 F_r = c$$

Therefore, the radial distances from the focus $F_r$ are

$$P_1 F_r = s - c \qquad P_2 F_r = s \qquad (6.111)$$

Now, from Kepler's equation with $e = 1$, the transfer time for the rectilinear ellipse is

$$\sqrt{\mu}(t_2 - t_1) = a^{\frac{3}{2}}[(E_2 - \sin E_2) - (E_1 - \sin E_1)] \qquad (6.112)$$

Also, from the equation of orbit for an ellipse with unit eccentricity,

$$PF_r = r = a(1 - \cos E) = 2a \sin^2 \tfrac{1}{2} E$$

Thus, the radial distances from the focus $F_r$ are

$$P_1 F_r = s - c = 2a \sin^2 \tfrac{1}{2} E_1 \qquad P_2 F_r = s = 2a \sin^2 \tfrac{1}{2} E_2 \qquad (6.113)$$

Therefore, when Eqs. (6.112) and (6.113) are compared with Eqs. (6.106) and (6.105), we see that the Lagrange parameters $\alpha$ and $\beta$ are simply the eccentric anomalies† of the respective end points $P_2$ and $P_1$ of the rectilinear orbit as illustrated in the first part of Fig. 6.20.

The situation is somewhat different when the elliptic arc from $P_1$ to $P_2$ has its vacant focus $\widetilde{F}^*$ along the upper part of the branch of the hyperbolic locus in Fig. 6.17. The transformation is illustrated in the right half of Fig. 6.19 for various stages as the focus $F$ moves to $F_r$ and $\widetilde{F}^*$ moves to $\widetilde{F}_r^*$. Since the orbit from $P_1$ and $P_2$ must always encircle the vacant focus, then, in the limit, the rectilinear ellipse is traversed from $P_1$ to $\widetilde{F}_r^*$ and back to $P_2$. The corresponding eccentric anomalies are shown in the second part of Fig. 6.20.

---

† This interpretation of the Lagrange parameters was made by John E. Prussing of the University of Illinois in a paper entitled "Geometrical Interpretation of the Angles $\alpha$ and $\beta$ in Lambert's Problem" which appeared in the *Journal of Guidance and Control*, Vol. 2, Sept.–Oct. 1979.

**284**    Two-Body Orbital Boundary-Value Problem    [Chap. 6

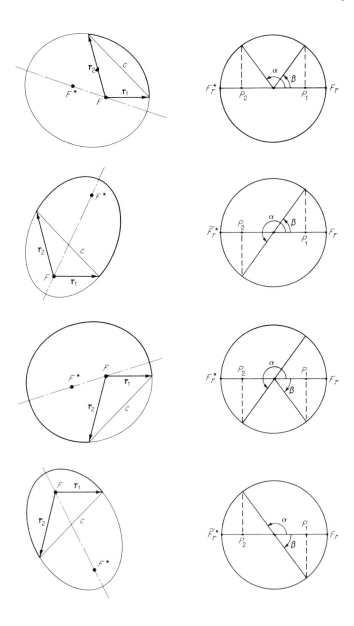

**Fig. 6.20:** Transformations of the four basic ellipses.

In the case for which the transfer angle $\theta$ exceeds 180 degrees, the quadrants for $\alpha$ and $\beta$ are determined as illustrated in the third and fourth parts of Fig. 6.20. In the first instance, the path from $P_1$ to $P_2$ must encircle the occupied focus $F_r$ as well as $F_r^*$. For the orbit corresponding to $\widetilde{F}^*$, the point $F_r$ is encircled but not $\widetilde{F}_r^*$.

It is clear, from the above consideration of all the possibilities, that we may adopt the following convention for assigning quadrants to the Lagrange parameters $\alpha$ and $\beta$

$$\begin{array}{llll} 0 \leq \alpha \leq 2\pi & 0 \leq \beta \leq \pi & \text{for} & \theta \leq \pi \\ 0 \leq \alpha \leq 2\pi & -\pi \leq \beta \leq 0 & \text{for} & \theta \geq \pi \end{array} \quad (6.114)$$

which will include all elliptic orbits.

◇ **Problem 6–28**

The Lagrange parameters $\alpha_m$ and $\beta_m$ for the minimum-energy orbit are

$$\alpha_m = \pi \qquad \cos\beta_m = \frac{3c - r_1 - r_2}{r_1 + r_2 + c}$$

and the transfer time is

$$\sqrt{\frac{\mu}{a_m^3}}(t_2 - t_1) = \pi - (\beta_m - \sin\beta_m)$$

◇ **Problem 6–29**

The Lagrange parameters $\alpha_F$ and $\beta_F$ for the fundamental ellipse are

$$\alpha_F = \pi - \beta_F \qquad \cos\beta_F = \frac{c}{r_1 + r_2}$$

and the transfer time is

$$\sqrt{\frac{\mu}{a_F^3}}(t_2 - t_1) = \pi - 2\beta_F$$

Obtain a corresponding expression for the transfer time for the conjugate of the fundamental ellipse.

### Transforming to a Rectilinear Hyperbola

Consider the hyperbola connecting $P_1$ and $P_2$ whose vacant focus $F^*$ lies along the upper part of the hyperbolic locus shown in Fig. 6.18. We again allow the eccentricity to approach unity in such a way that $a$, $r_1 + r_2$, and $c$ remain unchanged. The vacant focus $F^*$ moves clockwise on its locus while the occupied focus $F$ moves counterclockwise as before. In the limit, the orbit is a rectilinear hyperbola and the time to traverse the straight-line path from $P_1$ to $P_2$ is the same as for the original hyperbolic arc.

Just as in the case of the ellipse, the radial distances from the focus of the rectilinear hyperbola are again those given in Eq. (6.111). The equation of orbit, however, is

$$PF_r = r = a(1 - \cosh H)$$

so that $H_2$ and $H_1$ are identified with the parameters $\alpha$ and $\beta$ defined in Prob. 6–21.

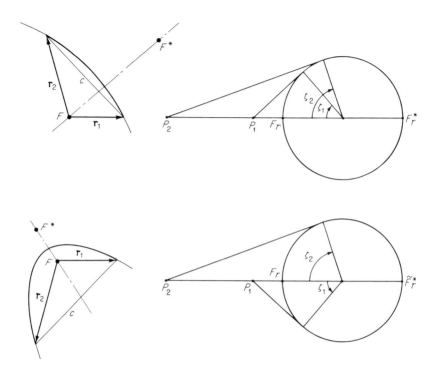

**Fig. 6.21:** Transformation of the two conjugate hyperbolas.

The Gudermannian of $H$ provides an appropriate geometrical representation in terms of an auxiliary circle. From Sect. 4.4 we recall the transformation
$$\sinh H = \tan \varsigma \qquad \cosh H = \sec \varsigma$$
so that $\alpha$ and $\beta$ correspond to the angles $\varsigma_2$ and $\varsigma_1$ illustrated in the first part of Fig. 6.21.

If we repeat the argument for the hyperbolic arc connecting $P_1$ and $P_2$ with focus at $\widetilde{F}^*$, we find that the path from $P_1$ to $P_2$ along the rectilinear path must encircle the focus $F_r$. Therefore, the angles $\varsigma_1$ and $\varsigma_2$ are as shown in the second part of Fig. 6.21. Since the orbit is traversed in the clockwise direction, the situation is identical to the problem of a counterclockwise orbit from $P_1$ to $P_2$ through a transfer angle of $2\pi - \theta$.

We may then adopt the following convention for values of $\alpha$ and $\beta$ (since $H$ and $\varsigma$ have the same signs)

$$\begin{array}{llll} 0 \leq \alpha & 0 \leq \beta & \text{for} & \theta \leq \pi \\ 0 \leq \alpha & \beta \leq 0 & \text{for} & \theta \geq \pi \end{array} \qquad (6.115)$$

which will encompass all hyperbolic orbits.

## Problem 6-30

The transfer time for an elliptic arc with vacant focus $F^*$ connecting points $P_1$ and $P_2$, after it has been transformed to a rectilinear ellipse, can be calculated from the one-dimensional form of the vis-viva integral

$$v^2 = \left(\frac{dr}{dt}\right)^2 = \mu\left(\frac{2}{r} - \frac{1}{a}\right)$$

(a) Derive the equation

$$\sqrt{\mu}(t_2 - t_1) = \int_{s-c}^{s} \frac{r\,dr}{\sqrt{2r - r^2/a}}$$

and carry out the integration using the following change of variable

$$r = a(1 - \cos x)$$

to obtain Lagrange's equation

$$\sqrt{\mu}(t_2 - t_1) = a^{\frac{3}{2}}[(\alpha - \sin\alpha) - (\beta - \sin\beta)]$$

where $\alpha$ and $\beta$ are angles in the upper half-plane.

(b) If the vacant focus of the elliptic arc is $\widetilde{F^*}$, the transfer time is increased by the amount

$$2\sqrt{\frac{a^3}{\mu}}[\pi - (\alpha - \sin\alpha)]$$

(c) If the original orbit is hyperbolic with vacant focus $F^*$, then the appropriate change of variable to evaluate the integral is

$$r = a(1 - \cosh x)$$

Hence, derive the hyperbolic form of Lagrange's equation

$$\sqrt{\mu}(t_2 - t_1) = (-a)^{\frac{3}{2}}[(\sinh\alpha - \alpha) - (\sinh\beta - \beta)]$$

(d) In the integral of part (a), let the semimajor axis of the orbit become infinite and evaluate the integral to obtain Euler's equation for the parabola.

## 6.8 Terminal Velocity Vector Diagrams

An elegant geometric interpretation of the invariants of Lambert's theorem is possible in the form of hodograph representations. By means of conventional compass and straight-edge techniques, the terminal velocity vectors can be readily constructed in a manner which explicitly displays these invariants.†

---

† These results were presented by the author at the American Astronautical Society Symposium on Unmanned Exploration of the Solar System held at Denver, Colorado in February 1965. They were published in *Advances in the Astronautical Sciences*, Vol. 19 as a paper entitled "Orbital Boundary-Value Problems."

Two expressions for the orbital parameter in terms of $\psi$ and $\phi$ are given in Sect. 6.6 as Eqs. (6.108) and (6.109). From them we obtain the following relations for the skewed velocity components

$$v_c = \sqrt{\frac{\mu}{a}} \frac{\sin\phi}{\cos\psi - \cos\phi} \qquad v_\rho = \sqrt{\frac{\mu}{a}} \frac{\sin\psi}{\cos\psi - \cos\phi} \qquad (6.116)$$

when substitution into Eqs. (6.15) is made. Furthermore, if $\psi$ and $\phi$ are replaced by $\alpha$ and $\beta$, according to Eqs. (6.104), we obtain

$$v_c = \sqrt{\frac{\mu}{4a}}(\cot \tfrac{1}{2}\beta + \cot \tfrac{1}{2}\alpha) \qquad v_\rho = \sqrt{\frac{\mu}{4a}}(\cot \tfrac{1}{2}\beta - \cot \tfrac{1}{2}\alpha) \quad (6.117)$$

Similar relations exist, of course, for hyperbolic orbits in terms of hyperbolic functions.

In the following, we shall, for convenience, restrict our discussion to the case for which the transfer angle $\theta$ does not exceed 180 degrees. Then we have, from Eq. (6.114),

$$0 \leq \tfrac{1}{2}\alpha \leq \pi \qquad 0 \leq \tfrac{1}{2}\beta \leq \tfrac{1}{2}\pi$$

so that $\cot \tfrac{1}{2}\beta$ is always positive, and $\cot \tfrac{1}{2}\alpha$ will be positive for those orbits with vacant focus $F^*$ and negative for those with vacant focus $\tilde{F}^*$. Thus, from the definition of $\alpha$ and $\beta$, we may write

$$v_c = P + Q \qquad v_\rho = P - Q \qquad (6.118)$$

where

$$P = \sqrt{\frac{\mu}{r_1 + r_2 - c} - \frac{\mu}{4a}} \qquad Q = \pm\sqrt{\frac{\mu}{r_1 + r_2 + c} - \frac{\mu}{4a}} \qquad (6.119)$$

with the choice of sign, plus or minus, depending, respectively, on whether the vacant focus is at $F^*$ or $\tilde{F}^*$.

### Elliptic Orbits

We consider first the construction of the terminal velocity vectors when the vacant focus is $F^*$. The technique is illustrated in Fig. 6.22 where all lengths indicated have the dimension of velocity.

1. Two parallel lines are constructed with a separation of $\sqrt{\mu/4a}$.
2. A compass is set to a radius of $\sqrt{\mu/(r_1 + r_2 + c)}$ and, with the point at $A$, the point $R$ is determined.
3. With the same radius, the compass point is set at $R$ and the point $D$ determined.
4. With a radius of $\sqrt{\mu/(r_1 + r_2 - c)}$, and the compass point at $R$, the point $B$ is located.
5. Two lines are drawn through point $B$ making angles $\phi_1$ and $\phi_2$ (see Fig. 6.3 ) with the line $AB$.

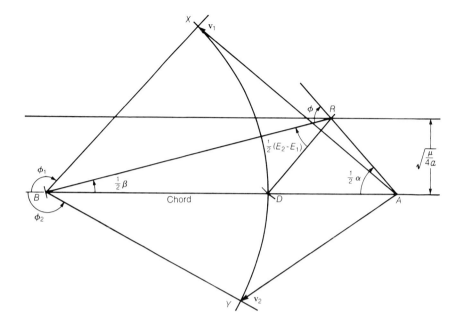

**Fig. 6.22**: Construction of elliptic ($F^*$) velocity vectors.

6. Finally, with the compass point at $B$, and radius equal to the distance from $B$ to $D$, the points $X$ and $Y$ are determined.

It is readily verified that the vectors from $A$ to $X$ and from $A$ to $Y$ are the velocity vectors $\mathbf{v}_1$ and $\mathbf{v}_2$ relative to the line $AB$ which itself is to be regarded as parallel to the chord connecting $P_1$ and $P_2$.

Several interesting by-products of the construction are the angles $\frac{1}{2}\alpha$ and $\frac{1}{2}\beta$, which are fundamental in the Lambert time equation, together with the angles $\psi = \frac{1}{2}(E_2 - E_1)$ and $\phi$, which is the angle whose cosine is $e\cos\frac{1}{2}(E_2 + E_1)$.

For the orbit having the same value of $a$ but with vacant focus $\widetilde{F}^*$, the construction is identical with the exception that the point $R$ is determined to the right of point $A$ instead of to the left. For this case, the construction of the terminal velocity vectors $\widetilde{\mathbf{v}}_1$ and $\widetilde{\mathbf{v}}_2$ is detailed in Fig. 6.23.

A few important observations can now be made. The lengths of the lines $RA$, $RD$, and $RB$ depend only on the dimensions of the triangle $\Delta FP_1P_2$ and not on the specific orbit. Therefore, if point $A$ is kept fixed and the semimajor axis allowed to vary, thereby altering the distance between the parallel lines, the motion of the points $R$, $D$, and $B$ can be readily visualized. When $a$ attains its minimum value $a_m$, the parallel lines are at their maximum separation, the points $A$ and $D$ coincide, and $\frac{1}{2}\alpha$

**290**   Two-Body Orbital Boundary-Value Problem   [Chap. 6

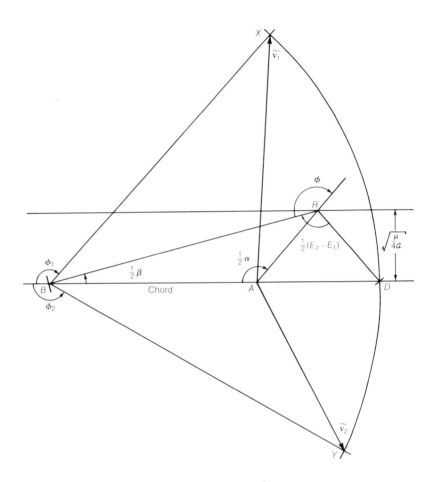

**Fig. 6.23:** Construction of elliptic ($\widetilde{F}^*$) velocity vectors.

becomes a right angle. This special case is illustrated in Fig. 6.24 and the resulting velocity vectors $\mathbf{v}_{m_1}$ and $\mathbf{v}_{m_2}$ are those for the minimum-energy path.

When the transfer angle $\theta$ is 180 degrees, none of the formulas for the velocity vectors given above is valid. In this case, we have

$$s = c = r_1 + r_2 \quad \text{and} \quad \mathbf{i}_{r_1} = -\mathbf{i}_c = -\mathbf{i}_{r_2}$$

so that the equations for $\mathbf{v}_1$ and $\mathbf{v}_2$ are meaningless. However, if we recall the appropriate equations (6.13), then the construction illustrated in Fig. 6.25 is readily validated.

Sect. 6.8]    Terminal Velocity Vector Diagrams    291

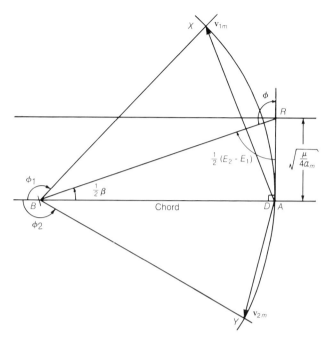

**Fig. 6.24:** Construction of minimum-energy velocity vectors.

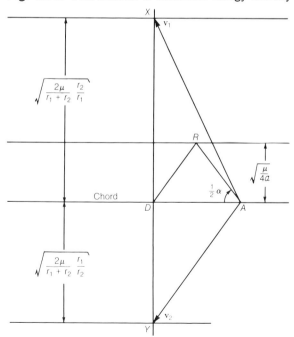

**Fig. 6.25:** Construction of elliptic ($\theta = 180°$) velocity vectors.

## Parabolic Orbit

The construction technique for the velocity vectors $\mathbf{v}_{p_1}$ and $\mathbf{v}_{p_2}$ for the single parabolic orbit connecting the points $P_1$ and $P_2$ for counterclockwise motion is illustrated in Fig. 6.26. The method is the same as described for elliptic orbits but, since $a$ is infinite for the parabola, the two parallel lines coincide. Clearly, the straight-edge and compass procedure is unaffected.

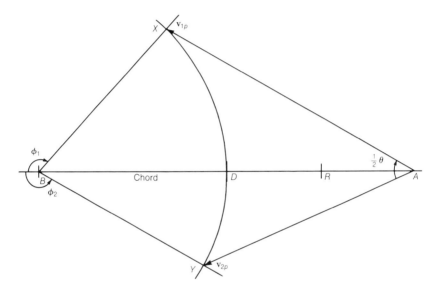

**Fig. 6.26:** Construction of parabolic velocity vectors.

## Hyperbolic Orbits

The expressions for $P$ and $Q$ given in Eqs. (6.119) are the same for the hyperbola connecting $P_1$ and $P_2$ except that $Q$ is always positive and the semimajor axis $a$ is negative. However, the construction technique must be altered. The method is outlined below and the operations are illustrated in Fig. 6.27.

1. Two parallel lines are drawn separated by $\sqrt{\mu/(-4a)}$.
2. The compass is set to a radius of $\sqrt{\mu/(r_1 + r_2 + c)}$ and, with the point set at $R$, the point $A'$ is determined.
3. With a new radius of $\sqrt{\mu/(r_1 + r_2 - c)}$ and with the point again set at $R$, the point $B'$ is located.
4. A perpendicular from $R$, and intersecting the line $AB$, determines the point $R'$.
5. With the compass point at $R'$, points $A$, $D$, and $B$ are located such that $R'A' = R'A = R'D$ and $R'B' = R'B$.

Sect. 6.8]    Terminal Velocity Vector Diagrams    293

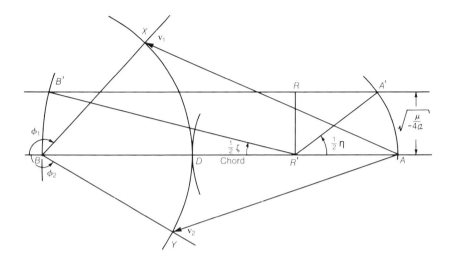

**Fig. 6.27**: Construction of hyperbolic velocity vectors.

6. Two lines are drawn through point $B$ making angles $\phi_1$ and $\phi_2$ with line $AB$.
7. With compass point at $B$, and radius equal to the distance from $B$ to $D$, the points $X$ and $Y$ are determined.

Finally, as before, the vectors from $A$ to $X$ and from $A$ to $Y$ are the velocity vectors $\mathbf{v}_1$ and $\mathbf{v}_2$ relative to the line $AB$ and, hence, to the chord connecting $P_1$ and $P_2$.

Again we note that the lengths of lines $RA'$ and $RB'$ do not depend on the orbit and, thus, $R$, $R'$, $A'$, and $B'$ remain fixed as the semimajor axis $a$ is varied. The variation of the other quantities can be visualized as before. In particular, we observe that there is no restriction on the size of $a$ which may vary from zero to minus infinity. In the limit, the same parabolic orbit is obtained as before.

In the construction diagram, the Lagrange parameters $\alpha$ and $\beta$ do not appear directly as by-products as they did for elliptic orbits. However, if we introduce the two auxiliary angles $\eta_1$ and $\eta_2$ defined as

$$\tan \tfrac{1}{2}\eta_2 = \sinh \tfrac{1}{2}\alpha \qquad \tan \tfrac{1}{2}\eta_1 = \sinh \tfrac{1}{2}\beta$$

then $\tfrac{1}{2}\eta_1$ and $\tfrac{1}{2}\eta_2$ have the geometrical significance indicated in Fig. 6.27.

Furthermore, the alternate form of the Lambert time equation, as a function of these new variables, is readily shown to be

$$\sqrt{\mu}(t_2 - t_1) = 2\sqrt{(-a)^3}\{[\tan\tfrac{1}{2}\eta_2 \sec\tfrac{1}{2}\eta_2 + \log\tfrac{1}{4}(\pi - \eta_2)]$$
$$- [\tan\tfrac{1}{2}\eta_1 \sec\tfrac{1}{2}\eta_1 + \log\tfrac{1}{4}(\pi - \eta_1)]\} \quad (6.120)$$

with $0 \le \eta_1 \le \eta_2 \le \pi$.

### Boundary Conditions at Infinity

An important orbital boundary-value problem is one for which a hyperbolic orbit is desired having a given velocity magnitude and direction at infinity. More specifically, let it be required to determine a velocity $\mathbf{v}_1$ corresponding to a given position vector $\mathbf{r}_1$ which will permit the attainment of a specified asymptotic hyperbolic velocity vector $\mathbf{v}_\infty$.

The equation for $\mathbf{v}_1$ can be derived as a limiting form of Eqs. (6.118) with $P$ and $Q$ defined by Eqs. (6.119). In particular, if we keep the transfer angle $\theta$ fixed and let $r_2$ become infinite, we readily see that

$$\lim_{\substack{\theta = \text{const.} \\ r_2 \to \infty}} \mathbf{i}_c = \mathbf{i}_\infty \quad \text{and} \quad \lim_{\substack{\theta = \text{const.} \\ r_2 \to \infty}} \frac{1}{s} = 0$$

where $\mathbf{i}_\infty$ is a unit vector in the direction of the asymptote. Furthermore, since $\lim s/r_2 = 1$, we have

$$\lim_{\substack{\theta = \text{const.} \\ r_2 \to \infty}} (s - c) = \lim_{\substack{\theta = \text{const.} \\ r_2 \to \infty}} \frac{r_1 r_2 (1 + \cos\theta)}{2s} = \frac{r_1}{2}(1 + \cos\theta)$$

Finally, from energy considerations, the magnitude of the asymptotic velocity is related to the semimajor axis $a$ according to

$$\frac{\mu}{-a} = v_\infty^2$$

Therefore, the formula for the velocity vector $\mathbf{v}_1$ is found to be

$$\mathbf{v}_1 = (D + \tfrac{1}{2}v_\infty)\mathbf{i}_\infty + (D - \tfrac{1}{2}v_\infty)\mathbf{i}_{r_1} \quad (6.121)$$

where

$$D = \sqrt{\frac{\mu}{r_1(1 + \cos\theta)} + \frac{v_\infty^2}{4}} \quad (6.122)$$

The graphical construction of the velocity vector $\mathbf{v}_1$ is the same as that for the general hyperbolic velocity vector with suitable account taken of the limiting processes described above. Referring to Fig. 6.27, we see that points $R$ and $A'$ coincide in the limit while the distance between $R$ and $B'$ becomes simply $\sqrt{\mu/r_1(1 + \cos\theta)}$. The reader should have no difficulty constructing the velocity vector diagram.

Chapter 7

# Solving Lambert's Problem

D ETERMINATION OF AN ORBIT, HAVING A SPECIFIED TRANSFER TIME and connecting two position vectors, is called *Lambert's Problem*. It is fundamental today as a means of targeting spacecraft and missiles. In the past, from the time of Euler and Lambert, it's solution was essential for obtaining the elements of the orbits of planets and comets from observations. Over the years a variety of techniques for solving this problem has been developed. Each is characterized by a particular form of the transfer-time equation and a particular choice of independent variable to be used in an iterative algorithm to determine the orbital elements. Some of these are the subject of the present chapter.

The first real progress in the solution of Lambert's Problem was made by Carl Friedrich Gauss in his book *Theoria Motus Corporum Coelestium in Sectionibus Conicis Solem Ambientium*—Theory of the Motion of the Heavenly Bodies Moving about the Sun in Conic Sections—which we have, heretofore, referred to simply as *Theoria Motus*. The story behind this book is a fascinating one and we allow Gauss to tell it in his own words by quoting from the Preface.†

> "*To determine the orbit of a heavenly body, without hypothetical assumption, from observations not embracing a great period of time, and not allowing a selection with a view to the application of special methods*, was almost wholly neglected up to the beginning of the present century; or, at least, not treated by any one in a manner worthy of its importance; since it assuredly commended itself to mathematicians by its difficulty and elegance, even if its great utility in practice were not apparent. An opinion had universally prevailed that a complete determination from observations embracing a short interval of time was impossible,—an ill-founded opinion,—for it is now clearly shown that the orbit of a heavenly body may be determined quite nearly from good

---

† Written in Göttingen on March 28, 1809. This work earned him the appointment as professor of astronomy and director of the observatory at Göttingen. Except for one trip to Berlin to attend a scientific meeting he remained at Göttingen for the remainder of his life.

observations embracing only a few days; and this without any hypothetical assumption.

"Some ideas occurred to me in the month of September of the year 1801, engaged at the time on a very different subject, which seemed to point to the solution of the great problem of which I have spoken. Under such circumstances we not unfrequently, for fear of being too much led away by an attractive investigation, suffer the associations of ideas, which, more attentively considered, might have proved most fruitful in results, to be lost from neglect. And the same fate might have befallen these conceptions, had they not happily occurred at the most propitious moment for their preservation and encouragement that could have been selected. For just about this time the report of the new planet, discovered on the first day of January of that year with the telescope at Palermo, was the subject of universal conversation; and soon afterwards the observations made by that distinguished astronomer Piazzi† from the above date to the eleventh of February were published. Nowhere in the annals of astronomy do we meet with so great an opportunity, and a greater one could hardly be imagined, for showing most strikingly, the value of this problem, than in this crisis and urgent necessity, when all hope of discovering in the heavens this planetary atom, among innumerable small stars after the lapse of nearly a year, rested solely upon a sufficiently approximate knowledge of its orbit to be based upon these very few observations. Could I ever have found a more seasonable opportunity to test the practical value of my conceptions, than now in employing them for the determination of the orbit of the planet Ceres, which during these forty-one days had described a geocentric arc of only three degrees, and after the lapse of a year must be looked for in a region of the heavens very remote from that in which it was last seen? This first application of the method was made in the month of October, 1801, and the first clear night, when the planet was sought for [by Baron Franz Xaver von Zach on December 7, 1801] as directed by the numbers deduced from it, restored the fugitive to observation. Three other new planets, subsequently discovered, furnished new opportunities for examining and verifying the efficiency and generality of the method."

Carl Friedrich Gauss (1777–1855) was born in Brunswick, Germany and seemed destined by tradition to a life of manual work with his father who was a mason. He taught himself to read and to calculate before he was three years old. In elementary school he displayed such extraordinary

---

† Giuseppe Piazzi (1746–1826).

intellect that Carl Wilhelm Ferdinand, the Duke of Brunswick, took a personal interest in his education—sending him in 1795 to the University of Göttingen where his research began in earnest. By the time he was eighteen he had invented the method of least squares and at nineteen he had made the first significant progress in Euclidean geometry in two thousand years by stating and proving the conditions which allow a regular polygon to be constructed with straightedge and compass. Until then he had been undecided on a career choice, philology or mathematics. Fortunately, his early successes in mathematics made the decision relatively easy. He often said later that such an overwhelming horde of new ideas stormed his mind before he was twenty that he could but record only a small fraction of them.

He transferred to the University of Helmstädt in 1798 and attracted the attention of Johann Friedrich Pfaff (1765–1825) who became both his teacher and friend. It was there in 1798 that he wrote his doctoral dissertation in which he gave the first proof of the fundamental theorem of algebra, that every $n^{th}$ degree polynomial has exactly $n$ roots. In 1801 he published his classic work in number theory *Disquisitiones Arithmeticae* just in time to tackle the problem of determining the orbital elements of Ceres, the largest of the asteroids.

Over seven years elapsed before Gauss published his method and he did not regret the delay.

> "For, the methods first employed have undergone so many and such great changes, that scarcely any trace of resemblance remains between the method in which the orbit of Ceres was first computed, and the form given in this work."

Indeed, thereafter, he was always slow to publish and preferred to polish his relatively few masterpieces rather than rush to print everything as was Euler's custom. His seal, a tree with but few fruits, bore the motto *Pauca sed matura* (Few, but ripe).

A list of Gauss' contributions to mathematics and to mathematical physics is almost endless and we cannot begin to enumerate them here. By the time of his death at the age of 78 he was hailed by his contemporaries as the "Prince of Mathematicians."

## 7.1 Formulations of the Transfer-Time Equation

We derive in this section two separate transfer-time relationships, the first utilizing Lagrange's equations from his proof of Lambert's theorem and the second from an adaptation of one of Gauss' equations from the *Theoria Motus*. Then, we combine the best features of the two formulations to derive a third.

For convenience, we summarize the relevant equations from Sect. 6.6 which are needed for this purpose. They are†

$$\sqrt{\mu}(t_2 - t_1) = 2a^{\frac{3}{2}}(\psi - \sin\psi\cos\phi) \tag{7.1}$$
$$r_1 + r_2 = 2a(1 - \cos\psi\cos\phi) \tag{7.2}$$
$$c = 2a\sin\psi\sin\phi \tag{7.3}$$
$$\lambda s = a(\cos\psi - \cos\phi) \tag{7.4}$$

For brevity, we have introduced the parameter $\lambda$ defined as

$$\lambda s = FS = \sqrt{r_1 r_2}\cos\tfrac{1}{2}\theta \tag{7.5}$$

Then, since

$$\lambda^2 = \frac{s-c}{s} \tag{7.6}$$

we see that $\lambda$ has a range $(-1, 1)$ and, of course, depends only on the geometric configuration of $P_1$ and $P_2$ with respect to the occupied focus.

## Lagrange's Equation

Recall from Sect. 6.6 that Lagrange defined two parameters

$$\alpha = \phi + \psi \qquad \beta = \phi - \psi$$

so that

$$\psi = \tfrac{1}{2}(\alpha - \beta) \qquad \phi = \tfrac{1}{2}(\alpha + \beta)$$

Then, from Eqs. (7.2) and (7.3), he obtained

$$\sin^2\tfrac{1}{2}\alpha = \frac{s}{2a} \qquad \sin^2\tfrac{1}{2}\beta = \frac{s-c}{2a} \tag{7.7}$$

Since Eq. (7.4) may be written as

$$\lambda s = 2a\sin\tfrac{1}{2}\alpha\sin\tfrac{1}{2}\beta$$

we may combine this with the first of Eqs. (7.7) and obtain the following relation between $\alpha$ and $\beta$

$$\sin\tfrac{1}{2}\beta = \lambda\sin\tfrac{1}{2}\alpha \tag{7.8}$$

Lagrange's form of the transfer-time equation for elliptic orbits is expressed in terms of the parameters $\alpha$ and $\beta$ as

$$\sqrt{\mu}(t_2 - t_1) = a^{\frac{3}{2}}\left[(\alpha - \sin\alpha) - (\beta - \sin\beta)\right] \tag{7.9}$$

---

† Recall that $\psi = \tfrac{1}{2}(E_2 - E_1)$ and $\cos\phi = e\cos\tfrac{1}{2}(E_2 + E_1)$ for elliptic orbits with similar relations for hyperbolic orbits.

Sect. 7.1]     Formulations of the Transfer-Time Equation     299

and, for fixed geometry, is a function only of the semimajor axis $a$. However, $a$ is not a convenient variable† for two important reasons: (1) the transfer time is a double-valued function of $a$—remember that each pair of conjugate orbits has the same semimajor axis—and (2) the derivative of the transfer time with respect to $a$,

$$\sqrt{\mu}\frac{d}{da}(t_2 - t_1) = \tfrac{3}{2}a^{\tfrac{1}{2}}[(\alpha - \sin\alpha) - (\beta - \sin\beta)]$$
$$- a^{-\tfrac{1}{2}}[s\tan\tfrac{1}{2}\alpha - (s - c)\tan\tfrac{1}{2}\beta]$$

is infinite for that value of $a = a_m = \tfrac{1}{2}s$ corresponding to the minimum-energy orbit for which $\alpha_m = \pi$.

Fortunately, we can transform Lagrange's equation to a much more convenient form‡ by using Eqs. (7.6) and (7.7) to write

$$\sqrt{\frac{\mu}{a_m^3}}(t_2 - t_1) = \frac{\alpha - \sin\alpha}{\sin^3\tfrac{1}{2}\alpha} - \lambda^3\frac{\beta - \sin\beta}{\sin^3\tfrac{1}{2}\beta} \qquad (7.10)$$

Similarly, for hyperbolic orbits, we obtain

$$\sqrt{\frac{\mu}{a_m^3}}(t_2 - t_1) = \frac{\sinh\alpha - \alpha}{\sinh^3\tfrac{1}{2}\alpha} - \lambda^3\frac{\sinh\beta - \beta}{\sinh^3\tfrac{1}{2}\beta} \qquad (7.11)$$

with $\alpha$ and $\beta$ given by

$$\sinh^2\tfrac{1}{2}\alpha = \frac{s}{-2a} \qquad \sinh^2\tfrac{1}{2}\beta = \frac{s - c}{-2a} \qquad (7.12)$$

and related according to

$$\sinh\tfrac{1}{2}\beta = \lambda\sinh\tfrac{1}{2}\alpha \qquad (7.13)$$

By defining a function $Q_\alpha$ as

$$Q_\alpha = \begin{cases} \dfrac{\alpha - \sin\alpha}{\sin^3\tfrac{1}{2}\alpha} \\[2ex] \dfrac{\sinh\alpha - \alpha}{\sinh^3\tfrac{1}{2}\alpha} \end{cases} \qquad (7.14)$$

then both equations, (7.10) and (7.11), are identical. Hence,

$$\sqrt{\frac{\mu}{a_m^3}}(t_2 - t_1) = Q_\alpha - \lambda^3 Q_\beta \qquad (7.15)$$

---

† Some authors have developed algorithms using the parameter $p$ as the iterated variable without mentioning that all orbits have the same parameter for a 180 degree transfer.

‡ This transformation is from the author's book *Astronautical Guidance* published by McGraw-Hill Book Co. in 1964 and first appeared in his *MIT Instrumentation Laboratory Report R-383* in Sept. 1962. The material comprising the present section and the next are from the author's paper "Lambert's Problem Revisited" which was published in the *AIAA Journal*, Vol. 15, May 1977, pp. 707–713.

As we shall see in the next section, $Q_\alpha$ is a hypergeometric function. Specifically,

$$Q_\alpha = \begin{cases} \frac{4}{3}F(3,1;\frac{5}{2};\sin^2\frac{1}{4}\alpha) \\ \frac{4}{3}F(3,1;\frac{5}{2};-\sinh^2\frac{1}{4}\alpha) \end{cases} \quad (7.16)$$

Therefore, if we define

$$x = \begin{cases} \cos\frac{1}{2}\alpha \\ \cosh\frac{1}{2}\alpha \end{cases} \quad \text{and} \quad y = \begin{cases} \cos\frac{1}{2}\beta \\ \cosh\frac{1}{2}\beta \end{cases} \quad (7.17)$$

then Eq. (7.15) becomes

$$\sqrt{\frac{\mu}{a_m^3}}(t_2 - t_1) = \frac{4}{3}F\left[3,1;\tfrac{5}{2};\tfrac{1}{2}(1-x)\right] - \frac{4}{3}\lambda^3 F\left[3,1;\tfrac{5}{2};\tfrac{1}{2}(1-y)\right] \quad (7.18)$$

with the positive quantity $y$ related to $x$ according to

$$y = \sqrt{1 - \lambda^2(1 - x^2)} \quad (7.19)$$

which is derived from Eqs. (7.8) and (7.13).

There is great advantage in regarding the transfer time as a function of the variable $x$, defined in the first of Eqs. (7.17) but also expressible as

$$x^2 = 1 - \frac{a_m}{a} \quad (7.20)$$

All of the problems anticipated with the semimajor axis $a$ used for this purpose have vanished. The graph of the transfer time as a function of $x$ for various values of $\lambda$, shown in Fig. 7.1, is single-valued, monotonic, and readily adapted to a Newton method of iterative solution. Further, we note that the variable $x$ has the following range and significance:

$$-1 < x < 1 \quad \text{elliptic orbits}$$
$$x = 1 \quad \text{parabolic orbit}$$
$$1 < x < \infty \quad \text{hyperbolic orbits}$$

and also that $x = 0$ corresponds to the transfer time for the minimum-energy path from $P_1$ to $P_2$.

## Gauss' Equation

Another form of the transfer-time equation for an elliptic orbit may be obtained by eliminating $\cos\phi$ between Eqs. (7.1) and (7.4). We have then,

$$\sqrt{\mu}(t_2 - t_1) = a^{\frac{3}{2}}(2\psi - \sin 2\psi) + 2\lambda s a^{\frac{1}{2}}\sin\psi$$

as first obtained by Gauss in his *Theoria Motus*. Similarly, for hyperbolic orbits,

$$\sqrt{\mu}(t_2 - t_1) = (-a)^{\frac{3}{2}}(\sinh 2\psi - 2\psi) + 2\lambda s(-a)^{\frac{1}{2}}\sinh\psi$$

## Formulations of the Transfer-Time Equation

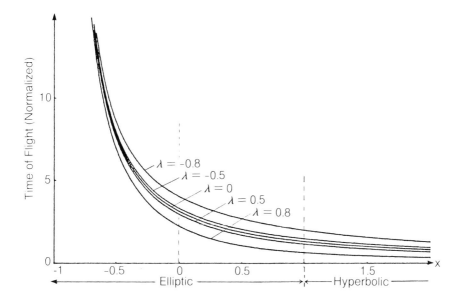

**Fig. 7.1:** Transfer time as a function of $x$.

We now depart from Gauss and define a positive quantity $\eta$ by

$$s\eta^2 = \begin{cases} 2a\sin^2\psi \\ -2a\sinh^2\psi \end{cases} \quad (7.21)$$

and write the transfer-time equation, in either case, in universal form

$$\sqrt{\frac{\mu}{a_m^3}}(t_2 - t_1) = \eta^3 Q_{2\psi} + 4\lambda\eta \quad (7.22)$$

where $Q_{2\psi}$ is defined by Eq. (7.14) with $\alpha$ replaced by $2\psi$.

Finally, by eliminating $\phi$ between Eqs. (7.2) and (7.4),

$$r_1 + r_2 = s\eta^2 + 2\lambda s \begin{cases} \cos\psi \\ \cosh\psi \end{cases}$$

so that, for fixed geometry, $\eta$ is a function only of $\psi$. Indeed, since

$$r_1 + r_2 = 2s - c = s(1 + \lambda^2)$$

we have

$$\eta^2 = \begin{cases} (1-\lambda)^2 + 4\lambda\sin^2\tfrac{1}{2}\psi \\ (1-\lambda)^2 - 4\lambda\sinh^2\tfrac{1}{2}\psi \end{cases} \quad (7.23)$$

The transfer time is thus a function only of the anomaly difference $E_2 - E_1$ (or $H_2 - H_1$). However, since (as shown in the following section)

$$Q_{2\psi} = \begin{cases} \frac{4}{3} F(3, 1; \frac{5}{2}; \sin^2 \frac{1}{2}\psi) \\ \frac{4}{3} F(3, 1; \frac{5}{2}; -\sinh^2 \frac{1}{2}\psi) \end{cases} \quad (7.24)$$

we may define

$$S_1 = \begin{cases} \sin^2 \frac{1}{2}\psi \\ -\sinh^2 \frac{1}{2}\psi \end{cases} \quad (7.25)$$

and express $t_2 - t_1$ more compactly as a function of $S_1$. Therefore,

$$\sqrt{\frac{\mu}{a_m^3}}(t_2 - t_1) = \frac{4}{3}\eta^3 F(3, 1; \frac{5}{2}; S_1) + 4\lambda\eta \quad (7.26)$$

where

$$\eta^2 = (1 - \lambda)^2 + 4\lambda S_1 \quad \text{with} \quad \eta \geq 0 \quad (7.27)$$

and

$$0 < S_1 < 1 \quad \text{elliptic orbits}$$
$$S_1 = 0 \quad \text{parabolic orbit}$$
$$-\infty < S_1 < 0 \quad \text{hyperbolic orbits}$$

A graph of the transfer time as a function of $-S_1$ for various values of $\lambda$ is given in Fig. 7.2. The curves are, indeed, monotonic but have little else to recommend them for a Newton method of iteration.

## Combined Equations

When we compare the two transfer-time formulations, as summarized in Figs. 7.1 and 7.2, it appears that the graph of $t_2 - t_1$ as a function of $x$ is more amenable to a mechanized iterative solution than is its graph as a function of $S_1$. The two sets of curves are identical for $\lambda = 0$, corresponding to a 180 degree transfer, but otherwise differ significantly in important characteristics.

On the other hand, if the basis of comparison is computation efficiency, then Gauss' equation is to be preferred since the evaluation of only one hypergeometric function is required rather than two. In both cases, one square root function is necessary.

It is possible to relate the independent variables $S_1$ and $x$ in a simple way so that the advantages of both formulations can be realized in a single expression. For this purpose, using Eq. (7.25) together with the relation

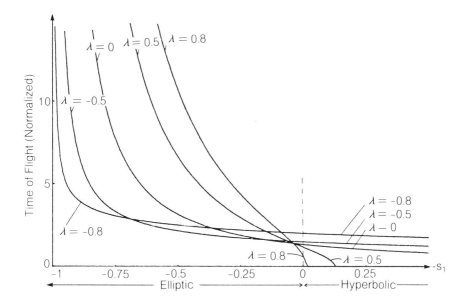

**Fig. 7.2:** Transfer time as a function of $-S_1$.

between $\psi$ and the Lagrange parameters $\alpha$ and $\beta$, we write

$$2S_1 = \begin{cases} 1 - \cos\psi = 1 - \cos\tfrac{1}{2}(\alpha - \beta) \\ 1 - \cosh\psi = 1 - \cosh\tfrac{1}{2}(\alpha - \beta) \end{cases}$$

$$= \begin{cases} 1 - \cos\tfrac{1}{2}\alpha\cos\tfrac{1}{2}\beta - \sin\tfrac{1}{2}\alpha\sin\tfrac{1}{2}\beta \\ 1 - \cosh\tfrac{1}{2}\alpha\cosh\tfrac{1}{2}\beta + \sinh\tfrac{1}{2}\alpha\sinh\tfrac{1}{2}\beta \end{cases}$$

Next, we employ Eqs. (7.8) and (7.13) to eliminate $\sin\tfrac{1}{2}\beta$ and $\sinh\tfrac{1}{2}\beta$. There results

$$2S_1 = \begin{cases} 1 - \cos\tfrac{1}{2}\alpha\cos\tfrac{1}{2}\beta - \lambda\sin^2\tfrac{1}{2}\alpha \\ 1 - \cosh\tfrac{1}{2}\alpha\cosh\tfrac{1}{2}\beta + \lambda\sinh^2\tfrac{1}{2}\alpha \end{cases}$$

$$= 1 - xy - \lambda(1 - x^2) \qquad (7.28)$$

Furthermore, the quantity $\eta^2$, as given in Eq. (7.27), is a much simpler expression in terms of $x$ and $y$. Thus,

$$\eta^2 = \begin{cases} 1 + \lambda^2 - 2\lambda\cos\tfrac{1}{2}(\alpha - \beta) \\ 1 + \lambda^2 - 2\lambda\cosh\tfrac{1}{2}(\alpha - \beta) \end{cases}$$

which we expand, using Eqs. (7.8) and (7.13) as before, to obtain

$$\eta^2 = \begin{cases} 1 - \lambda^2 - 2\lambda \cos \tfrac{1}{2}\alpha \cos \tfrac{1}{2}\beta + 2\lambda^2 \cos^2 \tfrac{1}{2}\alpha \\ 1 - \lambda^2 - 2\lambda \cosh \tfrac{1}{2}\alpha \cosh \tfrac{1}{2}\beta + 2\lambda^2 \cosh^2 \tfrac{1}{2}\alpha \end{cases}$$

$$= 1 - \lambda^2 - 2\lambda xy + 2\lambda^2 x^2$$

Then, from Eq. (7.19),

$$1 - \lambda^2 = y^2 - \lambda^2 x^2$$

so that

$$\eta^2 = (y - \lambda x)^2$$

Finally, since $y$ and $\eta$ are both positive and

$$y^2 - \lambda^2 x^2 \geq 0$$

it follows that

$$\eta = y - \lambda x \qquad (7.29)$$

Therefore, the transfer-time formulation which seems more appropriate than either Gauss' or Lagrange's alone, can be summarized. From the geometry of the problem, we first calculate

$$2a_m = s = \tfrac{1}{2}(r_1 + r_2 + c) \quad \text{and} \quad \lambda s = \sqrt{r_1 r_2} \cos \tfrac{1}{2}\theta$$

Then, starting with a suitable trial value of $x$, we compute

$$\begin{aligned} y &= \sqrt{1 - \lambda^2(1 - x^2)} \\ \eta &= y - \lambda x \\ S_1 &= \tfrac{1}{2}(1 - \lambda - x\eta) \\ Q &= \tfrac{4}{3} F(3, 1; \tfrac{5}{2}; S_1) \end{aligned} \qquad (7.30)$$

which are then used to obtain the transfer time from

$$\sqrt{\frac{\mu}{a_m^3}}(t_2 - t_1) = \eta^3 Q + 4\lambda \eta \qquad (7.31)$$

The process continues by systematically altering the value of $x$ until the required convergence is obtained. Note that only one square root and one hypergeometric function are required for each computation cycle.

In the next section, convenient derivative formulas for the transfer time are developed in the event that Newton's method is to be used for the iterative calculation of $x$.

Sect. 7.1]    Formulations of the Transfer-Time Equation

## Problem 7-1

In Lambert's paper *Insigniores orbitae Cometarum proprietates*, published in Augsburg in 1761, he derived the series

$$\sqrt{\mu}(t_2 - t_1) = \frac{\sqrt{2}}{3}[s^{\frac{3}{2}} \mp (s-c)^{\frac{3}{2}}]$$

$$+ \sum_{n=1}^{\infty} \frac{\sqrt{2}}{2n+3} \frac{(2n-1)!}{2^{3n-1} n! (n-1)!} [s^{n+\frac{3}{2}} \mp (s-c)^{n+\frac{3}{2}}] \frac{1}{a^n}$$

from the integral of Prob. 6–30 by expanding the integrand as a power series in $r$ and integrating term by term. This result can be obtained more easily from Eq. (7.18) by using the identity (1.18) derived in Sect. 1.1.

Determine the range of convergence of this series.

NOTE: The first term in the series is Euler's equation for the transfer time of a parabola.

*Johann Heinrich Lambert* 1761


### Multiple-Revolution Transfer Orbits

For elliptic orbits, we may wish to include the possibility of a number $N$ of complete orbits before termination at the point $P_2$. In this case, the transfer-time equation (7.26) is modified, using Eq. (7.20), as follows:

$$\sqrt{\frac{\mu}{a_m^3}}(t_2 - t_1) = \frac{2\pi N}{(1-x^2)^{\frac{3}{2}}} + \tfrac{4}{3}\eta^3 F(3, 1; \tfrac{5}{2}; S_1) + 4\lambda\eta \qquad (7.32)$$

When the transfer angle $\theta$ is less than 360 degrees ($N = 0$), the orbit connecting points $P_1$ and $P_2$ for a given transfer time is unique. However, if $\theta$ is greater than 360 degrees but less than 720 degrees ($N = 1$), $x$ is a double-valued function of the transfer time. Thus, corresponding to each value of $t_2 - t_1$ that is sufficiently large to ensure a solution, two orbits are obtainable. As $N$ increases so does the number of possible orbits for sufficiently large values of $t_2 - t_1$.

In Fig. 7.3 the complete family of solutions is illustrated for $N = 0$ and $N = 1$. Two interesting characteristics of these curves deserve comment: (1) The curve for $\lambda = 1$ and $N = 0$ terminates for $x = 0$. Since $\lambda = 1$ corresponds to a transfer angle of zero, the portion of the curve for negative $x$ corresponds, simply, to rectilinear orbits. (2) For $\lambda = -1$, that is a transfer angle of exactly 360 degrees, there is a discontinuity in the slope at that point on the curve corresponding to the minimum-energy orbit. However, $\lambda = -1$ for $N = 0$ is the same as $\lambda = 1$ for the case of a single multiple-revolution orbit ($N = 1$). Viewed from this perspective, the curve has a continuous derivative. This feature does suggest that, for $\lambda$ quite close in value to minus one, there will be a change in curvature.

**306**  Solving Lambert's Problem  [Chap. 7

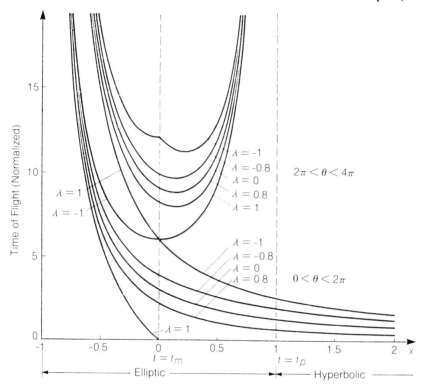

**Fig. 7.3:** Transfer time for multiple-revolution orbits.

### The Velocity Vector

The final step in the solution of the Lambert problem will, in many cases, be the calculation of the velocity vector $\mathbf{v}_1$ at the point $P_1$ in terms of that value of $x$ found to satisfy the transfer-time equation (7.31).

From Eqs. (6.2) and (6.3), the vector $\mathbf{v}_1$ may be written as

$$\mathbf{v}_1 = \frac{\sqrt{\mu}}{r_1}(\sigma_1 \, \mathbf{i}_{r_1} + \sqrt{p} \, \mathbf{i}_h \times \mathbf{i}_{r_1}) \qquad (7.33)$$

where $\mathbf{i}_{r_1}$ is the unit vector defining the direction of $P_1$ from the force center, $\mathbf{i}_h$ is the unit vector normal to the orbital plane, $p$ is the parameter of the orbit, and

$$\sigma_1 = \frac{\mathbf{r}_1 \cdot \mathbf{v}_1}{\sqrt{\mu}}$$

To complete the task, we must find convenient expressions for $p$ and $\sigma_1$.

First, from Eq. (6.108), in the previous chapter, we can derive

$$p = \frac{r_1 r_2 \sin^2 \tfrac{1}{2}\theta}{a \sin^2 \psi}$$

Sect. 7.2]  The Q Function  307

and, substituting from Eq. (7.21), obtain

$$p = \frac{r_1 r_2}{a_m \eta^2} \sin^2 \tfrac{1}{2}\theta \tag{7.34}$$

Thus, the parameter of the orbit is inversely proportional to $\eta^2$.

Second, by adapting Eq. (4.98), derived in Chapter 4 as an identity for the universal functions, we arrive at the following equation for $\sigma_1$:

$$\frac{\sigma_1}{\sqrt{p}} \sin \tfrac{1}{2}\theta = \cos \tfrac{1}{2}\theta - \sqrt{\frac{r_1}{r_2}} \begin{Bmatrix} \cos \psi \\ \cosh \psi \end{Bmatrix}$$

which, using Eqs. (7.34), (7.5), and (7.28), becomes

$$\sigma_1 = \frac{1}{\eta \sqrt{a_m}} [2\lambda a_m - r_1(\lambda + x\eta)] \tag{7.35}$$

Finally, Eq. (7.33) may be expressed as

$$\mathbf{v}_1 = \frac{1}{\eta} \sqrt{\frac{\mu}{a_m}} \left\{ \left[2\lambda \frac{a_m}{r_1} - (\lambda + x\eta)\right] \mathbf{i}_{r_1} + \sqrt{\frac{r_2}{r_1}} \sin \tfrac{1}{2}\theta\, \mathbf{i}_h \times \mathbf{i}_{r_1} \right\} \tag{7.36}$$

which is a most convenient form for computational purposes.

A different expression for the velocity vector $\mathbf{v}_1$ can be had which gives some geometric significance to the variables $x$ and $y$. This is the subject of the next problem.

### Problem 7–2

Define the unit vectors $\mathbf{j}_1$ and $\mathbf{j}_2$ in the directions of the minimum-energy velocity vectors at the initial and terminal points. Let $\mathbf{i}_1$ and $\mathbf{i}_2$ be defined to make the coordinate pairs $\mathbf{i}_1, \mathbf{j}_1$ and $\mathbf{i}_2, \mathbf{j}_2$ right-handed and orthogonal. Then, the velocity vectors at the two terminals may be written as

$$\mathbf{v}_1 = \sqrt{\frac{\mu}{a_m}} \left[ -x(\mathbf{i}_1 \cdot \mathbf{i}_{r_1})\mathbf{i}_1 + y\sqrt{\frac{r_2}{r_1}}(\mathbf{i}_2 \cdot \mathbf{i}_{r_2})\mathbf{j}_1 \right]$$

$$\mathbf{v}_2 = \sqrt{\frac{\mu}{a_m}} \left[ \phantom{-}x(\mathbf{i}_2 \cdot \mathbf{i}_{r_2})\mathbf{i}_2 + y\sqrt{\frac{r_1}{r_2}}(\mathbf{i}_1 \cdot \mathbf{i}_{r_1})\mathbf{j}_2 \right]$$

## 7.2 The Q Function

The function $Q_\alpha$, or simply, $Q$, defined in Eq. (7.14) as

$$Q = \frac{\alpha - \sin \alpha}{\sin^3 \tfrac{1}{2}\alpha} \tag{7.37}$$

for elliptic orbits can be shown to be a hypergeometric function of

$$z = \sin^2 \tfrac{1}{4}\alpha \tag{7.38}$$

To this end,† we differentiate $Q$ with respect to $z$

$$\sin^2 \tfrac{1}{2}\alpha \frac{dQ}{dz} + 6\cos \tfrac{1}{2}\alpha\, Q = 8$$

by using the chain rule and noting that

$$\frac{dz}{d\alpha} = \tfrac{1}{2}\sin \tfrac{1}{4}\alpha \cos \tfrac{1}{4}\alpha = \tfrac{1}{4}\sin \tfrac{1}{2}\alpha$$

Now, substituting for $\alpha$ from

$$\sin^2 \tfrac{1}{2}\alpha = 4z(1-z) \qquad \text{and} \qquad \cos \tfrac{1}{2}\alpha = 1 - 2z$$

we obtain

$$z(1-z)\frac{dQ}{dz} + (\tfrac{3}{2} - 3z)Q = 2 \tag{7.39}$$

Finally, differentiating a second time produces

$$z(1-z)\frac{d^2Q}{dz^2} + (\tfrac{5}{2} - 5z)\frac{dQ}{dz} - 3Q = 0 \tag{7.40}$$

which is Gauss' equation (1.12) with parameters 3, 1, and $\tfrac{5}{2}$. Since

$$\lim_{z \to 0} Q = \lim_{\alpha \to 0} Q = \tfrac{4}{3}$$

then the first part of Eq. (7.16) of the previous section is established. In a similar manner, we can verify the second part.

### Improving the Convergence

Since hypergeometric functions admit a wide variety of transformations, we are tempted to explore the possibility of improving their computational efficiency. In fact, we can develop a convenient recursion formula for this purpose which is a direct consequence of

1. Gauss' relation for contiguous functions‡

$$(\gamma - \alpha - \beta)F(\alpha, \beta; \gamma; z)$$
$$+ \alpha(1-z)F(\alpha+1, \beta; \gamma; z) - (\gamma - \beta)F(\alpha, \beta-1; \gamma; z) = 0$$

2. Quadratic transformation formula—Eq. (1.19)

$$F(\alpha, \beta; \alpha+\beta-\tfrac{1}{2}; z) = (1-z)^{-\alpha} F(2\alpha-1, 2\beta-1; \alpha+\beta-\tfrac{1}{2}; \tfrac{1}{2} - \tfrac{1}{2}\sqrt{1-z}\,)$$

From the relation for contiguous functions, with the parameters $\alpha$, $\beta$, and $\gamma$ chosen as $\alpha = 2$, $\beta = 1$, and $\gamma = \tfrac{5}{2}$, we obtain

$$2(1-z)F(3, 1; \tfrac{5}{2}; z) = \tfrac{3}{2} + \tfrac{1}{2}F(2, 1; \tfrac{5}{2}; z)$$

---

† The derivation is the same as was used in Sect. 4.7 for the same quantity expressed in terms of universal functions.

‡ This is identity number (3) in the subsection on contiguous functions in Sect. 1.1.

Then, with the same choice of parameters, the quadratic transformation gives
$$F(2,1; \tfrac{5}{2}; z) = (1-z)^{-2} F(3,1; \tfrac{5}{2}; \tfrac{1}{2} - \tfrac{1}{2}\sqrt{1-z})$$
By combining the two, we derive the recursive expression
$$F(3,1; \tfrac{5}{2}; S_n) = \frac{1}{4C_n}\left[3 + \frac{1}{\sqrt{C_n}} F(-\tfrac{1}{2},1; \tfrac{5}{2}; S_{n+1})\right] \qquad (7.41)$$
where
$$S_n = \begin{cases} \sin^2 \dfrac{\psi}{2^n} \\ -\sinh^2 \dfrac{\psi}{2^n} \end{cases} \qquad C_n = \begin{cases} \cos^2 \dfrac{\psi}{2^n} \\ \cosh^2 \dfrac{\psi}{2^n} \end{cases} \qquad (7.42)$$
and for which, with $n = 0, 1, 2, \ldots$, the following recursive relations hold:
$$S_{n+1} = \tfrac{1}{2}(1 - \sqrt{C_n}) \quad \text{and} \quad C_{n+1} = \tfrac{1}{2}(1 + \sqrt{C_n}) \qquad (7.43)$$

Since convergence of either the hypergeometric series or the continued fraction is enhanced when the argument is small, we may recursively use Eq. (7.41) to advantage in order to obtain for given $\psi$ as rapid convergence as might be desired. There is, of course, a penalty in that the expression for $Q$ becomes algebraically more complex. For example, if we apply the recursive identities successively, we generate the following sequence for $Q$:

$$Q = \frac{4}{3} F(3,1; \tfrac{5}{2}; S_1)$$
$$Q = \frac{1}{C_1}\left[1 + \frac{1}{3\sqrt{C_1}} F(3,1; \tfrac{5}{2}; S_2)\right] \qquad (7.44)$$
$$Q = \frac{1}{C_1}\left\{1 + \frac{1}{4C_2 \sqrt{C_1}}\left[1 + \frac{1}{3\sqrt{C_2}} F(3,1; \tfrac{5}{2}; S_3)\right]\right\} \qquad \text{etc.}$$

where
$$\begin{aligned} S_1 &= \tfrac{1}{2}(1 - \lambda - x\eta) & C_1 &= \tfrac{1}{2}(1 + \lambda + x\eta) \\ S_2 &= \tfrac{1}{2}(1 - \sqrt{C_1}) & C_2 &= \tfrac{1}{2}(1 + \sqrt{C_1}) \\ S_3 &= \tfrac{1}{2}(1 - \sqrt{C_2}) \text{ etc.} & C_3 &= \tfrac{1}{2}(1 + \sqrt{C_2}) \text{ etc.} \end{aligned}$$

Note that each time the recursion is applied an additional square root is required in the calculation of $Q$ while, at the same time, the magnitude of the argument of the hypergeometric function decreases as indicated in Eq. (7.42). A comparison of the number of levels necessary for evaluation of the continued fraction representations of the successive hypergeometric functions can be found in the previously cited paper "Lambert's Problem Revisited" and will not be repeated here.

◇ **Problem 7–3**

Establish the identity

$$F(3, 1; \tfrac{5}{2}; S_n) = \frac{1}{C_n} F(-\tfrac{1}{2}, 1; \tfrac{5}{2}; T_n)$$

from the linear transformation formula (1.15). By combining this with Eq. (7.41), derive the recursive expression

$$F(-\tfrac{1}{2}, 1; \tfrac{5}{2}; T_n) = \frac{1}{4}\left[3 + \frac{1}{C_{n+1}\sqrt{C_n}} F(-\tfrac{1}{2}, 1; \tfrac{5}{2}; T_{n+1})\right]$$

where

$$T_n = \begin{cases} -\tan^2 \dfrac{\psi}{2^n} \\ \tanh^2 \dfrac{\psi}{2^n} \end{cases}$$

Develop the following sequence for $Q$:

$$Q = \frac{4}{3C_1} F(-\tfrac{1}{2}, 1; \tfrac{5}{2}; T_1)$$

$$Q = \frac{1}{C_1}\left[1 + \frac{1}{3C_2\sqrt{C_1}} F(-\tfrac{1}{2}, 1; \tfrac{5}{2}; T_2)\right]$$

$$Q = \frac{1}{C_1}\left\{1 + \frac{1}{4C_2\sqrt{C_1}}\left[1 + \frac{1}{3C_3\sqrt{C_2}} F(-\tfrac{1}{2}, 1; \tfrac{5}{2}; T_3)\right]\right\} \quad \text{etc.}$$

where

$$T_1 = \frac{\lambda + x\eta - 1}{\lambda + x\eta + 1} \qquad T_2 = \frac{\sqrt{C_1} - 1}{\sqrt{C_1} + 1} \qquad T_3 = \frac{\sqrt{C_2} - 1}{\sqrt{C_2} + 1} \quad \text{etc.}$$

**Continued Fraction Representation**

The hypergeometric function $F(3, 1; \tfrac{5}{2}; z)$ satisfies the requirement necessary for expansion as a continued fraction. Therefore, according to the developments in Sect. 1.2 of Chapter 1, we have

$$F(3, 1; \tfrac{5}{2}; z) = \cfrac{1}{1 - \cfrac{\gamma_1 z}{1 - \cfrac{\gamma_2 z}{1 - \cfrac{\gamma_3 z}{1 - \cdots}}}} \qquad (7.45)$$

where

$$\gamma_n = \begin{cases} \dfrac{(n+2)(n+5)}{(2n+1)(2n+3)} & n \text{ odd} \\ \dfrac{n(n-3)}{(2n+1)(2n+3)} & n \text{ even} \end{cases} \qquad (7.46)$$

By using the continued fraction rather than the power series representation of the hypergeometric function, not only is the speed of convergence improved, for a given argument $z$, but the range of convergence is also expanded from $|z| < 1$ for the power series to $z < 1$ for the continued fraction—a range that encompasses the entire spectrum of arguments for Lambert's problem. (The convergence of this particular fraction was demonstrated in Chapter 1 in Prob. 1–19 for negative values of $z$ and, as an example in a subsection of Sect. 1.3 for positive $z$.)

A convenient technique for evaluating continued fractions from the top to the bottom was developed in Sect. 1.4 of Chapter 1. Applying this to the case at hand, the algorithm for determining $F(3, 1; \frac{5}{2}; z)$ can be summarized as follows.

Initialize:
$$\delta_1 = u_1 = \Sigma_1 = 1$$

Calculate:
$$\delta_{n+1} = \frac{1}{1 - \gamma_n z \delta_n}$$
$$u_{n+1} = u_n(\delta_{n+1} - 1) \qquad (7.47)$$
$$\Sigma_{n+1} = \Sigma_n + u_{n+1}$$

where $\gamma_n$ is given in Eq. (7.46). Repeated calculations, for $n = 1, 2, \ldots$, of these equations produces $F(3, 1; \frac{5}{2}; z)$ since

$$F(3, 1; \tfrac{5}{2}; z) = \lim_{n \to \infty} \Sigma_n$$

provided, of course, that $z < 1$.

◊ **Problem 7–4**
The continued fraction expansion of $F(-\frac{1}{2}, 1; \frac{5}{2}; z)$ is

$$F(-\tfrac{1}{2}, 1; \tfrac{5}{2}; z) = \cfrac{1}{1 - \cfrac{\omega_1 z}{1 - \cfrac{\omega_2 z}{1 - \cfrac{\omega_3 z}{1 - \cdots}}}}$$

where

$$\omega_n = \begin{cases} \dfrac{(n-2)(n+2)}{(2n+1)(2n+3)} & n \text{ odd} \\[2ex] \dfrac{n(n+4)}{(2n+1)(2n+3)} & n \text{ even} \end{cases}$$

Determine the range of convergence of this continued fraction.

## Derivative Formulas

When Newton's method is used to solve the transfer-time equation for $x$, the derivative of Eq. (7.31) is required. For this purpose, we use Eqs. (7.19) and (7.29) to obtain

$$\frac{d\eta}{dx} = -\frac{\lambda\eta}{y}$$

so that

$$\sqrt{\frac{\mu}{a_m^3}}\frac{d}{dx}(t_2 - t_1) = \eta^3\frac{dQ}{dx} - \frac{\lambda\eta}{y}(3\eta^2 Q + 4\lambda) \qquad (7.48)$$

The derivative of the $Q$ function implies differentiation of the hypergeometric function $F(3, 1; \frac{5}{2}; z)$ which is obtained from Eq. (7.39) as

$$\frac{d}{dz}F(3, 1; \tfrac{5}{2}; z) = \frac{3 + 3(2z - 1)F(3, 1; \tfrac{5}{2}; z)}{2z(1 - z)}$$

Unfortunately, this expression is indeterminate for $z = 0$, corresponding to the case of the parabolic orbit.

To resolve the indeterminacy, we write

$$F(3, 1; \tfrac{5}{2}; z) = \frac{1}{1 - \gamma_1 z G(z)}$$

from the continued fraction representation given in Eq. (7.45), with

$$G(z) = \cfrac{1}{1 - \cfrac{\gamma_2 z}{1 - \cfrac{\gamma_3 z}{1 - \cdots}}}$$

Then, we have

$$\frac{d}{dz}F(3, 1; \tfrac{5}{2}; z) = \frac{6 - 3\gamma_1 G(z)}{2(1 - z)[1 - \gamma_1 z G(z)]} \qquad (7.49)$$

The function $G(z)$ [instead of $F(3, 1; \tfrac{5}{2}; z)$] may be evaluated by a trivial modification of the algorithm summarized in Eqs. (7.47).

Finally, the derivative of $z$ with respect to $x$, where $z = S_n$, is easily obtained by noting that

$$\frac{dS_1}{dx} = -\frac{dC_1}{dx} = -C_1^2 \qquad (7.50)$$

When applying Newton's iteration, a note of caution is necessary. In the vicinity of the minimum energy orbit ($x = 0$) and for $\lambda$ in the range $-1.0 \leq \lambda \leq -0.97$ (close to a 360° transfer), the second derivative of the transfer-time function versus $x$ is negative.† Under these very specialized circumstances, a different iteration technique will be required.

◇ **Problem 7-5**
Demonstrate that

$$\frac{d}{dz}F(-\tfrac{1}{2},1;\tfrac{5}{2};z) = -\frac{1+3\omega_1 H(z)}{2(1-z)[1-\omega_1 z H(z)]}$$

where

$$H(z) = \cfrac{1}{1 - \cfrac{\omega_2 z}{1 - \cfrac{\omega_3 z}{1 - \cdots}}}$$

Also, verify that

$$\frac{dT_1}{dx} = -\frac{\eta^2}{2y}$$

NOTE: The formulas for the coefficients of this continued fraction are given in Prob. 7-4.

## 7.3 Gauss' Method

The insight and ingenuity of Gauss are aptly demonstrated in his approach to the orbit-determination problem. One of his goals was to formulate the transfer-time equation in such a manner as to render it totally insensitive to computational errors when the transfer angle $\theta$ is small—of the order of two or three degrees.

Consider first his treatment of the expression

$$a = \frac{r_1 + r_2 - 2\sqrt{r_1 r_2}\cos\tfrac{1}{2}\theta\cos\psi}{2\sin^2\psi} \tag{7.51}$$

for the semimajor axis, obtained by eliminating $\cos\phi$ between Eqs. (7.2) and (7.4). In this form the equation is not suitable for his use. The radii $r_1$ and $r_2$ are nearly equal and both $\theta$ and $\psi$ are small angles. Therefore, to compute $a$ from Eq. (7.51) requires calculating the small difference of two almost equal quantities and then dividing by a small quantity—totally unacceptable to be sure.

Instead, Gauss writes

$$a = \frac{2\sqrt{r_1 r_2}\cos\tfrac{1}{2}\theta\,(\ell + \sin^2\tfrac{1}{2}\psi)}{\sin^2\psi} \tag{7.52}$$

---

† This was first reported by E. R. Lancaster and R. C. Blanchard in a NASA TN D-5368 titled "A Unified Form of Lambert's Theorem" and published in Sept. 1969. This change in curvature was discussed in connection with Fig. 7.3 in the previous section.

where $\ell$ is defined by

$$\ell = \frac{r_1 + r_2}{4\sqrt{r_1 r_2}\cos\frac{1}{2}\theta} - \frac{1}{2} \quad (7.53)$$

If $\ell$ can be accurately determined, Eq. (7.52) presents no problem when used to calculate $a$. But, of course, the equation for $\ell$ is also completely inappropriate for the same reasons as before.

However, suppose we define a quantity $\omega$ by

$$\tan(\tfrac{1}{4}\pi + \omega) = \left(\frac{r_2}{r_1}\right)^{\frac{1}{4}} \quad (7.54)$$

so that

$$\sqrt{\frac{r_2}{r_1}} + \sqrt{\frac{r_1}{r_2}} = 2 + 4\tan^2 2\omega$$

Then, since Eq. (7.53) can be written as

$$\ell = \frac{\sqrt{\frac{r_2}{r_1}} + \sqrt{\frac{r_1}{r_2}}}{4\cos\frac{1}{2}\theta} - \frac{1}{2}$$

we have

$$\ell = \frac{\sin^2\frac{1}{4}\theta + \tan^2 2\omega}{\cos\frac{1}{2}\theta} \quad (7.55)$$

which is entirely insensitive to computational errors and is positive for $\theta$ less than 180 degrees.

From a different point of view, let us write

$$r_2 = r_1(1 + \epsilon) \quad (7.56)$$

so that $\epsilon$ is simply the fractional part of the quotient of $r_2$ and $r_1$. Then, since

$$\tan(\tfrac{1}{4}\pi + \omega) = (1 + \epsilon)^{\frac{1}{4}}$$

according to Eq. (7.54), we obtain

$$\tan^2 2\omega = \frac{\frac{1}{4}\epsilon^2}{\sqrt{\frac{r_2}{r_1}} + \frac{r_2}{r_1}\left(2 + \sqrt{\frac{r_2}{r_1}}\right)} \quad (7.57)$$

to be used in Eq. (7.55). This alternative seems much more useful for computational purposes.

Sect. 7.3]  Gauss' Method  315

The Classical Equations of Gauss

Turning now to the transfer-time equation in the form

$$\sqrt{\frac{\mu}{a^3}}(t_2 - t_1) = 2\psi - \sin 2\psi + \frac{2\sqrt{r_1 r_2}\cos\frac{1}{2}\theta}{a}\sin\psi$$

which is obtained by eliminating $\cos\phi$ between Eqs. (7.1) and (7.4), we substitute for $a$ from Eq. (7.52) and introduce a quantity $m$ defined by

$$m = \frac{\sqrt{\mu}(t_2 - t_1)}{(2\sqrt{r_1 r_2}\cos\frac{1}{2}\theta)^{\frac{3}{2}}} \qquad (7.58)$$

There results

$$\pm m \frac{\sin^3\psi}{(\ell + \sin^2\frac{1}{2}\psi)^{\frac{3}{2}}} = 2\psi - \sin 2\psi + \frac{\sin^3\psi}{\ell + \sin^2\frac{1}{2}\psi} \qquad (7.59)$$

where the choice of sign depends on whether $\sin\psi$ is positive or negative.† Finally, we observe that, for the case we are treating, $\theta < \pi$, the upper sign of Eq. (7.59) is to be adopted and by introducing a quantity $y$ defined by

$$y^2 = \frac{m^2}{\ell + \sin^2\frac{1}{2}\psi} \qquad (7.60)$$

the transfer-time equation takes the form

$$y^3 - y^2 = m^2 \frac{2\psi - \sin 2\psi}{\sin^3\psi} \qquad (7.61)$$

These equations, (7.60) and (7.61), are the classical equations of Gauss which are to be solved simultaneously for the variables $y$ and $\psi$. The quantities $\ell$ and $m$ are constants which depend only on the geometry, the transfer time $t_2 - t_1$, and the gravitational constant $\mu$. When $y$ and $\psi$ are found, Eq. (7.52) provides an error-free computation of the semimajor axis $a$. The orbital parameter could then be obtained from Eq. (6.108).

Before considering the solution of Gauss' equations, we will demonstrate that Gauss' quantity $y$ has a significant geometrical interpretation. From Eqs. (7.52), (6.108), and (7.60) we find that

$$p = \frac{r_1 r_2 \sin^2\frac{1}{2}\theta}{2\sqrt{r_1 r_2}\cos\frac{1}{2}\theta} \frac{y^2}{m^2} = \frac{r_1^2 r_2^2 y^2 \sin^2\theta}{\mu(t_2 - t_1)^2}$$

---

† Observe that Eq. (7.58) implies that $\cos\frac{1}{2}\theta$ is to be positive and non-zero. Later we shall modify the equations to account for the case of a negative value for $\cos\frac{1}{2}\theta$ but the 180 degree transfer is excluded—a significant limitation of Gauss' method which we shall address in the last section of this chapter.

Then, since $p = h^2/\mu$, we have

$$y = \frac{\frac{1}{2}h(t_2 - t_1)}{\frac{1}{2}r_1 r_2 \sin\theta} \tag{7.62}$$

the denominator of which is the area of the triangle $\Delta FP_1P_2$; the numerator, by Kepler's second law, is the area of the sector bounded by the radii $r_1$, $r_2$ and the arc of the orbit included between $P_1$ and $P_2$. Thus, *y is the ratio of the areas of the sector and the triangle.* (All too frequently, Gauss' equations are developed by postulating this area ratio as an essential variable. In so doing, the fundamental motivation of Gauss tends to be obscured.)

It is also interesting to observe that $m$ and $\ell$ are invariant under the transformation described in Sect. 6.7. Therefore, from either Eq. (7.60) or (7.61), the area-ratio $y$ must also be an invariant.

Solving Gauss' Equations

The classical memoir by Gauss on hypergeometric functions and their continued fraction expansions was published some four years after *Theoria Motus* so his development of the right-hand side of Eq. (7.61) appeared to be ad hoc and somewhat enigmatic. Later, he would have written

$$\frac{2\psi - \sin 2\psi}{\sin^3\psi} = \frac{4}{3}F(3, 1; \tfrac{5}{2}; \sin^2 \tfrac{1}{2}\psi)$$

which we called $Q$ in the previous section and developed in a continued fraction expansion. As a consequence of this relation, the quantity

$$x = \sin^2 \tfrac{1}{2}\psi \tag{7.63}$$

can replace $\psi$ as one of the two unknowns in Gauss' equations.

In the orbit-determination problem which Gauss was originally addressing, the transfer angle $\theta$ (and, therefore, $\psi$) was small. It is reasonable, then, to assume as a first approximation, that $\psi$ (and $x = \sin^2 \tfrac{1}{2}\psi$) are zero. With $Q(0) = \tfrac{4}{3}$, a corresponding value of $y$ is determined by solving the cubic equation

$$y^3 - y^2 = m^2 Q(x) \tag{7.64}$$

for $y$ (there being only one positive real root) and then obtaining a new value of $x$ from

$$x = \frac{m^2}{y^2} - \ell \tag{7.65}$$

The calculation is repeated until $x$ ceases to change within tolerable limits.

To improve the convergence, Gauss in typical fashion displayed remarkable ingenuity. The idea, similar to the one he used for Kepler's

Sect. 7.3]  Gauss' Method  317

equation described in Sect. 5.5, is to replace the cubic equation (7.64) by one that is less sensitive to changes in the variable $x$.

We have already seen that $Q(x)$ admits of a continued fraction expansion. Clearly, from Eqs. (7.45) and (7.46), it can be written as

$$Q = \frac{1}{\frac{3}{4} - \frac{9}{10}xX} \quad \text{where} \quad X = \frac{1}{1 + \frac{2}{35}xZ}$$

and

$$Z = \cfrac{1}{1 - \cfrac{\frac{40}{63}x}{1 - \cfrac{\frac{4}{99}x}{1 - \cdots}}}$$

Now, define a quantity $\xi$ as

$$\xi = x(1 - X)$$

so that $xX = x - \xi$ and, therefore,

$$Q = \frac{1}{\frac{3}{4} - \frac{9}{10}(x - \xi)} \tag{7.66}$$

The continued fraction representation for $\xi$ is found by noting that

$$1 - X = \frac{\frac{2}{35}xZ}{1 + \frac{2}{35}xZ} = \frac{\frac{2}{35}x}{\frac{2}{35}x + \frac{1}{Z}}$$

Hence,

$$\xi = \cfrac{\frac{2}{35}x^2}{1 + \frac{2}{35}x - \cfrac{\frac{40}{63}x}{1 - \cfrac{\frac{4}{99}x}{1 - \cfrac{\frac{70}{143}x}{1 - \cfrac{\frac{6}{65}x}{1 - \cfrac{\frac{36}{85}x}{1 - \cdots}}}}}} \tag{7.67}$$

so that if $x$ is, indeed, small, then $\xi$, which is of the order of $\frac{2}{35}x^2$, will be considerably smaller.

The next step is to write the cubic equation for $y$ as

$$y - 1 = \frac{m^2}{y^2 \left[\frac{3}{4} - \frac{9}{10}(x - \xi)\right]}$$

by substituting Eq. (7.66) into (7.64). The explicit dependence on $x$ is eliminated by using Eq. (7.65) to replace $y^2 x$ by $m^2 - y^2\ell$. A minor rearrangement of this result produces a cubic equation for $y$ in the form

$$y^3 - y^2 - hy - \frac{h}{9} = 0 \qquad (7.68)$$

where the coefficient $h$, defined as

$$h = \frac{m^2}{\frac{5}{6} + \ell + \xi} \qquad (7.69)$$

depends only on $\xi$ which we have already found to be of the order of $\sin^4 \frac{1}{4}(E_2 - E_1)$. Since $\ell$ is positive [as is evident from Eq. (7.55)] and $\xi$ is small, then $h$ is positive. Hence, the cubic equation for $y$ admits of exactly one positive real root.

In this way, his objective of designing a rapidly convergent algorithm was neatly accomplished. For a reasonably small transfer angle $\theta$ (and, hence, a correspondingly small value of $x$) we may first assume that $x = \xi = 0$. Then $h$ is determined from Eq. (7.69) and $y$ obtained as the positive real root of the cubic equation (7.68). Having now a trial value for $y$, a new value of $x$ is obtained from Eq. (7.65) with which an improved value of $h$ is found. The process is repeated until $y$ ceases to change by a preassigned amount—usually two or three iterations being sufficient. The method of successive substitutions, which he also used for solving Kepler's equation, obviously was a favorite technique of Gauss.

After $x$ and $y$ are determined, the semimajor axis $a$ and the orbital parameter $p$ are determined from the formulas

$$\frac{1}{a} = \frac{4r_1 r_2 y^2 \sin^2 \psi \cos^2 \frac{1}{2}\theta}{\mu(t_2 - t_1)^2} \quad \text{and} \quad p = \frac{r_1^2 r_2^2 y^2 \sin^2 \theta}{\mu(t_2 - t_1)^2} \qquad (7.70)$$

which involve only products and quotients, and as such are themselves error-free.

To determine the eccentricity which Gauss wrote in the form

$$e = \sin \phi = \frac{2 \tan \frac{1}{2}\phi}{1 + \tan^2 \frac{1}{2}\phi}$$

he also proceeded carefully and cleverly. Writing the parameter as

$$p = a(1 - e^2) = a \cos^2 \phi = \frac{r_1 r_2 \sin^2 \frac{1}{2}\theta}{a \sin^2 \psi}$$

[using Eq. (6.108) for this purpose], we have

$$\cos \phi = \frac{\sqrt{r_1 r_2} \sin \frac{1}{2}\theta}{a \sin \psi}$$

Now, substituting for $a$ from Eq. (7.52), gives

$$\cos \phi = \frac{\sin \psi \tan \frac{1}{2}\theta}{2(\ell + \sin^2 \frac{1}{2}\psi)}$$

Finally, by replacing $\ell$ with its equivalent from Eq. (7.55), we obtain

$$\tan^2 \tfrac{1}{2}\phi = \frac{1-\cos\phi}{1+\cos\phi} = \frac{\sin^2(\tfrac{1}{4}\theta - \tfrac{1}{2}\psi) + \tan^2 2\omega}{\sin^2(\tfrac{1}{4}\theta + \tfrac{1}{2}\psi) + \tan^2 2\omega} \qquad (7.71)$$

or, alternately,

$$\tan^2 \tfrac{1}{2}\phi = \frac{\sin^2 B}{\sin^2 A} = \frac{1+\cot^2 A}{1+\cot^2 B} \qquad (7.72)$$

where

$$\cot^2 A = \frac{\sin^2(\tfrac{1}{4}\theta - \tfrac{1}{2}\psi)}{\tan^2 2\omega} \quad \text{and} \quad \cot^2 B = \frac{\sin^2(\tfrac{1}{4}\theta + \tfrac{1}{2}\psi)}{\tan^2 2\omega}$$

The equations comprising the algorithm are universal as Gauss also demonstrated. By extending the definition of $x$ so that

$$x = \begin{cases} \sin^2 \tfrac{1}{4}(E_2 - E_1) & \text{ellipse} \\ 0 & \text{parabola} \\ -\sinh^2 \tfrac{1}{4}(H_2 - H_1) & \text{hyperbola} \end{cases} \qquad (7.73)$$

and allowing, thereby, values of $x$ to range from $-\infty$ to $+1$, all types of orbits are included. Furthermore, and fortunately, the continued fraction (7.67) converges over this range.

For $0 < \theta < \pi$ and $\pi < \theta < 2\pi$ Gauss did, however, choose to develop separate equations. On the other hand, if we change his notation slightly and use $m$ in place of $m^2$ so that $\ell$ and $m$ can be negative for $\theta > 180°$, we will be spared the unnecessary burden of addressing the two cases separately. But we must remember that, under these circumstances, $y$ can be negative.

Obviously, Gauss knew that his method was singular for $\theta = \pi$ but he judiciously avoided ever mentioning it, as if it were just a minor annoyance —a small flaw in an otherwise beautiful scheme. Indeed, he said *"The equations ... possess so much neatness, that there may seem nothing more to be desired."* This flaw, however, coupled with convergence difficulties when $\theta$ is not very small, has rendered the method impractical to the modern Astrodynamicist who is concerned with a more general range of orbit-determination problems than Gauss could have imagined.

◊ **Problem 7-6**
Derive the appropriate equations of Gauss' method for the case of hyperbolic orbits and verify that Gauss' equations are universal.

◊ **Problem 7-7**
The equation of the velocity vector $\mathbf{v}_1$ at $P_1$, using Gauss' parameters, is

$$\mathbf{v}_1 = \frac{4m^2\lambda s - y^2 r_1}{y r_1 (t_2 - t_1)} \mathbf{r}_1 + \frac{y}{t_2 - t_1} \mathbf{r}_2$$

# Problem 7-8

When the transfer angle $\theta$ is between 180 and 360 degrees (or, more generally, when $\cos\frac{1}{2}\theta$ is negative), Gauss defined the appropriate quantities

$$L = \frac{1}{2} - \frac{r_1 + r_2}{4\sqrt{r_1 r_2}\cos\frac{1}{2}\theta} = -\frac{\sin^2\frac{1}{4}\theta + \tan^2 2\omega}{\cos\frac{1}{2}\theta}$$

$$M = \frac{\sqrt{\mu}(t_2 - t_1)}{(-2\sqrt{r_1 r_2}\cos\frac{1}{2}\theta)^{\frac{3}{2}}}$$

$$Y = \frac{M}{\sqrt{L - \sin^2\frac{1}{2}\psi}}$$

The orbit-determination problem can then be solved iteratively using

$$H = \frac{M^2}{L - \frac{5}{6} - \xi} \qquad Y^3 + Y^2 - HY + \frac{H}{9} = 0 \qquad x = L - \frac{M^2}{Y^2}$$

Further, the semimajor axis and the orbital parameter are obtained from

$$\frac{1}{a} = \frac{4Y^2 r_1 r_2 \cos^2\frac{1}{2}\theta \sin^2\frac{1}{2}\psi}{\mu(t_2 - t_1)^2} \qquad p = \frac{Y^2 r_1^2 r_2^2 \sin^2\theta}{\mu(t_2 - t_1)^2}$$

Verify the assertion that replacing $m^2$ by $m$ will render Gauss' equations universal in so far as the question of quadrant is concerned.

## Solving Gauss' Cubic Equation

When the transfer angle $\theta$ is small, Hansen† devised a convenient method of solving Gauss' cubic equation (7.68) using continued fractions. For this purpose, we write the cubic in the form

$$y^2(y - 1) - \tfrac{1}{9}h(9y + 1) = 0$$

and, replacing $y$ by $z$ where $y = 1 + 10z$, obtain

$$z(1 + 10z)^2 - \tfrac{1}{9}h(1 + 9z) = z(1 + 9z)(1 + 11z) + z^3 - \tfrac{1}{9}h(1 + 9z) = 0$$

Hence

$$z(1 + 11z) = \frac{h}{9} - \frac{z^3}{1 + 9z}$$

If the $z^3$ term is neglected (and it will be small for small $\theta$), then

$$z = \frac{\frac{2}{9}h}{1 + \sqrt{1 + \frac{44}{9}h}} \tag{7.74}$$

---

† Peter Andreas Hansen (1795–1874) was the leading German theoretical astronomer of the mid-nineteenth century. In 1825 he was invited to succeed Johann Franz Encke as the director of the private observatory of the Duke of Mecklenburg at Seeberg, near Gotha. From then until the end of his life, Hansen's contributions to astronomy and celestial mechanics were so numerous and enriched so many branches of those fields that he was considered among the foremost astronomers of his time.

Sect. 7.3]  Gauss' Method  321

or, alternately,

$$z = \frac{\frac{1}{9}h}{1+11z} = \frac{\frac{1}{9}h}{1+\dfrac{\frac{11}{9}h}{1+\dfrac{\frac{11}{9}h}{1+\cdots}}} \tag{7.75}$$

If necessary, we can replace the $h$ by $h - 9z^3/(1+9z)$ and recalculate $z$ from either the quadratic or the continued fraction. Each time the process is repeated, the original value of $h$ is corrected *using the latest value of* $z$. When $z$ no longer changes, then the root of the original cubic is

$$y = 1 + 10z \tag{7.76}$$

In general, Eq. (7.68) can be reduced, first, to the canonical form

$$z^3 - 3(3h+1)z = 2(6h+1)$$

by the substitution $3y = z+1$, and next, to the form of Eq. (1.32)

$$w^3 - 3w = 2\frac{1+6h}{(1+3h)^{\frac{3}{2}}} \tag{7.77}$$

with the substitution $z = \sqrt{1+3h}\,w$. Then, according to the discussion in the last subsection of Sect. 1.2, $w$ may be obtained using continued fractions. With $w$ known, the solution of the original cubic equation is

$$y = \tfrac{1}{3}(1+\sqrt{1+3h}\,w) \tag{7.78}$$

◇ **Problem 7–9**
During a time interval of 0.008840956 year, the planet Mars moved through a central angle of $\theta = 2°$ from a radial distance of 1.397414 a.u. to a radial distance of 1.399588 a.u. from the sun. Use Gauss' method to determine the semimajor axis $a$, the mean daily motion $n$, and the eccentricity $e$ of the Martian orbit.
ANSWER: The exact values are $a = 1.523691$, $n = 0.524033$ deg/day, and $e = 0.093368$.

**Problem 7–10**
The solution of the cubic equation in Prob. 7–8 for $Y$ can be expressed as

$$Y = \tfrac{1}{3}(\sqrt{1+3H}\,W - 1) \quad \text{where} \quad W^3 - 3W + 2\frac{1+6H}{(1+3H)^{\frac{3}{2}}} = 0$$

By substituting

$$W = Z - \frac{1}{Z}$$

obtain

$$Z^6 - 2BZ^3 + 1 = 0$$

as a quadratic equation in $Z^3$. Show that the *discriminant* $D$ of the quadratic is

$$D = \frac{(1+6H)^2}{(1+3H)^3} - 1 = -\frac{27H}{(1+3H)^3}\left(H - \frac{\sqrt{5}+1}{6}\right)\left(H + \frac{\sqrt{5}-1}{6}\right)$$

Therefore, Gauss' cubic equation can have more than one real root. In fact,

$$0 < H < \tfrac{1}{6}(\sqrt{5}+1) \qquad \text{one real root}$$
$$H = \tfrac{1}{6}(\sqrt{5}+1) \qquad \text{three real roots (two are equal)}$$
$$H > \tfrac{1}{6}(\sqrt{5}+1) \qquad \text{three real roots (unequal)}$$

For a positive value of $H$, show that Gauss' cubic equation (if it has any positive real root) has one negative and two positive roots, and that the two positive roots will either be equal to $\tfrac{1}{6}(\sqrt{5}-1)$ or one will be greater and the other less than this limit. Furthermore, show that the largest of the three roots is always the desired one.

## 7.4 An Alternate Geometric Transformation

The geometric transformation of the orbital boundary-value problem described in Sect. 6.7 resulted in the coincidence of the major axis and the chord—hence, a rectilinear orbit. The algorithm developed in the next section requires a different transformation resulting in an orbit whose major axis is *perpendicular* to the chord. The geometry of this new orbit is illustrated in Fig. 7.4.

### Transforming the Mean Point to an Apse

It is not difficult to see that the fundamental ellipse becomes a circle under this transformation. For all other elliptic orbits, the transformed mean point will either coincide with pericenter or apocenter, depending on whether the original ellipse had its vacant focus $F^*$, respectively, either below or above $F_F^*$ on the hyperbolic locus illustrated in Fig. 6.17.

For discussion, consider an ellipse for which the mean point is the pericenter of the transformed orbit. The transfer time from pericenter of the new orbit to the point $P_2$ is just one-half of the transfer time of the original orbit from $P_1$ to $P_2$. The pericenter radius is $r_0$, the mean point of the original orbit, the terminal radius is $\tfrac{1}{2}(r_1+r_2)$, and the true anomaly $f$ is related to the original central angle $\theta$ according to

$$\sin f = \frac{\tfrac{1}{2}c}{\tfrac{1}{2}(r_1+r_2)} \qquad \text{and} \qquad \cos f = \frac{\sqrt{r_1 r_2}\cos\tfrac{1}{2}\theta}{\tfrac{1}{2}(r_1+r_2)} \qquad (7.79)$$

The transfer-time equation is now the elementary form of Kepler's equation

$$\tfrac{1}{2}\sqrt{\mu}(t_2 - t_1) = a^{\tfrac{3}{2}}(E - e_0 \sin E) \qquad (7.80)$$

### Sect. 7.4]  An Alternate Geometric Transformation

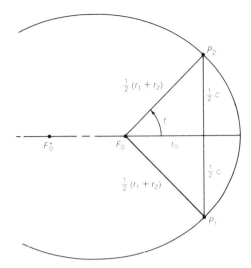

**Fig. 7.4:** Transformed ellipse with major axis perpendicular to the chord.

where $e_0$ is the eccentricity of the transformed orbit and $E$ is the eccentric anomaly corresponding to the true anomaly $f$. These quantities are simply related to the invariants $\psi$ and $\phi$ or $\alpha$ and $\beta$. Clearly,

$$E = \tfrac{1}{2}(E_2 - E_1) = \psi = \tfrac{1}{2}(\alpha - \beta) \tag{7.81}$$

Also, in general, from Eq. (6.74),

$$r_0 = a[1 - e\cos\tfrac{1}{2}(E_1 + E_2)] = a(1 - \cos\phi)$$

But, for the transformed orbit,

$$r_0 = a(1 - e_0)$$

so that

$$e_0 = \cos\phi = \cos\tfrac{1}{2}(\alpha + \beta) \tag{7.82}$$

Although the eccentricity is *not* an invariant, $e_0$ can, nevertheless, be expressed in terms of three of the basic length invariants of the boundary-value problem. Since

$$p = q(1 + e) = r(1 + e\cos f)$$

then

$$e = \frac{r - q}{q - r\cos f}$$

Therefore, in the present context, we have

$$e_0 = \frac{\tfrac{1}{2}(r_1 + r_2) - r_0}{r_0 - \sqrt{r_1 r_2}\cos\tfrac{1}{2}\theta} \tag{7.83}$$

If the vacant focus is above the vacant focus of the fundamental ellipse on the hyperbolic locus, then $r_0$ will be the apocenter distance. Equations

(7.82) and (7.83) would produce a negative value for $e_0$. Indeed, this case ($r_0$ being the apocenter radius) may be accounted for analytically by allowing the eccentricity $e_0$ to be negative.

For hyperbolic orbits traversed from $P_1$ to $P_2$, the above analysis is essentially the same. The expression for the eccentricity of the transformed hyperbolic orbit is exactly Eq. (7.83) and, of course, the problem of a negative value for $e_0$ will not arise.

### Relating $h$ and $\sigma$ to the Original Orbit

The angular momentum $h$ and the quantity $\sigma = \mathbf{r} \cdot \mathbf{v}/\sqrt{\mu}$ of the original orbit, each computed at the mean point, are

$$h = r_0 v_0 \sin \gamma_0 \quad \text{and} \quad \sqrt{\mu}\,\sigma_0 = r_0 v_0 \cos \gamma_0$$

where $\gamma_0$ is the flight-direction angle which, from Eq. (6.66), is

$$\cos \gamma_0 = e_F$$

where $e_F$ is the eccentricity of the fundamental ellipse.

The angular momentum $h_0$ of the transformed orbit is

$$h_0 = r_0 v_0$$

because of the invariance of $r_0$ and $v_0$. Thus, if $p_0$ is the parameter of the transformed orbit, we have

$$p = p_0(1 - e_F^2) \quad \text{and} \quad \sigma_0 = \sqrt{p_0}\, e_F$$

The quantities $\sigma$ and $\sigma_1$, corresponding respectively to the mean point and the initial point $P_1$ of the original orbit, are related according to Eq. (6.7). Therefore,

$$(r_0 \sigma_1 + r_1 \sigma_0) \tan \tfrac{1}{2}\delta = \sqrt{p}(r_0 - r_1)$$

where $\delta$, the transfer angle separating the mean point and the initial point, is determined from Eq. (6.69). Hence,

$$\sigma_1 = \frac{\sqrt{p_0}}{\tfrac{1}{2} c r_0}\{r_1[r_0 - \tfrac{1}{2}(r_1 + r_2)] + (r_0 - r_1)\sqrt{r_1 r_2} \cos \tfrac{1}{2}\theta\} \tag{7.84}$$

Further, since $p = p_0 \sin^2 \gamma_0 = p_0 \cos^2 \phi_F$, we have, from Eq. (6.48),

$$\frac{p}{p_m} = \frac{2p_0}{c} \tag{7.85}$$

The quantities $\sigma_1$ and $\sqrt{p}$ are precisely those needed to compute the velocity vector $\mathbf{v}_1$ at the initial point $P_1$ of the original orbit using Eq. (7.33).

## 7.5 Improving Gauss' Method

The simplicity of Gauss' method would certainly have been attractive to the modern Astrodynamicist except for two major flaws—the method is singular for a transfer angle of 180 degrees and the convergence rate is extremely slow when that angle is not very small. In this section, an algorithm is developed which represents a major improvement over Gauss' method and is made possible by exploiting a new principle, of which Gauss and his followers were probably not aware (even though it is a fundamental property of two-body orbits), together with a new wrinkle on an idea that Gauss himself invented in developing the iterative solution of Kepler's equation described in Sect. 5.5. The new principle is the invariance of the mean point (discussed in Sect. 6.4) used in conjunction with the geometric transformation of the boundary-value problem to bring the mean-point radius into coincidence with an orbital apse as described in the previous section. The result is that the transformed problem can then be simply described using the elementary form of Kepler's equation.

The second innovation is introducing a free parameter in Kepler's equation. The new twist is to choose this parameter, not to be a constant as Gauss did in his application, but rather to insure rapid convergence over the entire range of problems. Indeed, it is truly startling to observe just how rapid the convergence is when the iterated variable is anywhere near the solution value. It is not uncommon for this quantity to improve by as many as four or more significant decimal places *in a single iteration step*.

In this section, we derive an algorithm† which parallels exactly the elegant simplicity of the classical one but is completely devoid of the two basic faults of the original. In the process, we shall separately (1) remove the singularity at $\theta = \pi$ and (2) drastically improve the convergence for the entire range of transfer angles $0 < \theta < 2\pi$.

### Removing the Singularity

The singularity at $\theta = \pi$ in Gauss' method is removed using the transformation described in the previous section and the transfer-time equation is the elementary form of Kepler's equation (7.80). There are a number of invariants of this orbital transformation. The mean-point radius—the radius to that point in the orbit at which the tangent is parallel to the chord $P_1 P_2$—is one such invariant and is precisely the pericenter radius of the transformed orbit. The difference between the eccentric anomalies at $P_1$ and $P_2$ is also an invariant—half of this difference is just the eccentric

---

† The material in this section is from the paper "An Elegant Lambert Algorithm" by Richard H. Battin and Robin M. Vaughan published in the Nov.-Dec., 1984 issue of the *Journal of Guidance, Control, and Dynamics*. Several improvements have since been made which are incorporated here.

anomaly† $E$ in Kepler's equation (7.80). Obviously, the eccentricity $e_0$, given in Eq. (7.83), is *not* invariant even though the terms in that formula are each invariant.

Our objective is to convert Kepler's equation to a form resembling Gauss' equations (7.60) and (7.61). To this end, write (7.80) as

$$\frac{1}{2}\sqrt{\frac{\mu}{a^3}}(t_2 - t_1) = E - \sin E + (1 - e_0)\sin E \qquad (7.86)$$

and replace $1 - e_0$ by

$$1 - e_0 = \frac{r_0}{a} = \frac{r_{0p}}{a}\sec^2 \tfrac{1}{2}E$$

where we have used Eq. (6.79) for the second step. [Recall that $r_{0p}$ is the mean point radius (now, the pericenter radius) of the parabola connecting the terminals.] Then, we have

$$\frac{1}{2}\sqrt{\frac{\mu}{8r_{0p}^3}}(t_2 - t_1)\left(\frac{2r_{0p}}{a}\right)^{\frac{3}{2}} = E - \sin E + \left(\frac{2r_{0p}}{a}\right)\tan \tfrac{1}{2}E$$

which involves only $a$ and $E$. We can eliminate the semimajor axis by using the formula (4.32) which relates the true and eccentric anomalies. From

$$\tan^2 \tfrac{1}{2}f = \frac{1 + e_0}{1 - e_0}\tan^2 \tfrac{1}{2}E$$

we obtain

$$1 - e_0 = \frac{2\tan^2 \tfrac{1}{2}E}{\tan^2 \tfrac{1}{2}f + \tan^2 \tfrac{1}{2}E} \qquad (7.87)$$

Hence,

$$\frac{2r_{0p}}{a} = \frac{2(1 - e_0)}{1 + \tan^2 \tfrac{1}{2}E} = \frac{4\tan^2 \tfrac{1}{2}E}{(1 + \tan^2 \tfrac{1}{2}E)(\tan^2 \tfrac{1}{2}f + \tan^2 \tfrac{1}{2}E)} \qquad (7.88)$$

and, as a consequence, Kepler's equation takes the form

$$\sqrt{\frac{\mu}{8r_{0p}^3}}(t_2 - t_1)\frac{4\tan^3 \tfrac{1}{2}E}{\left[(\tan^2 \tfrac{1}{2}f + \tan^2 \tfrac{1}{2}E)(1 + \tan^2 \tfrac{1}{2}E)\right]^{\frac{3}{2}}} =$$

$$E - \sin E + \frac{4\tan^3 \tfrac{1}{2}E}{(\tan^2 \tfrac{1}{2}f + \tan^2 \tfrac{1}{2}E)(1 + \tan^2 \tfrac{1}{2}E)}$$

The analogy with Gauss' equations is now readily made. At the possible risk of confusing the reader, we will use the same notation as in Gauss'

---

† Recall that $E \equiv \psi$—the symbol used in the development of Gauss' method.

Sect. 7.5]  Improving Gauss' Method  327

method but the symbols will have different meanings. The advantage will be the ease of comparison of the two. Thus, if we define

$$x = \tan^2 \tfrac{1}{2} E \qquad \ell = \tan^2 \tfrac{1}{2} f \qquad m = \frac{\mu(t_2 - t_1)^2}{8 r_{0p}^3} \qquad (7.89)$$

the analog of the first equation of Gauss is had by defining $y$ as†

$$y^2 = \frac{m}{(\ell + x)(1 + x)} \qquad (7.90)$$

Substituting into Kepler's equation, we obtain the analog of the second of Gauss' equations

$$y^3 - y^2 = m \frac{E - \sin E}{4 \tan^3 \tfrac{1}{2} E} \qquad (7.91)$$

Just as in Gauss' method, the right-hand side of the second equation can be expressed using hypergeometric functions. Since,

$$\frac{E - \sin E}{4 \tan^3 \tfrac{1}{2} E} = \frac{1}{2 \tan^2 \tfrac{1}{2} E} \left( \frac{\tfrac{1}{2} E}{\tan \tfrac{1}{2} E} - \frac{1}{1 + \tan^2 \tfrac{1}{2} E} \right)$$

$$= \frac{1}{2x} \left( \frac{\arctan \sqrt{x}}{\sqrt{x}} - \frac{1}{1 + x} \right) = -\frac{d}{dx} \left( \frac{\arctan \sqrt{x}}{\sqrt{x}} \right)$$

we have, from Eq. (1.6),

$$\frac{E - \sin E}{4 \tan^3 \tfrac{1}{2} E} = -\frac{d}{dx} F(\tfrac{1}{2}, 1; \tfrac{3}{2}; -x) \qquad (7.92)$$

so that Eqs. (7.90) and (7.91) are also functions of $x$ and $y$ as was the case in Gauss' equations. The hypergeometric function satisfies the necessary requirement to be expanded as a continued fraction. Specifically,

$$F(\tfrac{1}{2}, 1; \tfrac{3}{2}; -x) = \cfrac{1}{1 + \cfrac{x}{3 + \cfrac{4x}{5 + \cfrac{9x}{7 + \cdots}}}} \qquad (7.93)$$

which should be compared with the results of Prob. 1–10.

Since we cannot, explicitly, differentiate continued fractions, an alternate form of Eq. (7.92) will be required. From the derivation of that equation, we see that

$$\frac{dF}{dx} = \frac{1}{2x} \left( \frac{1}{1 + x} - F \right) \qquad (7.94)$$

---

† Note that in this case $y$ is *not* the sector-triangle area ratio.

However, since $x$ can vanish, we must eliminate the indeterminacy somehow. For this purpose, define

$$F = \frac{1}{1+xG} \quad \text{where} \quad G = \cfrac{1}{3 + \cfrac{4x}{5 + \cfrac{9x}{7 + \cfrac{16x}{9 + \cfrac{25x}{11 + \cdots}}}}} \quad (7.95)$$

and obtain, therefrom,

$$\frac{dF}{dx} = -\frac{(1-G)F}{2(1+x)}$$

In the interest of simplifying the final result, it happens that we should also define $\xi$ by

$$G = \cfrac{1}{3 + \cfrac{4x}{\xi}} \quad \text{where} \quad \xi = 5 + \cfrac{9x}{7 + \cfrac{16x}{9 + \cfrac{25x}{11 + \cfrac{36x}{13 + \cdots}}}} \quad (7.96)$$

so that

$$\frac{dF}{dx} = -\frac{(2x+\xi)FG}{\xi(1+x)} = -\frac{2x+\xi}{(1+x)[4x+\xi(3+x)]} \quad (7.97)$$

Therefore, the analog of the second of Gauss' equations (7.91) can also be written as

$$y^3 - y^2 = \frac{m(2x+\xi)}{(1+x)[4x+\xi(3+x)]} \quad (7.98)$$

The reader should verify that these equations are also universal when the definition of $x$ is extended to include the other conics:

$$x = \begin{cases} \tan^2 \tfrac{1}{4}(E_2 - E_1) & \text{ellipse} \\ 0 & \text{parabola} \\ -\tanh^2 \tfrac{1}{4}(H_2 - H_1) & \text{hyperbola} \end{cases} \quad (7.99)$$

with its values ranging now from $-1$ to $+\infty$. Here again we should emphasize the importance of the continued fraction formulation. Unlike, any power series representation, *the continued fraction converges over the entire range of interest.*

## Computing $\ell$, $m$, and the Orbital Elements

From Eqs. (7.89) and (7.79), we have

$$\ell = \tan^2 \tfrac{1}{2} f = \frac{1 - \cos f}{1 + \cos f} = \frac{r_1 + r_2 - 2\sqrt{r_1 r_2} \cos \tfrac{1}{2}\theta}{r_1 + r_2 + 2\sqrt{r_1 r_2} \cos \tfrac{1}{2}\theta} \qquad (7.100)$$

However, just as was the case for Gauss' definition of $\ell$, this equation is not appropriate when $\theta$ is small; nor should it be used when $\theta$ is near 360 degrees. The same technique that Gauss employed is applicable here. Indeed, we easily deduce that

$$\ell = \begin{cases} \dfrac{\sin^2 \tfrac{1}{4}\theta + \tan^2 2\omega}{\sin^2 \tfrac{1}{4}\theta + \tan^2 2\omega + \cos \tfrac{1}{2}\theta} & \text{for } 0 < \theta < \pi \\[2ex] \dfrac{\cos^2 \tfrac{1}{4}\theta + \tan^2 2\omega - \cos \tfrac{1}{2}\theta}{\cos^2 \tfrac{1}{4}\theta + \tan^2 2\omega} & \text{for } \pi < \theta < 2\pi \end{cases} \qquad (7.101)$$

with $\tan^2 2\omega$ calculated from Eq. (7.57).

In like fashion, an error-free formula for computing the mean point radius of the parabola is possible. From Eq. (6.77), we have

$$r_{0p} = \tfrac{1}{4}(r_1 + r_2 + 2\sqrt{r_1 r_2} \cos \tfrac{1}{2}\theta) = \sqrt{r_1 r_2}\,(\cos^2 \tfrac{1}{4}\theta + \tan^2 2\omega) \qquad (7.102)$$

to be used in determining $m$ from the third of Eqs. (7.89).†

The orbital elements are as easily calculated and as error-free as they were in Gauss' method. Since

$$a(1 - e_0) = r_{0p}(1 + x) \qquad \text{and} \qquad 1 - e_0 = \frac{2x}{\ell + x}$$

we have

$$\frac{1}{a} = \frac{2x}{r_{0p}(\ell + x)(1 + x)} = \frac{2xy^2}{r_{0p} m}$$

so that we may compute the semimajor axis from

$$\frac{1}{a} = \frac{16 r_{0p}^2 xy^2}{\mu(t_2 - t_1)^2} = \frac{16 r_1 r_2 (\cos^2 \tfrac{1}{4}\theta + \tan^2 2\omega)^2 xy^2}{\mu(t_2 - t_1)^2} \qquad (7.103)$$

For the parameter, we first obtain

$$p_0 = a(1 - e_0^2) = \frac{\mu(t_2 - t_1)^2}{16 r_{0p}^2 xy^2} \times \frac{4\ell x}{(\ell + x)^2}$$

using Eq. (7.87) to derive the second factor. But, from Eq. (7.90),

$$y^2(\ell + x)^2 = \frac{m^2}{y^2(1 + x)^2} = \frac{\mu^2(t_2 - t_1)^4}{64 r_{0p}^6 y^2(1 + x)^2}$$

---

† As can be seen from the new definitions of $\ell$ and $m$, we have eliminated the singularity at $\theta = \pi$. However, there is now a singularity at $\theta = 2\pi$, which is more tolerable.

so that we have
$$p_0 = \frac{16\ell r_{0p}^4 y^2(1+x)^2}{\mu(t_2-t_1)^2}$$

Now recall, from the discussion of the last subsection of Sect. 7.4, that

$$p = p_0 \cos^2 \phi_F \quad \text{where} \quad \cos^2 \phi_F = \frac{2p_m}{c} = \frac{4r_1 r_2}{c^2} \sin^2 \tfrac{1}{2}\theta$$

Next, we show that $c^2$ can be expressed in terms of $\ell$ and $r_{0p}$. For this purpose, from Eq. (7.100),

$$1 + \ell = \frac{2(r_1+r_2)}{r_1+r_2+2\sqrt{r_1 r_2}\cos\tfrac{1}{2}\theta} = \frac{r_1+r_2}{2r_{0p}}$$

Also, from the first of Eqs. (7.79),

$$\sin^2 f = \frac{4\tan^2 \tfrac{1}{2}f}{(1+\tan^2 \tfrac{1}{2}f)^2} = \frac{4\ell}{(1+\ell)^2} = \frac{c^2}{(r_1+r_2)^2}$$

Then, from these last two equations, we obtain

$$c^2 = 16\ell r_{0p}^2 \tag{7.104}$$

to be used in the expression for $\cos^2 \phi_F$. Finally, then, the parameter may be calculated from

$$\begin{aligned}p &= \frac{4r_{0p}^2 r_1 r_2 y^2 (1+x)^2 \sin^2 \tfrac{1}{2}\theta}{\mu(t_2-t_1)^2} \\ &= \left[\frac{2r_1 r_2 (\cos^2 \tfrac{1}{4}\theta + \tan^2 2\omega) y(1+x)\sin\tfrac{1}{2}\theta}{\sqrt{\mu}(t_2-t_1)}\right]^2\end{aligned} \tag{7.105}$$

For an accurate determination of the eccentricity, we use the by now familiar notation

$$e = \sin\phi \quad \text{and} \quad e_0 = \sin\phi_0$$

so that the equation

$$p = p_0 \cos^2 \phi_F \quad \text{becomes} \quad \cos^2 \phi = \cos^2 \phi_0 \cos^2 \phi_F$$

or

$$1 - \sin^2 \phi = (1-\sin^2\phi_0)(1-\sin^2\phi_F)$$

Hence,

$$\sin^2 \phi = \sin^2 \phi_0 + \sin^2 \phi_F - \sin^2 \phi_0 \sin^2 \phi_F = \sin^2 \phi_F \cos^2 \phi_0 + \sin^2 \phi_0$$

Now, substitute for $\sin^2 \phi_F$ from

$$\sin^2 \phi_F = e_F^2 = \left(\frac{r_2 - r_1}{c}\right)^2 = \frac{r_1^2 \epsilon^2}{c^2}$$

where $\epsilon$ was defined in Eq. (7.56) as
$$r_2 = r_1(1+\epsilon)$$

Further, since,
$$c^2 = r_1^2 + r_2^2 - 2r_1 r_2 \cos\theta = r_1^2[2 + 2\epsilon + \epsilon^2 - 2(1+\epsilon)\cos\theta]$$

we have
$$\frac{c^2}{r_1^2} = \epsilon^2 + 4\frac{r_2}{r_1}\sin^2\tfrac{1}{2}\theta$$

Finally, using Eq. (7.87),
$$\sin^2\phi_0 = e_0^2 = \left(\frac{\ell - x}{\ell + x}\right)^2$$

and, as a result, the eccentricity is accurately computed from
$$e^2 = \frac{\epsilon^2 + 4\dfrac{r_2}{r_1}\sin^2\tfrac{1}{2}\theta \left(\dfrac{\ell-x}{\ell+x}\right)^2}{\epsilon^2 + 4\dfrac{r_2}{r_1}\sin^2\tfrac{1}{2}\theta} \tag{7.106}$$

In short, all of the precision-preserving techniques that Gauss so carefully crafted for his method exist also with the new formulation.

◇ **Problem 7-11**

Formulas for $\ell$, $m$, the semimajor axis, and the parameter can be expressed in terms of $\lambda$ defined in Eqs. (7.5) and (7.6). Specifically,

$$r_{0p} = \tfrac{1}{4}s(1+\lambda)^2 \qquad \ell = \left(\frac{1-\lambda}{1+\lambda}\right)^2 \qquad m = \frac{8\mu(t_2-t_1)^2}{s^3(1+\lambda)^6} = \frac{T^2}{(1+\lambda)^6}$$

and
$$\frac{1}{a} = \frac{8xy^2}{ms(1+\lambda)^2} \qquad p = \frac{2r_1 r_2 y^2(1+x)^2 \sin^2\tfrac{1}{2}\theta}{ms(1+\lambda)^2}$$

where
$$\lambda s = FS = \sqrt{r_1 r_2}\cos\tfrac{1}{2}\theta \quad \text{or} \quad \lambda = \pm\sqrt{\frac{s-c}{s}} \quad \text{and} \quad T = \sqrt{\frac{8\mu}{s^3}}(t_2 - t_1)$$

NOTE: This was the basis of the formulation used in the paper (previously cited) "An Elegant Lambert Algorithm." If $\theta$ is defined implicitly by the vectors $\mathbf{r}_1$ and $\mathbf{r}_2$, experience† has shown that these equations are preferred for computation.

---

† Allan Klumpp of the Jet Propulsion Laboratory made an extensive study of this algorithm in 1986—exploring the envelope of applicability and stress-testing it for all reasonable cases. He concluded that it "*offers the required compactness, speed, and reliability for manned and unmanned onboard guidance. The algorithm offers the accuracy and application range required for planetary orbit determination and theoretical astronomy.*"

## Improving the Convergence

In his *Theoria Motus*, Gauss developed an extremely efficient technique for solving the elementary form of Kepler's equation in the case of near parabolic orbits which was described in Sect. 5.5. The key was the introduction of a parameter specifically selected to accelerate the convergence of his successive substitution algorithm. Since our time equation is also the simple form of Kepler's equation, we are tempted to introduce a free parameter in this instance too.

For this purpose, with $\beta$ as yet unspecified, write Eq. (7.86) as

$$\frac{1}{2}\sqrt{\frac{\mu}{a^3}}(t_2 - t_1) = [1 + \beta(1 - e_0)]P + (1 - e_0)Q$$

where

$$P = E - \sin E \quad \text{and} \quad Q = \sin E - \beta P$$

Now, from Eqs. (7.87) and (7.92) together with a trigonometric identity for the sine function, we have

$$1 - e_0 = \frac{2x}{\ell + x} \qquad P = -4\tan^3 \tfrac{1}{2}E \, \frac{dF}{dx} \qquad \sin E = \frac{2\tan \tfrac{1}{2}E}{1 + x}$$

which are used to develop the expressions

$$[1 + \beta(1 - e_0)]P = -4\tan^3 \tfrac{1}{2}E \left[\frac{dF}{dx} + \frac{h_1}{(\ell + x)(1 + x)}\right]$$

$$(1 - e_0)Q = \frac{4\tan^3 \tfrac{1}{2}E}{(\ell + x)(1 + x)}(1 + h_1) = \frac{4y^2}{m}(1 + h_1)\tan^3 \tfrac{1}{2}E$$

with the quantity $h_1$ defined as

$$h_1 = 2\beta x(1 + x)\frac{dF}{dx} \tag{7.107}$$

Then, by combining the third of Eqs. (7.89) with Eq. (7.103), we also have

$$\frac{m}{2}\sqrt{\frac{\mu}{a^3}}(t_2 - t_1) = 4y^3 \tan^3 \tfrac{1}{2}E$$

so that Kepler's equation can be written as

$$y^3 - (1 + h_1)y^2 + m\left[\frac{dF}{dx} + \frac{h_1}{(\ell + x)(1 + x)}\right] = 0 \tag{7.108}$$

Clearly, if $\beta = 0$, then $h_1 = 0$ and Eq. (7.108) reduces to (7.91). Otherwise, $\beta$ can have any value whatsoever—*not necessarily constant*.

Now that we have this extra degree of freedom, how can we best use it? To decide, consider the general problem of the simultaneous solution of two equations by successive substitutions. In the top part of Fig. 7.5 we have plotted two arbitrary functions $y_1(x)$ and $y_2(x)$; the intersection of these two curves is the solution point. To find the solution by successive substitution, we start by choosing an initial value $x_0$ and then calculate the corresponding value of $y_1(x_0)$. Next, $y_1(x_0)$ is used to obtain a new value of $x$ by locating the point where $y_2(x) = y_1(x_0)$. The new value of $x$ is again used to calculate $y_1$ which, in turn, is used to find a new value for $x$. This process is represented in the figure by the horizontal and vertical dotted lines. Clearly, the curvature of the two functions greatly influences the number of iterations required to reach the solution. In the extreme, suppose that $y_1(x)$ is a constant as shown in the bottom part of the figure. Then the solution would be attained in just one iteration step since, for any $x_0$, we have $y_1(x_0)$ exactly equal to $y_2(x)$ at the solution point.

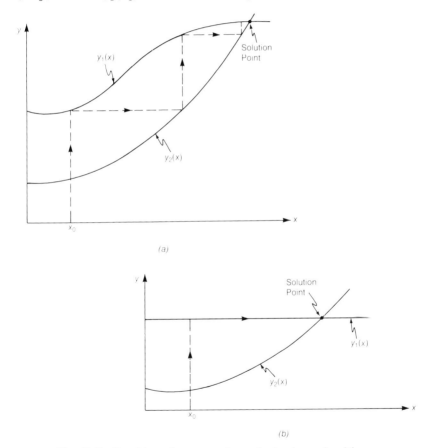

**Fig. 7.5:** Graphics of a successive-substitutions algorithm.

Using this argument, it appears that we should choose our free parameter $\beta$ so that $dy/dx$ will be zero at the solution point. Of course, we don't know the location of this point—if we did, no problem would exist at all. For the moment, we ignore this seemingly crucial matter, calculate the derivative of Eq. (7.108),

$$3y^2\frac{dy}{dx} - 2(1+h_1)y\frac{dy}{dx} - y^2\frac{dh_1}{dx} + m\frac{d^2F}{dx^2}$$
$$+ \frac{m}{(\ell+x)(1+x)}\frac{dh_1}{dx} + mh_1\frac{d}{dx}\frac{1}{(\ell+x)(1+x)} = 0$$

and examine the terms one by one.

At the solution point, we know that Eq. (7.90) is satisfied so that the terms involving $dh_1/dx$ cancel. Hence, for $dy/dx$ to be zero at this point, the sum of the fourth and sixth terms must vanish. Thus,

$$(\ell+x)^2(1+x)^2\frac{d^2F}{dx^2} - h_1(1+2x+\ell) = 0 \qquad (7.109)$$

and this is the equation we shall use to determine $\beta$. This parameter will be a function of $x$ which, of course, should be evaluated at the solution point wherever it may be. Since we don't know its location, we will use the function instead of its value—knowing full well that this is the right value at the solution point. It will be almost correct near that point and, hopefully, won't cause serious problems otherwise.†

To evaluate the free parameter $\beta$ from Eq. (7.109), we first note that since $F = F(\frac{1}{2}, 1; \frac{3}{2}; -x)$ is a hypergeometric function, it must satisfy Gauss' differential equation (1.12)

$$2x(1+x)\frac{d^2F}{dx^2} + (3+5x)\frac{dF}{dx} + F = 0$$

This can be used to eliminate the second derivative from Eq. (7.109). As a consequence, we obtain

$$(\ell+x)^2(1+x)\left[(3+5x)\frac{dF}{dx} + F\right] + 2x(1+2x+\ell)h_1 = 0 \qquad (7.110)$$

When we substitute for $dF/dx$ from Eq. (7.97) and make the appropriate reductions, we obtain the following equation‡ for $h_1$:

$$h_1 = \frac{(\ell+x)^2(1+3x+\xi)}{(1+2x+\ell)[4x+\xi(3+x)]} \qquad (7.111)$$

Finally, we have no difficulty in also deriving

$$h_2 = \frac{m(x-\ell+\xi)}{(1+2x+\ell)[4x+\xi(3+x)]} \qquad (7.112)$$

---

† Gauss achieved a similar flattening of his cubic equation (7.68) for values of $x$ near $x = 0$ since he was able to insure that his coefficient $h$ had only a second-order variation with $x$.

‡ Since $h_1$ is proportional to $\beta$, we have, in fact, determined the free parameter.

where we have introduced the notation $-h_2$ for the last term in Eq. (7.108).

In summary, then, the two analogs of Gauss' equations (7.60) and (7.61) are

$$x = \sqrt{\left(\frac{1-\ell}{2}\right)^2 + \frac{m}{y^2}} - \frac{1+\ell}{2} \qquad (7.113)$$

and

$$y^3 - y^2 - h_1 y^2 - h_2 = 0 \qquad (7.114)$$

with the coefficients determined from Eqs. (7.111) and (7.112).† The function $\xi(x)$ is calculated from the continued fraction

$$\xi = 5 + \cfrac{\frac{9}{7}x}{1 + \cfrac{\frac{16}{63}x}{1 + \cfrac{\frac{25}{99}x}{1 + \cfrac{\frac{36}{143}x}{1 + \cfrac{\frac{49}{195}x}{1 + \cfrac{\frac{64}{255}x}{1 + \cfrac{\frac{81}{323}x}{1 + \cdots}}}}}}} \qquad \text{for} \quad -1 \leq x < \infty \qquad (7.115)$$

which is equivalent to Eq. (7.96).

The mechanics of the algorithm are the same as for Gauss' method. From a trial value of $x$, the coefficients $h_1$ and $h_2$ are calculated. Then the cubic equation is solved for $y$ and a new value of $x$ determined from Eq. (7.113). The steps are repeated until convergence to an acceptable accuracy is obtained.

## Transforming the Function $\xi(x)$

In both Gauss' method and in the algorithm described here, the two operations which consume the most time are: (1) evaluating the continued fraction and (2) solving the cubic equation. In *Theoria Motus*, Gauss prepared tables for this purpose which span the region over which his method is useful and valid. In lieu of tables, we shall derive efficient computation techniques for these two aspects of our algorithm.

In Sect. 1.4 a top-down method is described for evaluating continued fractions which can be used to determine $\xi(x)$. It will not be considered further here. However, we must remark that, although the continued fraction representation of $\xi(x)$ in Eq. (7.115) converges rapidly for small values of $x$, the number of levels required increases significantly as $x$ becomes large.

---

† Equation (7.113) is obtained from Eq. (7.90) by solving for $x$ using the quadratic formula. The specified range $-1 < x < \infty$ governs the choice of sign for the radical.

A substantial improvement in the rate of convergence can be had at the expense of a little extra preliminary computation. For this purpose, we establish, using Prob. 1–10,

$$\frac{\tfrac{1}{2}E}{\tan \tfrac{1}{2}E} = F(1, \tfrac{1}{2}; \tfrac{3}{2}; -\tan^2 \tfrac{1}{2}E)$$

from which is deduced

$$\frac{\tfrac{1}{4}E}{\tan \tfrac{1}{4}E} = F(1, \tfrac{1}{2}; \tfrac{3}{2}; -\tan^2 \tfrac{1}{4}E)$$

Then, since

$$\tan^2 \tfrac{1}{4}E = \frac{\sec \tfrac{1}{2}E - 1}{\sec \tfrac{1}{2}E + 1} = \frac{\sqrt{1+x}-1}{\sqrt{1+x}+1} \equiv \eta$$

it follows that

$$F(x) \equiv F(1, \tfrac{1}{2}; \tfrac{3}{2}; -x) = \frac{2\tan \tfrac{1}{4}E}{\tan \tfrac{1}{2}E} F(1, \tfrac{1}{2}; \tfrac{3}{2}; -\eta) = \frac{2F(\eta)}{\sqrt{1+x}+1}$$

Therefore,

$$F(x) = \frac{1}{1+xG(x)} = \frac{2}{\sqrt{1+x}+1}\left(\frac{1}{1+\eta G(\eta)}\right) \qquad (7.116)$$

To solve for $G(x)$, this last equation is written as

$$2[1+xG(x)] = \sqrt{1+x}+1+(\sqrt{1+x}-1)G(\eta)$$
$$= \sqrt{1+x}+1+\frac{x}{\sqrt{1+x}+1}G(\eta)$$

leading to

$$2xG(x) - (\sqrt{1+x}-1) = 2xG(x) - \frac{x}{\sqrt{1+x}+1} = \frac{x}{\sqrt{1+x}+1}G(\eta)$$

Now $x$ may be cancelled as a common factor and we have

$$G(x) = \frac{1+G(\eta)}{2(\sqrt{1+x}+1)} \qquad (7.117)$$

Furthermore,

$$G(x) = \frac{1}{3+4x/\xi(x)} = \frac{1}{2(\sqrt{1+x}+1)}\left[1+\frac{1}{3+4\eta/\xi(\eta)}\right] \qquad (7.118)$$

and, in a similar manner, we determine

$$\xi(x) = \frac{8(\sqrt{1+x}+1)}{3+\dfrac{1}{\eta+\xi(\eta)}} \qquad (7.119)$$

Sect. 7.5]  Improving Gauss' Method  337

Continued Fraction Levels $[\xi(x), \xi(\eta)]$

| $\lambda\backslash T$ | 0.3 | 0.5 | 0.7 | 0.9 | 1.0 | 3.0 | 5.0 | 7.0 |
|---|---|---|---|---|---|---|---|---|
| $-0.9$ | 81, 19 | 50, 15 | 36, 13 | 27, 11 | 24, 10 | 12, 6 | 46, 9 | 81, 9 |
| $-0.7$ | 60, 17 | 37, 13 | 26, 10 | 19, 9 | 17, 8 | 16, 7 | 29, 8 | 38, 9 |
| $-0.5$ | 44, 14 | 26, 11 | 18, 9 | 13, 7 | 11, 6 | 15, 7 | 22, 8 | 26, 8 |
| $-0.3$ | 31, 12 | 19, 9 | 13, 7 | 10, 6 | 8, 5 | 12, 6 | 17, 7 | 20, 7 |
| $-0.1$ | 20, 9 | 13, 7 | 10, 6 | 8, 5 | 7, 5 | 11, 6 | 14, 7 | 17, 7 |
| 0.0 | 15, 8 | 11, 6 | 9, 6 | 7, 5 | 6, 5 | 10, 6 | 13, 6 | 15, 7 |
| 0.1 | 11, 6 | 9, 6 | 8, 5 | 7, 5 | 6, 4 | 9, 5 | 12, 6 | 14, 7 |
| 0.3 | 7, 5 | 7, 5 | 6, 4 | 5, 4 | 5, 4 | 8, 5 | 10, 6 | 12, 6 |
| 0.5 | 5, 4 | 5, 4 | 5, 4 | 4, 3 | 4, 3 | 7, 5 | 9, 5 | 11, 6 |
| 0.7 | 4, 3 | 4, 3 | 3, 3 | 3, 2 | 3, 3 | 6, 4 | 8, 5 | 9, 5 |
| 0.9 | 2, 2 | 3, 2 | 3, 3 | 3, 3 | 4, 3 | 6, 4 | 7, 5 | 8, 5 |

In summary, then, if we define

$$\eta = \frac{x}{(\sqrt{1+x}+1)^2} \quad \text{where} \quad -1 < \eta < 1 \tag{7.120}$$

we have shown that

$$\xi(x) = \cfrac{8(\sqrt{1+x}+1)}{3 + \cfrac{1}{5 + \eta + \cfrac{\frac{9}{7}\eta}{1 + \cfrac{\frac{16}{63}\eta}{1 + \cfrac{\frac{25}{99}\eta}{1 + \cfrac{\frac{36}{143}\eta}{1 + \cfrac{\frac{49}{195}\eta}{1 + \cdots}}}}}} \tag{7.121}$$

In the table† at the top of the page a comparison is made of the number of continued fraction levels required to compute $\xi(x)$ to eight significant digits for various values of

$$\lambda = \pm\sqrt{\frac{s-c}{s}} \quad \text{and} \quad T = \sqrt{\frac{8\mu}{s^3}}(t_2 - t_1) \tag{7.122}$$

using both Eqs. (7.115) and (7.121).

---

† The tables in this section are from the paper "An Elegant Lambert Algorithm."

## Solving the Cubic

Turning our attention now to solving the cubic equation (7.114), observe that Eq. (7.91) has only one positive real root since the right-hand side of that equation is positive for all orbits. Furthermore, that root must exceed unity in magnitude. This is the solution to our problem and, of course, must be a root of Eq. (7.114) also.

It is not difficult to show that $h_1(x)$, defined in Eq. (7.111), is always positive, but $h_2(x)$, defined in Eq. (7.112), can have either sign. Hence, there can be more than one positive real root of Eq. (7.114) and the question of which is the proper choice must be resolved.

It is easy to verify, for parabolic orbits with

$$\lambda > \frac{1 - \sqrt{5}}{1 + \sqrt{5}}$$

that $h_2$ is positive and, consequently, for that case Eq. (7.114) has exactly one positive real root. Therefore, if $\lambda$ and $T$ vary continuously, a simple continuity argument will suffice to prove that, when multiple roots appear, the largest is always the correct choice.

The classical explicit formulas for obtaining the roots of a cubic can, of course, be used for solving Eq. (7.114). However, there is an extremely attractive formula, utilizing continued fractions, which is guaranteed to produce always the correct root.

The transformation

$$y = \frac{2}{3}(1 + h_1)\left(\frac{b}{z} + 1\right)$$

where

$$b = \sqrt{\frac{27 h_2}{4(1 + h_1)^3} + 1}$$

will convert the cubic equation for $y$ to the canonical form

$$z^3 - 3z = 2b$$

which is just the one considered at the end of Sect. 1.2. The solution as a continued fraction is developed in that section.

In summary, to solve the cubic, we first calculate

$$u = \frac{B}{2(\sqrt{1 + B} + 1)} \quad \text{where} \quad B = \frac{27 h_2}{4(1 + h_1)^3} \quad (7.123)$$

Then, the largest positive real root of Eq. (7.114) is

$$y = \frac{1 + h_1}{3}\left(2 + \frac{\sqrt{1 + B}}{1 + 2u K^2(u)}\right) \quad (7.124)$$

Sect. 7.5]  Improving Gauss' Method  339

where $K(u)$ is calculated from the continued fraction

$$K(u) = \cfrac{\frac{1}{3}}{1 + \cfrac{\frac{4}{27}u}{1 + \cfrac{\frac{8}{27}u}{1 + \cfrac{\frac{2}{9}u}{1 + \cfrac{\frac{22}{81}u}{1 + \cfrac{\frac{208}{891}u}{1 + \cdots}}}}}} \qquad (7.125)$$

In general, the odd- and even-numbered coefficients of $u$ in the continued fraction, which we call $\gamma_{2n+1}$ and $\gamma_{2n}$, can be obtained from

$$\gamma_{2n+1} = \frac{2(3n+2)(6n+1)}{9(4n+1)(4n+3)} \quad \text{and} \quad \gamma_{2n} = \frac{2(3n+1)(6n-1)}{9(4n-1)(4n+1)}$$

The quantity $1 + B$ can be shown to be always positive so that no difficulty is encountered in the square root. Indeed, unless this term is positive, there will not exist a positive real root of Eq. (7.114).

Of course, we can always resort to Newton's method for finding the root but we must be careful that we converge to the correct one. A little analysis confirms that an appropriate starting value $y_0$ is either zero or $\frac{2}{3}(1+h_1)$ depending on whether $h_2/(1+h_1)^3$ is or is not less than $-\frac{4}{27}$, respectively. On each subsequent cycle in the iteration it makes sense to use the value of $y$ calculated during the previous cycle.

Finally, a few programming hints are appropriate. In calculating the coefficients of the cubic equation, $h_1(x)$ and $h_2(x)$, note that from Eq. (7.113), we can write

$$1 + x = \sqrt{L^2 + \frac{m}{y^2}} + L \quad \text{and} \quad \ell + x = \sqrt{L^2 + \frac{m}{y^2}} - L = \frac{m}{y^2(1+x)}$$

so that

$$1 + 2x + \ell = 2\sqrt{L^2 + \frac{m}{y^2}}$$

where

$$L = \frac{1-\ell}{2} = \begin{cases} \dfrac{\frac{1}{2}\cos\frac{1}{2}\theta}{\sin^2\frac{1}{4}\theta + \tan^2 2\omega + \cos\frac{1}{2}\theta} & \text{for } 0 < \theta < \pi \\[2ex] \dfrac{\frac{1}{2}\cos\frac{1}{2}\theta}{\cos^2\frac{1}{4}\theta + \tan^2 2\omega} & \text{for } \pi < \theta < 2\pi \end{cases}$$

## Comparing the Two Methods

While the derivations of Gauss' method and the new method for solving Lambert's problem are different, the final equations and mechanics of the algorithms are quite similar. The dimensionless parameters $\lambda$ and $T$ are the inputs to both methods. The new method requires somewhat more algebra for each iteration since there are two coefficients to be found for the cubic equation and Eq. (7.90) is quadratic in $x$. There is no need, however, to test the value of $\lambda$ as in Gauss' method since the new equations are valid for $-1 < \lambda \leq 1$. The efficiency of either procedure is measured by the number of iterations necessary to compute $x$ to a given accuracy.

In the table on the opposite page, the two methods are compared to contrast the number of iterations required to compute $x$ to eight significant digits. For Gauss' method, the initial values of $x$ selected to generate this table were

$$x_0 = \begin{cases} 0 & \text{parabola, hyperbola, i.e., } T \leq T_p = \tfrac{4}{3}(1-\lambda^3) \\ \dfrac{\ell}{1+2\ell} & \text{ellipse, i.e., } T > T_p \end{cases}$$

(For the ellipse, $x_0$ defines a circular orbit.) Note the rapid convergence of Gauss' method in the lower left-hand corner of the table. In this region, $x$ is nearly zero and the transfer angle $\theta$ is small. Gauss designed his method for problems of this type. The quantity $\xi(x)$ was ingeniously constructed to be of order $x^2$ so that it would be very small for small $x$—the result being that $h$, the coefficient in the cubic, is nearly independent of $x$ so that $y$ is almost constant.

The two major disadvantages are the singularity at $\lambda = 0$ and the convergence properties over the range of $\lambda$ and $T$ considered. Although, Gauss' method converged for all elliptic cases considered, it sometimes required more than 100 iterations to do so. For most of the hyperbolic cases, it did not converge at all.

On the other hand, the new method was designed to converge rapidly for any case independent of the value of $x$. The nearly uniform convergence behaviour of the new method is seen in the table. The initial value strategy chosen for $x$ was identical to that used for Gauss' method. In this case, those values are

$$x_0 = \begin{cases} 0 & \text{parabola, hyperbola} \\ \ell & \text{ellipse} \end{cases}$$

The rapid convergence in the lower left-hand corner of the table is retained using the new method and there is no significant difference in the rate of convergence for positive or negative values of $\lambda$. Although not shown in the table, a striking advantage of the new method is that *only one more iteration step is necessary to obtain four more significant figures in all the cases considered.*

Sect. 7.5]  Improving Gauss' Method  341

Number of Iterations (Gauss' Method, New Method)

| $\lambda \backslash T$ | 0.3 | 0.5 | 0.7 | 0.9 | 1.0 | 3.0 | 5.0 | 7.0 |
|---|---|---|---|---|---|---|---|---|
| −0.9 | †, 4 | †, 4 | †, 5 | †, 5 | †, 5 | 67, 8 | 210, 7 | 190, 6 |
| −0.7 | †, 4 | †, 5 | †, 5 | †, 5 | †, 5 | 44, 5 | 52, 5 | 52, 4 |
| −0.5 | †, 4 | †, 4 | †, 4 | †, 4 | †, 5 | 24, 4 | 26, 4 | 19, 3 |
| −0.3 | †, 4 | †, 4 | †, 4 | 29, 4 | 17, 4 | 14, 4 | 14, 3 | 19, 4 |
| −0.1 | †, 4 | 16, 4 | 10, 4 | 9, 4 | 8, 4 | 10, 4 | 12, 4 | 14, 5 |
| 0.0 | ‡, 4 | ‡, 4 | ‡, 4 | ‡, 4 | ‡, 4 | ‡, 4 | ‡, 4 | ‡, 5 |
| 0.1 | 5, 4 | 6, 4 | 6, 4 | 5, 4 | 5, 4 | 7, 4 | 9, 4 | 11, 5 |
| 0.3 | 3, 4 | 4, 4 | 4, 4 | 4, 4 | 4, 4 | 6, 4 | 8, 5 | 10, 5 |
| 0.5 | 3, 3 | 3, 3 | 3, 3 | 3, 3 | 3, 3 | 5, 4 | 6, 5 | 8, 5 |
| 0.7 | 3, 3 | 3, 3 | 3, 3 | 3, 3 | 3, 3 | 4, 4 | 5, 4 | 7, 5 |
| 0.9 | 2, 2 | 2, 2 | 2, 3 | 2, 3 | 3, 3 | 4, 4 | 5, 4 | 6, 5 |

† Gauss' method does not converge.
‡ Gauss' method is singular for $\theta = 180$ degrees.

◊ **Problem 7–12**

A vehicle in interplanetary space moves from $\mathbf{r}_1$ to $\mathbf{r}_2$ in a time interval of 0.010794065 year where

$$\mathbf{r}_1 = \begin{bmatrix} 0.159321004 \\ 0.579266185 \\ 0.052359607 \end{bmatrix} \text{ a.u.} \quad \text{and} \quad \mathbf{r}_2 = \begin{bmatrix} 0.057594337 \\ 0.605750797 \\ 0.068345246 \end{bmatrix} \text{ a.u.}$$

Use both Gauss's method and the new method to determine the velocity vector $\mathbf{v}_1$ at the position $\mathbf{r}_1$.

ANSWER:

$$\mathbf{v}_1 = \begin{bmatrix} -9.303603251 \\ 3.018641330 \\ 1.536362143 \end{bmatrix} \text{ a.u./year}$$

**Behaviour Near the Singularity**

Because the new method is singular, it is instructive to investigate its behaviour for values of $\lambda$ approaching $-1$. The last table shows the required number of iterations as $\lambda$ varies from $-0.90$ to $-0.99$. The uniformity of convergence persists except for a narrow region near $T = 5$. The increase in iteration steps is not the result of a poor initial guess for $x_0$; indeed, it turns out that the number of steps approaches a maximum when the solution of the problem approaches the minimum-energy orbit—that orbit for which the semimajor axis is

$$a_m = \tfrac{1}{2}s = \tfrac{1}{4}(r_1 + r_2 + c)$$

Number of Iterations for New Method near 360 degrees

| $\lambda \backslash T$ | 0.3 | 0.5 | 0.7 | 0.9 | 1.0 | 3.0 | 5.0 | 7.0 | 9.0 | 11.0 |
|---|---|---|---|---|---|---|---|---|---|---|
| $-0.99$ | 4 | 4 | 5 | 5 | 5 | 10 | 14 | 8 | 6 | 5 |
| $-0.98$ | 4 | 4 | 5 | 5 | 5 | 9 | 12 | 8 | 6 | 5 |
| $-0.97$ | 4 | 4 | 4 | 5 | 5 | 9 | 11 | 7 | 6 | 5 |
| $-0.96$ | 4 | 4 | 5 | 5 | 5 | 9 | 10 | 7 | 6 | 5 |
| $-0.95$ | 4 | 4 | 5 | 5 | 5 | 9 | 9 | 7 | 5 | 5 |
| $-0.94$ | 4 | 4 | 5 | 5 | 5 | 8 | 9 | 6 | 5 | 5 |
| $-0.92$ | 4 | 4 | 5 | 5 | 5 | 8 | 8 | 6 | 5 | 5 |
| $-0.90$ | 4 | 4 | 5 | 5 | 5 | 8 | 7 | 6 | 5 | 5 |

—when the transfer angle is close to 360 degrees. This same region also causes difficulties for Newton's method as noted at the end of Sect. 7.2. The transfer-time graph experiences a change in curvature, necessitating abandonment of the Newton iteration technique, when $\lambda$ is near $-1$. The new method takes a little longer than usual but it still converges without modification.

Appendix A

# Mathematical Progressions

In Gauss' algorithm for evaluating elliptic integrals, the arithmetic and geometric means play a fundamental role. Since these, as well as the harmonic mean, continually appear throughout our studies—sometimes in quite unexpected and surprising ways—we shall record here their precise definitions as well as some properties which the interested reader should be able to substantiate on his own.

## A.1  Arithmetic Progression

A series in which each term exceeds the preceding by a fixed quantity, called the *common difference*, is an *arithmetic series* or an *arithmetic progression*.

◊ **Problem A-1**
The sum of $n$ terms of an arithmetic progression is $n$ times the average of the first and last terms.

◊ **Problem A-2**
If $a$, $b$, $c$ are in arithmetic progression, then

$$b = \tfrac{1}{2}(a+c)$$

The quantity $b$ is called the *arithmetic mean* between $a$ and $c$.

## A.2  Geometric Progression

A series in which the ratio of each term to the preceding is a constant is called a *geometric series* or a *geometric progression* and the constant ratio is called the *common ratio*.

◊ **Problem A-3**
The sum of $n$ terms of a geometric progression is

$$\frac{a - rl}{1 - r}$$

where $a$ and $l$ are the first and last terms and $r$ is the common ratio.

## Problem A-4
If $a$, $b$, $c$ are positive quantities in geometric progression, then
$$b = \sqrt{ac}$$
Here, $b$ is referred to as the *geometric mean* between $a$ and $c$.

## Problem A-5
A point $P$ divides the diameter of a circle into two parts of lengths $a$ and $c$. The length of the line, drawn perpendicular to the diameter from $P$ to the point of intersection with the circle, is the geometric mean between $a$ and $c$.

## A.3 Harmonic Progression

A series of quantities, whose reciprocals form an arithmetic progression, is called a *harmonic series* or a *harmonic progression*.

## Problem A-6
If $a$, $b$, $c$ are three consecutive terms of a harmonic progression, then
$$\frac{1}{b} = \frac{1}{2}\left(\frac{1}{a} + \frac{1}{c}\right) \quad \text{or} \quad b = \frac{2ac}{a+c}$$
In this case, $b$ is called the *harmonic mean* between $a$ and $c$.

## Problem A-7
The geometric mean between two positive quantities $a$ and $c$ is the geometric mean between the arithmetic and harmonic means between $a$ and $c$.

## Problem A-8
The arithmetic, geometric, and harmonic means are in descending order of magnitude.

Appendix B

# Vector and Matrix Algebra

We assume that the reader is basically familiar with the techniques of vector and matrix algebra. Therefore, the following set of problems should be regarded as both a review and a ready reference of various identities and properties which will be used throughout this book.

## B.1 Vector Algebra

◇ **Problem B-1**
Derive the following identities

$$(\mathbf{a} \times \mathbf{b}) \times \mathbf{c} = (\mathbf{a} \cdot \mathbf{c})\mathbf{b} - (\mathbf{b} \cdot \mathbf{c})\mathbf{a}$$
$$\mathbf{a} \times (\mathbf{b} \times \mathbf{c}) = (\mathbf{a} \cdot \mathbf{c})\mathbf{b} - (\mathbf{a} \cdot \mathbf{b})\mathbf{c}$$
$$(\mathbf{a} \times \mathbf{b}) \cdot (\mathbf{c} \times \mathbf{d}) = (\mathbf{a} \cdot \mathbf{c})(\mathbf{b} \cdot \mathbf{d}) - (\mathbf{b} \cdot \mathbf{c})(\mathbf{a} \cdot \mathbf{d})$$
$$(\mathbf{a} \times \mathbf{b}) \times (\mathbf{c} \times \mathbf{d}) = (\mathbf{a} \cdot \mathbf{b} \times \mathbf{d})\mathbf{c} - (\mathbf{a} \cdot \mathbf{b} \times \mathbf{c})\mathbf{d}$$

◇ **Problem B-2**
If $\mathbf{a}$, $\mathbf{b}$, $\mathbf{p}$ are vectors from the origin to the points $A$, $B$, $P$ and

$$\mathbf{p} = \ell \mathbf{a} + m \mathbf{b}$$

where

$$1 = \ell + m$$

then the point $P$ lies on the line connecting points $A$ and $B$.

◇ **Problem B-3**
If $\mathbf{a}$, $\mathbf{b}$, $\mathbf{c}$, $\mathbf{p}$ are vectors from the origin to the points $A$, $B$, $C$, $P$ and

$$\mathbf{p} = \ell \mathbf{a} + m \mathbf{b} + n \mathbf{c}$$

where

$$1 = \ell + m + n$$

then the point $P$ lies in the plane determined by $A$, $B$, and $C$.

## Problem B-4

Consider four points in space $A$, $B$, $C$, $D$ so located that the line segments $AB$ and $CD$ are parallel. Let $\mathbf{a}$, $\mathbf{b}$, $\mathbf{c}$, $\mathbf{d}$ be vectors from the origin to the points $A$, $B$, $C$, $D$. If $m$ is the ratio of the lengths of $AB$ and $CD$, then

$$\mathbf{p} = \frac{\mathbf{b} - m\mathbf{d}}{1 - m} = \frac{\mathbf{a} - m\mathbf{c}}{1 - m}$$

is the vector from the origin to the intersection of $AC$ and $BD$.

## ◇ Problem B-5

If the vectors $\mathbf{a}$, $\mathbf{b}$, $\mathbf{c}$ are not parallel to a plane, then the solution of the equations

$$\mathbf{p} \cdot \mathbf{a} = \ell \qquad \mathbf{p} \cdot \mathbf{b} = m \qquad \mathbf{p} \cdot \mathbf{c} = n$$

can be written as

$$\mathbf{p} = \frac{\ell \, \mathbf{b} \times \mathbf{c} + m \, \mathbf{c} \times \mathbf{a} + n \, \mathbf{a} \times \mathbf{b}}{\mathbf{a} \cdot \mathbf{b} \times \mathbf{c}}$$

## ◇ Problem B-6

If the vectors $\mathbf{a}$ and $\mathbf{b}$ are perpendicular, then the solution of the equations

$$\mathbf{p} \times \mathbf{a} = \mathbf{b} \qquad \mathbf{p} \cdot \mathbf{a} = \ell$$

can be written as

$$\mathbf{p} = \frac{\mathbf{a} \times \mathbf{b} + \ell \mathbf{a}}{\mathbf{a} \cdot \mathbf{a}}$$

## Problem B-7

Consider four points in space $A$, $B$, $C$, $D$ so located that the line segments $AB$ and $CD$ are not parallel. Let $\mathbf{a}$, $\mathbf{b}$, $\mathbf{c}$, $\mathbf{d}$ be vectors from the origin to the points $A$, $B$, $C$, $D$. Then

$$\frac{|(\mathbf{c} - \mathbf{a}) \cdot (\mathbf{b} - \mathbf{a}) \times (\mathbf{d} - \mathbf{c})|}{|(\mathbf{b} - \mathbf{a}) \times (\mathbf{d} - \mathbf{c})|}$$

is the perpendicular distance between the lines $AB$ and $CD$.

## ◇ Problem B-8

By means of products express the condition that three vectors be parallel to a plane.

## ◇ Problem B-9

By means of products express the condition that the plane containing the vectors $\mathbf{a}$ and $\mathbf{b}$ be perpendicular to the plane containing $\mathbf{c}$ and $\mathbf{d}$.

## ◇ Problem B-10

By means of products find a vector which is in the plane of $\mathbf{b}$ and $\mathbf{c}$ and is perpendicular to $\mathbf{a}$.

Vector and Matrix Algebra       347

◇ **Problem B-11**
Consider a sphere of unit radius and three unit vectors $\mathbf{i}_A$, $\mathbf{i}_B$, $\mathbf{i}_C$, directed from its center to the three vertices of a spherical triangle on its surface. Let the vertex angles be $A$, $B$, $C$ and the opposite sides $a$, $b$, $c$, respectively. Then by properly interpreting the two sides of the vector identity

$$(\mathbf{i}_A \times \mathbf{i}_B) \cdot (\mathbf{i}_A \times \mathbf{i}_C) = (\mathbf{i}_B \cdot \mathbf{i}_C) - (\mathbf{i}_A \cdot \mathbf{i}_C)(\mathbf{i}_A \cdot \mathbf{i}_B)$$

deduce the *law of cosines* for spherical trigonometry

$$\cos a = \cos b \cos c + \sin b \sin c \cos A$$

## B.2 Matrix Algebra

◇ **Problem B-12**
For any square matrix $\mathbf{A}$, show that the products $\mathbf{A}\mathbf{A}^T$ and $\mathbf{A}^T\mathbf{A}$ are always symmetric.

◇ **Problem B-13**
If $\mathbf{A}$ is skew-symmetric, prove that $\mathbf{A}^2$ is symmetric.

◇ **Problem B-14**
If $\mathbf{A}$ and $\mathbf{B}$ are symmetric, then the product $\mathbf{AB}$ is symmetric if and only if $\mathbf{AB} = \mathbf{BA}$.

◇ **Problem B-15**
Determine the symmetry or skew-symmetry of $\mathbf{AB} - \mathbf{BA}$ in the cases for which
(a) both $\mathbf{A}$ and $\mathbf{B}$ are symmetric,
(b) both are skew-symmetric, and
(c) one is symmetric and the other is skew-symmetric.

◇ **Problem B-16**
Any square matrix $\mathbf{M}$ can be represented as the sum of a symmetric matrix and a skew-symmetric matrix. That is,

$$\mathbf{M} = \tfrac{1}{2}(\mathbf{M} + \mathbf{M}^T) + \tfrac{1}{2}(\mathbf{M} - \mathbf{M}^T)$$

◇ **Problem B-17**
If the matrix $\mathbf{B}$ is skew-symmetric and $\mathbf{I}$ is the identity matrix, show that the matrix $\mathbf{I} + \mathbf{B}$ is nonsingular. Then, demonstrate that

$$\mathbf{A} = (\mathbf{I} - \mathbf{B})(\mathbf{I} + \mathbf{B})^{-1}$$

is an orthogonal matrix.

*Georg Ferdinand Frobenius* 1894

◊ **Problem B-18**

Let the vectors **a**, **b**, **c** be the column vectors of a three-dimensional matrix **M**, i.e.,

$$\mathbf{M} = [\mathbf{a} \quad \mathbf{b} \quad \mathbf{c}]$$

Then,

$$\mathbf{M}^{-1} = \frac{1}{\mathbf{a} \cdot \mathbf{b} \times \mathbf{c}} [\mathbf{b} \times \mathbf{c} \quad \mathbf{c} \times \mathbf{a} \quad \mathbf{a} \times \mathbf{b}]^T$$

◊ **Problem B-19**

Verify the following identities for block partitioned matrices where the subscripts on the submatrices indicate the number of rows and the number of columns, respectively.

(1) $\begin{bmatrix} \mathbf{A}_{nn} & \mathbf{B}_{nm} \\ \mathbf{C}_{mn} & \mathbf{D}_{mm} \end{bmatrix} = \begin{bmatrix} \mathbf{A}_{nn} & \mathbf{O}_{nm} \\ \mathbf{O}_{mn} & \mathbf{I}_{mm} \end{bmatrix} \begin{bmatrix} \mathbf{I}_{nn} & \mathbf{A}_{nn}^{-1}\mathbf{B}_{nm} \\ \mathbf{C}_{mn} & \mathbf{D}_{mm} \end{bmatrix}$

(2) $\begin{bmatrix} \mathbf{A}_{nn} & \mathbf{B}_{nm} \\ \mathbf{C}_{mn} & \mathbf{D}_{mm} \end{bmatrix} = \begin{bmatrix} \mathbf{A}_{nn} & \mathbf{B}_{nm}\mathbf{D}_{mm}^{-1} \\ \mathbf{C}_{mn} & \mathbf{I}_{mm} \end{bmatrix} \begin{bmatrix} \mathbf{I}_{nn} & \mathbf{O}_{nm} \\ \mathbf{O}_{mn} & \mathbf{D}_{mm} \end{bmatrix}$

(3) $\begin{bmatrix} \mathbf{I}_{nn} & \mathbf{A}_{nm} \\ \mathbf{B}_{mn} & \mathbf{C}_{mm} \end{bmatrix} = \begin{bmatrix} \mathbf{I}_{nn} & \mathbf{O}_{nm} \\ \mathbf{B}_{mn} & \mathbf{I}_{mm} \end{bmatrix} \begin{bmatrix} \mathbf{I}_{nn} & \mathbf{A}_{nm} \\ \mathbf{O}_{mn} & \mathbf{C}_{mm} - \mathbf{B}_{mn}\mathbf{A}_{nm} \end{bmatrix}$

(4) $\begin{bmatrix} \mathbf{A}_{nn} & \mathbf{O}_{nm} \\ \mathbf{O}_{mn} & \mathbf{B}_{mm} \end{bmatrix} = \begin{bmatrix} \mathbf{A}_{nn} & \mathbf{O}_{nm} \\ \mathbf{O}_{mn} & \mathbf{I}_{mm} \end{bmatrix} \begin{bmatrix} \mathbf{I}_{nn} & \mathbf{O}_{nm} \\ \mathbf{O}_{mn} & \mathbf{B}_{mm} \end{bmatrix}$

(5) $\begin{bmatrix} \mathbf{I}_{nn} & \mathbf{O}_{nm} \\ \mathbf{A}_{mn} & \mathbf{B}_{mm} \end{bmatrix} = \begin{bmatrix} \mathbf{I}_{nn} & \mathbf{O}_{nm} \\ \mathbf{A}_{mn} & \mathbf{I}_{mm} \end{bmatrix} \begin{bmatrix} \mathbf{I}_{nn} & \mathbf{O}_{nm} \\ \mathbf{O}_{mn} & \mathbf{B}_{mm} \end{bmatrix}$

(6) $\begin{bmatrix} \mathbf{I}_{nn} & \mathbf{A}_{nm} \\ \mathbf{O}_{mn} & \mathbf{B}_{mm} \end{bmatrix} = \begin{bmatrix} \mathbf{I}_{nn} & \mathbf{O}_{nm} \\ \mathbf{O}_{mn} & \mathbf{B}_{mm} \end{bmatrix} \begin{bmatrix} \mathbf{I}_{nn} & \mathbf{A}_{nm} \\ \mathbf{O}_{mn} & \mathbf{I}_{mm} \end{bmatrix}$

(7) $\begin{bmatrix} \mathbf{I}_{nn} & \mathbf{O}_{nm} \\ \mathbf{O}_{mn} & \mathbf{A}_{mm} \end{bmatrix}^{-1} = \begin{bmatrix} \mathbf{I}_{nn} & \mathbf{O}_{nm} \\ \mathbf{O}_{mn} & \mathbf{A}_{mm}^{-1} \end{bmatrix}$

(8) $\begin{bmatrix} \mathbf{I}_{nn} & \mathbf{O}_{nm} \\ \mathbf{A}_{mn} & \mathbf{I}_{mm} \end{bmatrix}^{-1} = \begin{bmatrix} \mathbf{I}_{nn} & \mathbf{O}_{nm} \\ -\mathbf{A}_{mn} & \mathbf{I}_{mm} \end{bmatrix}$

(9) $\begin{bmatrix} \mathbf{I}_{nn} & \mathbf{A}_{nm} \\ \mathbf{O}_{mn} & \mathbf{I}_{mm} \end{bmatrix}^{-1} = \begin{bmatrix} \mathbf{I}_{nn} & -\mathbf{A}_{nm} \\ \mathbf{O}_{mn} & \mathbf{I}_{mm} \end{bmatrix}$

(10) $\begin{vmatrix} \mathbf{I}_{nn} & \mathbf{O}_{nm} \\ \mathbf{A}_{mn} & \mathbf{I}_{mm} \end{vmatrix} = 1$ $\quad \begin{vmatrix} \mathbf{I}_{nn} & \mathbf{A}_{nm} \\ \mathbf{O}_{mn} & \mathbf{I}_{mm} \end{vmatrix} = 1$

(11) $\begin{vmatrix} \mathbf{I}_{nn} & \mathbf{O}_{nm} \\ \mathbf{O}_{mn} & \mathbf{A}_{mm} \end{vmatrix} = |\mathbf{A}_{mm}|$ $\quad \begin{vmatrix} \mathbf{A}_{nn} & \mathbf{O}_{nm} \\ \mathbf{O}_{mn} & \mathbf{I}_{mm} \end{vmatrix} = |\mathbf{A}_{nn}|$

(12) $\begin{vmatrix} \mathbf{I}_{nn} & \mathbf{A}_{nm} \\ \mathbf{B}_{mn} & \mathbf{C}_{mm} \end{vmatrix} = |\mathbf{C}_{mm} - \mathbf{B}_{mn}\mathbf{A}_{nm}|$

## Problem B-20

Let $\mathbf{M}$ be a square matrix of dimension $n + m$ and partitioned as

$$\mathbf{M} = \begin{bmatrix} \mathbf{A}_{nn} & \mathbf{B}_{nm} \\ \mathbf{C}_{mn} & \mathbf{D}_{mm} \end{bmatrix}$$

where $\mathbf{A}_{nn}$ is an $n$-dimensional square matrix, $\mathbf{B}_{nm}$ is a rectangular matrix having $n$ rows and $m$ columns, etc. Utilize the identities established in the preceding problem to derive the following relationships.

(a) The matrix $\mathbf{M}$ can be expressed as

$$\mathbf{M} = \begin{bmatrix} \mathbf{A}_{nn} & \mathbf{O}_{nm} \\ \mathbf{O}_{mn} & \mathbf{I}_{mm} \end{bmatrix} \begin{bmatrix} \mathbf{I}_{nn} & \mathbf{O}_{nm} \\ \mathbf{C}_{mn} & \mathbf{I}_{mm} \end{bmatrix} \begin{bmatrix} \mathbf{I}_{nn} & \mathbf{A}_{nn}^{-1}\mathbf{B}_{nm} \\ \mathbf{O}_{mn} & \mathbf{E}_{mm} \end{bmatrix}$$

where

$$\mathbf{E}_{mm} = \mathbf{D}_{mm} - \mathbf{C}_{mn}\mathbf{A}_{nn}^{-1}\mathbf{B}_{nm}$$

with $\mathbf{I}$ and $\mathbf{O}$ the identity and zero matrices, respectively.

(b) The determinant of $\mathbf{M}$ is given by

$$|\mathbf{M}| = |\mathbf{A}_{nn}|\,|\mathbf{E}_{mm}|$$

(c) The inverse of $\mathbf{M}$ can be obtained from

$$\mathbf{M}^{-1} = \begin{bmatrix} \mathbf{I}_{nn} & -\mathbf{A}_{nn}^{-1}\mathbf{B}_{nm} \\ \mathbf{O}_{mn} & \mathbf{I}_{mm} \end{bmatrix} \begin{bmatrix} \mathbf{I}_{nn} & \mathbf{O}_{nm} \\ \mathbf{O}_{mn} & \mathbf{E}_{mm}^{-1} \end{bmatrix}$$

$$\begin{bmatrix} \mathbf{I}_{nn} & \mathbf{O}_{nm} \\ -\mathbf{C}_{mn} & \mathbf{I}_{mm} \end{bmatrix} \begin{bmatrix} \mathbf{A}_{nn}^{-1} & \mathbf{O}_{nm} \\ \mathbf{O}_{mn} & \mathbf{I}_{mm} \end{bmatrix}$$

provided that $\mathbf{E}_{mm}$ is nonsingular.

(d) An alternate factorization of $\mathbf{M}$ is

$$\mathbf{M} = \begin{bmatrix} \mathbf{F}_{nn} & \mathbf{B}_{nm}\mathbf{D}_{mm}^{-1} \\ \mathbf{O}_{mn} & \mathbf{I}_{mm} \end{bmatrix} \begin{bmatrix} \mathbf{I}_{nn} & \mathbf{O}_{nm} \\ \mathbf{C}_{mn} & \mathbf{I}_{mm} \end{bmatrix} \begin{bmatrix} \mathbf{I}_{nn} & \mathbf{O}_{nm} \\ \mathbf{O}_{mn} & \mathbf{D}_{mm} \end{bmatrix}$$

where

$$\mathbf{F}_{nn} = \mathbf{A}_{nn} - \mathbf{B}_{nm}\mathbf{D}_{mm}^{-1}\mathbf{C}_{mn}$$

(e) The determinant of $\mathbf{M}$ is then

$$|\mathbf{M}| = |\mathbf{D}_{mm}|\,|\mathbf{F}_{nn}|$$

(f) The inverse of $\mathbf{M}$ is then

$$\mathbf{M}^{-1} = \begin{bmatrix} \mathbf{I}_{nn} & \mathbf{O}_{nm} \\ \mathbf{O}_{mn} & \mathbf{D}_{mm}^{-1} \end{bmatrix} \begin{bmatrix} \mathbf{I}_{nn} & \mathbf{O}_{nm} \\ -\mathbf{C}_{mn} & \mathbf{I}_{mm} \end{bmatrix}$$

$$\begin{bmatrix} \mathbf{F}_{nn}^{-1} & \mathbf{O}_{nm} \\ \mathbf{O}_{mn} & \mathbf{I}_{mm} \end{bmatrix} \begin{bmatrix} \mathbf{I}_{nn} & -\mathbf{B}_{nm}\mathbf{D}_{mm}^{-1} \\ \mathbf{O}_{mn} & \mathbf{I}_{mm} \end{bmatrix}$$

provided that $\mathbf{F}_{nn}$ is nonsingular.

## Appendix B

◊ **Problem B-21**
Use the results of the previous problem to evaluate the determinant and to calculate the inverse of the matrix

$$\mathbf{M} = \begin{bmatrix} 2 & 1 & -1 & 4 \\ -2 & 3 & 2 & -5 \\ 1 & -2 & -3 & 2 \\ -4 & -3 & 2 & -2 \end{bmatrix}$$

◊ **Problem B-22**
As an exercise in quadratic forms, characteristic values, and characteristic vectors, consider the quadratic form

$$Q = 5x_1^2 - 2x_2^2 - 3x_3^2 + 12x_1x_2 - 8x_1x_3 + 20x_2x_3$$

(a) Find the matrix $\mathbf{A}$ such that $Q = \mathbf{x}^T \mathbf{A} \mathbf{x}$.
(b) Find the characteristic values and characteristic vectors of $\mathbf{A}$.
(c) Find the modal matrix $\mathbf{B}$ for which $\mathbf{B}^{-1} = \mathbf{B}^T$ and having the property that the transformation

$$\mathbf{x} = \mathbf{B}\mathbf{y}$$

will reduce the quadratic form $Q$ to canonical form, i.e., a sum of squares with no cross product terms.
(d) Write out the new quadratic form.

ANSWER:

(a) $$\mathbf{A} = \begin{bmatrix} 5 & 6 & -4 \\ 6 & -2 & 10 \\ -4 & 10 & -3 \end{bmatrix}$$

(b) $6, 9, -15$ and $\frac{1}{3}\begin{bmatrix} -2 \\ 1 \\ 2 \end{bmatrix}, \frac{1}{3}\begin{bmatrix} 2 \\ 2 \\ 1 \end{bmatrix}, \frac{1}{3}\begin{bmatrix} 1 \\ -2 \\ 2 \end{bmatrix}$

(c) $$\mathbf{B} = \begin{bmatrix} -\frac{2}{3} & \frac{2}{3} & \frac{1}{3} \\ \frac{1}{3} & \frac{2}{3} & -\frac{2}{3} \\ \frac{2}{3} & \frac{1}{3} & \frac{2}{3} \end{bmatrix}$$

(d) $$Q = 6y_1^2 + 9y_2^2 - 15y_3^2$$

Appendix C

# Power Series Manipulations

The formal manipulation of power series was a favorite tool of Newton and many of his followers. We summarize here some of the convenient algorithms for multiplying, dividing, extracting roots, and reversing power series. The reader should have little difficulty in verifying these for himself.

Consider the power series

$$A = \sum_{n=0}^{\infty} a_n x^n \qquad B = \sum_{n=0}^{\infty} b_n x^n \qquad C = \sum_{n=0}^{\infty} c_n x^n$$

with $a_0 = b_0 = c_0 = 1$.

◊ **Problem C–1**
If $C = AB$, then

$$c_n = \sum_{k=0}^{n} a_k b_{n-k}$$

◊ **Problem C–2**
If $C = A/B$, then

$$c_n = a_n - \sum_{k=1}^{n} b_k c_{n-k}$$

◊ **Problem C–3**
If $C = A^{\frac{1}{2}}$, then

$$2c_1 = a_1$$
$$2c_2 = a_2 - c_1^2$$
$$2c_3 = a_3 - 2c_1 c_2$$
$$2c_4 = a_4 - 2c_1 c_3 - c_2^2 \qquad \text{etc.}$$

## ◇ Problem C–4
If $C = A^{\frac{1}{3}}$, then

$$3c_1 = a_1$$
$$3c_2 = a_2 - 3c_1^2$$
$$3c_3 = a_3 - 6c_1 c_2 - c_1^3$$
$$3c_4 = a_4 - 3c_1^2 c_2 - 6c_1 c_3 - 3c_2^2 \qquad \text{etc.}$$

## ◇ Problem C–5
If
$$y = x + bx^2 + cx^3 + dx^4 + ex^5 + fx^6 + \cdots$$
then the formal *reversion of the series*
$$x = y + By^2 + Cy^3 + Dy^4 + Ey^5 + Fy^6 + \cdots$$
obtains if $B, C, \ldots$ are determined from

$$B = -b$$
$$C = -c + 2b^2$$
$$D = -d + 5bc - 5b^3$$
$$E = -e + 6bd - 21b^2 c + 3c^2 + 14b^4$$
$$F = -f + 7be - 28b^2 d - 28bc^2 + 84b^3 c - 42b^5 + 7cd \qquad \text{etc.}$$

Appendix D

# Linear Algebraic Systems

In the section on orbit determination—Sect. 3.7—we had occasion to use a fundamental property of linear algebraic equations. There are others, in addition, which we shall need from time to time. They are all stated here in the form of problems for the student.

◊ **Problem D-1**
Consider the system

$$a_{11}x_1 + a_{12}x_2 + \cdots + a_{1n}x_n + c_1 = 0$$
$$a_{21}x_1 + a_{22}x_2 + \cdots + a_{2n}x_n + c_2 = 0$$
$$\vdots$$
$$a_{n1}x_1 + a_{n2}x_2 + \cdots + a_{nn}x_n + c_n = 0$$

consisting of $n$ equations to be used to determine the $n$ variables $x_1$, $x_2$, ..., $x_n$. If, in addition to these equations, another, namely,

$$a_{n+1,1}x_1 + a_{n+1,2}x_2 + \cdots + a_{n+1,n}x_n + c_{n+1} = 0$$

is also specified, the system of $n+1$ equations thus obtained will, in general, be inconsistent. A necessary and sufficient condition for consistency is that

$$\begin{vmatrix} a_{11} & a_{12} & \cdots & a_{1n} & c_1 \\ a_{21} & a_{22} & \cdots & a_{2n} & c_2 \\ \vdots & \vdots & \ddots & \vdots & \vdots \\ a_{n1} & a_{n2} & \cdots & a_{nn} & c_n \\ a_{n+1,1} & a_{n+1,2} & \cdots & a_{n+1,n} & c_{n+1} \end{vmatrix} = 0$$

*Étienne Bezout*[†] 1764

---

[†] Étienne Bezout (1739-1783) published this result at the beginning of his paper *Sur le degré des équations résultantes de l'évanouissement des inconnues* in 1764. He also systematized the process of determining the signs of the terms of a determinant and made other significant contributions to the theory of equations.

## Problem D-2

Consider the $n^{\text{th}}$ order system of the previous problem with all of the $c$'s equal to zero. Then, we have $n$ *homogeneous* equations of the first degree in $n$ variables. The system can have a non-trivial solution, i.e., not all of the variables equal to zero, only if the determinant of the coefficients vanishes. Furthermore, in this case, we have

$$x_1 : x_2 : \cdots : x_n = A_{r1} : A_{r2} : \cdots : A_{rn}$$

where $r = 1, 2, \ldots, n$ and $A_{ij}$ is the cofactor of the element $a_{ij}$.

Thus, the ratios of the variables can be determined but their actual values are indeterminate.

## Problem D-3

The equation of a straight line passing through the points $x_1$, $y_1$ and $x_2$, $y_2$ may be written as a determinantal equation

$$\begin{vmatrix} x & y & 1 \\ x_1 & y_1 & 1 \\ x_2 & y_2 & 1 \end{vmatrix} = 0$$

## Problem D-4

As discussed in Appendix E, the general second-degree equation in rectangular coordinates

$$Ax^2 + Bxy + Cy^2 + Dx + Ey + F = 0$$

represents an ellipse, parabola, or hyperbola. The equation of such a curve passing through the five points $x_1, y_1, \ldots, x_5, y_5$ may be written in determinant form as

$$\begin{vmatrix} x^2 & xy & y^2 & x & y & 1 \\ x_1^2 & x_1 y_1 & y_1^2 & x_1 & y_1 & 1 \\ x_2^2 & x_2 y_2 & y_2^2 & x_2 & y_2 & 1 \\ x_3^2 & x_3 y_3 & y_3^2 & x_3 & y_3 & 1 \\ x_4^2 & x_4 y_4 & y_4^2 & x_4 & y_4 & 1 \\ x_5^2 & x_5 y_5 & y_5^2 & x_5 & y_5 & 1 \end{vmatrix} = 0$$

Appendix E

# Conic Sections

An equation of the second degree in rectangular coordinates has the form
$$Ax^2 + 2Bxy + Cy^2 + 2Dx + 2Ey + F = 0$$
By a rotation and translation of the coordinate axes, this equation can be reduced to one of the following forms:
$$A_1 x^2 + B_1 y^2 + C_1 = 0$$
$$y^2 + 2D_1 x = 0$$
If $C_1$ is not zero and if $A_1$, $B_1$, $C_1$ do not have the same algebraic sign, then the locus defined by the first form is a circle, ellipse, or hyperbola. The origin of coordinates is called the *center* of the conic, and the coordinate axes are called the *axes of symmetry*. On the other hand, if $D_1$ is not zero, the locus defined by the second form is a parabola. The origin of coordinates is called the *vertex*, and the $x$ axis is the *axis* of the parabola.

The set of problems, which constitute this Appendix, develop some of the many properties of the conic sections† and are entertaining exercises for the serious reader.

**Problem E-1**
The three quantities

$$A + C \qquad B^2 - AC \qquad \Delta \equiv \begin{vmatrix} A & B & D \\ B & C & E \\ D & E & F \end{vmatrix}$$

are invariant under a translation or rotation of coordinates.

For the parabola, $y^2 + 2D_1 x = 0$, we have
$$D_1 = -\sqrt{\frac{-\Delta}{(A+C)^3}}$$

---

† Menaechmus, who lived about 350 B.C., is the reputed discoverer of conic sections and the best Greek geometers, until the time of Pappus of Alexandria, devoted much attention to them. There followed a period of over a thousand years when they were almost completely forgotten. Then, in the seventeenth century, Kepler's work motivated a new interest in these important curves.

For the ellipse or hyperbola, $A_1 x^2 + B_1 y^2 + C_1 = 0$, the quantities
$$-\frac{A_1}{C_1} \quad \text{and} \quad -\frac{B_1}{C_1}$$
are obtainable as the roots of the quadratic equation
$$\lambda^2 - \frac{(B^2 - AC)(A + C)}{\Delta} \lambda + \frac{(B^2 - AC)^3}{\Delta^2} = 0$$

### Problem E-2
The sides of a triangle $\triangle A_1 A_2 A_3$ are represented by the equations
$$L_1 = a_1 x + b_1 y + c_1 = 0$$
$$L_2 = a_2 x + b_2 y + c_2 = 0$$
$$L_3 = a_3 x + b_3 y + c_3 = 0$$
If $k_1$, $k_2$, $k_3$ are constants, then
$$k_1 L_2 L_3 + k_2 L_3 L_1 + k_3 L_1 L_2 = 0$$
represents a conic section circumscribing the triangle.

### ◊ Problem E-3
When expanded, the equation of the preceding problem takes the form
$$Ax^2 + Bxy + Cy^2 + Dx + Ey + F = 0$$
If $k_1$, $k_2$, $k_3$ are so chosen that $B = 0$ and $A = C$, then the curve is the circle circumscribing the triangle $A_1 A_2 A_3$. In this way, determine the circle circumscribed about the triangle with sides
$$y - x = 0$$
$$x + y = 2$$
$$x + 3y + 4 = 0$$

### ◊ Problem E-4
The parametric equations of the ellipse and hyperbola can be written in algebraic form as
$$x = \frac{a(1 - w^2)}{1 + w^2} \quad y = \frac{2bw}{1 + w^2} \quad \text{ellipse}$$
$$x = -\frac{a(1 + w^2)}{1 - w^2} \quad y = \frac{2bw}{1 - w^2} \quad \text{hyperbola}$$
where
$$w = \begin{cases} \tan \tfrac{1}{2} E & \text{ellipse} \\ \tanh \tfrac{1}{2} H & \text{hyperbola} \end{cases}$$

## Problem E-5

An interesting construction of an ellipse is possible. Consider a rectangle of sides $2a$ and $2b$. Three sides of the rectangle are divided into an equal number of parts and the points of division connected to the opposite corners by straight lines as illustrated in the accompanying figure. The intersections of lines through like numbered points determine points on the ellipse whose semiaxes are $a$ and $b$.

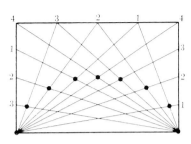

## Problem E-6

Two circles are represented by the equations

$$x^2 + y^2 + A_1 x + B_1 y + C_1 = 0$$
$$x^2 + y^2 + A_2 x + B_2 y + C_2 = 0$$

By subtracting these, the equation

$$(A_1 - A_2)x + (B_1 - B_2)y + C_1 - C_2 = 0$$

of a line is obtained. How is this line related to the two circles?

## ◊ Problem E-7

A circle is the locus of points the ratio of whose distances from two fixed points is a constant different from unity.

## ◊ Problem E-8

If $P(x_1, y_1)$ is a point outside the circle

$$x^2 + y^2 + Ax + By + C = 0$$

then

$$x_1^2 + y_1^2 + A x_1 + B y_1 + C$$

is the square of the length of the tangent from $P_1$ to the circle.

## ◊ Problem E-9

If a point moves so that the ratio of the lengths of the tangents from it to two fixed circles is constant, its locus is a circle or a straight line.

## ◊ Problem E-10

If the length of the tangent from a point $P$ to a fixed circle is equal to the distance from $P$ to a fixed straight line, the locus of $P$ is a parabola.

## Problem E-11

The locus of centers of all parallel chords of a parabola is a straight line.

## Problem E-12
If the normal to a parabola at point $P$ intersects the axis of the parabola at $N$, the projection of $PN$ on the axis is constant.

## Problem E-13
If the ends of a bar move on perpendicular lines, then a point $P$ on the bar at distances $a$ and $b$ from its ends describes an ellipse with semiaxes $a$ and $b$.

## Problem E-14
If a circle is deformed in such a way that the distances of its points from a fixed diameter are all changed in the same ratio, the resulting curve is an ellipse.

## Problem E-15
If the sum of the lengths of the tangents from a point $P$ to two fixed circles is constant, the locus of $P$ is an ellipse.

## Problem E-16
If a circle rolls without slipping so that it is always tangent to a fixed circle of twice its radius, then the locus of a point $P$ rigidly attached to the interior of the rolling circle is an ellipse. Further, the eccentricity of the ellipse is

$$e = \frac{2\sqrt{\lambda}}{1+\lambda}$$

where $\lambda$ is the ratio of the distance of $P$ from the center of the rolling circle to its radius.

*Philippe de La Hire* 1707

## Problem E-17
On a level plane find the locus of points where the crack of a rifle and the thud of the bullet against the target are heard simultaneously.

## Problem E-18
The two vertices of a triangle $A$ and $B$ are fixed while the vertex $C$ is allowed to vary. Find the locus of $C$ if $\angle ABC = 2\angle BAC$.

## Problem E-19
The ellipses and hyperbolas obtained by assigning values to the constant $\lambda$ in the equation

$$\frac{x^2}{a^2-\lambda} + \frac{y^2}{b^2-\lambda} = 1$$

are confocal, that is, all have the same foci. Two of these curves (an ellipse and a hyperbola) pass through each point $P(x,y)$, not on either axis, and their slopes $m$ at $P$ satisfy the equation

$$m^2 xy + m(x^2 - y^2 - a^2 + b^2) - xy = 0$$

Hence, they intersect at right angles so that the system of confocal curves is called *self-orthogonal*.

Appendix F

# Tschebycheff Approximations

One of the many uses to which Tschebycheff polynomials can be put is for the economization of power series. In particular, they can be employed to produce an efficient method for evaluating the function $C(A)$ needed in the extension of Gauss' method for solving Kepler's equation in the last section of Chapter 5.†

## F.1 Tschebycheff Polynomials

The geometric series
$$\sum_{k=0}^{n} z^k = \frac{1 - z^{n+1}}{1 - z}$$
with
$$z = y e^{i\phi} = y(\cos\phi + i \sin\phi)$$
can be used to develop the expansion
$$\sum_{k=0}^{n} y^k \cos k\phi = \frac{1 - y\cos\phi - y^{n+1}\cos(n+1)\phi + y^{n+2}\cos n\phi}{1 - 2y\cos\phi + y^2}$$
as Leonhard Euler found in 1760.‡ If $y < 1$, then we obtain the infinite expansion
$$\frac{1 - y\cos\phi}{1 - 2y\cos\phi + y^2} = \sum_{k=0}^{\infty} y^k \cos k\phi \qquad (\text{F.1})$$

The functions $\cos k\phi$ can be expanded by recursively applying the standard trigonometric identity
$$\cos(k+1)\phi = \cos k\phi \cos\phi - \sin k\phi \sin\phi$$
to obtain $\cos k\phi$ as a $k^{\text{th}}$ order polynomial in $\cos\phi$. These polynomials
$$T_k(\cos\phi) = \cos k\phi \qquad (\text{F.2})$$

---

† An example of such an economized series for a function related to $C(A)$ is given in the author's paper "Extension of Gauss' Method for the Solution of Kepler's Equation" coauthored with Tom Fill and referenced in Sect. 5.6.

‡ We have already encountered some of Euler's trigonometric series in Sect. 5.2 which were used in Lagrange's solution of Kepler's equation.

are called *Tschebycheff polynomials* after the man† who first discovered the various laws of approximations of functions by polynomials. Indeed, Tschebycheff discovered that these polynomials are the *best possible* choice in approximating a function over the interval $(-1, 1)$ to minimize the maximum error.

Now, if we define

$$x = \cos\phi \quad \text{so that} \quad -1 \leq x \leq 1$$

then the function $T(x, y)$, where

$$T(x, y) \equiv \frac{1 - xy}{1 - 2xy + y^2} = \sum_{n=0}^{\infty} T_n(x) y^n \tag{F.3}$$

is called the *generating function* for the Tschebycheff polynomials. Clearly, $|T_n(x)| \leq 1$ for $-1 \leq x \leq 1$ since $|\cos n\phi| \leq 1$ for $-\pi \leq \phi \leq \pi$.

We can derive the following recursion formula for the Tschebycheff polynomials:

$$T_{n+1}(x) - 2xT_n(x) + T_{n-1}(x) = 0 \tag{F.4}$$

with

$$T_0(x) = 1 \quad \text{and} \quad T_1(x) = x$$

from the trigonometric identity used earlier.

It is also easy to verify that $T_n(x)$ satisfies the differential equation

$$(1 - x^2)\frac{d^2 T_n}{dx^2} - x\frac{dT_n}{dx} + n^2 T_n = 0 \tag{F.5}$$

known as *Tschebycheff's equation*.

◊ **Problem F-1**

Tschebycheff polynomials can be expressed as hypergeometric functions. In particular, we have

$$T_n(x) = F[n, -n; \tfrac{1}{2}; \tfrac{1}{2}(1 - x)]$$

HINT: Use Eq. (1.14) of Sect. 1.1.

Of great importance is the fact that a function $f(x)$ can be expanded in the series

$$f(x) = \sum_{n=0}^{\infty} a_n T_n(x)$$

---

† Pafnuti L. Tschebycheff (1821–1894), a professor at the University of Petrograd, was one of the great Russian mathematicians and a leading engineer of the last century. He was led to the idea of the best uniform approximation from a purely practical problem—that of the construction of mechanisms producing a given trajectory of motion. Details of his calculations for mechanisms of this sort may be found in the publication "The Scientific Heritage of P. L. Tschebycheff," Volume II, Academy of Sciences of the USSR, 1945.

## Tschebycheff Approximations

over the range $-1 \leq x \leq 1$ where

$$a_0 = \frac{1}{\pi} \int_{-1}^{1} \frac{T_0(x)}{\sqrt{1-x^2}} f(x)\, dx \quad \text{and} \quad a_n = \frac{2}{\pi} \int_{-1}^{1} \frac{T_n(x)}{\sqrt{1-x^2}} f(x)\, dx$$

This property follows from the fact that the Tschebycheff polynomials are *orthogonal* over the interval $-1 \leq x \leq 1$ relative to the *weighting function* $(1-x^2)^{\frac{1}{2}}$; that is

$$\int_{-1}^{1} \frac{T_n(x) T_m(x)}{\sqrt{1-x^2}}\, dx = 0 \quad n \neq m$$

$$\int_{-1}^{1} \frac{T_n(x)^2}{\sqrt{1-x^2}}\, dx = \begin{cases} \frac{1}{2}\pi & n \neq 0 \\ \pi & n = 0 \end{cases}$$

Using the recursion formula (F.4), we list the first ten Tschebycheff polynomials:

$T_0(x) = 1$      $T_5(x) = 16x^5 - 20x^3 + 5x$
$T_1(x) = x$      $T_6(x) = 32x^6 - 48x^4 + 18x^2 - 1$
$T_2(x) = 2x^2 - 1$      $T_7(x) = 64x^7 - 112x^5 + 56x^3 - 7x$
$T_3(x) = 4x^3 - 3x$      $T_8(x) = 128x^8 - 256x^6 + 160x^4 - 32x^2 + 1$
$T_4(x) = 8x^4 - 8x^2 + 1$      $T_9(x) = 256x^9 - 576x^7 + 432x^5 - 120x^3 + 9x$

and, by reversing these series, obtain the Tschebycheff expansion of the first ten powers of $x$ in the form:

$1 = T_0$      $x^5 = \frac{1}{16}(10T_1 + 5T_3 + T_5)$
$x = T_1$      $x^6 = \frac{1}{32}(10T_0 + 15T_2 + 6T_4 + T_6)$
$x^2 = \frac{1}{2}(T_0 + T_2)$      $x^7 = \frac{1}{64}(35T_1 + 21T_3 + 7T_5 + T_7)$
$x^3 = \frac{1}{4}(3T_1 + T_3)$      $x^8 = \frac{1}{128}(35T_0 + 56T_2 + 28T_4 + 8T_6 + T_8)$
$x^4 = \frac{1}{8}(3T_0 + 4T_2 + T_4)$      $x^9 = \frac{1}{256}(126T_1 + 84T_3 + 36T_5 + 9T_7 + T_9)$

which we shall use for economizing power series.

### Problem F-2

Euler derived other expansions similar to Eq. (F.1). The following are of some interest:

$$\frac{y \sin \phi}{1 - 2y \cos \phi + y^2} = \sum_{k=1}^{\infty} y^k \sin k\phi$$

$$\frac{\cos \alpha - y \cos(\alpha - \phi)}{1 - 2y \cos \phi + y^2} = \sum_{k=0}^{\infty} y^k \cos(\alpha + k\phi)$$

$$\frac{\sin \alpha - y \sin(\alpha - \phi)}{1 - 2y \cos \phi + y^2} = \sum_{k=0}^{\infty} y^k \sin(\alpha + k\phi)$$

## F.2 Economization of Power Series

We conclude this appendix with an application to the economization of power series. If minimizing the maximum error is the governing criterion, then a satisfactory approximation may be afforded by fewer terms of the Tschebycheff series than by an ordinary power series in $x$.

As an example, suppose that $\sin x$ is represented by its series expansion

$$\sin x = x - \frac{x^3}{3!} + \frac{x^5}{5!} - \frac{x^7}{7!} + \cdots$$

and it is decided to use the truncated series

$$f(x) = x - \frac{x^3}{3!} + \frac{x^5}{5!}$$

to represent $\sin x$ over the interval $-1 \leq x \leq 1$.

The function $f(x)$ can be expressed in terms of Tschebycheff polynomials as

$$f(x) = T_1 - \tfrac{1}{24}(3T_1 + T_3) + \tfrac{1}{1920}(10T_1 + 5T_3 + T_5)$$

$$= \tfrac{169}{192}T_1 - \tfrac{5}{128}T_3 + \tfrac{1}{1920}T_5$$

Then, since $|T_n(x)| \leq 1$ in the interval $(-1, 1)$, the function

$$g(x) = \tfrac{169}{192}T_1 - \tfrac{5}{128}T_3$$

$$= \tfrac{383}{384}x - \tfrac{5}{32}x^3$$

will represent the function $\sin x$ with an error of less than $\tfrac{1}{1920}$ in the interval $-1 \leq x \leq 1$.

However, if we had simply dropped the $x^5$ term in $f(x)$, the error could have been as large as $\tfrac{1}{120}$.

◊ **Problem F-3**

Economize the power series

$$f(x) = 1 + \frac{x}{2} + \frac{x^2}{3} + \frac{x^3}{4} + \frac{x^4}{5} + \frac{x^5}{6}$$

if the desired accuracy requires the error to be less than 0.05.

NOTE: In the economized series, the highest power of $x$ will be the third.

Appendix G

# Plane Trigonometry

Many of the key results of Chapter 6 require trigonometric expressions in terms of the semiperimeter of a triangle. In a typical course in trigonometry, these formulas do not generally arise. Even if taught, they are, more than likely, rapidly forgotten through lack of use. In this appendix, the necessary formulas for the requirements of Chapter 6, and some other chapters as well, are presented in the form of problems for the reader.

Let $a$, $b$, $c$ denote the sides and $\alpha$, $\beta$, $\gamma$ the corresponding opposite angles of the plane triangle shown in the accompanying figure.

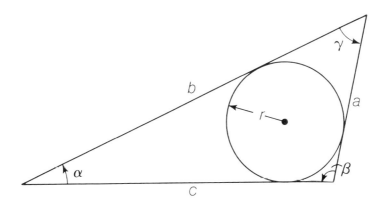

◊ **Problem G–1**
By inspection derive the *law of sines*

$$\frac{a}{\sin \alpha} = \frac{b}{\sin \beta} = \frac{c}{\sin \gamma}$$

◊ **Problem G–2**
From the law of sines derive the *law of tangents*

$$\frac{a+b}{a-b} = \frac{\tan \frac{1}{2}(\alpha+\beta)}{\tan \frac{1}{2}(\alpha-\beta)}$$

## Problem G-3
By inspection, show that
$$a = b\cos\gamma + c\cos\beta$$

## Problem G-4
From the result of the last problem, or using vector algebra, derive the *law of cosines*
$$a^2 = b^2 + c^2 - 2bc\cos\alpha$$

## Problem G-5
With $s$ denoting the semiperimeter of the triangle
$$s = \tfrac{1}{2}(a+b+c)$$
use the law of cosines to derive
$$\sin\tfrac{1}{2}\alpha = \sqrt{\frac{(s-b)(s-c)}{bc}} \qquad \cos\tfrac{1}{2}\alpha = \sqrt{\frac{s(s-a)}{bc}}$$
and obtain
$$\sin\alpha = \frac{2}{bc}\sqrt{s(s-a)(s-b)(s-c)}$$

## Problem G-6
Using the fact that the bisectors of the interior angles of a triangle meet at a point, which is the center of the inscribed circle whose radius we shall call $r$, show that
$$r = (s-a)\tan\tfrac{1}{2}\alpha = \sqrt{\frac{(s-a)(s-b)(s-c)}{s}}$$

## Problem G-7
Show that the area of the triangle is $rs$ and that the altitude on side $a$ is $2rs/a$.

## Problem G-8
Derive the identities
$$\begin{aligned}\sin\alpha + \sin\beta &= 2\sin\tfrac{1}{2}(\alpha+\beta)\cos\tfrac{1}{2}(\alpha-\beta)\\ \sin\alpha - \sin\beta &= 2\cos\tfrac{1}{2}(\alpha+\beta)\sin\tfrac{1}{2}(\alpha-\beta)\\ \cos\alpha + \cos\beta &= 2\cos\tfrac{1}{2}(\alpha+\beta)\cos\tfrac{1}{2}(\alpha-\beta)\\ \cos\alpha - \cos\beta &= -2\sin\tfrac{1}{2}(\alpha+\beta)\sin\tfrac{1}{2}(\alpha-\beta)\end{aligned}$$

## Problem G-9
Derive the identities
$$\begin{aligned}\sin\alpha\sin\beta &= \tfrac{1}{2}\cos(\alpha-\beta) - \tfrac{1}{2}\cos(\alpha+\beta)\\ \cos\alpha\cos\beta &= \tfrac{1}{2}\cos(\alpha-\beta) + \tfrac{1}{2}\cos(\alpha+\beta)\\ \sin\alpha\cos\beta &= \tfrac{1}{2}\sin(\alpha-\beta) + \tfrac{1}{2}\sin(\alpha+\beta)\end{aligned}$$

PART II

Chapter 8

# Non-Keplerian Motion

H ARMONICES MUNDI, WHICH WAS THE CULMINATION OF KEPLER'S revolutionary contribution to science and contained his third law of planetary motion, failed to account for the masses of the planets. Indeed, when we refer to *Keplerian orbits*, we are implicitly assuming that these masses are truly negligible, and that Kepler's so-called "laws" are exact. In fact, however, with the exception of two-body motion, the problems of celestial mechanics are, generally, incapable of exact mathematical solution. In many ways, this was fortunate for the development of science and engineering. (Indeed, if even the solution of Kepler's equation in the two-body problem had been simple to obtain in closed form, the history of mathematics might have been considerably altered.)

Celestial mechanics became the driving force which spurred the great mathematicians to incredible efforts to find useful methods of analyzing planetary motion. The elegant tools which they invented for this purpose had astonishing applicability in many diverse fields.

Sir Isaac Newton was the first to consider the attraction exerted by spheres and spheroids of uniform and varying density on a particle. In the *Principia*, Proposition 74, he showed that the attraction of a homogeneous sphere on a particle is the same as if the mass of the sphere were concentrated at its center. This was not an easy problem even for Newton. In 1684, almost 20 years after he began to apply his law of gravitation to planetary motion, his friend Edmund Halley urged him to publish his results. However, Newton lacked the proof that would eventually become Proposition 74. In a letter to Halley on June 20, 1686, he stated that until 1685 he suspected that the proposition was false. Then came the proof and the agreement to publish. Halley assisted Newton with the editorial work and even paid for the publication.

Newton also studied and obtained some approximate results for the general problem of three bodies which continues to be a major area of activity in Celestial Mechanics. He devoted so much time to the moon's motion—an important effort needed to improve the method of determining latitude—that he frequently complained it made his head ache.

The first particular solutions of the three-body problem were obtained by Lagrange in his prize memoir, *Essai sur le Problème des Trois Corps*, which was submitted to the Paris Academy in 1772. The solutions which

365

366                Non-Keplerian Motion                [Chap. 8

he found are precisely those given in this chapter. Referring to the colinear libration points of the Lagrange solution in *l'Exposition du Système du Monde*, Laplace remarked that if the moon had been given to the earth by Providence to illuminate the night, as some have maintained, the end sought has been only imperfectly attained; for if the moon were properly started in opposition to the sun, it would always remain there relatively, and the whole earth would have either the full moon or the sun always in view. (Actually, the configuration is unstable as was later proved by Joseph Liouville in the *Journal de Mathématiques* in 1845.)

We conclude this chapter with a spacecraft thrusting problem based on a paper by Hsue-shen Tsien written six years before the Russians launched their Sputnik. It is one of the very few such problems which has an exact solution—even though somewhat complex and involving elliptic integrals. At the time, Tsien was the Robert H. Goddard Professor of Jet Propulsion at the California Institute of Technology. He had been an extraordinarily talented engineering student from Shanghai who studied with Theodore von Kármán and taught at MIT. He had the misfortune of falling under false suspicion during the anti-Communist crusades led by Senator Joseph McCarthy. Tsien returned to the land of his birth where he has contributed much to its technological development.

## 8.1  Lagrange's Solution of the Three-Body Problem

Analytic solutions of the $n$-body problem do not exist in complete generality except for the case of $n = 2$. However, for the problem of three bodies, Lagrange discovered some particular solutions which merit our attention not only for the paucity of such results but also because they can be exploited for strategic location of earth satellites. In addition, they provide excellent examples of the advantages that often accrue when a problem is viewed from the perspective of a rotating coordinate system.

If the $n$ bodies are moving in coplanar circles around a common origin with a constant angular velocity $\omega$, then the position vectors $\mathbf{r}_i$ will all be constant in the rotating reference plane. From Eq. (2.54) we see that only the centripetal acceleration term will be nonzero and the equations of motion reduce to a set of algebraic equations

$$\mathbf{f}_i = -m_i \omega^2 \mathbf{r}_i \quad \text{for} \quad i = 1, 2, \ldots, n \quad (8.1)$$

where the force $\mathbf{f}_i$ is given by Eq. (2.37). We have already seen that the totality of these forces has a zero resultant so that by adding the $n$ vector equations, we have

$$\omega^2 (m_1 \mathbf{r}_1 + m_2 \mathbf{r}_2 + \cdots + m_n \mathbf{r}_n) = \mathbf{0} \quad (8.2)$$

Thus, the center of mass of the system must coincide with the axis of rotation if such motion is possible. Assuming this to be the case and noting

## Lagrange's Solution of the Three-Body Problem

that the position vectors are confined to a plane, then Eqs. (8.1) provide $2n$ nonlinear algebraic equations to be satisfied by the $2n$ components of the vectors $\mathbf{r}_1, \mathbf{r}_2, \ldots, \mathbf{r}_n$. For $n = 3$, the relevant equations are

$$\left(\frac{\omega^2}{G} - \frac{m_2}{r_{12}^3} - \frac{m_3}{r_{13}^3}\right)\mathbf{r}_1 + \frac{m_2}{r_{12}^3}\mathbf{r}_2 + \frac{m_3}{r_{13}^3}\mathbf{r}_3 = 0$$

$$\frac{m_1}{r_{21}^3}\mathbf{r}_1 + \left(\frac{\omega^2}{G} - \frac{m_1}{r_{21}^3} - \frac{m_3}{r_{23}^3}\right)\mathbf{r}_2 + \frac{m_3}{r_{23}^3}\mathbf{r}_3 = 0 \qquad (8.3)$$

$$m_1\mathbf{r}_1 + m_2\mathbf{r}_2 + m_3\mathbf{r}_3 = 0$$

where, for simplicity, we use the center of mass equation (8.2) instead of the third equation in (8.1).

### Equilateral Triangle Solution

There is a solution of these equations if the masses $m_1$, $m_2$, $m_3$ are located at the vertices of an equilateral triangle. To demonstrate, let $\rho$ be the constant length of the sides of the triangle so that

$$r_{12} = r_{13} = r_{21} = r_{23} = \rho$$

The equations are then linear and homogeneous in the components of $\mathbf{r}_1$, $\mathbf{r}_2$, $\mathbf{r}_3$ and will have a solution provided that the determinant of coefficients vanishes. It is easily seen that this condition is fulfilled if

$$\omega^2 = \frac{G}{\rho^3}(m_1 + m_2 + m_3) \qquad (8.4)$$

and, further, that all three of the equations in (8.3) are identical. Therefore, the equilateral triangle configuration, with appropriate initial conditions, is a particular solution of the three-body problem and any two of the three position vectors are arbitrary.

### Straight Line Solutions

There are also solutions of Eqs. (8.3) if the masses are colinear. Without loss of generality, the three masses are assumed to lie on the $\xi$ axis and are distributed so that $\xi_3 > \xi_2 > \xi_1$. Then we may write

$$\mathbf{r}_1 = \xi_1 \mathbf{i}_\xi$$
$$\mathbf{r}_2 = (\xi_1 + r_{12})\mathbf{i}_\xi$$
$$\mathbf{r}_3 = (\xi_1 + r_{12} + r_{23})\mathbf{i}_\xi$$

and the relations to be satisfied may be regarded as three scalar equations in the three unknowns $\omega^2$, $\xi_1$, and $r_{23}$ with $r_{12}$ as a parameter. Thus,

we have

$$\frac{\omega^2}{G} r_{12}^3 \frac{\xi_1}{r_{12}} + m_2 + \frac{1}{(1+\chi)^2} m_3 = 0$$

$$\frac{\omega^2}{G} r_{12}^3 \left(1 + \frac{\xi_1}{r_{12}}\right) - m_1 + \frac{1}{\chi^2} m_3 = 0$$

$$(m_1 + m_2 + m_3)\frac{\xi_1}{r_{12}} + m_2 + (1+\chi)m_3 = 0$$

where $\chi$ is defined as the ratio

$$\chi = \frac{r_{23}}{r_{12}}$$

The third and first equations give the coordinate of the mass $m_1$ relative to the center of rotation and the angular velocity in terms of $\chi$ as

$$\xi_1 = -r_{12} \frac{m_2 + (1+\chi)m_3}{m_1 + m_2 + m_3} \tag{8.5}$$

$$\omega^2 = \frac{G(m_1 + m_2 + m_3)}{r_{12}^3(1+\chi)^2} \frac{m_2(1+\chi)^2 + m_3}{m_2 + (1+\chi)m_3} \tag{8.6}$$

Substituting into the second equation, we obtain the *quintic equation of Lagrange* to be satisfied by $\chi$

$$(m_1 + m_2)\chi^5 + (3m_1 + 2m_2)\chi^4 + (3m_1 + m_2)\chi^3$$
$$- (m_2 + 3m_3)\chi^2 - (2m_2 + 3m_3)\chi - (m_2 + m_3) = 0 \tag{8.7}$$

as the condition for the existence of colinear solutions of the three-body problem.

The condition equation has one and only one positive root as can be seen from the fact that the coefficients change sign only once. However, a total of three different straight line solutions exist since two more can be obtained by a cyclic permutation of the order of the masses.

◊ **Problem 8–1**

Approximate solutions to Lagrange's quintic equation can be obtained when one of the three masses is of negligible size and the two finite masses are of different orders of magnitude.
(a) If $m_3 = 0$ and $m_2 < m_1$, then

$$\sigma^3 \equiv \frac{m_2}{3m_1} = \frac{\chi^3(3 + 3\chi + \chi^2)}{3(1-\chi^3)(1+\chi)^2} = \chi^3(1 - \chi + \frac{4}{3}\chi^2 - \frac{2}{3}\chi^3 + \cdots)$$

Obtain the cube root of this series as

$$\sigma = \chi - \tfrac{1}{3}\chi^2 + \tfrac{1}{3}\chi^3 + \tfrac{1}{81}\chi^4 + \cdots$$

### Sect. 8.1]  Lagrange's Solution of the Three-Body Problem

and, by series reversion, verify that

$$\chi = \sigma + \tfrac{1}{3}\sigma^2 - \tfrac{1}{9}\sigma^3 - \tfrac{31}{81}\sigma^4 + \cdots$$

(b) If $m_2 = 0$ and $m_1 < m_3$, show that

$$\varphi^3 \equiv \frac{m_1}{3m_3} = \frac{1 + 3\chi + 3\chi^2}{3\chi^3(3 + 3\chi + \chi^2)}$$

Then, with $\alpha$ defined as

$$\alpha = \frac{1}{1+\chi} \quad \text{or} \quad \chi = \frac{1-\alpha}{\alpha}$$

obtain

$$\varphi^3 = \frac{\alpha^3(3 - 3\alpha + \alpha^2)}{3(1-\alpha^3)(1-\alpha)^2} = \alpha^3(1 + \alpha + \tfrac{4}{3}\alpha^2 + \tfrac{8}{3}\alpha^3 + \cdots)$$

Hence, derive the expansion

$$\frac{1}{1+\chi} = \varphi - \tfrac{1}{3}\varphi^2 - \tfrac{1}{9}\varphi^3 - \tfrac{23}{81}\varphi^4 + \cdots$$

(c) If $m_1 = 0$ and $m_3 < m_2$, we have

$$\frac{m_3}{m_2} = \frac{(\chi^3 - 1)(1+\chi)^2}{3\chi^2 + 3\chi + 1}$$

Then, with $\beta$ defined as

$$\beta = \frac{1-\chi}{\chi} \quad \text{or} \quad \chi = \frac{1}{1+\beta}$$

obtain

$$\delta \equiv \frac{m_3}{m_2 + m_3} = -\frac{\beta(12 + 24\beta + 19\beta^2 + 7\beta^3 + \beta^4)}{7 + 14\beta + 13\beta^2 + 6\beta^3 + \beta^4}$$

$$= -\tfrac{12}{7}\beta + \tfrac{23}{49}\beta^3 - \tfrac{23}{49}\beta^4 + \cdots$$

Hence, derive the expansion

$$\frac{1-\chi}{\chi} = -\tfrac{7}{12}\delta - \tfrac{1,127}{20,736}\delta^3 - \tfrac{7,889}{248,832}\delta^4 + \cdots$$

(d) Obtain approximate numerical values for $\chi$ in the case for which the finite masses are the earth and the moon while the infinitesimal mass is a spacecraft. Compare with the exact values obtained by solving Lagrange's quintic equation using an appropriate method of numerical iteration.

NOTE: The ratio of the masses of the earth and the moon is 81.3007.

HINT: The formulas of Appendix C will be helpful.

◇ **Problem 8–2**
Develop the solution of Lagrange's equations (8.5)–(8.7) for the case $m_1 = m_2 = m_3$ from basic principles.

### Conic Section Solutions

Lagrange was able to obtain still a third kind of solution of the three-body problem in which the orbits are conic sections and include the equilateral triangle and straight line solutions as special cases. Again we confine ourselves to planar motion and utilize rotating coordinates, but this time the angular velocity and the radial distances are not constant.

Let $\mathbf{r}_i(t_0)$ for $i = 1, 2, 3$ be the (two-dimensional) position vectors of the three bodies at an initial time $t_0$ and let

$$\mathbf{r}_i(t) = \rho(t)\mathbf{r}_i(t_0) \tag{8.8}$$

be the position vectors as a function of time in a rotating coordinate system defined by the rotation matrix

$$\mathbf{R} = \begin{bmatrix} \cos\theta & -\sin\theta \\ \sin\theta & \cos\theta \end{bmatrix}$$

Since $\rho$ and $\theta$ are the same for the three bodies, the ratios of their mutual distances and the shape of the figure formed by the three bodies are unaltered with time.

From Eq. (2.37), the force vector acting on the $i^{\text{th}}$ mass is

$$\mathbf{f}_i(t) = G \sum_{j=1}^{3}{}' \frac{m_i m_j}{r_{ij}^3(t)}[\mathbf{r}_j(t) - \mathbf{r}_i(t)]$$

$$= \frac{1}{\rho^2}\mathbf{f}_i(t_0)$$

and from Eq. (2.51) the matrix $\mathbf{\Omega}$ is determined as†

$$\mathbf{\Omega} = \mathbf{R}\frac{d\mathbf{R}}{dt} = \begin{bmatrix} 0 & -1 \\ 1 & 0 \end{bmatrix}\frac{d\theta}{dt} = -\mathbf{J}\frac{d\theta}{dt}$$

Then, according to Eq. (2.53), the equations of motion are

$$m_i\left\{\left[\frac{d^2\rho}{dt^2} - \rho\left(\frac{d\theta}{dt}\right)^2\right]\mathbf{I} - \left[\frac{1}{\rho}\frac{d}{dt}\left(\rho^2\frac{d\theta}{dt}\right)\right]\mathbf{J}\right\}\mathbf{r}_i(t_0) = \frac{1}{\rho^2}\mathbf{f}_i(t_0) \tag{8.9}$$

where

$$\mathbf{I} = \begin{bmatrix} 1 & 0 \\ 0 & 1 \end{bmatrix} \quad \text{and} \quad \mathbf{J} = \begin{bmatrix} 0 & 1 \\ -1 & 0 \end{bmatrix}$$

Now, from the previous cases considered, we know that configurations of the three bodies exist for which

$$\mathbf{f}_i(t_0) = -m_i k^2 \mathbf{r}_i(t_0) \tag{8.10}$$

---

† The $\mathbf{J}$ matrix is fundamental in the definition of symplectic matrices which are introduced in Sect. 9.5.

where $k^2$ is a proportionality constant. For any of these configurations, the net force on each body is directed toward their mutual center of mass, i.e., along the radius vectors, so that we have, from Prob. 2-16,

$$r_i^2 \frac{d\theta}{dt} = c_i \quad \text{for} \quad i = 1, 2, 3 \tag{8.11}$$

where the $c_i$'s are constants. Therefore,

$$\frac{d}{dt}\left(\rho^2 \frac{d\theta}{dt}\right) = 0 \tag{8.12}$$

and the equations of motion will be satisfied if

$$\frac{d^2\rho}{dt^2} - \rho\left(\frac{d\theta}{dt}\right)^2 = -\frac{k^2}{\rho^2} \tag{8.13}$$

These last two differential equations for $\rho$ and $\theta$ are exactly the equations of motion in polar coordinates for the relative motion of two bodies as derived in Prob. 3-1.

## 8.2 The Restricted Problem of Three Bodies

When all of the masses are finite, the three-body problem admits of certain exact solutions in which the ratios of the mutual distances of the bodies are constant. However, if one of the masses is infinitesimal so that it has no appreciable effect on the motion of the other two, then the possible motions of the small mass are considerably expanded. This is the famous *restricted problem of three bodies* examples of which are approximated by a spacecraft in the earth-moon system or a planetary satellite in the planet-sun system.

Specifically, the problem is the description of the motion of an infinitesimal mass under the influence of two bodies of finite mass $m_1$ and $m_2$ which revolve around their common centroid in circular orbits. As before, let the origin of coordinates be at the center of mass of the system. The angular velocity of the finite masses

$$\omega^2 = \frac{G(m_1 + m_2)}{r_{12}^3} \tag{8.14}$$

is obtained from Eq. (8.4) where $r_{12}$ denotes the constant separation of $m_1$ and $m_2$.

To simplify the notation, let **r** denote the position vector of the infinitesimal mass relative to the center of mass of $m_1$ and $m_2$ and define

$$\boldsymbol{\rho}_1 = \mathbf{r} - \mathbf{r}_1 \qquad \boldsymbol{\rho}_2 = \mathbf{r} - \mathbf{r}_2$$

Then, the equation of motion of the small body in a coordinate system rotating with the constant angular velocity $\boldsymbol{\omega} = \omega\, \mathbf{i}_\zeta$ is

$$\frac{d^2\mathbf{r}}{dt^2} + 2\boldsymbol{\omega} \times \frac{d\mathbf{r}}{dt} + \boldsymbol{\omega} \times (\boldsymbol{\omega} \times \mathbf{r}) + \frac{Gm_1}{\rho_1^3}\boldsymbol{\rho}_1 + \frac{Gm_2}{\rho_2^3}\boldsymbol{\rho}_2 = \mathbf{0} \tag{8.15}$$

## Jacobi's Integral

Let the coordinates along the rotating axes be denoted by $\xi$, $\eta$, and $\varsigma$. In this system, the origin of coordinates will be at the center of mass of $m_1$ and $m_2$. We may, for convenience, assume that the finite masses are positioned always on the $\xi$-axis in the orbital plane, i.e., the $\xi, \eta$ plane. Then we have

$$\mathbf{r}_1 = \xi_1 \mathbf{i}_\xi \qquad \mathbf{r}_2 = \xi_2 \mathbf{i}_\xi \qquad \boldsymbol{\omega} = \omega \mathbf{i}_\varsigma$$

On the other hand, the infinitesimal mass has three degrees of freedom so that

$$\mathbf{r} = \xi \mathbf{i}_\xi + \eta \mathbf{i}_\eta + \varsigma \mathbf{i}_\varsigma$$

Therefore, the squares of its distance from $m_1$ and $m_2$ are given by

$$\rho_1^2 = (\xi - \xi_1)^2 + \eta^2 + \varsigma^2 \quad \text{and} \quad \rho_2^2 = (\xi - \xi_2)^2 + \eta^2 + \varsigma^2$$

The vector products in Eq. (8.15) can be expressed in component form as

$$\boldsymbol{\omega} \times \frac{d\mathbf{r}}{dt} = -\omega \frac{d\eta}{dt} \mathbf{i}_\xi + \omega \frac{d\xi}{dt} \mathbf{i}_\eta$$

$$\boldsymbol{\omega} \times (\boldsymbol{\omega} \times \mathbf{r}) = -\omega^2 (\xi \mathbf{i}_\xi + \eta \mathbf{i}_\eta)$$

Jacobi† obtained an integral for the equation of motion by defining a scalar function

$$J = \frac{\omega^2}{2}(\xi^2 + \eta^2) + \frac{Gm_1}{\rho_1} + \frac{Gm_2}{\rho_2} \tag{8.16}$$

with properties similar to the force function introduced in Sect. 2.4. Then, in terms of the gradient of $J$, Eq. (8.15) is written

$$\frac{d^2\mathbf{r}}{dt^2} + 2\boldsymbol{\omega} \times \frac{d\mathbf{r}}{dt} = \left[\frac{\partial J}{\partial \mathbf{r}}\right]^T \tag{8.17}$$

from which is obtained

$$\frac{d^2\mathbf{r}}{dt^2} \cdot \frac{d\mathbf{r}}{dt} = \frac{1}{2}\frac{d}{dt}\left(\frac{d\mathbf{r}}{dt} \cdot \frac{d\mathbf{r}}{dt}\right) = \frac{\partial J}{\partial \mathbf{r}}\frac{d\mathbf{r}}{dt} = \frac{dJ}{dt}$$

---

† Carl Gustav Jacob Jacobi (1804–1851) studied at the University of Berlin and became a professor at Königsberg in 1827, a post which he had to abandon fifteen years later because of ill health. The rest of his short life was spent in Berlin with a pension from the Prussian government. He was fortunate in that his fame was great in his lifetime and his students spread his ideas throughout Europe. Elliptic functions, functional determinants (called *Jacobians*), ordinary and partial differential equations, dynamics, celestial mechanics, fluid dynamics, and hyperelliptic integrals and functions were his major interests. His classic in dynamics, *Vorlesungen über Dynamik*, appeared posthumously in 1866.

Sect. 8.2]  The Restricted Problem of Three Bodies  373

This is a perfect differential and by integration we obtain the modified energy integral known as *Jacobi's integral*

$$v_{rel}^2 = \omega^2(\xi^2 + \eta^2) + \frac{2Gm_1}{\rho_1} + \frac{2Gm_2}{\rho_2} - C \tag{8.18}$$

where $C$ is a constant and $v_{rel}^2$ is the square of the magnitude of the observed velocity $d\mathbf{r}/dt$, i.e., the velocity relative to the rotating axes.

◇ **Problem 8–3**
Jacobi's integral can also be expressed in the form

$$v_{rel}^2 = -\mathbf{r} \cdot \boldsymbol{\omega} \times (\boldsymbol{\omega} \times \mathbf{r}) + \frac{2Gm_1}{\rho_1} + \frac{2Gm_2}{\rho_2} - C$$

**Rectilinear Oscillation of an Infinitesimal Mass**

As an example of the use of Jacobi's integral, consider the motion of the infinitesimal body along the straight line through the center of mass and perpendicular to the plane of rotation of the finite masses which, for simplicity, we assume to be equal, i.e., $m_1 = m_2 \equiv m$. Let $D$ and $\rho$ be the distances of each finite mass from the center and from the infinitesimal mass, respectively. Then, if $\varsigma$ is the distance of the small body from the center, we have

$$\rho^2 = D^2 + \varsigma^2 \quad \text{and} \quad \omega^2 = \frac{Gm}{4D^3}$$

Jacobi's integral, Eq. (8.18), is then

$$\left(\frac{d\varsigma}{dt}\right)^2 = \frac{4Gm}{\rho} - C \tag{8.19}$$

since $\xi = \eta = 0$, and the motion of the small mass is confined to the $\varsigma$ axis. If $v_0$ is its velocity when $\varsigma = 0$, then the constant $C$ is

$$C = \frac{4Gm}{D} - v_0^2$$

so that Jacobi's integral takes the form

$$\left(\frac{d\varsigma}{dt}\right)^2 = v_0^2 - 16\omega^2 D^2 \left(1 - \frac{D}{\rho}\right) \tag{8.20}$$

Introduce the angle $\theta$, for which $\varsigma = D \tan \theta$ and $\rho = D \sec \theta$, together with the quantity $B$ defined by

$$B = \frac{v_0^2}{16\omega^2 D^2} \tag{8.21}$$

and Jacobi's integral becomes

$$\left(\frac{d\theta}{dt}\right)^2 = 16\omega^2 \cos^4\theta [B - (1 - \cos\theta)] \tag{8.22}$$

Now, $d\theta/dt$ will vanish if and only if $B \leq 1$, in which case the motion of the small body will be oscillatory. Assume this to be the case and define $\theta_m$ to be that value of $\theta$ for which $d\theta/dt = 0$. Then, we have

$$B = 1 - \cos\theta_m \tag{8.23}$$

and the equation of motion takes the form

$$\left(\frac{d\theta}{dt}\right)^2 = 16\omega^2 \cos^4\theta (\cos\theta - \cos\theta_m) \tag{8.24}$$

Finally, define $x = \cos\theta$ and $x_m = \cos\theta_m$ to obtain

$$\left(\frac{dx}{dt}\right)^2 = 16\omega^2 x^4 (1 - x^2)(x - x_m) \tag{8.25}$$

as the required equation to be solved for $x$ as a function of the time $t$.

Let $T$ be the quarter period of the motion. Then,

$$4\omega T = \int_{x_m}^1 \frac{dx}{x^2 \sqrt{P(x)}} \quad \text{with} \quad P(x) = (1 - x^2)(x - x_m) \tag{8.26}$$

which is recognized as an elliptic integral according to Sect. 1.5.

Reducing this integral to the standard forms, which Legendre proved was always possible, involves some ad hoc techniques which are part of Legendre's proof.

1. By using a modification of the standard method of integration by parts, observe that

$$\frac{d}{dx}\frac{\sqrt{P(x)}}{x} = \frac{\frac{1}{2}xP'(x) - P(x)}{x^2\sqrt{P(x)}} = -\frac{x}{2\sqrt{P(x)}} - \frac{1}{2x\sqrt{P(x)}} + \frac{x_m}{x^2\sqrt{P(x)}}$$

which, when integrated, gives

$$4\omega x_m T = \left.\frac{\sqrt{P(x)}}{x}\right|_{x_m}^1 + \frac{1}{2}\int_{x_m}^1 \frac{x\,dx}{\sqrt{P(x)}} + \frac{1}{2}\int_{x_m}^1 \frac{dx}{x\sqrt{P(x)}}$$

and note that the integrated part vanishes at both limits.

2. Since $P(x)$ is of third degree, convert it to a fourth-degree polynomial with the substitution $x = 1 - z^2$. Hence,

$$4\omega x_m T = \int_0^\alpha \frac{(1-z^2)\,dz}{\sqrt{Q(z)}} + \int_0^\alpha \frac{dz}{(1-z^2)\sqrt{Q(z)}}$$

## Sect. 8.2] The Restricted Problem of Three Bodies

where

$$Q(z) = (2 - z^2)(\alpha^2 - z^2) \quad \text{and} \quad \alpha = \sqrt{1 - x_m}$$

3. Then, write

$$Q(z) = 2\alpha^2 \left(1 - \frac{z^2}{2}\right)\left(1 - \frac{z^2}{\alpha^2}\right) \equiv R(y) = (1 - y^2)(1 - k^2 y^2)$$

where we have defined $z = \alpha y$ and $k^2 = \frac{1}{2}\alpha^2$, to obtain

$$4\sqrt{2}\,\omega x_m T = \int_0^1 \frac{(1 - 2k^2 y^2)\,dy}{\sqrt{R(y)}} + \int_0^1 \frac{dy}{(1 - 2k^2 y^2)\sqrt{R(y)}}$$

Observe that the second integral is a complete elliptic integral of the third kind while the first integral can be written

$$\int_0^1 \frac{(1 - 2k^2 y^2)\,dy}{\sqrt{R(y)}} = \int_0^1 \frac{2(1 - k^2 y^2)\,dy}{\sqrt{R(y)}} - \int_0^1 \frac{dy}{\sqrt{R(y)}}$$

$$= 2\int_0^1 \sqrt{\frac{1 - k^2 y^2}{1 - y^2}}\,dy - \int_0^1 \frac{dy}{\sqrt{R(y)}}$$

which are complete elliptic integrals of the second and first kinds, respectively.

4. Finally, make the change of variable $y = \sin\phi$ to convert to Legendre's form [Eqs. (1.52) through (1.55)]. Hence,

$$4\sqrt{2}\,\omega x_m T = 2E(k) - K(k) + \Pi(2k^2, k, \tfrac{1}{2}\pi) \tag{8.27}$$

so that the period $4T$ is expressed as a linear combination of complete elliptic integrals of the first, second, and third kinds.

The complete elliptic integral of the third kind can be expressed in terms of complete and incomplete integrals of the first and second kinds. It would take us too far afield to prove this in general so we shall, instead, simply state the result.† In this instance, the characteristic is

$$n = B = 2\sin^2 \tfrac{1}{2}\theta_m = 2k^2 \quad \text{so that} \quad k^2 < n < 1$$

—just the requirement for the so-called "circular case." (Actually, four different cases are possible.) The formula for this one is

$$\Pi(n, \sin \tfrac{1}{2}\theta_m, \tfrac{1}{2}\pi) = K(\sin \tfrac{1}{2}\theta_m) + \tfrac{1}{2}\pi\delta_2 [1 - \Lambda_0(\epsilon, \sin \tfrac{1}{2}\theta_m)]$$

---

† See, for example, "Handbook of Mathematical Functions" published by Dover Publications, Inc. and edited by Milton Abramowitz and Irene A. Stegun.

where $\Lambda_0$ is *Heuman's Lambda function*†

$$\Lambda_0(\epsilon, \sin \tfrac{1}{2}\theta_m) = \frac{2}{\pi} K(\sin \tfrac{1}{2}\theta_m)[E(\epsilon, \cos \tfrac{1}{2}\theta_m)$$
$$- Q(\sin \tfrac{1}{2}\theta_m)F(\epsilon, \cos \tfrac{1}{2}\theta_m)] \quad (8.28)$$

The function $Q$ is defined in Eq. (1.76). The two parameters $\epsilon$, an angle in the first quadrant, and $\delta_2$ are defined as

$$\sin \epsilon = \sqrt{1-n} \sec \tfrac{1}{2}\theta_m \quad \text{and} \quad \delta_2 = \sqrt{\frac{n}{(1-n)(n-\sin^2 \tfrac{1}{2}\theta_m)}}$$

which, for the case at hand, are calculated from

$$\cos \epsilon = \tan \tfrac{1}{2}\theta_m \quad \text{and} \quad \delta_2 = \sqrt{2 \sec \theta_m}$$

The final result is then

$$4\omega x_m T = \sqrt{2} E(\sin \tfrac{1}{2}\theta_m) + \tfrac{1}{2}\pi\sqrt{\sec \theta_m}\, [1 - \Lambda_0(\epsilon, \sin \tfrac{1}{2}\theta_m)] \quad (8.29)$$

where $\cos \epsilon = \tan \tfrac{1}{2}\theta_m$ and $\Lambda_0$ is obtained from Eq. (8.28).

### Surfaces of Zero Relative Velocity

For various values of the constant $C$, Eq. (8.18) defines surfaces in the $\xi$, $\eta$, $\varsigma$ space on which the relative velocity $v_{rel}$ will be zero. Thus, with $C$ determined by the position and velocity of the small body at some instant of time, its subsequent motion will be confined to one side of the corresponding *surface of zero relative velocity*. Clearly, these surfaces are symmetrical with respect to the $\xi, \eta$ and $\xi, \varsigma$ planes.‡

Of particular interest are the curves formed by the intersection of these surfaces with the $\xi, \eta$ plane, an example§ of which is illustrated in Fig. 8.1. The equations for these curves are most conveniently represented by transforming to *bipolar coordinates*.

For this purpose, we note that

$$\rho_1^2 = (\xi - \xi_1)^2 + \eta^2 \qquad \rho_2^2 = (\xi - \xi_2)^2 + \eta^2$$

---

† C. Heuman, Tables of Complete Elliptic Integrals in the *Journal of Mathematical Physics*, Vol. 20, pp. 127–206, 1941.

‡ Recently, John Lundberg, Victor Szebehely, R. Steven Nerem, and Byron Beal used computer graphics to generate these three-dimensional surfaces. They are illustrated in their paper "Surfaces of Zero Velocity in the Restricted Problem of Three Bodies" which was published in the June 1985 issue of *Celestial Mechanics*, Vol. 36, pp. 191–205.

§ The curves marked $C_1$, $C_2$, ..., $C_5$ are in the order of decreasing values of the constant $C$. They are reproduced from the classical text *Celestial Mechanics* by Forest Ray Moulton. These contours were not drawn from numerical calculations, but were intended only to show, qualitatively, the character of the curves.

Sect. 8.2]   The Restricted Problem of Three Bodies

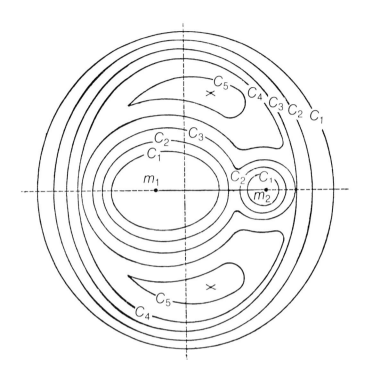

**Fig. 8.1:** Surfaces of zero relative velocity.

since $\zeta = 0$ in the $\xi$, $\eta$ plane. The coordinates $\xi_1$ and $\xi_2$ are determined from the equation of the centroid and the definition of $\rho \equiv r_{12}$:

$$m_1 \xi_1 + m_2 \xi_2 = 0 \qquad \xi_2 - \xi_1 = \rho$$

Hence,

$$m_1 \rho_1^2 + m_2 \rho_2^2 = (m_1 + m_2)(\xi^2 + \eta^2) + m_1 \xi_1^2 + m_2 \xi_2^2$$

$$= (m_1 + m_2)(\xi^2 + \eta^2) + \frac{m_1 m_2}{m_1 + m_2} \rho^2$$

so that

$$\xi^2 + \eta^2 = \frac{m_1 \rho_1^2 + m_2 \rho_2^2}{m_1 + m_2} - \frac{m_1 m_2}{(m_1 + m_2)^2} \rho^2$$

Since the second term on the right side of this last equation is a constant, we may use Eqs. (8.14) and (8.18) to express the zero relative velocity curves (8.16) as

$$2J(\rho_1, \rho_2) = C^* \tag{8.30}$$

where the function $J$ is defined by

$$J(\rho_1,\rho_2) = \frac{Gm_1}{2}\left(\frac{\rho_1^2}{\rho^3} + \frac{2}{\rho_1}\right) + \frac{Gm_2}{2}\left(\frac{\rho_2^2}{\rho^3} + \frac{2}{\rho_2}\right) \qquad (8.31)$$

and $C^*$ is a constant simply related to $C$.

### Problem 8-4
The problem of computing the locus of points in the $\xi, \eta$ plane for which $2J = C^*$ is easily managed in bipolar coordinates $\rho_1, \rho_2$.
(a) Verify that $2J = C^*$ may be written

$$x_1^3 - Ax_1 + 2 = 0$$

where

$$A = \frac{\rho C^*}{Gm_1} - \frac{m_2}{m_1}\left(x_2^2 + \frac{2}{x_2}\right)$$

with

$$x_1 = \frac{\rho_1}{\rho} > 0 \qquad x_2 = \frac{\rho_2}{\rho} > 0$$

and demonstrate that $A$ is always positive.
(b) The cubic equation for $x_1$ has always one negative root and either a pair of complex roots or a pair of real positive roots. Furthermore, no complex roots exist if $A \geq 3$.
(c) The requirement $A \geq 3$ is equivalent to

$$x_2^3 - Bx_2 + 2 \leq 0$$

where

$$B = 3 + \frac{\rho}{Gm_2}(C^* - 3\rho^2\omega^2)$$

and $B$ must be positive if $x_2$ is to be positive.
(d) If $C^* = 3\rho^2\omega^2$, the inequality for $x_2$ is

$$(x_2 - 1)^2(x_2 + 2) \leq 0$$

which is valid only for $x_2 = 1$. Further, the $x_2$ inequality is violated for any smaller value of $C^*$ if $x_2$ is required to be positive.
(e) The quantity $x_2$ must lie between the two positive real roots of

$$x^3 - Bx + 2 = 0$$

if both $x_1$ and $x_2$ are to be real and positive.
(f) When $C^* = 3\rho^2\omega^2$, then $\rho_1 = \rho_2 = \rho$ obtains corresponding to the equilateral triangle solution of the three-body problem.
(g) Finally, devise a computational procedure for determining the curves of zero relative velocity in the $\xi, \eta$ plane.

## Lagrangian Points

In terms of the $J$ function, the equations of motion of the infinitesimal body in the $\xi, \eta$ plane are, from Eq. (8.17),

$$\frac{d^2\xi}{dt^2} - 2\omega\frac{d\eta}{dt} = \frac{\partial J}{\partial \xi}$$
$$\frac{d^2\eta}{dt^2} + 2\omega\frac{d\xi}{dt} = \frac{\partial J}{\partial \eta} \tag{8.32}$$

Now consider points in the plane for which $\partial J/\partial \xi = \partial J/\partial \eta = 0$. If the small body is placed *at rest* at one of these points, it follows from the equations of motion that its acceleration will be zero. Thus, the body will remain relatively at rest forever unless acted upon by externally applied disturbing forces.

Just as in the case for which all three bodies have finite mass, we should expect five points of relative equilibrium corresponding to the vanishing of the gradient of the $J$ function. These five points are the *Lagrangian points* (also called *libration points*) and are usually labelled $L_1, L_2, \ldots, L_5$. The points $L_1, L_2, L_3$ lie along the straight line joining the two large masses while $L_4$ and $L_5$ are points in the plane of rotation which form an equilateral triangle with the two masses. In each case the three bodies are at rest when viewed in a coordinate system which rotates at the appropriate constant angular velocity.

Since

$$\frac{\partial J}{\partial \xi} = \frac{\partial J}{\partial \rho_1}\frac{\partial \rho_1}{\partial \xi} + \frac{\partial J}{\partial \rho_2}\frac{\partial \rho_2}{\partial \xi}$$
$$\frac{\partial J}{\partial \eta} = \frac{\partial J}{\partial \rho_1}\frac{\partial \rho_1}{\partial \eta} + \frac{\partial J}{\partial \rho_2}\frac{\partial \rho_2}{\partial \eta} \tag{8.33}$$

then, clearly, two points of relative equilibrium for the infinitesimal body can be determined from

$$\frac{\partial J}{\partial \rho_1} = \frac{\partial J}{\partial \rho_2} = 0 \tag{8.34}$$

which implies $\rho_1 = \rho_2 = \rho$. These correspond to the $L_4$ and $L_5$ Lagrangian points. Furthermore, since

$$\frac{\partial \rho_1}{\partial \eta} = \frac{\eta}{\rho_1} \qquad \frac{\partial \rho_2}{\partial \eta} = \frac{\eta}{\rho_2} \tag{8.35}$$

we see that $\partial J/\partial \eta$ vanishes identically for all points whose $\eta$ coordinate is zero. Hence, the $L_1, L_2, L_3$ points should correspond to those points on the $\xi$ axis for which $\partial J/\partial \xi = 0$ or

$$\frac{1}{\rho_1}\frac{\partial J}{\partial \rho_1}(\xi - \xi_1) + \frac{1}{\rho_2}\frac{\partial J}{\partial \rho_2}(\xi - \xi_2) = 0 \tag{8.36}$$

Consider the $L_1$ point for which the infinitesimal body lies between the two finite masses. Then

$$\xi - \xi_1 = \rho_1 \qquad \xi - \xi_2 = -\rho_2$$

and the point of relative equilibrium is determined from

$$L_1: \qquad \frac{\partial J}{\partial \rho_1} = \frac{\partial J}{\partial \rho_2} \qquad \text{and} \qquad \rho_1 + \rho_2 = \rho \qquad (8.37)$$

For the point $L_2$ to the right of $m_2$

$$\xi - \xi_1 = \rho_1 \qquad \xi - \xi_2 = \rho_2$$

so that the coordinate of $L_2$ is found from

$$L_2: \qquad \frac{\partial J}{\partial \rho_1} = -\frac{\partial J}{\partial \rho_2} \qquad \text{and} \qquad \rho_1 = \rho + \rho_2 \qquad (8.38)$$

Finally, for the point $L_3$ to the left of $m_1$, we have

$$\xi - \xi_1 = -\rho_1 \qquad \xi - \xi_2 = -\rho_2$$

and, therefore,

$$L_3: \qquad \frac{\partial J}{\partial \rho_1} = -\frac{\partial J}{\partial \rho_2} \qquad \text{and} \qquad \rho_2 = \rho + \rho_1 \qquad (8.39)$$

serves to specify the coordinates.

◊ **Problem 8-5**
The coordinates of the colinear Lagrangian points are determined from

$$L_1: \qquad \frac{m_2}{m_1} = \frac{\rho_2^2(3\rho^2 \rho_2 - 3\rho \rho_2^2 + \rho_2^3)}{(\rho^3 - \rho_2^3)(\rho - \rho_2)^2} \qquad \rho_1 = \rho - \rho_2$$

$$L_2: \qquad \frac{m_2}{m_1} = \frac{\rho_2^2(3\rho^2 \rho_2 + 3\rho \rho_2^2 + \rho_2^3)}{(\rho^3 - \rho_2^3)(\rho + \rho_2)^2} \qquad \rho_1 = \rho + \rho_2$$

$$L_3: \qquad \frac{m_1}{m_2} = \frac{\rho_1^2(3\rho^2 \rho_1 + 3\rho \rho_1^2 + \rho_1^3)}{(\rho^3 - \rho_1^3)(\rho + \rho_1)^2} \qquad \rho_2 = \rho + \rho_1$$

By appropriately relabeling the symbols, show that the quintic equations obtained are special cases of Lagrange's quintic equation derived in Sect. 8.1.

Sect. 8.2]   The Restricted Problem of Three Bodies   381

## Problem 8-6
For the restricted three-body problem, the force function $U$ and the kinetic energy $T$ are given by

$$U = \frac{Gm_1m_2}{\rho} + \frac{Gm_1m}{\rho_1} + \frac{Gm_2m}{\rho_2}$$

$$T = \tfrac{1}{2}m_1 r_1^2 \omega^2 + \tfrac{1}{2}m_2 r_2^2 \omega^2 + \tfrac{1}{2}m(\dot{x}^2 + \dot{y}^2 + \dot{z}^2)$$

where $m$ is the mass of the infinitesimal body.

Define the generalized coordinates $q_1$, $q_2$, $q_3$ to be the rectangular coordinates $\xi$, $\eta$, $\varsigma$ of the rotating coordinate frame, so that

$$x = q_1 \cos \omega t - q_2 \sin \omega t$$
$$y = q_1 \sin \omega t + q_2 \cos \omega t$$
$$z = \varsigma = q_3$$

In formulating the Lagrangian and Hamiltonian functions to be used in deriving the equations of motion, note that the constant terms in $T$ and $U$ may be omitted since they do not contribute when $T$ and $U$ are differentiated. Also note that $m$ will be a factor of the equations of motion and may, therefore, be ignored. Under these circumstances, the Lagrangian and Hamiltonian functions are

$$L = \tfrac{1}{2}[\dot{q}_1^2 + \dot{q}_2^2 + \dot{q}_3^2 + 2\omega(q_1 \dot{q}_2 - q_2 \dot{q}_1) + \omega^2(q_1^2 + q_2^2)] + U$$

$$H = \tfrac{1}{2}[p_1^2 + p_2^2 + p_3^2 - 2\omega(q_1 p_2 - q_2 p_1)] - U$$

where

$$U = G\left(\frac{m_1}{\rho_1} + \frac{m_2}{\rho_2}\right)$$

Finally, derive Lagrange's equations of motion

$$\frac{d^2 q_1}{dt^2} - 2\omega \frac{dq_2}{dt} = \omega^2 q_1 + \frac{\partial U}{\partial q_1}$$

$$\frac{d^2 q_2}{dt^2} + 2\omega \frac{dq_1}{dt} = \omega^2 q_2 + \frac{\partial U}{\partial q_2}$$

$$\frac{d^2 q_3}{dt^2} = \frac{\partial U}{\partial q_3}$$

and Hamilton's canonic equations

$$\frac{dq_1}{dt} = p_1 + \omega q_2 \qquad \frac{dp_1}{dt} = \omega p_2 + \frac{\partial U}{\partial q_1}$$

$$\frac{dq_2}{dt} = p_2 - \omega q_1 \qquad \frac{dp_2}{dt} = -\omega p_1 + \frac{\partial U}{\partial q_2}$$

$$\frac{dq_3}{dt} = p_3 \qquad \frac{dp_3}{dt} = \frac{\partial U}{\partial q_3}$$

## 8.3 Stability of the Lagrangian Points

Five particular solutions of the equations of motion for the infinitesimal body in the restricted problem of three bodies have been found. For each, the three bodies are at rest when viewed in a coordinate system which rotates at constant angular velocity about their common center of mass. In this section we investigate the stability of these solutions. Specifically, if the infinitesimal body is displaced slightly from its equilibrium position and given a small velocity, will it remain in the vicinity of the libration point or move rapidly away? In the first case, the point is said to be *stable* and in the second case, *unstable*.

The question of stability is resolved by studying the behavior of the linearized form of the equations of motion in the vicinity of each of the libration points. If $\mathbf{r}_0$ is the position vector of a particular libration point, then we write

$$\mathbf{r} = \mathbf{r}_0 + \delta\mathbf{r}$$
$$\mathbf{v} = \frac{d}{dt}(\delta\mathbf{r}) = \delta\mathbf{v} \quad (8.40)$$

where $\delta\mathbf{r}$ and $\delta\mathbf{v}$ are to be regarded as small increments in position and velocity—so small that products and powers of their components may be disregarded in the analysis.

As we shall see shortly, the linearized equations of motion will have the form

$$\frac{d\mathbf{x}}{dt} = \mathbf{M}\mathbf{x} \quad (8.41)$$

where the six-component vector $\mathbf{x}$ is partitioned as

$$\mathbf{x} = \begin{bmatrix} \delta\mathbf{r} \\ \delta\mathbf{v} \end{bmatrix} \quad (8.42)$$

and the six-dimensional matrix $\mathbf{M}$ is constant. The *characteristic equation* of $\mathbf{M}$ is determined by setting to zero the determinant of the matrix difference $\mathbf{M} - \lambda\mathbf{I}$, where $\mathbf{I}$ is the six-dimensional identity matrix and $\lambda$ is a parameter. Thus, the equation

$$|\mathbf{M} - \lambda\mathbf{I}| = 0 \quad (8.43)$$

is the sixth-order polynomial equation in $\lambda$ whose roots are called the *characteristic values* or *eigenvalues* of the matrix $\mathbf{M}$. It can be shown that the system defined by Eq. (8.41) is stable if none of the eigenvalues has a positive real part and if all multiple eigenvalues, i.e., repeated roots, have negative real parts.

## The Equilateral Libration Points

We first examine the stability of the Lagrangian points $L_4$ and $L_5$. For this purpose, let $\mathbf{f}(\mathbf{r})$ represent the gravitational force vector in Eq. (8.15) so that

$$\mathbf{f}(\mathbf{r}) = -\frac{Gm_1}{\rho_1^3}\boldsymbol{\rho}_1 - \frac{Gm_2}{\rho_2^3}\boldsymbol{\rho}_2$$

and expand this vector function of position in a Taylor series about the point $L_4$. We have

$$\mathbf{f}(\mathbf{r}) = \mathbf{f}(\mathbf{r}_0) + \left.\frac{\partial \mathbf{f}}{\partial \mathbf{r}}\right|_{\mathbf{r}=\mathbf{r}_0} \delta\mathbf{r} + \cdots \equiv \mathbf{f}_0 + \mathbf{F}_0 \delta\mathbf{r} + \cdots \quad (8.44)$$

and, as previously noted, higher-order terms in the expansion are to be neglected. The elements of the matrix $\mathbf{F}_0$ are gradients of the various components of the force vector $\mathbf{f}$ with respect to the position vector $\mathbf{r}$ evaluated at the point $L_4$.

To calculate the matrix $\mathbf{F}_0$, first note that

$$\frac{\partial \boldsymbol{\rho}_1}{\partial \mathbf{r}} = \frac{\partial \boldsymbol{\rho}_1}{\partial \boldsymbol{\rho}_1} = \mathbf{I} \quad \text{and} \quad \frac{\partial \rho_1}{\partial \mathbf{r}} = \frac{\partial \rho_1}{\partial \boldsymbol{\rho}_1} = \frac{1}{\rho_1}\boldsymbol{\rho}_1^T$$

with identical results obtaining for $\boldsymbol{\rho}_2$. Therefore,

$$\mathbf{F} = \frac{\partial \mathbf{f}}{\partial \mathbf{r}} = \frac{Gm_1}{\rho_1^5}(3\boldsymbol{\rho}_1\boldsymbol{\rho}_1^T - \rho_1^2 \mathbf{I}) + \frac{Gm_2}{\rho_2^5}(3\boldsymbol{\rho}_2\boldsymbol{\rho}_2^T - \rho_2^2 \mathbf{I})$$

Furthermore, since

$$\mathbf{r}_0 = \tfrac{1}{2}[(\xi_1 + \xi_2)\mathbf{i}_\xi + \sqrt{3}\,\rho\,\mathbf{i}_\eta]$$

it follows that

$$\boldsymbol{\rho}_1 = \tfrac{1}{2}\rho(\mathbf{i}_\xi + \sqrt{3}\,\mathbf{i}_\eta) \qquad \boldsymbol{\rho}_2 = \tfrac{1}{2}\rho(-\mathbf{i}_\xi + \sqrt{3}\,\mathbf{i}_\eta)$$

at the point $L_4$. Hence,

$$\mathbf{F}_0 = \frac{Gm_1}{4\rho^3}\begin{bmatrix} -1 & 3\sqrt{3} & 0 \\ 3\sqrt{3} & 5 & 0 \\ 0 & 0 & -4 \end{bmatrix} + \frac{Gm_2}{4\rho^3}\begin{bmatrix} -1 & -3\sqrt{3} & 0 \\ -3\sqrt{3} & 5 & 0 \\ 0 & 0 & -4 \end{bmatrix}$$

or, alternately,

$$\mathbf{F}_0 = \frac{\omega^2}{4}\begin{bmatrix} -1 & 0 & 0 \\ 0 & 5 & 0 \\ 0 & 0 & -4 \end{bmatrix} + \frac{\omega^2}{4}\left(\frac{m_1 - m_2}{m_1 + m_2}\right)\begin{bmatrix} 0 & 3\sqrt{3} & 0 \\ 3\sqrt{3} & 0 & 0 \\ 0 & 0 & 0 \end{bmatrix} \quad (8.45)$$

since

$$\omega^2 = \frac{G}{\rho^3}(m_1 + m_2)$$

In terms of the quantities $\delta\mathbf{r}$, $\delta\mathbf{v}$, and $\mathbf{F}_0$, the equations of motion (8.15) become

$$\frac{d}{dt}(\delta\mathbf{v}) + 2\boldsymbol{\omega} \times \delta\mathbf{v} + \boldsymbol{\omega} \times [\boldsymbol{\omega} \times (\mathbf{r}_0 + \delta\mathbf{r})] = \mathbf{f}_0 + \mathbf{F}_0\delta\mathbf{r} + \cdots$$

However, $\mathbf{r}_0$ is an equilibrium point so that $\boldsymbol{\omega} \times (\boldsymbol{\omega} \times \mathbf{r}_0)$ on the left side exactly cancels $\mathbf{f}_0$ on the right. Thus, we have

$$\frac{d}{dt}(\delta\mathbf{r}) = \delta\mathbf{v}$$
$$\frac{d}{dt}(\delta\mathbf{v}) + 2\boldsymbol{\omega} \times \delta\mathbf{v} + \boldsymbol{\omega} \times (\boldsymbol{\omega} \times \delta\mathbf{r}) = \mathbf{F}_0\delta\mathbf{r} \quad (8.46)$$

as a pair of vector differential equations for $\delta\mathbf{r}$ and $\delta\mathbf{v}$ valid in the vicinity of $\mathbf{r}_0$ and correct to first-order terms in the small quantities $\delta\mathbf{r}$ and $\delta\mathbf{v}$.

These equations can be written in the vector-matrix form of Eq. (8.41) by defining the matrix $\mathbf{M}$ as

$$\mathbf{M} = \begin{bmatrix} \mathbf{O} & \mathbf{I} \\ \mathbf{F}_0 - \boldsymbol{\Omega}\boldsymbol{\Omega} & -2\boldsymbol{\Omega} \end{bmatrix} \quad (8.47)$$

with $\mathbf{O}$ and $\mathbf{I}$ as the three-dimensional zero and identity matrices, and

$$\boldsymbol{\Omega} = \begin{bmatrix} 0 & -\omega & 0 \\ \omega & 0 & 0 \\ 0 & 0 & 0 \end{bmatrix} \quad (8.48)$$

The determination of the characteristic equation of $\mathbf{M}$, Eq. (8.43), is a routine and straightforward calculation. There results

$$|\mathbf{M} - \lambda\mathbf{I}| = (\lambda^2 + \omega^2)\left[\lambda^4 + \omega^2\lambda^2 + \frac{27}{4}\omega^4 \frac{m_1 m_2}{(m_1 + m_2)^2}\right] \quad (8.49)$$

so that the eigenvalues will all be imaginary provided

$$\omega^4 - 27\omega^4 \frac{m_1 m_2}{(m_1 + m_2)^2} \geq 0 \quad (8.50)$$

Therefore, the $L_4$ libration point (and, by symmetry, the $L_5$ point) will be stable if $m_1$ and $m_2$ are so related that

$$\frac{m_1}{m_2} + \frac{m_2}{m_1} \geq 25 \quad (8.51)$$

The masses of the sun and Jupiter satisfy this inequality and one might expect planets to exist approximating the equilateral triangle configurations. Such planets have been discovered and are known as the *Trojan asteroids*.

### The Colinear Libration Points

In a similar manner we can show that the Lagrangian points $L_1$, $L_2$, $L_3$ are unstable whatever might be the mass ratios. Since it is considerably easier to demonstrate this instability for motion in the $\xi, \eta$ plane rather than in three dimensions, we will utilize the equations of motion (8.32) for the infinitesimal body.

As before, define

$$\xi = \xi_0 + \delta\xi \qquad \eta = 0 + \delta\eta$$

where $(\xi_0, 0)$ are the coordinates of one of the colinear libration points. Then expand the right sides of Eqs. (8.32) in a Taylor series as

$$\frac{\partial J}{\partial \xi} = J_\xi + J_{\xi\xi}\delta\xi + J_{\eta\xi}\delta\eta + \cdots$$
$$\frac{\partial J}{\partial \eta} = J_\eta + J_{\xi\eta}\delta\xi + J_{\eta\eta}\delta\eta + \cdots \qquad (8.52)$$

with the subscript notation $J_\xi$, $J_{\xi\xi}$, etc. indicating the various partial derivatives of $J$ evaluated at the Lagrangian point. Again we neglect powers and products of $\delta\xi$ and $\delta\eta$.

The first derivatives of the $J$ function vanish at a Lagrangian point so that the constant terms in the series expansion are zero. Thus the linearized equations of motion are readily seen to be of the form of (8.41) where the $\mathbf{M}$ matrix is now four dimensional. Indeed, it is easy to show that

$$\mathbf{M} - \lambda\mathbf{I} = \begin{bmatrix} -\lambda & 0 & 1 & 0 \\ 0 & -\lambda & 0 & 1 \\ J_{\xi\xi} & J_{\eta\xi} & -\lambda & 2\omega \\ J_{\xi\eta} & J_{\eta\eta} & -2\omega & -\lambda \end{bmatrix} \qquad (8.53)$$

from which the characteristic equation of the matrix $\mathbf{M}$ is found to be

$$\lambda^4 + (4\omega^2 - J_{\xi\xi} - J_{\eta\eta})\lambda^2 + J_{\xi\xi}J_{\eta\eta} - J_{\xi\eta}^2 = 0 \qquad (8.54)$$

To resolve the question of stability, examine the signs of $J_{\xi\xi}$ and $J_{\eta\eta}$. Consider first

$$\frac{\partial^2 J}{\partial \xi^2} = \frac{\partial^2 J}{\partial \rho_1^2}\left(\frac{\partial \rho_1}{\partial \xi}\right)^2 + \frac{\partial J}{\partial \rho_1}\frac{\partial^2 \rho_1}{\partial \xi^2} + \frac{\partial^2 J}{\partial \rho_2^2}\left(\frac{\partial \rho_2}{\partial \xi}\right)^2 + \frac{\partial J}{\partial \rho_2}\frac{\partial^2 \rho_2}{\partial \xi^2}$$

Then, since

$$\rho_1 \frac{\partial \rho_1}{\partial \xi} = \xi - \xi_1 \qquad \rho_2 \frac{\partial \rho_2}{\partial \xi} = \xi - \xi_2$$

and

$$\left(\frac{\partial \rho_1}{\partial \xi}\right)^2 + \rho_1 \frac{\partial^2 \rho_1}{\partial \xi^2} = 1 \qquad \left(\frac{\partial \rho_2}{\partial \xi}\right)^2 + \rho_2 \frac{\partial^2 \rho_2}{\partial \xi^2} = 1$$

we have
$$\left(\frac{\partial \rho_1}{\partial \xi}\right)^2 = \left(\frac{\partial \rho_2}{\partial \xi}\right)^2 = 1 \qquad \frac{\partial^2 \rho_1}{\partial \xi^2} = \frac{\partial^2 \rho_2}{\partial \xi^2} = 0$$
at all points for which $\eta = \varsigma = 0$. Hence,
$$\frac{\partial^2 J}{\partial \xi^2} = \frac{\partial^2 J}{\partial \rho_1^2} + \frac{\partial^2 J}{\partial \rho_2^2} = Gm_1\left(\frac{1}{\rho^3} + \frac{2}{\rho_1^3}\right) + Gm_2\left(\frac{1}{\rho^3} + \frac{2}{\rho_2^3}\right)$$
is a positive quantity at any point on the $\xi$ axis.

Similarly, since
$$\rho_1 \frac{\partial \rho_1}{\partial \eta} = \eta \qquad \rho_2 \frac{\partial \rho_2}{\partial \eta} = \eta$$
and
$$\left(\frac{\partial \rho_1}{\partial \eta}\right)^2 + \rho_1 \frac{\partial^2 \rho_1}{\partial \eta^2} = 1 \qquad \left(\frac{\partial \rho_2}{\partial \eta}\right)^2 + \rho_2 \frac{\partial^2 \rho_2}{\partial \eta^2} = 1$$
we have, for $\eta = 0$,
$$\frac{\partial \rho_1}{\partial \eta} = \frac{\partial \rho_2}{\partial \eta} = 0 \qquad \frac{\partial^2 \rho_1}{\partial \eta^2} = \frac{1}{\rho_1} \qquad \frac{\partial^2 \rho_2}{\partial \eta^2} = \frac{1}{\rho_2}$$
so that
$$\frac{\partial^2 J}{\partial \eta^2} = \frac{1}{\rho_1}\frac{\partial J}{\partial \rho_1} + \frac{1}{\rho_2}\frac{\partial J}{\partial \rho_2}$$
at all points on the $\xi$ axis.

To address the question of the sign of $J_{\eta\eta}$ we must consider separately the three possibilities. For the $L_1$ point we have already established, in the previous section, that this point of relative equilibrium is determined from
$$L_1: \qquad \frac{\partial J}{\partial \rho_1} = \frac{\partial J}{\partial \rho_2} \qquad \text{and} \qquad \rho_1 + \rho_2 = \rho$$
Hence, at the point $L_1$, we have
$$\frac{\partial^2 J}{\partial \eta^2} = \left(\frac{1}{\rho_1} + \frac{1}{\rho_2}\right)\frac{\partial J}{\partial \rho_1} = Gm_1\rho_1\left(\frac{1}{\rho_1} + \frac{1}{\rho_2}\right)\left(\frac{1}{\rho^3} - \frac{1}{\rho_1^3}\right)$$
which is negative since $\rho_1 < \rho$. Similarly, for $L_2$ we have
$$\frac{\partial^2 J}{\partial \eta^2} = \left(\frac{1}{\rho_1} - \frac{1}{\rho_2}\right)\frac{\partial J}{\partial \rho_1} = Gm_1\rho_1\left(\frac{1}{\rho_1} - \frac{1}{\rho_2}\right)\left(\frac{1}{\rho^3} - \frac{1}{\rho_1^3}\right)$$
which is again negative since $\rho_2 < \rho_1$ and $\rho < \rho_1$. Finally, for $L_3$,
$$\frac{\partial^2 J}{\partial \eta^2} = \left(\frac{1}{\rho_2} - \frac{1}{\rho_1}\right)\frac{\partial J}{\partial \rho_2} = Gm_2\rho_2\left(\frac{1}{\rho_2} - \frac{1}{\rho_1}\right)\left(\frac{1}{\rho^3} - \frac{1}{\rho_2^3}\right)$$
is also negative since $\rho_1 < \rho_2$ and $\rho < \rho_2$.

Thus, in all cases, the constant term in the characteristic equation for **M** is negative at the colinear points so that at least one eigenvalue is real and positive. The motion is, therefore, unstable.

## 8.4 The Disturbing Function

The equations of motion of $n$ mass particles, interacting through their gravitational forces, were developed in Sect. 2.4. These equations can be reformulated as the relative motion of two bodies with the remaining $n-2$ bodies acting as disturbances which cause the resulting motion to deviate from a two-body orbit. This mathematical description of the problem will be most effective if the disturbing forces are small, for then the relative motion of the two bodies will be well approximated by conic or *Keplerian* orbits as they are sometimes called.

From Eqs. (2.37) and (2.38) with $i=1$ and 2, we write

$$\frac{d^2\mathbf{r}_1}{dt^2} = G\frac{m_2}{r_{12}^3}(\mathbf{r}_2 - \mathbf{r}_1) + G\sum_{j=3}^{n}\frac{m_j}{r_{1j}^3}(\mathbf{r}_j - \mathbf{r}_1)$$

$$\frac{d^2\mathbf{r}_2}{dt^2} = G\frac{m_1}{r_{21}^3}(\mathbf{r}_1 - \mathbf{r}_2) + G\sum_{j=3}^{n}\frac{m_j}{r_{2j}^3}(\mathbf{r}_j - \mathbf{r}_2)$$

as the equations of motion of $m_1$ and $m_2$ with respect to unaccelerated, i.e., inertial coordinate axes. The motion of $m_2$ relative to $m_1$ is obtained by subtracting the two differential equations. We have

$$\frac{d^2\mathbf{r}}{dt^2} + \frac{\mu}{r^3}\mathbf{r} = -G\sum_{j=3}^{n}m_j\left(\frac{1}{d_j^3}\mathbf{d}_j + \frac{1}{\rho_j^3}\boldsymbol{\rho}_j\right) \tag{8.55}$$

where, for convenience of notation, we have defined

$$\mathbf{r} = \mathbf{r}_2 - \mathbf{r}_1 \qquad \boldsymbol{\rho}_j = \mathbf{r}_j - \mathbf{r}_1 \qquad \mathbf{d}_j = \mathbf{r} - \boldsymbol{\rho}_j \tag{8.56}$$

and

$$\mu = G(m_1 + m_2)$$

Equation (8.55) describes the relative motion of $m_1$ and $m_2$ within a system of $n$ bodies. If $m_3, m_4, \ldots, m_n$ were nonexistent, the equation of motion would be exactly Eq. (3.2).

An alternate form of the right-hand side of the equation of relative motion is frequently convenient. It is readily verified that

$$\frac{1}{d_j^3}\mathbf{d}_j^T + \frac{1}{\rho_j^3}\boldsymbol{\rho}_j^T = -\frac{\partial}{\partial \mathbf{r}}\left(\frac{1}{d_j} - \frac{1}{\rho_j^3}\mathbf{r}\cdot\boldsymbol{\rho}_j\right)$$

Therefore, if we define

$$R_j = Gm_j\left(\frac{1}{d_j} - \frac{1}{\rho_j^3}\mathbf{r}\cdot\boldsymbol{\rho}_j\right) \tag{8.57}$$

we may write

$$\frac{d^2\mathbf{r}}{dt^2} + \frac{\mu}{r^3}\mathbf{r} = \sum_{j=3}^{n}\left[\frac{\partial R_j}{\partial \mathbf{r}}\right]^T \tag{8.58}$$

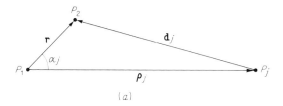

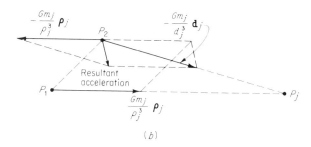

**Fig. 8.2:** Geometry of the disturbing acceleration.
(a) Position-vector diagram; (b) acceleration-vector diagram.

The scalar quantity $R_j$ is called the *disturbing function* associated with the disturbing body $m_j$.

Either Eq. (8.55) or (8.58) may be used to describe the motion of $m_2$ with respect to $m_1$. However, if $r$ is small compared to $\rho_j$, neither form is suitable for either analytical study or numerical integration. This point is clearly illustrated in Fig. 8.2. The disturbing effect of $m_j$ on the motion of $m_2$ relative to $m_1$ is seen to be calculated as the difference of two almost equal vectors. Several methods are available for circumventing this difficulty and preserving the significance of the results. Two of these are described below.

Explicit Calculation of the Disturbing Acceleration

The first method to be considered is a practical technique to alleviate the numerical troubles associated with the evaluation of the right-hand side of Eq. (8.55). We omit the identifier subscript $j$ and write

$$\frac{1}{d^3}\mathbf{d} + \frac{1}{\rho^3}\boldsymbol{\rho} = \frac{1}{d^3}\left[\mathbf{r} + \left(\frac{d^3}{\rho^3} - 1\right)\boldsymbol{\rho}\right]$$

Now it is clear that the potential difficulty arises in the evaluation of the quantity in parenthesis. Since

$$\frac{d^2}{\rho^2} = \frac{(\mathbf{r} - \boldsymbol{\rho}) \cdot (\mathbf{r} - \boldsymbol{\rho})}{\boldsymbol{\rho} \cdot \boldsymbol{\rho}}$$

## Sect. 8.4]   The Disturbing Function

this factor may be expressed as

$$\frac{d^3}{\rho^3} - 1 = f(q)$$

where $q$ and $f(q)$ are defined by

$$q = \frac{\mathbf{r} \cdot (\mathbf{r} - 2\boldsymbol{\rho})}{\boldsymbol{\rho} \cdot \boldsymbol{\rho}} \qquad f(q) = (1+q)^{\frac{3}{2}} - 1 \qquad (8.59)$$

A standard technique for evaluating $f(q)$ is to expand $(1+q)^{\frac{3}{2}}$ as a power series in $q$ so that

$$f(q) = \frac{3}{2}q\left(1 + \frac{1}{2^1 2!}q - \frac{1}{2^2 3!}q^2 + \frac{1\cdot 3}{2^3 4!}q^3 - \cdots\right)$$

However, a closed-form calculation is also possible. To this end write

$$f(q) = \frac{(1+q)^3 - 1}{1 + (1+q)^{\frac{3}{2}}}$$

Hence,

$$f(q) = q\frac{3 + 3q + q^2}{1 + (1+q)^{\frac{3}{2}}} \qquad (8.60)$$

and the evaluation of $f(q)$ is now clearly insensitive to the size of $q$ no matter how small.

Finally, then we have

$$\frac{d^2\mathbf{r}}{dt^2} + \frac{\mu}{r^3}\mathbf{r} = -G\sum_{j=3}^{n}\frac{m_j}{d_j^3}[\mathbf{r} + f(q_j)\boldsymbol{\rho}_j] \qquad (8.61)$$

where

$$q_j = \frac{\mathbf{r}\cdot(\mathbf{r} - 2\boldsymbol{\rho}_j)}{\boldsymbol{\rho}_j \cdot \boldsymbol{\rho}_j} = \frac{r}{\rho_j}\left(\frac{r}{\rho_j} - 2\cos\alpha_j\right) \qquad (8.62)$$

which describes the relative motion in a manner such that no loss of significance results in the calculation of the disturbing acceleration.

### Expansion of the Disturbing Function

The second method consists of expressing the disturbing function $R_j$ as a power series in $r/\rho_j$. For this purpose, we write Eq. (8.57) in the form (again, omitting the identifier subscript $j$)

$$R = \frac{Gm}{\rho}\left(\frac{\rho}{d} - \nu x\right)$$

where

$$x = \frac{r}{\rho} \qquad \text{and} \qquad \nu = \cos\alpha$$

with $\alpha$ as the angle between the vectors $\mathbf{r}$ and $\boldsymbol{\rho}$. It is also convenient to write $q$, previously defined in Eq. (8.59), as

$$q = x^2 - 2\nu x$$

Now,

$$\frac{\rho}{d} = (1+q)^{-\frac{1}{2}} = \sum_{i=0}^{\infty} \frac{(-1)^i (2i)!}{(2^i\, i!)^2} q^i$$

and, from the binomial theorem,

$$q^i = (x^2 - 2\nu x)^i = x^i \sum_{\ell=0}^{i} \frac{i!}{\ell!\,(i-\ell)!} x^\ell (-2\nu)^{i-\ell}$$

Hence,

$$(1+q)^{-\frac{1}{2}} = \sum_{i=0}^{\infty} \sum_{\ell=0}^{i} \frac{(-1)^\ell (2i)!}{2^{i+\ell}\, i!\,(i-\ell)!\,\ell!} x^{i+\ell} \nu^{i-\ell}$$

Since we are interested in the coefficients of the powers of $x$, we make a change in the summation indices by defining $k = i + \ell$ and replacing $i$ by $k - \ell$. Therefore, we obtain

$$(1+q)^{-\frac{1}{2}} = \sum_{k=0}^{\infty} \sum_{\ell=0}^{[\frac{1}{2}k]} \frac{(-1)^\ell (2k-2\ell)!}{2^k \ell!\,(k-\ell)!\,(k-2\ell)!} \nu^{k-2\ell} x^k$$

where the notation $[m]$ indicates the greatest integer contained in $m$. Thus,

$$\left[\tfrac{1}{2}k\right] = \begin{cases} \tfrac{1}{2}k & k \text{ even} \\ \tfrac{1}{2}(k-1) & k \text{ odd} \end{cases}$$

We see that the coefficients of $x^k$ are polynomials in $\nu$ which we symbolize as $P_k(\nu)$. Hence, we have shown that

$$(1 - 2\nu x + x^2)^{-\frac{1}{2}} = \sum_{k=0}^{\infty} P_k(\nu) x^k \tag{8.63}$$

where

$$P_k(\nu) = \sum_{\ell=0}^{[\frac{1}{2}k]} \frac{(-1)^\ell (2k-2\ell)!}{2^k \ell!\,(k-\ell)!\,(k-2\ell)!} \nu^{k-2\ell} \tag{8.64}$$

are known as *Legendre polynomials*. The first few Legendre polynomials are

$$P_0(\nu) = 1 \qquad P_3(\nu) = \tfrac{1}{2}(5\nu^3 - 3\nu)$$
$$P_1(\nu) = \nu \qquad P_4(\nu) = \tfrac{1}{8}(35\nu^4 - 30\nu^2 + 3)$$
$$P_2(\nu) = \tfrac{1}{2}(3\nu^2 - 1) \qquad P_5(\nu) = \tfrac{1}{8}(63\nu^5 - 70\nu^3 + 15\nu)$$

Sect. 8.4]  The Disturbing Function  391

From the subsection on Legendre polynomials later in this section, we have $|P_k(\cos \alpha)| \leq 1$. Thus, it is clear that the series (8.63) converges absolutely if $|x| < 1$. Therefore, the disturbing function may be expressed as the following convergent power series

$$R_j = \frac{Gm_j}{\rho_j} \left[ 1 + \sum_{k=2}^{\infty} \left(\frac{r}{\rho_j}\right)^k P_k(\cos \alpha_j) \right] \tag{8.65}$$

When $r/\rho_j$ is small, the series converges quite rapidly so that in many cases only a few terms are required for satisfactory accuracy. Finally, by substituting in Eq. (8.58), we have

$$\frac{d^2 \mathbf{r}}{dt^2} + \frac{\mu}{r^3} \mathbf{r} =$$

$$G \sum_{j=3}^{n} \frac{m_j}{\rho_j^2} \sum_{k=1}^{\infty} \left(\frac{r}{\rho_j}\right)^k \left[ P'_{k+1}(\cos \alpha_j) \mathbf{i}_{\rho_j} - P'_k(\cos \alpha_j) \mathbf{i}_r \right] \tag{8.66}$$

where $\mathbf{i}_r$ and $\mathbf{i}_{\rho_j}$ are unit vectors in the direction of $\mathbf{r}$ and $\boldsymbol{\rho}_j$, respectively. The prime on the Legendre polynomial indicates the derivative with respect to the argument $\cos \alpha_j$.

◊ **Problem 8–7**
Provide a detailed derivation of Eq. (8.66) by first showing that

$$\frac{\partial}{\partial \mathbf{r}} P_k(\cos \alpha) = P'_k(\cos \alpha) \frac{\partial}{\partial \mathbf{r}} \cos \alpha = \frac{1}{r} P'_k(\cos \alpha)(\mathbf{i}_\rho^T - \cos \alpha \, \mathbf{i}_r^T)$$

and then using the identity

$$k P_k(\cos \alpha) = P'_k(\cos \alpha) \cos \alpha - P'_{k-1}(\cos \alpha)$$

established in the subsection on Legendre polynomials later in this section.

**Problem 8–8**
For the expansion

$$(1 - 2x \cos \alpha + x^2)^{-\frac{1}{2}} = \sum_{k=0}^{\infty} a_k \cos k\alpha$$

the first two coefficients are

$$a_0 = \frac{2}{\pi} K(x) \quad \text{and} \quad a_1 = \frac{2}{\pi} \frac{K(x) - E(x)}{x}$$

where $K$ and $E$ are complete elliptic integrals of the first and second kinds, respectively. Derive Euler's recurrence formula

$$a_k = \frac{k-1}{k-\frac{1}{2}} \left( x + \frac{1}{x} \right) a_{k-1} - \frac{k-\frac{3}{2}}{k-\frac{1}{2}} a_{k-2}$$

for the coefficients.

### Jacobi's Expansion and Rodrigues' Formula

Jacobi invented a clever method of expanding the disturbing function, by using the Lagrange expansion theorem which also results in a derivation of Rodrigues' formula for Legendre polynomials. No motivation is provided but he might have reasoned as follows.

Assume that $(1 - 2\nu x + x^2)^{\frac{1}{2}}$ is the radical portion of the solution to the quadratic equation

$$ay^2 + by + c = 0$$

Then, of course, we must have

$$b^2 - 4ac = 1 - 2\nu x + x^2$$

Choose $b = 1$ and select the plus sign in the solution of the quadratic so that we will be dealing with the smaller of the two roots. Then

$$\frac{\partial y}{\partial \nu} = (1 - 2\nu x + x^2)^{-\frac{1}{2}}$$

results if we choose $a = -\frac{1}{2}x$. Hence, we must have $c = \frac{1}{2}(x - 2\nu)$.
The quadratic equation for $y$ can then be written

$$y = \nu + x \left[\tfrac{1}{2}(y^2 - 1)\right]$$

—exactly of the form of Eq. (5.9) with $x$ replacing $\alpha$ as the parameter. Therefore, we can apply Lagrange's expansion theorem and obtain

$$y = \nu + \sum_{k=1}^{\infty} \frac{x^k}{k!} \frac{d^{k-1}}{d\nu^{k-1}} [\tfrac{1}{2}(\nu^2 - 1)]^k$$

which represents that root of the quadratic equation which is equal to $\nu$ when $x = 0$. Differentiating one more time with respect to $\nu$ results in

$$(1 - 2\nu x + x^2)^{-\frac{1}{2}} = 1 + \sum_{k=1}^{\infty} \frac{x^k}{k!} \frac{d^k}{d\nu^k} [\tfrac{1}{2}(\nu^2 - 1)]^k$$

When this is compared with Eq. (8.63), we establish the identity

$$P_k(\nu) = \frac{1}{2^k k!} \frac{d^k}{d\nu^k} (\nu^2 - 1)^k \qquad (8.67)$$

known as *Rodrigues' formula*† for the Legendre polynomials.

---

† Olinde Rodrigues (1794–1851) published this basic formula in 1816. Other sets of orthogonal polynomials, such as the Tschebycheff polynomials, have similar formulas which are also called *Rodrigues' formulas* even though Rodrigues had absolutely nothing to do with them. For example,

$$T_n(x) = (-1)^n \frac{2^n n!}{(2n)!} \sqrt{1 - x^2} \frac{d^n}{dx^n} (1 - x^2)^{n - \frac{1}{2}}$$

is the one for the Tschebycheff polynomials.

## Legendre Polynomials

The function $\mathcal{L}(x, \nu)$, where

$$\mathcal{L}(x, \nu) = (1 - 2\nu x + x^2)^{-\frac{1}{2}} = \sum_{n=0}^{\infty} P_n(\nu) x^n \qquad (8.68)$$

is called the *generating function* for the Legendre polynomials $P_n(\nu)$ and can be used to derive some basic identities for these functions just as the generating function for Bessel functions was similarly used in Chapter 5. For example, from the identity $\mathcal{L}(x, -\nu) = \mathcal{L}(-x, \nu)$ we can deduce the property

$$P_n(-\nu) = (-1)^n P_n(\nu) \qquad (8.69)$$

It is also easy to verify that

$$P_n(1) = 1 \quad \text{and} \quad P_n(-1) = (-1)^n \qquad (8.70)$$

By differentiating the generating function, we develop the equation

$$(1 - 2\nu x + x^2) \frac{\partial \mathcal{L}}{\partial x} = (\nu - x) \mathcal{L}$$

Then, by substituting the power series for $\mathcal{L}$ and equating coefficients of $x^n$, we derive the recurrence formula

$$n P_n(\nu) - (2n - 1) \nu P_{n-1}(\nu) + (n - 1) P_{n-2}(\nu) = 0 \qquad (8.71)$$

Similarly, the differential equation

$$x \frac{\partial \mathcal{L}}{\partial x} = (\nu - x) \frac{\partial \mathcal{L}}{\partial \nu}$$

leads to the recurrence formula

$$\nu P_n'(\nu) - P_{n-1}'(\nu) = n P_n(\nu) \qquad (8.72)$$

By differentiating the recurrence formula (8.71) with respect to $\nu$ and using the formula Eq. (8.72) with $n$ replaced by $n - 1$, we can also show that

$$P_n'(\nu) - \nu P_{n-1}'(\nu) = n P_{n-1}(\nu) \qquad (8.73)$$

Other properties of the Legendre polynomials are developed in the problems to follow.

◇ **Problem 8–9**

Derive the following recurrence formula for the derivatives of the Legendre polynomials

$$(n - 1) P_n'(\nu) - (2n - 1) \nu P_{n-1}'(\nu) + n P_{n-2}'(\nu) = 0$$

## Problem 8–10
Use the identity
$$1 - 2x\cos\alpha + x^2 = (1 - xe^{i\alpha})(1 - xe^{-i\alpha})$$
to express $P_n(\cos\alpha)$ as the finite Fourier cosine series
$$P_n(\cos\alpha) = \sum_{k=0}^{n} \frac{1}{4^n} \binom{2k}{k}\binom{2n-2k}{n-k} \cos(n-2k)\alpha$$
With this result and Eq. (8.68) demonstrate that
$$|P_n(\cos\alpha)| \leq 1$$
HINT: All coefficients are positive and the maximum value occurs when $\alpha = 0$ for which $P_n(1) = 1$.

NOTE: Since $\cos n\alpha = T_n(\cos\alpha)$, we have also expressed $P_n(\cos\alpha)$ as a series of Tschebycheff polynomials.

## Problem 8–11
Calculate the $n^{\text{th}}$ derivative of the binomial expansion
$$(\nu^2 - 1)^n = \sum_{k=0}^{n} \frac{(-1)^k n!}{k!(n-k)!} \nu^{2n-2k}$$
to provide an alternate derivation of Rodrigues' formula.

## Problem 8–12
Using Rodrigues' formula and integration by parts, show that
$$\int_{-1}^{1} \nu^k P_n(\nu)\,d\nu = 0 \quad \text{for} \quad k = 0, 1, 2, \ldots, n-1$$
From this, deduce the orthogonality property of Legendre polynomials
$$\int_{-1}^{1} P_m(\nu) P_n(\nu)\,d\nu = 0$$
which holds when the integers $m$ and $n$ are unequal.

## Problem 8–13
By writing
$$\nu^2 - 1 = 2(\nu - 1)[1 + \tfrac{1}{2}(\nu - 1)]$$
and using the binomial theorem, obtain the expansion
$$(\nu^2 - 1)^n = \sum_{k=0}^{\infty} \frac{2^n n!}{2^k k!(n-k)!} (\nu - 1)^{n+k}$$
Then, calculate the $n^{\text{th}}$ derivative, to verify that
$$P_n(\nu) = F[-n, n+1; 1; \tfrac{1}{2}(1 - \nu)]$$
where $F$ denotes the hypergeometric function.

◊ **Problem 8–14**
Let $z = (\nu^2 - 1)^n$ so that

$$(1 - \nu^2)\frac{dz}{d\nu} + 2n\nu z = 0$$

Now differentiate $n + 1$ times, using Leibnitz's rule, to show that $y = P_n(\nu)$ is a solution of *Legendre's differential equation*

$$(1 - \nu^2)\frac{d^2y}{d\nu^2} - 2\nu\frac{dy}{d\nu} + n(n + 1)y = 0$$

NOTE: For Leibnitz's rule see Sect. 5.4.

## 8.5 The Sphere of Influence

When considering the disturbed motion of one body $m_2$ in the presence of two bodies $m_1$ and $m_3$, it is important for numerical computation to select the appropriate body to which the motion of $m_2$ is to be referred. More specifically, the question arises as to which of the following two descriptions of the motion is preferable and when a change of origin of coordinates should be made. The motion of $m_2$ relative to $m_1$ is described by

$$\frac{d^2\mathbf{r}}{dt^2} + \frac{G(m_1 + m_2)}{r^3}\mathbf{r} = -Gm_3\left(\frac{1}{d^3}\mathbf{d} + \frac{1}{\rho^3}\boldsymbol{\rho}\right)$$

while the motion of $m_2$ relative to $m_3$ is determined from

$$\frac{d^2\mathbf{d}}{dt^2} + \frac{G(m_2 + m_3)}{d^3}\mathbf{d} = -Gm_1\left(\frac{1}{r^3}\mathbf{r} - \frac{1}{\rho^3}\boldsymbol{\rho}\right)$$

According to Laplace, the advantage of either form depends on the ratio of the disturbing force to the corresponding central attraction. Whichever provides the smaller ratio is the one to be preferred. It happens that the surface boundary over which these two ratios are equal is almost spherical if $r$ is considerably smaller than $\rho$. For this reason, the boundary surface has been called the *sphere of influence*.†

For convenience in the analysis, we define four acceleration vectors $\mathbf{a}_{21}^p$, $\mathbf{a}_{21}^d$, $\mathbf{a}_{23}^p$, $\mathbf{a}_{23}^d$ with superscript labels distinguishing primary and disturbing acceleration components while subscript labels refer to the de-

---

† The concept of the *sphere of influence* originated with Pierre-Simon de Laplace when he was studying the motion of a comet which was about to pass near the planet Jupiter. In his orbit determination calculations he searched for a logical criterion to choose the origin of his coordinate system during various phases of the motion.

scriptions of motion—$m_2$ with respect to $m_1$ or $m_3$. Then we have

$$\mathbf{a}_{21}^p = -\frac{G(m_1 + m_2)}{r^2} \mathbf{i}_r$$

$$\mathbf{a}_{21}^d = \frac{Gm_3}{\rho^2} \sum_{k=1}^{\infty} x^k [P'_{k+1}(\nu) \mathbf{i}_\rho - P'_k(\nu) \mathbf{i}_r]$$

$$\mathbf{a}_{23}^p = -\frac{G(m_2 + m_3)}{d^2} \mathbf{i}_d$$

$$\mathbf{a}_{23}^d = \frac{Gm_1}{r^2}(x^2 \mathbf{i}_\rho - \mathbf{i}_r)$$

using the notation of Sect. 8.4.

The ratio of disturbing to primary acceleration for $m_2$ relative to $m_3$ is then exactly

$$\frac{a_{23}^d}{a_{23}^p} = \frac{m_1}{m_2 + m_3} \frac{1}{x^2}(1 - 2\nu x + x^2)\sqrt{1 - 2\nu x^2 + x^4}$$

while the corresponding ratio ($m_2$ relative to $m_1$) is

$$\frac{a_{21}^d}{a_{21}^p} = \frac{m_3}{m_1 + m_2} x^3 \sqrt{1 + 3\nu^2} + O(x^4)$$

if terms of order $x^4$ and higher are ignored.

We now equate these two ratios and assume that $x$ is small compared with unity. In this way, we obtain

$$x = \left[\frac{m_1(m_1 + m_2)}{m_3(m_2 + m_3)}\right]^{\frac{1}{5}} (1 + 3\nu^2)^{-\frac{1}{10}}$$

Also we note that $(1 + 3\nu^2)^{\frac{1}{10}}$ is at most equal to 1.15 and that, in many cases of interest, $m_2$ may be neglected in comparison with $m_1$ and $m_3$. Thus, we have approximately

$$\frac{r}{\rho} = \left(\frac{m_1}{m_3}\right)^{\frac{2}{5}} \tag{8.74}$$

as a valid result provided $m_1$ is much smaller than $m_3$.

Equation (8.74) defines a sphere about $m_1$ on the boundary of which the ratio of disturbing to primary accelerations is the same for either of the two descriptions of the relative motion of $m_2$. Inside this sphere, called *the sphere of influence of $m_1$ with respect to $m_3$*, it is appropriate to determine the motion of $m_2$ relative to $m_1$ as the origin, while outside we should use $m_3$ as the origin of coordinates.

A tabulation of the radii of the spheres of influence for the various planets of the solar system is given in the accompanying table. Here, of course, $m_1$ is the mass of the planet and $m_3$ is the mass of the sun.

Sect. 8.5]                The Sphere of Influence                397

Spheres of Influence of the Planets

| Planet  | Mean distance a.u. | Mass ratio planet/sun | Radius of sphere miles |
|---------|--------------------|-----------------------|------------------------|
| Mercury | 0.387099           | 0.00000017            | 70,000                 |
| Venus   | 0.723322           | 0.00000245            | 383,000                |
| Earth   | 1.000000           | 0.000002999           | 574,000                |
| Mars    | 1.523691           | 0.00000032            | 357,000                |
| Jupiter | 5.202803           | 0.000954786           | 29,937,000             |
| Saturn  | 9.538843           | 0.000285584           | 33,869,000             |
| Uranus  | 19.181951          | 0.000043727           | 32,152,000             |
| Neptune | 30.057779          | 0.000051776           | 53,904,000             |

◊ **Problem 8–15**
For the disturbed motion of $m_2$ with respect to $m_1$ in the presence of $m_3$

$$\frac{r}{\rho} = \left(\frac{m_1 + m_2}{m_3}\right)^{\frac{1}{3}} (1 + 3\cos^2\alpha)^{-\frac{1}{6}}$$

defines a surface about $m_1$ on the boundary of which the disturbing and primary accelerations are equal—assuming that $m_1 + m_2$ is much smaller than $m_3$.

**Problem 8–16**
When the three bodies $m_1$, $m_2$, and $m_3$ are, respectively, the moon, a spacecraft, and the earth, the mass ratio of the moon and earth is not small enough for Eq. (8.74) to be a good approximation of the boundary surface. Derive a better one by the following steps:
(a) Develop the following approximations for the two ratios:

$$\frac{a_{23}^d}{a_{23}^p} = \frac{m_1}{m_2 + m_3}\frac{1}{x^2}[1 - 2\nu x + O(x^2)]$$

$$\frac{a_{21}^d}{a_{21}^p} = \frac{m_3}{m_1 + m_2}x^3\sqrt{1 + 3\nu^2}\left[1 + \frac{6\nu^3}{1 + 3\nu^2}x + O(x^2)\right]$$

(b) By equating the two ratios, obtain

$$\frac{m_3(m_2 + m_3)}{m_1(m_1 + m_2)}x^5\sqrt{1 + 3\nu^2} = 1 - 2\nu\left(1 + \frac{3\nu^2}{1 + 3\nu^2}\right)x + O(x^2)$$

(c) By extracting the fifth root of both sides of the last equation and solving for $x$, show that

$$\frac{r}{\rho} = \left[\left(\frac{m_1(m_1 + m_2)}{m_3(m_2 + m_3)}\right)^{-\frac{1}{5}}(1 + 3\cos^2\alpha)^{\frac{1}{10}} + \frac{2}{5}\cos\alpha\left(\frac{1 + 6\cos^2\alpha}{1 + 3\cos^2\alpha}\right)\right]^{-1}$$

is the desired equation for the boundary surface.

(d) Considering this surface to be centered in the moon, the surface radius varies from about 32,400 miles in the earth direction to 41,000 miles at 90 degrees from the earth-moon line, to about 40,000 miles in the direction away from the earth. Further, the values of the acceleration ratios at these points are approximately 0.5, 0.4, and 0.6, respectively.

(e) For the sake of comparison, the radii of the corresponding surface about the earth for the sun-earth system are 499,000, 575,000, and 502,000 miles and the acceleration ratios are 0.1, 0.08, and 0.1, respectively.

*James S. Miller* 1962

## 8.6 The Canonical Coordinates of Jacobi

An alternate and symmetric form for the equations of relative motion of $n$ bodies is also possible using what are sometimes called *Jacobi coordinates*.† The $n$ mass particles $m_1$, $m_2$, ..., $m_n$ are ordered in any convenient sequence. Then the position of each body in the sequence is measured with respect to the center of mass of all bodies preceding it.

Specifically, define the position vectors $\boldsymbol{\rho}_1$, $\boldsymbol{\rho}_2$, ..., $\boldsymbol{\rho}_n$ as

$$\boldsymbol{\rho}_1 = \mathbf{r}_1$$

$$\boldsymbol{\rho}_2 = \mathbf{r}_2 - \mathbf{r}_1$$

$$\boldsymbol{\rho}_3 = \mathbf{r}_3 - \frac{m_1 \mathbf{r}_1 + m_2 \mathbf{r}_2}{m_1 + m_2}$$

...

$$\boldsymbol{\rho}_n = \mathbf{r}_n - \frac{m_1 \mathbf{r}_1 + m_2 \mathbf{r}_2 + \cdots + m_{n-1} \mathbf{r}_{n-1}}{m_1 + m_2 + \cdots + m_{n-1}}$$

or in a more compact notation with $k = 2, 3, \ldots, n$, as

$$\boldsymbol{\rho}_k = \mathbf{r}_k - \frac{1}{\sigma_{k-1}} \sum_{j=1}^{k-1} m_j \mathbf{r}_j \tag{8.75}$$

$$\sigma_k = m_1 + m_2 + \cdots + m_k \tag{8.76}$$

The problem is then to rewrite the equations of motion

$$m_j \frac{d^2 \mathbf{r}_j^T}{dt^2} = \frac{\partial U}{\partial \mathbf{r}_j} \quad \text{for} \quad j = 1, 2, \ldots, n \tag{8.77}$$

in terms of the vectors $\boldsymbol{\rho}_j$. [Refer to Sect. 2.4.]

---

† In the winter semester of 1842–43 Jacobi gave a course of lectures at the University of Königsberg on Dynamics which included some very important investigations on the integration of the differential equations which arise in Mechanics. His symmetric form for the equations of motion was published the following year in a memoir entitled *Sur l'élimination des nœuds dans le problème des trois corps*. Henri Poincaré (1854–1912) made general use of this system in his research in the problem of three bodies which appeared in his greatest work *Les Méthodes Nouvelles de la Mécanique Céleste* published in three volumes during the period 1892–1899.

Sect. 8.6]     The Canonical Coordinates of Jacobi     399

Substituting from Eqs. (8.75) into (8.77), produces

$$m_k \frac{d^2 \boldsymbol{\rho}_k^\mathrm{T}}{dt^2} = \frac{\partial U}{\partial \mathbf{r}_k} - \frac{m_k}{\sigma_{k-1}} \sum_{j=1}^{k-1} \frac{\partial U}{\partial \mathbf{r}_j} \tag{8.78}$$

and, using the property

$$\sum_{j=1}^{n} \frac{\partial U}{\partial \mathbf{r}_j} = \mathbf{0}^\mathrm{T}$$

derived in Sect. 2.4, we write the equations of motion as

$$m_k \frac{d^2 \boldsymbol{\rho}_k^\mathrm{T}}{dt^2} = \frac{\sigma_k}{\sigma_{k-1}} \frac{\partial U}{\partial \mathbf{r}_k} + \frac{m_k}{\sigma_{k-1}} \sum_{j=k+1}^{n-1} \frac{\partial U}{\partial \mathbf{r}_j} + \frac{m_k}{\sigma_{k-1}} \frac{\partial U}{\partial \mathbf{r}_n}$$

for $k = 2, 3, \ldots, n$.

To transform the gradients of the force function $U$, we note that variations in $\mathbf{r}_k$, holding all the other $\mathbf{r}$'s constant, produce changes only in $\boldsymbol{\rho}_k, \boldsymbol{\rho}_{k+1}, \ldots, \boldsymbol{\rho}_n$ and not in $\boldsymbol{\rho}_1, \boldsymbol{\rho}_2, \ldots, \boldsymbol{\rho}_{k-1}$. Hence,

$$\frac{\partial U}{\partial \mathbf{r}_k} = \frac{\partial U}{\partial \boldsymbol{\rho}_k} \frac{\partial \boldsymbol{\rho}_k}{\partial \mathbf{r}_k} + \frac{\partial U}{\partial \boldsymbol{\rho}_{k+1}} \frac{\partial \boldsymbol{\rho}_{k+1}}{\partial \mathbf{r}_k} + \cdots + \frac{\partial U}{\partial \boldsymbol{\rho}_n} \frac{\partial \boldsymbol{\rho}_n}{\partial \mathbf{r}_k}$$

$$= \frac{\partial U}{\partial \boldsymbol{\rho}_k} - \frac{m_k}{\sigma_k} \frac{\partial U}{\partial \boldsymbol{\rho}_{k+1}} - \frac{m_k}{\sigma_{k+1}} \frac{\partial U}{\partial \boldsymbol{\rho}_{k+2}} - \cdots - \frac{m_k}{\sigma_{n-1}} \frac{\partial U}{\partial \boldsymbol{\rho}_n}$$

or

$$\frac{\partial U}{\partial \mathbf{r}_k} = \frac{\partial U}{\partial \boldsymbol{\rho}_k} - m_k \sum_{j=k+1}^{n} \frac{1}{\sigma_{j-1}} \frac{\partial U}{\partial \boldsymbol{\rho}_j} \qquad \text{for} \quad k = 1, 2, \ldots, n-1$$

$$\frac{\partial U}{\partial \mathbf{r}_n} = \frac{\partial U}{\partial \boldsymbol{\rho}_n} \tag{8.79}$$

Substitute these in the equations of motion to obtain

$$\sigma_{k-1} \frac{d^2 \boldsymbol{\rho}_k^\mathrm{T}}{dt^2} = \frac{\sigma_k}{m_k} \frac{\partial U}{\partial \boldsymbol{\rho}_k} + \sum_{i=k+2}^{n} \left(1 - \frac{\sigma_k}{\sigma_{i-1}}\right) \frac{\partial U}{\partial \boldsymbol{\rho}_i} - \sum_{j=k+1}^{n-1} m_j \sum_{i=j+1}^{n} \frac{1}{\sigma_{i-1}} \frac{\partial U}{\partial \boldsymbol{\rho}_i}$$

after some obvious simplifications. Then, reverse the order of the double summation

$$\sum_{j=k+1}^{n-1} m_j \sum_{i=j+1}^{n} \frac{1}{\sigma_{i-1}} \frac{\partial U}{\partial \boldsymbol{\rho}_i} = \sum_{i=k+2}^{n} \frac{1}{\sigma_{i-1}} \frac{\partial U}{\partial \boldsymbol{\rho}_i} \sum_{j=k+1}^{i-1} m_j$$

and note that

$$\sum_{j=k+1}^{i-1} m_j = \sigma_{i-1} - \sigma_k$$

As a result, only the first term on the right-hand side remains.

Therefore, the equations of relative motion in the Jacobian coordinates are simply

$$m_k \frac{d^2 \boldsymbol{\rho}_k^T}{dt^2} = \frac{\sigma_k}{\sigma_{k-1}} \frac{\partial U}{\partial \boldsymbol{\rho}_k} \quad \text{for} \quad k = 2, 3, \ldots, n \quad (8.80)$$

which should be compared with the equivalent ones in Eqs. (8.58).

### Problem 8-17

In the lunar problem, the main concern is with the three bodies earth, moon, and sun. Let their respective masses be $m_E$, $m_M$, $m_S$ and let the point $B$ denote the center of gravity or *barycenter* of the earth and moon. Define the vector $\mathbf{r}_{EM}$ as the position vector of the moon relative to the earth with similar definitions for $\mathbf{r}_{ES}$, $\mathbf{r}_{MS}$, and $\mathbf{r}_{BS}$.

(a) Derive the equations of motion of the moon relative to the earth and the sun relative to the barycenter in the form

$$\frac{d^2 \mathbf{r}_{EM}}{dt^2} + \frac{G(m_E + m_M)}{r_{EM}^3} \mathbf{r}_{EM} = \frac{G m_S (m_E + m_M)}{m_E m_M} \left[ \frac{\partial F}{\partial \mathbf{r}_{EM}} \right]^T$$

$$\frac{d^2 \mathbf{r}_{BS}}{dt^2} = \frac{G(m_E + m_M + m_S)}{m_E + m_M} \left[ \frac{\partial F}{\partial \mathbf{r}_{BS}} \right]^T$$

where

$$F = \frac{m_M}{r_{MS}} + \frac{m_E}{r_{ES}}$$

(b) Obtain the expansion

$$F = \frac{m_E + m_S}{r_{BS}} + \frac{m_E m_M}{m_E + m_M} \frac{1}{r_{BS}} \left[ \left( \frac{r_{EM}}{r_{BS}} \right)^2 P_2(\cos \alpha) \right.$$

$$+ \frac{m_E - m_M}{m_E + m_M} \left( \frac{r_{EM}}{r_{BS}} \right)^3 P_3(\cos \alpha)$$

$$+ \frac{m_E^2 - m_E m_M + m_M^2}{(m_E + m_M)^2} \left( \frac{r_{EM}}{r_{BS}} \right)^4 P_4(\cos \alpha) + \cdots \right]$$

where

$$\cos \alpha = \frac{\mathbf{r}_{EM} \cdot \mathbf{r}_{BS}}{r_{EM} r_{BS}}$$

(c) The ratio of the second term to the first in the expansion of $F$ is approximately $8 \times 10^{-8}$. Hence, the motion of the sun relative to the earth-moon barycenter is essentially elliptic; i.e.,

$$\frac{d^2 \mathbf{r}_{BS}}{dt^2} + \frac{G(m_E + m_M + m_S)}{r_{BS}^3} \mathbf{r}_{BS} = 0$$

### ◊ Problem 8-18

In Jacobi coordinates, the energy integral is

$$\frac{1}{2} m_1 \frac{d\boldsymbol{\rho}_1}{dt} \cdot \frac{d\boldsymbol{\rho}_1}{dt} + \frac{1}{2} \sum_{k=2}^{n} \frac{m_k \sigma_{k-1}}{\sigma_k} \frac{d\boldsymbol{\rho}_k}{dt} \cdot \frac{d\boldsymbol{\rho}_k}{dt} - U = \text{constant}$$

## 8.7 Potential of Distributed Mass

Jacobi's canonical equations provide a simple heuristic means of developing the equations of motion of a mass particle in the gravitational field of a continuous distribution of mass. From Eq. (8.80), the motion of the $n^{\text{th}}$ particle of an $n$-body system is described by

$$m_n \frac{d^2 \mathbf{r}^T}{dt^2} = \frac{m_n + \sigma_{n-1}}{\sigma_{n-1}} \frac{\partial U}{\partial \mathbf{r}}$$

where

$$\mathbf{r} \equiv \boldsymbol{\rho}_n = \mathbf{r}_n - \mathbf{r}_{cm}$$

and $\mathbf{r}_{cm}$ is the center of mass of the first $n-1$ bodies. The force function, from Eq. (2.45), is

$$U = \frac{G}{2} \sum_{i=1}^{n} \sum_{j=1}^{n}{}' \frac{m_i m_j}{r_{ij}}$$

and since we are calculating the gradient of $U$ with respect to $\mathbf{r}$, only the terms for which $i$ or $j$ are equal to $n$ will contribute to the acceleration. As a consequence,

$$m_n \frac{d^2 \mathbf{r}^T}{dt^2} = \frac{m_n + \sigma_{n-1}}{\sigma_{n-1}} \frac{\partial}{\partial \mathbf{r}} \sum_{j=1}^{n-1} \frac{G m_j m_n}{r_{jn}}$$

Suppose now that $m_n$ is small compared to $\sigma_{n-1}$ which is the total mass of the other bodies. Then the equation of motion of $m_n$ is simply

$$\frac{d^2 \mathbf{r}^T}{dt^2} = \frac{\partial}{\partial \mathbf{r}} \sum_{j=1}^{n-1} \frac{G m_j}{r_{jn}} \qquad (8.81)$$

Finally, imagine that $n$ is a very large number and that the $n-1$ masses are densely clustered in some region of space apart from the $n^{\text{th}}$ mass particle. Then the summation is well approximated by an integration over the mass volume.

Therefore, if the mass particle is at the point $P(x, y, z)$ referenced to the center of mass of the aggregate, then

$$\frac{d^2 \mathbf{r}^T}{dt^2} = \frac{\partial}{\partial \mathbf{r}} \iiint \frac{G\, dm}{[(x-\xi)^2 + (y-\eta)^2 + (z-\varsigma)^2]^{\frac{1}{2}}} \qquad (8.82)$$

where $\xi$, $\eta$, $\varsigma$ are the coordinates of the mass element $dm$. The integral is the gravitational potential of the mass distribution at the point $P$ and is denoted by $V$.

An alternate expression for $V$ is more useful; namely,

$$V = \iiint \frac{G\, dm}{(r^2 + \rho^2 - 2r\rho \cos \gamma)^{\frac{1}{2}}} \qquad (8.83)$$

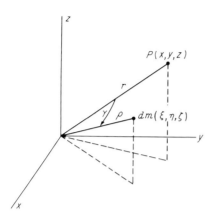

Fig. 8.3: Geometry of the potential function.

where $r$ and $\rho$ represent the radial distances of $P$ and $dm$ from the center of mass with $\gamma$ as the angle between the two radii illustrated in Fig. 8.3.

In calculating the potential function $V$ at point $P$ we assume that $P$ is at a greater radial distance from the origin of coordinates than is any part of the distributed mass. Furthermore, we will represent the reciprocal of the distance between $P$ and $dm$ as a series of Legendre polynomials and integrate this series term by term. Therefore,

$$V = \frac{G}{r} \sum_{k=0}^{\infty} \iiint \left(\frac{\rho}{r}\right)^k P_k(\cos\gamma)\, dm \qquad (8.84)$$

where the integration is taken over the mass volume. The components of force per unit mass at $P(x, y, z)$ are then obtained as the partial derivatives of $V$ with respect to the coordinates $x$, $y$, $z$.

### MacCullagh's Approximation

The first three terms in the power series expansion of the gravitational potential are directly related to the inertia properties of the attracting body. Indeed, if the distance between the point $P$ and the center of mass is large compared with the dimensions of the body, these terms

$$V = \frac{Gm}{r} + \frac{G}{r^2} \iiint \rho \cos\gamma\, dm + \frac{G}{2r^3} \iiint \rho^2 (3\cos^2\gamma - 1)\, dm$$

can serve as an adequate approximation to the true potential.

In the first term, $m$ is the total mass of the body. Because the origin of coordinates coincides with the center of mass, the second term will vanish since it is proportional to the first moment of mass with respect to a plane normal to the line joining the point $P$ and the center of mass. The third term involves the various inertia moments of the body.

## Potential of Distributed Mass

Specifically, let $A$, $B$, $C$ be the moments of inertia about axes which are assumed to coincide with the rectangular coordinate axes. Then

$$A = \iiint (\eta^2 + \varsigma^2)\,dm \quad B = \iiint (\xi^2 + \varsigma^2)\,dm \quad C = \iiint (\xi^2 + \eta^2)\,dm$$

so that

$$A + B + C = 2\iiint \rho^2\,dm$$

Further, define $I$ as the moment of inertia of the body about the line connecting the center of mass and the point $P$ at which the potential is computed. Therefore,

$$I = \iiint \rho^2 \sin^2 \gamma\,dm$$

Thus, we find that the relevant terms in the expansion of the potential are simply

$$V = \frac{Gm}{r} + \frac{G}{2r^3}(A + B + C - 3I) \tag{8.85}$$

This form of the gravitation potential is *MacCullagh's approximation*.†

The attracting force acting on a unit of mass located at $P$ is determined as the gradient of $V$. However, since $I$ depends on the location of $P$, it is necessary first to obtain an appropriate analytic form. For this purpose, write

$$I = \iiint \rho^2(1 - \cos^2 \gamma)\,dm = \iiint \left[\rho^2 - \left(\frac{\xi x + \eta y + \varsigma z}{r}\right)^2\right] dm$$

Then, if the coordinate axes coincide with the principal axes of inertia, we have

$$I = \frac{1}{r^2}(Ax^2 + By^2 + Cz^2) \tag{8.86}$$

since the products of inertia will then be zero.

The gradient is now readily calculated by employing the two vector identities

$$\frac{\partial r}{\partial \mathbf{r}} = \mathbf{i}_r^T \quad \text{and} \quad \frac{\partial}{\partial \mathbf{r}}\left(\frac{x}{r}\right) = \frac{1}{r}\left(\mathbf{i}_x^T - \frac{x}{r}\mathbf{i}_r^T\right)$$

The final result is expressed in the form

$$\left[\frac{\partial V}{\partial \mathbf{r}}\right]^T = -\frac{Gm}{r^2}\mathbf{i}_r - \frac{3G}{2r^4}(A + B + C - 5I)\mathbf{i}_r$$
$$- \frac{3G}{r^5}(Ax\,\mathbf{i}_x + By\,\mathbf{i}_y + Cz\,\mathbf{i}_z) \tag{8.87}$$

---

† James MacCullagh (1809–1847) was a professor of mathematics and natural philosophy at Trinity College, Dublin, Ireland. His many contributions during his short life were recognized by the Royal Irish Academy with the award of its first gold medal.

The shape of the moon is often approximated by the triaxial ellipsoid and the experimentally determined values of the moments of inertia are

$$\frac{C-A}{C} = 0.0006313$$

$$\frac{B-A}{C} = 0.0002278$$

$$C = 0.392 \text{ (moon mass)} \times \text{(moon radius)}^2$$

◇ **Problem 8–19**
Because of the particular form in which the moment of inertia data for the moon is presented, a more convenient expression is generally used. Show that

$$V = \frac{Gm}{r} + \frac{G}{2r^3}\left\{(B-A)[1 - 3(\mathbf{i}_r \cdot \mathbf{i}_y)^2] + (C-A)[1 - 3(\mathbf{i}_r \cdot \mathbf{i}_z)^2]\right\}$$

and calculate the gradient $\partial V/\partial \mathbf{r}$.

### Expansion as a Series of Legendre Functions

The higher-order terms in the series expansion of the potential function cannot be interpreted in a simple fashion. In order to further the analysis, it is convenient to utilize spherical coordinates with $r$, $\phi$, $\theta$ specifying the location of $P$ and $\rho$, $\beta$, $\lambda$ similarly used for the mass element $dm$ as illustrated in Fig. 8.4. Let $D(\rho, \beta, \lambda)$ denote the mass density so that the

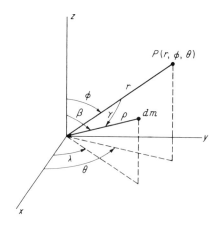

**Fig. 8.4:** Spherical coordinates.

element of mass $dm$ can be expressed as

$$dm = D(\rho, \beta, \lambda)\rho^2 \sin\beta \, d\rho \, d\beta \, d\lambda$$

Finally, the angle $\gamma$ is related to the spherical coordinates through the cosine law of spherical trigonometry

$$\cos\gamma = \cos\phi\cos\beta + \sin\phi\sin\beta\cos(\theta - \lambda) \quad (8.88)$$

## Sect. 8.7]   Potential of Distributed Mass

The further development of the potential function is expedited by means of the so-called *addition theorem for Legendre polynomials*† which involves the associated Legendre functions as well as the Legendre polynomials. The former are defined as

$$P_k^j(\nu) = (1 - \nu^2)^{\frac{1}{2}j} \frac{d^j}{d\nu^j} P_k(\nu) \tag{8.89}$$

and called the *associated Legendre functions of the first kind of degree $k$ and order $j$*. For example,

$$P_1^1(\cos\phi) = \sin\phi \qquad P_2^1(\cos\phi) = 3\sin\phi\cos\phi \qquad P_2^2(\cos\phi) = 3\sin^2\phi$$

If the cosine law (8.88) is rewritten as

$$P_1(\cos\gamma) = P_1(\cos\phi)P_1(\cos\beta) + P_1^1(\cos\phi)P_1^1(\cos\beta)\cos(\theta - \lambda)$$

the result is precisely the addition theorem for Legendre polynomials of the first degree. For the second degree, recall that

$$P_2(\cos\gamma) = \tfrac{1}{2}(3\cos^2\gamma - 1)$$

into which we substitute for $\cos\gamma$ from Eq. (8.88) and proceed to derive

$$P_2(\cos\gamma) = P_2(\cos\phi)P_2(\cos\beta) + \tfrac{1}{3}P_2^1(\cos\phi)P_2^1(\cos\beta)\cos(\theta - \lambda)$$
$$+ \tfrac{1}{12}P_2^2(\cos\phi)P_2^2(\cos\beta)\cos 2(\theta - \lambda)$$

In this way, the plausibility of the general result

$$P_k(\cos\gamma) = P_k(\cos\phi)P_k(\cos\beta)$$
$$+ 2\sum_{j=1}^{k} \frac{(k-j)!}{(k+j)!} P_k^j(\cos\phi)P_k^j(\cos\beta)\cos j(\theta - \lambda) \tag{8.90}$$

is established.

With the use of the addition theorem, the expansion of the potential function $V$ given in Eq. (8.84) may be written

$$V(r,\phi,\theta) = \frac{Gm}{r} + \sum_{k=1}^{\infty} \frac{1}{r^{k+1}} \Big\{ A_k P_k(\cos\phi)$$
$$+ \sum_{j=1}^{k} \big[ B_k^j P_k^j(\cos\phi)\cos j\theta + C_k^j P_k^j(\cos\phi)\sin j\theta \big] \Big\} \tag{8.91}$$

---

† The reader is acquainted with "addition theorems" for the more familiar functions. For example, we have

$$\sin(A+B) = \sin A\cos B + \cos A\sin B \qquad \text{and} \qquad \exp(A+B) = \exp(A)\exp(B)$$

where the constant coefficients are obtained from

$$A_k = G \iiint \rho^{k+2} D(\rho, \beta, \lambda) P_k(\cos \beta) \sin \beta \, d\rho \, d\beta \, d\lambda$$

$$B_k^j = 2G \frac{(k-j)!}{(k+j)!} \iiint \rho^{k+2} D(\rho, \beta, \lambda) P_k^j(\cos \beta) \cos j\lambda \sin \beta \, d\rho \, d\beta \, d\lambda$$

$$C_k^j = 2G \frac{(k-j)!}{(k+j)!} \iiint \rho^{k+2} D(\rho, \beta, \lambda) P_k^j(\cos \beta) \sin j\lambda \sin \beta \, d\rho \, d\beta \, d\lambda$$

As before, the quantity $m$ in the first term of Eq. (8.91) is simply the total mass of the attracting matter, i.e.,

$$m = \iiint D(\rho, \beta, \lambda) \rho^2 \sin \beta \, d\rho \, d\beta \, d\lambda$$

A tremendous simplification occurs if the mass distribution is symmetric about the $z$ axis. In this important case the density is a function only of $\rho$ and $\beta$, so that the integration with respect to $\lambda$ may be performed independently. Since

$$\int_0^{2\pi} \sin j\lambda \, d\lambda = \int_0^{2\pi} \cos j\lambda \, d\lambda = 0$$

for $j = 1, 2, \ldots$, the coefficients $B_k^j$ and $C_k^j$ vanish identically.

With both axial symmetry and the fact that the origin of coordinates coincides with the center of mass, the constant $A_1$ is identically zero. For a proof observe that

$$A_1 = G \iiint \rho \cos \beta \, dm$$

and thus is proportional to the first moment of mass $m$ with respect to the $x, y$ plane.

Finally, if the mass is distributed in homogeneous concentric layers, then $A_k$ vanishes identically for all $k$. For this case the density is a function of $\rho$ only and

$$A_k = 2\pi G \int_0^R \rho^{k+2} D(\rho) \, d\rho \int_0^\pi P_k(\cos \beta) \sin \beta \, d\beta$$

where $R$ is the radius of the spherical-shaped mass. The second integral is zero as can be deduced from the results of Prob. 8–12. Only the first term remains in Eq. (8.91) for the potential, and we conclude that the net effect at point $P$ is the same as if all the mass were acting from a point at the center of the body.

Sect. 8.7]     Potential of Distributed Mass     407

For many practical applications the assumption of axial symmetry for a body in the solar system is reasonable. With $r_{eq}$ denoting the equatorial radius of the body, the conventional form of the external potential is then

$$V(r,\phi) = \frac{Gm}{r}\left[1 - \sum_{k=2}^{\infty} J_k \left(\frac{r_{eq}}{r}\right)^k P_k(\cos\phi)\right] \quad (8.92)$$

The coefficients $J_k$ are readily identified with $A_k$; however, an explicit numerical determination by integration is clearly impossible. Their values must instead be empirically obtained using suitable experiments such as observations of satellite orbits. The odd-order terms are antisymmetric about the equatorial plane and will be zero for a symmetrically shaped body.

Values of these coefficients for the earth are

$$J_2 = \phantom{-}0.00108263$$
$$J_3 = -0.00000254$$
$$J_4 = -0.00000161$$

◇ **Problem 8–20**

The force per unit mass at a point external to an axially symmetric mass distribution is given by

$$-\frac{Gm}{r^2}\left\{\mathbf{i}_r - \sum_{k=2}^{\infty} J_k \left(\frac{r_{eq}}{r}\right)^k \left[P'_{k+1}(\cos\phi)\,\mathbf{i}_r - P'_k(\cos\phi)\,\mathbf{i}_z\right]\right\}$$

(a) The radial component of this force is

$$-\frac{Gm}{r^2}\left[1 - \sum_{k=2}^{\infty} J_k \left(\frac{r_{eq}}{r}\right)^k (k+1)P_k(\cos\phi)\right]$$

(b) The circumferential component (perpendicular to the radius and in the plane containing the axis of symmetry) is

$$-\frac{Gm}{r^2}\sum_{k=2}^{\infty} J_k \left(\frac{r_{eq}}{r}\right)^k P_k^1(\cos\phi)$$

(c) The axial component is

$$-\frac{Gm}{r^2}\left[\cos\phi - \sum_{k=2}^{\infty} J_k \left(\frac{r_{eq}}{r}\right)^k (k+1)P_{k+1}(\cos\phi)\right]$$

NOTE: Deriving an expression for the axial component of force was the original problem which led Adrien-Marie Legendre to the discovery of his polynomials. It appeared in his paper *Recherches sur l'attraction des sphéroïds* written in 1782 but not published until 1785. Legendre was a professor at the École Militaire and until his death in 1833 at the age of 81 he worked diligently in many areas of mathematics. His name lives on in a great number of theorems.

## Problem 8–21

An oblate spheroid is an axially symmetric body whose meridian section is an ellipse. Show that the gravitational potential of a solid homogeneous oblate spheroid at a point $P(x, y, z)$, remote compared to the dimensions of the body, may be calculated as a power series in $e$ of the form

$$V = \frac{Gm}{r}\left[1 + \frac{b^2}{10r^4}(x^2 + y^2 - 2z^2)e^2 + \cdots\right]$$

where

$$m = \tfrac{4}{3}\pi D a^2 b$$

is the mass of the spheroid, $D$ is the constant density, and $a$, $b$, $e$ are the semimajor axis, semiminor axis, and eccentricity of the elliptical cross section.

HINT: Use the polar equation of the ellipse with the origin of coordinates at the center developed in Prob. 4.8.

## Problem 8–22

The associated Legendre function of the first kind of degree $n$ and order $m$ is defined by

$$P_n^m(x) = (1 - x^2)^{\frac{1}{2}m}\frac{d^m}{dx^m}P_n(x)$$

where $P_n(x)$ is the $n^{\text{th}}$ order Legendre polynomial. Using the result of Prob. 8–13, which expresses $P_n(x)$ as a hypergeometric function, show that

$$P_n^m(\cos\phi) = \frac{(n+m)!}{(n-m)!}\tan^m \tfrac{1}{2}\phi\, F(-n, n+1; m+1; \sin^2 \tfrac{1}{2}\phi)$$

## 8.8 Spacecraft Motion Under Continuous Thrust

One of the possible disturbing accelerations affecting a spacecraft is the thrust acceleration produced by the vehicle's engines. Trajectory determination under these circumstances, as in almost all problems of disturbed motion, generally requires the application of special numerical techniques. However, there are several examples of some practical interest for which considerable analysis can be made. Specifically, if the thrust acceleration is constant in magnitude and directed radially, tangentially, or circumferentially, then it is possible to obtain, at least partially, some quite interesting mathematical results.[†]

Constant thrust acceleration is, of course, somewhat of a fiction. Nevertheless, the scarcity of mathematically tractable examples in this field

---

† The key results of this section are from two papers by Hsue-shen Tsien and David J. Benney, now professor of mathematics at MIT. The first, by Tsien, is from the *Journal of the American Rocket Society*, vol. 23, July–August, 1953, pp. 233–236 titled "Take-off from Satellite Orbit" and the second, by Benney, from *Jet Propulsion*, vol. 28, March, 1958, pp. 167–169 titled "Escape from a Circular Orbit Using Tangential Thrust."

### Constant Radial Acceleration

A vehicle is initially in a circular orbit of radius $r_0$ and at time $t = t_0$ a constant radial thrust acceleration is applied until the vehicle attains parabolic velocity—frequently called *escape velocity*. If $a_{Tr}$ is the rocket thrust acceleration per unit mass in the radial direction, then the equations of motion in polar coordinates are

$$\frac{d^2 r}{dt^2} - r\left(\frac{d\theta}{dt}\right)^2 + \frac{\mu}{r^2} = a_{Tr}$$

$$\frac{d}{dt}\left(r^2 \frac{d\theta}{dt}\right) = 0$$

The second equation is integrated at once to give

$$r^2 \frac{d\theta}{dt} = \sqrt{\mu r_0} \tag{8.93}$$

with the constant determined from the initial value of the circumferential velocity which is simply $\sqrt{\mu/r_0}$. An integral of the first equation is obtained by substituting for $d\theta/dt$ and observing that

$$\frac{1}{2}\frac{d}{dr}\left(\frac{dr}{dt}\right)^2 = \frac{d^2 r}{dt^2}$$

Therefore, if the thrust acceleration is constant, the radial velocity as a function of the radius is found to be

$$\left(\frac{dr}{dt}\right)^2 = (r - r_0)\left[2a_{Tr} - \frac{\mu}{r_0 r^2}(r - r_0)\right] \tag{8.94}$$

The vehicle will reach escape velocity when

$$\frac{1}{2}\left[\left(\frac{dr}{dt}\right)^2 + \left(r\frac{d\theta}{dt}\right)^2\right] - \frac{\mu}{r} = 0$$

which occurs when the radial distance becomes

$$r_e = r_0\left(1 + \frac{\mu}{2r_0^2 a_{Tr}}\right) \tag{8.95}$$

Then, by combining Eqs. (8.94) and (8.95), we may write the expression for the radial velocity as

$$\left(\frac{dr}{dt}\right)^2 = \frac{2a_{Tr}}{r^2}(r - r_0)[r^2 - (r_e - r_0)(r - r_0)] \tag{8.96}$$

From this last equation we find that the radial velocity will vanish if the thrust acceleration is not sufficiently large so that escape conditions can be attained before the radial distance exceeds five times the radius of the initial circular orbit. For if $r_e > 5r_0$, then the second factor on the right side of Eq. (8.96) will vanish when

$$r = \tfrac{1}{2}(r_e - r_0) - \tfrac{1}{2}\sqrt{(r_e - r_0)(r_e - 5r_0)}$$

This distance is easily seen to be less than the escape radius $r_e$. Therefore, we conclude from Eq. (8.95) that $a_{Tr} > \mu/8r_0^2$ must obtain if the vehicle is to attain escape velocity under a constant radial acceleration.

It is convenient to introduce the dimensionless parameter $\beta$ defined by

$$\beta = \sqrt{\frac{\mu}{8r_0^2 a_{Tr}}} \qquad (8.97)$$

Then the condition for escape is $\beta < 1$.

Assuming that $a_{Tr}$ exceeds the critical value, i.e., $\beta < 1$, then the time required to reach escape velocity, denoted by $t_e - t_0$, is computed from

$$\sqrt{\frac{\mu}{r_0^3}}(t_e - t_0) = \frac{2\beta}{\sqrt{r_0}} \int_{r_0}^{r_e} \frac{r\,dr}{\sqrt{(r - r_0)[r^2 - 4\beta^2 r_0(r - r_0)]}} \qquad (8.98)$$

where

$$r_e = r_0(1 + 4\beta^2) \qquad (8.99)$$

According to Sect. 1.5, this is an elliptic integral.

Transforming the Integral to Normal Form

The reduction of the integral of Eq. (8.98) is neither simple nor straightforward and depends on a number of ad hoc substitutions, contained in Legendre's proof,† which are outlined below.

1. The radicand of Eq. (8.98) is of the third degree and has one real root $r = r_0$. Make the change of variable

$$r = r_0 + z^2$$

so that the new radicand will be of the fourth degree and involve no odd powers of the variable $z$. Using the notation $I$ for the right-side of Eq. (8.98), we obtain

$$I = \frac{4\beta}{\sqrt{r_0}} \int_0^{z_1} \frac{r_0 + z^2}{\sqrt{P(z)}}\,dz$$

---

† Refer, for example, to Philip Franklin's book *A Treatise on Advanced Calculus* published by John Wiley & Sons, Inc. in 1940.

Sect. 8.8]    Spacecraft Motion Under Continuous Thrust    411

where we have defined
$$P(z) = (z^2 - z_1 z + r_0)(z^2 + z_1 z + r_0) \quad \text{and} \quad z_1 = 2\beta\sqrt{r_0}$$

2. The radicand $P(z)$ is next reduced to a form with no linear terms in the two quadratic factors. This is accomplished by a *linear fractional transformation*, also called a *bilinear* or *Möbius*† *transformation*. Let
$$y = \frac{\sqrt{r_0} - z}{\sqrt{r_0} + z} \quad \text{or} \quad z = \frac{1-y}{1+y}\sqrt{r_0}$$
and obtain
$$P(z) = \frac{4r_0^2}{(1+y)^4} Q(y)$$
where
$$Q(y) = [(1+\beta)y^2 + (1-\beta)][(1-\beta)y^2 + (1+\beta)]$$
The integral then becomes
$$I = 8\beta \int_{y_0}^{1} \frac{1+y^2}{(1+y)^2} \times \frac{dy}{\sqrt{Q(y)}}$$
and the lower limit of integration
$$y_0 = \frac{1 - 2\beta}{1 + 2\beta} \quad \text{is such that} \quad -1 < y_0 < 1$$

3. The rational part of the new integrand can be decomposed into a number of partial fractions resulting in
$$I = 8\beta \int_{y_0}^{1} \left[ 1 - \frac{2}{1+y} + \frac{2}{(1+y)^2} \right] \frac{dy}{\sqrt{Q(y)}}$$
To reduce these integrals, observe that
$$\frac{d}{dy} \frac{\sqrt{Q(y)}}{1+y} = \frac{\frac{1}{2} Q'(y)(1+y) - Q(y)}{(1+y)^2 \sqrt{Q(y)}} = \frac{R(y)}{(1+y)^2 \sqrt{Q(y)}}$$
and expand $R(y)$ as a fourth-order polynomial in $1+y$. Thus,
$$R(y) = (1 - \beta^2)(1+y)^4 - 2(1 - \beta^2)(1+y)^3 + 4(1+y) - 4$$
and we have
$$\frac{4}{\sqrt{Q(y)}} \left[ \frac{1}{1+y} - \frac{1}{(1+y)^2} \right] = \frac{d}{dy} \frac{\sqrt{Q(y)}}{1+y} - \frac{(1-\beta^2)(y^2-1)}{\sqrt{Q(y)}}$$

---

† August Ferdinand Möbius (1790–1868), Professor of Astronomy at Leipzig University, studied theoretical astronomy with Gauss and mathematics with Pfaff. He is most frequently remembered for his discovery of the one-sided surface called the *Möbius strip* in September, 1858. Johann Benedict Listing (1806–1882) discovered the same surface in July, 1858 but published in 1861. Therefore, both Listing and Möbius should share the credit for this delightful mathematical oddity.

The integral is now

$$I = 4\beta(1+\beta^2)\int_{y_0}^{1}\frac{dy}{\sqrt{Q(y)}} + 4\beta(1-\beta^2)\int_{y_0}^{1}\frac{y^2}{\sqrt{Q(y)}}dy + 4\beta C_1$$

with the integrated part

$$C_1 = -\int_{y_0}^{1}\frac{d}{dy}\frac{\sqrt{Q(y)}}{1+y}dy = \frac{\sqrt{Q(y_0)}}{1+y_0} - \frac{\sqrt{Q(1)}}{2} = \frac{\sqrt{1+8\beta^2}}{1+2\beta} - 1$$

4. Next develop real substitutions to transform the radicand $Q(y)$ to the standard form

$$(1-x^2)(1-k^2x^2)$$

where $k < 1$. For this purpose, write

$$Q(y) = (1-\beta^2)\left(\frac{1+\beta}{1-\beta}y^2 + 1\right)\left(\frac{1-\beta}{1+\beta}y^2 + 1\right)$$

Then, the substitution

$$u^2 = 1 + \frac{1-\beta}{1+\beta}y^2$$

will result in

$$\frac{dy}{\sqrt{Q(y)}} = \frac{du}{(1+\beta)\sqrt{U(u)}} \quad \text{and} \quad \frac{y^2 dy}{\sqrt{Q(y)}} = \frac{(u^2-1)\,du}{(1-\beta)\sqrt{U(u)}}$$

where

$$U(u) = (u^2-1)(u^2-k^2) \quad \text{and} \quad k^2 = \frac{4\beta}{(1+\beta^2)}$$

The integral now takes the form

$$I = 4\beta(1+\beta)\int_{u_0}^{u_1}\frac{u^2 - \tfrac{1}{2}k^2}{\sqrt{U(u)}}du + 4\beta C_1$$

with the new limits of integration

$$u_0 = \frac{\sqrt{1+8\beta^2}}{1+2\beta}u_1 \quad \text{and} \quad u_1 = \sqrt{\frac{2}{1+\beta}}$$

falling in the range $1 < u_0 < u_1 < \sqrt{2}$.

## Sect. 8.8] Spacecraft Motion Under Continuous Thrust

5. The final substitution

$$x = \frac{1}{u}$$

gives

$$I = 4\beta(1+\beta) \int_{x_0}^{x_1} \left( \frac{1}{x^2} - \frac{k^2}{2} \right) \frac{dx}{\sqrt{X(x)}} + 4\beta C_1$$

and the radicand

$$X(x) = (1 - x^2)(1 - k^2 x^2)$$

has the required form. The limits of integration

$$x_0 = \frac{1}{u_1} \quad \text{and} \quad x_1 = \frac{1}{u_0}$$

are such that

$$\frac{1}{\sqrt{2}} < x_0 < x_1 < 1$$

Then, since

$$\frac{d}{dx} \frac{\sqrt{X(x)}}{x} = \frac{\frac{1}{2} x X'(x) - X(x)}{x^2 \sqrt{X(x)}} = k^2 \frac{x^2}{\sqrt{X(x)}} - \frac{1}{x^2 \sqrt{X(x)}}$$

$$= -\frac{1 - k^2 x^2}{\sqrt{X(x)}} + \frac{1}{\sqrt{X(x)}} - \frac{1}{x^2 \sqrt{X(x)}}$$

we have

$$I = 4\beta(1+\beta) \left[ (1 - \tfrac{1}{2} k^2) \int_{x_0}^{x_1} \frac{dx}{\sqrt{(1-x^2)(1-k^2 x^2)}} \right.$$

$$\left. - \int_{x_0}^{x_1} \sqrt{\frac{1-k^2 x^2}{1-x^2}} \, dx \right] + 4\beta C_1 + 4\beta(1+\beta) C_2$$

and the new integrated part is

$$C_2 = -\int_{x_0}^{x_1} \frac{d}{dx} \frac{\sqrt{X(x)}}{x} dx = \frac{1-\beta}{1+\beta} \left[ 1 - \frac{1-2\beta}{1+2\beta} \frac{1}{\sqrt{1+8\beta^2}} \right]$$

The two integral terms of $I$ are the Jacobian forms of the elliptic integrals of the first and second kind. Transforming to the Legendre forms gives the following expression for the time interval $t_e - t_0$:

$$\sqrt{\frac{\mu}{r_0^3}} (t_e - t_0) = 4\beta \frac{1+\beta^2}{1+\beta} \int_{\phi_0}^{\phi_1} \frac{d\phi}{\sqrt{1 - k^2 \sin^2 \phi}}$$

$$- 4\beta(1+\beta) \int_{\phi_0}^{\phi_1} \sqrt{1 - k^2 \sin^2 \phi} \, d\phi + 4\beta^2 \left[ \frac{3}{\sqrt{1+8\beta^2}} - 1 \right] \quad (8.100)$$

The modulus of the two elliptic integrals is

$$k = \frac{2\sqrt{\beta}}{1+\beta}$$

and their amplitudes are the angles

$$\phi_0 = \arcsin\sqrt{\frac{1+\beta}{2}}$$

and

$$\phi_1 = \arcsin\left(\frac{1+2\beta}{\sqrt{1+8\beta^2}}\sin\phi_0\right)$$

which are both in the first quadrant with $0 < \phi_0 < \phi_1 < \tfrac{1}{2}\pi$.

Landen's transformation of Sect. 1.5 will convert the result to a simpler and more symmetric form with the parameter $\beta$ as the modulus of the elliptic integrals. The appropriate equations are (1.61) and (1.69), with the angles $\phi_0$ and $\phi_1$ replaced by

$$\theta_0 = \frac{\pi}{2} \quad \text{and} \quad \theta_1 = \arcsin\left(\frac{|1-4\beta^2|}{\sqrt{1+8\beta^2}}\right)$$

The angle $\theta_1$ is in the second quadrant.

Using the identities of Prob. 1–27 in Sect. 1.5, we have

$$F(\beta,\theta_1) = F[\beta,\pi+(\theta_1-\pi)] = 2K(\beta) + F(\beta,\theta_1-\pi)$$
$$= 2K(\beta) - F(\beta,\pi-\theta_1)$$

and

$$E(\beta,\theta_1) = 2E(\beta) - E(\beta,\pi-\theta_1)$$

so that the final form of Eq. (8.100) is expressed as†

$$\sqrt{\frac{\mu}{r_0^3}}(t_e - t_0) = 4\beta\bigg\{[K(\beta)-E(\beta)] - [F(\beta,\psi)-E(\beta,\psi)]$$
$$+ \frac{3\beta}{\sqrt{1+8\beta^2}} - \beta\sin\psi\bigg\} \quad (8.101)$$

where $\psi = \pi - \theta_1$ is an angle in the first quadrant determined from

$$\sin\psi = \frac{|1-4\beta^2|}{\sqrt{1+8\beta^2}}$$

Values of the elliptic integrals can be numerically computed using the methods developed in Sect. 1.5.

---

† In Tsien's paper, he credits Dr. Y. T. Wu for obtaining his result in terms of elliptic integrals but no details are supplied. Unfortunately, Dr. Wu's expression seems to differ from the one derived here by the present author.

## Problem 8-23

If the radial-thrust acceleration $a_{Tr}$ is so small that $\beta$ exceeds one, the vehicle, starting from a circular orbit of radius $r_0$, cannot reach escape conditions. Instead, it will spiral out to a maximum altitude

$$r_a = \frac{2\beta}{\beta + \sqrt{\beta^2 - 1}} r_0$$

in the time interval $t_a - t_0$ calculated from

$$\sqrt{\frac{\mu}{r_0^3}}(t_a - t_0) = \frac{4\beta}{\sqrt{k}}[(1+k)K(k) - E(k)]$$

where

$$k = \frac{r_a}{r_0} - 1 = \frac{1}{\left(\beta + \sqrt{\beta^2 - 1}\right)^2}$$

If the radial-thrust acceleration suddenly ceases at the moment when the maximum altitude is achieved, the resulting orbit will be characterized by the orbital elements

$$a = \frac{r_0(1+k)^2}{1+2k} \qquad e = \frac{k}{1+k} \qquad p = r_0$$

NOTE: In 1985, one of the author's students Bill Kromydas made a careful analysis of this problem by solving the equations of motion and comparing the results with analytically determined values. From an initial radius $r_0 = 8000\,\text{km}$ and $\beta = 1.1$ the maximum altitude $r_a = 11294.6668\,\text{km}$ is achieved in the time $6372.083$ sec. using a value of $\mu = 398,600\,\text{km}^3/\text{sec}^2$. If the constant radial-thrust acceleration is maintained, the vehicle spirals back to the initial radius $r_0$ in the same time interval but the outbound and return orbits are quite different as seen from Fig. 8.5.

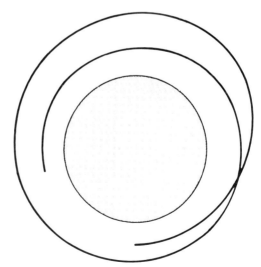

**Fig. 8.5:** Constant radial-thrust acceleration orbit.

## Constant Tangential Acceleration

A vehicle is initially in a circular orbit of radius $r_0$ and at time $t = t_0$ a constant tangential-thrust acceleration is applied. If $a_{Tt}$ is the rocket-thrust acceleration per unit mass in the direction tangent to the orbit, then the equations of motion, from Prob. 2-19, can be written†

$$\frac{dv}{dt}\mathbf{i}_t + \frac{v^2}{\rho}\mathbf{i}_n = a_{Tt}\mathbf{i}_t - \frac{\mu}{r^2}\mathbf{i}_r$$

In terms of components in the tangential and normal directions, we have

$$\frac{dv}{dt} = a_{Tt} - \frac{\mu}{r^2}\cos\gamma \quad \text{and} \quad \frac{v^2}{\rho} = \frac{\mu}{r^2}\sin\gamma \qquad (8.102)$$

where $\gamma$ is the flight-direction angle.

With $s$ used to denote the arc length of the orbit, i.e., the distance travelled by the vehicle, then, clearly,

$$(ds)^2 = (dr)^2 + (r\,d\theta)^2$$

and

$$r\frac{d\theta}{ds} = \sin\gamma \quad \text{and} \quad \frac{dr}{ds} = \cos\gamma$$

so that the equations of motion can be expressed as

$$v\frac{dv}{ds} = a_{Tt} - \frac{\mu}{r^2}\frac{dr}{ds} \quad \text{and} \quad \frac{v^2}{\rho} = \frac{\mu}{r}\frac{d\theta}{ds} \qquad (8.103)$$

Finally, substitute for the curvature $1/\rho$ from Prob. 2-19 part (5) to obtain the equations of motion in the form

$$\frac{1}{2}\frac{dv^2}{ds} + \frac{\mu}{r^2}\frac{dr}{ds} = a_{Tt}$$

$$rv^2\frac{d^2r}{ds^2} + \left(v^2 - \frac{\mu}{r}\right)\left[\left(\frac{dr}{ds}\right)^2 - 1\right] = 0 \qquad (8.104)$$

with the initial conditions

$$r(t_0) = r_0 \qquad v^2(t_0) = v_0^2 = \frac{\mu}{r_0} \qquad \left.\frac{dr}{ds}\right|_{t=t_0} = 0$$

The first of Eqs. (8.104) can be integrated exactly provided the thrust acceleration is constant. Indeed, we obtain

$$v^2 = 2sa_{Tt} + \mu\left(\frac{2}{r} - \frac{1}{r_0}\right) \qquad (8.105)$$

---

† Recall that $v$ is the magnitude of the velocity vector and $\rho$ is the instantaneous radius of curvature of the orbit.

Sect. 8.8]    Spacecraft Motion Under Continuous Thrust    417

Also, if the thrust acceleration is quite small so that $d^2r/ds^2$ is essentially zero, then the second of Eqs. (8.104) requires that $v^2 - \mu/r = 0$. In other words, the vehicle's orbit is always very nearly circular. Hence, replacing $v^2$ by $\mu/r$ in Eq. (8.105) and solving for $r$ produces

$$r = \frac{r_0}{1 - \dfrac{2sa_{Tt}}{v_0^2}} \qquad (8.106)$$

When the vehicle attains escape velocity $(v^2 = 2\mu/r)$, the second of Eqs. (8.104) requires that

$$2r\frac{d^2r}{ds^2} = 1 - \left(\frac{dr}{ds}\right)^2 \qquad (8.107)$$

must prevail. Then, by substituting for $r$ from Eq. (8.106) and solving for $s$, we find

$$s_{esc} = \frac{v_0^2}{2a_{Tt}}\left[1 - \frac{1}{v_0}(20a_{Tt}^2 r_0^2)^{\frac{1}{4}}\right] \qquad (8.108)$$

where $s_{esc}$ denotes the distance travelled by the vehicle before reaching escape speed. The corresponding radial distance is

$$r_{esc} = \frac{r_0 v_0}{(20a_{Tt}^2 r_0^2)^{\frac{1}{4}}} \qquad (8.109)$$

From Eqs. (8.105) and (8.106), it follows that

$$v^2 = v_0^2 - 2sa_{Tt}$$

Then, since $v = ds/dt$, we readily calculate the time

$$t_{esc} - t_0 = \frac{v_0}{a_{Tt}}\left[1 - \left(\frac{20a_{Tt}^2 r_0^2}{v_0^4}\right)^{\frac{1}{8}}\right] \qquad (8.110)$$

to reach escape conditions. Furthermore, it is of interest to note that

$$N_{esc} = \frac{1}{2\pi}\int_0^{s_e}\frac{ds}{r} = \frac{v_0^2}{8\pi a_{Tt} r_0}\left(1 - \frac{\sqrt{20}a_{Tt} r_0}{v_0^2}\right) \qquad (8.111)$$

is the approximate number of revolutions of the planet before escape.

## ◊ Problem 8-24

A vehicle is initially in a circular orbit of radius $r_0$ and at time $t = t_0$ a constant circumferential-thrust acceleration is applied. The equations of motion are

$$\frac{d^2r}{dt^2} - r\left(\frac{d\theta}{dt}\right)^2 = -\frac{\mu}{r^2}$$

$$\frac{d}{dt}\left(r^2\frac{d\theta}{dt}\right) = ra_{T\theta}$$

For very small values of $a_{T\theta}$, the radial acceleration will be quite small so that the centripetal acceleration will be approximately balanced by gravity. With this assumption, the radius as a function of time is

$$r = \frac{r_0}{\left[1 - \frac{(t-t_0)a_{T\theta}}{v_0}\right]^2}$$

and the time to reach escape velocity is

$$t_{esc} - t_0 = \frac{v_0}{a_{T\theta}}\left[1 - \left(\frac{2a_{T\theta}r_0}{v_0^2}\right)^{\frac{1}{4}}\right]$$

Escape occurs at a radial distance of

$$r_{esc} = \frac{r_0 v_0}{\sqrt{2a_{T\theta}r_0}}$$

after, approximately,

$$N_{esc} = \frac{v_0^2}{8\pi a_{T\theta}r_0}\left(1 - \frac{2a_{T\theta}r_0}{v_0^2}\right)$$

revolutions.

Compare the efficiency of thrusting tangentially versus circumferentially.

Chapter 9

# Patched-Conic Orbits and Perturbation Methods

BASIC TO THE DETERMINATION OF PRECISION SPACECRAFT ORBITS IS an appropriate first approximation by a sequence of two-body orbits. For example, the initial portion of the orbit for a free-return, flyby, interplanetary voyage may be approximated by an ellipse whose focus is at the center of the sun. When the spacecraft is within the sphere of influence of the planet, the orbit is then essentially hyperbolic with the planet at the focus. Again for the return trip, the trajectory is approximated by an ellipse. For each of the three parts, the assumption is made that only one gravitational center is active at a time. The resulting orbit is an amazingly good representation of the actual motion and can be utilized for many important problems.

Although the *patched-conic approximation*, as it is frequently called, is not adequate as a precise reference orbit, it does afford a convenient means of exploring a variety of initial and boundary conditions at earth and the target planet in an efficient manner. Indeed, one can expect to achieve significant economies in computation time without compromising the essential ingredients of the problem.

When a precision orbit is obtained based on the conic approximations, certain quantities can be regarded as invariant: the total time of flight and the position vectors at the time of insertion into orbit and the return perigee are possible invariants. Thus, the approximate patched-conic solution can relate to the precise orbit in important and fundamental ways. Precision-orbit determination is accomplished by making slight adjustments in both the orbit insertion and the return velocities.

Perturbation methods can be used for both the problem of determining precision orbits and the problem of insuring that a spacecraft in flight will meet certain specified boundary conditions. In celestial mechanics it is customary to distinguish between the two classes—*general perturbations* and *special perturbations*. In the first class are included methods of generalizing the expressions for simple two-body motion of a planet about the sun to include the disturbing effects of the other planets by utilizing infinite trigonometric series expansions and term-by-term integration; the resulting expressions are known as general perturbations. In the second class fall all numerical methods for deriving the disturbed orbit by direct integration of either the rectangular coordinates or a set of *osculating orbital elements*. The latter method, devised by and named for Johann Franz Encke, is the subject of Sect. 9.4.

It is beyond our scope to consider all or even a significant number of the many techniques which have been developed for specialized application. Except for some examples in Chapter 10, we avoid the subject of general perturbations entirely. Specifically, the discussion here is limited to special perturbation techniques which have been found particularly useful in space trajectory calculations.

In contrast to the perturbation methods of celestial mechanics, the method of *linearized perturbations* does not provide an exact description of the motion but is an enormously valuable tool for our purposes. Basically, the approach is to linearize the equations of motion by a series expansion about a *nominal* or *reference orbit* in which only first-order terms are retained. For the results to remain valid it is, of course, necessary to restrict the magnitude of the deviations from the nominal orbit. When applicable, many advantages accrue from the linearized method of analysis. First of all, the resulting equations are far simpler. Of even greater importance, however, is the fact that superposition techniques are possible. In fact, all the tools of linear analysis can be exploited to obtain solutions to a wide variety of problems. The material developed in this chapter does, indeed, form the basis of the navigation theories presented in Chapters 13 and 14.

The so-called *perturbation matrices* introduced in Sects. 9.5 and 9.6 are frequently referred to as *sensitivity coefficients* in that they provide a convenient description of the manner in which errors propagate along a reference orbit. Thus, these matrices are useful, not only for navigation in the vicinity of a reference orbit, but also to assist in the preparation of the reference orbit itself.

Linear perturbation methods are particularly advantageous for designing spacecraft orbits to achieve certain boundary conditions. A specific application of the method is given in Sect. 9.8 to the problem of determining precise circumlunar trajectories. For this problem, as well as the guidance problems discussed in Chapter 11, the concept of the so-called *method of adjoints* is fundamental. One set of perturbation equations describes the propagation of errors in the forward direction along the orbit. The adjoint equations, on the other hand, describe the propagation of errors in the backward direction, that is, corresponding to motion which would result if the orbit were traversed in the opposite direction. There exists an entire body of mathematics relative to the theory of adjoint differential equations, and we shall have occasion to draw upon this knowledge.

## 9.1 Approach Trajectories Near a Target Planet

When in the vicinity of a planet, a vehicle in a solar orbit experiences velocity perturbations which depend on the relative velocity between the vehicle and the planet and the distance separating the two at the point of closest approach. If only the gravitational field of the planet affected the motion of the spacecraft, the vehicle would make its approach along a hyperbolic path. Actually, the period of time for which the planet's gravitation is significant is small when compared with the total time of the mission. Furthermore, during this time, the distance between the planet and the spacecraft is small when compared with its distance from the sun. As a consequence, for the brief period of contact, solar gravity affects both the planet and the vehicle in essentially the same way. Therefore, in the discussion of planetary approach, solar gravity may be ignored with the assurance that its effects would not alter the results significantly. In this section we shall consider separately the problems of a close pass and a surface impact.

### Close Pass of a Target Planet

We can view the effect of a planetary contact as an impulsive change in the velocity of a vehicle in a solar orbit. At a sufficiently great distance the motion of the space vehicle with respect to a target planet is essentially along the asymptote of the approach hyperbola. Refer to Fig. 9.1 and define $\nu$ as the angle between the asymptote and the conjugate axis of the hyperbolic path of approach. The vertex is, of course, the point of closest approach of the vehicle and the planet. Clearly, the total effect on the velocity of the spacecraft, after contact with the planet, is simply a rotation in the plane of motion of the inbound relative velocity vector $\mathbf{v}_{\infty i}$ by an amount $2\nu$. The direction of rotation can increase or decrease the solar orbital velocity depending on the orientation of the plane of relative motion. But the magnitude of the outbound relative velocity vector $\mathbf{v}_{\infty o}$ is the same as the inbound magnitude.

Let $e$ and $a$ be the eccentricity and semimajor axis of the hyperbolic orbit with $r_m$ denoting the distance between the vertex and the focus. The vertex is, of course, the point of closest approach of the spacecraft to the planet, and we have the relationship

$$r_m = a(1-e) = \frac{\mu}{v_\infty^2}(e-1)$$

where $a$ is determined from the vis-viva integral and $\mu$ is the gravitational constant of the planet. Solving for $e$ and noting that

$$e = \csc \nu$$

**422**  Patched-Conic Orbits and Perturbation Methods  [Chap. 9

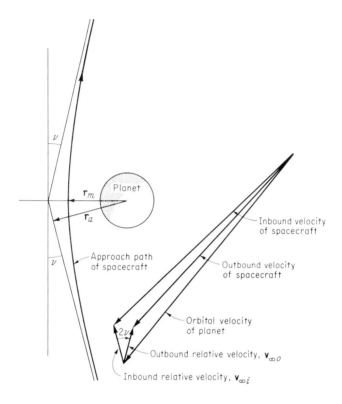

**Fig. 9.1:** Motion of spacecraft in the vicinity of a target planet.

we obtain

$$\sin \nu = \frac{1}{1 + r_m v_\infty^2 / \mu} \tag{9.1}$$

For navigation purposes, the vector $\mathbf{r}_a$ shown in Fig. 9.1 has more significance than the minimum passing distance. It is a vector from the focus of the hyperbolic orbit and perpendicular to the velocity vector $\mathbf{v}_{\infty i}$. One might think of the terminus of the vector $\mathbf{r}_a$ as the *point of aim* for the approach. Since

$$r_a = -ae\cos\nu = -a\cot\nu$$

and is, therefore, equal to the semiminor axis of the hyperbola, we have the following alternate expression for the turn angle $\nu$ in terms of the aim-point distance $r_a$:

$$\tan \nu = \frac{\mu}{r_a v_\infty^2} \tag{9.2}$$

Sect. 9.1]    Approach Trajectories Near a Target Planet    423

Also, by eliminating $\nu$ between Eqs. (9.1) and (9.2), we obtain

$$r_a = r_m \sqrt{1 + \frac{2\mu}{r_m v_\infty^2}} \quad (9.3)$$

as a means of determining the aim point for a specific passing distance.

◊ **Problem 9-1**
For the purpose of an error analysis it is desirable to determine the variations to be expected in the magnitude of the velocity change during planetary contact as a function of variations in the point of aim, $r_a$. Assume for simplicity that the magnitude of the approach velocity $v_\infty$ is unaffected by variations in $r_a$ and derive the expression

$$\frac{d|\mathbf{v}_{\infty o} - \mathbf{v}_{\infty i}|}{dr_a} = -\frac{v_\infty}{r_m}(1 - \sin \nu) \sin 2\nu$$

### Tisserand's Criterion

When a comet passes close to a planet, the elements of its orbit can be so drastically altered that the identity of the comet can be questionable. To resolve this problem, Tisserand[†] established, in 1889, a relationship among the comet elements which remains essentially unaltered by the perturbations. This same relationship, which we now will derive, can be used to analyze the effect of planetary contact on a spacecraft.

Tisserand's contribution is a particular interpretation of Jacobi's integral derived in Sect. 8.2. The first step in the development is to rewrite that integral from the form given in Prob. 8–3 involving rotating coordinates to the corresponding one in fixed coordinates. For this purpose, using the notation of Sect. 2.5, we solve Eq. (2.50)

$$\mathbf{v}^* = \mathbf{R}(\mathbf{v} + \mathbf{\Omega r}) = \mathbf{R}(\mathbf{v} + \mathbf{\Omega R}^T \mathbf{r}^*)$$

for $\mathbf{v}$ to obtain

$$\mathbf{v} = \mathbf{R}^T(\mathbf{v}^* - \mathbf{R \Omega R}^T \mathbf{r}^*) = \mathbf{R}^T(\mathbf{v}^* + \mathbf{\Omega}^* \mathbf{r}^*)$$

The second form of this equation follows from Prob. 2–17. Next we calculate

$$\mathbf{v}^T \mathbf{v} = v^{*2} - \mathbf{r}^T \mathbf{\Omega v} + \mathbf{v}^T \mathbf{\Omega r} - \mathbf{r}^T \mathbf{\Omega \Omega r}$$
$$= v^{*2} + 2\boldsymbol{\omega} \cdot \mathbf{r} \times \mathbf{v} - \mathbf{r} \cdot \boldsymbol{\omega} \times (\boldsymbol{\omega} \times \mathbf{r})$$
$$= v^{*2} + 2\boldsymbol{\omega}^* \cdot \mathbf{r}^* \times \mathbf{v}^* + r^{*2} \omega^{*2} - (\boldsymbol{\omega}^* \cdot \mathbf{r}^*)^2$$

---

[†] François Félix Tisserand (1845–1896) was a professor of astronomy at the University of Toulouse before he became the director of the Paris observatory in 1892. Publication of his greatest work *Traité de Mécanique Céleste* began in 1889 and was completed a few months before his death. The four volumes constituted an updated version of LaPlace's *Mécanique Céleste*.

Recall that the product of the matrix $\Omega^*$ and any vector is equivalent to the vector product of $\boldsymbol{\omega}$ and that vector.

Then, since
$$\boldsymbol{\omega}^* = -\boldsymbol{\omega} \qquad r^{*2} = r^2 \qquad (\boldsymbol{\omega}^* \cdot \mathbf{r}^*)^2 = (\boldsymbol{\omega} \cdot \mathbf{r})^2$$
we have
$$v^2 = v^{*2} - 2\boldsymbol{\omega} \cdot \mathbf{r}^* \times \mathbf{v}^* - \mathbf{r} \cdot \boldsymbol{\omega} \times (\boldsymbol{\omega} \times \mathbf{r})$$
which, when compared with the equation of Prob. 8–3, produces
$$v^{*2} = 2\boldsymbol{\omega} \cdot \mathbf{r}^* \times \mathbf{v}^* + \frac{2Gm_1}{\rho_1} + \frac{2Gm_2}{\rho_2} - C \qquad (9.4)$$
as the desired result.

If $m_1$ and $m_2$ are the masses of the sun and a planet, respectively, then $m_2 \ll m_1$. Therefore, when the comet (or spacecraft) is not close to the planet, we may discard the term $2Gm_2/\rho_2$ in Jacobi's integral. Also, from Eq. (8.14), we have

$$\omega^2 = \frac{G(m_1 + m_2)}{\rho^3} \approx \frac{Gm_1}{\rho^3} = \frac{\mu}{\rho^3} \quad \text{where we have defined} \quad \rho \equiv r_{12}$$

Furthermore, $\mathbf{r}^* \times \mathbf{v}^*$ is just the angular momentum vector $\mathbf{h}$ of the small body with respect to the sun so that

$$\boldsymbol{\omega} \cdot \mathbf{r}^* \times \mathbf{v}^* = \omega h \cos i = \omega \sqrt{\mu a (1 - e^2)} \cos i$$

where $i$ is the inclination angle of the body's orbital plane with respect to the ecliptic; $a$ and $e$ are, of course, the semimajor axis and the eccentricity of the small body's orbit. In addition, $v^{*2}$ may be replaced by its equivalent from the vis-viva integral

$$v^{*2} = \mu \left( \frac{2}{\rho_1} - \frac{1}{a} \right)$$

When these are substituted in Jacobi's integral (9.4), we obtain

$$\frac{1}{a} + 2\sqrt{\frac{a(1 - e^2)}{\rho^3}} \cos i = \text{constant}$$

or, equivalently,

$$\frac{1}{a_1} + 2\sqrt{\frac{a_1(1 - e_1^2)}{\rho^3}} \cos i_1 = \frac{1}{a_2} + 2\sqrt{\frac{a_2(1 - e_2^2)}{\rho^3}} \cos i_2 \qquad (9.5)$$

where $a_1$, $e_1$, $i_1$ are the semimajor axis, eccentricity and orbital inclination prior to the planetary contact and $a_2$, $e_2$, $i_2$ are the orbital elements after contact. As previously noted, the distance between the sun and the planet is $\rho$. Equation (9.5) is generally referred to as *Tisserand's criterion for the identification of comets*.

Sect. 9.1]    Approach Trajectories Near a Target Planet    425

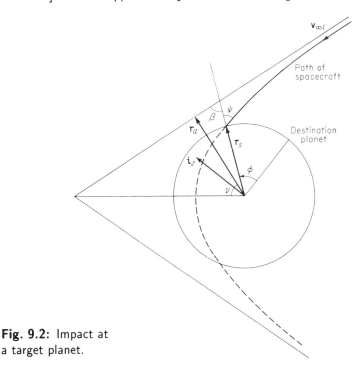

**Fig. 9.2:** Impact at a target planet.

### Surface Impact at a Target Planet

Consider now the problem of pointing a vehicle in a direction to impact a planetary surface at a specified point. For simplicity, the following analysis assumes the point of impact to lie in the plane formed by the planet's polar axis and the direction of the relative velocity vector—i.e., we are addressing the problem of impacting at a specified latitude. Generally, small adjustments in the orbit can alter the time of arrival to accommodate a desired longitude of impact.

Referring to Fig. 9.2 we see that the choice of latitude $\phi$ together with the inbound relative velocity vector $\mathbf{v}_{\infty i}$ serves to determine the angle $\beta$ and the point of impact $\mathbf{r}_s$. If $\mathbf{v}_{\infty i}$ is expressed in a planetocentric equatorial system of coordinates, then

$$\sin(\beta - \phi) = \frac{\mathbf{i}_z \cdot \mathbf{v}_{\infty i}}{v_\infty} \qquad (9.6)$$

where $\mathbf{i}_z$ is the unit vector in the direction of the planet's north polar axis.

In order to determine the point of aim, we first note that the parameter of the hyperbolic orbit is

$$p = a(1 - e^2) = \frac{\mu \cot^2 \nu}{v_\infty^2}$$

Thus, we have

$$r_s = \frac{p}{1 + e\cos(\frac{1}{2}\pi - \beta + \nu)}$$

$$= \frac{\mu \cot^2 \nu / v_\infty^2}{1 + \cot\nu \sin\beta - \cos\beta}$$

Using Eq. (9.2), we obtain the following quadratic equation for determining $r_a$ in terms of $\beta$ and $v_\infty$:

$$r_a^2 - r_a r_s \sin\beta - \frac{\mu}{v_\infty^2} r_s(1 - \cos\beta) = 0 \qquad (9.7)$$

The angle of incidence $\psi$, shown in Fig. 9.2, is important for the atmospheric entry problem and may be determined (using the results of Prob. 3–23) from

$$\tan\psi = \frac{p}{r_s e \sin f}$$

where the true anomaly $f$ is $\frac{1}{2}\pi - \beta + \nu$. Thus, we have

$$\tan\psi = \frac{r_a^2}{r_s(r_a \cos\beta + \mu \sin\beta / v_\infty^2)} \qquad (9.8)$$

◇ **Problem 9–2**

The quantity $r_s(d\phi/dr_a)$ can be interpreted as a linear miss ratio for a spacecraft entering a planet's atmosphere. It is, in effect, the magnification factor which must be applied to an error in the magnitude of the point of aim $r_a$ to produce a corresponding error in the impact. Derive the expression

$$\frac{d\phi}{dr_a} = \frac{1}{r_a}\left(2 - \frac{r_s}{r_a}\sin\beta\right)\tan\psi$$

in which we assume that $\mathbf{v}_\infty$ does not vary with $r_a$.

◇ **Problem 9–3**

A spacecraft is returning from Mars and the approach velocity relative to the earth is

$$\mathbf{v}_\infty = 10,619\,\mathbf{i}_x + 9,682\,\mathbf{i}_y + 6,493\,\mathbf{i}_z \text{ fps}$$

expressed in a geocentric ecliptic coordinate system. If it is desired to impact in the general area of the Gulf of Mexico, compute the magnitude of the point of aim $r_a$, the incidence angle $\psi$, and the linear miss ratio defined in Prob. 9–2. The latitude of the Gulf of Mexico may be taken as 28°.

## 9.2 Interplanetary Orbits

The path of an interplanetary spacecraft is, of course, completely determined by the initial conditions, i.e., the velocity vector of the vehicle at the time of departure from earth. Prior to injection into orbit, the spacecraft has a velocity with respect to the sun of just under 100,000 fps, which is the same as the orbital velocity of the earth. The problem then is to determine the impulse in velocity needed to attain a suitable interplanetary orbit so that the spacecraft will intersect the orbit of the destination planet at a predetermined point in space and time. In order to stay within the realm of two-body analysis, it is necessary that the velocity impulse occur at a point sufficiently far removed from the gravitational field of the earth that only solar attraction is important. However, for simplicity, we shall assume that the impulse takes place at a point on the earth's orbit. Then, if the travel-time to the destination planet is specified, the two-body orbit may be determined using the methods of Chapter 7.

It is possible to establish an orbit to the planets Mars or Venus with a departure or excess hyperbolic velocity from the earth which is only slightly larger than the minimum escape velocity. The greater part of the voyage is made in free flight under the action of solar gravity—the periods of acceleration and of proximity to planets being insignificant compared with the total duration of the flight. For the most part the influence of the various planets on the path of the spacecraft is almost negligible. Therefore, by far the more substantial portion of the trip is made in a nearly true elliptic orbit.

The Hohmann† transfer orbit for a Martian voyage is an ellipse, with the sun at one focus, whose perihelion is the point of tangency with the Mars orbit as shown in Fig. 9.3. If the planetary orbits were coplanar circles, then this path would require the least expenditure of fuel for the transfer as shown in Sect. 11.3. However, the orbits of the planets are not coplanar, and although the angle between the orbital planes of earth and Mars is only 1.85°, the effect in terms of required velocity of departure is not a minor one. For a vehicle moving solely under the influence of solar gravity, the trajectory plane must include the position of earth at departure, the position of the destination planet at arrival, and the sun as the center of attraction. If the launch and arrival positions are nearly 180° apart, as measured with the sun at the vertex, then the trajectory plane can and generally will be inclined at a large angle to the ecliptic. Such an orbit involves a relative velocity between the spacecraft and earth which is comparable to the earth's own velocity about the sun. These

---

† Walter Hohmann (1880–1945), a German engineer, first published this result in his paper *Erreichbarkeit der Himmelskörper* in Munich, Germany in 1925. An English translation was, subsequently, published by NASA in 1960 when America first became interested in space travel.

**428**  Patched-Conic Orbits and Perturbation Methods  [Chap. 9

orbits therefore involve an impractically large expenditure of energy at departure, despite the fact that in the simplified two-dimensional model they are optimum in this respect.

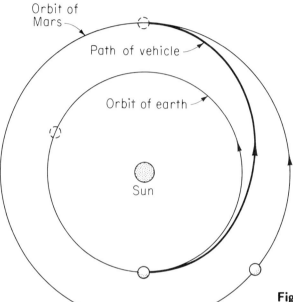

**Fig. 9.3:** Earth-to-Mars Hohmann orbit.

Apart from the three-dimensional effect described above, the cotangential transfer ellipse, if continued past the destination planet, would not provide a suitable return trajectory to earth. For a one-way trip this is not relevant; however, for a spacecraft which is to be recovered or for a manned mission, this consideration is important. The outbound trip to Mars along the Hohmann orbit consumes between 8 and 9 months. If the vehicle continued its flight with no extra propulsion, it would return to the original point of departure in space only to find the earth nearly on the opposite side of the sun. Therefore, either the vehicle must wait in the vicinity of Mars until the time is right for a return voyage or the original trajectory must be revised so that the vehicle will, indeed, encounter the earth when it returns to the earth's orbit. It requires a significant velocity change to enter an orbit about Mars and, subsequently, to depart from the planet for the return trip. However, if no stopover is required, the vehicle can, in principle, require no extra fuel for the round-trip mission.

## Planetary Flyby Orbits

For purpose of discussion, consider the problem of placing a spacecraft in an orbit that passes within a few thousand miles of another planet and subsequently returns to earth. Determining a suitable one-way trajectory is straightforward and the added complication of requiring the vehicle to return to earth without additional propulsion (except that needed to correct for navigation inaccuracies) would not contribute significantly to the difficulty of obtaining a solution were it not for the orbital deflection caused by the gravitation field of the planet as the spacecraft passes. However, with the material developed in the preceding section as background, a computation procedure needed to determine round-trip, flyby interplanetary orbits may be formulated as follows.

The outbound portion of the round-trip trajectory is determined as for the one-way case. The spacecraft velocity vector can then be calculated, and the velocity relative to the destination planet determined. Since the gravitation field of the planet can only rotate this velocity vector, the spacecraft must leave the planet for a return trip to earth with a known relative velocity and at a known time. The problem of establishing a return trajectory is solved basically by an iteration. The procedure consists of making, and systematically revising, an estimate of the time required for the return trip. For each such estimate a new trajectory is calculated until one is obtained which matches the relative velocity magnitude at the target planet.

It is, of course, possible that no return path exists corresponding to the required departure time and the relative velocity magnitude. However, when a matching pair of trajectories, outbound and return, has been found, one final step remains. It is necessary to determine if the velocity change at the destination planet can be effected during the period of contact solely by the planet's gravitation. The required turn angle $2\nu$ is readily computed from the inbound and outbound relative velocity vectors. Thus

$$\sin 2\nu = \frac{|\mathbf{v}_{\infty o} \times \mathbf{v}_{\infty i}|}{v_\infty^2} \tag{9.9}$$

The minimum passing distance can then be determined from

$$r_m = \frac{\mu(\csc \nu - 1)}{v_\infty^2} \tag{9.10}$$

If $r_m$ is of reasonable magnitude, the solution is complete and a satisfactory round-trip path has been found.

◇ **Problem 9-4**

The vector point of aim $\mathbf{r}_a$ can be determined from

$$\mathbf{r}_a = \frac{\mu}{2v_\infty^2 \sin^2 \nu} \, \mathbf{i}_{\infty i} \times (\mathbf{i}_{\infty i} \times \mathbf{i}_{\infty o})$$

## Impulse Control of Flyby Altitude

If $\mathbf{v}_{\infty i}$ and $\mathbf{v}_{\infty o}$ of equal magnitude have been determined but do not result in an acceptable passing distance of the planet, we can, in fact, select a specific value for $r_m$ and calculate an appropriate velocity impulse to be applied to achieve a desired altitude.

For a given $r_m$ and $\mathbf{v}_{\infty i}$ we can determine the turn angle $\nu$ from Eq. (9.1) and the point of aim radius from Eq. (9.3). The inbound and outbound hyperbolas will be identical in size and shape but one is rotated with respect to the other, about their common focus, through the angle $2\vartheta$ which we must determine. The intersection of these two hyperbolas $\mathbf{r}_s$ is the point at which the velocity impulse is to be applied. Clearly, from symmetry, the direction of the intersection point from the focus is simply

$$\mathbf{i}_{r_s} = \frac{\mathbf{v}_{\infty i} - \mathbf{v}_{\infty o}}{|\mathbf{v}_{\infty i} - \mathbf{v}_{\infty o}|} \tag{9.11}$$

so that the velocity impulse $\Delta \mathbf{v}$ is calculated from

$$\Delta \mathbf{v} = -2 v_s \sin \vartheta \, \mathbf{i}_{r_s}$$

where

$$v_s^2 = \frac{2\mu}{r_s} + v_\infty^2$$

To determine $\vartheta$ and $r_s$, we define the angle $\delta$ such that

$$\cos(\pi + 2\delta) = \mathbf{i}_{\infty i} \cdot \mathbf{i}_{\infty o} \tag{9.12}$$

and note that the angle $\beta$, from Fig. 9.2, is $\pi - \delta$. Then, since

$$\delta + \nu + \vartheta = \tfrac{1}{2}\pi$$

we have

$$\vartheta = \tfrac{1}{2}\pi - \nu - \delta$$

Finally, from Eq. (9.7), we obtain the radius

$$r_s = \frac{r_a^2}{r_a \sin \delta + \mu(1 + \cos \delta)/v_\infty^2} \tag{9.13}$$

at which the velocity impulse

$$\Delta \mathbf{v} = -2 \cos(\nu + \delta) \sqrt{\frac{2\mu}{r_s} + v_\infty^2} \, \mathbf{i}_{r_s} \tag{9.14}$$

is to be applied.

## Examples of Free-Return, Flyby Orbits

The simplest possible round-trip trajectory would be an orbit whose period is a multiple of the earth's period. Consider first the possibilities of a spacecraft orbit with a period of 1 year which intersects the orbits of both earth and the destination planet. It is shown in the first problem at the end of this section that the minimum required departure velocity for a Mars mission is more than 50,000 fps after the vehicle has escaped from the earth's gravitational field. However, for the Venus trip, the short outbound time of flight and the large gravitational pull together make possible conditions which more nearly approximate those required for a 1 year round-trip trajectory. Normally, the period of the outbound orbit will be slightly less than 0.8 year while the return path will have a period of approximately 1 year. A typical round trip requires roughly 1.2 years, of which about 0.4 year is spent from earth to Venus and 0.8 year in return.

With the severe propulsion requirements ruling out the 1-year round-trip to Mars, an alternate possibility is a space vehicle orbit with a 2-year period. An example† of such a trajectory is illustrated in Fig. 9.4. The departure velocity for the Mars mission is 18,200 fps after escape and 1.5293 years are required for the outbound trip. After passing 7,892 miles from the surface with a relative excess hyperbolic velocity of 28,852 fps, the vehicle returns to earth 0.3673 year after contact.

Unfortunately, the 2-year round-trip to Mars has a somewhat tight restriction with respect to times of launch. Although we may expect this class of trajectories approximately to recur with the Martian synodical period of 780 days, the duration of the time for favorable launch conditions with reasonable velocities and passing conditions at Mars is roughly one month.

The tolerances on the 1-year Venus and the 3-year Mars trajectories are much less severe. For the 3-year Martian reconnaissance trajectory, the space vehicle makes two circuits about the sun while the earth makes three. Thus, either the earth to Mars trajectory or the return trajectory, but not both, will be characterized by a heliocentric angle of travel which exceeds a full revolution.

A typical round-trip Venusian reconnaissance trajectory is illustrated in Fig. 9.5. For the example shown, the vehicle velocity relative to earth after escape is 15,000 fps. After 0.3940 year the spacecraft passes 5,932 miles from the surface of the planet with a relative approach velocity of 25,100 fps and returns to earth 0.8635 year later, entering the atmosphere with a velocity of 50,738 fps. The motion of the space vehicle relative to Venus during the period of contact is illustrated in Fig. 9.6. The direction

---

† The example trajectories used in this section are taken from the author's book *Astronautical Guidance*.

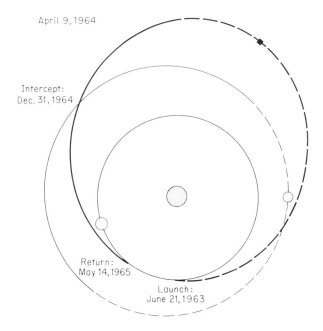

**Fig. 9.4:** Two-year Martian flyby trajectory.

of motion of the planet is shown, together with the hyperbolic contact trajectory of the spacecraft.

The trajectory chosen to illustrate the 3-year round-trip Martian reconnaissance mission is in the next three figures. The earth-to-Mars orbit is diagrammed in Fig. 9.7 and the return path in Fig. 9.8. The departure velocity is 12,000 fps, and 1.1970 year is consumed to reach Mars. After passing 4,903 miles from the surface with a relative approach velocity of 21,567 fps, the vehicle makes one complete orbit of the sun and returns to earth 1.9131 years after contact with a reentry velocity of 39,921 fps. The relative motion of the spacecraft during the planetary contact is diagrammed in Fig. 9.9. In this example, the Martian gravity field alone has the effect of quadrupling the out-of-plane component of the vehicle velocity and, thereby, causing a rotation of approximately 30° in the orbital line of nodes.

Returning to the Venusian reconnaissance trajectory shown in Fig. 9.5, it is of interest to note that the increased velocity introduced at Venus is sufficient to carry the spacecraft on the return trip to a distance of about 1.35 astronomical units from the sun. Since Mars at perihelion is only at a distance of 1.38 astronomical units, the interesting possibility arises of a dual contact with both planets and a total time of flight for the round trip just in excess of 1 year. This would clearly be an improvement over the

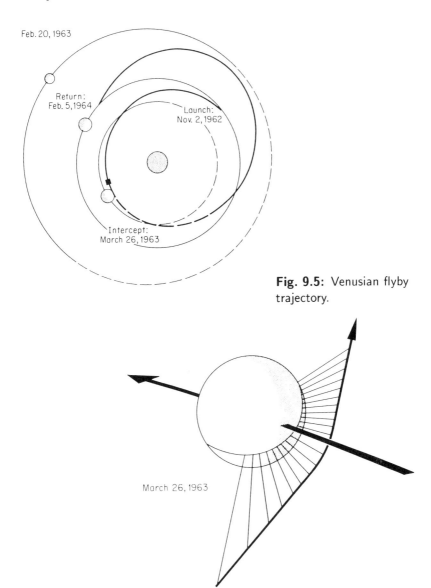

**Fig. 9.5:** Venusian flyby trajectory.

**Fig. 9.6:** Orientation during Venusian contact.

3.2-year round trip to Mars alone. The principal drawback to such a double reconnaissance is the infrequency of possible launch dates. The synodical periods for Venus and Mars are 584 and 780 days, respectively. Therefore, one can expect favorable conditions for round-trip missions to each planet individually to recur with the corresponding synodical frequency. On the

**434** Patched-Conic Orbits and Perturbation Methods [Chap. 9

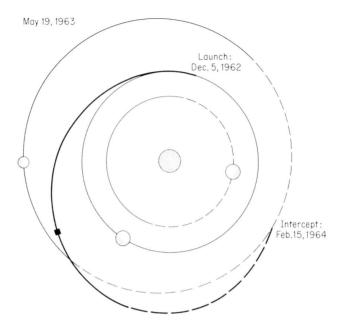

**Fig. 9.7:** Three-year Martian flyby trajectory, outbound.

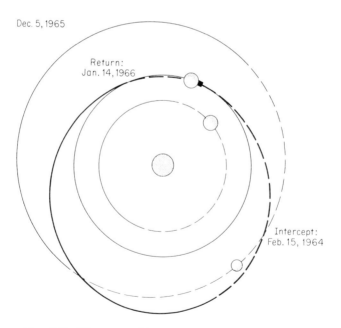

**Fig. 9.8:** Three-year Martian flyby trajectory, inbound.

Sect. 9.2]  Interplanetary Orbits  435

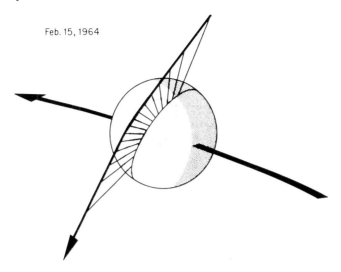

Fig. 9.9: Orientation during Martian contact.

other hand, roughly 2,340 days are required before any particular configuration of the three planets, earth, Venus and Mars, will be approximately repeated. Even then the likelihood of a configuration existing at all which would admit of the dual mission seems, at first, to be remote.

Nevertheless, on June 9, 1972, the ideal circumstances did prevail. On that date a vehicle in a parking orbit from Cape Canaveral on the 110° launch azimuth course could have been injected into just such a trajectory at the geographical location of 5° west and 18° south and with an injection velocity of 39,122 fps. After escape the vehicle would have had a velocity relative to the earth of 15,000 fps. The first planet encountered is Venus after a trip lasting 0.4308 year. The vehicle passes 4,426 miles from the surface of the planet and receives, from the Venusian gravity field alone, a velocity impulse sending it in the direction of Mars. The second portion of the trip consumes 0.3949 year, and the spacecraft contacts Mars passing at a minimum distance of 1,538 miles from the surface. The trip from Mars to earth takes an additional 0.4348 year, and the vehicle returns on September 13, 1973. This truly remarkable trajectory is illustrated in Fig. 6 of the Introduction to this book.

It might be expected from previous remarks that similar conditions would have existed approximately $6\frac{1}{2}$ years earlier. Indeed, the trajectory shown in Fig. 9.10 was possible on February 6, 1966, and is similar in all respects but one. With a departure velocity of 16,500 fps the vehicle contacts Venus after 0.4196 year and Mars 0.5454 year later with respective passing distances of 1,616 and 7,515 miles. Now, however, the encounter with Mars occurs quite far from the Martian perihelion. Thus, in order to catch up with the earth, the vehicle must once again pass inside the earth's orbit with the result that the return trip from Mars requires 0.8950 year.

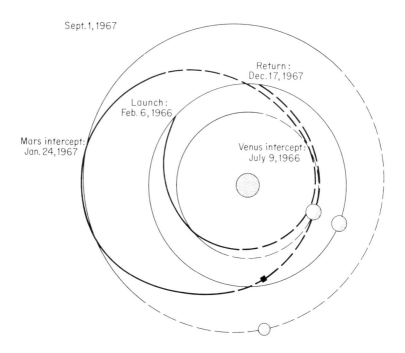

**Fig. 9.10:** Double flyby trajectory.

◊ **Problem 9-5**

Determine the minimum departure velocity from earth for an orbit to Mars with a period of 1 year. Show that the orbit is an ellipse tangent to the orbit of Mars and that departure from earth occurs at an extremity of the minor axis of the transfer orbit.

NOTE: In this problem and the next assume a simplified model of the solar system, i.e., circular, coplanar planetary orbits.

◊ **Problem 9-6**

A vehicle is in a circular orbit about the sun at a distance of one astronomical unit. A velocity impulse is applied to place the vehicle in a transfer orbit with a period of 1.5 years which will intersect the orbit of Mars after traversing a heliocentric angle of 140°. Assuming that the closest approach to the surface of Mars is 3,000 miles and that the vehicle passes ahead of the planet, determine the period of the new orbit.

## 9.3 Circumlunar Trajectories

The motion of a spacecraft in cislunar space is governed primarily by the gravitational fields of the earth and the moon. The effects of solar gravity and the perturbations arising from the nonspherical shape of the attracting bodies are important in a final analysis but are neglected in obtaining the conic approximation.

The calculation of circumlunar trajectories is more difficult than the corresponding interplanetary problem because the time spent within the lunar sphere of influence is a significant fraction of the total mission time. Thus, it would be out of the question to regard the effect of the moon as simply an impulsive change in the vehicle velocity as was done in the computations of the previous sections.

An adequate approximate trajectory may be had† by matching in both position and velocity at the junction points (1) an ellipse from earth to the sphere of influence of the moon whose focus is at the center of the earth, (2) a hyperbola around the moon, and (3) an ellipse from the sphere of influence back to earth. The simplified problem, though itself fairly complex, is tractable when the relevant parameters and independent variables are identified. Clearly, an analogous procedure could be used for interplanetary trajectories if it is desired to obtain a better approximation than would result from the simplified treatment described earlier.

In our analysis the following parameters are found to be a convenient choice for the independent variables:

1. $r_m$, the perilune altitude or minimum passing distance. This parameter is directly related to the total time of flight.
2. $t_A$, the time of arrival at the sphere of influence of the moon on the outbound trajectory. This is specified as a Julian date.‡ It was decided to fix this time, rather than the time of injection, since the time of flight from injection to the sphere of influence is a parameter which will be varied during the iteration process. Thus, since the position of the moon does not change with time of flight, there is no need during that iteration for continual redetermination of the position of the moon.
3. $i_L$, the angle of inclination of the outbound trajectory plane with respect to the equatorial plane of the earth. This parameter cannot

---

† The calculations described in this section were originally published in April, 1962 as an *MIT Instrumentation Laboratory Report R-353* entitled "Circumlunar Trajectory Calculations" authored by Richard H. Battin and James S. Miller.

‡ It is conventional in astronomical calculations to number consecutively the astronomical days, beginning at Greenwich noon, from January 1 of the year 4713 B.C.. The number assigned to a day is called the *Julian day number* and denotes the number of days that have elapsed, at Greenwich noon on the day assigned, since the epoch. The Julian year consists of exactly 365.25 Julian days and the Julian century of 36,525 Julian days. In the nautical almanac the correspondence is made to relate the Julian day number to the ordinary calendar day. For example, the epoch 1960 Jan. 1.5 ET (meaning ephemeris time) corresponds to the Julian day number 2,436,935.

**438** Patched-Conic Orbits and Perturbation Methods  [Chap. 9

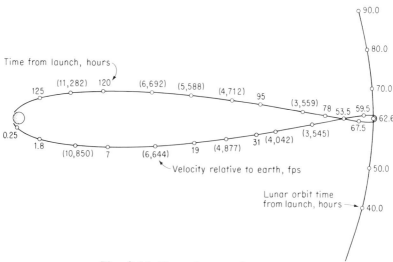

**Fig. 9.11** Circumlunar trajectory.

be freely chosen, since it depends somewhat on the latitude of the launch point. For example, if the parking orbit plane contains the latitude of Cape Canaveral, the inclination angle of the plane cannot be less than this latitude.

4. $i_R$, the angle of inclination of the return trajectory plane with respect to the equatorial plane of the earth. This parameter primarily affects the latitude of the re-entry point.

5. $r_L$, the perigee distance of the outbound trajectory. For these circumlunar calculations, the injection is assumed to place the spacecraft at the perigee of the outbound ellipse.

6. $r_R$, the vacuum perigee distance of the return trajectory. This parameter, which is the perigee that the return ellipse would have in the absence of an atmosphere of the earth, affects the re-entry flight-path angle and, therefore, cannot be freely chosen.

With these six quantities specified, the trajectory is completely determined except for a fourfold ambiguity in the orientation of the outbound and return orbital planes. This problem will be discussed later in detail.

In Fig. 9.11, a precise circumlunar trajectory is shown for which the independent variables in the conic approximation have the values:†

$r_m = 1,180$ miles  $\qquad i_R = 35.0°$
$t_A = 555.125$ Julian days  $\qquad r_L = 4,077$ miles
$i_L = 28.3°$  $\qquad r_R = 4,008$ miles

---

† The time $t_A$ is given in Julian days from the midnight preceding December 31, 1966, ET.

Sect. 9.3]   Circumlunar Trajectories   439

The trajectory is plotted to scale, projected into the plane of the moon's orbit in earth-centered nonrotating coordinates. Time measured in hours from launch and velocity relative to the earth measured in feet per second are indicated at several points along the path. Figure 9.12 shows a portion of the trajectory near the earth together with a portion of the hyperbolic path relative to the moon. In the latter case the velocities shown are measured relative to the moon.

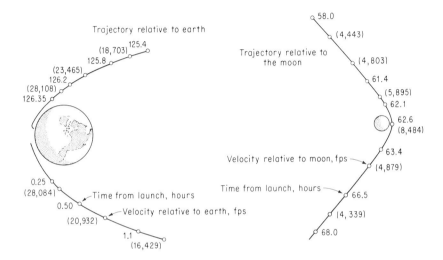

**Fig. 9.12:** Expanded views of trajectory near earth and moon.

The general approach taken here in the development of a calculation technique is first to obtain two earth-centered elliptic orbits, one outbound and one returning, which satisfy the desired end conditions and which, at the sphere of influence, have relative-velocity vectors aligned with the center of the moon. By adjusting the times of flight on the two trajectories, it is possible to cause these vectors both to assume a given magnitude. According to the two-body assumptions concerning vehicle motion within the sphere of influence, the effect of lunar gravity is simply to rotate the inbound relative-velocity vector in the plane of relative motion. The two relative-velocity vectors determine the plane of this motion. Thus, the possibility exists of establishing a realistic hyperbolic pass at the moon by translating these vectors in their common plane to obtain the proper offset from the moon so that the vehicle will indeed pass the moon at a distance which is compatible with the magnitude and the angle between the relative-velocity vectors.

An outline of the overall computational procedure will now be given. Some of the more important mathematical details will be described in the following subsection.

1. With $t_A$, $i_L$, and $r_L$ specified, a point on the sphere of influence of the moon and a time of flight $t_{FL}$ from injection to this point are selected. Let $\mathbf{r}_T$ be the vector from the center of the earth to the selected point on the sphere. From these values an outbound elliptic trajectory may be calculated and the position and velocity vectors $\mathbf{r}_{TM}$ and $\mathbf{v}_{TM}$ relative to the moon at the sphere of influence are determined.
2. The vectors $\mathbf{r}_{TM}$ and $\mathbf{v}_{TM}$ completely specify a hyperbolic trajectory with the moon at the focus. Thus, the perilune distance $r_m$ and the perpendicular distance $r_a$ measured from the center of the moon to the asymptote of the hyperbola are obtained. At the sphere of influence the motion of the vehicle relative to the moon is essentially along this asymptote, so that $r_a$ is the distance at which the vehicle would pass the moon's center if the lunar gravitation were not present. The detailed geometry is shown in Fig. 9.13.

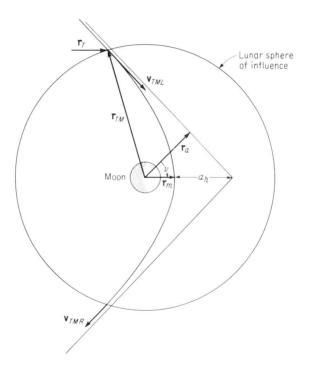

**Fig. 9.13:** Hyperbolic contact at the moon.

3. When $\mathbf{r}_T$ is varied on the sphere of influence and steps 1 and 2 are repeated, a trajectory is obtained for which $r_a$ is zero and the relative-velocity vector $\mathbf{v}_{TM}$ is directed at the center of the moon. The iteration is accomplished by first varying $\mathbf{r}_T$ systematically over the sphere of influence until the calculated $r_m$ is less than some preassigned value. (20,000 miles is satisfactory to initiate the second phase of the iteration.) Then, since moving the position vector $\mathbf{r}_T$ over the sphere of influence by a small amount does not essentially alter the magnitude or direction of $\mathbf{v}_{TM}$, an improved value of $\mathbf{r}_T$ may be obtained from

$$\mathbf{r}_T = \mathbf{r}_{EM} - r_S \frac{\mathbf{v}_{TM}}{v_{TM}}$$

where $\mathbf{r}_{EM}$ is the position vector of the moon relative to the earth and $r_S$ is the radius of the sphere of influence. Usually only four or five cycles are required to reduce $r_a$ to less than one mile.

4. With the use of the calculated value of the magnitude $v_{TM}$ and the original specified passing distance $r_m$, the time interval $t_S$ that the vehicle would spend within the sphere of influence if the direction and magnitude of $\mathbf{v}_{TM}$ were compatible with $r_m$ may be calculated. Then $t_A + t_S$ is the time at which the vehicle leaves the sphere of influence on the return trip.

The time interval $t_S$ is most easily computed as follows: The semimajor axis $a_h$ of the hyperbola is determined from the vis-viva integral as

$$a_h = \left( \frac{v_{TM}^2}{\mu_M} - \frac{2}{r_S} \right)^{-1} \quad (9.15)$$

where $\mu_M$ is the gravitational constant of the moon. The required turn angle $2\nu$ may then be calculated from

$$\sin \nu = \frac{1}{1 + r_m/a_h} \quad (9.16)$$

as shown in Sect. 9.1. Then, since the eccentricity of the hyperbola is simply $\csc \nu$, it follows from the hyperbolic form of Kepler's equation that

$$t_S = 2 \sqrt{\frac{a_h^3}{\mu_M}} \left( \csc \nu \sinh H - H \right) \quad (9.17)$$

where the argument $H$ is determined from

$$\cosh H = \left( 1 + \frac{r_S}{a_h} \right) \sin \nu \quad (9.18)$$

5. Steps 1, 2, and 3 are repeated with $t_A + t_S$, $i_R$, and $r_R$ specified and a selected value of the time of flight $t_{FR}$ from the sphere to the return perigee.

6. The value of $t_{FR}$ is systematically adjusted and step 5 is repeated until the magnitudes of the two vectors $\mathbf{v}_{TML}$ and $\mathbf{v}_{TMR}$ are equal. The angle $2\nu$ between the two vectors is computed from

$$\sin 2\nu = \frac{|\mathbf{v}_{TML} \times \mathbf{v}_{TMR}|}{v_{TM}^2} \qquad (9.19)$$

and the passing distance $r_m$ from†

$$r_m = a_h(\csc \nu - 1) \qquad (9.20)$$

7. The value of $t_{FL}$ is systematically adjusted and steps 1 to 6 are repeated until the calculated value of $r_m$ agrees with the desired value.
8. The vectors $\mathbf{r}_{TML}$ and $\mathbf{r}_{TMR}$ are changed in the plane determined by $\mathbf{v}_{TML}$ and $\mathbf{v}_{TMR}$ to offset each relative-velocity vector by the amount $r_a$ where

$$r_a = a_h \cot \nu \qquad (9.21)$$

This step also involves an iteration. Although $\mathbf{v}_{TML}$ and $\mathbf{v}_{TMR}$ change only slightly in direction as $\mathbf{r}_{TML}$ and $\mathbf{r}_{TMR}$ are displaced, the effect on $r_a$ is greater than can be tolerated. However, the change in the relative-velocity magnitude is less than one foot per second.
9. For each of the newly established velocity vectors, $r_m$ is recalculated as a final check on the validity of step 8. In every instance, experience has shown that the mismatch in $r_m$ is less than one mile.

### Calculating the Conic Arcs

The details required to mechanize this procedure are, for the most part, straightforward. However, certain portions of the calculations are, perhaps, not immediately self-evident.

The basic problem described in step 1 of the outline is to determine the position and velocity vectors $\mathbf{r}_L$ and $\mathbf{v}_L$ associated with an injection at perigee which will produce an elliptic arc whose plane is inclined at an angle $i_L$ to the equatorial plane, having a perigee distance $r_L$ and requiring a time $t_{FL}$ to reach a given position $\mathbf{r}_T$.

In general, as seen in Fig. 9.14, there are two planes which satisfy the required conditions with two exceptions: (1) for a 90° inclination angle only one such plane is defined, and (2) no solution is possible if the desired inclination angle is smaller than the latitude of the target position relative to the earth's equatorial plane. Let $L$ and $\lambda$ be the latitude and longitude

---

† The reader should compare this result with Eq. (9.10). Although we are assuming, in the present case, that the motion takes place essentially along the asymptote of the approach hyperbola, the results would be grossly in error if the velocity at infinity were used to calculate $r_m$ as was done for interplanetary orbits.

Sect. 9.3]  Circumlunar Trajectories  443

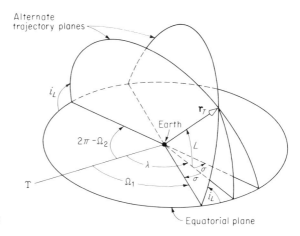

**Fig. 9.14:** Geometry of the trajectory planes.

of $\mathbf{r}_T$. When two planes exist, the longitudes of their ascending nodes, $\Omega_1$ and $\Omega_2$, are determined from

$$\Omega_1 = \lambda - \sigma \quad \text{and} \quad \Omega_2 = \lambda + \sigma + \pi$$

where

$$\sin \sigma = \frac{\tan L}{\tan i_L}$$

Either plane may be suitable for an outbound trajectory. Since the same conditions hold for the return trajectory, there are potentially four different circumlunar orbits satisfying the conditions of the problem.

In addition to the geometrical limitation imposed on the trajectory inclination, there is another constraint which must be examined. Since only the perigee radius $r_L$ is specified, the central angle $\theta$, through which the vehicle moves from $\mathbf{r}_L$ to $\mathbf{r}_T$, must be selected to coincide with the given time of flight. The time $t_{FL}$ and the corresponding angle $\theta$ cannot be freely chosen if the outbound and return orbits are to be elliptic.

The trajectory from $\mathbf{r}_L$ to $\mathbf{r}_T$ was called a tangent ellipse in Sect. 6.3, and the semimajor axis $a$, obtained from the results of that section, is

$$a = \frac{r_L(r_L - r_T \cos \theta)}{2r_L - r_T(1 + \cos \theta)} \tag{9.22}$$

in the current notation. Hence, for an elliptic orbit $(0 < a < \infty)$, the condition

$$1 + \cos \theta < \frac{2r_L}{r_T}$$

must be fulfilled.

**444**  **Patched-Conic Orbits and Perturbation Methods**  [Chap. 9]

When the denominator of the equation for $a$ vanishes, i.e., at $\theta = \theta_p$, where

$$\cos \theta_p = \frac{2r_L}{r_T} - 1$$

the required conic is a parabola and the corresponding time of flight $t_{FL(p)}$, as calculated from the formula derived in Prob. 4-3, is given by

$$t_{FL(p)} = \frac{1}{3}\sqrt{\frac{2(r_T - r_L)}{\mu_E}}(r_T + 2r_L)$$

The specified value of $t_{FL}$ must exceed $t_{FL(p)}$ if the trajectory is to be an ellipse.

In summary, therefore, the outbound trajectory problem is solvable only if

$$i_L > L \quad \text{and} \quad t_{FL} > t_{FL(p)}$$

Identical arguments apply for the return trajectory.

With the orientation of the trajectory plane determined, there remains the problem of calculating the central angle $\theta$, following which the orbital elements $a$ and $e$ are obtained from Eq. (9.22) and

$$e = 1 - \frac{r_L}{a} \tag{9.23}$$

One method of procedure is to use Lagrange's form of the time of flight for an elliptic arc as given in Sect. 6.6 and written in the form

$$t_{FL} = \sqrt{\frac{a^3}{\mu_E}}\left[(\alpha - \sin \alpha) - (\beta - \sin \beta)\right] \tag{9.24}$$

where

$$\sin \tfrac{1}{2}\alpha = \sqrt{\frac{s}{2a}} \quad \text{and} \quad \sin \tfrac{1}{2}\beta = \sqrt{\frac{s-c}{2a}}$$

The semiperimeter $s$ is

$$s = \tfrac{1}{2}(r_L + r_T + c)$$

and $c$ is the linear distance from perigee to $\mathbf{r}_T$; that is,

$$c^2 = r_L^2 + r_T^2 - 2r_L r_T \cos \theta$$

From Eq. (9.22) $a$ is given as a function of $\theta$ so that Eq. (9.24) expresses the time of flight $t_{FL}$ in terms of the single variable $\theta$.

Because of the transcendental nature of Eq. (9.24) it is not possible to express $\theta$ in terms of $t$ using a finite number of elementary functions. However, a simple Newton iteration will converge quite rapidly.

## Circumlunar Trajectories

For this purpose, the following derivative $dt_{FL}/d\theta$ is required

$$\frac{dt_{FL}}{d\theta} = \frac{3}{2}\frac{t_{FL}}{a}\frac{da}{d\theta} + \sqrt{\frac{a}{\mu_E}}\left[\tan\tfrac{1}{2}\alpha\left(\frac{ds}{d\theta} - \frac{s}{a}\frac{da}{d\theta}\right)\right.$$
$$\left. + \tan\tfrac{1}{2}\beta\left(\frac{ds}{d\theta} + \frac{s-c}{a}\frac{da}{d\theta}\right)\right]$$

where

$$\frac{ds}{d\theta} = \frac{r_L r_T}{2c}\sin\theta$$

and

$$\frac{1}{a}\frac{da}{d\theta} = -\frac{2r_T(a - r_L)\sin\theta}{r_L^2 + c^2 - r_T^2}$$

Unfortunately, the right side of the expression for the derivative is indeterminate for $\theta = \pi$. However, it can be shown that

$$\lim_{\theta\to\pi}\frac{\sin\theta}{\sin\alpha} = \sqrt{\frac{r_L}{r_T}}\left(1 + \frac{r_L}{r_T}\right)$$

so that

$$\left.\frac{dt_{FL}}{d\theta}\right|_{\theta=\pi} = \sqrt{\frac{r_T^3(r_L + r_T)}{2\mu_E r_L}}$$

For step 2 of the procedural outline the position and velocity vectors $\mathbf{r}_{TM}$ and $\mathbf{v}_{TM}$ relative to the moon at the sphere of influence are required. These may be calculated in the following manner.

Equation (2.6), with $i = i_L$ and $\Omega = \Omega_1$ or $\Omega_2$, provides the direction cosines of the normal $\mathbf{i}_\varsigma$ to the elliptic trajectory plane. Thus

$$\mathbf{i}_\varsigma = \begin{bmatrix} \sin\Omega\sin i_L \\ -\cos\Omega\sin i_L \\ \cos i_L \end{bmatrix}$$

Then the velocity vector at $\mathbf{r}_T$ relative to the earth is

$$\mathbf{v}_T = \frac{1}{r_T}\left(\sqrt{\frac{\mu_E}{p}}\,e\sin\theta\,\mathbf{r}_T + \frac{\sqrt{\mu_E p}}{r_T}\mathbf{i}_\varsigma\times\mathbf{r}_T\right)$$

where $p$ is the parameter of the trajectory. Note that the position vector at injection $\mathbf{r}_L$ is easily obtained as

$$\mathbf{r}_L = \frac{r_L}{r_T}(\cos\theta\,\mathbf{r}_T - \sin\theta\,\mathbf{i}_\varsigma\times\mathbf{r}_T)$$

Finally, if $\mathbf{r}_{EM}$ and $\mathbf{v}_{EM}$ are the position and velocity vectors of the moon relative to the earth at time $t_A$, then

$$\mathbf{r}_{TM} = \mathbf{r}_T - \mathbf{r}_{EM}$$

$$\mathbf{v}_{TM} = \mathbf{v}_T - \mathbf{v}_{EM}$$

In order to determine the passing distance at the moon, it is sufficiently accurate for the purpose to assume that the velocity vector $\mathbf{v}_{TM}$ lies along the asymptote of the approach hyperbola. Then the unit vector $\mathbf{i}_a$, lying in the plane of the hyperbola and normal to the asymptote, is computed from

$$\mathbf{i}_a = \frac{\mathbf{v}_{TM} \times (\mathbf{r}_{TM} \times \mathbf{v}_{TM})}{|\mathbf{v}_{TM} \times (\mathbf{r}_{TM} \times \mathbf{v}_{TM})|}$$

and the offset or aiming point distance $r_a$ is

$$r_a = \mathbf{r}_{TM} \cdot \mathbf{i}_a$$

The minimum passing distance $r_m$ is computed using Eq. (9.20) with the angle $\nu$ determined from Eq. (9.21). The perilune position vector $\mathbf{r}_m$ is then readily found to be

$$\mathbf{r}_m = r_m \left( \cos \nu \, \mathbf{i}_a + \frac{\sin \nu}{v_{TM}} \mathbf{v}_{TM} \right)$$

◇ **Problem 9-7**

For the calculation described in the subsection above, Kepler's rather than Lagrange's equation could have been used

$$t_{FL} = \sqrt{\frac{a^3}{\mu_E}} \left[ E - \left( 1 - \frac{r_L}{a} \right) \sin E \right]$$

where $E$ is the eccentric anomaly of the target location $\mathbf{r}_T$.

Verify the relation

$$a = \frac{r_T - r_L \cos E}{1 - \cos E}$$

so that Kepler's equation can be written as

$$t_{FL} = \sqrt{\frac{r_T - r_L \cos E}{\mu_E (1 - \cos E)^3}} \left[ (r_T - r_L \cos E) E - (r_T - r_L) \sin E \right]$$

in terms of the single variable $E$. The appropriate derivative needed for the Newton iteration is then

$$\frac{dt_{FL}}{dE} = t_{FL} \left[ \frac{1}{2} \frac{r_L \sin E}{r_T - r_L \cos E} + \frac{r_T(1 - \cos E) + r_L E \sin E}{(r_T - r_L \cos E) E - (r_T - r_L) \sin E} - \frac{3}{2} \frac{\sin E}{1 - \cos E} \right]$$

◇ **Problem 9-8**

The sign of the radial component of the velocity vector $\mathbf{v}_1$ for a transfer orbit connecting $\mathbf{r}_1$ and $\mathbf{r}_2$ through a central angle $\theta$ is the same as the sign of

$$\sin \theta \left[ \left( 1 - \frac{p}{r_2} \right) - \left( 1 - \frac{p}{r_1} \right) \mathbf{r}_1 \cdot \mathbf{r}_2 \right]$$

where $p$ is the parameter of the orbit.

## 9.4 The Osculating Orbit and Encke's Method

The equation of the disturbed relative motion of two bodies, as developed in Sect. 8.4, is of the form

$$\frac{d^2\mathbf{r}}{dt^2} + \frac{\mu}{r^3}\mathbf{r} = \mathbf{a}_d \qquad (9.25)$$

where $\mathbf{r}$ is the vector position of one mass with respect to the other and $\mathbf{a}_d$ is the vector acceleration arising from the presence of the disturbing bodies. Actually, of course, the interpretation of $\mathbf{a}_d$ can be more general and indeed may include all relevant forces which tend to prevent the relative motion from being precisely a Keplerian orbit. For example, consider the motion of a space vehicle in the vicinity of the earth. Then, apart from the inverse-square central gravitational force, the other forces which influence the motion to varying degrees will include: (1) the nonspherical shape of the earth, (2) atmospheric lift and drag, (3) the sun, moon and other planets of the solar system, (4) solar radiation pressure, and (5) thrust acceleration from the vehicle's engines.

The most straightforward method for determining the position and velocity, $\mathbf{r}(t)$ and $\mathbf{v}(t)$, when the orbit is not a conic is a direct numerical integration of the equations of motion in rectangular coordinates known in celestial mechanics as *Cowell's method*.† The integration formulas used in the Cowell method actually were first given by Carl Friedrich Gauss and were well adapted to the computation techniques available at the end of the last century. Today, when Eq. (9.25) is integrated numerically in rectangular coordinates by any technique whatsoever, the method is still referred to as Cowell's method.

The choice of integrating the complete equations of motion (9.25) is reasonable if the disturbing acceleration magnitude $a_d$ is of the same or higher order as that due to the central force field. On the other hand, if $a_d$ is small, the method can be inefficient. Cowell's method may then require relatively small interval lengths independent of the size of $a_d$ to ensure a given accuracy. However, if only the differential accelerations instead of the total acceleration are integrated, considerable accuracy can be obtained with a larger interval when $a_d$ is small. This procedure, which

---

† Philip Herbert Cowell (1870–1949) graduated from Trinity College, Cambridge, England after having displayed unusual ability in mathematics. His two positions, first as chief assistant at the Royal Observatory at Greenwich in 1896 and superintendent of the *Nautical Almanac* in 1910, did not really provide him with the scope for theoretical research as would have been possible with a Cambridge professorship. Nevertheless, Cowell made important contributions to the theory of the motion of the moon for which he was elected a fellow of the Royal Society in 1906. His name will be remembered for the step-by-step numerical integration method of the planetary equations of motion in rectangular coordinates. He first applied his method to the newly discovered eighth moon of Jupiter and then to predict the return of Halley's comet in 1910.

will now be described in detail, is known as *Encke's method*† even though it was proposed two years earlier in 1849 by that famous Harvard University father-son team William Cranch Bond (1789–1859) and George Phillips Bond (1825–1865)—the latter often referred to as "the father of celestial photography."

If at a particular instant of time $t_0$, all the effects embodied in the vector $\mathbf{a}_d$ ceased to exercise any influence on the motion, the resulting orbit would be a conic and the position and velocity vectors would be exactly computable from the two-body formulas. Expressed differently, at any instant of time $t_0$, the position and velocity vectors of the relative motion may be used to define a two-body orbit. The terminology *osculating orbit* is used to describe this instantaneous conic path associated with the time $t_0$. Of course, the true motion never actually takes place along the osculating orbit; however, if the disturbing forces are small compared with the central body force, then over short intervals of time, the actual position in orbit will differ from the associated position in the osculating orbit by a correspondingly small amount.

The concept of the osculating orbit can be successfully exploited in the calculation of perturbed orbits. Specifically, let $\mathbf{r}(t)$, $\mathbf{v}(t)$ and $\mathbf{r}_{osc}(t)$, $\mathbf{v}_{osc}(t)$ represent, respectively, the position and velocity in the true orbit and the osculating orbit as functions of time. At time $t_0$, we have

$$\mathbf{r}(t_0) = \mathbf{r}_{osc}(t_0) \qquad \mathbf{v}(t_0) = \mathbf{v}_{osc}(t_0)$$

so that at a later time $t = t_0 + \Delta t$, we can write

$$\mathbf{r}(t) = \mathbf{r}_{osc}(t) + \boldsymbol{\delta}(t) \qquad \mathbf{v}(t) = \mathbf{v}_{osc}(t) + \boldsymbol{\nu}(t)$$

The vector difference $\boldsymbol{\delta}(t)$ is easily seen to satisfy the following differential equation:

$$\frac{d^2 \boldsymbol{\delta}}{dt^2} + \frac{\mu}{r_{osc}^3} \boldsymbol{\delta} = \frac{\mu}{r_{osc}^3} \left( 1 - \frac{r_{osc}^3}{r^3} \right) \mathbf{r} + \mathbf{a}_d \qquad (9.26)$$

subject to the initial conditions

$$\boldsymbol{\delta}(t_0) = \mathbf{0} \qquad \text{and} \qquad \left. \frac{d\boldsymbol{\delta}}{dt} \right|_{t=t_0} = \boldsymbol{\nu}(t_0) = \mathbf{0}$$

---

† Johann Franz Encke (1791–1865), the eighth child of a Lutheran preacher, studied mathematics and astronomy at the University of Göttingen. His education, though twice interrupted by military service during the Wars of Liberation, was guided by Gauss who procured for him in 1816 a post at the small Seeberg observatory near Gotha. As a result of his work on the computation of the orbit of a comet, which had the unusual period of scarcely four years, he was promoted, first to director of that observatory, and later in 1825, to a professorship at the Academy of Sciences in Berlin and the director of the Berlin observatory. The comet which brought him fame was later called Encke's comet. In Berlin, he also became editor of the *Berliner astronomisches Jahrbuch* in which many of his contributions to orbit determination and perturbation computations were published.

Sect. 9.4]    The Osculating Orbit and Encke's Method    449

The numerical difficulties which would arise from the evaluation of the coefficient of **r** in Eq. (9.26) may be avoided by employing the technique described in Sect. 8.4. Since

$$\mathbf{r}(t) = \mathbf{r}_{osc}(t) + \boldsymbol{\delta}(t)$$

it follows that

$$1 - \frac{r_{osc}^3}{r^3} = -f(q) = 1 - (1+q)^{\frac{3}{2}}$$

where

$$q = \frac{\boldsymbol{\delta} \cdot (\boldsymbol{\delta} - 2\mathbf{r})}{\mathbf{r} \cdot \mathbf{r}}$$

As before, the computation of $f(q)$ is expedited by writing

$$f(q) = q \frac{3 + 3q + q^2}{1 + (1+q)^{\frac{3}{2}}}$$

Encke's method may now be summarized as follows:

1. Position and velocity in the osculating orbit are calculated using the Lagrangian coefficients

$$\mathbf{r}_{osc}(t) = F\mathbf{r}(t_0) + G\mathbf{v}(t_0)$$
$$\mathbf{v}_{osc}(t) = F_t\mathbf{r}(t_0) + G_t\mathbf{v}(t_0)$$

   The functions $F$, $G$, ... are calculated using any of the appropriate formulas developed in Chapters 3 and 4. Of course, the calculation of these coefficients can be accomplished only by first solving the appropriate form of Kepler's equation.

2. Deviations from the osculating orbit are then obtained by a numerical integration of

$$\frac{d^2\boldsymbol{\delta}}{dt^2} + \frac{\mu}{r_{osc}^3}\boldsymbol{\delta} = -\frac{\mu}{r_{osc}^3}f(q)\mathbf{r}(t) + \mathbf{a}_d \quad (9.27)$$

where $\mathbf{r} = \mathbf{r}_{osc} + \boldsymbol{\delta}$. At any time the true position and velocity vectors are obtained by simply adding the computed deviations $\boldsymbol{\delta}$ and $\boldsymbol{\nu}$ to the osculating quantities $\mathbf{r}_{osc}$ and $\mathbf{v}_{osc}$.

The various terms in Eq. (9.27) must remain small, i.e., of the same order as $a_d(t)$, if the method is to be efficient. As the deviation vector $\boldsymbol{\delta}$ grows in magnitude, the various acceleration terms will eventually increase in size. Therefore, in order to maintain the efficiency, a new osculating orbit should then be defined using the computed values of the true position and velocity vectors. The process of selecting a new conic orbit from which to calculate deviations is called *rectification*. When rectification occurs, the initial conditions for the $\boldsymbol{\delta}$ differential equation are again zero and the only nonzero driving accelerations immediately following rectification are simply the disturbing accelerations $\mathbf{a}_d$.

It is worthwhile emphasizing the role that Encke's method plays in maintaining control of numerical errors. The determination of position and velocity in the osculating orbit is subject only to round-off errors and is independent of the particular numerical technique used to perform the integration. The accuracy in the computation of $\boldsymbol{\delta}$ and $\boldsymbol{\nu}$, the deviations from the osculating orbit, is limited by both round-off and truncation—the effects of the latter will propagate from step to step along the integrated orbit. The integrated quantities themselves are small and, when added to the osculating quantities, will have little effect on the determination of the true orbit. Before the errors in the deviations can grow in size sufficient to have a detrimental effect, a new osculating orbit is selected through the process of rectification.

◇ **Problem 9-9**

Derive the appropriate differential equation for Encke's method when the universal anomaly $\chi$ (defined in Chapter 4) rather than $t$ is the independent variable. Discuss the possible advantage of such a formulation of the integration process.

◇ **Problem 9-10**

For Encke's method,

$$1 - \frac{r_{osc}^3}{r^3} = 1 - (1+q)^{-\frac{3}{2}} = q\frac{3 + 3q + q^2}{(1+q)^{\frac{3}{2}} + (1+q)^3}$$

with

$$q = \frac{(\boldsymbol{\delta} + 2\mathbf{r}_{osc}) \cdot \boldsymbol{\delta}}{r_{osc}^2}$$

provides an alternate means of calculating the coefficient of $\mathbf{r}$ in Eq. (9.26). Further, this coefficient may be expressed as a power series in $q$ of the form

$$1 - \frac{r_{osc}^3}{r^3} = 3\frac{q}{2}\left[1 - \frac{5}{2}\left(\frac{q}{2}\right) + \frac{5 \cdot 7}{2 \cdot 3}\left(\frac{q}{2}\right)^2 - \frac{5 \cdot 7 \cdot 9}{2 \cdot 3 \cdot 4}\left(\frac{q}{2}\right)^3 + \cdots\right]$$

Determine the range of values of $q$ for which the series will converge.

NOTE: This power series expansion is the classical method used for calculating $f(q)$ in Encke's method.

## 9.5 Linearization and the State Transition Matrix

The osculating orbit introduced by Encke may be regarded as a *nominal* or *reference orbit* against which deviations are computed to establish the true orbit in the problem of disturbed motion. The Encke perturbation equation is not an approximation since it contains exactly the same information as the original equations of motion. Approximations are introduced only when a particular numerical integration procedure is selected to produce numerical solutions.

## Sect. 9.5]   Linearization and the State Transition Matrix   451

In this section we are concerned with a somewhat different kind of perturbation problem. In contrast with Encke's method, the method of *linearized perturbations* does not provide an exact description of the motion. Basically, the approach is to linearize the equations of motion by a series expansion about a reference orbit in which only first-order terms are retained. For the results to remain valid it is, of course, necessary to restrict the magnitude of the deviations from the nominal path. When applicable, many advantages accrue from the linearized method of analysis—the principal one being, of course, that the process of superposition may be applied.

Paralleling the development of Encke's method, let the vectors $\mathbf{r}(t)$, $\mathbf{v}(t)$ and $\mathbf{r}_{ref}(t)$, $\mathbf{v}_{ref}(t)$ represent the position and velocity along the actual orbit and the corresponding quantities along a reference path which need *not* be a conic. We write

$$\mathbf{r}(t) = \mathbf{r}_{ref}(t) + \boldsymbol{\delta}(t) \qquad \mathbf{v}(t) = \mathbf{v}_{ref}(t) + \boldsymbol{\nu}(t) \qquad (9.28)$$

as before and note that both the actual and reference quantities satisfy the same basic equations of motion

$$\frac{d\mathbf{r}}{dt} = \mathbf{v} \qquad \frac{d\mathbf{v}}{dt} = \mathbf{g}(\mathbf{r}) \qquad (9.29)$$

The vector $\mathbf{g}$ includes all relevant gravitational effects under consideration.

Since $\mathbf{g}$ is a function of $\mathbf{r}$, we may expand $\mathbf{g}(\mathbf{r})$ in a Taylor series about the point $\mathbf{r}_{ref}$

$$\mathbf{g}(\mathbf{r}) = \mathbf{g}(\mathbf{r}_{ref}) + \mathbf{G}(\mathbf{r}_{ref})\boldsymbol{\delta} + O(\delta^2)$$

where

$$\mathbf{G}(\mathbf{r}_{ref}) = \left.\frac{\partial \mathbf{g}}{\partial \mathbf{r}}\right|_{\mathbf{r}=\mathbf{r}_{ref}} \qquad (9.30)$$

is referred to as the *gravity gradient matrix*. Then, substituting in the equations of motion, we have

$$\frac{d\mathbf{r}_{ref}}{dt} + \frac{d\boldsymbol{\delta}}{dt} = \mathbf{v}_{ref} + \boldsymbol{\nu} \qquad \frac{d\mathbf{v}_{ref}}{dt} + \frac{d\boldsymbol{\nu}}{dt} = \mathbf{g}(\mathbf{r}_{ref}) + \mathbf{G}\boldsymbol{\delta}$$

if all terms of order $\delta^2$ and higher are neglected. But the reference quantities satisfy Eq. (9.29), so that

$$\frac{d\boldsymbol{\delta}}{dt} = \boldsymbol{\nu} \quad \text{and} \quad \frac{d\boldsymbol{\nu}}{dt} = \mathbf{G}(\mathbf{r}_{ref})\boldsymbol{\delta} \qquad (9.31)$$

are obtained as the linearized differential equations for the deviation vectors $\boldsymbol{\delta}$ and $\boldsymbol{\nu}$. Since the $\mathbf{G}$ matrix depends only upon the reference orbit, it may be regarded as a known function of time.

At this point it is convenient to introduce a six-dimensional deviation vector $\mathbf{x}$ called the *state vector* and defined by

$$\mathbf{x} = \begin{bmatrix} \boldsymbol{\delta} \\ \boldsymbol{\nu} \end{bmatrix} \qquad (9.32)$$

so that the linearized equations of motion can be simply written

$$\frac{d\mathbf{x}}{dt} = \mathbf{F}(t)\mathbf{x} \tag{9.33}$$

The six-dimensional coefficient matrix $\mathbf{F}(t)$ is partitioned as

$$\mathbf{F}(t) = \begin{bmatrix} \mathbf{O} & \mathbf{I} \\ \mathbf{G}(t) & \mathbf{O} \end{bmatrix} \tag{9.34}$$

where $\mathbf{O}$ and $\mathbf{I}$ are the three-dimensional zero and identity matrices.

The sixth-order system of linear differential equations admits six linearly independent solutions $\mathbf{x}_1(t)$, $\mathbf{x}_2(t)$, ..., $\mathbf{x}_6(t)$ which may be regarded as the columns of a six-dimensional matrix $\boldsymbol{\Phi}$. If the initial conditions are prescribed at a time $t_0$ in such a manner that $\mathbf{x}_j(t_0)$ has all components zero except the $j^{th}$ which is unity, then the matrix $\boldsymbol{\Phi}(t,t_0)$ will be a function of both $t$ and $t_0$ satisfying the matrix differential equation

$$\frac{d}{dt}\boldsymbol{\Phi}(t,t_0) = \mathbf{F}(t)\boldsymbol{\Phi}(t,t_0) \tag{9.35}$$

subject to the initial conditions

$$\boldsymbol{\Phi}(t_0,t_0) = \mathbf{I} \tag{9.36}$$

where $\mathbf{I}$ is now the six-dimensional identity matrix.

The matrix $\boldsymbol{\Phi}(t,t_0)$ is frequently referred to as the *state transition matrix* (or *fundamental matrix*). Indeed, if the deviation or state vector $\mathbf{x}$ is known at a time $t_0$, then its value at time $t$ is obtained simply from the product

$$\mathbf{x}(t) = \boldsymbol{\Phi}(t,t_0)\mathbf{x}(t_0) \tag{9.37}$$

[Clearly, $\mathbf{x}(t)$ satisfies the system equation (9.33) since $\boldsymbol{\Phi}(t,t_0)$ satisfies Eq. (9.35) and has the proper value at time $t_0$ according to Eq. (9.36).]

Solution of the Forced Linear System

If the basic equations of motion (9.29) include an additive disturbing acceleration $\mathbf{a}_d(t)$, then the system equations (9.33) will have the form

$$\frac{d\mathbf{x}}{dt} = \mathbf{F}(t)\mathbf{x} + \mathbf{K}\mathbf{a}_d(t) \tag{9.38}$$

where the $6 \times 3$ compatibility matrix $\mathbf{K}$ is defined by

$$\mathbf{K} = \begin{bmatrix} \mathbf{O} \\ \mathbf{I} \end{bmatrix}$$

The additional term is often referred to as the *forcing function*.

Sect. 9.5]   Linearization and the State Transition Matrix   453

The solution of the inhomogeneous system (9.38) can be written explicitly using the state transition matrix of the homogeneous system (9.33). Indeed, we have

$$\mathbf{x}(t) = \mathbf{\Phi}(t,t_0)\mathbf{x}(t_0) + \mathbf{\Phi}(t,t_0)\int_{t_0}^{t}\mathbf{\Phi}^{-1}(\tau,t_0)\mathbf{K}\mathbf{a}_d(\tau)\,d\tau \qquad (9.39)$$

which is readily verified by direct substitution into Eq. (9.38) and using the fact that $\mathbf{\Phi}$ satisfies Eq. (9.35).

The inverse of the matrix $\mathbf{\Phi}$ appearing in the integrand of Eq. (9.39) is easily seen to be

$$\mathbf{\Phi}^{-1}(t,t_0) = \mathbf{\Phi}(t_0,t) \qquad (9.40)$$

By interchanging $t$ and $t_0$ in Eq. (9.37), we have

$$\mathbf{x}(t_0) = \mathbf{\Phi}(t_0,t)\mathbf{x}(t)$$

and, therefore,

$$\mathbf{x}(t) = \mathbf{\Phi}(t,t_0)\mathbf{\Phi}(t_0,t)\mathbf{x}(t)$$

Hence, $\mathbf{\Phi}(t,t_0)\mathbf{\Phi}(t_0,t)$ must be the identity matrix.

Symplectic Property of the Transition Matrix

The transition matrix is an example of a class called *symplectic matrices*. An even-dimensional matrix $\mathbf{A}$ is said to be symplectic if

$$\mathbf{A}^T\mathbf{J}\mathbf{A} = \mathbf{J} \qquad (9.41)$$

where

$$\mathbf{J} = \begin{bmatrix} \mathbf{O} & \mathbf{I} \\ -\mathbf{I} & \mathbf{O} \end{bmatrix}$$

Since $\mathbf{J}^2 = -\mathbf{I}$, the $\mathbf{J}$ matrix is analogous to the pure imaginary $\sqrt{-1}$ in complex algebra.

From the definition (9.41) it is seen that a symplectic matrix bears the same relationship to the matrix $\mathbf{J}$ that an orthogonal matrix bears to the identify matrix $\mathbf{I}$. Thus, if $\mathbf{P}$ is an orthogonal matrix, then

$$\mathbf{P}^T\mathbf{I}\mathbf{P} = \mathbf{I}$$

The importance of identifying the transition matrix as symplectic lies in the ease with which the inverse may be obtained. Postmultiply Eq. (9.41) by $\mathbf{A}^{-1}$ and premultiply by $\mathbf{J}$, to obtain

$$\mathbf{A}^{-1} = -\mathbf{J}\mathbf{A}^T\mathbf{J} \qquad (9.42)$$

so that the inverse of a symplectic matrix is found by a simple rearrangement of the elements. By comparison, the inverse of an orthogonal matrix is equal to its transpose—again, by rearrangement of the elements.

To show that $\boldsymbol{\Phi}(t,t_0)$ is symplectic we first note that

$$\boldsymbol{\Phi}^\mathrm{T}(t_0,t_0)\mathbf{J}\boldsymbol{\Phi}(t_0,t_0) = \mathbf{J}$$

since $\boldsymbol{\Phi}(t_0,t_0)$ is just the identity matrix. Therefore, to complete the proof, we need show only that

$$\frac{d}{dt}[\boldsymbol{\Phi}^\mathrm{T}(t,t_0)\mathbf{J}\boldsymbol{\Phi}(t,t_0)] = \mathbf{O}$$

For this purpose, we use Eq. (9.35) and write

$$\frac{d}{dt}[\boldsymbol{\Phi}^\mathrm{T}(t,t_0)\mathbf{J}\boldsymbol{\Phi}(t,t_0)] = \boldsymbol{\Phi}^\mathrm{T}(t,t_0)[\mathbf{F}^\mathrm{T}(t)\mathbf{J} + \mathbf{J}\mathbf{F}(t)]\boldsymbol{\Phi}(t,t_0)$$

$$= \boldsymbol{\Phi}^\mathrm{T}(t,t_0)\begin{bmatrix} \mathbf{G}(t) - \mathbf{G}^\mathrm{T}(t) & \mathbf{O} \\ \mathbf{O} & \mathbf{I} - \mathbf{I} \end{bmatrix}\boldsymbol{\Phi}(t,t_0)$$

$$= \mathbf{O}$$

The last step follows from the fact that $\mathbf{G}(t) = \mathbf{G}^\mathrm{T}(t)$; that is, the gravity gradient is a symmetric matrix.

Finally, if the transition matrix is partitioned as

$$\boldsymbol{\Phi}(t,t_0) = \begin{bmatrix} \boldsymbol{\Phi}_1(t,t_0) & \boldsymbol{\Phi}_2(t,t_0) \\ \boldsymbol{\Phi}_3(t,t_0) & \boldsymbol{\Phi}_4(t,t_0) \end{bmatrix} \quad (9.43)$$

then the inverse is directly obtained from

$$\boldsymbol{\Phi}^{-1}(t,t_0) = \boldsymbol{\Phi}(t_0,t) = \begin{bmatrix} \boldsymbol{\Phi}_4^\mathrm{T}(t,t_0) & -\boldsymbol{\Phi}_2^\mathrm{T}(t,t_0) \\ -\boldsymbol{\Phi}_3^\mathrm{T}(t,t_0) & \boldsymbol{\Phi}_1^\mathrm{T}(t,t_0) \end{bmatrix} \quad (9.44)$$

which can be demonstrated using Eq. (9.42).

◇ **Problem 9–11**

For two-body motion the vector $\mathbf{g}$ in Eq. (9.29) is simply

$$\mathbf{g}(\mathbf{r}) = -\frac{\mu}{r^3}\mathbf{r}$$

For this case, the gravity gradient matrix along the reference orbit is

$$\mathbf{G} = \frac{\mu}{r_{ref}^5}(3\mathbf{r}_{ref}\mathbf{r}_{ref}^\mathrm{T} - r_{ref}^2\mathbf{I}) = \frac{\mu}{r_{ref}^5}\begin{bmatrix} 3x^2 - r^2 & 3xy & 3xz \\ 3yx & 3y^2 - r^2 & 3yz \\ 3zx & 3zy & 3z^2 - r^2 \end{bmatrix}_{ref}$$

◇ **Problem 9–12**

Verify that the determinant of a symplectic matrix is $\pm 1$ and use this result to deduce

$$|\boldsymbol{\Phi}(t,t_0)| = 1$$

## Problem 9-13
A matrix $\mathbf{A}$ satisfies the matrix differential equation

$$\frac{d\mathbf{A}}{dt} = \mathbf{B}(t)\mathbf{A}$$

Demonstrate that the determinant $|\mathbf{A}|$ satisfies the scalar differential equation

$$\frac{d|\mathbf{A}|}{dt} = \operatorname{tr}(\mathbf{B})\,|\mathbf{A}|$$

and use the result to give an alternate proof of the fact that the determinant of the transition matrix is unity.

## Problem 9-14
If $\mathbf{\Phi}(t)$ satisfies the differential equation

$$\frac{d\mathbf{\Phi}}{dt} = \mathbf{F}\mathbf{\Phi}$$

then $\mathbf{\Phi}^{-1}(t)$ satisfies

$$\frac{d\mathbf{\Phi}^{-1}}{dt} = -\mathbf{\Phi}^{-1}\mathbf{F}$$

Hence, deduce that

$$\mathbf{\Phi}(t, t_0) = \mathbf{\Phi}(t)\mathbf{\Phi}^{-1}(t_0)$$

regardless of the initial conditions assigned to $\mathbf{\Phi}(t)$.

## Problem 9-15
If $\mathbf{A}$ and $\mathbf{B}$ are symplectic matrices, then $\mathbf{C} = \mathbf{AB}$ is symplectic. Further, the inverse and transpose of a symplectic matrix are both symplectic.

## Problem 9-16
Any two-dimensional matrix whose determinant is unity is symplectic.

## Problem 9-17
A matrix $\mathbf{A}(t)$ of even dimension $2n$ satisfies the differential equation

$$\frac{d\mathbf{A}}{dt} = \mathbf{B}(t)\mathbf{A} \quad \text{with} \quad \mathbf{A}(0) = \mathbf{I}$$

If the coefficient matrix $\mathbf{B}$ is partitioned into $n$-dimensional blocks as

$$\mathbf{B} = \begin{bmatrix} \mathbf{B}_1 & \mathbf{B}_2 \\ \mathbf{B}_3 & \mathbf{B}_4 \end{bmatrix}$$

find the necessary and sufficient conditions that these partitions must satisfy for $\mathbf{A}$ to be a symplectic matrix.

## 9.6 Fundamental Perturbation Matrices

Consider a vehicle launched into orbit and moving under the influence of one or more gravity fields to reach a target point. Let $\mathbf{r}_{ref}(t)$ and $\mathbf{v}_{ref}(t)$ be the position and velocity vectors at time $t$ for a vehicle in a reference orbit connecting the initial and terminal points. Because of errors from any of a number of sources, the vehicle will fail to follow the exact reference path so that the true position and velocity vectors $\mathbf{r}(t)$ and $\mathbf{v}(t)$ will deviate from the associated reference quantities. It will be assumed that these deviations from the reference path are always small so that linearization techniques are applicable.

At any time $t$ later than $t_0$, the position and velocity vectors will be a function, not only of time, but also of the position and velocity that the vehicle had at the earlier time $t_0$. Thus, we may expand $\mathbf{r}[t, \mathbf{r}(t_0), \mathbf{v}(t_0)]$ in a Taylor series about the reference quantities to obtain

$$\mathbf{r}[t, \mathbf{r}(t_0), \mathbf{v}(t_0)] = \mathbf{r}_{ref}[t, \mathbf{r}_{ref}(t_0), \mathbf{v}_{ref}(t_0)]$$
$$+ \left.\frac{\partial \mathbf{r}(t)}{\partial \mathbf{r}(t_0)}\right|_{ref} [\mathbf{r}(t_0) - \mathbf{r}_{ref}(t_0)] + \left.\frac{\partial \mathbf{r}(t)}{\partial \mathbf{v}(t_0)}\right|_{ref} [\mathbf{v}(t_0) - \mathbf{v}_{ref}(t_0)] + \cdots$$

Similarly, for the velocity vector,

$$\mathbf{v}[t, \mathbf{r}(t_0), \mathbf{v}(t_0)] = \mathbf{v}_{ref}[t, \mathbf{r}_{ref}(t_0), \mathbf{v}_{ref}(t_0)]$$
$$+ \left.\frac{\partial \mathbf{v}(t)}{\partial \mathbf{r}(t_0)}\right|_{ref} [\mathbf{r}(t_0) - \mathbf{r}_{ref}(t_0)] + \left.\frac{\partial \mathbf{v}(t)}{\partial \mathbf{v}(t_0)}\right|_{ref} [\mathbf{v}(t_0) - \mathbf{v}_{ref}(t_0)] + \cdots$$

These expansions may be written more briefly as

$$\mathbf{r}(t) = \mathbf{r}_{ref}(t) + \left.\frac{\partial \mathbf{r}}{\partial \mathbf{r}_0}\right|_{ref} \delta\mathbf{r}_0 + \left.\frac{\partial \mathbf{r}}{\partial \mathbf{v}_0}\right|_{ref} \delta\mathbf{v}_0 + \cdots$$

$$\mathbf{v}(t) = \mathbf{v}_{ref}(t) + \left.\frac{\partial \mathbf{v}}{\partial \mathbf{r}_0}\right|_{ref} \delta\mathbf{r}_0 + \left.\frac{\partial \mathbf{v}}{\partial \mathbf{v}_0}\right|_{ref} \delta\mathbf{v}_0 + \cdots$$

or, in vector-matrix notation, as

$$\begin{bmatrix} \delta\mathbf{r} \\ \delta\mathbf{v} \end{bmatrix} = \Phi(t, t_0) \begin{bmatrix} \delta\mathbf{r}_0 \\ \delta\mathbf{v}_0 \end{bmatrix} \quad \text{where} \quad \Phi(t, t_0) = \begin{bmatrix} \dfrac{\partial \mathbf{r}}{\partial \mathbf{r}_0} & \dfrac{\partial \mathbf{r}}{\partial \mathbf{v}_0} \\ \dfrac{\partial \mathbf{v}}{\partial \mathbf{r}_0} & \dfrac{\partial \mathbf{v}}{\partial \mathbf{v}_0} \end{bmatrix}_{ref} \quad (9.45)$$

The matrix $\Phi(t, t_0)$ is the state transition matrix introduced in the previous section.

## Sect. 9.6]  Fundamental Perturbation Matrices  457

Equation (9.45) can also be written in the form

$$\begin{bmatrix} \delta \mathbf{r}_0 \\ \delta \mathbf{v}_0 \end{bmatrix} = \mathbf{\Phi}(t_0, t) \begin{bmatrix} \delta \mathbf{r} \\ \delta \mathbf{v} \end{bmatrix} \quad \text{where} \quad \mathbf{\Phi}(t_0, t) = \begin{bmatrix} \dfrac{\partial \mathbf{r}_0}{\partial \mathbf{r}} & \dfrac{\partial \mathbf{r}_0}{\partial \mathbf{v}} \\ \dfrac{\partial \mathbf{v}_0}{\partial \mathbf{r}} & \dfrac{\partial \mathbf{v}_0}{\partial \mathbf{v}} \end{bmatrix}_{ref} \quad (9.46)$$

so that

$$\mathbf{\Phi}(t, t_0) \mathbf{\Phi}(t_0, t) = \mathbf{I}$$

Here we shall explore other properties of $\mathbf{\Phi}$ and its partitions. Later in Chapter 11, applications to the problems of midcourse velocity corrections needed to fulfill particular mission objectives will be developed.

### Partitions of the Transition Matrix

Define the partitions† of the transition matrix to be

$$\mathbf{\Phi}(t, t_0) = \begin{bmatrix} \widetilde{\mathbf{R}}(t) & \mathbf{R}(t) \\ \widetilde{\mathbf{V}}(t) & \mathbf{V}(t) \end{bmatrix} \quad (9.47)$$

and, since

$$\frac{d\mathbf{\Phi}(t, t_0)}{dt} = \begin{bmatrix} \mathbf{O} & \mathbf{I} \\ \mathbf{G} & \mathbf{O} \end{bmatrix} \mathbf{\Phi}(t, t_0) \quad \text{with} \quad \mathbf{\Phi}(t_0, t_0) = \mathbf{I} \quad (9.48)$$

then

$$\frac{d\widetilde{\mathbf{R}}}{dt} = \widetilde{\mathbf{V}} \quad \frac{d\widetilde{\mathbf{V}}}{dt} = \mathbf{G}\widetilde{\mathbf{R}} \qquad \frac{d\mathbf{R}}{dt} = \mathbf{V} \quad \frac{d\mathbf{V}}{dt} = \mathbf{G}\mathbf{R}$$
$$\widetilde{\mathbf{R}}(t_0) = \mathbf{I} \quad \widetilde{\mathbf{V}}(t_0) = \mathbf{O} \qquad \mathbf{R}(t_0) = \mathbf{O} \quad \mathbf{V}(t_0) = \mathbf{I} \quad (9.49)$$

Thus, the fundamental perturbation matrices can be generated as solutions of two pairs of uncoupled matrix differential equations.

The elements of the $\mathbf{R}$ and $\mathbf{V}$ matrices represent deviations in position and velocity from the reference path as a result of certain specific deviations in the launch velocity from its reference value. For example,

---

† The matrices $\mathbf{R}$ and $\mathbf{V}$, together with their adjoints $\mathbf{R}^*$ and $\mathbf{V}^*$, were first introduced in the paper "A Navigation Theory for Round-trip Reconnaissance Missions to Venus and Mars" authored by Richard H. Battin and J. Halcombe Laning, Jr. and presented at the *Fourth Air Force Ballistic Missile Division and Space Technology Laboratory Symposium* held at the University of California Los Angeles in August 1959. It was subsequently published in *Planetary Space Science*, vol. 7 by Pergamon Press in 1961.

At that time, the concept of state space, state vectors, and the state transition matrix was in its infancy and certainly not familiar to the authors. It was only recently that the author introduced the matrices $\widetilde{\mathbf{R}}$, $\widetilde{\mathbf{V}}$, $\widetilde{\mathbf{R}}^*$, and $\widetilde{\mathbf{V}}^*$ which add, significantly, to the elegance of the expressions.

the first columns of these matrices are the vector deviations at time $t$ due to a unit change in the first component of the velocity at time $t_0$. Corresponding interpretations may be ascribed to the other columns as well. Similar interpretations, of course, apply to $\tilde{\mathbf{R}}$ and $\tilde{\mathbf{V}}$ also.

Since $\Phi(t, t_0)$ is symplectic, we have

$$\Phi^{-1}(t, t_0) = \begin{bmatrix} \mathbf{V}^T & -\mathbf{R}^T \\ -\tilde{\mathbf{V}}^T & \tilde{\mathbf{R}}^T \end{bmatrix}$$

Therefore,

$$\begin{bmatrix} \tilde{\mathbf{R}} & \mathbf{R} \\ \tilde{\mathbf{V}} & \mathbf{V} \end{bmatrix} \begin{bmatrix} \mathbf{V}^T & -\mathbf{R}^T \\ -\tilde{\mathbf{V}}^T & \tilde{\mathbf{R}}^T \end{bmatrix} = \begin{bmatrix} \mathbf{V}^T & -\mathbf{R}^T \\ -\tilde{\mathbf{V}}^T & \tilde{\mathbf{R}}^T \end{bmatrix} \begin{bmatrix} \tilde{\mathbf{R}} & \mathbf{R} \\ \tilde{\mathbf{V}} & \mathbf{V} \end{bmatrix} = \begin{bmatrix} \mathbf{I} & \mathbf{O} \\ \mathbf{O} & \mathbf{I} \end{bmatrix}$$

from which the following identities are readily obtained:

$$\begin{array}{ll} \tilde{\mathbf{R}}\mathbf{V}^T - \mathbf{R}\tilde{\mathbf{V}}^T = \mathbf{I} & \mathbf{V}^T\tilde{\mathbf{R}} - \mathbf{R}^T\tilde{\mathbf{V}} = \mathbf{I} \\ -\tilde{\mathbf{R}}\mathbf{R}^T + \mathbf{R}\tilde{\mathbf{R}}^T = \mathbf{O} & \mathbf{V}^T\mathbf{R} - \mathbf{R}^T\mathbf{V} = \mathbf{O} \\ \tilde{\mathbf{V}}\mathbf{V}^T - \mathbf{V}\tilde{\mathbf{V}}^T = \mathbf{O} & -\tilde{\mathbf{V}}^T\tilde{\mathbf{R}} + \tilde{\mathbf{R}}^T\tilde{\mathbf{V}} = \mathbf{O} \\ -\tilde{\mathbf{V}}\mathbf{R}^T + \mathbf{V}\tilde{\mathbf{R}}^T = \mathbf{I} & -\tilde{\mathbf{V}}^T\mathbf{R} + \tilde{\mathbf{R}}^T\mathbf{V} = \mathbf{I} \end{array} \quad (9.50)$$

These identities can be used to establish the symmetry of several products of the partitions of the transition matrix. Indeed, four of these relations show immediately that

$$\mathbf{R}\tilde{\mathbf{R}}^T \quad \mathbf{V}\tilde{\mathbf{V}}^T \quad \mathbf{R}^T\mathbf{V} \quad \tilde{\mathbf{R}}^T\tilde{\mathbf{V}} \quad \text{are symmetric.} \quad (9.51)$$

Furthermore, the matrix $\mathbf{VR}^{-1}$ is also symmetric as can be seen from

$$\mathbf{VR}^{-1} = \mathbf{R}^{-T}\mathbf{R}^T\mathbf{VR}^{-1} = \mathbf{R}^{-T}\mathbf{V}^T\mathbf{RR}^{-1} = \mathbf{R}^{-T}\mathbf{V}^T$$
$$= (\mathbf{VR}^{-1})^T$$

Similarly, we can demonstrate the symmetry of three additional matrix products. In summary,

$$\mathbf{VR}^{-1} \quad \mathbf{R}^{-1}\tilde{\mathbf{R}} \quad \mathbf{V}^{-1}\tilde{\mathbf{V}} \quad \tilde{\mathbf{V}}\tilde{\mathbf{R}}^{-1} \quad \text{are symmetric} \quad (9.52)$$

and, since the inverses of symmetric matrices are symmetric, then

$$\mathbf{RV}^{-1} \quad \tilde{\mathbf{R}}^{-1}\mathbf{R} \quad \tilde{\mathbf{V}}^{-1}\mathbf{V} \quad \tilde{\mathbf{R}}\tilde{\mathbf{V}}^{-1} \quad \text{are symmetric.} \quad (9.53)$$

From the set of eight identities derived above involving the identity matrix $\mathbf{I}$, only one is unique as can be shown using the symmetry properties just developed. The first and last identities in each column of Eqs. (9.50) are obviously the same. The last equations in each column can be written as

$$\mathbf{V} = \tilde{\mathbf{R}}^{-T} + \tilde{\mathbf{V}}\mathbf{R}^T\tilde{\mathbf{R}}^{-T} \quad \text{and} \quad \mathbf{V} = \tilde{\mathbf{R}}^{-T} + \tilde{\mathbf{R}}^{-T}\tilde{\mathbf{V}}^T\mathbf{R}$$

But

$$\tilde{\mathbf{R}}^{-T}\tilde{\mathbf{V}}^T\mathbf{R} = \tilde{\mathbf{V}}\tilde{\mathbf{R}}^{-1}\mathbf{R} = \tilde{\mathbf{V}}\mathbf{R}^T\tilde{\mathbf{R}}^{-T}$$

with the result that
$$\mathbf{V}\tilde{\mathbf{R}}^T - \tilde{\mathbf{V}}\mathbf{R}^T = \mathbf{I} \tag{9.54}$$
is the only one of the four identities with which we need be concerned. Indeed, this identity shows that the partitions of the transition matrix are *not* independent. If any three are known, the fourth may be determined from Eq. (9.54).

The two matrices $\tilde{\mathbf{C}}$ and $\mathbf{C}$ defined as
$$\tilde{\mathbf{C}} = \tilde{\mathbf{V}}\tilde{\mathbf{R}}^{-1} \quad \text{and} \quad \mathbf{C} = \mathbf{V}\mathbf{R}^{-1} \tag{9.55}$$
are symmetric and have special significance. If $\mathbf{v}_0$ is constant, then
$$\tilde{\mathbf{C}}\tilde{\mathbf{R}}\,\delta\mathbf{r}_0 = \tilde{\mathbf{V}}\,\delta\mathbf{r}_0 \quad \text{so that} \quad \tilde{\mathbf{C}}\,\delta\mathbf{r} = \delta\mathbf{v}$$
Similarly, if $\mathbf{r}_0$ is constant, then
$$\mathbf{C}\mathbf{R}\,\delta\mathbf{v}_0 = \mathbf{V}\,\delta\mathbf{v}_0 \quad \text{so that} \quad \mathbf{C}\,\delta\mathbf{r} = \delta\mathbf{v}$$
Therefore,
$$\tilde{\mathbf{C}} = \left.\frac{\partial \mathbf{v}}{\partial \mathbf{r}}\right|_{\mathbf{v}_0=\text{constant}} \quad \text{and} \quad \mathbf{C} = \left.\frac{\partial \mathbf{v}}{\partial \mathbf{r}}\right|_{\mathbf{r}_0=\text{constant}} \tag{9.56}$$

Further note, in terms of $\tilde{\mathbf{C}}$ and $\mathbf{C}$, Eq. (9.54) may be written either as
$$\mathbf{V} = \tilde{\mathbf{C}}\mathbf{R} + \tilde{\mathbf{R}}^{-T} \quad \text{or} \quad \tilde{\mathbf{V}} = \mathbf{C}\tilde{\mathbf{R}} - \mathbf{R}^{-T} \tag{9.57}$$

We can also show that the matrices $\tilde{\mathbf{C}}$ and $\mathbf{C}^{-1}$ satisfy so-called *matrix Ricatti equations*.† By differentiating the identity $\tilde{\mathbf{C}}\tilde{\mathbf{R}} = \tilde{\mathbf{V}}$, we have
$$\frac{d\tilde{\mathbf{C}}}{dt}\tilde{\mathbf{R}} + \tilde{\mathbf{C}}\frac{d\tilde{\mathbf{R}}}{dt} = \frac{d\tilde{\mathbf{V}}}{dt}$$
But $d\tilde{\mathbf{R}}/dt = \tilde{\mathbf{V}}$ and $d\tilde{\mathbf{V}}/dt = \mathbf{G}\tilde{\mathbf{R}}$ so that
$$\frac{d\tilde{\mathbf{C}}}{dt}\tilde{\mathbf{R}} + \tilde{\mathbf{C}}\tilde{\mathbf{V}} = \mathbf{G}\tilde{\mathbf{R}}$$
which can be written as
$$\frac{d\tilde{\mathbf{C}}}{dt} + \tilde{\mathbf{C}}^2 = \mathbf{G} \quad \text{with} \quad \tilde{\mathbf{C}}(t_0) = \mathbf{O} \tag{9.58}$$

---

† The nonlinear equation
$$\frac{dy}{dx} = a_0(x) + a_1(x)y + a_2(x)y^2$$
was important to the early history of ordinary differential equations. It acquired its significance when it was introduced in 1724 by Jacopo Francesco, Count Ricatti of Venice (1676–1754), in his work in acoustics.

Jean Le Rond D'Alembert (1717–1783) was the first to consider the general form of the equation and to use the term "Ricatti equation."

## 460  Patched-Conic Orbits and Perturbation Methods  [Chap. 9

Similarly,

$$\frac{d\mathbf{C}^{-1}}{dt} + \mathbf{C}^{-1}\mathbf{G}\mathbf{C}^{-1} = \mathbf{I} \quad \text{with} \quad \mathbf{C}^{-1}(t_0) = \mathbf{O} \qquad (9.59)$$

These differential equations also provide independent demonstrations that the matrices $\widetilde{\mathbf{C}}$ and $\mathbf{C}$ are symmetric. The initial condition $\mathbf{O}$ is a symmetric matrix and since $\mathbf{G}$ and $\mathbf{I}$ are symmetric, the differential equations are also symmetric. Hence, $\widetilde{\mathbf{C}}$ and $\mathbf{C}^{-1}$ are initially symmetric and must remain so thereafter.

◊ **Problem 9-18**

Let $\delta \mathbf{r}_1$ and $\delta \mathbf{r}_2$ be the deviations in position at two distinct times $t_1$ and $t_2$. Then the position and velocity deviations from reference values at any time $t$ are given by

$$\delta \mathbf{r}(t) = \widetilde{\mathbf{R}}(t)\mathbf{c} + \mathbf{R}(t)\mathbf{d}$$
$$\delta \mathbf{v}(t) = \widetilde{\mathbf{V}}(t)\mathbf{c} + \mathbf{V}(t)\mathbf{d}$$

where

$$\mathbf{c} = \widetilde{\mathbf{R}}_1^{-1}(\widetilde{\mathbf{R}}_2\widetilde{\mathbf{R}}_1^{-1} - \mathbf{R}_2\mathbf{R}_1^{-1})^{-1}(\delta\mathbf{r}_2 - \mathbf{R}_2\mathbf{R}_1^{-1}\delta\mathbf{r}_1)$$
$$\mathbf{d} = \mathbf{R}_1^{-1}(\mathbf{R}_2\mathbf{R}_1^{-1} - \widetilde{\mathbf{R}}_2\widetilde{\mathbf{R}}_1^{-1})^{-1}(\delta\mathbf{r}_2 - \widetilde{\mathbf{R}}_2\widetilde{\mathbf{R}}_1^{-1}\delta\mathbf{r}_1)$$

NOTE: For convenience, we are using the notation $\widetilde{\mathbf{R}}_1 \equiv \widetilde{\mathbf{R}}(t_1)$, etc.

◊ **Problem 9-19**

By differentiating the identity

$$\mathbf{\Phi}(t,t_0)\mathbf{\Phi}(t_0,t) = \mathbf{I}$$

and using Eq. (9.48), derive

$$\frac{d\mathbf{\Phi}^T(t_0,t)}{dt} = -\begin{bmatrix} \mathbf{O} & \mathbf{G} \\ \mathbf{I} & \mathbf{O} \end{bmatrix} \mathbf{\Phi}^T(t_0,t)$$

which is called the *adjoint system* of the system of equations (9.48). Since

$$\mathbf{\Phi}(t_0,t) = \mathbf{\Phi}^{-1}(t,t_0)$$

then the matrix $\mathbf{\Phi}^{-T}(t,t_0)$ is the adjoint of the matrix $\mathbf{\Phi}(t,t_0)$.

NOTE: In general, the *adjoint* of the matrix system of differential equations

$$\frac{d\mathbf{X}}{dt} = \mathbf{F}(t)\mathbf{X} \quad \text{is the system} \quad \frac{d\mathbf{Y}}{dt} = -\mathbf{F}^T(t)\mathbf{Y}$$

and the matrix $\mathbf{Y}$ is the *adjoint* of $\mathbf{X}$. The system is called *self-adjoint* if $\mathbf{F}$ is a skew-symmetric matrix, i.e., $\mathbf{F} = -\mathbf{F}^T$.

## The Adjoint Matrices

For a time $t_1$ later than time $t_0$ let us define the partitions of the transition matrix $\Phi(t, t_1)$ as

$$\Phi(t, t_1) = \begin{bmatrix} \widetilde{\mathbf{R}}^\star(t) & \mathbf{R}^\star(t) \\ \widetilde{\mathbf{V}}^\star(t) & \mathbf{V}^\star(t) \end{bmatrix} \quad (9.60)$$

Then, using the arguments of the last subsection, we have

$$\frac{d\widetilde{\mathbf{R}}^\star}{dt} = \widetilde{\mathbf{V}}^\star \quad \frac{d\widetilde{\mathbf{V}}^\star}{dt} = \mathbf{G}\widetilde{\mathbf{R}}^\star \quad \frac{d\mathbf{R}^\star}{dt} = \mathbf{V}^\star \quad \frac{d\mathbf{V}^\star}{dt} = \mathbf{G}\mathbf{R}^\star$$

$$\widetilde{\mathbf{R}}^\star(t_1) = \mathbf{I} \quad \widetilde{\mathbf{V}}^\star(t_1) = \mathbf{O} \quad \mathbf{R}^\star(t_1) = \mathbf{O} \quad \mathbf{V}^\star(t_1) = \mathbf{I} \quad (9.61)$$

Solutions of these differential equations for $t_0 \leq t \leq t_1$ are generated by integrating backwards from $t_1$ to $t_0$. They are the same equations as (9.49) with initial conditions prescribed at time $t_1$ instead of $t_0$. The differential equations for the matrix partitions of the transition matrix $\Phi$ are, for this reason, said to be *self-adjoint* even though the coefficient matrix $\mathbf{F}$ of the six-dimensional system is not skew-symmetric.

Needless to say, all of the symmetry properties and other relations derived for the unstarred matrices in the previous subsection hold also for the starred matrices. In particular, the fundamental matrices $\widetilde{\mathbf{C}}^\star$ and $\mathbf{C}^\star$ defined as

$$\widetilde{\mathbf{C}}^\star = \widetilde{\mathbf{V}}^\star \widetilde{\mathbf{R}}^{\star -1} \quad \text{and} \quad \mathbf{C}^\star = \mathbf{V}^\star \mathbf{R}^{\star -1} \quad (9.62)$$

are readily seen to be symmetric. Furthermore,

$$\widetilde{\mathbf{C}}^\star = \left.\frac{\partial \mathbf{v}}{\partial \mathbf{r}}\right|_{\mathbf{v}_1 = \text{constant}} \quad \text{and} \quad \mathbf{C}^\star = \left.\frac{\partial \mathbf{v}}{\partial \mathbf{r}}\right|_{\mathbf{r}_1 = \text{constant}} \quad (9.63)$$

Thus, for example, $\mathbf{C}^\star \delta \mathbf{r} = \delta \mathbf{v}$ which is the velocity deviation required at time $t$ if the vehicle is to pass through the reference point at time $t_1$. For this and other reasons, the matrices $\Phi(t, t_0)$ and $\Phi(t_1, t)$ are sometimes referred to as the *navigation matrix* and the *guidance matrix*, respectively.

The guidance matrix at time $t_0$ is simply related to the navigation matrix at time $t_1$. Since

$$\Phi(t, t_1)\big|_{t=t_0} = \Phi^{-1}(t, t_0)\big|_{t=t_1}$$

we have the following matrix identity†

$$\begin{bmatrix} \widetilde{\mathbf{R}}^\star(t_0) & \mathbf{R}^\star(t_0) \\ \widetilde{\mathbf{V}}^\star(t_0) & \mathbf{V}^\star(t_0) \end{bmatrix} = \begin{bmatrix} \mathbf{V}^{\mathrm{T}}(t_1) & -\mathbf{R}^{\mathrm{T}}(t_1) \\ -\widetilde{\mathbf{V}}^{\mathrm{T}}(t_1) & \widetilde{\mathbf{R}}^{\mathrm{T}}(t_1) \end{bmatrix} \quad (9.64)$$

---

† The relation $\mathbf{R}^\star(t_0) = -\mathbf{R}^{\mathrm{T}}(t_1)$ was first discovered accidently by William F. Marscher in 1961 from the print-out of a file of the fundamental perturbation matrices. It was not exactly obvious since the two matrices $\mathbf{R}^\star(t_0)$ and $\mathbf{R}(t_1)$ appeared on the first and last sheets of a three-inch stack of paper.

There are yet other relations, but holding for all time $t$, which can be derived. From Eqs. (9.49) and (9.61), we have

$$\frac{d\mathbf{V}}{dt} - \mathbf{GR} = \mathbf{O} \quad \text{and} \quad \frac{d\mathbf{V}^{\star\mathrm{T}}}{dt} - \mathbf{R}^{\star\mathrm{T}}\mathbf{G} = \mathbf{O}$$

Now, premultiply the first equation by $\mathbf{R}^{\star\mathrm{T}}$, postmultiply the second by $\mathbf{R}$, and subtract to obtain

$$\mathbf{R}^{\star\mathrm{T}}\frac{d\mathbf{V}}{dt} - \frac{d\mathbf{V}^{\star\mathrm{T}}}{dt}\mathbf{R} = \mathbf{O}$$

which can be rewritten as

$$\frac{d}{dt}(\mathbf{R}^{\star\mathrm{T}}\mathbf{V} - \mathbf{V}^{\star\mathrm{T}}\mathbf{R}) = \mathbf{O}$$

Hence,

$$\mathbf{R}^{\star\mathrm{T}}(t)\mathbf{V}(t) - \mathbf{V}^{\star\mathrm{T}}(t)\mathbf{R}(t) = \mathbf{K} = \text{constant} \qquad (9.65)$$

The constant matrix $\mathbf{K}$ can be determined from the initial conditions at either $t_0$ or $t_1$. Thus,

$$\mathbf{K} = \mathbf{R}^{\star\mathrm{T}}(t_0) = -\mathbf{R}(t_1) \qquad (9.66)$$

In a similar manner, we obtain

$$\begin{aligned}\widetilde{\mathbf{R}}^{\star\mathrm{T}}(t)\mathbf{V}(t) - \widetilde{\mathbf{V}}^{\star\mathrm{T}}(t)\mathbf{R}(t) &= \widetilde{\mathbf{R}}^{\star\mathrm{T}}(t_0) = \mathbf{V}(t_1)\\ \widetilde{\mathbf{R}}^{\star\mathrm{T}}(t)\widetilde{\mathbf{V}}(t) - \widetilde{\mathbf{V}}^{\star\mathrm{T}}(t)\widetilde{\mathbf{R}}(t) &= -\widetilde{\mathbf{V}}^{\star}(t_0) = \widetilde{\mathbf{V}}^{\mathrm{T}}(t_1) \qquad (9.67)\\ \widetilde{\mathbf{R}}^{\mathrm{T}}(t)\mathbf{V}^{\star}(t) - \widetilde{\mathbf{V}}^{\mathrm{T}}(t)\mathbf{R}^{\star}(t) &= \mathbf{V}^{\star}(t_0) = \widetilde{\mathbf{R}}^{\mathrm{T}}(t_1)\end{aligned}$$

◇ **Problem 9-20**

The matrix $\mathbf{\Lambda}(t)$ satisfies the equation

$$\mathbf{R}^{\star\mathrm{T}}\mathbf{\Lambda} = \mathbf{K} \quad \text{where} \quad \mathbf{\Lambda} = (\mathbf{C}^{\star} - \mathbf{C})\mathbf{R}$$

with $\mathbf{K}$ defined in Eq. (9.65). Furthermore, $\mathbf{\Lambda}(t)$ is the solution of

$$\frac{d\mathbf{\Lambda}}{dt} + \mathbf{C}^{\star}\mathbf{\Lambda} = \mathbf{O} \quad \text{with} \quad \mathbf{\Lambda}(t_0) = -\mathbf{I}$$

Demonstrate that the matrix $\mathbf{\Lambda}(t)$ relates the deviation in velocity at time $t_0$ to the velocity change required at time $t$ in order to intercept the reference point at time $t_1$.

◇ **Problem 9-21**

The state transition matrix can be expressed as either

$$\mathbf{\Phi}(t, t_0) = \begin{bmatrix} \mathbf{I} & \mathbf{O} \\ \mathbf{C}(t) & \mathbf{I} \end{bmatrix} \begin{bmatrix} \mathbf{V}^{\star\mathrm{T}}(t_0) & -\mathbf{R}^{\star\mathrm{T}}(t_0) \\ \mathbf{R}^{\star-1}(t_0) & \mathbf{O} \end{bmatrix}$$

or

$$\mathbf{\Phi}(t, t_0) = \begin{bmatrix} \mathbf{R}(t) & \mathbf{O} \\ \mathbf{V}(t) & -\mathbf{R}^{-\mathrm{T}}(t) \end{bmatrix} \begin{bmatrix} -\mathbf{C}^{\star}(t_0) & \mathbf{I} \\ -\mathbf{I} & \mathbf{O} \end{bmatrix}$$

## Problem 9-22
The state transition matrix can also be expressed as

$$\Phi_{t,t_0} = \begin{bmatrix} \mathbf{R} & \mathbf{R}^\star \\ \mathbf{V} & \mathbf{V}^\star \end{bmatrix} \begin{bmatrix} (\mathbf{C}_0^{\star-1}\mathbf{V}_0 - \mathbf{R}_0)^{-1} & \mathbf{O} \\ \mathbf{O} & (\mathbf{C}_0^{-1}\mathbf{V}_0^\star - \mathbf{R}_0^\star)^{-1} \end{bmatrix} \begin{bmatrix} -\mathbf{I} & \mathbf{C}_0^{\star-1} \\ -\mathbf{I} & \mathbf{C}_0^{-1} \end{bmatrix}$$

using the notation $\mathbf{R} \equiv \mathbf{R}(t)$ and $\mathbf{R}_0 \equiv \mathbf{R}(t_0)$, etc.

## Problem 9-23
Suppose that a reference trajectory, together with a complete set of perturbation matrices, has been established between launch and target points at times $t_0$ and $t_1$, respectively. It is desired to change the target point to an earlier time $t_1'$ along the same reference path. The $\mathbf{R}$ and $\mathbf{V}$ matrices will, of course, be unchanged, since they are related to time $t_0$. However, the new starred matrices $\mathbf{R}^{\star\prime}$ and $\mathbf{V}^{\star\prime}$ can be obtained from

$$\mathbf{R}^{\star\prime}(t) = -\mathbf{R}(t)\mathbf{\Lambda}^{-1}(t_1') - \mathbf{R}^\star(t)\mathbf{\Lambda}^{\star-1}(t_1')$$
$$\mathbf{V}^{\star\prime}(t) = -\mathbf{V}(t)\mathbf{\Lambda}^{-1}(t_1') - \mathbf{V}^\star(t)\mathbf{\Lambda}^{\star-1}(t_1')$$

NOTE: The matrix $\mathbf{\Lambda}$ is defined in Prob. 9–20. The matrix $\mathbf{\Lambda}^\star$ is defined by interchanging the starred and unstarred matrices; specifically,

$$\mathbf{\Lambda} = (\mathbf{C}^\star - \mathbf{C})\mathbf{R} \quad \text{and} \quad \mathbf{\Lambda}^\star = (\mathbf{C} - \mathbf{C}^\star)\mathbf{R}^\star$$

## Problem 9-24
Let $\delta \mathbf{r}_1$ and $\delta \mathbf{v}_1$ be the deviations in position and velocity at some time $t_1$ —not necessarily the terminal time. Then, the position and velocity deviations at any other time $t$ can be determined from

$$\delta \mathbf{r}(t) = \mathbf{R}(t)\mathbf{c} + \mathbf{R}^\star(t)\mathbf{c}^\star$$
$$\delta \mathbf{v}(t) = \mathbf{V}(t)\mathbf{c} + \mathbf{V}^\star(t)\mathbf{c}^\star$$

where

$$\mathbf{c} = \mathbf{\Lambda}_1^{-1}(\mathbf{C}_1^\star \delta\mathbf{r}_1 - \delta\mathbf{v}_1)$$
$$\mathbf{c}^\star = \mathbf{\Lambda}_1^{\star-1}(\mathbf{C}_1 \delta\mathbf{r}_1 - \delta\mathbf{v}_1)$$

NOTE: The matrices $\mathbf{\Lambda}$ and $\mathbf{\Lambda}^\star$ are defined in Prob. 9–23.

## 9.7 Calculating the Perturbation Matrices

The fundamental perturbation matrices $\mathbf{R}$, $\mathbf{R}^\star$, $\mathbf{V}$, $\mathbf{V}^\star$ were shown, in the previous section, to satisfy certain matrix differential equations. However, the initial conditions for the $\mathbf{R}$ and $\mathbf{V}$ matrices are given at the initial time $t_0$ while those for the $\mathbf{R}^\star$ and $\mathbf{V}^\star$ matrices are given at the terminal time $t_1$. Therefore, if one wished to determine values of these matrices at some time $t$ in the interval $t_0 < t < t_1$ by solving differential equations, it would be necessary to solve 18 simultaneous equations for $\mathbf{R}$ and $\mathbf{V}$ starting from $t_0$ and integrating forward to time $t$ and then repeat the operation for $\mathbf{R}^*$ and $\mathbf{V}^*$ starting from $t_1$ and integrating backward to time $t$. In addition, since values of the reference position and velocity vectors are needed in this computation, six more differential equations would have to be solved in both the forward and backward directions. This brings the total to 48 simultaneous differential equations which must be integrated to determine the required information. (Of course, the backward integration is unnecessary if the matrices $\widetilde{\mathbf{R}}$ and $\widetilde{\mathbf{V}}$ are used instead of $\mathbf{R}^*$ and $\mathbf{V}^*$.) In this section explicit formulas are derived for the special case of two-body orbits which do not involve the solution of differential equations.

For the two-body problem, the position and velocity vectors $\mathbf{r}$, $\mathbf{v}$ at time $t$ are related to their values $\mathbf{r}_0$, $\mathbf{v}_0$ at some epoch time $t_0$ according to

$$\mathbf{r} = F\mathbf{r}_0 + G\mathbf{v}_0 \qquad \mathbf{r}_0 = G_t\mathbf{r} - G\mathbf{v}$$
$$\mathbf{v} = F_t\mathbf{r}_0 + G_t\mathbf{v}_0 \qquad \mathbf{v}_0 = -F_t\mathbf{r} + F\mathbf{v} \tag{9.68}$$

where the Lagrangian coefficients are expressed in terms of the universal functions $U_n(\chi;\alpha)$ as†

$$F = 1 - \frac{U_2}{r_0} = \frac{1}{r_0}(rU_0 - \sigma U_1)$$

$$\sqrt{\mu}\,G = r_0 U_1 + \sigma_0 U_2 = rU_1 - \sigma U_2 = \sqrt{\mu}(t - t_0) - U_3$$

$$F_t = -\frac{\sqrt{\mu}\,U_1}{rr_0} \tag{9.69}$$

$$G_t = 1 - \frac{U_2}{r} = \frac{1}{r}(r_0 U_0 + \sigma_0 U_1)$$

where

$$\sqrt{\mu}\,\sigma = \mathbf{r}^T\mathbf{v} \qquad \text{and} \qquad \alpha = \frac{2}{r_0} - \frac{v_0^2}{\mu} = \frac{2}{r} - \frac{v^2}{\mu}$$

or, in terms of the true anomaly difference between $\mathbf{r}_0$ and $\mathbf{r}$, as

$$F = 1 - \frac{r}{p}(1 - \cos\theta) \qquad \sqrt{\mu}\,G = \frac{rr_0}{\sqrt{p}}\sin\theta$$

$$F_t = \frac{\sqrt{\mu}}{rr_0^2}[r_0\sigma_0(1 - F) - \sqrt{\mu}\,G] \qquad G_t = 1 - \frac{r_0}{r}(1 - F) \tag{9.70}$$

and the magnitude of the position vector $\mathbf{r}$ from

$$r = \frac{pr_0}{r_0 + (p - r_0)\cos\theta - \sqrt{p}\,\sigma_0 \sin\theta} \tag{9.71}$$

---

† The relevant equations of Sects. 4.5 and 4.6 are summarized here for reference. The variant forms of these equations, which also are needed, are easily obtained and left for the reader to verify.

Sect. 9.7]  Calculating the Perturbation Matrices

The universal form of Kepler's equation is

$$\sqrt{\mu}(t - t_0) = r_0 U_1 + \sigma_0 U_2 + U_3 = rU_1 - \sigma U_2 + U_3 \tag{9.72}$$

and the following relations for $r$ and $\sigma$ also obtain:

$$\begin{aligned} r &= r_0 U_0 + \sigma_0 U_1 + U_2 & r_0 &= rU_0 - \sigma U_1 + U_2 \\ \sigma &= \sigma_0 U_0 + (1 - \alpha r_0) U_1 & \sigma_0 &= \sigma U_0 - (1 - \alpha r) U_1 \end{aligned} \tag{9.73}$$

Certain identities, which exist among the functions $U_n$, are

$$U_n + \alpha U_{n+2} = \frac{\chi^n}{n!}$$

$$U_0^2 + \alpha U_1^2 = 1 \qquad U_1^2 - U_0 U_2 = U_2$$

$$U_0 U_3 - U_1 U_2 = U_3 - \chi U_2 \qquad U_1 U_3 - U_2^2 = 2U_4 - \chi U_3$$

along with the differential relations

$$\frac{\partial U_0}{\partial \chi} = -\alpha U_1 \qquad \frac{\partial U_n}{\partial \chi} = U_{n-1} \quad \text{for} \quad n = 1, 2, \ldots$$

and

$$\frac{\partial U_n}{\partial \alpha} = \tfrac{1}{2}(nU_{n+2} - \chi U_{n+1})$$

The derivation of the following differentials and resulting matrices is straightforward but tedious.† We report certain intermediate milestones as an aid to the serious reader.

First we establish

$$dU_0 = -\alpha U_1 \, d\varsigma - \tfrac{1}{2} U_1^2 \, d\alpha$$

$$dU_1 = U_0 \, d\varsigma - \tfrac{1}{2} U_1 U_2 \, d\alpha$$

$$dU_2 = U_1 \, d\varsigma - \tfrac{1}{2} U_2^2 \, d\alpha$$

$$dU_3 = U_2 \, d\varsigma - \tfrac{1}{2}(U_2 U_3 - 3U_5 + \chi U_4) \, d\alpha$$

where $d\varsigma$ is defined by

$$d\varsigma = d\chi + \tfrac{1}{2} U_3 \, d\alpha$$

Using these, we calculate the differentials for $r_0$ and $\sqrt{\mu}(t - t_0)$ as

$$dr_0 = -\sigma_0 \, d\varsigma - \tfrac{1}{2}(r_0 + r) U_2 \, d\alpha - U_1 \, d\sigma + U_0 \, dr$$

$$\sqrt{\mu} \, d(t - t_0) = 0 = r_0 \, d\varsigma + \tfrac{1}{2} \sqrt{\mu} \, C \, d\alpha - U_2 \, d\sigma + U_1 \, dr$$

---

† These results, in part, first appeared in the AIAA Professional Study Series book *Space Guidance and Navigation* used for a two day course conducted by the author and Donald C. Fraser at the University of California Santa Barbara in 1970. They were published in the *Journal of Guidance and Control*, Vol. 1, September-October, 1978 as a part of the paper entitled "The Epoch State Navigation Filter" by Richard H. Battin, Steven R. Croopnick, and Joan E. Lenox.

where the symbol $C$, introduced for notational convenience, is defined by†

$$\sqrt{\mu}\, C = 3U_5 - \chi U_4 - \sqrt{\mu}(t - t_0)U_2 \tag{9.74}$$

The differentials of the vectors $\mathbf{r}_0$ and $\mathbf{v}_0$ are obtained after some extensive manipulation and can be expressed in the form

$$d\mathbf{r}_0 = -\frac{r_0}{\sqrt{\mu}}\mathbf{v}_0\, d\varsigma + \tfrac{1}{2}U_2(\mathbf{r} - \mathbf{r}_0)\, d\alpha + \frac{U_2}{\sqrt{\mu}}\mathbf{v}\, d\sigma$$
$$+ \left(\frac{U_2}{r^2}\mathbf{r} - \frac{U_1}{\sqrt{\mu}}\mathbf{v}\right) dr + G_t\, d\mathbf{r} - G\, d\mathbf{v} \tag{9.75}$$

$$d\mathbf{v}_0 = \frac{\sqrt{\mu}}{r_0^2}\mathbf{r}_0\, d\varsigma - \frac{1}{r_0}(\mathbf{v} - \mathbf{v}_0)(\tfrac{1}{2}rU_2\, d\alpha + U_1\, d\sigma)$$
$$+ \frac{1}{r}[U_0(\mathbf{v} - \mathbf{v}_0) + F_t\mathbf{r}]\, dr - F_t\, d\mathbf{r} + F\, d\mathbf{v} \tag{9.76}$$

The matrices

$$\mathbf{R}_0^\star \equiv \mathbf{R}^\star(t_0) = \frac{\partial \mathbf{r}_0}{\partial \mathbf{v}} \qquad \text{and} \qquad \mathbf{V}_0^\star \equiv \mathbf{V}^\star(t_0) = \frac{\partial \mathbf{v}_0}{\partial \mathbf{v}}$$

are determined by first observing

$$\frac{\partial \alpha}{\partial \mathbf{v}} = -\frac{2}{\mu}\mathbf{v}^T \qquad \frac{\partial \sigma}{\partial \mathbf{v}} = \frac{1}{\sqrt{\mu}}\mathbf{r}^T$$
$$\frac{\partial \varsigma}{\partial \mathbf{v}} = \frac{1}{\sqrt{\mu}\, r_0}(C\mathbf{v}^T + U_2\mathbf{r}^T) \tag{9.77}$$

and remembering that $\mathbf{r}$ is constant. Then, we have

$$\mathbf{R}_0^\star = \frac{r_0}{\mu}(1 - F)[(\mathbf{v} - \mathbf{v}_0)\mathbf{r}^T - (\mathbf{r} - \mathbf{r}_0)\mathbf{v}^T] - \frac{C}{\mu}\mathbf{v}_0\mathbf{v}^T - G\mathbf{I} \tag{9.78}$$

$$\mathbf{V}_0^\star = \frac{r}{\mu}(\mathbf{v} - \mathbf{v}_0)(\mathbf{v} - \mathbf{v}_0)^T + \frac{1}{r_0^3}[r_0(1 - F)\mathbf{r}_0\mathbf{r}^T + C\mathbf{r}_0\mathbf{v}^T] + F\mathbf{I} \tag{9.79}$$

Explicit formulas for the matrices

$$\widetilde{\mathbf{R}}_0^\star \equiv \widetilde{\mathbf{R}}^\star(t_0) = \frac{\partial \mathbf{r}_0}{\partial \mathbf{r}} \qquad \text{and} \qquad \widetilde{\mathbf{V}}_0^\star \equiv \widetilde{\mathbf{V}}^\star(t_0) = \frac{\partial \mathbf{v}_0}{\partial \mathbf{r}}$$

can also be derived. First, as the analog of Eqs. (9.77), we have

$$\frac{\partial \alpha}{\partial \mathbf{r}} = -\frac{2}{r^3}\mathbf{r}^T \qquad \frac{\partial \sigma}{\partial \mathbf{r}} = \frac{1}{\sqrt{\mu}}\mathbf{v}^T \qquad \frac{\partial r}{\partial \mathbf{r}} = \frac{1}{r}\mathbf{r}^T \tag{9.80}$$

and

$$\frac{\partial \varsigma}{\partial \mathbf{r}} = \frac{1}{\sqrt{\mu}}\left[\frac{\mu C}{r_0 r^3}\mathbf{r}^T + (\mathbf{v} - \mathbf{v}_0)^T\right] \tag{9.81}$$

---

† In Sect. 4.7, continued fraction evaluations of the $U$-functions through fifth order were given which can be conveniently used to calculate the quantity $C$.

Then, remembering that $\mathbf{v}$ is constant, use Eqs. (9.75) and (9.76) to obtain

$$\tilde{\mathbf{R}}_0^\star = \frac{r_0}{\mu}(\mathbf{v} - \mathbf{v}_0)(\mathbf{v} - \mathbf{v}_0)^T + \frac{1}{r^3}[r_0(1-F)\mathbf{r}_0\mathbf{r}^T - C\mathbf{v}_0\mathbf{r}^T] + G_t\mathbf{I} \quad (9.82)$$

$$\tilde{\mathbf{V}}_0^\star = \frac{1}{r^2}(\mathbf{v} - \mathbf{v}_0)\mathbf{r}^T + \frac{1}{r_0^2}\mathbf{r}_0(\mathbf{v} - \mathbf{v}_0)^T$$

$$+ F_t\left[\frac{1}{r^2}\mathbf{r}\mathbf{r}^T - \frac{1}{\mu r}(\mathbf{v} - \mathbf{v}_0)\mathbf{r}^T(\mathbf{v}\mathbf{r}^T - \mathbf{r}\mathbf{v}^T) - \mathbf{I}\right] + \frac{\mu C}{r^3 r_0^3}\mathbf{r}_0\mathbf{r}^T \quad (9.83)$$

The matrices

$$\mathbf{R} \equiv \mathbf{R}(t) = \frac{\partial \mathbf{r}}{\partial \mathbf{v}_0} \quad \text{and} \quad \mathbf{V} \equiv \mathbf{V}(t) = \frac{\partial \mathbf{v}}{\partial \mathbf{v}_0}$$

together with

$$\tilde{\mathbf{R}} \equiv \tilde{\mathbf{R}}(t) = \frac{\partial \mathbf{r}}{\partial \mathbf{r}_0} \quad \text{and} \quad \tilde{\mathbf{V}} \equiv \tilde{\mathbf{V}}(t) = \frac{\partial \mathbf{v}}{\partial \mathbf{r}_0}$$

can be calculated in a similar manner. On the other hand, they may be obtained using the identities (9.64). We have

$$\mathbf{R} = \frac{r_0}{\mu}(1-F)[(\mathbf{r} - \mathbf{r}_0)\mathbf{v}_0^T - (\mathbf{v} - \mathbf{v}_0)\mathbf{r}_0^T] + \frac{C}{\mu}\mathbf{v}\mathbf{v}_0^T + G\mathbf{I} \quad (9.84)$$

$$\mathbf{V} = \frac{r_0}{\mu}(\mathbf{v} - \mathbf{v}_0)(\mathbf{v} - \mathbf{v}_0)^T + \frac{1}{r^3}[r_0(1-F)\mathbf{r}\mathbf{r}_0^T - C\mathbf{r}\mathbf{v}_0^T] + G_t\mathbf{I} \quad (9.85)$$

$$\tilde{\mathbf{R}} = \frac{r}{\mu}(\mathbf{v} - \mathbf{v}_0)(\mathbf{v} - \mathbf{v}_0)^T + \frac{1}{r_0^3}[r_0(1-F)\mathbf{r}\mathbf{r}_0^T + C\mathbf{v}\mathbf{r}_0^T] + F\mathbf{I} \quad (9.86)$$

$$\tilde{\mathbf{V}} = -\frac{1}{r_0^2}(\mathbf{v} - \mathbf{v}_0)\mathbf{r}_0^T - \frac{1}{r^2}\mathbf{r}(\mathbf{v} - \mathbf{v}_0)^T$$

$$+ F_t\left[\mathbf{I} - \frac{1}{r^2}\mathbf{r}\mathbf{r}^T + \frac{1}{\mu r}(\mathbf{r}\mathbf{v}^T - \mathbf{v}\mathbf{r}^T)\mathbf{r}(\mathbf{v} - \mathbf{v}_0)^T\right] - \frac{\mu C}{r^3 r_0^3}\mathbf{r}\mathbf{r}_0^T \quad (9.87)$$

## 9.8 Precision Orbit Determination

With the background established in the earlier sections of this chapter, we are prepared to complete the description of a method for determining exact interplanetary and circumlunar orbits. The patched-conic approximation provides the first step in an iteration procedure which will result in a precise orbit that takes into account all relevant disturbing forces.

### Precision Orbits for Lambert's Problem

Consider the now familiar problem of establishing an orbit between two position vectors $\mathbf{r}_1$ and $\mathbf{r}_2$ requiring a time $t_2 - t_1$ to traverse the arc which we have called Lambert's problem. In absence of all except the single central force, the orbit will be a conic, the calculation of which has been throughly treated in Chapter 7. We may now start at position $\mathbf{r}_1$ with the computed conic velocity vector $\mathbf{v}_1$ and integrate the equations of motion, subject to all relevant perturbations, by any of the methods described earlier in this chapter. At the end of a time interval $t_2 - t_1$ the position vector will, of course, fail to coincide with $\mathbf{r}_2$. However, if the perturbation forces are, indeed, small, we may expect the difference $\delta\mathbf{r}_2^{(1)}$, that is, the new terminus minus the original position $\mathbf{r}_2$, to be small.

The next step is to replace $\mathbf{r}_2$ by $\mathbf{r}_2^{(1)} = \mathbf{r}_2 - \delta\mathbf{r}_2^{(1)}$ and repeat the process. A conic arc requiring a time $t_2 - t_1$ for traversal is computed to connect $\mathbf{r}_1$ and the new location $\mathbf{r}_2^{(1)}$. Using the new conic velocity at $\mathbf{r}_1$ as an initial condition, we again integrate the complete equations of motion for a time interval $t_2 - t_1$. This time the difference $\delta\mathbf{r}_2^{(2)}$ between the new terminus and the original $\mathbf{r}_2$ should be smaller in magnitude. If it is still too large to be acceptable, we set $\mathbf{r}_2^{(2)} = \mathbf{r}_2^{(1)} - \delta\mathbf{r}_2^{(2)}$ and continue the procedure until acceptable accuracy is attained. Usually, three or four iteration steps are sufficient for most problems.

### Precision Free-Return Orbits

The technique described above, together with the perturbation matrices introduced earlier in this chapter, can be exploited to great advantage for determining a precise circumlunar trajectory from the patched-conic approximation developed in Sect. 9.3. The following quantities will remain invariant during the course of the iteration process: (1) $\mathbf{r}_L$, the injection position vector; (2) $t_L$, the time of launch; (3) $t_A$, the time of arrival at the lunar sphere of influence; (4) $t_D = t_A + t_S$, the time of departure from the lunar sphere of influence; (5) $\mathbf{r}_R$, the return-position vector; and (6) $t_R$, the time of return.

Let $\mathbf{r}_{TA}$ and $\mathbf{r}_{TD}$ be vectors from the center of the earth to the points on the lunar sphere of influence where the trajectory arrives and departs, respectively. During the course of the calculation, these vectors will be altered, but initially, they are taken from the patched-conics approximation. Then, with $\mathbf{r}_L$ and $\mathbf{r}_{TA}$ fixed, a precise trajectory connecting these points in the time $t_{FL} = t_A - t_L$ is readily obtained using the procedure just described. In like manner, two additional precise arcs are determined connecting $\mathbf{r}_R$ and $\mathbf{r}_{TD}$ in the time interval $t_{FR} = t_R - t_D$ and connecting $\mathbf{r}_{TML}$ and $\mathbf{r}_{TMR}$, the position vectors relative to the moon at the sphere of influence, in the time interval $t_S$ spent within the sphere.

## Precision Orbit Determination

As a result, a piecewise continuous precise circumlunar trajectory is obtained having velocity discontinuities at the sphere of influence. These two velocity mismatches can be eliminated, using perturbation matrices, in a manner now to be described.

Suppose $r_{TA}$ and $r_{TD}$ are shifted by the amounts $\delta r_A$ and $\delta r_D$, respectively. The velocity vectors at and exterior to the sphere must change by the amounts

$$\delta \mathbf{v}_L(t_A) = \mathbf{C}_L(t_A)\,\delta \mathbf{r}_A$$

$$\delta \mathbf{v}_R(t_D) = \mathbf{C}_R^\star(t_D)\,\delta \mathbf{r}_D$$

if $\mathbf{r}_L$, $\mathbf{r}_R$, and the times of flight are to remain invariant. Here, $\mathbf{C}_L(t_A)$ is the $\mathbf{C}$ matrix for the outbound trajectory leg evaluated at time $t_A$. Similarly, $\mathbf{C}_R^\star(t_D)$ is the $\mathbf{C}^\star$ matrix for the return trajectory leg at time $t_D$. The velocity vectors at and interior to the sphere will also change. The problem is, of course, to determine $\delta \mathbf{r}_A$ and $\delta \mathbf{r}_D$ so that the resulting velocity changes will exactly cancel the original velocity differences at the points of discontinuity.

For this purpose, consider the hyperbolic arc within the sphere of influence from $t_A$ to $t_D$. Let $\mathbf{R}_h$, $\mathbf{R}_h^\star$, $\mathbf{C}_h$, $\mathbf{C}_h^\star$ be the perturbation matrices associated with this trajectory. Then the velocity vectors at and interior to the sphere will change by

$$\delta \mathbf{v}_h(t_A) = \mathbf{R}_h^{-1}(t_D)\,\delta \mathbf{r}_D + \mathbf{C}_h^\star(t_A)\,\delta \mathbf{r}_A$$

$$\delta \mathbf{v}_h(t_D) = \mathbf{C}_h(t_D)\,\delta \mathbf{r}_D + \mathbf{R}_h^{\star -1}(t_A)\,\delta \mathbf{r}_A$$

if the time of flight is invariant.

Initially, the velocity mismatch at $\mathbf{r}_{TA}$ is

$$\Delta \mathbf{v}_A = \mathbf{v}_{TML} - \mathbf{v}_h(t_A)$$

and at $\mathbf{r}_{TD}$ the mismatch is

$$\Delta \mathbf{v}_D = \mathbf{v}_h(t_D) - \mathbf{v}_{TMR}$$

Thus, $\delta \mathbf{r}_A$ and $\delta \mathbf{r}_D$ must be chosen such that

$$\delta \mathbf{v}_L(t_A) - \delta \mathbf{v}_h(t_A) + \Delta \mathbf{v}_A = 0$$

$$\delta \mathbf{v}_h(t_D) - \delta \mathbf{v}_R(t_D) + \Delta \mathbf{v}_D = 0$$

The solution may be written as

$$\delta \mathbf{r}_A = -\mathbf{R}_h^\star(t_A)[\mathbf{B}(t_D) + \mathbf{R}_h(t_D)\mathbf{A}^{-1}(t_A)\mathbf{R}_h^\star(t_A)]^{-1}\mathbf{R}_h(t_D)\,\Delta \mathbf{v}_A$$

$$- \mathbf{A}(t_A)[\mathbf{A}(t_A) + \mathbf{R}_h^\star(t_A)\mathbf{B}^{-1}(t_D)\mathbf{R}_h(t_D)]^{-1}\mathbf{R}_h^\star(t_A)\,\Delta \mathbf{v}_D \quad (9.88)$$

$$\delta \mathbf{r}_D = \mathbf{B}(t_D)[\mathbf{B}(t_D) + \mathbf{R}_h(t_D)\mathbf{A}^{-1}(t_A)\mathbf{R}_h^\star(t_A)]^{-1}\mathbf{R}_h(t_D)\,\Delta \mathbf{v}_A$$

$$- \mathbf{R}_h(t_D)[\mathbf{A}(t_A) + \mathbf{R}_h^\star(t_A)\mathbf{B}^{-1}(t_D)\mathbf{R}_h(t_D)]^{-1}\mathbf{R}_h^\star(t_A)\,\Delta \mathbf{v}_D \quad (9.89)$$

where, for notational convenience, the matrices $\mathbf{A}(t_A)$ and $\mathbf{B}(t_D)$ are defined as

$$\mathbf{A}(t_A) = \mathbf{C}_h^{\star-1}(t_A)[\mathbf{C}_h^{\star-1}(t_A) - \mathbf{C}_L^{-1}(t_A)]^{-1}\mathbf{C}_L^{-1}(t_A)$$

$$\mathbf{B}(t_D) = \mathbf{C}_R^{\star-1}(t_D)[\mathbf{C}_R^{\star-1}(t_D) - \mathbf{C}_h^{-1}(t_D)]^{-1}\mathbf{C}_h^{-1}(t_D)$$

With $\delta\mathbf{r}_A$ and $\delta\mathbf{r}_D$ calculated from Eqs. (9.88) and (9.89), the position vectors $\mathbf{r}_{TA}$ and $\mathbf{r}_{TD}$ are replaced by $\mathbf{r}_{TA} + \delta\mathbf{r}_A$ and $\mathbf{r}_{TD} + \delta\mathbf{r}_D$. Three new precision trajectories connecting the terminus pairs

(1) $\mathbf{r}_L$ and $\mathbf{r}_{TA} + \delta\mathbf{r}_A$
(2) $\mathbf{r}_{TML} + \delta\mathbf{r}_A$ and $\mathbf{r}_{TMR} + \delta\mathbf{r}_D$
(3) $\mathbf{r}_{TD} + \delta\mathbf{r}_D$ and $\mathbf{r}_R$

are then computed. Velocity mismatches at the sphere of influence will still exist but should be smaller in magnitude than those produced by the initial trial. The entire process can be repeated until satisfactory velocity continuity is obtained.

Chapter 10

# Variation of Parameters

ANALYTICAL DEVELOPMENT OF THE VARIATION OF PARAMETERS WAS first given by Leonhard Euler in a series of memoirs on the mutual perturbations of Jupiter and Saturn for which he received the prizes of the French Academy in the years 1748 and 1752. The method is also called *the variation of orbital elements* or *the variation of constants*—the latter referring to integration constants. Euler's treatment of the method of variation of parameters was not entirely general since he did not consider the orbital elements as being simultaneously variable. It is noteworthy, however, that the first steps in the expansion of the disturbing function were made by Euler in those papers.

Joseph-Louis Lagrange wrote his first memoir on the perturbations of Jupiter and Saturn in 1766 in which he made further advances in the variation of parameters method. His final equations were still incorrect because he regarded the major axes and the times of perihelion passage as constants. However, his expressions for the angle of inclination, the longitude of the ascending node, and the argument of perihelion were all perfectly correct. Later, in 1782, he developed completely and for the first time the method of the variation of parameters in a prize memoir on the perturbations of comets moving in elliptical orbits. One of the objectives in this chapter is the derivation of Lagrange's planetary equations.

The most dramatic application of the method was made independently and almost simultaneously by the Englishman John Couch Adams (1819–1892) and the Frenchman Urbain-Jean-Joseph Le Verrier (1811–1877). Each predicted the existence and apparent position of the planet Neptune from the otherwise unexplained irregularities in the motion of Uranus.† The story is one of the most fascinating in the history of astronomy and is an impressive example of the precision which can be achieved using variational methods.

The planet Uranus was discovered on March 13, 1781 by Sir William Herschel shortly before Euler's death in 1783. The other planets had been known since ancient times and Herschel's findings opened the door to an

---

† A mathematical account of the procedures used by Adams and Le Verrier is given in William Marshall Smart's book *Celestial Mechanics* published in 1953 by Longmans, Green and Co.

era of astronomical discoveries of major importance. Since Uranus is almost visible to the unaided eye, the German astronomer Johann Elbert Bode† (1747–1826) suspected that it might have been mistaken for a star in the past. When the orbital elements had been determined with sufficient accuracy, a search of the old catalogs revealed that Uranus had been observed at least 19 different times—the earliest by the first Astronomer Royal, John Flamsteed, in the year 1690.

When the French astronomer Alexis Bouvard attempted to reconcile the new observations with the old during his preparation of tables for Jupiter, Saturn, and Uranus which were published in 1821, he was forced to abandon the earlier data because of serious and unexplained discrepancies. Even so, the planet began to deviate more and more from Bouvard's predicted positions and, by 1846, the error in longitude was almost two minutes of arc. In 1842, Friedrich Wilhelm Bessel, suspecting the presence of an ultra-Uranian planet, announced his intention of investigating the motion of Uranus. Unfortunately, he died before much could be accomplished.

On July 3, 1841, an undergraduate at St. John's College in Cambridge, England wrote in his journal

> "Formed a design in the beginning of this week of investigating, as soon as possible after taking my degree, the irregularities in the motion of Uranus ... in order to find out whether they may be attributed to the action of an undiscovered planet beyond it ..."

True to his word, by 1845 John Couch Adams had obtained a solution

---

† In 1766 the German astronomer Johann Daniel Titius (1729–1796) of Whittenberg found an empirical formula for the distances of the planets from the sun—a "solution" of the problem to which Kepler devoted so much misplaced energy. According to Titius, the formula for the mean distance

$$a_n = \tfrac{1}{10}\left(4 + 3 \times [2^{n-2}]\right) \text{ a.u.}$$

with $n = 1, 2, 3, 4$ holds for the planets Mercury, Venus, earth, Mars and with $n = 6, 7$ for Jupiter and Saturn. (The symbol $[x]$ denotes the greatest integer contained in $x$.) The approximation is, indeed, remarkably good. When Uranus was found to conform to the rule for $n = 8$, the formula took on greater significance. The empty space, corresponding to $n = 5$, inspired Johann Bode, director of the Berlin Observatory, to declare

> "Is it not highly probable that a planet actually revolves in the orbit which the finger of the Almighty has drawn for it? Can we believe the Creator of the world has left this space empty? Certainly not!"

An association of European astronomers was formed to search for the missing planet. When Ceres was discovered by Giuseppe Piazzi on the first day of January, 1801 at approximately 2.8 a.u., Titius' rule became *Bode's law*. (It is, of course, not a law and, ironically, its association with Titius is almost forgotten.) It is, therefore, not surprising that both men, Adams and Le Verrier, used Bode's law to estimate the mean distance of Neptune as 38.8 a.u.—but it was, in fact, the first planet to violate the rule. (Neptune's mean distance is actually 30.1 astronomical units.)

and in September of that year he gave the results of his computations, on where the new planet could be found, to James Challis, director of the Cambridge observatory. Challis expressed little interest. The next month, Adams contacted the Astronomer Royal, Sir George Biddell Airy, who also reacted with a similar lack of enthusiasm.

Meanwhile, in that same year, the French astronomer Urbain-Jean-Joseph Le Verrier turned his attention to the Uranus problem and *published* his results on June 1, 1846. When Airy saw the close agreement with Adams' calculations, he suggested to Challis on July 9, 1846 that he search for the planet. Indeed, Challis did observe Neptune on August 4 but failed to recognize it as the object of his quest. He had neglected to reconcile his observations with those of the previous night—an unforgivable blunder for a man of his experience.

On September 18, 1846, Le Verrier requested the German astronomer Johann Gottfried Galle to look for the planet with the hope that it could be distinguished from a star by its disk-like appearance. Then on September 23, 1846, after only an hour, Galle found the planet Neptune within one degree of the position computed by Le Verrier.

The reader can imagine the controversy between the English and the French over who deserved the credit. But justice prevailed and when the battle subsided, it was universally agreed that both Adams and Le Verrier would share equally in the glory.

## 10.1 Variational Methods for Linear Equations

The first application of the method of variation of parameters was made by John Bernoulli† in 1697 to solve the linear differential equation of the first order. The most general such equation is

$$\frac{dy}{dt} + f(t)y = g(t) \tag{10.1}$$

For the solution, consider first the homogeneous linear equation

$$\frac{dy}{dt} + f(t)y = 0 \tag{10.2}$$

---

† After Newton and Leibnitz, the Bernoulli brothers, James and John, were the two most important founders of the calculus. James Bernoulli (1655–1705) trained for the ministry at the urging of his father but managed to teach himself mathematics. In 1686 he turned to mathematics exclusively and became a professor at the University of Basle. His younger brother John (1667–1748) was steered into business by his father but turned, instead, to medicine and learned mathematics on the side from his brother. Mathematics again won out and he became a professor at Groningen in Holland and, later, succeeded his brother at Basle. His most famous student at the university was Leonhard Euler who completed his studies there at the age of fifteen. It was through the assistance of the younger Bernoullis, Nicholas (1695–1726) and Daniel (1700–1782), both sons of John and both accomplished mathematicians, that Euler in 1733 secured an appointment at the St. Petersburg Academy in Russia.

The Bernoulli family was, indeed, a unique source of mathematical talent.

474                         Variation of Parameters                [Chap. 10

The variables are separable

$$\frac{dy}{y} = -f(t)\,dt$$

and we have

$$y = ce^{-\int f(t)\,dt} \qquad (10.3)$$

where $c$ is a constant.

Suppose now that we allow $c$ to be a function of $t$ and determine the relation that $c(t)$ must satisfy if Eq. (10.3) is to be a solution of the inhomogeneous equation (10.1). By direct substitution we find

$$\frac{dc}{dt} e^{-\int f(t)\,dt} = g(t)$$

Hence

$$c(t) = C + \int g(t) e^{\int f(t)\,dt}\,dt$$

where $C$ is a constant. Thus, the general solution of Eq. (10.1) is

$$y = Ce^{-\int f\,dt} + e^{-\int f\,dt}\int g e^{\int f\,dt}\,dt \qquad (10.4)$$

and involves two quadratures.

◊ **Problem 10-1**
   Obtain the general solutions of

(1) $$\frac{dy}{dt} - ay = e^{at} \qquad (a = \text{constant})$$

(2) $$\frac{dy}{dt}\cos t + y\sin t = 1$$

using the method of variation of parameters.

Lagrange, in 1774, extended the method to the general $n^{\text{th}}$ order linear differential equation

$$L(y) = g(t) \qquad (10.5)$$

where the operator $L(y)$ is defined by

$$L(y) \equiv \frac{d^n y}{dt^n} + f_1(t)\frac{d^{n-1}y}{dt^{n-1}} + \cdots + f_{n-1}(t)\frac{dy}{dt} + f_n(t)y$$

It is convenient to convert Eq. (10.5) to a system of $n$ first-order equations written in vector-matrix form. For this purpose, define

$$\mathbf{y}^T = \begin{bmatrix} y & \dfrac{dy}{dt} & \dfrac{d^2 y}{dt^2} & \cdots & \dfrac{d^{n-2}y}{dt^{n-2}} & \dfrac{d^{n-1}y}{dt^{n-1}} \end{bmatrix}$$

$$\mathbf{g}^T = \begin{bmatrix} 0 & 0 & 0 & \cdots & 0 & g \end{bmatrix}$$

## Sect. 10.1] Variational Methods for Linear Equations

$$\mathbf{F} = \begin{bmatrix} 0 & 1 & 0 & \cdots & 0 & 0 \\ 0 & 0 & 1 & \cdots & 0 & 0 \\ \vdots & \vdots & \vdots & \ddots & \vdots & \vdots \\ 0 & 0 & 0 & \cdots & 0 & 1 \\ -f_n & -f_{n-1} & -f_{n-2} & \cdots & -f_2 & -f_1 \end{bmatrix}$$

so that the scalar differential equation (10.5) is equivalent to

$$\frac{d\mathbf{y}}{dt} = \mathbf{Fy} + \mathbf{g} \tag{10.6}$$

Suppose that $n$ linearly independent solutions of the homogeneous equation

$$L(y) = 0 \tag{10.7}$$

are known. Call them $\mathbf{y}_1(t)$, $\mathbf{y}_2(t)$, ..., $\mathbf{y}_n(t)$ and form a matrix with these vectors as the columns of the array. This is the *Wronskian matrix* $\mathbf{W}$ defined as

$$\mathbf{W} = \begin{bmatrix} \mathbf{y}_1 & \mathbf{y}_2 & \cdots & \mathbf{y}_n \end{bmatrix}$$

Clearly then, $\mathbf{W}$ satisfies the matrix differential equation

$$\frac{d\mathbf{W}}{dt} = \mathbf{FW} \tag{10.8}$$

and the general solution of the homogeneous equation is

$$\mathbf{y}_h = \mathbf{Wc} \tag{10.9}$$

where the components of the vector $\mathbf{c}$ are $n$ arbitrary constants.

As before we allow the elements of the vector $\mathbf{c}$ to be functions of $t$ and require that

$$\mathbf{y} = \mathbf{Wc}(t) \tag{10.10}$$

be a solution of Eq. (10.6). That is,

$$\frac{d\mathbf{W}}{dt}\mathbf{c} + \mathbf{W}\frac{d\mathbf{c}}{dt} = \mathbf{FWc} + \mathbf{g}$$

But $\mathbf{W}$ is a solution of Eq. (10.8), so that the differential equation for $\mathbf{c}(t)$ is reduced to

$$\mathbf{W}\frac{d\mathbf{c}}{dt} = \mathbf{g} \tag{10.11}$$

Now, the functions $\mathbf{y}_1(t)$, $\mathbf{y}_2(t)$, ..., $\mathbf{y}_n(t)$ were given as linearly independent so that the Wronskian determinant is not zero. Therefore, the matrix $\mathbf{W}$ is not singular so that

$$\frac{d\mathbf{c}}{dt} = \mathbf{W}^{-1}\mathbf{g} \tag{10.12}$$

which is solved by quadratures for the elements of the vector $\mathbf{c}$.

◇ **Problem 10–2**
Obtain the general solutions of

(1) $$\frac{d^2y}{dt^2} + y = \sec t$$

(2) $$\frac{d^3y}{dt^3} + 4\frac{dy}{dt} = 4\cot 2t$$

(3) $$\frac{d^2y}{dt^2} + 2\frac{dy}{dt} - 3y = te^{-t}$$

◇ **Problem 10–3**
For the second-order linear differential operator

$$L(y) = \frac{d^2y}{dt^2} - \frac{6}{t^2}y$$

show that

$$y_1(t) = t^3 \quad \text{and} \quad y_2(t) = \frac{1}{t^2}$$

are linearly independent solutions of $L(y) = 0$. Then use the method of variation of parameters to obtain the general solution of

$$L(y) = t \log t$$

## 10.2 Lagrange's Planetary Equations

The method of the variation of parameters, as originally developed by Lagrange, was to study the disturbed motion of two bodies in the form

$$\frac{d\mathbf{r}}{dt} = \mathbf{v} \qquad \frac{d\mathbf{v}}{dt} + \frac{\mu}{r^3}\mathbf{r} = \left[\frac{\partial R}{\partial \mathbf{r}}\right]^T \tag{10.13}$$

where $R$ is the disturbing function defined in Sect. 8.4. The solution of the undisturbed or two-body motion is known and may be expressed functionally in the form

$$\mathbf{r} = \mathbf{r}(t, \boldsymbol{\alpha}) \qquad \mathbf{v} = \mathbf{v}(t, \boldsymbol{\alpha}) \tag{10.14}$$

where the components of the vector $\boldsymbol{\alpha}$ are the six constants of integration (orbital elements). As in the previous section, we allow $\boldsymbol{\alpha}$ to be a time dependent quantity and require that the two-body solution (10.14) exactly satisfy the equations (10.13) for the disturbed motion.

A set of differential equations for $\boldsymbol{\alpha}(t)$ will result as before; however, they will not be solvable by quadrature. The new set of equations will, in fact, be a transformation of the dependent variables of the problem from the original position and velocity vectors $\mathbf{r}(t)$ and $\mathbf{v}(t)$ to the time-varying orbital elements $\boldsymbol{\alpha}(t)$. Although the differential equations for $\boldsymbol{\alpha}(t)$ will be as complex as the original version, they will have advantages similar

Sect. 10.2]   Lagrange's Planetary Equations   477

to those encountered in Encke's method, i.e., only the disturbing and not the total acceleration will effect changes in $\boldsymbol{\alpha}(t)$. Indeed, one may regard the method of variation of orbital elements as a form of Encke's method in which rectification of the osculating orbit is performed continuously rather than at discrete and widely separated instants of time.

To obtain the variational equations, we substitute Eqs. (10.14) into Eqs. (10.13) and use the fact that

$$\frac{\partial \mathbf{r}}{\partial t} = \mathbf{v} \qquad \frac{\partial \mathbf{v}}{\partial t} + \frac{\mu}{r^3}\mathbf{r} = \mathbf{0} \qquad (10.15)$$

Here, the partial derivatives serve to emphasize that when the vector $\boldsymbol{\alpha}$ is considered to be constant, then Eqs. (10.14) are solutions of the equations which describe the undisturbed motion.

For the disturbed motion

$$\frac{d\mathbf{r}}{dt} = \frac{\partial \mathbf{r}}{\partial t} + \frac{\partial \mathbf{r}}{\partial \boldsymbol{\alpha}}\frac{d\boldsymbol{\alpha}}{dt}$$

and, paralleling the arguments used in the previous section, we have

$$\frac{\partial \mathbf{r}}{\partial \boldsymbol{\alpha}}\frac{d\boldsymbol{\alpha}}{dt} = \mathbf{0} \qquad (10.16)$$

as the condition to be imposed on $\boldsymbol{\alpha}(t)$. Physically, this means we are requiring the velocity vectors of both the disturbed and undisturbed motion to be identical. Similarly,

$$\frac{d\mathbf{v}}{dt} = \frac{\partial \mathbf{v}}{\partial t} + \frac{\partial \mathbf{v}}{\partial \boldsymbol{\alpha}}\frac{d\boldsymbol{\alpha}}{dt}$$

and, using the second of Eqs. (10.15), we find that

$$\frac{\partial \mathbf{v}}{\partial \boldsymbol{\alpha}}\frac{d\boldsymbol{\alpha}}{dt} = \left[\frac{\partial R}{\partial \mathbf{r}}\right]^T \qquad (10.17)$$

must obtain if Eqs. (10.13) are to be satisfied. Equations (10.16) and (10.17) are the required six scalar differential equations to be satisfied by the vector of orbital elements $\boldsymbol{\alpha}(t)$.

### The Lagrange Matrix and Lagrangian Brackets

The two matrix-vector variational equations can be combined to produce a more convenient and compact form. For this purpose, we first multiply Eq. (10.16) by $[\partial \mathbf{v}/\partial \boldsymbol{\alpha}]^T$. Then, multiply Eq. (10.17) by $[\partial \mathbf{r}/\partial \boldsymbol{\alpha}]^T$ and subtract the two. The result is expressed as

$$\mathbf{L}\frac{d\boldsymbol{\alpha}}{dt} = \left[\frac{\partial R}{\partial \boldsymbol{\alpha}}\right]^T \qquad (10.18)$$

478  Variation of Parameters  [Chap. 10

where the matrix

$$\mathbf{L} = \left[\frac{\partial \mathbf{r}}{\partial \boldsymbol{\alpha}}\right]^\mathrm{T} \frac{\partial \mathbf{v}}{\partial \boldsymbol{\alpha}} - \left[\frac{\partial \mathbf{v}}{\partial \boldsymbol{\alpha}}\right]^\mathrm{T} \frac{\partial \mathbf{r}}{\partial \boldsymbol{\alpha}} \qquad (10.19)$$

is six-dimensional and skew-symmetric. The form of the right-hand side of Eq. (10.18) follows from the *chain rule* of partial differentiation

$$\frac{\partial R}{\partial \boldsymbol{\alpha}} = \frac{\partial R}{\partial \mathbf{r}} \frac{\partial \mathbf{r}}{\partial \boldsymbol{\alpha}}$$

The element in the $i^\mathrm{th}$ row and $j^\mathrm{th}$ column of the *Lagrange matrix* $\mathbf{L}$ is denoted by $[\alpha_i, \alpha_j]$ and will be referred to as a *Lagrangian bracket*. From Eq. (10.19) we have

$$\begin{aligned}[] [\alpha_i, \alpha_j] &= \frac{\partial \mathbf{r}}{\partial \alpha_i} \cdot \frac{\partial \mathbf{v}}{\partial \alpha_j} - \frac{\partial \mathbf{r}}{\partial \alpha_j} \cdot \frac{\partial \mathbf{v}}{\partial \alpha_i} \\ &= \frac{\partial \mathbf{r}^\mathrm{T}}{\partial \alpha_i} \frac{\partial \mathbf{v}}{\partial \alpha_j} - \frac{\partial \mathbf{r}^\mathrm{T}}{\partial \alpha_j} \frac{\partial \mathbf{v}}{\partial \alpha_i} = \frac{\partial \mathbf{v}^\mathrm{T}}{\partial \alpha_j} \frac{\partial \mathbf{r}}{\partial \alpha_i} - \frac{\partial \mathbf{v}^\mathrm{T}}{\partial \alpha_i} \frac{\partial \mathbf{r}}{\partial \alpha_j} \end{aligned} \qquad (10.20)$$

An important property of the Lagrange matrix $\mathbf{L}$ is displayed when we calculate the partial derivative of the Lagrangian bracket with respect to $t$. Thus,

$$\frac{\partial}{\partial t}[\alpha_i, \alpha_j] = \frac{\partial}{\partial \alpha_j}\left(\frac{\partial \mathbf{v}^\mathrm{T}}{\partial t}\right) \frac{\partial \mathbf{r}}{\partial \alpha_i} + \frac{\partial \mathbf{v}^\mathrm{T}}{\partial \alpha_j} \frac{\partial \mathbf{v}}{\partial \alpha_i} - \frac{\partial}{\partial \alpha_i}\left(\frac{\partial \mathbf{v}^\mathrm{T}}{\partial t}\right) \frac{\partial \mathbf{r}}{\partial \alpha_j} - \frac{\partial \mathbf{v}^\mathrm{T}}{\partial \alpha_i} \frac{\partial \mathbf{v}}{\partial \alpha_j}$$

and, clearly, the second and fourth terms cancel immediately. Using the gravitational potential function $V = \mu/r$, the second one of Eqs. (10.15) becomes

$$\frac{\partial \mathbf{v}^\mathrm{T}}{\partial t} = \frac{\partial V}{\partial \mathbf{r}}$$

so that

$$\begin{aligned} \frac{\partial}{\partial t}[\alpha_i, \alpha_j] &= \frac{\partial}{\partial \alpha_j}\left(\frac{\partial V}{\partial \mathbf{r}}\right) \frac{\partial \mathbf{r}}{\partial \alpha_i} - \frac{\partial}{\partial \alpha_i}\left(\frac{\partial V}{\partial \mathbf{r}}\right) \frac{\partial \mathbf{r}}{\partial \alpha_j} \\ &= \frac{\partial}{\partial \mathbf{r}}\left(\frac{\partial V}{\partial \alpha_j}\right) \frac{\partial \mathbf{r}}{\partial \alpha_i} - \frac{\partial}{\partial \mathbf{r}}\left(\frac{\partial V}{\partial \alpha_i}\right) \frac{\partial \mathbf{r}}{\partial \alpha_j} \\ &= \frac{\partial^2 V}{\partial \alpha_j \partial \alpha_i} - \frac{\partial^2 V}{\partial \alpha_i \partial \alpha_j} = 0 \end{aligned}$$

In view of this discussion, we can summarize the properties of the Lagrangian brackets as

(1)  $[\alpha_i, \alpha_i] = 0$
(2)  $[\alpha_i, \alpha_j] = -[\alpha_j, \alpha_i]$
(3)  $\dfrac{\partial}{\partial t}[\alpha_i, \alpha_j] = 0$

Sect. 10.2]     Lagrange's Planetary Equations     479

or, equivalently, for the Lagrange matrix,

$$\mathbf{L}^T = -\mathbf{L} \quad \text{and} \quad \frac{\partial \mathbf{L}}{\partial t} = \mathbf{O} \tag{10.21}$$

The fact that the matrix $\mathbf{L}$ is *not an explicit function of* $t$ will be exploited to great advantage in determining the elements of the Lagrange matrix.

### Problem 10–4
Consider the case for which the position and velocity vectors $\mathbf{r}_0$ and $\mathbf{v}_0$ at some instant of time $t_0$ are used as orbital elements, i.e.,

$$\boldsymbol{\alpha}^T = [\,\mathbf{r}_0^T \quad \mathbf{v}_0^T\,]$$

Using the fact that the Lagrangian matrix $\mathbf{L}$ is independent of time, show that

$$\mathbf{L} = \mathbf{J}$$

where the matrix $\mathbf{J}$ is defined in Sect. 9.5. Then show that the variational equations are

$$\frac{d\mathbf{r}_0^T}{dt} = -\frac{\partial R}{\partial \mathbf{v}_0} \qquad \frac{d\mathbf{v}_0^T}{dt} = \frac{\partial R}{\partial \mathbf{r}_0}$$

which are in the so-called *canonical form* and should be compared to Hamilton's canonical form of the equations of motion developed in Prob. 2–13.

### Computing the Lagrangian Brackets

To compute the Lagrangian brackets we first select an appropriate set of orbital elements. The classical choice is

$$\boldsymbol{\alpha}^T = [\,\Omega \quad i \quad \omega \quad a \quad e \quad \lambda\,] \tag{10.22}$$

where $\Omega$, $i$, $\omega$ are the three Euler angles, $a$ is the semimajor axis, $e$ is the eccentricity and

$$\lambda = -n\tau \tag{10.23}$$

where $n$ is the mean motion

$$n = \sqrt{\frac{\mu}{a^3}}$$

and $\tau$ is the time of pericenter passage.

The position and velocity vectors, expressed in reference coordinates as functions of the orbital elements, are then

$$\mathbf{r} = \begin{bmatrix} l_1 & l_2 & l_3 \\ m_1 & m_2 & m_3 \\ n_1 & n_2 & n_3 \end{bmatrix} \begin{bmatrix} a(\cos E - e) \\ b \sin E \\ 0 \end{bmatrix} \tag{10.24}$$

$$\mathbf{v} = \begin{bmatrix} l_1 & l_2 & l_3 \\ m_1 & m_2 & m_3 \\ n_1 & n_2 & n_3 \end{bmatrix} \begin{bmatrix} -an \sin E/(1 - e \cos E) \\ bn \cos E/(1 - e \cos E) \\ 0 \end{bmatrix} \tag{10.25}$$

**480**                    Variation of Parameters                  [Chap. 10

[The components of $\mathbf{r}$ and $\mathbf{v}$ in orbital plane coordinates are obtained from Eqs. (4.40) and the transformation to reference coordinates via the rotation matrix from Eq. (2.3).] Thus, $\mathbf{r}$ and $\mathbf{v}$ are functions of the Euler angles through the direction cosine elements of the rotation matrix as given in Eqs. (2.9). The elements $a$ and $e$ enter the relations explicitly through Eqs. (10.24) and (10.25), and implicitly through the mean motion $n$, the semiminor axis

$$b = a\sqrt{1 - e^2}$$

and through the eccentric anomaly $E$ in Kepler's equation

$$E - e \sin E = nt + \lambda \tag{10.26}$$

We begin by calculating the partial derivatives of $\mathbf{r}$ and $\mathbf{v}$ with respect to each of the orbital elements. Since the Lagrangian brackets are not explicit functions of time, we may set $t$ equal to any convenient value after the differentiation. The expressions will be simplest at pericenter for which $t = \tau$, $r = q = a(1 - e)$, and $E = 0$.

Consider first the partial derivative of $\mathbf{r}$ with respect to $\Omega$. From Eq. (2.9), we have

$$\frac{\partial l_1}{\partial \Omega} = \frac{\partial}{\partial \Omega}(\cos \Omega \cos \omega - \sin \Omega \sin \omega \cos i)$$
$$= -\sin \Omega \cos \omega - \cos \Omega \sin \omega \cos i$$
$$= -m_1$$

and, similarly,

$$\frac{\partial m_1}{\partial \Omega} = l_1 \qquad \frac{\partial n_1}{\partial \Omega} = 0$$

Now since,

$$\begin{bmatrix} a(\cos E - e) \\ b \sin E \\ 0 \end{bmatrix} = \begin{bmatrix} q \\ 0 \\ 0 \end{bmatrix}$$

at pericenter, the derivatives of the other direction cosines do not enter in the calculation of $\partial \mathbf{r}/\partial \Omega$. The result appears in the next equation set.

The derivatives of $\mathbf{r}$ with respect to $i$ and $\omega$ are entirely similar and in exactly the same way we compute the derivatives of $\mathbf{v}$. Therefore, at pericenter,

$$\frac{\partial \mathbf{r}}{\partial \Omega} = q \begin{bmatrix} -m_1 \\ l_1 \\ 0 \end{bmatrix} \qquad \frac{\partial \mathbf{r}}{\partial i} = q \sin \omega \begin{bmatrix} l_3 \\ m_3 \\ n_3 \end{bmatrix} \qquad \frac{\partial \mathbf{r}}{\partial \omega} = q \begin{bmatrix} l_2 \\ m_2 \\ n_2 \end{bmatrix}$$

$$\tag{10.27}$$

$$\frac{\partial \mathbf{v}}{\partial \Omega} = \frac{nab}{q} \begin{bmatrix} -m_2 \\ l_2 \\ 0 \end{bmatrix} \qquad \frac{\partial \mathbf{v}}{\partial i} = \frac{nab \cos \omega}{q} \begin{bmatrix} l_3 \\ m_3 \\ n_3 \end{bmatrix} \qquad \frac{\partial \mathbf{v}}{\partial \omega} = -\frac{nab}{q} \begin{bmatrix} l_1 \\ m_1 \\ n_1 \end{bmatrix}$$

## Lagrange's Planetary Equations

To calculate $\partial \mathbf{r}/\partial a$, we first determine

$$\frac{\partial}{\partial a}[a(\cos E - e)] = \cos E - e - a \sin E \frac{\partial E}{\partial a}$$

$$\frac{\partial}{\partial a}(b \sin E) = \sqrt{1-e^2} \sin E + b \cos E \frac{\partial E}{\partial a}$$

Then, by differentiating Kepler's equation (10.26)

$$\frac{\partial E}{\partial a} - e \cos E \frac{\partial E}{\partial a} = \frac{\partial n}{\partial a} t = -\frac{3n}{2a} t$$

we obtain

$$\frac{\partial E}{\partial a} = -\frac{3nt}{2r}$$

so that at pericenter

$$\frac{\partial}{\partial a}[a(\cos E - e)] = \frac{q}{a} \quad \text{and} \quad \frac{\partial}{\partial a}(b \sin E) = -\frac{3bn\tau}{2q}$$

Therefore,

$$\frac{\partial \mathbf{r}}{\partial a} = \frac{q}{a} \begin{bmatrix} l_1 \\ m_1 \\ n_1 \end{bmatrix} - \frac{3bn\tau}{2q} \begin{bmatrix} l_2 \\ m_2 \\ n_2 \end{bmatrix} \qquad (10.28)$$

and, similarly,

$$\frac{\partial \mathbf{v}}{\partial a} = \frac{3a^2 n^2 \tau}{2q^2} \begin{bmatrix} l_1 \\ m_1 \\ n_1 \end{bmatrix} - \frac{bn}{2q} \begin{bmatrix} l_2 \\ m_2 \\ n_2 \end{bmatrix} \qquad (10.29)$$

To calculate the derivatives with respect to $e$ and $\lambda$, note that

$$\frac{\partial E}{\partial e} - \sin E - e \cos E \frac{\partial E}{\partial e} = 0 \quad \text{and} \quad \frac{\partial E}{\partial \lambda} - e \cos E \frac{\partial E}{\partial \lambda} = 1$$

Hence, for $E = 0$, we have

$$\frac{\partial E}{\partial e} = 0 \quad \text{and} \quad \frac{\partial E}{\partial \lambda} = \frac{a}{q}$$

and the rest of the derivation is as before. There obtains

$$\frac{\partial \mathbf{r}}{\partial e} = -a \begin{bmatrix} l_1 \\ m_1 \\ n_1 \end{bmatrix} \qquad \frac{\partial \mathbf{r}}{\partial \lambda} = \frac{ab}{q} \begin{bmatrix} l_2 \\ m_2 \\ n_2 \end{bmatrix}$$

$$\frac{\partial \mathbf{v}}{\partial e} = \frac{na^3}{bq} \begin{bmatrix} l_2 \\ m_2 \\ n_2 \end{bmatrix} \qquad \frac{\partial \mathbf{v}}{\partial \lambda} = -\frac{na^3}{q^2} \begin{bmatrix} l_1 \\ m_1 \\ n_1 \end{bmatrix} \qquad (10.30)$$

all of which are, of course, valid only at pericenter.

With all of the derivatives evaluated, it is a simple task to calculate the Lagrangian brackets defined in Eq. (10.20). For example,

$$[i, \Omega] = \frac{\partial \mathbf{r}}{\partial i} \cdot \frac{\partial \mathbf{v}}{\partial \Omega} - \frac{\partial \mathbf{r}}{\partial \Omega} \cdot \frac{\partial \mathbf{v}}{\partial i}$$
$$= nab[(l_2 m_3 - l_3 m_2) \sin \omega - (l_1 m_3 - l_3 m_1) \cos \omega]$$
$$= nab(n_1 \sin \omega + n_2 \cos \omega)$$
$$= nab \sin i$$

and again

$$[\lambda, a] = \frac{\partial \mathbf{r}}{\partial \lambda} \cdot \frac{\partial \mathbf{v}}{\partial a} - \frac{\partial \mathbf{r}}{\partial a} \cdot \frac{\partial \mathbf{v}}{\partial \lambda}$$
$$= -\frac{ab^2 n}{2q^2} + \frac{na^3 q}{q^2 a} = \frac{na}{2}\left(\frac{2aq - b^2}{q^2}\right)$$
$$= \tfrac{1}{2} na$$

Because of the skew-symmetry of the matrix $\mathbf{L}$, there are just 15 distinct brackets to evaluate and only six of these turn out to be different from zero. The results are summarized as follows:

$[i, \Omega] = nab \sin i$

$[\omega, \Omega] = 0 \qquad [\omega, i] = 0$

$[a, \Omega] = -\tfrac{1}{2} nb \cos i \quad [a, i] = 0 \quad [a, \omega] = -\tfrac{1}{2} nb$

$[e, \Omega] = \dfrac{na^3 e}{b} \cos i \quad [e, i] = 0 \quad [e, \omega] = \dfrac{na^3 e}{b} \quad [e, a] = 0$

$[\lambda, \Omega] = 0 \qquad [\lambda, i] = 0 \quad [\lambda, \omega] = 0 \qquad [\lambda, a] = \tfrac{1}{2} na \quad [\lambda, e] = 0$

With the elements of the Lagrange matrix determined, Eq. (10.18) may be written in component form as

$$-nab \sin i \frac{di}{dt} + \frac{nb}{2} \cos i \frac{da}{dt} - \frac{na^3 e}{b} \cos i \frac{de}{dt} = \frac{\partial R}{\partial \Omega}$$

$$nab \sin i \frac{d\Omega}{dt} = \frac{\partial R}{\partial i}$$

$$\frac{nb}{2} \frac{da}{dt} - \frac{na^3 e}{b} \frac{de}{dt} = \frac{\partial R}{\partial \omega}$$

$$-\frac{nb}{2} \cos i \frac{d\Omega}{dt} - \frac{nb}{2} \frac{d\omega}{dt} - \frac{na}{2} \frac{d\lambda}{dt} = \frac{\partial R}{\partial a}$$

$$\frac{na^3 e}{b} \cos i \frac{d\Omega}{dt} + \frac{na^3 e}{b} \frac{d\omega}{dt} = \frac{\partial R}{\partial e}$$

$$\frac{na}{2} \frac{da}{dt} = \frac{\partial R}{\partial \lambda}$$

These are easily solved for the derivatives of the orbital elements to produce the classical form of *Lagrange's planetary equations*:

$$\begin{aligned}
\frac{d\Omega}{dt} &= \frac{1}{nab\sin i}\frac{\partial R}{\partial i} \\
\frac{di}{dt} &= -\frac{1}{nab\sin i}\frac{\partial R}{\partial \Omega} + \frac{\cos i}{nab\sin i}\frac{\partial R}{\partial \omega} \\
\frac{d\omega}{dt} &= -\frac{\cos i}{nab\sin i}\frac{\partial R}{\partial i} + \frac{b}{na^3 e}\frac{\partial R}{\partial e} \\
\frac{da}{dt} &= \frac{2}{na}\frac{\partial R}{\partial \lambda} \\
\frac{de}{dt} &= -\frac{b}{na^3 e}\frac{\partial R}{\partial \omega} + \frac{b^2}{na^4 e}\frac{\partial R}{\partial \lambda} \\
\frac{d\lambda}{dt} &= -\frac{2}{na}\frac{\partial R}{\partial a} - \frac{b^2}{na^4 e}\frac{\partial R}{\partial e}
\end{aligned} \quad (10.31)$$

Equations (10.31) demonstrate explicitly that the matrix $\mathbf{L}$ is nonsingular so long as the eccentricity $e$ is neither zero nor one and the inclination angle $i$ is not zero. It should be remarked that a different choice of orbital elements will alleviate these annoying singularities as seen in a later section of this chapter.

◇ **Problem 10–5**
The Lagrangian brackets are the sum of three Jacobians

$$[\alpha_i, \alpha_j] = \frac{\partial(x, \dot{x})}{\partial(\alpha_i, \alpha_j)} + \frac{\partial(y, \dot{y})}{\partial(\alpha_i, \alpha_j)} + \frac{\partial(z, \dot{z})}{\partial(\alpha_i, \alpha_j)}$$

where $\dot{x}$, $\dot{y}$, and $\dot{z}$ denote the time derivatives of $x$, $y$, and $z$.
NOTE: The *Jacobian* is a determinant defined by

$$\frac{\partial(u, v)}{\partial(\alpha, \beta)} = \begin{vmatrix} \dfrac{\partial u}{\partial \alpha} & \dfrac{\partial u}{\partial \beta} \\ \dfrac{\partial v}{\partial \alpha} & \dfrac{\partial v}{\partial \beta} \end{vmatrix}$$

**Problem 10–6**
Consider a new set of orbital elements $[\boldsymbol{\alpha}^T \quad \boldsymbol{\beta}^T]$ where

$$\boldsymbol{\alpha} = \begin{bmatrix} -\tfrac{1}{2}n^2 a^2 \\ nab \\ nab\cos i \end{bmatrix} \quad \boldsymbol{\beta} = \begin{bmatrix} -\tau \\ \omega \\ \Omega \end{bmatrix}$$

so that in Eqs. (10.31) the disturbing function $R = R(\Omega, i, \omega, a, e, \lambda)$ is to be replaced by

$$R = R^*(\boldsymbol{\alpha}, \boldsymbol{\beta})$$

**484**  Variation of Parameters  [Chap. 10]

First, verify the relations

$$\frac{\partial R}{\partial \Omega} = \frac{\partial R^*}{\partial \beta_3}$$

$$\frac{\partial R}{\partial i} = -nab\sin i \frac{\partial R^*}{\partial \alpha_3}$$

$$\frac{\partial R}{\partial \omega} = \frac{\partial R^*}{\partial \beta_2}$$

$$\frac{\partial R}{\partial a} = \frac{n^2 a}{2}\frac{\partial R^*}{\partial \alpha_1} + \frac{nb}{2}\frac{\partial R^*}{\partial \alpha_2} + \frac{nb}{2}\cos i \frac{\partial R^*}{\partial \alpha_3} + \frac{3\lambda}{2na}\frac{\partial R^*}{\partial \beta_1}$$

$$\frac{\partial R}{\partial e} = -\frac{na^3 e}{b}\frac{\partial R^*}{\partial \alpha_2} - \frac{na^3 e}{b}\cos i \frac{\partial R^*}{\partial \alpha_3}$$

$$\frac{\partial R}{\partial \lambda} = \frac{1}{n}\frac{\partial R^*}{\partial \beta_1}$$

and then show that Lagrange's planetary equations, in terms of the alternate set of orbital elements, are in canonical form; i.e.,

$$\frac{d\boldsymbol{\alpha}}{dt}^{\mathrm{T}} = \frac{\partial R^*}{\partial \boldsymbol{\beta}} \quad \text{and} \quad \frac{d\boldsymbol{\beta}}{dt}^{\mathrm{T}} = -\frac{\partial R^*}{\partial \boldsymbol{\alpha}}$$

The partitioned vector elements $\boldsymbol{\alpha}$ and $\boldsymbol{\beta}$ are said to be *canonically conjugate*.
NOTE: The orbital elements $\alpha_1$, $\alpha_2$, $\alpha_3$ are, respectively, the total energy, the angular momentum, and the component of the angular momentum vector along the reference $z$ axis.

## 10.3  Gauss' Form of the Variational Equations

Although Lagrange's variational equations were derived for the special case in which the disturbing acceleration was represented as the gradient of the disturbing function, this restriction is wholly unnecessary. If the disturbed relative motion of two bodies is formulated as in Sect. 9.4 according to

$$\frac{d^2\mathbf{r}}{dt^2} + \frac{\mu}{r^3}\mathbf{r} = \mathbf{a}_d \qquad (10.32)$$

then it is readily seen that the derivation in the previous section leading to Eq. (10.18) is still valid with the result now expressed as

$$\mathbf{L}\frac{d\boldsymbol{\alpha}}{dt} = \left[\frac{\partial \mathbf{r}}{\partial \boldsymbol{\alpha}}\right]^{\mathrm{T}} \mathbf{a}_d \qquad (10.33)$$

The elements of the Lagrange matrix, i.e., the Lagrangian brackets, are as calculated in the previous section. However, the matrix coefficient of the disturbing acceleration vector $\mathbf{a}_d$ in Eq. (10.33) is needed to obtain the appropriate variational equations. (The reader should understand that although $\partial \mathbf{r}/\partial \boldsymbol{\alpha}$ was computed in the previous section as a part of the determination of the Lagrangian brackets, the derivatives so obtained were valid only at the instantaneous pericenter.) We now derive the variational

## Gauss' Form of the Variational Equations

equations appropriate to various choices for component resolutions of the disturbing acceleration vector.

All of the equations derived in the following subsection and attributed to Gauss can also be obtained more simply from the equations of Sect. 10.5. Because of the complexity of the results, it is useful that they be derived using two different methods. It is left for the reader to verify that the two sets of variational equations so obtained are, indeed, equivalent.

### Gauss' Equations in Polar Coordinates

The rotation matrix

$$\mathbf{R} = \begin{bmatrix} l_1 & l_2 & l_3 \\ m_1 & m_2 & m_3 \\ n_1 & n_2 & n_3 \end{bmatrix}$$

will affect an orthogonal transformation of vector components from osculating orbital plane coordinates $\mathbf{i}_e$, $\mathbf{i}_p$, $\mathbf{i}_h$ to the reference coordinates $\mathbf{i}_x$, $\mathbf{i}_y$, $\mathbf{i}_z$. The direction cosine elements of $\mathbf{R}$ are related to the Euler angles through Eqs. (2.9).

Using the asterisk to distinguish a vector resolved along reference axes from the same vector resolved along osculating axes, we have

$$\mathbf{r}^* = \begin{bmatrix} x \\ y \\ z \end{bmatrix} \qquad \mathbf{r} = \begin{bmatrix} r \\ 0 \\ 0 \end{bmatrix} \qquad \mathbf{a}_d^* = \begin{bmatrix} a_{dx} \\ a_{dy} \\ a_{dz} \end{bmatrix} \qquad \mathbf{a}_d = \begin{bmatrix} a_{dr} \\ a_{d\theta} \\ a_{dh} \end{bmatrix}$$

so that

$$\mathbf{r}^* = \mathbf{R}\mathbf{R}_f \mathbf{r} \qquad \text{and} \qquad \mathbf{a}_d^* = \mathbf{R}\mathbf{R}_f \mathbf{a}_d \qquad (10.34)$$

The rotation matrix

$$\mathbf{R}_f = \begin{bmatrix} \cos f & -\sin f & 0 \\ \sin f & \cos f & 0 \\ 0 & 0 & 1 \end{bmatrix} = \begin{bmatrix} (a/r)(\cos E - e) & -(b/r)\sin E & 0 \\ (b/r)\sin E & (a/r)(\cos E - e) & 0 \\ 0 & 0 & 1 \end{bmatrix}$$

provides the necessary transformation from local osculating polar coordinates $\mathbf{i}_r$, $\mathbf{i}_\theta$, $\mathbf{i}_h$ to the orbital plane coordinates $\mathbf{i}_e$, $\mathbf{i}_p$, $\mathbf{i}_h$.

Let $\alpha$ be any one of the six orbital elements. Then, to derive the variational equations in terms of the osculating polar components of the disturbing acceleration, we may calculate

$$\frac{\partial \mathbf{r}^{*\,T}}{\partial \alpha} \mathbf{a}_d^* \equiv \mathbf{a}_d^{*\,T} \frac{\partial \mathbf{r}^*}{\partial \alpha}$$

and replace the term $\partial R/\partial \alpha$ with this quantity in the Lagrange planetary equations (10.31).

For the three Euler angle elements, we first obtain

$$\frac{\partial \mathbf{R}}{\partial \Omega} = \begin{bmatrix} -m_1 & -m_2 & -m_3 \\ l_1 & l_2 & l_3 \\ 0 & 0 & 0 \end{bmatrix}$$

$$\frac{\partial \mathbf{R}}{\partial i} = \begin{bmatrix} l_3 \sin \omega & l_3 \cos \omega & \sin \Omega \cos i \\ m_3 \sin \omega & m_3 \cos \omega & -\cos \Omega \cos i \\ n_3 \sin \omega & n_3 \cos \omega & -\sin i \end{bmatrix}$$

$$\frac{\partial \mathbf{R}}{\partial \omega} = \begin{bmatrix} l_2 & -l_1 & 0 \\ m_2 & -m_1 & 0 \\ n_2 & -n_1 & 0 \end{bmatrix}$$

Then,

$$\frac{\partial R}{\partial \alpha} = \mathbf{a}_d^{*\mathrm{T}} \frac{\partial \mathbf{r}^*}{\partial \alpha} = \mathbf{a}_d^{\mathrm{T}} \mathbf{R}_f^{\mathrm{T}} \mathbf{R}^{\mathrm{T}} \frac{\partial \mathbf{R}}{\partial \alpha} \mathbf{R}_f \mathbf{r} \quad (10.35)$$

is evaluated for $\alpha = \Omega$, $i$, $\omega$. We have

$$\frac{\partial R}{\partial \Omega} = \mathbf{a}_d^{\mathrm{T}} \begin{bmatrix} 0 \\ r \cos i \\ -r \cos \theta \sin i \end{bmatrix}$$

$$\frac{\partial R}{\partial i} = \mathbf{a}_d^{\mathrm{T}} \begin{bmatrix} 0 \\ 0 \\ r \sin \theta \end{bmatrix} \qquad \frac{\partial R}{\partial \omega} = \mathbf{a}_d^{\mathrm{T}} \begin{bmatrix} 0 \\ r \\ 0 \end{bmatrix}$$

(10.36)

where

$$\theta = \omega + f$$

is the argument of latitude defined in Sect. 3.4.

The vector $\mathbf{r}^*$ depends on the remaining three elements through the vector

$$\mathbf{R}_f \mathbf{r} = \begin{bmatrix} a(\cos E - e) \\ b \sin E \\ 0 \end{bmatrix}$$

The derivatives with respect to $a$, $e$, $\lambda$ (following the arguments used in computing the Lagrangian brackets) are obtained as follows:

$$\frac{\partial}{\partial a}(\mathbf{R}_f \mathbf{r}) = \begin{bmatrix} \cos E - e + (3ant/2r) \sin E \\ (b/a) \sin E - (3ant/2r) \cos E \\ 0 \end{bmatrix}$$

$$= \begin{bmatrix} (r/a) \cos f + (3ant/2b) \sin f \\ (r/a) \sin f - (3ant/2b)(e + \cos f) \\ 0 \end{bmatrix}$$

Sect. 10.3] Gauss' Form of the Variational Equations 487

$$\frac{\partial}{\partial e}(\mathbf{R}_f \mathbf{r}) = \begin{bmatrix} -(a^2/r)\sin^2 E - a \\ -(a^2 e/b)\sin E + (ab/r)\sin E \cos E \\ 0 \end{bmatrix}$$

$$= \begin{bmatrix} -(a^2 r/b^2)\sin^2 f - a \\ (a^2 r/b^2)\sin f \cos f \\ 0 \end{bmatrix}$$

$$\frac{\partial}{\partial \lambda}(\mathbf{R}_f \mathbf{r}) = \begin{bmatrix} -(a^2/r)\sin E \\ (ab/r)\cos E \\ 0 \end{bmatrix} = \begin{bmatrix} -(a^2/b)\sin f \\ (a^2/b)(e + \cos f) \\ 0 \end{bmatrix}$$

Then, for $\alpha = a$, $e$, $\lambda$, we evaluate

$$\frac{\partial R}{\partial \alpha} = \mathbf{a}_d^{*\mathrm{T}} \frac{\partial \mathbf{r}^*}{\partial \alpha} = \mathbf{a}_d \mathbf{R}_f^{\mathrm{T}} \mathbf{R}^{\mathrm{T}} \mathbf{R} \frac{\partial}{\partial \alpha}(\mathbf{R}_f \mathbf{r}) = \mathbf{a}_d \mathbf{R}_f^{\mathrm{T}} \frac{\partial}{\partial \alpha}(\mathbf{R}_f \mathbf{r}) \qquad (10.37)$$

and obtain

$$\frac{\partial R}{\partial a} = \mathbf{a}_d^{\mathrm{T}} \begin{bmatrix} (r/a) - (3ant/2b)e\sin f \\ -(3ant/2b)(1 + e\cos f) \\ 0 \end{bmatrix}$$

(10.38)

$$\frac{\partial R}{\partial e} = \mathbf{a}_d^{\mathrm{T}} \begin{bmatrix} -a\cos f \\ [1 + (ar/b^2)]a\sin f \\ 0 \end{bmatrix} \qquad \frac{\partial R}{\partial \lambda} = \mathbf{a}_d^{\mathrm{T}} \begin{bmatrix} (a^2/b)e\sin f \\ (a^2/b)(1 + e\cos f) \\ 0 \end{bmatrix}$$

Eliminating the Secular Term

Before writing out the complete set of variational equations, let us address an undesirable complication caused by the presence of the linear function of time $t$ in the expression for $\partial R/\partial a$. If we examine the Lagrange planetary equations (10.31), we see that $\partial R/\partial a$ appears in, and only in, the equation for the time rate of change of the element $\lambda$. An element exhibiting such behavior is clearly inconvenient at best when large values of $t$ are to be considered and should, therefore, be avoided if at all possible. Fortunately, the difficulty can be overcome in the following manner.

Differentiate the mean anomaly

$$M = nt + \lambda$$

to obtain

$$\frac{dM}{dt} = n - \frac{3nt}{2a}\frac{da}{dt} + \frac{d\lambda}{dt}$$

since the derivative of the mean motion $n = \sqrt{\mu/a^3}$ is simply

$$\frac{dn}{dt} = -\frac{3n}{2a}\frac{da}{dt}$$

Then, using Eqs. (10.31), we have

$$\frac{dM}{dt} = n - \frac{2}{na}\left(\frac{3nt}{2a}\frac{\partial R}{\partial \lambda} + \frac{\partial R}{\partial a}\right) - \frac{b^2}{na^4 e}\frac{\partial R}{\partial e}$$

It is apparent from Eqs. (10.38) that the parenthesized factor in this last equation does not contain $t$ explicitly because of the cancellation. Therefore, an effective artifice for avoiding the difficulty associated with the choice of $\lambda$ as an orbital element is to replace the variational equation for $\lambda$ in Lagrange's equations by

$$\frac{dM}{dt} = n + \frac{d\beta}{dt} \qquad (10.39)$$

where

$$\frac{d\beta}{dt} = -\frac{3t}{a^2}\frac{\partial R}{\partial \lambda} - \frac{2}{na}\frac{\partial R}{\partial a} - \frac{b^2}{na^4 e}\frac{\partial R}{\partial e} \qquad (10.40)$$

The quantity $\beta$ is then to be regarded as the sixth orbital element instead of $\lambda = -n\tau$.

## Summary of Gauss' Equations

Finally, we are ready to summarize the complete set of variational equations. By substituting Eqs. (10.36) and (10.38) into Lagrange's planetary equations (noting that $p = b^2/a$ and $h = nab$), we obtain

$$\begin{aligned}
\frac{d\Omega}{dt} &= \frac{r \sin \theta}{h \sin i} a_{dh} \\
\frac{di}{dt} &= \frac{r \cos \theta}{h} a_{dh} \\
\frac{d\omega}{dt} &= \frac{1}{he}[-p \cos f\, a_{dr} + (p+r)\sin f\, a_{d\theta}] - \frac{r \sin \theta \cos i}{h \sin i} a_{dh} \\
\frac{da}{dt} &= \frac{2a^2}{h}\left(e \sin f\, a_{dr} + \frac{p}{r} a_{d\theta}\right) \\
\frac{de}{dt} &= \frac{1}{h}\{p \sin f\, a_{dr} + [(p+r)\cos f + re]a_{d\theta}\} \\
\frac{dM}{dt} &= n + \frac{b}{ahe}[(p \cos f - 2re)a_{dr} - (p+r)\sin f\, a_{d\theta}]
\end{aligned} \qquad (10.41)$$

(It should be noted that variational equations for either the eccentric or true anomaly may be used in place of the sixth equation above for the mean anomaly. The appropriate equations are the subject of a problem later in this section.)

If initial conditions are specified for $\Omega$, $i$, $\omega$, $a$, $e$, $M$, these differential equations may be integrated by any convenient numerical method. Needless to say, as a part of the integration process, Kepler's equation must be solved for the osculating eccentric anomaly and the osculating true anomaly determined from an appropriate identity; specifically,

$$M = E - e \sin E \quad \text{and} \quad \tan \tfrac{1}{2} f = \sqrt{\frac{1+e}{1-e}} \tan \tfrac{1}{2} E$$

Generally, when the disturbing acceleration is small, a relatively large integration step can be employed. On the other hand, it is necessary to point out that, for this particular choice of orbital elements, the advantage of the variational method is lost for orbits of low inclination or small eccentricity. In these singular cases, the rates of change of $\Omega$, $\omega$ and/or $\beta$ will be large despite the fact that the disturbing acceleration is small. Particular techniques for avoiding these difficulties are treated in later sections.

◊ **Problem 10–7**

Let $a_{dt}$ and $a_{dn}$ be the components of the disturbing acceleration in the plane of the osculating orbit along the velocity vector and perpendicular to it. Show that

$$\begin{bmatrix} a_{dr} \\ a_{d\theta} \end{bmatrix} = \frac{h}{pv} \begin{bmatrix} e \sin f & -(1 + e \cos f) \\ 1 + e \cos f & e \sin f \end{bmatrix} \begin{bmatrix} a_{dt} \\ a_{dn} \end{bmatrix}$$

and then derive the variational equations in the form

$$\frac{d\Omega}{dt} = \frac{r \sin \theta}{h \sin i} a_{dh}$$

$$\frac{di}{dt} = \frac{r \cos \theta}{h} a_{dh}$$

$$\frac{d\omega}{dt} = \frac{1}{ev} \left[ 2 \sin f \, a_{dt} + \left( 2e + \frac{r}{a} \cos f \right) a_{dn} \right] - \frac{r \sin \theta \cos i}{h \sin i} a_{dh}$$

$$\frac{da}{dt} = \frac{2a^2 v}{\mu} a_{dt}$$

$$\frac{de}{dt} = \frac{1}{v} \left[ 2(e + \cos f) a_{dt} - \frac{r}{a} \sin f \, a_{dn} \right]$$

$$\frac{dM}{dt} = n - \frac{b}{eav} \left[ 2 \left( 1 + \frac{e^2 r}{p} \right) \sin f \, a_{dt} + \frac{r}{a} \cos f \, a_{dn} \right]$$

◊ **Problem 10–8**

The variational equations for the eccentric and the true anomalies are, in polar coordinates,

$$\frac{dE}{dt} = \frac{na}{r} + \frac{1}{nae} \left[ (\cos f - e) a_{dr} - \left( 1 + \frac{r}{a} \right) \sin f \, a_{d\theta} \right]$$

$$\frac{df}{dt} = \frac{h}{r^2} + \frac{1}{eh} [p \cos f \, a_{dr} - (p + r) \sin f \, a_{d\theta}]$$

and, in tangential–normal coordinates,

$$\frac{dE}{dt} = \frac{na}{r} - \frac{1}{ebv} [2a \sin f \, a_{dt} + r(e + \cos f) a_{dn}]$$

$$\frac{df}{dt} = \frac{h}{r^2} - \frac{1}{ev} \left[ 2 \sin f \, a_{dt} + \left( 2e + \frac{r}{a} \cos f \right) a_{dn} \right]$$

◇ **Problem 10-9**

The disturbing function for the constant radial thrust acceleration problem of Sect. 8.8 is simply
$$R = r a_{Tr}$$
Derive the variational equations (10.41) for this case directly from the Lagrange planetary equations.

## 10.4 Nonsingular Elements

For orbits of zero inclination angle, the line of nodes does not exist. For orbits of zero eccentricity the line of apsides is meaningless. Therefore, it is not surprising to find singularities in the variational equations for those elements associated with the node or pericenter. These are the longitude of the node $\Omega$, the argument of pericenter $\omega$, the time of pericenter passage $\tau$ (or $\lambda = -n\tau$), and any of the anomalies which are measured from pericenter.

To find variational equations which are nonsingular, we must search for combinations of the usual orbital elements which do not depend on either the line of nodes or the apsidal line. For example, if we add the variational equations for $\Omega$ and $\omega$, the resulting equation exhibits no singularity for vanishing inclination angle $i$. Specifically, from the first and third of Eqs. (10.41), we have

$$\frac{d\varpi}{dt} = \frac{1}{he}[-p\cos f\, a_{dr} + (p+r)\sin f\, a_{d\theta}] + \frac{r}{h}\sin\theta \tan \tfrac{1}{2}i\, a_{dh}$$

where
$$\varpi = \Omega + \omega \tag{10.42}$$
is the longitude of pericenter as defined in Sect. 3.4.

The singularity due to zero eccentricity is still present so that $\varpi$ itself is not a suitable nonsingular orbital element. However, by adding together the variational equations, for $\varpi$ and $M$, we obtain an equation devoid of either singularity. Since

$$\frac{b}{ahe} - \frac{1}{he} = \frac{b-a}{ahe} = \frac{b^2-a^2}{ahe(a+b)} = -\frac{ae}{h(a+b)}$$

it follows that

$$\frac{dl}{dt} = n - \frac{ae}{h(a+b)}[p\cos f\, a_{dr} - (p+r)\sin f\, a_{d\theta}] - \frac{2br}{ah}a_{dr} + \frac{r\sin\theta \tan \tfrac{1}{2}i}{h}a_{dh}$$

where
$$l = \varpi + M \tag{10.43}$$
is the mean longitude defined in Sect. 4.3.

Clearly, $l$ should replace $M$ in our set of nonsingular variables, but the equation just obtained is not yet suitable since it involves the true anomaly

## Sect. 10.4]  Nonsingular Elements

$f$ which is referenced to pericenter. To pursue the question further, let us examine the augmented form of Kepler's equation

$$l = \varpi + M = \varpi + E - e\sin E$$

If we define

$$K = \varpi + E \tag{10.44}$$

as the *eccentric longitude*, corresponding to the mean longitude $l$, then Kepler's equation becomes

$$l = K + e\sin\varpi \cos K - e\cos\varpi \sin K$$

Furthermore, the equation of orbit may be written either in terms of $K$ or in terms of the true longitude

$$L = \varpi + f \tag{10.45}$$

also defined in Sect. 4.3. We have

$$r = a(1 - e\cos E) = a(1 - e\sin\varpi \sin K - e\cos\varpi \cos K)$$

or

$$r = \frac{p}{1 + e\cos f} = \frac{p}{1 + e\sin\varpi \sin L + e\cos\varpi \cos L}$$

Observe that both in the equation of orbit and in Kepler's equation the eccentricity $e$ and the longitude of pericenter $\varpi$ appear only in the combinations $e\sin\varpi$ and $e\cos\varpi$. These functions are, therefore, promising candidates for new elements to replace $e$ and $\varpi$.

Therefore, define $P_1$ and $P_2$ as orbital elements, where

$$P_1 = e\sin\varpi \quad \text{and} \quad P_2 = e\cos\varpi \tag{10.46}$$

and obtain variational equations by differentiating and using the variational equations already obtained for $e$ and $\varpi$. Hence,

$$\frac{dP_1}{dt} = e\cos\varpi \frac{d\varpi}{dt} + \sin\varpi \frac{de}{dt}$$

$$= -\frac{1}{h}[p\cos L\, a_{dr} - (p+r)\sin L\, a_{d\theta} - rP_1 a_{d\theta}] + \frac{r\sin\theta \tan\frac{1}{2}i}{h} P_2 a_{dh}$$

with a similar expression for $P_2$.

Although these equations are nonsingular, the argument of latitude $\theta$ needs to be expressed in terms of the true longitude $L$. For this purpose, we write

$$\theta = \omega + f = L - \Omega$$

so that

$$\sin\theta = \sin L \cos\Omega - \cos L \sin\Omega$$

Now, we know that $\Omega$ is not itself a nonsingular element. However, $\sin\theta$ appears in the variational equation for $P_1$ multiplied by $\tan\frac{1}{2}i$ suggesting that the functions $\tan \frac{1}{2} i \sin \Omega$ and $\tan \frac{1}{2} i \cos \Omega$ would be suitable candidates for new elements to replace $\Omega$ and $i$.

Again, we are led to define $Q_1$ and $Q_2$ as orbital elements, where

$$Q_1 = \tan \tfrac{1}{2} i \sin \Omega \quad \text{and} \quad Q_2 = \tan \tfrac{1}{2} i \cos \Omega \qquad (10.47)$$

and obtain

$$\begin{aligned}\frac{dQ_1}{dt} &= \tan \tfrac{1}{2} i \cos \Omega \frac{d\Omega}{dt} + \frac{1}{2} \sec^2 \tfrac{1}{2} i \sin \Omega \frac{di}{dt} \\ &= \frac{r}{2h} \sec^2 \tfrac{1}{2} i (\sin \theta \cos \Omega + \cos \theta \sin \Omega) a_{dh} \\ &= \frac{r}{2h}(1 + Q_1^2 + Q_2^2) \sin L \, a_{dh}\end{aligned}$$

with a similar result for $Q_2$. The element set is now complete.

Finally, we note that the classical elements are easily recoverable from the new elements. For example,

$$e^2 = P_1^2 + P_2^2 \qquad \tan^2 \tfrac{1}{2} i = Q_1^2 + Q_2^2$$
$$\tan \varpi = \frac{P_1}{P_2} \qquad \tan \Omega = \frac{Q_1}{Q_2}$$

provided, of course, that $P_2$ and $Q_2$ are not zero.

We now summarize the variational equations for the elements $a$, $P_1$, $P_2$, $Q_1$, $Q_2$, $l$ which have recently been named the *equinoctial variables* by Professor Roger A. Broucke of the University of Texas. They are, indeed, nonsingular except for the rectilinear orbit $h = 0$ and for the orbit whose inclination angle $i = \pi$. (These singularities can also be eliminated but we will not pursue the question further.)

With $P_1$, $P_2$, $Q_1$, and $Q_2$ chosen to replace the classical elements $e$, $\Omega$, $i$, and $\omega$ and defined as†

$$P_1 = e \sin \varpi \qquad Q_1 = \tan \tfrac{1}{2} i \sin \Omega$$
$$P_2 = e \cos \varpi \qquad Q_2 = \tan \tfrac{1}{2} i \cos \Omega$$

1. The equation for the semimajor axis is

$$\frac{da}{dt} = \frac{2a^2}{h} \left[ (P_2 \sin L - P_1 \cos L) a_{dr} + \frac{p}{r} a_{d\theta} \right] \qquad (10.48)$$

2. The equations for $P_1$, $P_2$, $Q_1$, and $Q_2$ are

$$\frac{dP_1}{dt} = \frac{r}{h} \left\{ -\frac{p}{r} \cos L \, a_{dr} + \left[ P_1 + \left(1 + \frac{p}{r}\right) \sin L \right] a_{d\theta} \right.$$
$$\left. - P_2(Q_1 \cos L - Q_2 \sin L) a_{dh} \right\} \qquad (10.49)$$

---

† Lagrange first introduced this element set (using $i$ instead of $\frac{1}{2} i$) in 1774 for his study of secular variations. His notation for the four elements was $h$, $l$, $p$, and $q$.

$$\frac{dP_2}{dt} = \frac{r}{h}\left\{\frac{p}{r}\sin L\, a_{dr} + \left[P_2 + \left(1 + \frac{p}{r}\right)\cos L\right]a_{d\theta}\right.$$
$$\left. + P_1(Q_1\cos L - Q_2\sin L)a_{dh}\right\} \quad (10.50)$$

$$\frac{dQ_1}{dt} = \frac{r}{2h}(1 + Q_1^2 + Q_2^2)\sin L\, a_{dh} \quad (10.51)$$

$$\frac{dQ_2}{dt} = \frac{r}{2h}(1 + Q_1^2 + Q_2^2)\cos L\, a_{dh} \quad (10.52)$$

3. The equation for the mean longitude is

$$\frac{dl}{dt} = n - \frac{r}{h}\left\{\left[\frac{a}{a+b}\left(\frac{p}{r}\right)(P_1\sin L + P_2\cos L) + \frac{2b}{a}\right]a_{dr}\right.$$
$$+ \frac{a}{a+b}\left(1 + \frac{p}{r}\right)(P_1\cos L - P_2\sin L)a_{d\theta}$$
$$\left. + (Q_1\cos L - Q_2\sin L)a_{dh}\right\} \quad (10.53)$$

where

$$b = a\sqrt{1 - P_1^2 - P_2^2} \qquad h = nab$$

$$\frac{p}{r} = 1 + P_1\sin L + P_2\cos L \qquad \frac{r}{h} = \frac{h}{\mu(1 + P_1\sin L + P_2\cos L)}$$

4. The true longitude $L$ is obtained from the mean longitude $l$ by first solving Kepler's equation

$$l = K + P_1\cos K - P_2\sin K$$

for the eccentric longitude $K$ and determining $r$ from the equation of orbit

$$r = a(1 - P_1\sin K - P_2\cos K)$$

Then, $L$ is calculated from the eccentric longitude according to the easily derived relations

$$\sin L = \frac{a}{r}\left[\left(1 - \frac{a}{a+b}P_2^2\right)\sin K + \frac{a}{a+b}P_1P_2\cos K - P_1\right]$$
$$\cos L = \frac{a}{r}\left[\left(1 - \frac{a}{a+b}P_1^2\right)\cos K + \frac{a}{a+b}P_1P_2\sin K - P_2\right]$$

where

$$\frac{a}{a+b} = \frac{1}{1 + \sqrt{1 - e^2}} = \frac{1}{1 + \sqrt{1 - P_1^2 - P_2^2}}$$

or alternately expressed as

$$\frac{a}{a+b} = \frac{\beta}{e}$$

in terms of the parameter $\beta$ defined in Prob. 4–7.

Verification of the validity of these equations is left as an exercise for the reader.

## Problem 10–10

The *equinoctial coordinate axes* are defined with respect to the reference axes as follows:

(1) a positive rotation about the vector $\mathbf{i}_z$ through an angle $\Omega$ to establish the direction of the ascending node $\mathbf{i}_n$,
(2) a positive rotation about the vector $\mathbf{i}_n$ through an angle $i$ to establish the direction of $\mathbf{i}_h$, and
(3) a negative rotation about the vector $\mathbf{i}_h$ through an angle $\Omega$.

The position and velocity vectors are expressed in components along the equinoctial axes as

$$\mathbf{r} = r \begin{bmatrix} \cos L \\ \sin L \\ 0 \end{bmatrix} \quad \text{and} \quad \mathbf{v} = \frac{h}{p} \begin{bmatrix} -P_1 - \sin L \\ P_2 + \cos L \\ 0 \end{bmatrix}$$

and the rotation matrix, to transform from equinoctial coordinates to reference coordinates, in terms of the equinoctial elements is

$$\mathbf{R} = \frac{1}{1 + Q_1^2 + Q_2^2} \begin{bmatrix} 1 - Q_1^2 + Q_2^2 & 2Q_1 Q_2 & 2Q_1 \\ 2Q_1 Q_2 & 1 + Q_1^2 - Q_2^2 & -2Q_2 \\ -2Q_1 & 2Q_2 & 1 - Q_1^2 - Q_2^2 \end{bmatrix}$$

## ◇ Problem 10–11

The equations of motion for the constant radial thrust problem of Sect. 8.8 can be written as the following set of nonsingular variational equations:

$$\frac{dP_1}{dt} = -\frac{h}{\mu} \cos\theta\, a_{Tr}$$

$$\frac{dP_2}{dt} = \frac{h}{\mu} \sin\theta\, a_{Tr}$$

$$\frac{da}{dt} = \frac{2a^2}{h}(P_2 \sin\theta - P_1 \cos\theta) a_{Tr}$$

$$\frac{d\theta}{dt} = \frac{\mu^2}{h^3}(P_1 \sin\theta + P_2 \cos\theta)^2$$

where

$$h = \sqrt{\mu a(1 - P_1^2 - P_2^2)}$$

with the initial conditions at $t = t_0$ obtained from

$$P_1 = P_2 = \theta = 0$$

and

$$a = r_0$$

## 10.5 The Poisson Matrix and Vector Variations

The Lagrange matrix $\mathbf{L}$ defined in Eq. (10.19) can be written in a more compact form which renders obvious the proper expression for its inverse. Define a six-dimensional state vector $\mathbf{s}$ whose partitions are the position and velocity vectors $\mathbf{r}$ and $\mathbf{v}$:

$$\mathbf{s} = \begin{bmatrix} \mathbf{r} \\ \mathbf{v} \end{bmatrix} \quad (10.54)$$

Then, we can readily show that

$$\mathbf{L} = \left[\frac{\partial \mathbf{s}}{\partial \boldsymbol{\alpha}}\right]^T \mathbf{J} \frac{\partial \mathbf{s}}{\partial \boldsymbol{\alpha}} \quad (10.55)$$

where the $6 \times 6$ matrix $\mathbf{J}$, introduced in Sect. 9.5, is defined by

$$\mathbf{J} = \begin{bmatrix} \mathbf{O} & \mathbf{I} \\ -\mathbf{I} & \mathbf{O} \end{bmatrix} \quad (10.56)$$

The vector $\mathbf{s}$ is, of course, a function of both the time $t$ and the orbital elements $\boldsymbol{\alpha}$. Therefore,

$$\mathbf{s} = \mathbf{s}(t, \boldsymbol{\alpha}) \quad (10.57)$$

may be thought of as a transformation from element space to state space. The matrix $\partial \mathbf{s}/\partial \boldsymbol{\alpha}$ is called the *Jacobian matrix* of the transformation.

The inverse transformation

$$\boldsymbol{\alpha} = \boldsymbol{\alpha}(t, \mathbf{s}) \quad (10.58)$$

certainly exists and the associated Jacobian matrix $\partial \boldsymbol{\alpha}/\partial \mathbf{s}$ is the inverse of $\partial \mathbf{s}/\partial \boldsymbol{\alpha}$. Indeed, the matrix $\partial \boldsymbol{\alpha}/\partial \mathbf{s}$ is frequently called the *matrizant* of the two-body problem. Therefore,

$$\frac{\partial \mathbf{s}}{\partial \boldsymbol{\alpha}} \frac{\partial \boldsymbol{\alpha}}{\partial \mathbf{s}} = \mathbf{I} \quad \text{and} \quad \frac{\partial \boldsymbol{\alpha}}{\partial \mathbf{s}} \frac{\partial \mathbf{s}}{\partial \boldsymbol{\alpha}} = \mathbf{I}$$

Since $\mathbf{J}^2 = -\mathbf{I}$, it is now trivial to construct the inverse of the Lagrange matrix from the form given in Eq. (10.55).

### The Poisson Matrix

The matrix

$$\mathbf{P} = \frac{\partial \boldsymbol{\alpha}}{\partial \mathbf{s}} \mathbf{J} \left[\frac{\partial \boldsymbol{\alpha}}{\partial \mathbf{s}}\right]^T \quad (10.59)$$

is called the *Poisson matrix*.† Clearly, we have

$$\mathbf{LP} = \mathbf{PL} = -\mathbf{I}$$

---

† Siméon-Denis Poisson (1781–1840) was one of the greatest of the nineteenth century analysts and mathematical physicists. Although he was urged by his father to study medicine, he entered the Ecole Polytechnique first as a student and then as a professor of mathematics. He was one of the founders of the mathematical theory of elasticity and a major contributor to the theories of heat conduction and water waves.

so that
$$P = -L^{-1}$$
Since $L$ is skew-symmetric, so also is $P$ and, further, the transpose of the Poisson matrix is the inverse of the Lagrange matrix

$$P^T = L^{-1} \tag{10.60}$$

The element in the $i^{th}$ row and $j^{th}$ column of the matrix $P$ is denoted by $(\alpha_i, \alpha_j)$ and called a *Poisson bracket*. Now, from the expanded form of Eq. (10.59)

$$P = \frac{\partial \alpha}{\partial r}\left[\frac{\partial \alpha}{\partial v}\right]^T - \frac{\partial \alpha}{\partial v}\left[\frac{\partial \alpha}{\partial r}\right]^T \tag{10.61}$$

it follows that the Poisson brackets are obtained from

$$(\alpha_i, \alpha_j) = \frac{\partial \alpha_i}{\partial r}\left[\frac{\partial \alpha_j}{\partial v}\right]^T - \frac{\partial \alpha_i}{\partial v}\left[\frac{\partial \alpha_j}{\partial r}\right]^T \tag{10.62}$$

Furthermore, they have properties identical to those demonstrated for the Lagrangian brackets in Sect. 10.2.

Using the Poisson matrix, we may write Lagrange's variational equation (10.18) as

$$\frac{d\alpha}{dt} = P^T \left[\frac{\partial R}{\partial \alpha}\right]^T \tag{10.63}$$

in terms of the disturbing function $R$ or as

$$\frac{d\alpha}{dt} = P^T \left[\frac{\partial r}{\partial \alpha}\right]^T a_d \tag{10.64}$$

in terms of the disturbing acceleration vector $a_d$. Now, substitute for $P^T$ from Eq. (10.61), and in the first case,

$$\frac{d\alpha}{dt} = \frac{\partial \alpha}{\partial v}\left[\frac{\partial R}{\partial r}\right]^T - \frac{\partial \alpha}{\partial r}\left[\frac{\partial R}{\partial v}\right]^T$$

But $R$ is a function only of position, so that the result is simply

$$\frac{d\alpha}{dt} = \frac{\partial \alpha}{\partial v}\left[\frac{\partial R}{\partial r}\right]^T \tag{10.65}$$

In the second case,

$$\frac{d\alpha}{dt} = \frac{\partial \alpha}{\partial v}\left[\frac{\partial r}{\partial r}\right]^T a_d - \frac{\partial \alpha}{\partial r}\left[\frac{\partial r}{\partial v}\right]^T a_d$$

and since the state-vector components are to be regarded as independent variables, then

$$\frac{\partial r}{\partial r} = I \quad \text{and} \quad \frac{\partial r}{\partial v} = 0$$

Hence, the result
$$\frac{d\alpha}{dt} = \frac{\partial \alpha}{\partial \mathbf{v}} \mathbf{a}_d \qquad (10.66)$$

This last equation is particularly useful in that it provides a direct method for determining variational equations of vector orbital elements as well as scalar elements in vector form—as such they will be independent of the coordinate system in which the components of the disturbing acceleration vector $\mathbf{a}_d$ might be expressed.

### Variation of the Semimajor Axis

We begin with the energy or vis-viva integral, defined in Eq. (3.17), which is written as
$$\mu\left(\frac{2}{r} - \frac{1}{a}\right) = v^2 = \mathbf{v} \cdot \mathbf{v} = \mathbf{v}^T \mathbf{v}$$

Then, we calculate the partial derivative with respect to the vector $\mathbf{v}$ and obtain
$$\frac{\mu}{a^2} \frac{\partial a}{\partial \mathbf{v}} = 2\mathbf{v}^T$$

According to Eq. (10.66), we have
$$\frac{da}{dt} = \frac{\partial a}{\partial \mathbf{v}} \mathbf{a}_d$$

so that the variational equation for the semimajor axis $a$ is simply
$$\frac{da}{dt} = \frac{2a^2}{\mu} \mathbf{v} \cdot \mathbf{a}_d \qquad (10.67)$$

### Variation of the Angular Momentum Vector

According to Eq. (3.12), the angular momentum vector is defined as
$$\mathbf{h} = \mathbf{r} \times \mathbf{v}$$

Paralleling the arguments used in Sect. 2.2, we replace the vector product by the matrix-vector product
$$\mathbf{h} = \mathbf{S}_r \mathbf{v} \qquad (10.68)$$

where the skew-symmetric matrix $\mathbf{S}_r$ is defined as
$$\mathbf{S}_r = \begin{bmatrix} 0 & -z & y \\ z & 0 & -x \\ -y & x & 0 \end{bmatrix}$$

Then, we calculate
$$\frac{\partial \mathbf{h}}{\partial \mathbf{v}} = \mathbf{S}_r \frac{\partial \mathbf{v}}{\partial \mathbf{v}} = \mathbf{S}_r \mathbf{I} = \mathbf{S}_r$$

so that Eq. (10.66) gives

$$\frac{d\mathbf{h}}{dt} = \frac{\partial \mathbf{h}}{\partial \mathbf{v}}\mathbf{a}_d = \mathbf{S}_r \mathbf{a}_d$$

Thus, the variational equation for the vector angular momentum is

$$\frac{d\mathbf{h}}{dt} = \mathbf{r} \times \mathbf{a}_d \tag{10.69}$$

There are two possible vector forms for the variation of the scalar angular momentum $h$. On the one hand, if we write

$$h^2 = \mathbf{h} \cdot \mathbf{h} = \mathbf{h}^T \mathbf{h}$$

then we have

$$2h\frac{\partial h}{\partial \mathbf{v}} = 2\mathbf{h}^T \frac{\partial \mathbf{h}}{\partial \mathbf{v}} = 2\mathbf{h}^T \mathbf{S}_r$$

so that

$$\frac{dh}{dt} = \mathbf{i}_h \cdot \mathbf{r} \times \mathbf{a}_d = \mathbf{i}_h \times \mathbf{r} \cdot \mathbf{a}_d \tag{10.70}$$

or, alternately,

$$\frac{dh}{dt} = r\,\mathbf{i}_\theta \cdot \mathbf{a}_d \tag{10.71}$$

On the other hand,

$$h^2 = (\mathbf{r} \times \mathbf{v}) \cdot (\mathbf{r} \times \mathbf{v})$$

$$= (\mathbf{r} \cdot \mathbf{r})(\mathbf{v} \cdot \mathbf{v}) - (\mathbf{r} \cdot \mathbf{v})(\mathbf{r} \cdot \mathbf{v})$$

$$= \mathbf{r}^T \mathbf{r}\mathbf{v}^T \mathbf{v} - \mathbf{r}^T \mathbf{v}\mathbf{r}^T \mathbf{v}$$

so that

$$2h\frac{\partial h}{\partial \mathbf{v}} = 2\mathbf{r}^T \mathbf{r}\mathbf{v}^T - 2\mathbf{r}^T \mathbf{v}\mathbf{r}^T$$

Hence,

$$\frac{dh}{dt} = \frac{1}{h}\mathbf{r}^T(\mathbf{r}\mathbf{v}^T - \mathbf{v}\mathbf{r}^T)\mathbf{a}_d \tag{10.72}$$

or, alternately,

$$\frac{dh}{dt} = \frac{1}{h}[r^2(\mathbf{v} \cdot \mathbf{a}_d) - (\mathbf{r} \cdot \mathbf{v})(\mathbf{r} \cdot \mathbf{a}_d)] \tag{10.73}$$

Sect. 10.5]   The Poisson Matrix and Vector Variations   499

Variation of the Eccentricity Vector

The eccentricity or Laplace vector was defined in Eq. (3.14) and can be written in any of the possible forms

$$\mu \mathbf{e} = \mathbf{v} \times \mathbf{h} - \mu \mathbf{i}_r$$
$$= -\mathbf{S}_h \mathbf{v} - \mu \mathbf{i}_r = \mathbf{S}_v \mathbf{h} - \mu \mathbf{i}_r$$

where the matrices $\mathbf{S}_h$ and $\mathbf{S}_v$ are constructed in the same manner as the matrix $\mathbf{S}_r$ used for the angular momentum derivation. Again we have

$$\mu \frac{\partial \mathbf{e}}{\partial \mathbf{v}} = -\mathbf{S}_h \frac{\partial \mathbf{v}}{\partial \mathbf{v}} + \mathbf{S}_v \frac{\partial \mathbf{h}}{\partial \mathbf{v}}$$

so that

$$\mu \frac{d\mathbf{e}}{dt} = -\mathbf{h} \times \mathbf{a}_d + \mathbf{v} \times \frac{d\mathbf{h}}{dt}$$

Thus, the variation of the eccentricity vector can be expressed in any of the following forms:

$$\mu \frac{d\mathbf{e}}{dt} = \mathbf{a}_d \times (\mathbf{r} \times \mathbf{v}) + (\mathbf{a}_d \times \mathbf{r}) \times \mathbf{v} \qquad (10.74)$$

$$\mu \frac{d\mathbf{e}}{dt} = 2(\mathbf{v} \cdot \mathbf{a}_d)\mathbf{r} - (\mathbf{r} \cdot \mathbf{a}_d)\mathbf{v} - (\mathbf{r} \cdot \mathbf{v})\mathbf{a}_d \qquad (10.75)$$

$$\mu \frac{d\mathbf{e}}{dt} = (2\mathbf{r}\mathbf{v}^T - \mathbf{v}\mathbf{r}^T - \mathbf{r}^T\mathbf{v}\mathbf{I})\mathbf{a}_d \qquad (10.76)$$

By now it should be apparent that each of the variational equations derived thus far can be obtained formally according to the rule:

*Apply the usual rules of differentiation to any two-body identity. Treat* $\mathbf{r}$ *as constant, orbital elements as variables, and replace the time rate of change of* $\mathbf{v}$ *by* $\mathbf{a}_d$.

This convenient rule has general validity.

For example, to obtain the variational equation for the eccentricity, begin with the expression

$$p = h^2/\mu = a(1 - e^2)$$

defining the parameter. Then,

$$2h\frac{dh}{dt} = \mu(1 - e^2)\frac{da}{dt} - 2\mu ae\frac{de}{dt}$$

and, substituting from Eqs. (10.67) and (10.73), yields

$$\frac{de}{dt} = \frac{1}{\mu ae}[(\mathbf{r} \cdot \mathbf{v})(\mathbf{r} \cdot \mathbf{a}_d) + (pa - r^2)(\mathbf{v} \cdot \mathbf{a}_d)] \qquad (10.77)$$

### Variation of the Inclination and Longitude of the Node

The angular momentum vector **h** is, of course, normal to the plane of the osculating orbit and may be expressed in terms of components along reference axes as

$$\mathbf{h} = h\,\mathbf{i}_h = h(\sin\Omega \sin i\,\mathbf{i}_x - \cos\Omega \sin i\,\mathbf{i}_y + \cos i\,\mathbf{i}_z)$$

since the unit vector $\mathbf{i}_h$ is identical to the vector $\mathbf{i}_\zeta$ of Eq. (2.6). Applying the formal rule, i.e., calculating the ordinary time derivative of this two-body identity, results in

$$\frac{d\mathbf{h}}{dt} = h\sin i\,\frac{d\Omega}{dt}\mathbf{i}_n - h\frac{di}{dt}\mathbf{i}_m + \frac{dh}{dt}\mathbf{i}_h$$

where $\mathbf{i}_n$ is a unit vector in the direction of the ascending node and $\mathbf{i}_m$ is in the plane of the osculating orbit and normal to $\mathbf{i}_n$ such that $\mathbf{i}_n$, $\mathbf{i}_m$, $\mathbf{i}_h$ form an orthogonal triad. Expressions for $\mathbf{i}_n$ and $\mathbf{i}_m$ in terms of components along the reference axes are given in Eqs. (2.5) and (2.8).

The appropriate variational equations for the longitude of the node $\Omega$ and the inclination angle $i$ are obtained by calculating the scalar product of the last equation with $\mathbf{i}_n$ and $\mathbf{i}_m$, respectively. We have

$$\frac{d\Omega}{dt} = \frac{1}{h\sin i}\mathbf{i}_n \times \mathbf{r}\cdot\mathbf{a}_d = \frac{r\sin\theta}{h\sin i}\mathbf{i}_h\cdot\mathbf{a}_d \tag{10.78}$$

$$\frac{di}{dt} = -\frac{1}{h}\mathbf{i}_m \times \mathbf{r}\cdot\mathbf{a}_d = \frac{r\cos\theta}{h}\mathbf{i}_h\cdot\mathbf{a}_d \tag{10.79}$$

where $\theta = \omega + f$ is the argument of latitude. Note that a third scalar product with $\mathbf{i}_h$ produces the same variational equation for $h$ as obtained previously in Eq. (10.70).

### Variation of the Argument of Pericenter

The argument of latitude $\theta$ is defined as the angle between the position vector and the ascending node. Thus, from

$$\mathbf{i}_n = \cos\Omega\,\mathbf{i}_x + \sin\Omega\,\mathbf{i}_y$$

it follows that

$$\cos\theta = \mathbf{i}_n\cdot\mathbf{i}_r = \cos\Omega\,(\mathbf{i}_x\cdot\mathbf{i}_r) + \sin\Omega\,(\mathbf{i}_y\cdot\mathbf{i}_r)$$

Hence,

$$-\sin\theta\,\frac{\partial\theta}{\partial\mathbf{v}} = [-\sin\Omega\,(\mathbf{i}_x\cdot\mathbf{i}_r) + \cos\Omega\,(\mathbf{i}_y\cdot\mathbf{i}_r)]\frac{\partial\Omega}{\partial\mathbf{v}}$$

Next, from the results of Prob. 3–21,

$$\mathbf{i}_x\cdot\mathbf{i}_r = \cos\Omega\cos\theta - \sin\Omega\sin\theta\cos i$$
$$\mathbf{i}_y\cdot\mathbf{i}_r = \sin\Omega\cos\theta + \cos\Omega\sin\theta\cos i$$

### The Poisson Matrix and Vector Variations

so that, after substitution and cancellation, we obtain

$$\frac{\partial \theta}{\partial \mathbf{v}} = -\cos i \frac{\partial \Omega}{\partial \mathbf{v}}$$

or

$$\frac{\partial \theta}{\partial \mathbf{v}} \mathbf{a}_d = -\cos i \frac{d\Omega}{dt} \qquad (10.80)$$

This last expression is the perturbative derivative of $\theta$, i.e., the change in $\theta$ due to the change in $\mathbf{i}_n$ from which the angle $\theta$ is measured. The total time rate of change of $\theta$ is the sum

$$\frac{d\theta}{dt} = \frac{\partial \theta}{\partial t} + \frac{\partial \theta}{\partial \mathbf{v}} \mathbf{a}_d$$

where $\partial \theta / \partial t$ represents the change in $\theta$ due to ordinary two-body motion with constant orbital elements as specified by Kepler's second law. Thus,

$$\frac{d\theta}{dt} = \frac{h}{r^2} - \cos i \frac{d\Omega}{dt} \qquad (10.81)$$

with $\partial \Omega / dt$ obtained from Eq. (10.78).

Finally, since $\theta = \omega + f$, we can use Eq. (10.80) to write the variational equation for the argument of pericenter as

$$\frac{d\omega}{dt} = -\frac{\partial f}{\partial \mathbf{v}} \mathbf{a}_d - \cos i \frac{d\Omega}{dt} \qquad (10.82)$$

which involves the perturbative derivative of the true anomaly $f$. This we calculate in the next subsection.

### Variations of the Anomalies

By differentiating the equation of orbit

$$r(1 + e \cos f) = \frac{h^2}{\mu}$$

obtain

$$re \sin f \frac{\partial f}{\partial \mathbf{v}} = r \cos f \frac{\partial e}{\partial \mathbf{v}} - \frac{2h}{\mu} \frac{\partial h}{\partial \mathbf{v}} \qquad (10.83)$$

Also, from Eq. (3.31), we establish

$$\frac{\mu}{h} re \sin f = \mathbf{r} \cdot \mathbf{v}$$

which, when differentiated, yields

$$re \cos f \frac{\partial f}{\partial \mathbf{v}} = -r \sin f \frac{\partial e}{\partial \mathbf{v}} + \frac{\mathbf{r} \cdot \mathbf{v}}{\mu} \frac{\partial h}{\partial \mathbf{v}} + \frac{h}{\mu} \mathbf{r}^T \qquad (10.84)$$

Now multiply Eq. (10.83) by $\sin f$, Eq. (10.84) by $\cos f$, and add the two. After some fairly straightforward reduction, we obtain

$$reh\frac{\partial f}{\partial \mathbf{v}} = p\cos f \mathbf{r}^T - (p+r)\sin f \frac{\partial h}{\partial \mathbf{v}}$$

Then, substitution for $\partial h/\partial \mathbf{v}$ produces

$$\frac{\partial f}{\partial \mathbf{v}} = \frac{r}{h^2 e}\left\{\left[\frac{h}{p}(\cos f + e) + \frac{h}{r}\right]\mathbf{r}^T - (p+r)\sin f \mathbf{v}^T\right\} \qquad (10.85)$$

as the perturbative derivative of the true anomaly.

This last expression for $\partial f/\partial \mathbf{v}$ can be used in Eq. (10.82) to complete the variational equation for the argument of pericenter $\omega$. It may also be used to obtain the total time rate of change of the true anomaly as

$$\frac{df}{dt} = \frac{h}{r^2} + \frac{\partial f}{\partial \mathbf{v}}\mathbf{a}_d \qquad (10.86)$$

Thus far, the formulas in this section are equally valid for hyperbolic as well elliptic osculating orbits. In the remainder of this subsection, we consider the eccentric and mean anomalies which, of course, apply only to the ellipse and leave as an exercise the parallel arguments for the hyperbola.

From the identity relating the eccentric and true anomalies

$$\cos E = \frac{\cos f + e}{1 + e\cos f}$$

we obtain, in the usual manner,

$$b\frac{\partial E}{\partial \mathbf{v}} = r\frac{\partial f}{\partial \mathbf{v}} - \frac{ra}{p}\sin f \frac{\partial e}{\partial \mathbf{v}}$$

which, after substitution and reduction, results in

$$\frac{\partial E}{\partial \mathbf{v}} = \frac{r}{\mu b e}\left[\frac{h}{p}(\cos f + e)\mathbf{r}^T - (r+a)\sin f \mathbf{v}^T\right] \qquad (10.87)$$

Similarly, from Kepler's equation

$$M = E - e\sin E$$

we obtain

$$\frac{\partial M}{\partial \mathbf{v}} = \frac{r}{a}\frac{\partial E}{\partial \mathbf{v}} - \sin E\frac{\partial e}{\partial \mathbf{v}}$$

or, in reduced form,

$$\frac{\partial M}{\partial \mathbf{v}} = \frac{rb}{ha^2 e}\left[\cos f \mathbf{r}^T - \frac{a}{h}(r+p)\sin f \mathbf{v}^T\right] \qquad (10.88)$$

Sect. 10.6]  Applications of the Variational Method  503

The total time derivatives of the eccentric and mean anomalies are then

$$\frac{dE}{dt} = \frac{na}{r} + \frac{\partial E}{\partial \mathbf{v}} \mathbf{a}_d$$
$$\frac{dM}{dt} = n + \frac{\partial M}{\partial \mathbf{v}} \mathbf{a}_d \qquad (10.89)$$

## 10.6 Applications of the Variational Method

In this section we consider several interesting and important applications of the concepts thus far developed in this chapter. The first example utilizes the Lagrange planetary equations to study the average effect of the $J_2$ term in the earth's gravitational potential on the motion of an earth orbiting satellite. The second example is an application of Gauss' form of the variational equations to analyze the effect of atmospheric drag on the orbital elements of a satellite in earth orbit.

### Effect of $J_2$ on Satellite Orbits

The disturbing function associated with the $J_2$ term in the earth's gravitational field

$$R = -\frac{Gm}{r} J_2 \left(\frac{r_{eq}}{r}\right)^2 P_2(\cos\phi) \qquad (10.90)$$

is obtained from Eq. (8.92). The colatitude angle $\phi$ is related to the orbital elements and calculated from

$$\cos\phi = \mathbf{i}_r \cdot \mathbf{i}_z = \sin(\omega + f)\sin i$$

using the results of Prob. 3–21. Hence, the Legendre polynomial $P_2(\cos\phi)$ is expressed as

$$P_2(\cos\phi) = \tfrac{1}{2}[3\sin^2(\omega + f)\sin^2 i - 1]$$

so that the disturbing function assumes the form

$$R = -\frac{GmJ_2 r_{eq}^2}{2p^3}(1 + e\cos f)^3[3\sin^2(\omega + f)\sin^2 i - 1] \qquad (10.91)$$

where $r$ has been replaced by the equation of orbit.

The disturbing function can be expanded as a Fourier series in the mean anomaly $M$ using the technique of Sect. 5.3. The constant term in the series is simply the average value of $R$ over one orbit, i.e.,

$$\overline{R} = \frac{1}{2\pi}\int_0^{2\pi} R\, dM$$

Since $dM = n\,dt$ and $r^2 df = h\,dt$, then clearly,

$$\overline{R} = \frac{1}{2\pi}\int_0^{2\pi} \frac{n}{h} R r^2 df$$

Substituting from Eq. (10.91) and performing the integration yields

$$\overline{R} = \frac{n^2 J_2 r_{eq}^2}{4(1-e^2)^{\frac{3}{2}}}(2 - 3\sin^2 i) \tag{10.92}$$

Thus, the average value of the disturbing function depends only on the three orbital elements $a$, $e$, and $i$.

When $\overline{R}$ is used for $R$ in Lagrange's planetary equations (10.31), we have, immediately, expressions for the average rates of change of the satellite orbital elements during a single revolution. For example, since $\overline{R}$ is not a function of $\Omega$, $\omega$, or $\lambda$, we see that

$$\frac{\overline{da}}{dt} = 0 \qquad \frac{\overline{de}}{dt} = 0 \tag{10.93}$$

On the other hand, we obtain for the longitude of the node

$$\frac{\overline{d\Omega}}{dt} = -\frac{3}{2} J_2 \left(\frac{r_{eq}}{p}\right)^2 n \cos i \tag{10.94}$$

Thus, the plane of the orbit rotates about the earth's polar axis in a direction opposite to that of the motion of the satellite with a mean rate of rotation given by Eq. (10.94). This phenomenon is referred to as the *regression of the node*.

In a similar manner, we obtain for the mean rate of rotation of the line of apsides

$$\frac{\overline{d\omega}}{dt} = \frac{3}{4} J_2 \left(\frac{r_{eq}}{p}\right)^2 n(5\cos^2 i - 1) \tag{10.95}$$

It is apparent that there exists a *critical inclination angle*

$$i_{crit} = 63°\,26'.1$$

such that, if $i$ exceeds $i_{crit}$, the line of apsides will regress while, if $i$ is smaller than $i_{crit}$, the apsidal line will advance.

◇ **Problem 10-12**

For an earth orbiting satellite, show that

$$\frac{\overline{d\Omega}}{dt} = -9.96 \left(\frac{r_{eq}}{a}\right)^{3.5} (1-e^2)^{-2} \cos i \quad \text{degrees/day}$$

$$\frac{\overline{d\omega}}{dt} = 5.0 \left(\frac{r_{eq}}{a}\right)^{3.5} (1-e^2)^{-2}(5\cos^2 i - 1) \quad \text{degrees/day}$$

using appropriate values for the physical data of the earth.

Sect. 10.6]   Applications of the Variational Method   505

Effect of Atmospheric Drag on Satellite Orbits

For a satellite in an elliptic orbit around a nonrotating spherical earth, consider the effect of atmospheric drag on the eccentricity of the orbit. If the drag acceleration is proportional to the product of the atmospheric density $\rho$ and the square of the magnitude of the velocity vector and if it acts in a direction opposite to $\mathbf{v}$, then the disturbing acceleration vector is given by

$$\mathbf{a}_d = -c\rho v^2\, \mathbf{i}_t \qquad (10.96)$$

where the proportionality constant $c$ is called the *ballistic coefficient*.

Assume an isothermal atmosphere for which the atmospheric density at a distance $r$ from the center of the earth may be approximated by

$$\rho(r) = \rho_0 \exp\left(-\frac{r-q}{H}\right) \qquad (10.97)$$

where $\rho_0$ is the density at the pericenter radius $q$ and $H$ is the density scale height of the atmosphere.

The appropriate variational equation for our purpose is

$$\frac{de}{dt} = \frac{2}{v}(e + \cos f)a_{dt}$$

according to the results of Prob. 10–7. For the particular application, this equation becomes

$$\frac{de}{dt} = -2c\rho v(e + \cos f) \qquad (10.98)$$

Now, using Eqs. (4.29) and (4.40), the true anomaly and the velocity may be expressed in terms of the eccentric anomaly as

$$e + \cos f = \frac{p}{r}\cos E \quad \text{and} \quad v^2 = n^2 a^2 \frac{1 + e\cos E}{1 - e\cos E}$$

Also, the density model is written in terms of $E$ as

$$\rho = \rho_0 \exp[-\nu(1 - \cos E)]$$

where the constant $\nu$ is defined as

$$\nu = \frac{ae}{H}$$

In this way, Eq. (10.98) is written as

$$\frac{de}{dt} = -2c\rho_0 \frac{pna}{r} e^{-\nu} e^{\nu \cos E} \cos E \sqrt{\frac{1 + e\cos E}{1 - e\cos E}} \qquad (10.99)$$

The last two factors in Eq. (10.99) can be developed as a Fourier cosine series in $E$ using (1) the series manipulations developed in Appendix C, and (2) the trigonometric expansions developed in Sect. 5.2. As a result, we obtain

$$\cos E \sqrt{\frac{1 + e\cos E}{1 - e\cos E}} = \sum_{k=0}^{\infty} A_k \cos kE$$

where the coefficients $A_0, A_1, \ldots$ are power series in the eccentricity. Through terms of order $e^4$, these coefficients are

$$\begin{aligned}
A_0 &= \tfrac{1}{2}e(1 + \tfrac{3}{8}e^2) & A_3 &= \tfrac{1}{8}e^2(1 + \tfrac{15}{16}e^2) \\
A_1 &= 1 + \tfrac{3}{8}e^2 + \tfrac{15}{64}e^4 & A_4 &= \tfrac{1}{16}e^3 \\
A_2 &= \tfrac{1}{2}e(1 + \tfrac{1}{2}e^2) & A_5 &= \tfrac{3}{128}e^4
\end{aligned} \qquad (10.100)$$

Suppose now that we are interested in the average value of $de/dt$ over one orbit. Then,

$$\overline{\frac{de}{dt}} = \frac{1}{2\pi} \int_0^{2\pi} \frac{de}{dt} dM = \frac{1}{2\pi} \int_0^{2\pi} \frac{r}{a} \frac{de}{dt} dE$$

so that

$$\overline{\frac{de}{dt}} = -2c\rho_0 pne^{-\nu} \sum_{k=0}^{\infty} A_k I_k(\nu) \qquad (10.101)$$

where

$$I_k(\nu) = \frac{1}{\pi} \int_0^{\pi} e^{\nu \cos E} \cos kE\, dE \qquad (10.102)$$

is the *modified Bessel function of the first kind of order* $k$. Methods for obtaining numerical values for these special functions are developed in the following subsection.

A similar analysis can be used to study the average rate of change of the other orbital elements which are affected by atmospheric drag.

◊ **Problem 10-13**

Expand the coefficients $A_0, A_1, \ldots$ as power series in the eccentricity $e$, analogous to Eqs. (10.100), through terms of order $e^5$.

◊ **Problem 10-14**

Determine an expression for the average value of the time rate of change of the argument of pericenter $\omega$ for a satellite due to atmospheric drag.

◊ **Problem 10-15**

Calculate the average change in eccentricity of an earth orbiting satellite during 30,000 orbits assuming that $q = 4200$ miles, $H = 70$ miles, $e = \tfrac{1}{4}$, $\rho_0 = 2 \times 10^{-14}$ slugs/cu. ft. and $c = 1.0$ ft.$^2$/slug.

Sect. 10.6]     Applications of the Variational Method     507

Modified Bessel Functions

The function $\mathcal{I}(x,\nu)$, where

$$\mathcal{I}(x,\nu) = \exp\left[\frac{\nu}{2}\left(x + \frac{1}{x}\right)\right] = \sum_{n=-\infty}^{\infty} I_n(\nu)x^n \qquad (10.103)$$

is called the *generating function* for the modified Bessel functions $I_n(\nu)$. By paralleling the arguments of Sect. 5.3, we can use the generating function to develop $I_n(\nu)$ as a power series in $\nu$

$$I_n(\nu) = \sum_{j=0}^{\infty} \frac{(\frac{1}{2}\nu)^{n+2j}}{j!\,(n+j)!} \qquad (10.104)$$

which, when compared with Eq. (5.22), shows that

$$J_n(i\nu) = i^n I_n(\nu) \quad \text{where} \quad i = \sqrt{-1} \qquad (10.105)$$

This last identity, which is the basis for the alternate name, "Bessel functions with imaginary arguments," can be utilized for establishing the recurrence formula

$$nI_n(\nu) = \tfrac{1}{2}\nu\bigl[I_{n-1}(\nu) - I_{n+1}(\nu)\bigr] \qquad (10.106)$$

directly from the corresponding formula (5.27) for $J_n(\nu)$.

Similarly, the continued fraction

$$\frac{I_1(\nu)}{I_0(\nu)} = \cfrac{\tfrac{1}{2}\nu}{1 + \cfrac{(\tfrac{1}{2}\nu)^2}{2 + \cfrac{(\tfrac{1}{2}\nu)^2}{3 + \cfrac{(\tfrac{1}{2}\nu)^2}{4 + \cfrac{(\tfrac{1}{2}\nu)^2}{5 + \cdots}}}}} \qquad (10.107)$$

can be obtained directly from the results of Prob. 5–17.

When the argument $\nu$ of the modified Bessel function is small, the series expansion (10.104) is appropriate for calculating $I_0(\nu)$. However, for large $\nu$, the following *asymptotic series* can be used more effectively:†

$$I_0(\nu) \simeq (2\pi\nu)^{-\frac{1}{2}} e^\nu S(\nu) \qquad (10.108)$$

where

$$S(\nu) = 1 + \frac{1^2}{1!\,8\nu} + \frac{1^2 \cdot 3^2}{2!\,(8\nu)^2} + \frac{1^2 \cdot 3^2 \cdot 5^2}{3!\,(8\nu)^3} + \cdots$$

---

† Here the symbol $\simeq$ means "equal with small percentage error when $\nu$ is large and positive." The theory of asymptotic expansions is too extensive to be developed here. Suffice it to say that the series for $S(\nu)$ is usually divergent. Such divergent expansions, with properties like this one, are known as *asymptotic* series. When $\nu$ is large, the use of the asymptotic series involves less computation than the convergent series.

Knowing the value of $I_0(\nu)$, the function $I_1(\nu)$ can be calculated from the continued fraction expansion (10.107). The recursion formula (10.106) can then be used to calculate modified Bessel functions of higher order.

◇ **Problem 10–16**
By expanding

$$e^{\nu \cos \theta} = A_0 + \sum_{k=1}^{\infty} A_k \cos k\theta$$

in a Fourier cosine series, verify that $I_k(\nu) = \frac{1}{2} A_k$. Hence, obtain the integral form of the modified Bessel function of the first kind of order $n$

$$I_n(\nu) = \frac{1}{\pi} \int_0^{\pi} e^{\nu \cos \theta} \cos n\theta \, d\theta$$

which is equivalent to Eq. (10.102).
HINT: The derivation is identical to that used for $J_n(\nu)$ in Sect. 5.3.

◇ **Problem 10–17**
The differential equation

$$\nu^2 \frac{d^2 y}{d\nu^2} + \nu \frac{dy}{d\nu} - (\nu^2 + n^2) y = 0$$

is satisfied by $y = I_n(\nu)$.

◇ **Problem 10–18**
Derive the identities

$$\frac{d}{d\nu}[\nu^n I_n(\nu)] = \nu^n I_{n-1}(\nu) \qquad \frac{d}{d\nu}[\nu^{-n} I_n(\nu)] = \nu^{-n} I_{n+1}(\nu)$$

$$\frac{dI_n}{d\nu} = \tfrac{1}{2}(I_{n-1} + I_{n+1})$$

## 10.7 Variation of the Epoch State Vector

A frequently useful set of orbital elements is the pair of position and velocity vectors $\mathbf{r}_0$ and $\mathbf{v}_0$ at some epoch time $t_0$. Variational equations for these elements have the form

$$\frac{d\mathbf{r}_0}{dt} = \frac{\partial \mathbf{r}_0}{\partial \mathbf{v}} \mathbf{a}_d \qquad \text{and} \qquad \frac{d\mathbf{v}_0}{dt} = \frac{\partial \mathbf{v}_0}{\partial \mathbf{v}} \mathbf{a}_d \qquad (10.109)$$

according to the general result in Eq. (10.66). (The alternate form of these equations, in terms of the disturbing function $R$, was the result of Prob. 10–4.) By solving these equations numerically, the actual disturbed position and velocity $\mathbf{r}$, $\mathbf{v}$ at time $t$ for the orbit can be had from the standard closed-form, two-body equations.

## Relation to the Perturbation Matrices

We recognize that the coefficient matrices in Eqs. (10.109) are simply the adjoint perturbation matrices introduced in Sect. 9.6 and expressed explicitly in Sect. 9.7 by Eqs. (9.78) and (9.79). Therefore, the variational equations for $\mathbf{r}_0$ and $\mathbf{v}_0$ can be written as

$$\frac{d\mathbf{r}_0}{dt} = \mathbf{R}_0^\star \mathbf{a}_d \quad \text{and} \quad \frac{d\mathbf{v}_0}{dt} = \mathbf{V}_0^\star \mathbf{a}_d \qquad (10.110)$$

It is instructive to calculate directly the Lagrange and Poisson matrices for the case at hand—namely, $\boldsymbol{\alpha}^T = [\mathbf{r}_0^T \quad \mathbf{v}_0^T]$. Since

$$\frac{\partial \mathbf{r}}{\partial \boldsymbol{\alpha}} = \left[\frac{\partial \mathbf{r}}{\partial \mathbf{r}_0} \quad \frac{\partial \mathbf{r}}{\partial \mathbf{v}_0}\right] = [\tilde{\mathbf{R}} \quad \mathbf{R}]$$

$$\frac{\partial \mathbf{v}}{\partial \boldsymbol{\alpha}} = \left[\frac{\partial \mathbf{v}}{\partial \mathbf{r}_0} \quad \frac{\partial \mathbf{v}}{\partial \mathbf{v}_0}\right] = [\tilde{\mathbf{V}} \quad \mathbf{V}]$$

then, from Eq. (10.19), we have

$$\mathbf{L} = \begin{bmatrix} \tilde{\mathbf{R}}^T \\ \mathbf{R}^T \end{bmatrix} [\tilde{\mathbf{V}} \quad \mathbf{V}] - \begin{bmatrix} \tilde{\mathbf{V}}^T \\ \mathbf{V}^T \end{bmatrix} [\tilde{\mathbf{R}} \quad \mathbf{R}]$$

$$= \begin{bmatrix} \tilde{\mathbf{R}}^T\tilde{\mathbf{V}} & \tilde{\mathbf{R}}^T\mathbf{V} \\ \mathbf{R}^T\tilde{\mathbf{V}} & \mathbf{R}^T\mathbf{V} \end{bmatrix} - \begin{bmatrix} \tilde{\mathbf{V}}^T\tilde{\mathbf{R}} & \tilde{\mathbf{V}}^T\mathbf{R} \\ \mathbf{V}^T\tilde{\mathbf{R}} & \mathbf{V}^T\mathbf{R} \end{bmatrix}$$

Using the identities (9.50) from Sect. 9.6, we see that

$$\mathbf{L} = \begin{bmatrix} \mathbf{O} & \mathbf{I} \\ -\mathbf{I} & \mathbf{O} \end{bmatrix} = \mathbf{J}$$

which agrees with the result of Prob. 10–4.

Therefore, the identities (9.50), which followed from the symplectic property of the state transition matrix $\boldsymbol{\Phi}$, could also be established using the property that the Lagrange matrix $\mathbf{L}$ is not an explicit function of time.

Similarly, since

$$\begin{bmatrix} \partial \mathbf{r}_0/\partial \mathbf{r} \\ \partial \mathbf{v}_0/\partial \mathbf{r} \end{bmatrix} = \begin{bmatrix} \tilde{\mathbf{R}}_0^\star \\ \tilde{\mathbf{V}}_0^\star \end{bmatrix} \quad \text{and} \quad \begin{bmatrix} \partial \mathbf{r}_0/\partial \mathbf{v} \\ \partial \mathbf{v}_0/\partial \mathbf{v} \end{bmatrix} = \begin{bmatrix} \mathbf{R}_0^\star \\ \mathbf{V}_0^\star \end{bmatrix}$$

then, from Eq. (10.61), the Poisson matrix is obtained as

$$\mathbf{P} = \begin{bmatrix} \tilde{\mathbf{R}}_0^\star \\ \tilde{\mathbf{V}}_0^\star \end{bmatrix} [\mathbf{R}_0^{\star T} \quad \mathbf{V}_0^{\star T}] - \begin{bmatrix} \mathbf{R}_0^\star \\ \mathbf{V}_0^\star \end{bmatrix} [\tilde{\mathbf{R}}_0^{\star T} \quad \tilde{\mathbf{V}}_0^{\star T}]$$

$$= \begin{bmatrix} \tilde{\mathbf{R}}_0^\star \mathbf{R}_0^{\star T} & \tilde{\mathbf{R}}_0^\star \mathbf{V}_0^{\star T} \\ \tilde{\mathbf{V}}_0^\star \mathbf{R}_0^{\star T} & \tilde{\mathbf{V}}_0^\star \mathbf{V}_0^{\star T} \end{bmatrix} - \begin{bmatrix} \mathbf{R}_0^\star \tilde{\mathbf{R}}_0^{\star T} & \mathbf{R}_0^\star \tilde{\mathbf{V}}_0^{\star T} \\ \mathbf{V}_0^\star \tilde{\mathbf{R}}_0^{\star T} & \mathbf{V}_0^\star \tilde{\mathbf{V}}_0^{\star T} \end{bmatrix} = \begin{bmatrix} \mathbf{O} & \mathbf{I} \\ -\mathbf{I} & \mathbf{O} \end{bmatrix} = \mathbf{J}$$

Hence,

$$\frac{d\alpha}{dt} = \mathbf{P}^\mathrm{T} \left[\frac{\partial \mathbf{r}}{\partial \alpha}\right]^\mathrm{T} \mathbf{a}_d = \mathbf{J}^\mathrm{T} \begin{bmatrix} \widetilde{\mathbf{R}}^\mathrm{T} \\ \mathbf{R}^\mathrm{T} \end{bmatrix} \mathbf{a}_d = \begin{bmatrix} -\mathbf{R}^\mathrm{T} \\ \widetilde{\mathbf{R}}^\mathrm{T} \end{bmatrix} \mathbf{a}_d = \begin{bmatrix} \mathbf{R}_0^\star \\ \mathbf{V}_0^\star \end{bmatrix} \mathbf{a}_d$$

and is the same as Eqs. (10.110).

Avoiding Secular Terms

It is desirable from the point of view of numerical integration to avoid secular terms in the variational equations of motion. These arise due to the presence of the quantity $C$, defined in Sect. 9.7 as

$$\sqrt{\mu}\, C = 3U_5 - \chi U_4 - \sqrt{\mu}(t - t_0)U_2 \tag{10.111}$$

in the expressions for $\mathbf{R}_0^\star$ and $\mathbf{V}_0^\star$ given in Eqs. (9.78) and (9.79). They can be eliminated by allowing the epoch time $t_0$ to vary rather than remain constant.† In order to obtain the proper form of the matrix coefficients for Eqs. (10.109), we must modify the derivation of Sect. 9.7.

For this purpose, the differential of Kepler's equation is now

$$-\sqrt{\mu}\, dt_0 = r_0\, d\varsigma + \tfrac{1}{2}\sqrt{\mu}\, C\, d\alpha - U_2\, d\sigma + U_1\, dr$$

so that

$$\frac{\partial \varsigma}{\partial \mathbf{v}} = \frac{1}{\sqrt{\mu}\, r_0}\left(C\mathbf{v}^\mathrm{T} + U_2 \mathbf{r}^\mathrm{T} - \mu\frac{\partial t_0}{\partial \mathbf{v}}\right) \tag{10.112}$$

Then, we have

$$\frac{\partial \mathbf{r}_0}{\partial \mathbf{v}} = \frac{r_0}{\mu}(1 - F)[\mathbf{v}\mathbf{r}^\mathrm{T} - (\mathbf{r} - \mathbf{r}_0)\mathbf{v}^\mathrm{T}] - G\mathbf{I}$$

$$+ \frac{1}{\mu}\mathbf{v}_0\left[\mu\frac{\partial t_0}{\partial \mathbf{v}} - r_0(1 - F)\mathbf{r}^\mathrm{T} - C\mathbf{v}^\mathrm{T}\right] \tag{10.113}$$

$$\frac{\partial \mathbf{v}_0}{\partial \mathbf{v}} = \frac{r}{\mu}(\mathbf{v} - \mathbf{v}_0)(\mathbf{v} - \mathbf{v}_0)^\mathrm{T} + F\mathbf{I}$$

$$- \frac{1}{r_0^3}\mathbf{r}_0\left[\mu\frac{\partial t_0}{\partial \mathbf{v}} - r_0(1 - F)\mathbf{r}^\mathrm{T} - C\mathbf{v}^\mathrm{T}\right] \tag{10.114}$$

By electing to have $t_0$ vary in such a manner that

$$\mu\frac{\partial t_0}{\partial \mathbf{v}} = r_0(1 - F)\mathbf{r}^\mathrm{T} + C\mathbf{v}^\mathrm{T} \tag{10.115}$$

---

† This was suggested to the author by George H. Born. The reader is referred to the paper "Special Perturbations Employing Osculating Reference States" by G. H. Born, E. J. Christensen, and L. K. Seversike published in *Celestial Mechanics*, Vol. 9, March 1974, pp. 41–53.

then the matrix coefficients reduce to†

$$\frac{\partial \mathbf{r}_0}{\partial \mathbf{v}} = \frac{r_0}{\mu}(1-F)[\mathbf{v}\mathbf{r}^T - (\mathbf{r}-\mathbf{r}_0)\mathbf{v}^T] - G\mathbf{I}$$
$$\frac{\partial \mathbf{v}_0}{\partial \mathbf{v}} = \frac{r}{\mu}(\mathbf{v}-\mathbf{v}_0)(\mathbf{v}-\mathbf{v}_0)^T + F\mathbf{I}$$
(10.116)

Equation (10.115) implies an extra variational equation to be solved numerically in addition to Eqs. (10.109). Again the presence of the term involving $C$ in Eq. (10.115) makes it undesirable to determine $t_0$ using numerical integration.

### Variation of the True Anomaly Difference

Although somewhat in violation of the spirit of the variational method, the problem can be avoided by solving, instead, the differential equation for the true anomaly difference $\theta$ between $\mathbf{r}$ and $\mathbf{r}_0$. An additional advantage accrues in that the necessity of repeatedly solving Kepler's equation also is avoided. The penalty paid, however, is that the equations are not valid for rectilinear motion, since $\theta$ is not defined for $p = 0$.

For the variation in the true anomaly difference, we first obtain the differential for $\sigma_0$ as

$$d\sigma_0 = -(1-\alpha r_0)\, d\varsigma + \tfrac{1}{2}(r_0+r)U_1\, d\alpha + U_0\, d\sigma$$

using the notation and equations of Sect. 9.7. Then, by comparing Eqs. (9.69) and (9.70), observe that

$$pU_1 = r\sqrt{p}\sin\theta - r\sigma_0(1-\cos\theta)$$
$$pU_2 = rr_0(1-\cos\theta)$$

The variation with respect to $\mathbf{v}$ is calculated for each of these last two equations using the differentials of $U_1$ and $U_2$ obtained in Sect. 9.7. There result two expressions for $\partial\varsigma/\partial\mathbf{v}$, the first having as its coefficient $F$ and the second the coefficient $G$.

Solve these two equations for $\partial\varsigma/\partial\mathbf{v}$ by multiplying the first by $G_t$, the second by $F_t$, subtracting, and using the identity

$$FG_t - F_tG = 1$$

After considerable simplification, using, in particular, the relation

$$p = 2r_0 - \alpha r_0^2 - \sigma_0^2$$

---

† It is no longer proper to label these matrices as $\mathbf{R}_0^*$ and $\mathbf{V}_0^*$. This was erroneously done in the paper "The Epoch State Navigation Filter" which was previously cited in Sect. 9.7. They are not partitions of a symplectic transition matrix so that the results in the section *State Transition Matrix* of the paper which exploit that property are not valid. Fortunately, the simulation results quoted in the paper are correct. They were taken from the thesis of Joan (Edwards) Lenox, the third author, who did not obtain her results in the manner described in that paper.

we obtain

$$\frac{\partial \varsigma}{\partial \mathbf{v}} = \frac{r_0}{\sqrt{p}} \frac{\partial \theta}{\partial \mathbf{v}} - \frac{1}{2r}\left[\frac{1}{r_0}U_1 U_2 + \frac{1}{p}(r_0 U_1 + \sigma_0 U_2)\right]\frac{\partial p}{\partial \mathbf{v}}$$
$$- \frac{r}{2r_0}U_1 U_2 \frac{\partial \alpha}{\partial \mathbf{v}} - \frac{U_2}{rr_0}(r_0 + \sigma U_1)\frac{\partial \sigma}{\partial \mathbf{v}}$$

Finally, substituting from Eqs. (10.112) and applying the readily derived variational derivative for $p$,

$$\mu \frac{\partial p}{\partial \mathbf{v}} = 2\mathbf{r}^{\mathrm{T}}(\mathbf{r}\mathbf{v}^{\mathrm{T}} - \mathbf{v}\mathbf{r}^{\mathrm{T}})$$

yields

$$\frac{\partial \theta}{\partial \mathbf{v}} = \frac{r}{\sqrt{\mu p}}\left(\frac{1}{r}\mathbf{r}^{\mathrm{T}} - \frac{1}{r_0}\mathbf{r}_0^{\mathrm{T}}\right) \qquad (10.117)$$

It is interesting to note that the final forms of the variational equations do not involve the universal functions at all.

### Variational Equations of Motion

We now summarize the complete set of equations describing the relative motion of two bodies under the influence of a disturbing acceleration vector $\mathbf{a}_d$. The two vector and one scalar differential equations are

$$\frac{d\mathbf{r}_0}{dt} = \frac{rr_0}{\mu}(1 - \cos\theta)[(\mathbf{r}\cdot\mathbf{a}_d)\mathbf{v} - (\mathbf{v}\cdot\mathbf{a}_d)(\mathbf{r} - \mathbf{r}_0)] - \frac{rr_0}{\sqrt{\mu p}}\sin\theta\,\mathbf{a}_d \quad (10.118)$$

$$\frac{d\mathbf{v}_0}{dt} = \frac{r}{\mu}[(\mathbf{v} - \mathbf{v}_0)\cdot\mathbf{a}_d](\mathbf{v} - \mathbf{v}_0) + \left[1 - \frac{r}{p}(1 - \cos\theta)\right]\mathbf{a}_d \qquad (10.119)$$

$$\frac{d\theta}{dt} = \frac{\sqrt{\mu p}}{r^2} + \frac{r}{\sqrt{\mu p}}\left[\frac{1}{r}(\mathbf{r}\cdot\mathbf{a}_d) - \frac{1}{r_0}(\mathbf{r}_0\cdot\mathbf{a}_d)\right] \qquad (10.120)$$

where

$$\mathbf{r} - \mathbf{r}_0 = \frac{r}{p}(1 - \cos\theta)\mathbf{r}_0 + \frac{rr_0}{\sqrt{\mu p}}\sin\theta\,\mathbf{v}_0$$

$$\mathbf{v} - \mathbf{v}_0 = \frac{1}{pr_0}[\sqrt{\mu}\,\sigma_0(1 - \cos\theta) - \sqrt{\mu p}\sin\theta]\mathbf{r}_0 - \frac{r_0}{p}(1 - \cos\theta)\mathbf{v}_0$$

and

$$r = \frac{pr_0}{r_0 + (p - r_0)\cos\theta - \sqrt{p}\,\sigma_0\sin\theta}$$

$$\mu p = (\mathbf{r}_0 \times \mathbf{v}_0)\cdot(\mathbf{r}_0 \times \mathbf{v}_0)$$

$$\sqrt{\mu}\,\sigma_0 = \mathbf{r}_0 \cdot \mathbf{v}_0$$

Sect. 10.7]   Variation of the Epoch State Vector   513

### Problem 10–19
Five of the six orbital elements are contained in the vector expressions for angular momentum and eccentricity and variational equations have already been derived for these quantities. In order to compute $\mathbf{r}$ and $\mathbf{v}$, we may use the instantaneous ascending node as a reference point. This is, frequently, to be preferred over the choice of pericenter since the latter point does not exist for circular orbits.

(a) From the vectors $\mathbf{h}$ and $\mathbf{e}$ show that the calculations

$$\mathbf{i}_n = \frac{\mathbf{i}_z \times \mathbf{h}}{|\mathbf{i}_z \times \mathbf{h}|}$$

$$\mathbf{r}_0 = \frac{h^2}{\mu + \mu \mathbf{e} \cdot \mathbf{i}_n} \mathbf{i}_n$$

$$\mathbf{v}_0 = \frac{1}{h^2}\{\mu \mathbf{h} \times \mathbf{i}_n + \mu \mathbf{e} \times [(\mathbf{h} \times \mathbf{i}_n) \times \mathbf{i}_n]\}$$

will produce the initial position and velocity vectors $\mathbf{r}_0$ and $\mathbf{v}_0$.

(b) The vectors $\mathbf{r}$ and $\mathbf{v}$ can then be calculated using the Lagrange coefficients after first solving the universal form of Kepler's equation. The parameters in Kepler's equation

$$\sqrt{\mu}(t - t_0) = r_0 U_1(\chi; \alpha) + \sigma_0 U_2(\chi; \alpha) + U_3(\chi; \alpha)$$

can be obtained from

$$r_0 = \frac{h^2}{\mu + \mu \mathbf{e} \cdot \mathbf{i}_n}$$

$$\sigma_0 = \frac{\mathbf{r}_0 \cdot \mathbf{v}_0}{\sqrt{\mu}} = \frac{r_0}{h^2\sqrt{\mu}} \mu \mathbf{e} \cdot \mathbf{i}_n \times \mathbf{h}$$

$$\alpha = \frac{\mu}{h^2}(1 - e^2)$$

(c) The sixth orbital element is $t_0$ and its variational equation is

$$\frac{dt_0}{dt} = \left[\frac{r_0^2}{h^2}\left(\cot i \, \mathbf{i}_n + \frac{G}{rr_0}\mathbf{h}\right) \times \mathbf{r} + \left(1 + \frac{r_0}{r}\right)\frac{U_2}{\mu}\mathbf{r} + \frac{c}{\mu}\mathbf{v}\right] \cdot \mathbf{a}_d$$

or, alternately,

$$\frac{dt_0}{dt} = \left\{\frac{r_0^2}{h^2}\left[\frac{(\mathbf{i}_z \cdot \mathbf{h})(\mathbf{i}_z \times \mathbf{h})}{(\mathbf{i}_z \times \mathbf{h}) \cdot (\mathbf{i}_z \times \mathbf{h})} + \frac{G}{rr_0}\mathbf{h}\right] \times \mathbf{r} + \left(1 + \frac{r_0}{r}\right)\frac{U_2}{\mu}\mathbf{r} + \frac{c}{\mu}\mathbf{v}\right\} \cdot \mathbf{a}_d$$

NOTE: These variational equations for $t_0$ were first published in the AIAA Professional Study Series *Space Guidance and Navigation* previously referenced in Sect. 9.7. It was used in a two day AIAA short course presented by the author and Donald C. Fraser at the University of California Santa Barbara in 1970.

514                    Variation of Parameters                    [Chap. 10

◊ **Problem 10-20**

As noted earlier in this section, we may bypass the problem of the sixth orbital element if we are willing to compromise, somewhat, the spirit of the variational method. This is made possible by solving the following differential equation

$$\frac{d\theta}{dt} = \frac{h}{r^2} - \frac{1}{h\tan i}\mathbf{i}_n \times \mathbf{r} \cdot \mathbf{a}_d$$

directly for $\theta$. When coupled with the two vector differential equations

$$\frac{d\mathbf{h}}{dt} = \mathbf{r} \times \mathbf{a}_d$$

$$\mu\frac{d\mathbf{e}}{dt} = \mathbf{a}_d \times (\mathbf{r} \times \mathbf{v}) + (\mathbf{a}_d \times \mathbf{r}) \times \mathbf{v}$$

and the algebraic equations

$$\mathbf{i}_m = \frac{1}{h}\mathbf{h} \times \mathbf{i}_n$$

$$r = \frac{h^2}{\mu + \mu\mathbf{e}\cdot\mathbf{i}_n\cos\theta + \mu\mathbf{e}\cdot\mathbf{i}_m\sin\theta}$$

$$\mathbf{r} = r\cos\theta\,\mathbf{i}_n + r\sin\theta\,\mathbf{i}_m$$

$$\mathbf{v} = \frac{1}{h^2}\mathbf{h} \times \mu\mathbf{e} - \frac{\mu}{h}(\sin\theta\,\mathbf{i}_n - \cos\theta\,\mathbf{i}_m)$$

we have a complete system for propagating the orbital position and velocity vectors.

Chapter 11

# Two-Body Orbital Transfer

THE ALTERATION OF A SPACE VEHICLE'S ORBIT BY ONE OR MORE discrete changes in velocity for the purpose of fulfilling certain mission objectives forms the subject of the present chapter. We first consider ideal impulsive velocity changes which, although physically unrealizable, are, nevertheless, sufficiently good approximations for many purposes if the engine burn time is a small portion of the total mission time. Then, in the last two sections, attention is given to the more realistic problem of powered-flight guidance, i.e., directing the rocket engine thrust vector during the powered maneuver. First, a general-purpose guidance technique is developed, which is applicable to a variety of missions and is based on the velocity-to-be-gained concept. Then, for problems involving more general terminal constraints, some elementary optimal guidance laws are derived.

Many orbital transfer problems involve a minimization criterion to be satisfied. Usually, the objective is to minimize the sum of the required velocity impulses, sometimes referred to as the *characteristic velocity*. For example, one might require the smallest velocity impulse at a given position in orbit to transfer to a new orbit which will intersect a fixed point. The problem might be expanded to include another velocity change at the target point to place the vehicle in still another orbit. The requirement would then be to choose the transfer orbit which minimizes the sum of the two velocity changes. Of some interest, also, is an original proof of the optimality of the Hohmann transfer—a surprisingly difficult result to prove.

Problems of this nature often require solutions of algebraic equations of rather high degree which can be solved only by tedious numerical methods. However, in some cases, certain geometrical properties of the solution are readily perceived which enhance the interest of the subject and frequently lead to a better insight into the underlying mechanisms. In the selection of topics for this chapter, the author has been primarily attracted by those problems for which a geometrical description is possible.

The first few sections of the chapter are devoted to one- and two-impulse transfer problems of the type just described. They find application in a variety of missions such as satellite rendezvous, satellite interception, ballistic missiles, and determination of interplanetary trajectories. Because certain of these problems are conveniently solved or interpreted in the hodograph plane, a section on hodograph analysis has been included.

Injection from a circular coasting orbit is the subject of another section. Here the problem is to determine the velocity required at a point in orbit to achieve specified hyperbolic velocity conditions. The optimization problem reduces to finding the appropriate point in orbit to apply the injection impulse. In the present treatment full consideration is given to realistic constraints on the initial coasting orbit.

## 11.1 The Envelope of Accessibility

In Sect. 6.1 it was shown that an elliptic orbit connecting two points $P_1$ and $P_2$ with focus at $F$ is possible only if the semimajor axis $a$ is not smaller than a minimum value $a_m$. The vacant focus $F_m^*$, corresponding to the orbit with minimum $a$, lies on the chord $P_1 P_2$ and is located so that

$$P_2 F_m^* = s - r_2 \qquad F_m^* P_1 = s - r_1$$

where $r_1$ and $r_2$ are the respective lengths of the radii $FP_1$ and $FP_2$ and $s$ is the semiperimeter of the triangle $\Delta F P_1 P_2$. The magnitude of the velocity vector at $P_1$ for the orbit is

$$v_{1m}^2 = 2\mu \left( \frac{1}{r_1} - \frac{1}{s} \right) \tag{11.1}$$

If a body is projected from point $P_1$ with an initial speed $v_1$ which is less than $v_{1m}$, then regardless of the direction in which it is projected, the body cannot reach the point $P_2$. An important problem, which we can now solve, is to determine the locus of the points $P_2$ which are just accessible from $P_1$ with the fixed initial speed $v_1$.

For any point $P_2$ to be barely accessible from the fixed point $P_1$, the speed $v_1$ must correspond to the minimum energy path from $P_1$ to $P_2$. The implication of Eq. (11.1), is that $P_2$ must be so located that the semiperimeter $s$ is fixed. Obviously, then

$$P_2 P_1 + P_2 F = 2s - r_1 = \text{constant}$$

so that the locus of $P_2$ is an ellipse, as shown in Fig. 11.1, with foci at the gravitational center $F$ and the initial point $P_1$, and with semimajor axis $a_\ell$, eccentricity $e_\ell$, and parameter $p_\ell$ given by

$$2a_\ell = \frac{1+\alpha}{1-\alpha} r_1 \qquad e_\ell = \frac{1-\alpha}{1+\alpha} \qquad p_\ell = \frac{2\alpha r_1}{1-\alpha^2} \tag{11.2}$$

in terms of the dimensionless quantity $\alpha$ defined by

$$\alpha = \frac{r_1 v_1^2}{2\mu}$$

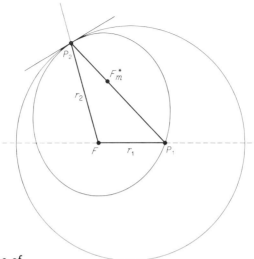

**Fig. 11.1:** Envelope of accessibility.

The tangent of the elliptical locus at $P_2$ bisects the angle between $P_1 P_2$ and $F P_2$ extended according to the property of ellipses demonstrated in Sect. 4.1. On the other hand, from the discussion of Sect. 6.1, the orbital tangent of the minimum-energy ellipse bisects the same angle. Thus, the locus of the barely accessible points $P_2$ and the associated minimum-energy orbits from $P_1$ to $P_2$ are tangent at $P_2$. Hence, the elliptical locus is the envelope of all possible orbits from $P_1$ and its interior includes all points accessible from $P_1$ with an initial speed $v_1$.†

◊ **Problem 11–1**
A ballistic missile launched from the surface of the earth has a velocity capability $v_1$. Assume that the earth is a nonrotating sphere of radius $r_1$ and neglect the effects of atmospheric drag. Then the maximum range attainable, expressed as arc length along a great circle, is

$$r_1 \theta = 2 r_1 \arcsin\left(\frac{\alpha}{1 - \alpha}\right) \qquad \text{where} \qquad \alpha = \frac{r_1 v_1^2}{2\mu}$$

Further, to attain this range, the angle $\gamma$ of the velocity vector measured from the vertical must be determined from

$$\gamma = \tfrac{1}{4}(\pi + \theta)$$

where $\theta$ is the range angle and $\mu$ is the gravitational constant of the earth.

---

† The envelope of accessibility is well known in the field of exterior ballistics and the author F. L. Beckner has written at length on the subject in his paper "Regions Accessible to a Ballistic Weapon" in the *Proceedings of the Fifth AFBMD/STL Aerospace Symposium*, vol. III, pp. 317–366, Academic Press, New York, 1960.

## 11.2 Optimum Single-Impulse Transfer

The notion of an envelope of accessibility, as discussed in the previous section, applies only if the initial point $P_1$ is not moving with respect to the center of attraction at $F$. However, in almost all problems in space-flight mechanics the vehicle or spacecraft does possess an initial velocity with respect to the center of attraction, and it is sensible, therefore, to confine our attention to the additional velocity which must be supplied to accomplish a given objective.

The simplest problem of this kind is one for which the initial velocity is given with the requirement that a velocity increment be determined which will place the vehicle in a new orbit to intersect a fixed point in space. For an application one might postulate a simplified model of the solar system in which the planets describe coplanar elliptic orbits about the sun. A vehicle departing from the earth will have an initial velocity with respect to the sun that is the same as the earth's orbital velocity. Furthermore, in order to stay within the realm of two-body analysis, we need to assume that the velocity impulse occurs at a point sufficiently far removed from the gravitational field of the earth that only solar attraction is important. At first, this model may seem to be simplified to the point of being meaningless. However, many of the essential ingredients of interplanetary flight remain, and the results of the analysis, when properly interpreted, will provide insight to the more realistic problem.

Let $\mathbf{v}_0$ be the initial velocity of the vehicle at $P_1$ so that

$$\Delta \mathbf{v}_1 = \mathbf{v}_1 - \mathbf{v}_0 \qquad (11.3)$$

is the velocity impulse required. Then, the problem of determining the minimum impulsive velocity magnitude $\Delta v_1$ is one of selecting a point on the hyperbolic locus of Fig. 6.4 such that the corresponding vector $\Delta \mathbf{v}_1$ is perpendicular to the hyperbola. It should be remarked that the case for which the initial velocity vector $\mathbf{v}_0$ does not lie in the transfer orbital plane presents no difficulties. The out-of-plane component of the $\mathbf{v}_0$ vector must be canceled by an out-of-plane component of $\Delta \mathbf{v}_1$. The in-plane components can then be handled as just described.

Let $\mathbf{v}_1$ be resolved into chordal and radial components as described in Sect. 6.1 and write

$$\mathbf{v}_1 = \mathbf{v}_{\rho_1} + \mathbf{v}_{c_1}$$

For this velocity vector, the tangent to the hyperbolic locus of velocity vectors is parallel to $\mathbf{v}_{\rho_1} - \mathbf{v}_{c_1}$ according to the discussion on asymptotic coordinates at the end of Sect. 4.4. Therefore, since the shortest distance from a point to a curve is perpendicular to the tangent of the curve, the optimum $\mathbf{v}_1$ is determined from

$$(\mathbf{v}_{\rho_1} - \mathbf{v}_{c_1}) \cdot (\mathbf{v}_{\rho_1} + \mathbf{v}_{c_1} - \mathbf{v}_0) = 0 \qquad (11.4)$$

Sect. 11.2]   Optimum Single-Impulse Transfer   519

which may be written

$$\left(\frac{v_{\rho_1}}{v_{c_1}}\right)^2 - 1 - \frac{1}{v_{c_1}}\left[\frac{v_{\rho_1}}{v_{c_1}}(\mathbf{v}_0 \cdot \mathbf{i}_{r_1}) - (\mathbf{v}_0 \cdot \mathbf{i}_c)\right] = 0 \qquad (11.5)$$

According to Sect. 6.1, the orbital parameter is related to the ratio of the chordal and radial velocity components. Therefore,

$$x^2 = \frac{v_{\rho_1}}{v_{c_1}} = \frac{p_m}{p} \quad \text{where} \quad p_m = \frac{2r_1 r_2}{c}\sin^2 \tfrac{1}{2}\theta \qquad (11.6)$$

Also we have, from Eq. (6.15)

$$\frac{1}{v_{c_1}} = \frac{r_1 r_2 \sin\theta}{c\sqrt{\mu p}} = \sqrt{\frac{2r_1 r_2}{\mu c}}\, x \cos \tfrac{1}{2}\theta$$

where $x > 0$, so that the orthogonality condition takes the form

$$x^4 - \sqrt{\frac{2r_1 r_2}{\mu c}}\cos \tfrac{1}{2}\theta(\mathbf{v}_0 \cdot \mathbf{i}_{r_1})\, x^3 + \sqrt{\frac{2r_1 r_2}{\mu c}}\cos \tfrac{1}{2}\theta(\mathbf{v}_0 \cdot \mathbf{i}_c)\, x - 1 = 0 \qquad (11.7)$$

which is a quartic equation to be solved for a positive real $x$.

Using Descartes' rule of signs,† we can easily see that, for $0 < \theta < \pi$,

1. If $\mathbf{v}_0 \cdot \mathbf{i}_{r_1} \leq 0$, it follows that $\mathbf{v}_0 \cdot \mathbf{i}_c > 0$ and the quartic has one positive real root.
2. If $\mathbf{v}_0 \cdot \mathbf{i}_c \leq 0$, it follows that $\mathbf{v}_0 \cdot \mathbf{i}_{r_1} > 0$ and again the quartic has one positive real root.

On the other hand, for $\pi < \theta < 2\pi$,

1. If $\mathbf{v}_0 \cdot \mathbf{i}_{r_1} \geq 0$, it follows that $\mathbf{v}_0 \cdot \mathbf{i}_c < 0$ and the quartic has one positive real root.
2. If $\mathbf{v}_0 \cdot \mathbf{i}_c \geq 0$, it follows that $\mathbf{v}_0 \cdot \mathbf{i}_{r_1} < 0$ and again the quartic has one positive real root.

There remains to be examined the pie-shaped region defined by the lines $\mathbf{v}_0 \cdot \mathbf{i}_{r_1} = 0$ and $\mathbf{v}_0 \cdot \mathbf{i}_c = 0$. For $\theta < \pi$, we can write‡

$$\mathbf{v}_0 \cdot \mathbf{i}_{r_1} = v_0 \cos\gamma_0 = \frac{\sqrt{\mu p_0}}{r_1}\cot\gamma_0$$

$$\mathbf{v}_0 \cdot \mathbf{i}_c = v_0 \cos(\phi_1 - \gamma_0) = \frac{\sqrt{\mu p_0}}{r_1}(\cos\phi_1 \cot\gamma_0 + \sin\phi_1)$$

---

† Descartes' rule states that the number of positive real roots of an equation with real coefficients is either equal to the number of its variations of sign or is less than that number by a positive even integer. A root of multiplicity $m$ is here counted as $m$ roots.

‡ The same argument can be used when the transfer angle $\theta > \pi$ with the same conclusions substantiated.

where $p_0$ and $\gamma_0$ are the parameter and flight direction angle of the initial orbit at $P_1$. Then define

$$P = \sqrt{\frac{2r_2 p_0}{r_1 c}} \cos \tfrac{1}{2}\theta \cot \gamma_0$$

$$Q = \sqrt{\frac{2r_2 p_0}{r_1 c}} \cos \tfrac{1}{2}\theta (\cos \phi_1 \cot \gamma_0 + \sin \phi_1)$$

so that the quartic may be written as

$$x^4 - Px^3 + Qx - 1 = 0 \tag{11.8}$$

Now, for the minimum-energy orbit, $P = Q$ and the velocity vector divides the pie-shaped region exactly in half. Then, the quartic can be factored as

$$(x-1)[x^3 + (1-P)x^2 + (1-P)x + 1] = 0$$

Therefore, $x = 1$ is the only positive real root provided that $P \leq 1$. However, if $P > 1$ so that $R \equiv P - 1 > 0$, then the cubic equation

$$x^3 - Rx^2 - Rx + 1 = 0$$

will have one negative real root and either two positive real roots or two conjugate complex roots depending on whether or not $R \geq 1$. Indeed, if $R = 1$, then the cubic has the roots $-1, 1, 1$. In short, if $P$ exceeds 2, we will have extraneous positive roots of the quartic with which to deal. A general analysis of the pie-shaped region is not practical; however, since multiple positive roots are possible, our only course is to be wary when $\mathbf{v}_0$ lies within this region.

### Problem 11-2

Consider a single-impulse transfer of a vehicle from a circular orbit to a new trajectory which will intercept a fixed point in space. Let the radius of the circular orbit be $r_1$ and the radius of the target point be $r_2$. Further, let $\phi$ be the latitude of the target point above the original circular orbit plane.

(a) Derive the relations

$$(\Delta \nu_1)^2 = 1 + \nu_{\theta_1}^2 - 2\nu_{\theta_1} \cos i + \nu_{r_1}^2$$

$$\sin i = \frac{\sin \phi}{\sin \theta}$$

$$\frac{1}{R_{12}} = \frac{\nu_{\theta_1}^2}{1 + (\nu_{\theta_1}^2 - 1)\cos\theta - \nu_{\theta_1} \nu_{r_1} \sin\theta}$$

where $i$ is the angle of inclination between the circular orbit plane and the transfer orbit plane, $\theta$ is the central transfer angle,

$$R_{12} = \frac{r_1}{r_2} \qquad \Delta\nu_1 = \frac{\Delta v_1}{v_0} \qquad v_0 = \sqrt{\frac{\mu}{r_1}}$$

and

$$\nu_{r_1} = \frac{v_{r_1}}{v_0} \qquad \nu_{\theta_1} = \frac{v_{\theta_1}}{v_0}$$

(b) The three equations of part (a) define $\Delta\nu_1$ as a function of the two quantities $\nu_{\theta_1}$ and $\theta$. Derive algebraic relations which define $\nu_{\theta_1}$ and $\theta$ corresponding to the minimum value of $\Delta\nu_1$.

(c) If $\phi = 90°$, then the optimum $\Delta\nu_1$ occurs for $\theta = 90°$ with

$$\nu_{\theta_1} = (1 + R_{12}^2)^{-\frac{1}{4}} \qquad (\Delta\nu_1)^2 = 2(1 + R_{12}^2)^{\frac{1}{2}} + 1 - 2R_{12}$$

Wayne Tempelman† 1961

◊ **Problem 11–3**

A vehicle is in orbit about a point $F$ and has a velocity $\mathbf{v}_0$ when at $P_1$. At this point the minimum velocity impulse $\Delta \mathbf{v}_1$ is applied to place the vehicle on a transfer orbit to intersect the point $P_2$ so located that $FP_1$ is perpendicular to $P_1P_2$. If the velocity immediately following the impulse is $\mathbf{v}_1$, then the sum of the direction angles that $\mathbf{v}_1$ and $\Delta \mathbf{v}_1$ make with the line $FP_1$ extended is $90°$.

HINT: Interpret Eq. (11.4) geometrically.

## Optimum Transfer from a Circular Orbit

When the initial orbit is circular, Eq. (11.8) reduces to

$$x^4 + Qx - 1 = 0 \tag{11.9}$$

with

$$Q^2 = \left(\frac{r_2}{c}\right)^3 (1 + \cos\theta) \sin^2\theta \tag{11.10}$$

Completing the square on the quartic results in

$$(x^2 + \xi)^2 - \eta(x + \varsigma)^2 = x^4 + Qx - 1 \tag{11.11}$$

provided that

$$2\xi - \eta = 0 \qquad -2\eta\varsigma = Q \qquad \xi^2 - \eta\varsigma^2 = -1$$

Hence, we must have

$$\xi = \frac{\eta}{2} \qquad \varsigma = -\frac{Q}{2\eta}$$

and $\eta$ as the solution of the cubic equation

$$\eta^3 + 4\eta = Q^2 \tag{11.12}$$

By writing $\eta = \frac{2}{3}\sqrt{3}\, y$, the cubic is transformed to

$$y^3 + 3y = \frac{3\sqrt{3}}{8} Q^2 \equiv 2B$$

---

† "Minimum-energy Intercepts Originating from a Circular Orbit" published in the *Journal of the Aerospace Sciences*, vol. 28, December 1961, pp. 924–929.

which is exactly Barker's equation with $B = \frac{3}{16}\sqrt{3}\,Q^2$. Thus, we can express the solution of Eq. (11.12) as

$$\eta = \frac{\frac{3}{4}AQ^2}{1 + A + A^2} \quad \text{where} \quad A = \left(\sqrt{1+B^2} + B\right)^{\frac{2}{3}} \tag{11.13}$$

To obtain the solution of the quartic equation, factor the left side of Eq. (11.11) so that

$$\left[x^2 + \sqrt{\eta}\,x - \tfrac{1}{2}\left(\sqrt{\eta^2+4} - \eta\right)\right]\left[x^2 - \sqrt{\eta}\,x + \tfrac{1}{2}\left(\sqrt{\eta^2+4} + \eta\right)\right] = 0$$

Note that the cubic equation has been used to write

$$\frac{Q}{\sqrt{\eta}} = \sqrt{\eta^2 + 4}$$

Hence, the only positive real root of the quartic equation is

$$x = \tfrac{1}{2}\left(\sqrt{2\sqrt{\eta^2+4} - \eta} - \sqrt{\eta}\right)$$

corresponding to a parameter value of

$$p = \frac{2p_m}{\sqrt{\eta^2+4} - \sqrt{2\eta\sqrt{\eta^2+4} - \eta^2}} \tag{11.14}$$

Alternately, following a little algebra, we derive

$$p = \frac{p_m}{8}\left(\frac{\eta^2 + 8 + \eta\sqrt{\eta^2+4}}{3\eta^2 + 16}\right)^2 \left(\eta + 2\sqrt{\eta^2+4}\right)$$

$$\times \left[2\eta^2 + 8 + \eta\sqrt{\eta^2+4} + \sqrt{(3\eta^2+16)(\eta^2 + 2\eta\sqrt{\eta^2+4})}\right] \tag{11.15}$$

which is free of numerical problems for all possible orbits and geometry.

◇ **Problem 11–4**

The necessary condition for an optimum single impulse transfer from a circular orbit can be expressed as

$$\sin\nu - \tan\nu = \frac{4}{Q} = \frac{4(c/r_2)^{\frac{3}{2}}}{\sin\theta\sqrt{1+\cos\theta}}$$

where the angle $\nu$, introduced in Sect. 6.5, is given by

$$x^2 = \frac{p_m}{p} = \cot^2 \tfrac{1}{2}\nu$$

and $0 \leq \nu \leq \pi$. In this form, the equation can be solved for $\nu$ almost by inspection using a table of trigonometric functions.

## Sect. 11.2]      Optimum Single-Impulse Transfer      523

◊ **Problem 11-5**
Using ordinary polar coordinates to represent the velocity vector $\mathbf{v}_1$ together with the results of Prob. 4-19, the necessary condition for the optimum single impulse transfer from a circular orbit can be expressed as

$$v_1 \sin(2\gamma_1 - \phi_1) = \sqrt{\frac{\mu}{r_1}} \sin \phi_1 \sin \gamma_1$$

With the findings of Prob. 6-2, write this necessary condition as

$$\sin^2(2\gamma_1 - \phi_1) \tan \tfrac{1}{2}\theta + \sin^3 \gamma_1 \sin(\gamma_1 - \phi_1) \sin \phi_1 = 0$$

—a form which depends only on the flight direction angle.

### Sufficient Condition for an Optimum Elliptic Transfer

Consider again the problem of the optimum single-impulse transfer from a circular orbit with the objective of determining the conditions for which the transfer orbit will be an ellipse. For this analysis it is convenient to use the angle $\nu$ defined in Sect. 6.5.

Consider elliptical orbits whose vacant foci $F^*$ are below the chord and for which the transfer angle $\theta$ is less than 180 degrees. In this case $\nu$, as defined in Sect. 6.5, will be an angle in the second quadrant and, from Eqs. (6.90), it is apparent that we must have

$$\sin \nu \geq \sqrt{\frac{c}{s}} \quad \text{and} \quad -\tan \nu \geq \sqrt{\frac{c}{s-c}} \quad (11.16)$$

with the equal signs obtaining for parabolic orbits. Then, according to the result of Prob. 11-4, a sufficient condition for the orbit to be an ellipse is that

$$\sqrt{\frac{c}{s}} + \sqrt{\frac{c}{s-c}} < \frac{4(c/r_2)^{\frac{3}{2}}}{\sin\theta\sqrt{1+\cos\theta}} \quad (11.17)$$

Therefore, with the exercise of a little algebra, this can be converted to

$$\sqrt{s} + \sqrt{s-c} < 2\sqrt{2r_1} \times \frac{c}{r_2 \sin\theta} = \frac{2\sqrt{2r_1}}{\sin\phi_1}$$

where $\phi_1$ is an exterior angle of the triangle $\Delta P_1 F P_2$. But

$$(\sqrt{s} + \sqrt{s-c})^2 = r_1 + r_2 + 2r_1 r_2 \cos \tfrac{1}{2}\theta \leq (\sqrt{r_1} + \sqrt{r_2})^2$$

so that it is sufficient to require

$$(\sqrt{r_1} + \sqrt{r_2}) \sin \phi_1 < 2\sqrt{2r_1}$$

or, even more conservatively,

$$1 + \sqrt{\frac{r_2}{r_1}} < 2\sqrt{2} \quad (11.18)$$

In the context of interplanetary voyages from the earth, according to Eq. (11.18) an optimum elliptic path to an inner planet always exists. Furthermore, this condition holds also for a trip from earth to Mars. On the other hand, for the remainder of the outer planets, the inequality does not hold and one might expect to find values of $\theta$ for which the optimum trajectory is hyperbolic. As a matter of fact, this situation does prevail for Jupiter and the planets beyond.

One should note, however, that the condition of Eq. (11.17) will always be satisfied if $\sin \theta$ is small enough. Thus, for the outer planets there are sectors near $\theta = 0$ and $\theta = \pi$ for which optimum elliptic trajectories exist. However, the farther from the sun, the smaller these sectors become.

## 11.3 Two-Impulse Transfer between Coplanar Orbits

The ideas presented in the previous section are particularly well suited for the analysis of the optimum two-impulse transfer between two given coplanar orbits. As before, at the point $P_1$ the velocity of the vehicle prior to the velocity impulse is $\mathbf{v}_0$. Immediately following the velocity change, the velocity is $\mathbf{v}_1$ and the vehicle is now in orbit to intersect a point $P_2$. Upon arrival at $P_2$ the vehicle velocity is $\mathbf{v}_2$ and it is now desired to make a second velocity change so that the velocity will then be $\mathbf{v}_3$. The problem is to choose the transfer orbit in such a way that the sum of the magnitudes of the velocity increments is a minimum.

If we resolve $\mathbf{v}_0$ and $\mathbf{v}_3$ into chordal and radial components as before, the velocity increment at $P_1$ will be

$$\Delta v_1 = |\mathbf{v}_1 - \mathbf{v}_0| = [(v_c - v_{c_0})^2 + (v_\rho - v_{\rho_0})^2 + 2(v_c - v_{c_0})(v_\rho - v_{\rho_0}) \cos \phi_1]^{\frac{1}{2}}$$

and similarly at $P_2$

$$\Delta v_2 = |\mathbf{v}_3 - \mathbf{v}_2| = [(v_{c_3} - v_c)^2 + (v_{\rho_3} - v_\rho)^2 + 2(v_{c_3} - v_c)(v_{\rho_3} - v_\rho) \cos \phi_2]^{\frac{1}{2}}$$

The advantage of this particular approach is immediately apparent. The only variables are $v_c$ and $v_\rho$ and these in turn are simply related. The optimum transfer is then found by setting to zero the derivative of $\Delta v_1 + \Delta v_2$ with respect to either $v_c$ or $v_\rho$ and again forming the quotient of $v_c$ and $v_\rho$. Unfortunately, the resulting algebraic equation is of the eighth degree.

Complicated though the solution may be, nevertheless, the optimum transfer is again capable of simple geometric interpretation which could be a tremendous aid to assist in the necessary numerical work. The desired property results from the determination of a necessary condition for a given transfer to be optimum.

Assume that $v_c$ and $v_\rho$ are the chordal and radial components of the optimum transfer orbit. Let the optimum orbit be changed by a small amount and denote the changes in the terminal velocity vectors by $\delta\mathbf{v}_1$ and $\delta\mathbf{v}_2$. Then since

$$\delta[(\Delta\mathbf{v}_1 \cdot \Delta\mathbf{v}_1)^{\frac{1}{2}} + (\Delta\mathbf{v}_2 \cdot \Delta\mathbf{v}_2)^{\frac{1}{2}}] = 0$$

it follows that

$$\frac{(\mathbf{v}_1 - \mathbf{v}_0) \cdot \delta\mathbf{v}_1}{\Delta v_1} = \frac{(\mathbf{v}_3 - \mathbf{v}_2) \cdot \delta\mathbf{v}_2}{\Delta v_2} \qquad (11.19)$$

Hence, $\delta\mathbf{v}_1$ and $\delta\mathbf{v}_2$ must have equal projections on the directions of $\Delta\mathbf{v}_1$ and $\Delta\mathbf{v}_2$, respectively. On the other hand, for $\delta\mathbf{v}_1$ and $\delta\mathbf{v}_2$ to be admissible† changes in the terminal velocities, we must have $\delta\mathbf{v}_1$ and $\delta\mathbf{v}_2$ tangent to their respective hyperbolic loci of the possible transfer velocity vectors as well as

$$\frac{\delta v_c}{v_c} = -\frac{\delta v_\rho}{v_\rho}$$

which is obtained from Eq. (6.16). The situation is illustrated in Fig. 11.2. The change $\delta\mathbf{v}_1$, to be an admissible variation, must be parallel to $\mathbf{v}_{\rho_1} - \mathbf{v}_{c_1}$ or, alternately, parallel to $\sin(\phi_1 - \gamma_1)\,\mathbf{i}_{r_1} - \sin\gamma_1\,\mathbf{i}_c$. Likewise, $\delta\mathbf{v}_2$ must be parallel to $\mathbf{v}_{\rho_2} - \mathbf{v}_{c_2}$ as well as $-\sin(\phi_2 + \gamma_2)\,\mathbf{i}_{r_2} - \sin\gamma_2\,\mathbf{i}_c$. Thus,

$$\delta\mathbf{v}_1 = \frac{\mathbf{v}_{\rho_1} - \mathbf{v}_{c_1}}{|\mathbf{v}_{\rho_1} - \mathbf{v}_{c_1}|} \delta v_1 \qquad \text{and} \qquad \delta\mathbf{v}_2 = \frac{\mathbf{v}_{\rho_2} - \mathbf{v}_{c_2}}{|\mathbf{v}_{\rho_2} - \mathbf{v}_{c_2}|} \delta v_2$$

and, using similar triangle arguments,

$$\frac{\delta v_1}{|\mathbf{v}_{\rho_1} - \mathbf{v}_{c_1}|} = \frac{\delta v_{\rho_1}}{v_{\rho_1}} = -\frac{\delta v_{c_1}}{v_{c_1}} \qquad \text{and} \qquad \frac{\delta v_2}{|\mathbf{v}_{\rho_2} - \mathbf{v}_{c_2}|} = \frac{\delta v_{\rho_2}}{v_{\rho_2}} = -\frac{\delta v_{c_2}}{v_{c_2}}$$

But, $v_{c_1} = v_{c_2}$ and $\delta v_{c_1} = \delta v_{c_2}$, so that

$$\frac{\delta v_1}{|\mathbf{v}_{\rho_1} - \mathbf{v}_{c_1}|} = \frac{\delta v_2}{|\mathbf{v}_{\rho_2} - \mathbf{v}_{c_2}|}$$

with the consequence that the necessary condition (11.19) may be written as

$$\frac{\Delta\mathbf{v}_1}{\Delta v_1} \cdot (\mathbf{v}_{\rho_1} - \mathbf{v}_{c_1}) = \frac{\Delta\mathbf{v}_2}{\Delta v_2} \cdot (\mathbf{v}_{\rho_2} - \mathbf{v}_{c_2}) \qquad (11.20)$$

The geometrical characteristic of the optimum transfer orbit is now readily perceived. The vector difference between the chordal and radial components of the terminal velocities for an optimum transfer must have equal projections on the corresponding velocity increment vectors.

Since the chordal and radial components are meaningless for a transfer through 180°, an alternate form of the necessary condition is desirable. Refer to Fig. 11.2 and denote by $A$ the side of the triangle whose other

---

† Admissible changes in the orbit are, of course, required to satisfy the compatibility conditions of Sect. 6.1.

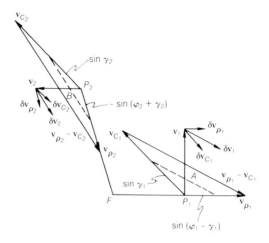

**Fig. 11.2:** Admissible variations in terminal velocity vectors.

sides are $\sin\gamma_1$ and $\sin(\phi_1 - \gamma_1)$. Similarly, let $B$ be the third side of the corresponding triangle at $P_2$. Then,

$$\delta \mathbf{v}_1 = \frac{1}{A}[\ \sin(\phi_1 - \gamma_1)\,\mathbf{i}_{r_1} - \sin\gamma_1\,\mathbf{i}_c]\,\delta v_1$$

$$\delta \mathbf{v}_2 = \frac{1}{B}[-\sin(\phi_2 + \gamma_2)\,\mathbf{i}_{r_2} - \sin\gamma_2\,\mathbf{i}_c]\,\delta v_2$$

By similar triangles,

$$\frac{A}{\sin\gamma_1} = \frac{|\mathbf{v}_{\rho_1} - \mathbf{v}_{c_1}|}{v_{c_1}} \quad \text{and} \quad \frac{B}{\sin\gamma_2} = \frac{|\mathbf{v}_{\rho_2} - \mathbf{v}_{c_2}|}{v_{c_2}}$$

so that we have

$$\frac{\sin\gamma_1}{v_{c_1}}\frac{\delta v_1}{A} = \frac{\delta v_1}{|\mathbf{v}_{\rho_1} - \mathbf{v}_{c_1}|} \quad \text{and} \quad \frac{\sin\gamma_2}{v_{c_2}}\frac{\delta v_2}{B} = -\frac{\delta v_2}{|\mathbf{v}_{\rho_2} - \mathbf{v}_{c_2}|}$$

Therefore, the necessary condition for the two-impulse transfer to be optimum can also be expressed as

$$\frac{\Delta\mathbf{v}_1}{\Delta v_1} \cdot \left(\frac{p_m}{p}\mathbf{i}_{r_1} - \mathbf{i}_c\right) = \frac{\Delta\mathbf{v}_2}{\Delta v_2} \cdot \left(\frac{p_m}{p}\mathbf{i}_{r_2} - \mathbf{i}_c\right) \qquad (11.21)$$

if we recall, from Eqs. (6.26) and (6.27), that

$$\frac{p_m}{p} = \frac{\sin(\phi_1 - \gamma_1)}{\sin\gamma_1} = -\frac{\sin(\phi_2 + \gamma_2)}{\sin\gamma_2}$$

## Cotangential Transfer Orbits

As a first guess it would seem reasonable to suppose that the optimum transfer orbit would be tangent to both the initial orbit through $P_1$ and the final orbit through $P_2$. When the transfer angle $\theta$ is $180°$ and the initial and final orbits are circles, the optimum transfer orbit is, indeed, doubly cotangential as was originally shown by Walter Hohmann and considered in the next subsection. This characteristic of the optimum transfer prevails even if the orbits are noncircular provided that their apsidal lines coincide and that the transfer is between pericenter and apocenter. However, in general, for $\theta \neq 180°$, the doubly cotangential orbit is not optimum as can be readily demonstrated from the necessary condition just derived.

For this purpose, assume that the optimum $\mathbf{v}_1$ and $\mathbf{v}_0$ are parallel and similarly for $\mathbf{v}_2$ and $\mathbf{v}_3$. Then the projections of the vectors $\mathbf{v}_{p_1} - \mathbf{v}_{c_1}$ and $\mathbf{v}_{p_2} - \mathbf{v}_{c_2}$ on the velocity increment vectors are the same as the projections on the optimum transfer velocities $\mathbf{v}_1$ and $\mathbf{v}_2$. Therefore, we must have

$$\frac{\mathbf{v}_1}{v_1} \cdot (\mathbf{v}_{p_1} - \mathbf{v}_{c_1}) = \frac{\mathbf{v}_2}{v_2} \cdot (\mathbf{v}_{p_2} - \mathbf{v}_{c_2})$$

or simply

$$\frac{v_p^2 - v_c^2}{v_1} = \frac{v_p^2 - v_c^2}{v_2}$$

It follows then that the doubly cotangential transfer can be optimum only if $v_1 = v_2$—that is, only if $r_1 = r_2$.

The same conclusion holds if the transfer angle $\theta = 180°$ provided that the axes of the initial and final orbits do not coincide. This follows from the basic form of the necessary condition (11.19). Under these circumstances, we have

$$\frac{\mathbf{v}_1}{v_1} \cdot \delta\mathbf{v}_1 = \frac{\mathbf{v}_2}{v_2} \cdot \delta\mathbf{v}_2$$

for the cotangential transfer. But admissible variations are in the radial direction only since the circumferential components of all transfer orbit velocity vectors are the same. Furthermore, from Eq. (6.5) we must have

$$\delta v_{r_1} = -\delta v_{r_2}$$

so that the cotangential transfer is optimum only if $\mathbf{v}_1$ and $\mathbf{v}_2$ are parallel. Alternately, since $\mathbf{i}_c = -\mathbf{i}_{r_1} = \mathbf{i}_{r_2}$, then Eq. (11.19) takes the form

$$\frac{\Delta\mathbf{v}_1}{\Delta v_1} \cdot \mathbf{i}_{r_1} = \frac{\Delta\mathbf{v}_2}{\Delta v_2} \cdot \mathbf{i}_{r_1}$$

or, simply,

$$\left(\frac{\mathbf{v}_1}{v_1} + \frac{\mathbf{v}_2}{v_2}\right) \cdot \mathbf{i}_{r_1} = 0$$

and the conclusion is the same.

Even though the cotangential transfer orbit is rarely optimum, there may be mission objectives which make such transfers desirable. For example, during the final stages of a manned orbital rendezvous, it might be advantageous for the two vehicles involved in the maneuver to be moving in the same direction. Then the task of nulling the relative velocity would be dramatically simplified. In the following two problems we show how such orbits can be found.

### Problem 11-6

Consider three coplanar, confocal ellipses with semimajor axes and distances from center to focus denoted, respectively, by $a_0$, $a_1$, $a_2$ and $c_0$, $c_1$, $c_2$. Let $\gamma_{ij}$ be the orientation angle between the axes of any pair of ellipses labeled $i$ and $j$.

(a) The distance between the centers of any pair of ellipses is

$$d_{ij} = \sqrt{c_i^2 - 2c_i c_j \cos \gamma_{ij} + c_j^2}$$

(b) The ellipse pairs $0, 1$ and $0, 2$ will each have a single point of tangency if and only if

$$d_{10}^2 = (a_1 - a_0)^2 \quad \text{and} \quad d_{20}^2 = (a_2 - a_0)^2$$

HINT: Derive the condition that two ellipses have only one point in common.

(c) The locus of the centers of all possible elliptic transfer orbits, which are tangent to ellipses 1 and 2 (assumed to be nonintersecting), is itself an ellipse whose foci are the centers of the two terminal ellipses and whose semimajor axis $a$ and eccentricity $e$ are given by

$$a = \frac{|a_1 - a_2|}{2} \quad \text{and} \quad e = \frac{d_{12}}{|a_1 - a_2|}$$

W. Li-Shu Wen† 1961

### Problem 11-7

A vehicle is in an elliptic orbit with semimajor axis $a_0$ about a center of attraction at $F$ with the vacant focus $F_0^*$. A target vehicle is in a nonintersecting elliptic orbit about the same center with its vacant focus at $F_2^*$ and with semimajor axis $a_2$. A velocity impulse is initiated at point $P_1$ on the first orbit to intercept the target at point $P_2$. If the transfer orbit is an ellipse tangent to the initial orbit at $P_1$ and the final orbit at $P_2$, show that its vacant focus $F_1^*$ is located on an ellipse whose foci are at $F_0^*$ and $F_2^*$ and whose semimajor axis is $|a_2 - a_0|$. Develop a graphical construction technique for determining the point $P_2$ when $P_1$ is given.

Geza S. Gedeon‡ 1958

---

† "A Study of Cotangential Elliptical Transfer Orbits in Space Flight," *Journal of the Aerospace Sciences*, vol. 28, May, 1961, pp. 411–417.

‡ "Orbital Mechanics of Satellites," in the *Proceedings of the American Astronautical Society*, paper 19, August, 1958.

## The Hohmann Transfer Orbit

Consider the problem of the two-impulse transfer between circular orbits. From Eqs. (6.2) and (6.3), the velocity increments can be written as

$$\Delta \mathbf{v}_1 = \mathbf{v}_1 - \mathbf{v}_0 = \frac{\sqrt{\mu p}}{r_1} [\cot \gamma_1 \, \mathbf{i}_{r_1} + (1 - A_1) \, \mathbf{i}_{\theta_1}]$$

$$\Delta \mathbf{v}_2 = \mathbf{v}_3 - \mathbf{v}_2 = -\frac{\sqrt{\mu p}}{r_2} [\cot \gamma_2 \, \mathbf{i}_{r_2} - (1 - A_2) \, \mathbf{i}_{\theta_2}]$$

where, using Eqs. (6.28), (6.29), and (6.19), we have defined

$$A_1^2 = \frac{r_1}{p} = \frac{r_2 \cot \gamma_1 \sin \theta + r_1 - r_2 \cos \theta}{r_2 (1 - \cos \theta)}$$

$$A_2^2 = \frac{r_2}{p} = \frac{r_2 - r_1 \cos \theta - r_1 \cot \gamma_2 \sin \theta}{r_1 (1 - \cos \theta)}$$

It will be convenient to write the trigonometric functions of $\theta$ in terms of $x = \cot \tfrac{1}{2}\theta$, so that

$$\sin \theta = \frac{2x}{1+x^2} \qquad \cos \theta = -\frac{1-x^2}{1+x^2} \qquad 1 - \cos \theta = \frac{2}{1+x^2}$$

Indeed, all of the equations will be simpler if we further introduce symbols for the cotangents of the direction angles. Therefore, define

$$x = \cot \tfrac{1}{2}\theta \qquad x_1 = \cot \gamma_1 \qquad x_2 = \cot \gamma_2$$

so that $A_1$ and $A_2$ can be expressed as

$$2r_2 A_1^2 = \frac{2r_2 r_1}{p} = r_1(1+x^2) + r_2(1-x^2) + 2r_2 x_1 x$$

$$2r_1 A_2^2 = \frac{2r_1 r_2}{p} = r_2(1+x^2) + r_1(1-x^2) - 2r_1 x_2 x$$

The minimization of $\Delta v_1 + \Delta v_2$ can be formulated as a constrained optimization problem. Specifically, we desire to minimize the function

$$J(x, x_1, x_2) = \frac{1}{r_1}\sqrt{x_1^2 + (1 - A_1)^2} + \frac{1}{r_2}\sqrt{x_2^2 + (1 - A_2)^2} \qquad (11.22)$$

subject to two constraints—the first being the necessary condition, Eq. (11.21), for the optimum two-impulse transfer for fixed transfer angle $\theta$ written as

$$F(x, x_1, x_2) = \frac{x_1[r_1(1+x^2) + r_2(1-x^2) + 2r_2 x_1 x] - r_2(1 - A_1)x}{\sqrt{x_1^2 + (1-A_1)^2}}$$

$$+ \frac{x_2[r_2(1+x^2) + r_1(1-x^2) - 2r_1 x_2 x] + r_1(1 - A_2)x}{\sqrt{x_2^2 + (1-A_2)^2}} = 0 \qquad (11.23)$$

The second is the relation between the direction angles $\gamma_1$ and $\gamma_2$ from Eq. (6.8) which can be conveniently expressed as

$$G(x, x_1, x_2) = r_1 A_2^2 - r_2 A_1^2 = 0 \tag{11.24}$$

With the introduction of the Lagrange multipliers $\lambda_1$ and $\lambda_2$, the constrained optimization problem is equivalent to minimizing the function

$$J(x, x_1, x_2) - \lambda_1 F(x, x_1, x_2) - \lambda_2 G(x, x_1, x_2)$$

when $x$, $x_1$ and $x_2$ are unrestricted. The appropriate necessary conditions for this minimum are

$$\frac{\partial J}{\partial x} - \lambda_1 \frac{\partial F}{\partial x} - \lambda_2 \frac{\partial G}{\partial x} = 0$$

$$\frac{\partial J}{\partial x_1} - \lambda_1 \frac{\partial F}{\partial x_1} - \lambda_2 \frac{\partial G}{\partial x_1} = 0 \qquad \frac{\partial J}{\partial x_2} - \lambda_1 \frac{\partial F}{\partial x_2} - \lambda_2 \frac{\partial G}{\partial x_2} = 0$$

which, together with the two constraint equations $F = G = 0$, provide five equations to be solved for $x$, $x_1$, $x_2$, $\lambda_1$ and $\lambda_2$.

It is easy to verify that all of the partial derivatives, as well as the constraints, vanish for $x = x_1 = x_2 = 0$ with the exception of $\partial F/\partial x$ which is then equal to $r_1 - r_2$. Therefore, the complete solution to the optimization problem is

$$x = x_1 = x_2 = \lambda_1 = 0 \quad \text{and} \quad \lambda_2 = \text{arbitrary constant}$$

Thus, we have shown that the optimum two-impulse transfer between circular orbits is tangent to both the initial and final orbits with a transfer angle of 180 degrees. This was first recognized by Walter Hohmann in his paper published in Munich, Germany in the year 1925 and, ever after, such orbits have been known as *Hohmann orbits*.

◇ **Problem 11-8**

Consider the optimum transfer problem between two circular orbits of radii $r_1$ and $r_2$ with $r_1 < r_2$.

(a) For the Hohmann transfer consisting of two velocity impulses $\Delta \mathbf{v}_1$ and $\Delta \mathbf{v}_2$ applied tangentially to the initial and final orbits and separated by a central angle of $180°$, we have

$$\frac{\Delta v_1 + \Delta v_2}{v_0} = \left(1 - \frac{1}{R_{21}}\right)\sqrt{\frac{2R_{21}}{1+R_{21}}} + \frac{1}{\sqrt{R_{21}}} - 1$$

where

$$R_{21} = \frac{r_2}{r_1} \quad \text{and} \quad v_0 = \sqrt{\frac{\mu}{r_1}}$$

(b) A *bielliptical transfer* consists of the three velocity impulses $\Delta \mathbf{v}_1$, $\Delta \mathbf{v}_i$, $\Delta \mathbf{v}_2$ applied tangentially in the following order: (1) $\Delta \mathbf{v}_1$ applied at the initial orbit to attain, after a $180°$ transfer, an intermediate point located on a circle of radius $r_i > r_2$ with zero radial velocity; (2) $\Delta \mathbf{v}_i$ applied to attain, again after a $180°$

transfer, a point located on the final orbit; and (3) $\Delta \mathbf{v}_2$ applied to match the terminal velocities. Then

$$\frac{\Delta v_1 + \Delta v_i + \Delta v_2}{v_0} = \sqrt{\frac{2R_{i1}}{1+R_{i1}}} - 1$$
$$+ \sqrt{\frac{2}{R_{i1}}} \left( \sqrt{\frac{R_{21}}{R_{21}+R_{i1}}} - \frac{1}{\sqrt{1+R_{i1}}} \right) + \frac{1}{\sqrt{R_{21}}} \left( \sqrt{\frac{2R_{i1}}{R_{i1}+R_{21}}} - 1 \right)$$

where

$$R_{i1} = \frac{r_i}{r_1}$$

(c) If the ratio $R_{21}$ is sufficiently large, it is always possible to select $R_{i1}$ such that the bielliptical transfer will be more economical than the Hohmann transfer.

*Rudolf F. Hoelker* and *Paul S. Silber*† 1961

## 11.4 Orbit Transfer in the Hodograph Plane

Many orbital problems can be solved graphically by representing two-body motion in the hodograph plane—a concept which was introduced in Sect. 3.5. Although clearly limited in numerical accuracy, nevertheless these graphical techniques not only serve as convenient checks on analytical computations but also can provide real insight as to the underlying principles involved. We shall now discuss several applications of the hodograph method to orbital transfer problems. Then, at the end of the section, a number of exercises are provided for the reader to test his grasp of the technique.

### Single Velocity Impulse

In order to develop a convenient graphical solution to the problem of trajectory modification following an impulsive velocity change, consider the following two vector identities:

$$\frac{h_1}{\mu} \mathbf{v}_1 = \frac{h_1}{h_0} \left( \frac{h_0}{\mu} \mathbf{v}_0 + \frac{h_0}{\mu} \Delta \mathbf{v}_1 \right)$$
$$\frac{h_0}{\mu} \mathbf{v}_0 = \frac{h_0}{h_1} \left( \frac{h_1}{\mu} \mathbf{v}_1 - \frac{h_1}{\mu} \Delta \mathbf{v}_1 \right)$$

where $\mathbf{v}_0$ is the initial velocity and $\mathbf{v}_1$ is the velocity immediately following the incremental change $\Delta \mathbf{v}_1$. The first identity shows that the vector $h_1 \mathbf{v}_1 / \mu$ is determined in two steps: an ordinary vector addition of $h_0 \mathbf{v}_0 / \mu$ and $h_0 \Delta \mathbf{v}_1 / \mu$ followed by a scalar multiplication of the resulting vector

---

† "The Bi-elliptical Transfer between Coplanar Circular Orbits," in the *Proceedings of the Fourth AFBMD/STL Symposium*, vol. 3, Pergamon Press, New York, 1961, pp. 164–175.

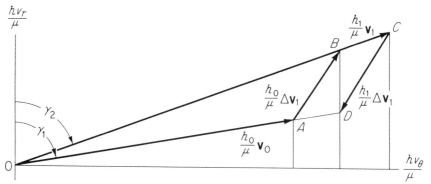

**Fig. 11.3:** Hodograph interpretation of velocity impulse.

sum by the factor $h_1/h_0$. These two operations are interpreted graphically in Fig. 11.3. The initial velocity vector terminus is at point $A$; the vector addition places the vector terminus at point $B$; and the scale-factor change places it finally at point $C$.

On the other hand, if we begin with the velocity $\mathbf{v}_1$ and subtract the increment $\Delta \mathbf{v}_1$, we are again back at $\mathbf{v}_0$. The second identity allows us to interpret this graphically as an ordinary vector subtraction of $h_1 \mathbf{v}_1/\mu$ and $h_1 \Delta \mathbf{v}_1/\mu$, which carries the vector terminus from point $C$ to point $D$, followed by a scale-factor change $h_0/h_1$, which places the terminus back at the point $A$.

Now since $rv_\theta = h$, we have

$$\frac{h_0}{\mu} v_{\theta_1} = \frac{h_0 h_1}{\mu r_1} \quad \text{and} \quad \frac{h_1}{\mu} v_{\theta_0} = \frac{h_1 h_0}{\mu r_1} \qquad (11.25)$$

so that the points $B$ and $D$ have the same abscissa. Furthermore, since

$$\begin{aligned} \text{Abscissa of } A &= \frac{p_0}{r_1} = \frac{h_0^2}{\mu r_1} \\ \text{Abscissa of } C &= \frac{p_1}{r_1} = \frac{h_1^2}{\mu r_1} \end{aligned} \qquad (11.26)$$

it follows that

Abscissa of $B$ = abscissa of $D$ =
$$\sqrt{(\text{abscissa of } A)(\text{abscissa of } C)} \qquad (11.27)$$

Thus, the abscissa of either $B$ or $D$ is seen to be the geometric mean between the abscissas of $A$ and $C$.

The following construction will produce the vector $h_1 \mathbf{v}_1/\mu$ from the vectors $h_0 \mathbf{v}_0/\mu$ and $\Delta \mathbf{v}_1$:

1. Perform the vector addition of $h_0 \mathbf{v}_0/\mu$ and $h_0 \Delta \mathbf{v}_1/\mu$ to obtain the intermediate point $B$.

2. Drop a perpendicular from $B$ to intersect the line $OA$ extended at the point $D$.
3. Draw a line through $D$ and parallel to the line $AB$ to intersect at $C$ the line $OB$ extended.

The values of the new orbital elements are immediately evident. The new angular momentum is determined from

$$h_1 = h_0 \frac{\text{abscissa of } B}{\text{abscissa of } A} \tag{11.28}$$

and the rotation of the line of apsides is just the difference between the true anomalies $f_0$ and $f_1$.

The construction must be modified for the case in which the increment $\Delta\mathbf{v}_1$ takes place in the original direction of motion $\mathbf{v}_0$. The point $B$ is determined as before; however, the scale change to locate the final point $C$ must be made numerically using the angular momentum relation given in Eq. (11.28).

## Transfer to a Specified Orbit

Consider now the problem of transferring at a given position $P$ from an initial orbit with elements $e_0$ and $h_0$ to a new orbit with elements $e_1$ and $h_1$. In this case the point $A$ is known and the point $C$ is determined, since it must lie on the circle of radius $e_1$ with an abscissa

$$\text{Abscissa of } C = \frac{h_1^2}{h_0^2}(\text{abscissa of } A) \tag{11.29}$$

The points $B$ and $D$ are then determined from Eq. (11.27) as a geometric mean. With $B$ determined, the velocity change $\Delta\mathbf{v}_1$ required for the transfer is obtained.

## Transfer from a Circular to a Hyperbolic Orbit

Suppose that a vehicle is initially in a circular satellite orbit of a planet and that a tangential velocity impulse $\Delta\mathbf{v}_1$ is applied of such a magnitude that the vehicle moves away from the planet along a hyperbolic path with an ultimate speed $v_\infty$ attained asymptotically with increasing distance. The velocity $v_\infty$ is frequently referred to as the *excess hyperbolic velocity*, that is, excess over the final value of zero velocity that would result from a parabolic *escape* from the planet.

From the vis-viva integral, it follows that $v_\infty$ is related to the semi-major axis of the hyperbola according to

$$v_\infty^2 = -\frac{\mu}{a}$$

so that immediately following the velocity impulse $\Delta \mathbf{v}_1$ we have

$$\frac{r_1 v_1^2}{\mu} = 2 + \frac{r_1 v_\infty^2}{\mu}$$

Then since the original orbit was circular and the increment was applied tangentially, it follows that

$$\frac{h_1 v_{\theta_1}}{\mu} = 2 + \frac{r_1 v_\infty^2}{\mu}$$

The solution of the problem can now be obtained graphically with the following construction:

1. The terminus of the initial velocity vector $h_0 \mathbf{v}_0/\mu$ has an abscissa of unity. Thus, the point $A$ coincides with the center of concentric circles of constant eccentricity as shown in Fig. 11.4. The terminus $C$ of the velocity vector $h_1 \mathbf{v}_1/\mu$ has an abscissa of $2 + r_1 v_\infty^2/\mu$ and a zero ordinate. Since $r_1$ and $v_\infty$ are known, the point $C$ is determined.
2. Describe a circle with center at $A$ and radius $1 + r_1 v_\infty^2/\mu$ to determine the intercept with the vertical axis. The circle represents conditions along the hyperbolic orbit.
3. The ordinate of the intercept is $h_1 v_\infty/\mu$, so that the angular momentum $h_1$ of the hyperbola is determined. The intermediate point $B$, which is the terminus of the vector $h_0 \Delta \mathbf{v}_1/\mu$ and lies on the horizontal axis, is determined as the geometric mean between the points $A$ and $C$. We have

$$\text{Abscissa of } B = 1 + \frac{h_0 \Delta v_{\theta_1}}{\mu} = \sqrt{2 + \frac{r_1 v_\infty^2}{\mu}}$$

The required velocity increment $\Delta v_1$ is thus determined. The point $P_1$ at which the velocity impulse is applied is frequently referred to as the *point of injection*. The angle $\theta$ through which the position vector turns from injection until asymptotic conditions are achieved is also immediately evident in the hodograph diagram.

◇ **Problem 11-9**
Discuss the solution in the hodograph plane of a single impulse transfer from an initial orbit with elements $e_0$ and $h_0$ to a new orbit with elements $e_1$ and $h_1$ in such a manner that no rotation of the apsidal line occurs. In particular, show how the position is obtained at which the impulse is to be made.

◇ **Problem 11-10**
Use the hodograph plane to show that, for the two-point boundary-value problem, the bisector of the transfer angle is perpendicular to the difference $\mathbf{v}_2 - \mathbf{v}_1$ of the terminal velocity vectors.

## Sect. 11.4]  Orbit Transfer in the Hodograph Plane

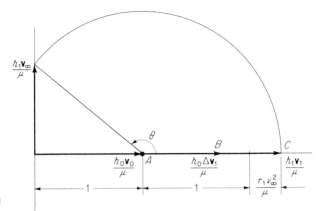

**Fig. 11.4:** Hodograph interpretation of orbital injection.

◊ **Problem 11-11**

Consider the problem of the Hohmann transfer between two circular orbits of radii $r_1 = 1$ and $r_2 = 2$ and assume the gravitational constant $\mu = 1$ for simplicity.
(a) What is the eccentricity $e_1$ of the transfer orbit?
(b) If the vehicle is initially in the circular orbit of radius $r_1$, what is the angular momentum $h_0$? What will be the angular momentum $h_2$ when the maneuver is completed?
(c) The first velocity impulse $\Delta \mathbf{v}_1$ is applied tangentially. Using the hodograph method, calculate the angular momentum $h_1$ of the transfer orbit and the magnitude of the velocity impulse.
(d) The second impulse $\Delta \mathbf{v}_2$ is applied tangentially when the radius $r = r_2$ is attained. Again using the hodograph, calculate the magnitude $\Delta v_2$.

◊ **Problem 11-12**

Determine the point in an elliptic orbit where a velocity impulse, made at right angles to the velocity vector and in the plane of motion, will result in the greatest instantaneous change in the eccentricity.

◊ **Problem 11-13**

A vehicle is in a circular orbit about a center of attraction when a velocity impulse is suddenly made resulting in a new orbit whose angular momentum is a factor of 1.5 times the original. If the new flight path direction immediately following the impulse is $60°$, find the eccentricity of the new orbit.

◊ **Problem 11-14**

Illustrate the effect in the hodograph plane of a velocity impulse applied in the radial direction. Label carefully the points $A$, $B$, $C$, and $D$. At what point in an orbit will a velocity change made in the radial direction cause the greatest change in eccentricity and how will the true anomaly be effected?

## ◇ Problem 11–15

A vehicle in a parabolic orbit with unit angular momentum is moving away from a planet whose gravitational constant $\mu$ is also unity. At a point $45°$ from pericenter, the radial component of velocity is suddenly decreased by $1/\sqrt{2}$ units. What is the eccentricity of the new orbit and where is the vehicle relative to the new pericenter?

## ◇ Problem 11–16

At a certain point in a circular orbit, an instantaneous change in the vehicle's course is made in such a fashion that the angular momentum is reduced by a factor of two but the period is unchanged. What is the eccentricity of the new orbit? What is the true anomaly $f$ and flight direction angle $\gamma$ immediately following the course change?

## ◇ Problem 11–17

A vehicle in an elliptic orbit is to transfer to a parabolic orbit by a single velocity impulse without a change in the angular momentum $h_0$.
(a) In what direction must the impulse be applied?
(b) If the velocity impulse has a magnitude of $\Delta v_1 = \mu/h_0\sqrt{3}$, where in the orbit should it be applied to rotate the line of apsides clockwise by $30°$?
(c) If no rotation of the line of apsides is to be permitted, where should the impulse be applied?

## ◇ Problem 11–18

A vehicle in a parabolic orbit moves from point $P_1$ to $P_2$ through a central angle $\theta$. Using the hodograph, find the angle between the velocity vectors at $P_1$ and $P_2$, i.e., the angle through which the velocity vector rotates during the motion.

HINT: Don't forget the rotation of $\mathbf{i}_r$.

## ◇ Problem 11–19

At a point in a hyperbolic orbit, a spacecraft's flight direction angle $\gamma$ is $30°$. If its speed is twice the ultimate speed $v_\infty$, what is the eccentricity of the orbit and through what central angle will it move before attaining asymptotic conditions? Assume that $h$ is numerically equal to $\mu$.

## 11.5   Injection from Circular Orbits

For a typical interplanetary mission, a spacecraft is launched from Cape Canaveral into a nearly circular earth satellite orbit. Then, at an appropriate point on the trajectory, an engine restart is initiated and the vehicle moves away from the earth along an essentially hyperbolic path relative to the earth. The asymptotic value of the relative velocity vector is the departure velocity of the vehicle with respect to the earth. The sphere of influence of the earth extends to a distance of half a million miles, beyond

Sect. 11.5]  Injection from Circular Orbits  537

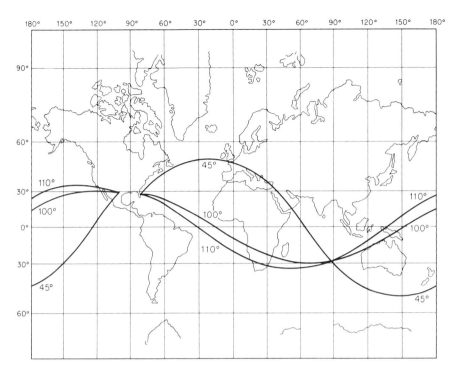

**Fig. 11.5:** Loci of points of injection.

which the effect of earth gravity diminishes rapidly. Then solar gravity provides the only significant force field to govern the flight of the vehicle.

Figure 11.5 shows a map of the world upon which are plotted three permissible coasting orbits having azimuth directions of 45, 100, and 110°. Completely arbitrary azimuths are restricted by range safety requirements and geographic restrictions might also limit the choice of injection points.

Consider the problem of a vehicle in a circular coasting orbit established from a fixed launch point on the surface of the earth. Assume that an interplanetary orbit from earth to a destination planet has been determined and it is desired to find the point on the coasting orbit where the minimum impulsive change in velocity can be made so that the vehicle will move away from the earth along a hyperbola whose asymptotic velocity vector is $\mathbf{v}_\infty$. For simplicity, and as an excellent first approximation, it will be assumed that the nominal time of injection occurs when the interplanetary orbit intersects the orbit of the earth and that the velocity $\mathbf{v}_\infty$ is the velocity of the spacecraft relative to the earth at this instant.

We shall postulate that the coasting orbit is established by a launch from a point on the earth's surface having a latitude $\phi_L$ and that the azimuth of the firing angle measured from north is $\alpha_L$. These two quan-

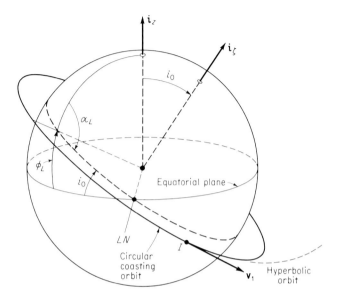

**Fig. 11.6:** Geometry of the coasting orbit.

tities determine the inclination angle $i_0$ of the coasting orbit plane to the equatorial plane. The relation is†

$$\cos i_0 = \cos \phi_L \sin \alpha_L \qquad (11.30)$$

as can be seen from Fig. 11.6. The time of launch determines the longitude of the node. We shall assume that the actual time of launch can vary by plus or minus 12 hours from its nominal value without seriously affecting the interplanetary orbit parameters. This assumption will permit the circular coasting orbit to be rotated arbitrarily about the earth's polar axis, thereby permitting an extra degree of freedom needed to optimize the injection velocity impulse. Once the point of injection along the coasting orbit has been located, it is then a simple matter to determine where this point lies geographically relative to a launch point fixed to the surface of the earth.

With the preceding discussion as background, we shall now consider an analysis of the injection problem. If the radius of the circular coasting orbit is $r_1$ and $\mu$ is the gravitational constant of the earth, then the initial orbital speed is

$$v_0 = \sqrt{\frac{\mu}{r_1}}$$

---

† Since $i_0$ and $\pi - \alpha_L$ are the two interior angles of a spherical right triangle with the side opposite $i_0$ being $\phi_L$, this relation follows immediately from a standard identity of spherical trigonometry.

## Injection from Circular Orbits

From the interplanetary orbit calculations the asymptotic relative velocity vector $\mathbf{v}_\infty$ is determined. Hence, from the vis-viva integral, the magnitude of the velocity vector immediately following the injection impulse is

$$v_1 = \sqrt{\frac{2\mu}{r_1} + v_\infty^2}$$

Since $\mathbf{v}_0$ and $\mathbf{v}_1$ are fixed in magnitude, the velocity change $\Delta\mathbf{v}_1 = \mathbf{v}_1 - \mathbf{v}_0$ is minimized by making the angle $\psi$ between them as small as possible. Clearly, if a point on the coasting orbit can be found such that $\mathbf{r}_1$, $\mathbf{v}_0$, and $\mathbf{v}_\infty$ are coplanar, then the optimum point of injection occurs at the perigee of the escape hyperbola—the point at which the angle $\psi$ will be zero.

### Optimum Injection

The problem is more complex if no such point exists. In general, we assume that injection occurs at an arbitrary point $\mathbf{r}_1 = r_1 \mathbf{i}_{r_1}$, so that the velocity vectors $\mathbf{v}_0$ and $\mathbf{v}_1$ just prior to and subsequent to the impulse are given by

$$\mathbf{v}_0 = \sqrt{\frac{\mu}{r_1}} \mathbf{i}_{\theta_1}$$

$$\mathbf{v}_1 = \tfrac{1}{2} v_\infty [(D+1) \mathbf{i}_\infty + (D-1) \mathbf{i}_{r_1}]$$

where

$$D = \sqrt{1 + \frac{4\mu}{r_1 v_\infty^2 (1 + \mathbf{i}_\infty \cdot \mathbf{i}_{r_1})}}$$

as derived in Sect. 6.8.

Also, we have

$$\mathbf{i}_{r_1} = \begin{bmatrix} \cos\Omega\cos\theta - \sin\Omega\sin\theta\cos i_0 \\ \sin\Omega\cos\theta + \cos\Omega\sin\theta\cos i_0 \\ \sin\theta\sin i_0 \end{bmatrix}$$

$$\mathbf{i}_{\theta_1} = \begin{bmatrix} -\cos\Omega\sin\theta - \sin\Omega\cos\theta\cos i_0 \\ -\sin\Omega\sin\theta + \cos\Omega\cos\theta\cos i_0 \\ \cos\theta\sin i_0 \end{bmatrix}$$

which are obtained from the results of Prob. 3–21. Then the quantity to be minimized is

$$|\mathbf{v}_1 - \mathbf{v}_0|^2 = \frac{3\mu}{r_1} + v_\infty^2 - (D+1)\mathbf{v}_\infty \cdot \mathbf{v}_0$$

which is equivalent to maximizing $J$ defined by

$$J = (D+1)\mathbf{i}_\infty \cdot \mathbf{i}_{\theta_1} \tag{11.31}$$

540    Two-Body Orbital Transfer    [Chap. 11

For convenience, choose the direction of the $x$ axis so that the $\mathbf{i}_\infty$ vector lies in the $xz$ plane and let $\beta$ be the angle it makes with the $x$ axis. Then, the two scalar products, of which $J$ is composed, are

$$\mathbf{i}_\infty \cdot \mathbf{i}_{r_1} = \sin\theta \sin i_0 \cos\beta + (\cos\Omega\cos\theta - \sin\Omega\sin\theta\cos i_0)\sin\beta$$
$$\mathbf{i}_\infty \cdot \mathbf{i}_{\theta_1} = \cos\theta \sin i_0 \cos\beta - (\cos\Omega\sin\theta + \sin\Omega\cos\theta\cos i_0)\sin\beta$$
(11.32)

so that $\Omega$ and $\theta$ are the quantities at our disposal for maximizing $J$.

If $\mathbf{r}_1$, $\mathbf{v}_0$, and $\mathbf{v}_\infty$ are not coplanar, we can still inject in the horizontal plane. The angle $\psi$ will not necessarily be zero of course, but will instead be the angle between the planes of the circular and hyperbolic orbits.

For this case, let $\mathbf{i}_\varsigma$ be the unit vector normal to the coasting orbit plane and let $\nu$ be the turn angle—i.e., the angle between $\mathbf{v}_1$ and $\mathbf{v}_\infty$. Then, since

$$\mathbf{i}_\infty \cdot \mathbf{i}_{r_1} = \cos(\tfrac{1}{2}\pi + \nu) = -\sin\nu$$

we find that

$$D = \frac{1 + \sin\nu}{1 - \sin\nu}$$

Therefore, the problem of maximizing $J$ is accomplished by maximizing

$$\mathbf{i}_\infty \cdot \mathbf{i}_{\theta_1} = \mathbf{i}_\infty \cdot \mathbf{i}_\varsigma \times \mathbf{i}_{r_1}$$

where

$$\mathbf{i}_\varsigma = \begin{bmatrix} \sin\Omega \sin i_0 \\ -\cos\Omega \sin i_0 \\ \cos i_0 \end{bmatrix} \qquad \mathbf{i}_\infty = \begin{bmatrix} \sin\beta \\ 0 \\ \cos\beta \end{bmatrix}$$

The triple scalar product is maximized if the three unit vectors $\mathbf{i}_\infty$, $\mathbf{i}_\varsigma$, and $\mathbf{i}_{r_1}$ are as nearly orthogonal as possible. But

$$\mathbf{i}_\infty \cdot \mathbf{i}_{r_1} = -\sin\nu \qquad \text{and} \qquad \mathbf{i}_\varsigma \cdot \mathbf{i}_{r_1} = 0$$

so that, in fact, we want to make $\mathbf{i}_\varsigma \cdot \mathbf{i}_\infty$ as small as possible.

### Tangential Injection from Perigee ($\beta + i_0 \geq 90°$)

If

$$\mathbf{i}_\varsigma \cdot \mathbf{i}_\infty = \sin\Omega \sin i_0 \sin\beta + \cos i_0 \cos\beta = 0$$

then

$$\sin\Omega = -\frac{\cot\beta}{\tan i_0} \tag{11.33}$$

which will not exceed unity in magnitude provided that $\beta + i_0 \geq \tfrac{1}{2}\pi$. When this is the case, there are two distinct circular orbital planes which contain $\mathbf{v}_\infty$ and are shown in Fig. 11.7. The longitudes of their respective

Sect. 11.5] Injection from Circular Orbits 541

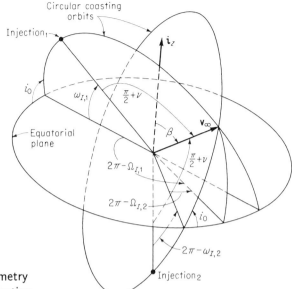

**Fig. 11.7:** Geometry of tangential injection.

ascending nodes, measured from the projection of $\mathbf{v}_\infty$ onto the equatorial plane, are then†

$$\Omega_{I,1} = \pi + \arcsin\left(\frac{\cot\beta}{\tan i_0}\right) \qquad \Omega_{I,2} = 2\pi - \arcsin\left(\frac{\cot\beta}{\tan i_0}\right) \quad (11.34)$$

To obtain the arguments of the injection points measured from their respective ascending nodes, we rewrite Eqs. (11.32) as

$$\sin\theta \sin i_0 \cos\beta + (\cos\Omega\cos\theta - \sin\Omega\sin\theta\cos i_0)\sin\beta = -\sin\nu$$
$$\cos\theta \sin i_0 \cos\beta - (\cos\Omega\sin\theta + \sin\Omega\cos\theta\cos i_0)\sin\beta = \cos\nu \quad (11.35)$$

Then, multiply the first by $\sin\theta$, the second by $\cos\theta$, and add to obtain

$$\sin i_0 \cos\beta - \sin\Omega\cos i_0 \sin\beta = \cos(\theta + \nu)$$

Finally, with Eq. (11.33) used to eliminate $\sin\Omega$ and with $\theta$ replaced by $\omega_I$, we have

$$\cos(\omega_I + \nu) = \frac{\cos\beta}{\sin i_0}$$

Hence

$$\omega_{I,1} = \arccos\left(\frac{\cos\beta}{\sin i_0}\right) - \nu \qquad \omega_{I,2} = -\arccos\left(\frac{\cos\beta}{\sin i_0}\right) - \nu \quad (11.36)$$

---

† Here and in succeeding equations of this section, principal values of all inverse trigonometric functions are postulated.

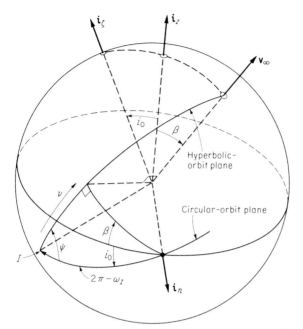

**Fig. 11.8:** Geometry of nontangential injection.

**Nontangential Injection from Perigee** $(\beta + i_0 < 90°)$

When $\mathbf{i}_\varsigma \cdot \mathbf{i}_\infty$ cannot be zero, the angle between the planes of the circular and hyperbolic orbits is nonzero as illustrated in Fig. 11.8. This angle $\psi$ will be as small as possible if $\mathbf{i}_\varsigma$, $\mathbf{i}_z$, and $\mathbf{i}_\infty$ are coplanar. Under these circumstances,

$$\cos(\beta + i_0) = \mathbf{i}_\varsigma \cdot \mathbf{i}_\infty$$
$$= \sin\Omega \sin i_0 \sin\beta + \cos i_0 \cos\beta$$

so that $\sin\Omega$ must equal $-1$ or, equivalently,

$$\Omega_I = 270° \qquad (11.37)$$

Also, from the first of Eqs. (11.35),

$$\mathbf{i}_\infty \cdot \mathbf{i}_{r_1} = \sin\theta \sin(\beta + i_0) = -\sin\nu$$

so that, replacing $\theta$ by $\omega_I$, we have the argument of the injection point measured from the ascending node in the plane of the coasting orbit

$$\omega_I = 2\pi - \arcsin\left(\frac{\sin\nu}{\sin(\beta + i_0)}\right) \qquad (11.38)$$

Clearly, horizontal injection is not possible unless $\beta + i_0 \geq \nu$.

Finally, since $\psi$ and $\beta + i_0$ are the interior angles of a right spherical triangle with the side opposite $\beta + i_0$ being $\nu$, then we can calculate the angle between the initial and final orbital planes from

$$\psi = \arcsin\left(\frac{\cos(\beta + i_0)}{\cos \nu}\right) \tag{11.39}$$

by using a standard identity for right spherical triangles.

## 11.6 Midcourse Orbit Corrections

Guidance and navigation techniques of spacecraft in interplanetary or cislunar space are often based on the method of linearized perturbations introduced in Sects. 9.5 and 9.6. In this case, it is assumed that the spacecraft does not deviate substantially from a selected reference orbit. In interplanetary space, deviations as large as one percent of an astronomical unit are about the maximum which could be expected; generally, they would be much smaller. When in proximity to a planet, it is necessary to keep deviations from course to within a percent or so of the distance to the planet in order to avoid the use of unduly large velocity corrections. Under these circumstances, perturbation techniques can be used to calculate these deviations and associated velocity corrections.

Explicit guidance techniques can also be employed using conic arcs suitably modified to account for small noncentral force field effects. In either case, both fixed- and variable-time-of-arrival velocity corrections can be calculated by methods which shall now be described.

### Fixed-Time-of-Arrival Orbit Corrections

Suppose that at time $t$ a vehicle is found to deviate from the reference path by an amount $\delta \mathbf{r}(t)$ in position and $\delta \mathbf{v}(t)$ in velocity. We wish to determine what the velocity deviation should be for that particular position deviation so that the vehicle will arrive at the target point at the predetermined or reference time $t_A$. For this purpose we can use the matrices defined in Sect. 9.6 to write

$$\begin{bmatrix} \tilde{\mathbf{R}}^\star(t) & \mathbf{R}^\star(t) \\ \tilde{\mathbf{V}}^\star(t) & \mathbf{V}^\star(t) \end{bmatrix} \begin{bmatrix} \mathbf{0} \\ \delta \mathbf{v}(t_A) \end{bmatrix} = \begin{bmatrix} \delta \mathbf{r}(t) \\ \delta \mathbf{v}(t) \end{bmatrix}$$

Therefore, we have

$$\delta \mathbf{r}(t) = \mathbf{R}^\star(t)\,\delta \mathbf{v}(t_A) \quad \text{and} \quad \delta \mathbf{v}(t) = \mathbf{V}^\star(t)\,\delta \mathbf{v}(t_A)$$

and, eliminating $\delta \mathbf{v}(t_A)$, we obtain

$$\delta \mathbf{v}(t) = \mathbf{V}^\star(t)\mathbf{R}^{\star\,-1}(t)\,\delta \mathbf{r}(t) = \mathbf{C}^\star(t)\,\delta \mathbf{r}(t) \tag{11.40}$$

The velocity vector $\mathbf{v}$, whose gradient with respect to $\mathbf{r}$ is $\mathbf{C}^*$, is that velocity required at $\mathbf{r}$ to reach the target point.

If a velocity correction $\Delta \mathbf{v}(t)$ is to be made at this time, it can be expressed as

$$\Delta \mathbf{v}(t) = \delta \mathbf{v}(t^+) - \delta \mathbf{v}(t^-) = \mathbf{C}^*(t)\,\delta \mathbf{r}(t) - \delta \mathbf{v}(t^-) \qquad (11.41)$$

where the superscripts $-$ and $+$ are used to distinguish the velocity just prior to a correction from the velocity immediately following the correction.

For these calculations to remain valid, it is necessary, of course, to restrict the magnitude of the deviations from the corresponding nominal values. Alternately, we could target any intermediate point $\mathbf{r}_T$—such as the point on the planet's sphere of influence through which the reference orbit passes at the reference time $t_T$. Then, if $\mathbf{r}$ and $\mathbf{v}$ are the position and velocity of the vehicle at the time the correction is to be made, these vectors can be extrapolated to the time $t_T$, using an orbital integration technique such as Encke's method, in order to determine the point $\mathbf{r}'_T$ at which the spacecraft would be found at the reference time if no corrective action were taken. By calculating the conic arc connecting the position vectors $\mathbf{r}$ and $\mathbf{r}'_T$ with a transfer time of $t_T - t$ (i.e., solving Lambert's problem using the methods of Chapter 7) the conic velocity $\mathbf{v}_{C_1}$ at $\mathbf{r}$ is determined. The difference between the conic velocity and the vehicle's actual velocity is a good measure of the effect of the disturbing perturbations. A second conic arc connecting the spacecraft position vector $\mathbf{r}$ and the desired target point $\mathbf{r}_T$ produces the conic velocity vector $\mathbf{v}_{C_2}$. If this velocity is corrected for the effect of perturbations, the velocity necessary to reach the desired target from position $\mathbf{r}$ is obtained. Thus, an excellent approximation to the required velocity correction is just the difference between the two conic velocities; specifically,

$$\Delta \mathbf{v} = \mathbf{v}_{C_2} - \mathbf{v}_{C_1} \qquad (11.42)$$

The computation may, of course, be repeated iteratively to achieve any desired degree of convergence. However, in practice, one computation cycle is usually sufficient.

### Variable-Time-of-Arrival Orbit Corrections

A reduction in fuel requirements can be accomplished if we permit the time of contact with the target planet to be a variable chosen in such a way that the velocity correction will have the smallest possible magnitude. Of course, we assume that the spacecraft is controlled in the vicinity of a reference interplanetary orbit just as for the linearized fixed-time-of-arrival guidance scheme.

To calculate the variable-time-of-arrival velocity correction, consider the effect of changing the arrival time $t_A$ by a small amount $\delta t$. Let $\mathbf{r}_p(t)$

and $\mathbf{v}_p(t)$ be, respectively, the position and velocity vectors of the target planet. Then the new point of contact will be

$$\mathbf{r}_p(t_A + \delta t) = \mathbf{r}_p(t_A) + \mathbf{v}_p(t_A)\delta t$$

if the linearization assumptions remain valid. At this time the spacecraft position will be

$$\mathbf{r}(t_A + \delta t) = \mathbf{r}(t_A) + \mathbf{v}(t_A)\delta t$$

Thus, if we require that $\mathbf{r}(t_A + \delta t) = \mathbf{r}_p(t_A + \delta t)$, then we have

$$\delta \mathbf{r}(t_A) = \mathbf{r}(t_A) - \mathbf{r}_p(t_A) = -\mathbf{v}_r(t_A)\delta t$$

where

$$\mathbf{v}_r(t_A) = \mathbf{v}(t_A) - \mathbf{v}_p(t_A) \qquad (11.43)$$

is the velocity of the spacecraft relative to the target planet at the nominal time of arrival. In this manner, the objective of nulling position errors would be ultimately attained but not at the reference arrival time.

To determine the variable-time-of-arrival correction, we may write, as before,

$$\begin{bmatrix} \widetilde{\mathbf{R}}^\star(t) & \mathbf{R}^\star(t) \\ \widetilde{\mathbf{V}}^\star(t) & \mathbf{V}^\star(t) \end{bmatrix} \begin{bmatrix} \delta\mathbf{r}(t_A) \\ \delta\mathbf{v}(t_A) \end{bmatrix} = \begin{bmatrix} \delta\mathbf{r}(t) \\ \delta\mathbf{v}(t) \end{bmatrix}$$

which is multiplied by the matrix $\begin{bmatrix} -\mathbf{C}^\star(t) & \mathbf{I} \end{bmatrix}$ to obtain

$$[-\mathbf{C}^\star(t)\widetilde{\mathbf{R}}^\star(t) + \widetilde{\mathbf{V}}^\star(t)]\delta\mathbf{r}(t_A) = -\mathbf{C}^\star(t)\delta\mathbf{r}(t) + \delta\mathbf{v}(t)$$

Then, using the starred form of the second of Eqs. (9.57), we have

$$\delta\mathbf{v}(t) = \mathbf{C}^\star(t)\delta\mathbf{r}(t) + \mathbf{R}^{\star-\mathrm{T}}(t)\mathbf{v}_r(t_A)\delta t \qquad (11.44)$$

as the required deviation in velocity at time $t$ if we are to arrive at the new target point at the time $t_A + \delta t$.

If we define $\Delta\mathbf{v}'(t)$ as the velocity correction to be applied at time $t$, we may write

$$\Delta\mathbf{v}'(t) = \Delta\mathbf{v}(t) + \mathbf{w}(t)\delta t \qquad (11.45)$$

where $\Delta\mathbf{v}(t)$ is the fixed-time-of-arrival correction and $\mathbf{w}$ is defined by

$$\mathbf{w}(t) = \mathbf{R}^{\star-\mathrm{T}}(t)\mathbf{v}_r(t_A) \qquad (11.46)$$

With the objective of selecting $\delta t$ so as to minimize the magnitude of $\Delta\mathbf{v}'$, clearly the best choice is that which will render the velocity correction vector normal to $\mathbf{w}$. Calling this value $\delta t_A$, we have, from Eq. (11.45),

$$\delta t_A = -\frac{\Delta\mathbf{v}\cdot\mathbf{w}}{\mathbf{w}\cdot\mathbf{w}} \qquad (11.47)$$

As a consequence, the velocity correction $\Delta\mathbf{v}'$ of smallest magnitude which will accomplish the mission is simply related to $\Delta\mathbf{v}$ by

$$\mathrm{Min}\,\Delta\mathbf{v}' = \mathbf{M}\,\Delta\mathbf{v} \qquad (11.48)$$

The matrix $\mathbf{M}$ is an example of a *projection operator* and is defined by

$$\mathbf{M} = \mathbf{I} - \frac{\mathbf{w}\mathbf{w}^T}{\mathbf{w}^T\mathbf{w}} \tag{11.49}$$

Thus, the variable-time-of-arrival correction is only a component of the fixed-time-of-arrival correction.

### Problem 11-20

A vehicle is approaching a target planet whose gravitation constant is $\mu$.
(a) The velocity vector may be expressed in the form

$$\mathbf{v} = \sqrt{\frac{\mu}{q(1+e)}} [-\sin f \, \mathbf{i}_e + (e + \cos f) \, \mathbf{i}_p]$$

where $q$ is the pericenter distance.

(b) If it is desired to make small variations $\delta e$ and $\delta f$ in the eccentricity $e$ and the true anomaly $f$, use the equation of orbit in the form

$$r = \frac{q(1+e)}{1 + e \cos f}$$

to show that $\delta e$ and $\delta f$ must be related by

$$(1 - \cos f)\, \delta e + e(1+e) \sin f \, \delta f = 0$$

in order that $r$ and $q$ remain unchanged.

(c) The corresponding change $\delta \mathbf{v}$ in the velocity vector to insure the invariance of $q$ is determined from

$$\delta \mathbf{v} = \frac{e}{2(1 - \cos f)} \sqrt{\frac{\mu}{q(1+e)}} [(1 - \cos f)^2 \, \mathbf{i}_e - (2 + e - \cos f) \sin f \, \mathbf{i}_p] \, \delta f$$

Hence, to a first approximation, the direction along which a velocity change may be made without altering the altitude at pericenter is the same as the direction of the vector

$$(1 - \cos f)^2 \, \mathbf{i}_e - (2 + e - \cos f) \sin f \, \mathbf{i}_p$$

HINT: To a first-order approximation

$$\delta \mathbf{i}_e = -\mathbf{i}_p \, \delta f \quad \text{and} \quad \delta \mathbf{i}_p = \mathbf{i}_e \, \delta f$$

*James E. Potter 1963*

### Problem 11-21
For the final velocity correction it may be desirable to select a new time of arrival in such a way as to minimize the sum of the magnitudes of the correction and the velocity deviation at the time of arrival at the target. The change $\delta t$ in the nominal time of arrival is then determined as the solution of

$$\frac{\mathbf{w}^T(\mathbf{w}\,\delta t + \Delta \mathbf{v})}{\sqrt{\Delta \mathbf{v} \cdot \Delta \mathbf{v} + \mathbf{w}^T(\mathbf{w}\,\delta t + 2\Delta \mathbf{v})\,\delta t}} + \frac{\boldsymbol{\lambda}^T(\boldsymbol{\lambda}\,\delta t - \mathbf{R}^{*-1}\,\delta \mathbf{r})}{\sqrt{\delta \mathbf{r}^T \mathbf{R}^{*-T}\mathbf{R}^{*-1}\,\delta \mathbf{r} + \boldsymbol{\lambda}^T(\boldsymbol{\lambda}\,\delta t - 2\mathbf{R}^{*-1}\,\delta \mathbf{r})\,\delta t}} = 0$$

where

$$\boldsymbol{\lambda} = (\mathbf{V}_A \boldsymbol{\Lambda}^{-1} + \boldsymbol{\Lambda}^{*-1})\mathbf{w}$$

and $\Delta \mathbf{v}$ is the fixed-time-of-arrival velocity correction.

### Pericenter Guidance

When a velocity correction is made in the vicinity of a planet, the arrival time at pericenter may be permitted to vary thereby reducing substantially the required velocity correction as well as the terminal velocity deviation from its nominal value. Specifically, let the desired terminal conditions at the target planet be a specified altitude at pericenter and a fixed plane in which the pericenter vector is to lie. Again, as before, the orbit is extrapolated forward in time to locate the pericenter vector $\mathbf{r}'_p$ which would result in the absence of a velocity correction. A conic arc, with $\mathbf{r}'_p$ as pericenter and connecting the position vector $\mathbf{r}$, is then determined to obtain a measure of the gravitational perturbation. The desired pericenter vector $\mathbf{r}_p$ is calculated from $\mathbf{r}'_p$ by scaling its length to correspond to the required pericenter distance and then rotating it into the required plane while, at the same time, keeping the transfer angle $\theta$ fixed. A second conic arc with $\mathbf{r}_p$ as pericenter is calculated and the difference between the two conic velocities again provides an excellent approximation to the necessary velocity correction.

Theoretically, the desired plane should not be fixed in space, but should rotate with the planet. However, the change in pericenter arrival time, combined with the planet's own rotation, generally leads to terminal deviations which are smaller than the navigation uncertainties. Hence, it is sufficiently accurate to aim for a fixed plane when approaching pericenter.

Pericenter guidance may be summarized as follows:

1. The conic velocity $\mathbf{v}_{C_1}$ required at $\mathbf{r}$ to attain pericenter at $\mathbf{r}'_p$ is computed from

$$\mathbf{v}_{C_1} = \frac{\sqrt{\mu p}}{rr'_p \sin \theta} \left\{ \mathbf{r}'_p - \left[ 1 - \frac{r_p}{p}(1 - \cos \theta) \right] \mathbf{r} \right\}$$

**548**  Two-Body Orbital Transfer  [Chap. 11]

where $\theta$ is the angle between $\mathbf{r}$ and $\mathbf{r}'_p$. The parameter $p$ of the conic, called a tangent ellipse in Sect. 6.3, is given by

$$p = \frac{rr'_p(1 - \cos\theta)}{r'_p - r\cos\theta}$$

2. The pericenter vector $\mathbf{r}'_p$ is rotated into the desired plane and scaled to the desired length $r_p$ by means of

$$\mathbf{r}_p = r_p \left[ \sqrt{1 - \beta\cos\theta}\, \text{Unit}(\mathbf{i}_n \times \mathbf{i}_r) - \beta\, \mathbf{i}_n \times (\mathbf{i}_n \times \mathbf{i}_r) \right]$$

where

$$\beta = \frac{\cos\theta}{1 - (\mathbf{i}_n \cdot \mathbf{i}_r)^2}$$

The unit vector $\mathbf{i}_n$ is normal to the desired plane in the direction of the angular momentum vector.

3. The conic velocity $\mathbf{v}_{C_2}$, required to attain pericenter at $\mathbf{r}_p$, is then calculated by repeating the first step with $\mathbf{r}_p$ in place of $\mathbf{r}'_p$.

4. The magnitude of the required velocity correction may be further reduced by noting there is a direction along which a velocity change can be made without altering the altitude at pericenter according to the results of Prob. 11–20. If the component of velocity correction along this insensitive direction is deducted from the total correction, the effect will be simply a small rotation of the pericenter vector $\mathbf{r}_p$. This *direction of insensitivity* is computed from

$$\mathbf{i}_d = \text{Unit}\left[-(1 - \cos\theta)^2\, \mathbf{i}_{r_p} + \sin\theta\left(1 - \cos\theta + \frac{p}{r_p}\right) \mathbf{i}_r \times \mathbf{i}_n\right]$$

where $\mathbf{i}_{r_p}$ is a unit vector in the direction of $\mathbf{r}_p$.

5. The velocity correction is then given by the component of the vector $\mathbf{v}_{C_2} - \mathbf{v}_{C_1}$ in the plane perpendicular to $\mathbf{i}_d$ and is calculated from

$$\Delta\mathbf{v} = (\mathbf{I} - \mathbf{i}_d \mathbf{i}_d^T)(\mathbf{v}_{C_2} - \mathbf{v}_{C_1}) \qquad (11.50)$$

It is not appropriate to aim for a fixed plane when making a velocity correction if the desired terminal conditions are a vacuum pericenter distance (which is equivalent to an entry angle) and a landing site fixed to the planet. The plane must be determined so that the spacecraft will be directed to the desired landing site.

◊ **Problem 11-22**

Derive the formula given in Step 2 of the above summary for rotating and scaling the pericenter vector $\mathbf{r}_p$.

Sect. 11.6]     Midcourse Orbit Corrections     549

◊ **Problem 11-23**
A vehicle is approaching a planet along a hyperbolic path. Let $v_\infty$ and $r_0$ be the velocity at infinity and the minimum distance from the center of the planet, respectively. At the instant the vehicle is at the point of closest approach, a velocity impulse $\Delta v$ is applied in the direction opposite to the motion in order to place the vehicle in an elliptic orbit of eccentricity $e$ about the planet. Then

$$\Delta v = \sqrt{v_\infty^2 + \frac{2\mu}{r_0}} - \sqrt{\frac{\mu(1+e)}{r_0}}$$

where $\mu$ is the gravitational constant of the planet.

◊ **Problem 11-24**
If $P$ is the orbital period of a spacecraft, then a small increase $\delta a$ in the semimajor axis $a$ will produce an increase of $(3P/2a)\,\delta a$ in the period.

◊ **Problem 11-25**
A satellite is in an elliptic orbit about the earth with apogee and perigee denoted by $r_a$ and $r_p$. If at apogee a small impulse $\delta v$ in velocity is suddenly made, then the perigee will be increased by the amount

$$\delta r_p = \frac{2P}{\pi}\sqrt{\frac{r_p}{r_a}}\,\delta v$$

where $P$ is the period of the satellite.

◊ **Problem 11-26**
A satellite is in an elliptic orbit. A small impulse $\delta v$ in velocity in the tangential direction is suddenly made. The eccentricity will be changed by the amount

$$\delta e = \frac{2p}{rv}\cos E\,\delta v$$

so that the eccentricity is increased in the first and fourth quadrants and it is decreased in the second and third quadrants.
The line of apsides will be rotated by the amount

$$\delta\omega = \frac{2}{ev}\sin f\,\delta v$$

so that the rotation is in the forward direction when $v_r$ is positive and backward when $v_r$ is negative.
HINT: Use the method of variation of parameters.

◊ **Problem 11-27**
At a certain point in an elliptic orbit a small error in the gravitation constant $\mu$ has no effect on the accuracy of determining the orbital eccentricity. The point at which this can occur must necessarily be at an extremity of the minor axis.
HINT: Use the vis-viva integral and the formula $p = a(1 - e^2)$.

## 11.7 Powered Orbital Transfer Maneuvers

The orbit transfer maneuvers considered thus far in this chapter have been accomplished by ideal impulsive velocity changes. It was assumed that the velocity required to achieve certain mission objectives could be attained instantaneously. This assumption is frequently justified for preliminary mission analysis but is clearly inadequate to solve the general powered-flight guidance problem. However, the concept of an impulsive velocity change can be exploited to provide an excellent rocket engine steering law which is applicable for a wide variety of orbit transfers.

The guidance and control problem, as it is considered in this section, is not directly concerned with the design or response characteristics of the physical components of the inflight guidance system. It is postulated that the system includes inertial instruments capable of measuring thrust acceleration along three mutually orthogonal axes which are nonrotating. The measured acceleration vector $\mathbf{a}_T$ of the vehicle is defined to be the acceleration resulting from the sum of rocket thrust and aerodynamic forces, if any, and would be zero if the vehicle moved under the action of gravity alone. The sum of $\mathbf{a}_T$ and $\mathbf{g}$, the gravitation vector, represents the total vehicle acceleration with respect to an inertial frame of reference.

Let a velocity vector $\mathbf{v}_r$ be defined as the instantaneous velocity, corresponding to the present vehicle location $\mathbf{r}$, required to satisfy a set of stated mission objectives. The velocity difference

$$\mathbf{v}_g = \mathbf{v}_r - \mathbf{v} \tag{11.51}$$

where $\mathbf{v}$ is the present vehicle velocity, is then the instantaneous velocity-to-be-gained. Since

$$\frac{d\mathbf{v}}{dt} = \mathbf{g}(\mathbf{r}) + \mathbf{a}_T$$

the rate of change of the velocity-to-be-gained $\mathbf{v}_g$ can be expressed as

$$\frac{d\mathbf{v}_g}{dt} = \frac{d\mathbf{v}_r}{dt} - \mathbf{g}(\mathbf{r}) - \mathbf{a}_T \tag{11.52}$$

The required velocity $\mathbf{v}_r$ is a function of both time $t$ and position $\mathbf{r}(t)$ so that

$$\frac{d\mathbf{v}_r}{dt} = \frac{\partial \mathbf{v}_r}{\partial t} + \frac{\partial \mathbf{v}_r}{\partial \mathbf{r}} \frac{d\mathbf{r}}{dt} = \frac{\partial \mathbf{v}_r}{\partial t} + \frac{\partial \mathbf{v}_r}{\partial \mathbf{r}} \mathbf{v}$$

$$= \frac{\partial \mathbf{v}_r}{\partial t} + \frac{\partial \mathbf{v}_r}{\partial \mathbf{r}} (\mathbf{v}_r - \mathbf{v}_g)$$

$$= \mathbf{g}(\mathbf{r}) - \frac{\partial \mathbf{v}_r}{\partial \mathbf{r}} \mathbf{v}_g \tag{11.53}$$

following the motion of the vehicle. Substituting in Eq. (11.52), gives

$$\frac{d\mathbf{v}_g}{dt} = -\frac{\partial \mathbf{v}_r}{\partial \mathbf{r}}\mathbf{v}_g - \mathbf{a}_T$$

Therefore, since $\mathbf{C}^\star = \partial \mathbf{v}_r/\partial \mathbf{r}$, as defined in Sect. 9.6, we have the following differential equation for the velocity-to-be-gained vector†

$$\frac{d\mathbf{v}_g}{dt} = -\mathbf{C}^\star \mathbf{v}_g - \mathbf{a}_T \tag{11.54}$$

### Constant Gravity Field Example

It is instructive to examine the velocity-to-be-gained equations for the special case in which the gravity vector $\mathbf{g}$ is a constant. The solution of the linearized equations of motion

$$\frac{d\mathbf{r}}{dt} = \mathbf{v} \quad \text{and} \quad \frac{d\mathbf{v}}{dt} = \mathbf{g} \quad \text{with} \quad \mathbf{r}(t_0) = \mathbf{r}_0 \quad \text{and} \quad \mathbf{v}(t_0) = \mathbf{v}_0$$

is readily seen to be

$$\mathbf{r}(t) = \mathbf{r}_0 + (t - t_0)\mathbf{v}_0 + \tfrac{1}{2}(t - t_0)^2 \mathbf{g}$$

To adapt this solution to our current notation, we replace $\mathbf{r}_0$ by $\mathbf{r}(t)$, $\mathbf{v}_0$ by $\mathbf{v}_r(t)$, and $\mathbf{r}(t)$ by $\mathbf{r}(t_1) = \mathbf{r}_1$. Thus, the required velocity is determined to be

$$\mathbf{v}_r(t) = \frac{1}{t_1 - t}[\mathbf{r}_1 - \mathbf{r}(t) - \tfrac{1}{2}(t_1 - t)^2 \mathbf{g}]$$

Hence,

$$\mathbf{C}^\star(t) = \frac{\partial \mathbf{v}_r}{\partial \mathbf{r}} = -\frac{1}{t_1 - t}\mathbf{I}$$

so that the equation for the velocity-to-be-gained, Eq. (11.54), is simply

$$\frac{d\mathbf{v}_g}{dt} = \frac{1}{t_1 - t}\mathbf{v}_g - \mathbf{a}_T \tag{11.55}$$

Alternatively, we could have differentiated the equation for $\mathbf{v}_r(t)$ to obtain

$$(t_1 - t)\frac{d\mathbf{v}_r}{dt} - \mathbf{v}_r = -\mathbf{v} + (t_1 - t)\mathbf{g}$$

or

$$(t_1 - t)\frac{d\mathbf{v}_r}{dt} = \mathbf{v}_g + (t_1 - t)\mathbf{g}$$

Then, by subtracting the equation for the vehicle velocity written in the form

$$(t_1 - t)\frac{d\mathbf{v}}{dt} = (t_1 - t)(\mathbf{g} + \mathbf{a}_T)$$

---

† For an alternate derivation of this basic equation, see pps. 8–9 of the Introduction.

we have
$$(t_1 - t)\frac{d\mathbf{v}_g}{dt} = \mathbf{v}_g - (t_1 - t)\mathbf{a}_T$$
which is the same as Eq. (11.55).

One of the advantages of exploring the constant gravity model is the ease with which an optimum control law for driving the velocity-to-be-gained vector to zero can be derived. For this purpose, form the scalar product of Eq. (11.55) and the vector $\mathbf{v}_g$ to obtain

$$\frac{d}{dt}(\mathbf{v}_g \cdot \mathbf{v}_g) = \frac{d}{dt}v_g^2 = \frac{2}{t_1 - t}v_g^2 - 2\mathbf{a}_T \cdot \mathbf{v}_g$$

or

$$(t_1 - t)\frac{d}{dt}v_g^2 = 2v_g^2 - 2(t_1 - t)\mathbf{a}_T \cdot \mathbf{v}_g$$

which can be integrated by parts from the present time $t$ to the time of engine cutoff $t_{co}$. Thus,

$$(t_1 - t)v_g^2(t)\Big|_t^{t_{co}} + \int_t^{t_{co}} v_g^2(t)\, dt = \int_t^{t_{co}} [2v_g^2(t) - 2(t_1 - t)\mathbf{a}_T \cdot \mathbf{v}_g]\, dt$$

Since the velocity-to-be-gained is zero at $t = t_{co}$, the integrated part is simply $-(t_1 - t)v_g^2(t)$. Therefore,

$$\int_t^{t_{co}} [2(t_1 - t)\mathbf{a}_T \cdot \mathbf{v}_g - v_g^2]\, dt = (t_1 - t)v_g^2(t) \qquad (11.56)$$

Now, for any particular time $t$, the right-hand side of Eq. (11.56) is determined. Thus, to minimize the integration interval $t_{co} - t$, the remaining engine burn time, we should maximize $\mathbf{a}_T \cdot \mathbf{v}_g$. This is accomplished by aligning the thrust direction with the $\mathbf{v}_g$ vector. The steering law thus obtained is optimum and, which is most important, independent of the time history of the thrust acceleration vector $\mathbf{a}_T(t)$.

### Cross-Product Steering

In the general case, for any gravity field, a convenient and, in fact, efficient guidance law can be developed by recognizing that an effective way to drive all three components of $\mathbf{v}_g$ to zero simultaneously is to align the time rate of change of the $\mathbf{v}_g$ vector with the vector itself. Mathematically, we require a direction of $\mathbf{a}_T$ to be chosen such that

$$\frac{d\mathbf{v}_g}{dt} \times \mathbf{v}_g = 0$$

In particular, we can verify from Eq. (11.55) that, for the constant gravity example, this law, which is called *cross-product steering*, is the same as thrusting in the direction of the velocity-to-be-gained.

## Sect. 11.7] Powered Orbital Transfer Maneuvers

For convenience, we introduce the notation

$$\mathbf{p}(t) = -\mathbf{C}^\star(t)\mathbf{v}_g(t) = \frac{d\mathbf{v}_r}{dt} - \mathbf{g}(\mathbf{r}) \qquad (11.57)$$

so that cross-product steering is equivalent to choosing the direction of the thrust acceleration vector such that

$$\mathbf{a}_T \times \mathbf{v}_g = \mathbf{p} \times \mathbf{v}_g \qquad (11.58)$$

Then a vector postmultiplication of Eq. (11.58) by $\mathbf{v}_g$ yields

$$(\mathbf{a}_T \cdot \mathbf{v}_g)\mathbf{v}_g - v_g^2 \mathbf{a}_T = (\mathbf{p} \cdot \mathbf{v}_g)\mathbf{v}_g - v_g^2 \mathbf{p}$$

or

$$\mathbf{a}_T = \mathbf{p} + (q - \mathbf{i}_{v_g} \cdot \mathbf{p})\,\mathbf{i}_{v_g} \qquad (11.59)$$

where $\mathbf{i}_{v_g}$ is the unit vector in the direction of $\mathbf{v}_g$. The scalar quantity

$$q = \mathbf{a}_T \cdot \mathbf{i}_{v_g}$$

can be calculated by squaring both sides of Eq. (11.59). We find that

$$q = \sqrt{a_T^2 - p^2 + (\mathbf{i}_{v_g} \cdot \mathbf{p})^2} \qquad (11.60)$$

Since $a_T$ is measurable in the vehicle using, for example, inertially oriented accelerometers, the direction of $\mathbf{a}_T$ may be calculated from Eq. (11.59).

As can be seen from Eq. (11.60), if $a_T$ is not sufficiently large, it will not be possible to align the vector $\mathbf{v}_g$ with its derivative. With typical chemical rockets, for which the burn time is relatively short, no difficulty is encountered with this guidance logic. When this is not the case, we can always resort to thrusting in the direction of the velocity-to-be-gained.

### Estimation of Burn Time

A rough estimation† of the rocket burn time can be had if we consider $\mathbf{C}^\star$ to be a constant matrix. Writing the fundamental differential equation for velocity-to-be-gained in terms of the vector $\mathbf{p}$ defined in Eq. (11.57), we have

$$\frac{d\mathbf{v}_g}{dt} = -\mathbf{C}^\star \mathbf{v}_g - \mathbf{a}_T = \mathbf{p} - \mathbf{a}_T$$

For a steer law which renders $\mathbf{v}_g$ irrotational, the vector $\mathbf{p}$ will have a fixed direction and will be proportional to $\mathbf{v}_g$ assuming $\mathbf{C}^\star$ to be a constant. Let $A\mathbf{v}_g$ and $B\mathbf{v}_g$ be the components of $\mathbf{p}$ along and perpendicular to $\mathbf{v}_g$, respectively. Then

$$A = \frac{\mathbf{p} \cdot \mathbf{v}_g}{\mathbf{v}_g \cdot \mathbf{v}_g} \qquad \text{and} \qquad B = \sqrt{\frac{\mathbf{p} \cdot \mathbf{p}}{\mathbf{v}_g \cdot \mathbf{v}_g} - A^2} \qquad (11.61)$$

---

† This analysis was made by Edward M. Copps in the midsixties during the Apollo days.

Now

$$\frac{d\mathbf{v}_g}{dt} \cdot \frac{d\mathbf{v}_g}{dt} = \left(\frac{dv_g}{dt}\right)^2 = p^2 + a_T^2 - 2\mathbf{p} \cdot \mathbf{a}_T = a_T^2 + 2\mathbf{p} \cdot \frac{d\mathbf{v}_g}{dt} - p^2$$

$$= a_T^2 + 2Av_g \frac{dv_g}{dt} - v_g^2(A^2 + B^2)$$

so that

$$\frac{dv_g}{dt} = -a_T\sqrt{1 - \frac{B^2}{a_T^2}v_g^2} + Av_g$$

$$= -a_T\left(1 - \frac{B^2}{2a_T^2}v_g^2 + \cdots\right) + Av_g$$

By neglecting the higher-order terms in this expansion, we obtain a Ricatti equation for $v_g$ which can then be transformed into the linear second-order differential equation

$$\frac{d^2y}{dt^2} + \left(\frac{1}{a_T}\frac{da_T}{dt} - A\right)\frac{dy}{dt} - \tfrac{1}{2}B^2 y = 0$$

using the change of variable

$$\frac{1}{y}\frac{dy}{dt} = -\frac{B^2}{2a_T}v_g$$

Note that the thrust acceleration term is the only time-varying coefficient in the equation.

Consider now a constant thrust rocket engine for which

$$a_T = \frac{F}{m_0 - \dot{m}t} = \frac{F}{m_0}\left[1 + \frac{\dot{m}}{m_0}t + \left(\frac{\dot{m}}{m_0}\right)^2 t^2 + \cdots\right]$$

where we have used the notation $\dot{m}$ for the time rate of change of the rocket mass and $F$ for the constant rocket force. Assume further that the ratio of the time rate of change of the thrust acceleration to the thrust acceleration has a constant value $D$ so that we may write

$$a_T = a_T(0)\left(1 + Dt + \tfrac{1}{2}D^2 t^2 + \cdots\right)$$

Therefore, by comparison, we have

$$\frac{1}{a_T}\frac{da_T}{dt} = \frac{1}{m_0}\frac{dm}{dt}$$

so that the acceleration profile will match that expected from a rocket to the first order in time and the second-order differential equation in question will then have constant coefficients. The solution is simply

$$y = c_1 e^{\lambda_1 t} + c_2 e^{\lambda_2 t}$$

where
$$2\lambda_1, \ 2\lambda_2 = -\frac{\dot{m}}{m_0} + A \pm \sqrt{A^2 + 2B^2 + \frac{\dot{m}}{m_0}\left(\frac{\dot{m}}{m_0} - 2A\right)} \quad (11.62)$$

In terms of the original variable $v_g$, we obtain
$$\frac{B^2}{2a_T}v_g = -\frac{\lambda_1 e^{\lambda_1 t} + c\lambda_2 e^{\lambda_2 t}}{e^{\lambda_1 t} + c e^{\lambda_2 t}}$$
where the constant $c$ is to be determined using the fact that $t = 0$ at the time of ignition. Thus,
$$c = \frac{w - \lambda_1}{w - \lambda_2} \quad \text{if we define} \quad w = \frac{B^2 v_g(0)}{2a_T(0)} \quad (11.63)$$

Finally, the estimate of the burn time is calculated from the fact that the velocity-to-be-gained $v_g$ vanishes at cutoff. Hence,
$$\text{Burn time} \approx \frac{1}{\lambda_2 - \lambda_1} \log \frac{\lambda_2(\lambda_1 - w)}{\lambda_1(\lambda_2 - w)} \quad (11.64)$$

### Hyperbolic Injection Guidance

In the remainder of this section, we shall consider as examples several specific mission objectives and calculate the corresponding $\mathbf{C}^*$ matrix for each.

For example, the velocity required to establish a hyperbolic orbit from position $\mathbf{r}$ to attain, ultimately, a velocity $\mathbf{v}_\infty$ is given by
$$\mathbf{v}_r = \tfrac{1}{2}v_\infty[(D+1)\mathbf{i}_\infty + (D-1)\mathbf{i}_r]$$
where
$$v_\infty^2(r + \mathbf{i}_\infty^T \mathbf{r})(D^2 - 1) = 4\mu$$
which we have used several times before and was derived in Sect. 6.8. To obtain the $\mathbf{C}^*$ matrix, we calculate
$$\frac{\partial \mathbf{v}_r}{\partial \mathbf{r}} = \tfrac{1}{2}v_\infty(\mathbf{i}_\infty + \mathbf{i}_r)\frac{\partial D}{\partial \mathbf{r}} + \tfrac{1}{2}v_\infty(D-1)\frac{\partial \mathbf{i}_r}{\partial \mathbf{r}}$$
Now
$$\frac{\partial \mathbf{i}_r}{\partial \mathbf{r}} = \frac{\partial}{\partial \mathbf{r}}\left(\frac{\mathbf{r}}{r}\right) = \frac{1}{r}\frac{\partial \mathbf{r}}{\partial \mathbf{r}} - \frac{1}{r^2}\mathbf{r}\frac{\partial r}{\partial \mathbf{r}} = \frac{1}{r}(\mathbf{I} - \mathbf{i}_r \mathbf{i}_r^T)$$
and
$$\frac{\partial D}{\partial \mathbf{r}} = \frac{(D^2-1)^2}{8\mu D}(\mathbf{i}_\infty + \mathbf{i}_r)^T$$
so that
$$\mathbf{C}^* = \frac{v_\infty(D-1)}{2r}(\mathbf{I} - \mathbf{i}_r \mathbf{i}_r^T) - \frac{v_\infty^3(D^2-1)^2}{16\mu D}(\mathbf{i}_\infty + \mathbf{i}_r)(\mathbf{i}_\infty + \mathbf{i}_r)^T \quad (11.65)$$

## Circular-Orbit Insertion Guidance

Consider the problem of guiding a vehicle into a circular orbit of a planet by a rocket braking maneuver initiated on an approach trajectory. The velocity vector $\mathbf{v}_r$ may be defined as the velocity the vehicle should have to be in a circular orbit at distance $r$ from the planet and in a plane whose unit normal is $\mathbf{i}_n$. Thus, if $\mu$ is the gravitational constant of the planet and $\mathbf{i}_r$ is the unit vector in the direction or $\mathbf{r}$, then

$$\mathbf{v}_r = \sqrt{\frac{\mu}{r}}\, \mathbf{i}_n \times \mathbf{i}_r \tag{11.66}$$

By driving the $\mathbf{v}_g$ vector to zero, we are able to control the shape and orientation of the final orbit, but no direct control of the radius of the orbit is possible. However, an empirical relationship between the final radius and the pericenter of the approach trajectory can readily be determined, so that a desired radius can be established by proper selection of the pericenter of the approach orbit and the ignition time.

The corresponding $\mathbf{C}^*$ matrix is readily calculated if we write the expression for the required velocity vector, Eq. (11.66), in the form

$$\mathbf{v}_r = \mathbf{S}_n \mathbf{r} \sqrt{\frac{\mu}{r^3}}$$

with the matrix $\mathbf{S}_n$ defined as

$$\mathbf{S}_n = \begin{bmatrix} 0 & -n_z & n_y \\ n_z & 0 & -n_x \\ -n_y & n_x & 0 \end{bmatrix}$$

where $n_x$, $n_y$, $n_z$ are the direction cosines of the unit vector $\mathbf{i}_n$. Indeed, it is easy to obtain

$$\mathbf{C}^* = \sqrt{\frac{\mu}{r^3}}\, \mathbf{S}_n (\mathbf{I} - \tfrac{3}{2} \mathbf{i}_r \mathbf{i}_r^T) \tag{11.67}$$

Note that the $\mathbf{C}^*$ of Eq. (11.67) is not symmetric while the one derived in the previous subsection is symmetric. Actually, the only $\mathbf{C}^*$ matrix which we proved to be symmetric was that for the time-constrained, two-point, boundary-value problem—i.e., Lambert's problem. Indeed, we are taking some liberties in even using the notation $\mathbf{C}^*$ for cases other than Lambert's problem.

## Sect. 11.7]    Powered Orbital Transfer Maneuvers    557

◇ **Problem 11–28**

The required velocity to achieve a specified flight direction angle $\gamma_T$ at the target position $\mathbf{r}_T$ is given by

$$\mathbf{v}_r = B\left[\frac{1}{D}(\mathbf{r}_T - \mathbf{r}) + D\,\mathbf{i}_r\right]$$

where $B$ and $D$ are the positive roots of

$$B^2 = \frac{\mu}{rr_T(1+\cos\theta)} = \frac{\mu}{r_T r + \mathbf{r}_T^T \mathbf{r}}$$

and

$$D^2 = r_T - r\frac{\sin(\gamma_T + \theta)}{\sin\gamma_T} = r_T - r(\cos\theta + \cot\gamma_T \sin\theta)$$

$$= r_T - \mathbf{i}_{r_T}^T \mathbf{r} - r\cot\gamma_T \sin\theta$$

with $\theta$ as the transfer angle between the present position $\mathbf{r}$ and the target $\mathbf{r}_T$.

(a) Derive the partial derivatives

$$\frac{\partial r}{\partial \mathbf{r}} = \mathbf{i}_r^T \qquad \frac{\partial \mathbf{r}}{\partial \mathbf{r}} = \mathbf{I} \qquad \frac{\partial \mathbf{i}_r}{\partial \mathbf{r}} = \frac{1}{r}(\mathbf{I} - \mathbf{i}_r \mathbf{i}_r^T) \qquad \sin\theta\,\frac{\partial\theta}{\partial \mathbf{r}} = -\mathbf{i}_{r_T}^T \frac{\partial \mathbf{i}_r}{\partial \mathbf{r}}$$

(b) Use these results to verify that

$$\frac{\partial B}{\partial \mathbf{r}} = -\frac{B}{2r(1+\cos\theta)}(\mathbf{i}_r + \mathbf{i}_{r_T})^T$$

and

$$\frac{\partial D}{\partial \mathbf{r}} = \frac{\cos(\gamma_T+\theta)}{2D\sin\gamma_T \sin\theta}\mathbf{i}_{r_T}^T - \frac{\cos\gamma_T}{2D\sin\gamma_T \sin\theta}\mathbf{i}_r^T$$

(c) Calculate the $\mathbf{C}^*$ matrix and determine if it is symmetric.

◇ **Problem 11–29**

The required velocity to attain an elliptic orbit of a specified semimajor axis and eccentricity can be determined from

$$\mathbf{v}_r = \pm\sqrt{\frac{\mu}{p}\left[e^2 - \left(\frac{p}{r}-1\right)^2\right]}\,\mathbf{i}_r + \frac{\sqrt{\mu p}}{r}\,\mathbf{i}_h \times \mathbf{i}_r$$

where $\mathbf{i}_h$ is the unit vector in the direction of the desired angular momentum vector and $p$ is the parameter of the desired orbit. The matrix $\mathbf{C}^*$ is obtained from

$$\mathbf{C}^* = \pm\frac{\sqrt{\mu p}}{r^2}\left[\left(\frac{re}{p-r}\right)^2 - 1\right]^{-\frac{1}{2}}\mathbf{i}_r \mathbf{i}_r^T$$

$$\pm\left\{\frac{\mu}{p}\left[e^2 - \left(\frac{p}{r}-1\right)^2\right]\right\}^{\frac{1}{2}}(\mathbf{I} - \mathbf{i}_r\mathbf{i}_r^T) - \frac{\sqrt{\mu p}}{r^2}\mathbf{S}_h(\mathbf{I} - 2\,\mathbf{i}_r\mathbf{i}_r^T)$$

where the matrix $\mathbf{S}_h$ has the property that $\mathbf{S}_h \mathbf{i}_r = \mathbf{i}_h \times \mathbf{i}_r$.

## ◇ Problem 11-30

The required velocity to reach the target position $\mathbf{r}_T$ with specified energy or, equivalently, fixed semimajor axis, is

$$\mathbf{v}_r = P(\mathbf{i}_c + \mathbf{i}_r) + Q(\mathbf{i}_c - \mathbf{i}_r)$$

where

$$P = \sqrt{\frac{\mu}{r_1 + r_2 - c} - \frac{\mu}{4a}} \qquad Q = \pm\sqrt{\frac{\mu}{r_1 + r_2 + c} - \frac{\mu}{4a}}$$

according to the introductory material in Sect. 6.8.

The associated $\mathbf{C}^\star$ matrix is

$$\mathbf{C}^\star = -\frac{\mu}{8P(s-c)^2}(\mathbf{i}_c + \mathbf{i}_r)(\mathbf{i}_c + \mathbf{i}_r)^T + \frac{\mu}{8Qs^2}(\mathbf{i}_c - \mathbf{i}_r)(\mathbf{i}_c - \mathbf{i}_r)^T$$
$$-\frac{P+Q}{c}(\mathbf{I} - \mathbf{i}_c\mathbf{i}_c^T) + \frac{P-Q}{r}(\mathbf{I} - \mathbf{i}_r\mathbf{i}_r^T)$$

## Problem 11-31

Repeat the derivation of the $\mathbf{C}^\star$ matrix using the expression for the required velocity given in Prob. 11-30 but constraining the transfer time $t_T - t$ instead of the semimajor axis. In other words, determine the $\mathbf{C}^\star$ for Lambert's problem in the form

$$\mathbf{C}^\star = \left(\frac{P-Q}{r} - \frac{P+Q}{c}\right)\mathbf{I} + \left(\frac{\mu Q}{16aP\Delta} - \frac{\mu}{8P(s-c)^2}\right)(\mathbf{i}_r + \mathbf{i}_c)(\mathbf{i}_r + \mathbf{i}_c)^T$$
$$+ \left(\frac{\mu P}{16aQ\Delta} + \frac{\mu}{8Qs^2}\right)(\mathbf{i}_c - \mathbf{i}_r)(\mathbf{i}_c - \mathbf{i}_r)^T - \left(\frac{P-Q}{r} + \frac{\mu}{8a\Delta}\right)\mathbf{i}_r\mathbf{i}_r^T$$
$$+ \left(\frac{P+Q}{r} + \frac{\mu}{8a\Delta}\right)\mathbf{i}_c\mathbf{i}_c^T$$

where

$$\Delta = 3PQ(t_T - t) + (s - c)Q - sP$$

*Frederick H. Martin*† 1966

## 11.8 Optimal Guidance Laws

The velocity-to-be-gained guidance technique developed in the previous section, is workable if it is possible to define, at each instant of thrusting, a required velocity to meet mission objectives which is a function only of current position. However, for such missions as inserting a spacecraft into an orbit of a specified size and orientation, soft-landing a vehicle on the surface of the moon, and orbital rendezvous, this requirement cannot be met.

---

† Fred Martin's derivation in his thesis had some errors. The corrected version is courtesy of William M. Robertson of the Charles Stark Draper Laboratory.

## Optimal Guidance Laws

When the spacecraft is propelled by an engine whose thrust magnitude and direction can be controlled, a variety of terminal conditions can be met. However, to avoid excessive fuel expenditure, appropriate guidance laws must be developed which are optimum, or nearly so, using techniques of the Calculus of Variations. It is beyond the scope of this book to develop fully this field; nevertheless, we can design some practical guidance techniques using elementary variational principles.

### Terminal State Vector Control

The development of an explicit steering equation for a controllable thrust engine, which will guide a vehicle to a desired set of terminal conditions, is based on the solution of a simple variational problem. Let it be required to find the acceleration program $\mathbf{a}(t)$ which will minimize the functional

$$J = \int_{t_0}^{t_1} a(t)^2 \, dt = \int_{t_0}^{t_1} \mathbf{a}^T(t)\mathbf{a}(t) \, dt \qquad (11.68)$$

If $\mathbf{a}(t)$ is the total acceleration vector, then the equations of motion are

$$\frac{d\mathbf{r}}{dt} = \mathbf{v} \quad \text{and} \quad \frac{d\mathbf{v}}{dt} = \mathbf{a} \qquad (11.69)$$

subject to

$$\begin{aligned} \mathbf{r}(t_0) &= \mathbf{r}_0 \\ \mathbf{v}(t_0) &= \mathbf{v}_0 \end{aligned} \quad \text{and} \quad \begin{aligned} \mathbf{r}(t_1) &= \mathbf{r}_1 \\ \mathbf{v}(t_1) &= \mathbf{v}_1 \end{aligned} \qquad (11.70)$$

This minimization problem is readily solved using the Calculus of Variations. In general, there exist infinitely many sets of functions which satisfy the differential equations and the boundary conditions. To each of these sets corresponds a particular value of $J$. Among these sets, we shall suppose that there is one which generates a minimum value for $J$. This minimal set will be denoted by $\mathbf{r}_m(t)$ and $\mathbf{v}_m(t)$ produced by the acceleration program $\mathbf{a}_m(t)$.

Consider a set of functions $\boldsymbol{\delta}(t)$, $\boldsymbol{\nu}(t)$, and $\boldsymbol{\zeta}(t)$ which satisfy

$$\frac{d\boldsymbol{\delta}}{dt} = \boldsymbol{\nu} \quad \text{and} \quad \frac{d\boldsymbol{\nu}}{dt} = \boldsymbol{\zeta} \qquad (11.71)$$

subject to the boundary conditions

$$\boldsymbol{\delta}(t_0) = \boldsymbol{\delta}(t_1) = \mathbf{0} \qquad \boldsymbol{\nu}(t_0) = \boldsymbol{\nu}(t_1) = \mathbf{0} \qquad \boldsymbol{\zeta}(t_0) = \boldsymbol{\zeta}(t_1) = \mathbf{0} \qquad (11.72)$$

Then, form the one-parameter family of so-called *admissible functions*

$$\begin{aligned} \mathbf{r}(t, \alpha) &= \mathbf{r}_m(t) + \alpha \boldsymbol{\delta}(t) \\ \mathbf{v}(t, \alpha) &= \mathbf{v}_m(t) + \alpha \boldsymbol{\nu}(t) \\ \mathbf{a}(t, \alpha) &= \mathbf{a}_m(t) + \alpha \boldsymbol{\zeta}(t) \end{aligned} \qquad (11.73)$$

which includes the minimal set corresponding to $\alpha = 0$.

The functional $J$ will then be the following function of $\alpha$:

$$J(\alpha) = \int_{t_0}^{t_1} \mathbf{a}_m^T \mathbf{a}_m \, dt + 2\alpha \int_{t_0}^{t_1} \mathbf{a}_m^T \boldsymbol{\zeta} \, dt + \alpha^2 \int_{t_0}^{t_1} \boldsymbol{\zeta}^T \boldsymbol{\zeta} \, dt \qquad (11.74)$$

so that a necessary condition for $J(\alpha)$ to have a minimum is

$$\left.\frac{dJ}{d\alpha}\right|_{\alpha=0} = 0 = 2 \int_{t_0}^{t_1} \mathbf{a}_m^T \, d\boldsymbol{\nu} \qquad (11.75)$$

The right side of Eq. (11.75) is integrated by parts

$$\int_{t_0}^{t_1} \mathbf{a}_m^T \, d\boldsymbol{\nu} = -\int_{t_0}^{t_1} \frac{d\mathbf{a}_m^T}{dt} \boldsymbol{\nu}(t) \, dt$$

and the integrated part vanishes since $\boldsymbol{\nu}(t_0) = \boldsymbol{\nu}(t_1) = \mathbf{0}$. Therefore, a necessary condition for $J(\alpha)$ to have a minimum is that

$$\int_{t_0}^{t_1} \frac{d\mathbf{a}_m^T}{dt} \frac{d\boldsymbol{\delta}}{dt} \, dt = 0 \qquad (11.76)$$

for every admissible variation $\boldsymbol{\delta}(t)$ satisfying the boundary conditions. It follows from the *Fundamental Lemma* of the Calculus of Variations that

$$\frac{d\mathbf{a}_m}{dt} = \mathbf{c}_1$$

where $\mathbf{c}_1$ is a vector constant.

To prove the lemma, note that Eq. (11.76) can be written as

$$\int_{t_0}^{t_1} \left(\frac{d\mathbf{a}_m}{dt} - \mathbf{c}_1\right)^T \frac{d\boldsymbol{\delta}}{dt} \, dt = 0$$

which must hold for all $\boldsymbol{\delta}(t)$. In particular, it must be true for

$$\boldsymbol{\delta}(t) = \int_{t_0}^{t} \frac{d\mathbf{a}_m}{dt} \, dt - \mathbf{c}_1(t - t_0)$$

with $\mathbf{c}_1$ chosen so that $\boldsymbol{\delta}(t_1) = \mathbf{0}$. Hence,

$$\int_{t_0}^{t_1} \left(\frac{d\mathbf{a}_m}{dt} - \mathbf{c}_1\right)^T \left(\frac{d\mathbf{a}_m}{dt} - \mathbf{c}_1\right) dt = 0$$

and this is possible only if $d\mathbf{a}_m/dt = \mathbf{c}_1$. Hence,

$$\mathbf{a}_m(t) = \mathbf{c}_1 t + \mathbf{c}_2 \qquad (11.77)$$

Therefore, the optimum acceleration program (if there is one) must be a linear function of time. The constants of integration $\mathbf{c}_1$ and $\mathbf{c}_2$ can be chosen so that $\mathbf{r}_m(t)$ and $\mathbf{v}_m(t)$ satisfy the boundary conditions. The

Sect. 11.8]  Optimal Guidance Laws  561

final result is simply

$$\mathbf{a}_m(t) = \frac{4}{t_{go}}[\mathbf{v}_1 - \mathbf{v}(t)] + \frac{6}{t_{go}^2}\{\mathbf{r}_1 - [\mathbf{r}(t) + \mathbf{v}_1 t_{go}]\} \quad (11.78)$$

where

$$t_{go} = t_1 - t$$

denotes the time-to-go before thrust termination.

For a guidance maneuver, the total acceleration $\mathbf{a}(t)$ is the sum of the thrust acceleration $\mathbf{a}_T(t)$ and the gravity acceleration $\mathbf{g}(\mathbf{r})$. If the gravity vector were, indeed, a constant, then the exact solution to the guidance problem would be

$$\mathbf{a}_T(t) = \frac{4}{t_{go}}[\mathbf{v}_1 - \mathbf{v}(t)] + \frac{6}{t_{go}^2}\{\mathbf{r}_1 - [\mathbf{r}(t) + \mathbf{v}_1 t_{go}]\} - \mathbf{g}[\mathbf{r}(t)] \quad (11.79)$$

In problems of practical interest, the vector $\mathbf{g}$ is not constant and the integral-square criterion of Eq. (11.68) is not appropriate for fuel minimization. However, it happens that Eq. (11.79) does provide a nearly optimum steer law for a wide variety of problems and, in fact, was the basis of the lunar-landing guidance method for the Apollo missions.

All of the quantities in the steer law can be either measured onboard the spacecraft or calculated in the spacecraft computer. As the terminal conditions are approached, time-to-go approaches zero and the computation clearly becomes unstable. The difficulty is avoided, with only slight loss in potential performance, by holding the time-to-go factor constant in Eq. (11.79) when it is less than some preassigned amount. Engine cutoff can then be commanded when the actual time-to-go reaches zero.

◊ **Problem 11-32**
A more general functional to be minimized is

$$J = \int_{t_0}^{t_1} F(t, \mathbf{x}, \mathbf{x}') \, dt$$

where the prime indicates differentiation with respect to time. By writing

$$\mathbf{x}(t, \alpha) = \mathbf{x}_m(t) + \alpha \boldsymbol{\epsilon}(t) \quad \text{with} \quad \boldsymbol{\epsilon}(t_0) = \boldsymbol{\epsilon}(t_1) = \mathbf{0}$$

where $\boldsymbol{\epsilon}(t)$ is an admissible function, then $J$ will be a function of $\alpha$. A necessary condition for $dJ/d\alpha$ to be zero when $\alpha = 0$ is†

$$\frac{\partial F}{\partial \mathbf{x}} - \frac{d}{dt}\frac{\partial F}{\partial \mathbf{x}'} = \mathbf{0}^T$$

---

† This famous differential equation is sometimes called the Euler-Lagrange equation but it should be mentioned that Lagrange was only eight years old when Euler first obtained the result. Its solutions have been named *extremals* because they are the only functions which can give $J$ a maximum or a minimum value.

NOTE: An essential part of the derivation consists of proving that if

$$\int_{t_0}^{t_1} \mathbf{y}(t) \cdot \boldsymbol{\epsilon}(t)\, dt = 0$$

for all admissible functions $\boldsymbol{\epsilon}(t)$, then the vector $\mathbf{y}$ is identically zero.

*Leonhard Euler* 1744

### Problem 11-33
An integral of Euler's equation exists if neither $t$ nor $\mathbf{x}$ appears explicitly as an argument of $F(t, \mathbf{x}, \mathbf{x}')$. In particular, the extremals for

$$J = \int_{t_0}^{t_1} F(\mathbf{x}, \mathbf{x}')\, dt \qquad \text{satisfy} \qquad F - \frac{\partial F}{\partial \mathbf{x}'} \mathbf{x}' = \text{constant}$$

HINT: Verify the relation

$$\frac{d}{dt}\left(F - \frac{\partial F}{\partial \mathbf{x}'} \mathbf{x}'\right) = \left(\frac{\partial F}{\partial \mathbf{x}} - \frac{d}{dt}\frac{\partial F}{\partial \mathbf{x}'}\right) \mathbf{x}'$$

### Problem 11-34
The curve $y = y(x)$, connecting the two points $x_1, y_1$ and $x_2, y_2$, generates a surface of revolution when rotated about the $x$ axis. If the surface area is to be a minimum, the curve must be a *catenary*; that is, $y$ will be a hyperbolic cosine function of $x$.

HINT: Find the extremals for the functional

$$S = 2\pi \int_{x_1}^{x_2} y\sqrt{1 + y'^2}\, dx$$

Then, to solve Euler's equation, try the substitution $w = dy/dx$ so that

$$\frac{d^2y}{dx^2} = w\frac{dw}{dy}$$

## The Linear-Tangent Law

Consider the problem of guiding a rocket launched from the surface of the earth in such a manner as to maximize its total energy per unit mass. For simplification, we shall ignore the small part of the trajectory within the sensible atmosphere and assume that no external forces other than gravity are influencing the course of the vehicle. Furthermore, to keep the problem manageable, we presume a given time history of thrust acceleration magnitude and assume a flat-earth approximation so that the gravity vector will be a constant. The direction of the thrust vector is the quantity we are to control to solve the optimization problem.

Sect. 11.8]     Optimal Guidance Laws

Under these circumstances, the equations of motion of a vehicle in the $x, y$ plane are

$$\frac{d\mathbf{x}}{dt} = \frac{d}{dt}\begin{bmatrix} x \\ y \\ v_x \\ v_y \end{bmatrix} = \mathbf{f}[\mathbf{x}(t), \beta(t)] = \begin{bmatrix} v_x \\ v_y \\ a_T \cos\beta \\ a_T \sin\beta - g \end{bmatrix} \quad (11.80)$$

where $\mathbf{x}(t)$ is the four-dimensional state vector and $\beta(t)$ is the control variable of the system. The initial conditions for the state are $\mathbf{x}(t_0) = \mathbf{0}$ and the thrust acceleration $a_T(t)$ is a known function of time. [No initial value will be specified for the control variable $\beta$; in fact, $\beta(0)$ will be determined as a part of the optimal solution.] The quantity to be maximized is

$$J = gy(t_1) + \tfrac{1}{2}[v_x^2(t_1) + v_y^2(t_1)] = gx_2(t_1) + \tfrac{1}{2}[x_3^2(t_1) + x_4^2(t_1)] \quad (11.81)$$

As before, we write

$$\begin{aligned} \mathbf{x}(t) &= \mathbf{x}_m(t) + \alpha\boldsymbol{\epsilon}(t) \quad \text{with} \quad \boldsymbol{\epsilon}(t_0) = \mathbf{0} \\ \beta(t) &= \beta_m(t) + \alpha\gamma(t) \end{aligned} \quad (11.82)$$

so that a necessary condition for $J$ to have a maximum is

$$\left.\frac{dJ}{d\alpha}\right|_{\alpha=0} = g\epsilon_2(t_1) + x_{3m}(t_1)\epsilon_3(t_1) + x_{4m}(t_1)\epsilon_4(t_1) = 0 \quad (11.83)$$

This is the so-called *Mayer form* of the general optimization problem† —so called because the specification of the optimum is entirely in terms of the end conditions.

In order to include the dynamics of the rocket in the problem, we first derive the differential equation to be satisfied by the admissible functions $\epsilon(t)$ and $\gamma(t)$. Since the state vector is a function of both $t$ and $\alpha$, we differentiate the state equation (11.80) partially with respect to $\alpha$ and obtain

$$\frac{\partial}{\partial\alpha}\left(\frac{d\mathbf{x}}{dt}\right) = \frac{d\boldsymbol{\epsilon}}{dt} = \frac{\partial\mathbf{f}}{\partial\mathbf{x}}\frac{\partial\mathbf{x}}{\partial\alpha} + \frac{\partial\mathbf{f}}{\partial\beta}\frac{\partial\beta}{\partial\alpha} = \frac{\partial\mathbf{f}}{\partial\mathbf{x}}\boldsymbol{\epsilon}(t) + \frac{\partial\mathbf{f}}{\partial\beta}\gamma(t)$$

Thus, the admissible variations must always be such that the equation

$$\frac{d\boldsymbol{\epsilon}}{dt} - \frac{\partial\mathbf{f}}{\partial\mathbf{x}}\boldsymbol{\epsilon} - \frac{\partial\mathbf{f}}{\partial\beta}\gamma = 0 \quad (11.84)$$

is satisfied. This condition still holds if the scalar product with an arbitrary function $\lambda(t)$ is taken and the product integrated between the limits $t_0$

---

† Christian Gustav Adolph Mayer (1839–1908) was born in Leipzig, Germany in a family of wealthy merchants. Mayer chose mathematics and physics as a way of life and his entire career was spent as a professor at the University of Heidelberg where he enjoyed great respect from his students and colleagues. He achieved important results in both partial differential equations and optimization criteria in variational calculus.

and $t_1$. Therefore,

$$I \equiv \int_{t_0}^{t_1} \left( \boldsymbol{\lambda}^T(t) \frac{d\boldsymbol{\epsilon}}{dt} + \frac{\partial F}{\partial \mathbf{x}} \boldsymbol{\epsilon} + \frac{\partial F}{\partial \beta} \gamma \right) dt = 0 \qquad (11.85)$$

where the function $F$ is defined by

$$F = -\boldsymbol{\lambda}^T(t) \mathbf{f}[\mathbf{x}(t), \beta(t)] \qquad (11.86)$$

This is the technique first used by Lagrange for finding the maximum or minimum of a function of several variables subject to certain constraints. The function $F$ is called the *Lagrange expression* and the components of $\boldsymbol{\lambda}$ are the *Lagrange multipliers*.

The second term in Eq. (11.85) can be integrated by parts

$$\int_{t_0}^{t_1} \frac{\partial F}{\partial \mathbf{x}} \boldsymbol{\epsilon}(t) \, dt = \mathbf{b}^T(t_1) \boldsymbol{\epsilon}(t_1) - \int_{t_0}^{t_1} \mathbf{b}^T(t) \frac{d\boldsymbol{\epsilon}}{dt} \, dt$$

where we have defined

$$\mathbf{b}^T(t) = \int_{t_0}^{t} \frac{\partial F}{\partial \mathbf{x}} \, dt = -\int_{t_0}^{t} \boldsymbol{\lambda}^T(t) \frac{\partial \mathbf{f}}{\partial \mathbf{x}} \, dt \qquad (11.87)$$

Hence,

$$I = \mathbf{b}^T(t_1)\boldsymbol{\epsilon}(t_1) + \int_{t_0}^{t_1} [\boldsymbol{\lambda}(t) - \mathbf{b}(t)]^T \frac{d\boldsymbol{\epsilon}}{dt} \, dt + \int_{t_0}^{t_1} \frac{\partial F}{\partial \beta} \gamma(t) \, dt$$

and, by choosing $\boldsymbol{\lambda}$ to be equal to $\mathbf{b}$, we have

$$I = \boldsymbol{\lambda}^T(t_1)\boldsymbol{\epsilon}(t_1) + \int_{t_0}^{t_1} \frac{\partial F}{\partial \beta} \gamma(t) \, dt$$

$$= \boldsymbol{\lambda}^T(t_1)\boldsymbol{\epsilon}(t_1) - \int_{t_0}^{t_1} \boldsymbol{\lambda}^T(t) \frac{\partial \mathbf{f}}{\partial \beta} \gamma(t) \, dt = 0 \qquad (11.88)$$

When $I$, whose value must be identically zero since $\boldsymbol{\epsilon}$ and $\gamma$ are admissible variations, is added to Eq. (11.83), the terminal values of the Lagrange multipliers, which are still at our disposal, can be so chosen that the necessary condition for a maximum is simply

$$\left.\frac{dJ}{d\alpha}\right|_{\alpha=0} + I = \int_{t_0}^{t_1} \boldsymbol{\lambda}^T(t) \frac{\partial \mathbf{f}}{\partial \beta} \gamma(t) \, dt = 0 \qquad (11.89)$$

The appropriate values of the components of $\boldsymbol{\lambda}(t_1)$ for this purpose are

$$\begin{aligned} \lambda_1(t_1) &= 0 \\ \lambda_2(t_1) &= g \\ \lambda_3(t_1) &= x_{3m}(t_1) = v_{xm}(t_1) \\ \lambda_4(t_1) &= x_{4m}(t_1) = v_{ym}(t_1) \end{aligned} \qquad (11.90)$$

The only circumstance under which the necessary condition (11.89) can be satisfied for all admissible variations $\gamma(t)$ is that

$$\lambda^{T}(t)\frac{\partial \mathbf{f}}{\partial \beta} = 0 \tag{11.91}$$

which can be established using the variation of the Fundamental Lemma stated in the note of Prob. 11–32. This requirement is called the *optimality condition* and the vector $\lambda$ is, frequently, called the *co-state*. Furthermore, from Eq. (11.87) with $\lambda = \mathbf{b}$, we see that the vector-matrix differential equation for the co-state is

$$\frac{d\lambda^{T}}{dt} = -\lambda^{T}\frac{\partial \mathbf{f}}{\partial \mathbf{x}} \tag{11.92}$$

For the case at hand, the partial derivatives of $\mathbf{f}(\mathbf{x}, \beta)$ are

$$\frac{\partial \mathbf{f}}{\partial \mathbf{x}} = \begin{bmatrix} 0 & 0 & 1 & 0 \\ 0 & 0 & 0 & 1 \\ 0 & 0 & 0 & 0 \\ 0 & 0 & 0 & 0 \end{bmatrix} \quad \text{and} \quad \frac{\partial \mathbf{f}}{\partial \beta} = \begin{bmatrix} 0 \\ 0 \\ -a_T \sin\beta \\ a_T \cos\beta \end{bmatrix}$$

so that

$$\lambda_1(t) = c_1 = 0$$
$$\lambda_2(t) = c_2 = g$$
$$\lambda_3(t) = c_1 t + c_3 = v_{xm}(t_1)$$
$$\lambda_4(t) = c_2 t + c_4 = g(t_1 - t) + v_{ym}(t_1)$$

Also, from the optimality condition (11.91)

$$-\lambda_3(t)\sin\beta_m + \lambda_4(t)\cos\beta_m = 0$$

Therefore, the optimum program for $\beta(t)$ is

$$\tan\beta_m(t) = \frac{\lambda_4(t)}{\lambda_3(t)} = \frac{g(t_1 - t) + v_{ym}(t_1)}{v_{xm}(t_1)} \tag{11.93}$$

called the *linear-tangent law*. At the final time $t_1$, the thrust direction is tangent to the vehicle's velocity vector.

The two-dimensional vector

$$\lambda_p(t) = \begin{bmatrix} \lambda_3(t) \\ \lambda_4(t) \end{bmatrix}$$

which has the direction of the optimal acceleration vector, was called the *primer vector* by Derek F. Lawden who authored the monograph *Optimal Trajectories for Space Navigation* in 1963.† The tip of the primer vector describes a straight-line locus in the $x, y$ plane during the time of thrusting.

---

† Published in London by Butterworth & Co. Ltd.

The linear-tangent law can be used as an approximate steer law in a realistic gravity field. For even greater simplicity, one might represent the angle $\beta$ itself as a linear function of time. This latter technique is the basis of the so-called "iterative guidance mode" which was used successfully to guide the Saturn launch vehicle as it carried the Apollo spacecraft into an earth parking orbit prior to its voyage to the moon.

◊ **Problem 11–35**
Show that the optimal control variable $\beta$ will be a constant if the problem is to maximize just the kinetic energy rather than the total energy of the rocket.

In general, it is not a simple matter to obtain a solution of the complete set of variational equations. The equations of the state and the co-state, (11.80) and (11.92), together with any constraints on the variables, form a large set of nonlinear first-order differential equations with boundary conditions specified at each end. Special iterative methods, which form an entire subject in themselves, are required for most practical problems of performance optimization.

Chapter 12

# Numerical Integration of Differential Equations

THE POPULAR METHODS BEFORE THE ADVENT OF MODERN DIGITAL computers for the step-by-step integration of differential equations had an essential feature in common. At each step of the process, use was made of the function values already obtained in the previous steps. Thus, if we had arrived at the value $y_n$, then to determine $y_{n+1}$ these methods required the use of the values for $y_{n-1}, y_{n-2}, \ldots$, the number of which depended on the desired accuracy and on the particular method employed. They were based on simple finite-difference formulas which were easy to apply using manual methods. However, the disadvantages were that they required special start-up procedures and were not readily amenable to changing the size of the integration interval.

The Runge-Kutta methods do not utilize preceding function values and so were frequently used by hand computers for starting an integration process. Then, the switch was made to finite-difference methods because the Runge-Kutta formulas were too difficult to continue by hand. However, in programming a method for the digital computer, it is inconvenient to use special instructions for a starting process. Furthermore, constant shifting of data is required which is difficult and time-consuming in a digital program—the manual computer operator does this simply by moving his eyes down the page. On the other hand, the more complicated formulas of the Runge-Kutta methods are easily programmed and are, today, frequently preferred to the more complex logic required for classical finite-difference methods. Step-size changes are also easily implemented. For these reasons, and the fact that most books on numerical methods give much attention to finite differences, Runge-Kutta processes are treated exclusively in this chapter.

The integration methods developed by Nyström in 1925 are especially appropriate for Cowell's and Encke's formulation of the equations of motion in orbital mechanics. He adapted Runge-Kutta techniques to the special class of second-order differential equations whose right-hand side is not an explicit function of the derivative of the dependent variable. Nyström found it possible to achieve a higher order of agreement with the Taylor series expansion of the solution for a given number of evaluations of the right-hand side than could be expected for the general case. Thus, with two evaluations he could achieve agreement with an error of fourth order, of fifth order with three evaluations, and of sixth order with four. Nyström developed special algorithms for these three cases.†

The interesting speculation as to whether or not this computational advantage persists for the higher-order methods remained unanswered for thirty years. Then, in 1955, Julius Albrecht published a special "symmetric" (i.e., equal subdivisions of the integration interval) algorithm for a sixth-order method with five evaluations.‡

Much of this chapter is devoted to developing general solutions of the condition equations through eighth order using the fewest number of right-hand side evaluations or *stages* as possible.§ (To the author's knowledge, this task has not been previously undertaken in any systematic way.) The phenomenon, first observed by Nyström, of $m - 1$ stages for $m^{th}$-order accuracy, does not seem to prevail beyond $m = 6$. Indeed, an interesting proof of the nonexistence of a seventh-order, six-stage algorithm is given in the author's paper. However, general methods are achieved for seventh and eighth order by increasing the number of stages by one.

In this chapter, we also develop explicit Runge-Kutta methods applied to general first-order differential equations which are appropriate for the variation of parameters formulation of the equations of motion. In this case, the number of stages required is the same as the order of the algorithm for $m \leq 4$. Furthermore, it has been shown by Butcher¶ that for $m \geq 5$, an order $m$ algorithm of this type can be achieved only if the number of stages $n$ is greater than $m$ and for $m \geq 7$, we must have $n > m + 1$. The existence of particular methods shows that his are the best possible results up to order seven. For order eight, at least ten stages are necessary but no method has been published requiring fewer than eleven.

The task of developing efficient higher-order algorithms in either case— R-K or R-K-N—is complicated by the fact that the number of condition equations for the parameters, many of which are nonlinear, increases with increasing order considerably faster than the number of parameters to be determined. By increasing the number of stages, the set of parameters is also enlarged. The number of equations is unchanged but the efficiency of the algorithm is, of course, adversely affected.

---

† Nyström, E. J., "Über die Numerische Integration von Differentialgleichungen," *Acta Societatis Scientiarum Ferricæ*, Vol. 50, No. 13, 1925, pp. 1–55.

‡ Albrecht, J., "Beiträge zum Runge-Kutta-Verfahren," *Zeitschrift für Angewandte Mathematik und Mechanik*, Vol. 35, March 1955, pp. 100–110.

§ From the author's paper "Resolution of Runge-Kutta-Nyström Condition Equations through Eighth Order," *AIAA Journal*, Vol. 14, August 1976, pp. 1012–1021. See also the author's comment in the *AIAA Journal*, Vol. 15, May 1977, p. 763.

¶ Butcher, J. C., "On the Attainable Order of Runge-Kutta Methods," *Mathematics of Computation*, Vol. 19, 1965, pp. 408–417.

## 12.1 Fundamental Considerations

We consider a method of numerically integrating the special class of second-order vector differential equations

$$\frac{d\mathbf{x}}{dt} = \mathbf{y} \qquad \frac{d\mathbf{y}}{dt} = \mathbf{f}(\mathbf{x}) \qquad (12.1)$$

where $\mathbf{f}$ is *not* an explicit function of $\mathbf{y}$. If we adopt the notation

$$\mathbf{x}(t_0) = \mathbf{x}_0 \qquad \mathbf{y}(t_0) = \mathbf{y}_0 \qquad \mathbf{f}(\mathbf{x}_0) = \mathbf{f}_0$$

and let $h$ denote the time interval

$$h = t - t_0 \qquad (12.2)$$

then a first-order Taylor series expansion

$$\begin{aligned}\mathbf{x} &= \mathbf{x}_0 + h\mathbf{y}_0 + O(h^2) \\ \mathbf{y} &= \mathbf{y}_0 + h\mathbf{f}_0 + O(h^2)\end{aligned} \qquad (12.3)$$

will give values for $\mathbf{x}$ and $\mathbf{y}$ at time $t$ in terms of their values at time $t_0$ with an error of order $h^2$ as indicated by the notation $O(h^2)$.

A second-order Taylor series

$$\begin{aligned}\mathbf{x} &= \mathbf{x}_0 + h\mathbf{y}_0 + \tfrac{1}{2}h^2\mathbf{f}_0 + O(h^3) \\ \mathbf{y} &= \mathbf{y}_0 + h\mathbf{f}_0 + \tfrac{1}{2}h^2\mathbf{f}_0' + O(h^3)\end{aligned}$$

has an error of order $h^3$. The derivative of the vector $\mathbf{f}(\mathbf{x})$ is obtained from

$$\mathbf{f}_0' = \mathbf{F}_0 \mathbf{y}_0 \qquad \text{where} \qquad \mathbf{F}_0 = \left.\frac{\partial \mathbf{f}}{\partial \mathbf{x}}\right|_{t=t_0}$$

and, fortunately, the evaluation of the matrix $\mathbf{F}_0$ can be avoided. Consider the Taylor series expansion

$$\mathbf{f}(\mathbf{x}_0 + hp\mathbf{y}_0) = \mathbf{f}_0 + hp\mathbf{F}_0\mathbf{y}_0 + O(h^2)$$

where $p$ is a constant to be specified. It is clear that equivalent accuracy in the computation of $\mathbf{y}$ may be had if we replace

$$h\mathbf{f}_0' = h\mathbf{F}_0\mathbf{y}_0 \qquad \text{by} \qquad \frac{1}{p}[\mathbf{f}(\mathbf{x}_0 + hp\mathbf{y}_0) - \mathbf{f}_0]$$

Thus, we have

$$\begin{aligned}\mathbf{x} &= \mathbf{x}_0 + h\mathbf{y}_0 + \tfrac{1}{2}h^2\mathbf{f}_0 + O(h^3) \\ \mathbf{y} &= \mathbf{y}_0 + h\left(1 - \frac{1}{2p}\right)\mathbf{f}_0 + \frac{1}{2p}h\mathbf{f}(\mathbf{x}_0 + hp\mathbf{y}_0) + O(h^3)\end{aligned}$$

which is a much more convenient computation.

It appears that two values of the vector $\mathbf{f}$ are required in the equations for $\mathbf{x}$ and $\mathbf{y}$. However, by choosing $p = \tfrac{1}{2}$ and noting that $\mathbf{f}_0$ differs from

$\mathbf{f}(\mathbf{x}_0 + h p \mathbf{y}_0)$ by terms of order $h$, we may write the integration formulas more simply as

$$\mathbf{x} = \mathbf{x}_0 + h\mathbf{y}_0 + \tfrac{1}{2}h^2 \mathbf{f}(\mathbf{x}_0 + \tfrac{1}{2}h\mathbf{y}_0) + O(h^3)$$
$$\mathbf{y} = \mathbf{y}_0 + h\mathbf{f}(\mathbf{x}_0 + \tfrac{1}{2}h\mathbf{y}_0) + O(h^3)$$

Thus, we have a second-order method requiring only one evaluation of the function $\mathbf{f}(\mathbf{x})$ for the value $\mathbf{x} = \mathbf{x}_0 + \tfrac{1}{2}h\mathbf{y}_0$.

The derivation may be formalized in a manner which lends itself more readily to higher-order methods. Suppose we seek equations of the form

$$\begin{aligned}\mathbf{x} &= \mathbf{x}_0 + h\mathbf{y}_0 + h^2 a \mathbf{k} + O(h^3) \\ \mathbf{y} &= \mathbf{y}_0 + hb\mathbf{k} + O(h^3)\end{aligned} \tag{12.4}$$

where

$$\mathbf{k} = \mathbf{f}(\mathbf{x}_0 + h p \mathbf{y}_0) \tag{12.5}$$

which, with appropriate values for $a$, $b$, and $p$, can be made to agree, through terms of order $h^2$, with the Taylor series expansions

$$\begin{aligned}\mathbf{x} &= \mathbf{x}_0 + h\mathbf{y}_0 + \tfrac{1}{2}h^2 \boldsymbol{\alpha}_0 + O(h^3) \\ \mathbf{y} &= \mathbf{y}_0 + h(\boldsymbol{\alpha}_0 + \tfrac{1}{2}h\boldsymbol{\alpha}_1) + O(h^3)\end{aligned} \tag{12.6}$$

where the vectors $\boldsymbol{\alpha}_0$ and $\boldsymbol{\alpha}_1$ are defined by

$$\boldsymbol{\alpha}_0 \equiv \mathbf{f}_0 \quad \text{and} \quad \boldsymbol{\alpha}_1 \equiv \mathbf{f}_0' = \mathbf{F}_0 \mathbf{y}_0$$

By expanding $\mathbf{f}(\mathbf{x}_0 + p h \mathbf{y}_0)$ in a Taylor series, we have

$$\mathbf{k} = \boldsymbol{\alpha}_0 + h p \boldsymbol{\alpha}_1 + O(h^2) \tag{12.7}$$

Then, by substituting Eq. (12.7) into Eqs. (12.4), there results

$$\begin{aligned}\mathbf{x} &= \mathbf{x}_0 + h\mathbf{y}_0 + h^2 a \boldsymbol{\alpha}_0 + O(h^3) \\ \mathbf{y} &= \mathbf{y}_0 + hb(\boldsymbol{\alpha}_0 + h p \boldsymbol{\alpha}_1) + O(h^3)\end{aligned} \tag{12.8}$$

The corresponding coefficients $\boldsymbol{\alpha}_0$ and $\boldsymbol{\alpha}_1$ in Eqs. (12.6) and (12.8) must be equal so that the following equations for $a$, $b$, and $p$ are obtained:

$$(\alpha) \qquad a = \tfrac{1}{2} \qquad \begin{bmatrix}1 \\ p\end{bmatrix} b = \begin{bmatrix}1 \\ \tfrac{1}{2}\end{bmatrix}$$

Thus, we have $a = p = \tfrac{1}{2}$, $b = 1$, and the derivation is complete.

In summary, then, the second-order integration algorithm is

$$\begin{aligned}\mathbf{x} &= \mathbf{x}_0 + h\mathbf{y}_0 + \tfrac{1}{2}h^2 \mathbf{k} + O(h^3) \\ \mathbf{y} &= \mathbf{y}_0 + h\mathbf{k} + O(h^3)\end{aligned} \tag{12.9}$$

where

$$\mathbf{k} = \mathbf{f}(\mathbf{x}_0 + \tfrac{1}{2}h\mathbf{y}_0) \tag{12.10}$$

◇ **Problem 12–1**
For the differential equations

$$\frac{d\mathbf{x}}{dt} = \mathbf{y} \qquad \frac{d\mathbf{y}}{dt} = \mathbf{f}(t, \mathbf{x})$$

it is possible to dispense with the special role played by the independent variable $t$ by augmenting the original system of equations (12.1). If we write

$$\frac{d}{dt}\begin{bmatrix} \mathbf{x} \\ t \end{bmatrix} = \begin{bmatrix} \mathbf{y} \\ 1 \end{bmatrix} \qquad \frac{d}{dt}\begin{bmatrix} \mathbf{y} \\ 1 \end{bmatrix} = \begin{bmatrix} \mathbf{f}\left(\begin{bmatrix} \mathbf{x} \\ t \end{bmatrix}\right) \\ 0 \end{bmatrix}$$

then

$$\mathbf{k} = \begin{bmatrix} \mathbf{f}\left(\begin{bmatrix} \mathbf{x}_0 \\ t_0 \end{bmatrix} + \frac{1}{2}h\begin{bmatrix} \mathbf{y}_0 \\ 1 \end{bmatrix}\right) \\ 0 \end{bmatrix}$$

and

$$\begin{bmatrix} \mathbf{x} \\ t \end{bmatrix} = \begin{bmatrix} \mathbf{x}_0 \\ t_0 \end{bmatrix} + h\begin{bmatrix} \mathbf{y}_0 \\ 1 \end{bmatrix} + \frac{1}{2}h^2\mathbf{k} + O(h^3)$$

$$\begin{bmatrix} \mathbf{y} \\ 1 \end{bmatrix} = \begin{bmatrix} \mathbf{y}_0 \\ 1 \end{bmatrix} + h\mathbf{k} + O(h^3)$$

Therefore, for the differential equations having $\mathbf{f}$ as an explicit function of both $t$ and $\mathbf{x}$, the algorithm is the same as Eqs. (12.9) with $\mathbf{k}$ computed from

$$\mathbf{k} = \mathbf{f}(t_0 + \tfrac{1}{2}h, \mathbf{x}_0 + \tfrac{1}{2}h\mathbf{y}_0)$$

## 12.2 Third-Order R-K-N Algorithms

For higher-order methods it is convenient to introduce an *indicial notation* for vectors and their derivatives. As an example, for three-dimensional vectors, we define

$$x^i = \begin{bmatrix} x^1 \\ x^2 \\ x^3 \end{bmatrix} \qquad y^i = \begin{bmatrix} y^1 \\ y^2 \\ y^3 \end{bmatrix} \qquad f^i = \begin{bmatrix} f^1 \\ f^2 \\ f^3 \end{bmatrix}$$

and

$$\frac{\partial f^i}{\partial x^j} = f^i_j = \begin{bmatrix} f^1_1 & f^1_2 & f^1_3 \\ f^2_1 & f^2_2 & f^2_3 \\ f^3_1 & f^3_2 & f^3_3 \end{bmatrix}$$

The differential equations (12.1) are then written as

$$\frac{dx^i}{dt} = y^i \qquad \frac{dy^i}{dt} = f^i(x^i)$$

and the higher-order derivatives of the vector $f^i$ are as follows

$$\frac{df^i}{dt} = \sum_{j=1}^{3} \frac{\partial f^i}{\partial x^j} \frac{dx^j}{dt} = \sum_{j=1}^{3} f_j^i y^j \equiv f_j^i y^j$$

$$\frac{d^2 f^i}{dt^2} = \sum_{j=1}^{3} \frac{\partial f^i}{\partial x^j} \frac{dy^j}{dt} + \sum_{k=1}^{3}\sum_{j=1}^{3} \left[ \frac{\partial}{\partial x^k} \left( \frac{\partial f^i}{\partial x^j} \right) \frac{dx^k}{dt} \right] y^j$$

$$= \sum_{j=1}^{3} f_j^i f^j + \sum_{j=1}^{3}\sum_{k=1}^{3} f_{jk}^i y^j y^k \equiv f_j^i f^j + f_{jk}^i y^j y^k$$

where, for convenience of notation, we are using the so-called *summation convention*, i.e., an index occurring both as a subscript and as a superscript in a single term implies summation on that index. It is important to realize that while $f_j^i$ is a matrix, the quantity $f_{jk}^i$ is a three-dimensional array which can not be represented as a matrix.†

Adopting the formalism of the second-order method, we define the following vectors:

$$\alpha_0^i \equiv f^i \qquad \alpha_1^i \equiv f_j^i y^j \qquad \alpha_2^i \equiv f_{jk}^i y^j y^k \qquad \beta_2^i \equiv f_j^i f^j \qquad (12.11)$$

where it is understood that the function $f^i$ and all of its derivatives are evaluated for $x^i = x^i(t_0)$. Then we may express the third-order Taylor series as

$$\begin{aligned}\mathbf{x} &= \mathbf{x}_0 + h\mathbf{y}_0 + h^2(\tfrac{1}{2}\boldsymbol{\alpha}_0 + \tfrac{1}{6}h\boldsymbol{\alpha}_1) + O(h^4) \\ \mathbf{y} &= \mathbf{y}_0 + h[\boldsymbol{\alpha}_0 + \tfrac{1}{2}h\boldsymbol{\alpha}_1 + \tfrac{1}{6}h^2(\boldsymbol{\alpha}_2 + \boldsymbol{\beta}_2)] + O(h^4)\end{aligned} \qquad (12.12)$$

We seek solutions of the form

$$\begin{aligned}\mathbf{x} &= \mathbf{x}_0 + h\mathbf{y}_0 + h^2(a_0 \mathbf{k}_0 + a_1 \mathbf{k}_1) + O(h^4) \\ \mathbf{y} &= \mathbf{y}_0 + h(b_0 \mathbf{k}_0 + b_1 \mathbf{k}_1) + O(h^4)\end{aligned} \qquad (12.13)$$

where

$$\begin{aligned}\mathbf{k}_0 &= \mathbf{f}(\mathbf{x}_0 + h p_0 \mathbf{y}_0) \\ \mathbf{k}_1 &= \mathbf{f}(\mathbf{x}_0 + h p_1 \mathbf{y}_0 + h^2 q_1 \mathbf{k}_0)\end{aligned} \qquad (12.14)$$

which will agree with Eqs. (12.12) through terms of order $h^3$ by proper choice of the constants $a_0$, $a_1$, $b_0$, $b_1$, $p_0$, $p_1$, and $q_1$.

---

† It is a useful exercise for the student to calculate these derivatives for the special case of two-body motion; that is,

$$\mathbf{f}(\mathbf{r}) = -\frac{\mu}{r^3}\mathbf{r}$$

## Taylor's Expansion of $f^i(x^i + \delta^i)$

We also require the Taylor series for a *vector function of a vector* in order to expand the **k** vectors. For the derivation, we write

$$\mathbf{x} = \mathbf{x}_0 + \tau\, \boldsymbol{\delta}$$

where the vectors $\mathbf{x}_0$ and $\boldsymbol{\delta}$ are to be regarded as constants. Then the function $\mathbf{f}(\mathbf{x})$ will be a function of the single variable $\tau$. Specifically,

$$\mathbf{f}(\mathbf{x}) = \mathbf{f}(\mathbf{x}_0 + \tau\, \boldsymbol{\delta}) \equiv \mathbf{g}(\tau)$$

The ordinary Taylor series for the function $\mathbf{g}(\tau)$ is

$$\mathbf{g}(\tau) = \mathbf{g}(0) + \mathbf{g}'(0)\tau + \mathbf{g}''(0)\frac{\tau^2}{2!} + \mathbf{g}'''(0)\frac{\tau^3}{3!} + \cdots$$

But we have

$$\mathbf{g}'(\tau) = \frac{d\mathbf{f}}{d\tau} = \frac{\partial \mathbf{f}}{\partial \mathbf{x}}\frac{d\mathbf{x}}{d\tau} = \frac{\partial \mathbf{f}}{\partial \mathbf{x}} \boldsymbol{\delta} = \left( \boldsymbol{\delta} \cdot \left[ \frac{\partial}{\partial \mathbf{x}} \right]^T \right) \mathbf{f}(\mathbf{x})$$

and, similarly,

$$\mathbf{g}''(\tau) = \left( \boldsymbol{\delta} \cdot \left[ \frac{\partial}{\partial \mathbf{x}} \right]^T \right) \frac{\partial \mathbf{f}}{\partial \mathbf{x}} \boldsymbol{\delta} = \left( \boldsymbol{\delta} \cdot \left[ \frac{\partial}{\partial \mathbf{x}} \right]^T \right)^2 \mathbf{f}(\mathbf{x})$$

$$\mathbf{g}'''(\tau) = \left( \boldsymbol{\delta} \cdot \left[ \frac{\partial}{\partial \mathbf{x}} \right]^T \right)^3 \mathbf{f}(\mathbf{x})$$

and so on for the higher derivatives, since each application of the operator $\boldsymbol{\delta} \cdot [\partial/\partial \mathbf{x}]^T$ on any function of $\mathbf{x} = \mathbf{x}_0 + \tau\, \boldsymbol{\delta}$ is equivalent to differentiation with respect to $\tau$.

Next, put $\tau = 1$ in the Taylor series for $\mathbf{g}(\tau)$ to obtain

$$\mathbf{g}(1) = \mathbf{g}(0) + \mathbf{g}'(0) + \mathbf{g}''(0)\frac{1}{2!} + \mathbf{g}'''(0)\frac{1}{3!} + \cdots$$

But when $\tau = 0$, we have $\mathbf{x} = \mathbf{x}_0$. Also, with $\tau = 1$, then $\mathbf{x} = \mathbf{x}_0 + \boldsymbol{\delta}$. Hence, the series for $\mathbf{g}(1)$ gives the desired result

$$\mathbf{f}(\mathbf{x}+\boldsymbol{\delta}) = \mathbf{f}(\mathbf{x}_0) + \left( \boldsymbol{\delta} \cdot \left[ \frac{\partial}{\partial \mathbf{x}} \right]^T \right) \mathbf{f}(\mathbf{x}) \bigg|_{\mathbf{x}=\mathbf{x}_0}$$
$$+ \frac{1}{2!} \left( \boldsymbol{\delta} \cdot \left[ \frac{\partial}{\partial \mathbf{x}} \right]^T \right)^2 \mathbf{f}(\mathbf{x}) \bigg|_{\mathbf{x}=\mathbf{x}_0} + \cdots$$

In our indicial notation, this expansion can be written much more compactly as

$$f^i(x^i + \delta^i) = f^i + f^i_j \delta^j + \tfrac{1}{2} f^i_{jk} \delta^j \delta^k + \tfrac{1}{6} f^i_{jk\ell} \delta^j \delta^k \delta^\ell + \cdots \quad (12.15)$$

where it is understood that after the differentiations are carried out, we are to replace $x^i$ by $x^i(t_0)$.

## Deriving the Condition Equations

The Taylor series expansions of Eqs. (12.14) are obtained from Eq. (12.15). First, with

$$\delta^i = hp_0 y^i$$

we have

$$k_0^i = f^i + f^i_j hp_0 y^j + \tfrac{1}{2} f^i_{jk}(hp_0 y^j)(hp_0 y^k) + O(h^3)$$

Then, with

$$\delta^i = hp_1 y^i + h^2 q_1 k_0^i = hp_1 y^i + h^2 q_1 f^i + O(h^3)$$

we obtain

$$k_1^i = f^i + f^i_j(hp_1 y^j + h^2 q_1 f^j) + \tfrac{1}{2} f^i_{jk}(hp_1 y^j)(hp_1 y^k) + O(h^3)$$

Finally, using the definitions (12.11), we write

$$\mathbf{k}_0 = \boldsymbol{\alpha}_0 + hp_0 \boldsymbol{\alpha}_1 + \tfrac{1}{2} h^2 p_0^2 \boldsymbol{\alpha}_2 + O(h^3)$$
$$\mathbf{k}_1 = \boldsymbol{\alpha}_0 + hp_1 \boldsymbol{\alpha}_1 + h^2(\tfrac{1}{2} p_1^2 \boldsymbol{\alpha}_2 + q_1 \boldsymbol{\beta}_2) + O(h^3) \tag{12.16}$$

Then, by substituting Eqs. (12.16) in Eqs. (12.13) and comparing the result with Eqs. (12.12), the required equations are found to be

$$(\boldsymbol{\alpha}) \quad \begin{bmatrix} 1 & 1 \\ p_0 & p_1 \end{bmatrix} \begin{bmatrix} a_0 \\ a_1 \end{bmatrix} = \begin{bmatrix} \tfrac{1}{2} \\ \tfrac{1}{6} \end{bmatrix} \qquad \begin{bmatrix} 1 & 1 \\ p_0 & p_1 \\ p_0^2 & p_1^2 \end{bmatrix} \begin{bmatrix} b_0 \\ b_1 \end{bmatrix} = \begin{bmatrix} 1 \\ \tfrac{1}{2} \\ \tfrac{1}{3} \end{bmatrix}$$

$$(\boldsymbol{\beta}) \qquad\qquad\qquad\qquad\qquad\qquad\qquad q_1 b_1 = \tfrac{1}{2}(\tfrac{1}{3})$$

Equations ($\boldsymbol{\alpha}$) and ($\boldsymbol{\beta}$) are the *condition equations* for the algorithm of third order but their solution for the parameters is not unique.

## Solving the Condition Equations

The first of the condition equations ($\boldsymbol{\alpha}$) consists of two sets of equations for the coefficients $a_0$, $a_1$ and $b_0$, $b_1$ in terms of $p_0$ and $p_1$. In the second set, if we subtract the second equation from the first and the third equation from the second, we have

$$\begin{bmatrix} 1 & 1 \\ p_0 & p_1 \end{bmatrix} \begin{bmatrix} (1-p_0)b_0 \\ (1-p_1)b_1 \end{bmatrix} = \begin{bmatrix} \tfrac{1}{2} \\ \tfrac{1}{6} \end{bmatrix}$$

which is identical to the first set. Therefore, $a_0$ and $a_1$ can be calculated from

$$a_0 = (1-p_0)b_0 \quad \text{and} \quad a_1 = (1-p_1)b_1 \tag{12.17}$$

after the other quantities have been determined.

## Sect. 12.2]    Third-Order R-K-N Algorithms

The original set of equations for $b_0$, $b_1$ in terms of $p_0$, $p_1$ will be consistent, according to the results of Prob. D–1 in Appendix D, if and only if the determinant

$$D_3 = \begin{vmatrix} 1 & 1 & 1 \\ p_0 & p_1 & \frac{1}{2} \\ p_0^2 & p_1^2 & \frac{1}{3} \end{vmatrix} \tag{12.18}$$

vanishes. By simple row and column operations, we have

$$D_3 = \begin{vmatrix} 1 & 0 & 1 \\ p_0 & p_1 - p_0 & \frac{1}{2} \\ p_0^2 & p_1^2 - p_0^2 & \frac{1}{3} \end{vmatrix} = (p_1 - p_0) \begin{vmatrix} 1 & 0 & 0 \\ p_0 & 1 & \frac{1}{2} - p_0 \\ p_0^2 & p_0 + p_1 & \frac{1}{3} - p_0^2 \end{vmatrix}$$

$$= (p_1 - p_0) \begin{vmatrix} 1 & \frac{1}{2} - p_0 \\ p_0 + p_1 & \frac{1}{3} - p_0^2 \end{vmatrix} = (p_1 - p_0) \begin{vmatrix} 1 & \frac{1}{2} - p_0 \\ p_1 & \frac{1}{3} - \frac{1}{2} p_0 \end{vmatrix}$$

$$= (p_1 - p_0) \begin{vmatrix} 1 & \frac{1}{2} - p_0 \\ 0 & \frac{1}{3} - \frac{1}{2}(p_0 + p_1) + p_0 p_1 \end{vmatrix}$$

Therefore,

$$D_3 = (p_1 - p_0)[\tfrac{1}{3} - \tfrac{1}{2}(p_0 + p_1) + p_0 p_1] \tag{12.19}$$

The first factor of $D_3$ is a second-order *Vandermonde determinant*

$$V_2 = \begin{vmatrix} 1 & 1 \\ p_0 & p_1 \end{vmatrix} = (p_1 - p_0) \tag{12.20}$$

and the second factor

$$L_3(p_0, p_1) = \tfrac{1}{3} - \tfrac{1}{2}(p_0 + p_1) + p_0 p_1 \tag{12.21}$$

is called the *constraint function*. An interesting interpretation of the constraint function is possible using the identity for the determinant of block partitioned matrices given in Prob. B–20 in Appendix B:

$$\begin{vmatrix} \mathbf{A}_{nn} & \mathbf{B}_{nm} \\ \mathbf{C}_{mn} & \mathbf{D}_{mm} \end{vmatrix} = |\mathbf{A}_{nn}||\mathbf{D}_{mm} - \mathbf{C}_{mn} \mathbf{A}_{nn}^{-1} \mathbf{B}_{nm}|$$

By choosing

$$\mathbf{A}_{22} = \mathbf{V}_2 \quad \mathbf{B}_{21} = \begin{bmatrix} 1 \\ \frac{1}{2} \end{bmatrix} \quad \mathbf{C}_{12} = [p_0^2 \ p_1^2] \quad \mathbf{D}_{11} = [\tfrac{1}{3}]$$

and noting that

$$\begin{bmatrix} b_0 \\ b_1 \end{bmatrix} = \mathbf{V}_2^{-1} \begin{bmatrix} 1 \\ \frac{1}{2} \end{bmatrix}$$

we have

$$D_3 = V_2[\tfrac{1}{3} - (p_0^2 b_0 + p_1^2 b_1)] = V_2 L_3(p_0, p_1) \tag{12.22}$$

so that $L_3$ is simply the residue of the third equation for $b_0$, $b_1$. Thus, a necessary and sufficient condition for the consistency of Eqs. ($\alpha$) is that

$$L_3(p_0, p_1) = 0 \tag{12.23}$$

The complete solution of the condition equations is then

$$b_0 = \frac{\frac{1}{2} - p_1}{p_0 - p_1} \qquad b_1 = \frac{\frac{1}{2} - p_0}{p_1 - p_0} \qquad q_1 = \frac{p_1 - p_0}{3 - 6p_0} \tag{12.24}$$

We may choose $p_0$ and $p_1$ arbitrarily, subject to the constraint $L_3 = 0$, leaving us with one free parameter. The coefficients $a_0$ and $a_1$ are then calculated from Eqs. (12.17).

In his original paper, Nyström chose $p_0 = 0$ and gave the algorithm

$$\begin{aligned} \mathbf{x} &= \mathbf{x}_0 + h\mathbf{y}_0 + \tfrac{1}{4}h^2(\mathbf{k}_0 + \mathbf{k}_1) + O(h^4) \\ \mathbf{y} &= \mathbf{y}_0 + \tfrac{1}{4}h(\mathbf{k}_0 + 3\mathbf{k}_1) + O(h^4) \end{aligned} \tag{12.25}$$

where†

$$\begin{aligned} \mathbf{k}_0 &= \mathbf{f}(t_0, \mathbf{x}_0) \\ \mathbf{k}_1 &= \mathbf{f}(t_0 + \tfrac{2}{3}h, \mathbf{x}_0 + \tfrac{2}{3}h\mathbf{y}_0 + \tfrac{2}{9}h^2\mathbf{k}_0) \end{aligned} \tag{12.26}$$

◊ **Problem 12-2**

Construct a second-order algorithm from Eqs. (12.12)–(12.14) and (12.16) by deriving the condition equations

$$(\alpha) \qquad \begin{bmatrix} 1 & 1 \end{bmatrix} \begin{bmatrix} a_0 \\ a_1 \end{bmatrix} = \tfrac{1}{2} \qquad \begin{bmatrix} 1 & 1 \\ p_0 & p_1 \end{bmatrix} \begin{bmatrix} b_0 \\ b_1 \end{bmatrix} = \begin{bmatrix} 1 \\ \tfrac{1}{2} \end{bmatrix}$$

There are now three free parameters $p_0$, $p_1$, and $a_0$ at our disposal. By choosing

$$p_0 = 0 \qquad p_1 = 1 \qquad a_0 = b_0$$

develop the second-order algorithm

$$\begin{aligned} \mathbf{x} &= \mathbf{x}_0 + h\mathbf{y}_0 + \tfrac{1}{2}h^2\mathbf{k}_0 + O(h^3) \\ \mathbf{y} &= \mathbf{y}_0 + \tfrac{1}{2}h(\mathbf{k}_0 + \mathbf{k}_1) + O(h^3) \end{aligned}$$

where

$$\begin{aligned} \mathbf{k}_0 &= \mathbf{f}(t_0, \mathbf{x}_0) \\ \mathbf{k}_1 &= \mathbf{f}(t_0 + h, \mathbf{x}_0 + h\mathbf{y}_0 + \tfrac{1}{2}h^2\mathbf{k}_0) \end{aligned}$$

In this case, although there are apparently two stages, this is true for the first step only. Thereafter, the last computed value for $\mathbf{k}_1$ is the same as the value of $\mathbf{k}_0$ required for the next step.

---

† Henceforth, we shall assume that $\mathbf{f} = \mathbf{f}(t, \mathbf{x})$ with $t$ being incremented in accord with the argument developed in Prob. 12–1.

## 12.3 Fourth-Order R-K-N Algorithms

Paralleling the arguments of the preceding section, the vector $f^i$ and its first three derivatives may be expressed as

$$f^i = \alpha_0^i \qquad \frac{d^2 f^i}{dt^2} = \alpha_2^i + \beta_2^i$$
$$\frac{df^i}{dt} = \alpha_1^i \qquad \frac{d^3 f^i}{dt^3} = \alpha_3^i + 3\beta_3^i + \gamma_3^i \qquad (12.27)$$

where we have defined

$$\begin{aligned} \alpha_0^i &\equiv f^i \\ \alpha_1^i &\equiv f_j^i y^j \\ \alpha_2^i &\equiv f_{jk}^i y^j y^k & \beta_2^i &\equiv f_j^i \alpha_0^j \\ \alpha_3^i &\equiv f_{jk\ell}^i y^j y^k y^\ell & \beta_3^i &\equiv f_{jk}^i \alpha_0^j y^k & \gamma_3^i &\equiv f_j^i \alpha_1^j \end{aligned} \qquad (12.28)$$

Therefore, the fourth-order Taylor series expansions can be written as

$$\begin{aligned} \mathbf{x} &= \mathbf{x}_0 + h\mathbf{y}_0 + h^2[\tfrac{1}{2}\boldsymbol{\alpha}_0 + \tfrac{1}{6}h\boldsymbol{\alpha}_1 + \tfrac{1}{24}h^2(\boldsymbol{\alpha}_2 + \boldsymbol{\beta}_2)] + O(h^5) \\ \mathbf{y} &= \mathbf{y}_0 + h[\boldsymbol{\alpha}_0 + \tfrac{1}{2}h\boldsymbol{\alpha}_1 + \tfrac{1}{6}h^2(\boldsymbol{\alpha}_2 + \boldsymbol{\beta}_2) \\ &\qquad + \tfrac{1}{24}h^3(\boldsymbol{\alpha}_3 + 3\boldsymbol{\beta}_3 + \boldsymbol{\gamma}_3)] + O(h^5) \end{aligned} \qquad (12.29)$$

Again we seek solutions of the form

$$\begin{aligned} \mathbf{x} &= \mathbf{x}_0 + h\mathbf{y}_0 + h^2(a_0 \mathbf{k}_0 + a_1 \mathbf{k}_1 + a_2 \mathbf{k}_2) + O(h^5) \\ \mathbf{y} &= \mathbf{y}_0 + h(b_0 \mathbf{k}_0 + b_1 \mathbf{k}_1 + b_2 \mathbf{k}_2) + O(h^5) \end{aligned} \qquad (12.30)$$

with

$$\begin{aligned} \mathbf{k}_0 &= \mathbf{f}(\mathbf{x}_0 + hp_0 \mathbf{y}_0) \\ \mathbf{k}_1 &= \mathbf{f}(\mathbf{x}_0 + hp_1 \mathbf{y}_0 + h^2 c_{10} \mathbf{k}_0) \\ \mathbf{k}_2 &= \mathbf{f}[\mathbf{x}_0 + hp_2 \mathbf{y}_0 + h^2(c_{20} \mathbf{k}_0 + c_{21} \mathbf{k}_1)] \end{aligned} \qquad (12.31)$$

where, for notational convenience in our later equations, we define

$$c_{10} = q_1 \qquad \text{and} \qquad c_{20} = q_2 - c_{21}$$

As before, we set $\delta^i = hp_0 y^i$ in Eq. (12.15) and obtain

$$k_0^i = f^i + f_j^i hp_0 y^j + \tfrac{1}{2} f_{jk}^i hp_0 y^j hp_0 y^k + \tfrac{1}{6} f_{jk\ell}^i hp_0 y^j hp_0 y^k hp_0 y^\ell + O(h^4)$$

Then, with

$$\delta^i = hp_1 y^i + h^2 c_{10} k_0^i = hp_1 y^i + h^2 q_1 \alpha_0^i + h^3 p_0 q_1 \alpha_1^i + O(h^4)$$

we have

$$\begin{aligned} k_1^i &= f^i + f_j^i(hp_1 y^j + h^2 q_1 \alpha_0^j + h^3 p_0 q_1 \alpha_1^j) \\ &\quad + \tfrac{1}{2} f_{jk}^i (hp_1 y^j + h^2 q_1 \alpha_0^j)(hp_1 y^k + h^2 q_1 \alpha_0^k) \\ &\quad + \tfrac{1}{6} f_{jk\ell}^i hp_1 y^j hp_1 y^k hp_1 y^\ell + O(h^4) \end{aligned}$$

**578**  Numerical Integration of Differential Equations  [Chap. 12]

Finally, with

$$\delta^i = hp_2 y^i + h^2(c_{20} k_0^i + c_{21} k_1^i)$$
$$= hp_2 y^i + h^2[c_{20}(\alpha_0^i + hp_0 \alpha_1^i) + c_{21}(\alpha_0^i + hp_1 \alpha_1^i)] + O(h^4)$$
$$= hp_2 y^i + h^2 q_2 \alpha_0^i + h^3[p_0 q_2 + (p_1 - p_0)c_{21}]\alpha_1^i + O(h^4)$$

we obtain

$$k_2^i = f^i + f_j^i \{hp_2 y^j + h^2 q_2 \alpha_0^j + h^3[p_0 q_2 + (p_1 - p_0)c_{21}]\alpha_1^j\}$$
$$+ \tfrac{1}{2} f_{jk}^i (hp_2 y^j + h^2 q_2 \alpha_0^j)(hp_2 y^k + h^2 q_2 \alpha_0^k)$$
$$+ \tfrac{1}{6} f_{jk\ell}^i hp_2 y^j hp_2 y^k hp_2 y^\ell + O(h^4)$$

Hence, using the definitions (12.28), the following Taylor series expansions of Eqs. (12.31) result:

$$\mathbf{k}_0 = \boldsymbol{\alpha}_0 + hp_0 \boldsymbol{\alpha}_1 + \tfrac{1}{2} h^2 p_0^2 \boldsymbol{\alpha}_2 + \tfrac{1}{6} h^3 p_0^3 \boldsymbol{\alpha}_3 + O(h^4)$$
$$\mathbf{k}_1 = \boldsymbol{\alpha}_0 + hp_1 \boldsymbol{\alpha}_1 + h^2(\tfrac{1}{2} p_1^2 \boldsymbol{\alpha}_2 + q_1 \boldsymbol{\beta}_2)$$
$$\qquad + h^3(\tfrac{1}{6} p_1^3 \boldsymbol{\alpha}_3 + p_1 q_1 \boldsymbol{\beta}_3 + p_0 q_1 \boldsymbol{\gamma}_3) + O(h^4)$$
$$\mathbf{k}_2 = \boldsymbol{\alpha}_0 + hp_2 \boldsymbol{\alpha}_1 + h^2(\tfrac{1}{2} p_2^2 \boldsymbol{\alpha}_2 + q_2 \boldsymbol{\beta}_2)$$
$$\qquad + h^3 \{\tfrac{1}{6} p_2^3 \boldsymbol{\alpha}_3 + p_2 q_2 \boldsymbol{\beta}_3 + [p_0 q_2 + (p_1 - p_0)c_{21}]\boldsymbol{\gamma}_3\} + O(h^4)$$

When these expressions for the **k**'s are substituted in Eqs. (12.30) and the results compared with Eqs. (12.29), we obtain the following condition equations for the fourth-order method:

$$(\boldsymbol{\alpha}) \quad \begin{bmatrix} 1 & 1 & 1 \\ p_0 & p_1 & p_2 \\ p_0^2 & p_1^2 & p_2^2 \end{bmatrix} \begin{bmatrix} a_0 \\ a_1 \\ a_2 \end{bmatrix} = \begin{bmatrix} \tfrac{1}{2} \\ \tfrac{1}{6} \\ \tfrac{1}{12} \end{bmatrix} \qquad \begin{bmatrix} 1 & 1 & 1 \\ p_0 & p_1 & p_2 \\ p_0^2 & p_1^2 & p_2^2 \\ p_0^3 & p_1^3 & p_2^3 \end{bmatrix} \begin{bmatrix} b_0 \\ b_1 \\ b_2 \end{bmatrix} = \begin{bmatrix} 1 \\ \tfrac{1}{2} \\ \tfrac{1}{3} \\ \tfrac{1}{4} \end{bmatrix}$$

$$(\boldsymbol{\beta}) \quad \begin{bmatrix} 1 & 1 \end{bmatrix} \begin{bmatrix} q_1 a_1 \\ q_2 a_2 \end{bmatrix} = \tfrac{1}{2}(\tfrac{1}{12}) \qquad \begin{bmatrix} 1 & 1 \\ p_1 & p_2 \end{bmatrix} \begin{bmatrix} q_1 b_1 \\ q_2 b_2 \end{bmatrix} = \tfrac{1}{2} \begin{bmatrix} \tfrac{1}{3} \\ \tfrac{1}{4} \end{bmatrix}$$

$$(\boldsymbol{\gamma}) \qquad\qquad\qquad\qquad\qquad (p_1 - p_0)c_{21} b_2 = \tfrac{1}{6}(\tfrac{1}{4} - p_0)$$

Again, the equations for the coefficients $a_i$ are redundant with those for $b_i$ if $a_i = (1 - p_i)b_i$.

## Vandermonde Matrices and Constraint Functions

Equations ($\alpha$) contain four linear equations for the three coefficients $b_0$, $b_1$, $b_2$ with the matrix of coefficients being a rectangular *Vandermonde matrix*.† These matrices are fundamental to our work in this chapter so that we now digress for the moment to explore some of their properties.

Square Vandermonde matrices have simple explicit determinants and inverses. Consider, for example, the third-order matrix

$$\mathbf{V}_3 = \begin{bmatrix} 1 & 1 & 1 \\ p_0 & p_1 & p_2 \\ p_0^2 & p_1^2 & p_2^2 \end{bmatrix}$$

From the fundamental properties of determinants, we see that the determinant of $\mathbf{V}_3$ is a second-order polynomial in $p_2$ with roots $p_0$ and $p_1$. Therefore, $V_3$ is proportional to $(p_2 - p_0)(p_2 - p_1)$. Furthermore, the coefficient of $p_2^2$ is the cofactor

$$V_2 = \begin{vmatrix} 1 & 1 \\ p_0 & p_1 \end{vmatrix}$$

so that we have

$$V_3 = (p_1 - p_0)(p_2 - p_0)(p_2 - p_1)$$

Clearly, the scheme can be generalized—for if we consider the Vandermonde matrix of order $\ell$

$$\mathbf{V}_\ell = \|p_j^i\| \qquad (i,j = 0, 1, \ldots, \ell - 1) \qquad (12.32)$$

then the determinant $V_\ell$ is obtained either recursively from

$$V_\ell = V_{\ell-1} \prod_{j=0}^{\ell-2} (p_{\ell-1} - p_j) \qquad (12.33)$$

or, explicitly, from

$$V_\ell = \prod_{i=j+1}^{\ell-1} \prod_{j=0}^{\ell-2} (p_i - p_j) \qquad (12.34)$$

---

† Alexandre-Théophile Vandermonde (1735–1796) was encouraged to pursue a musical career by his physician father but a friend stimulated an interest in mathematics. He was elected to the Académie des Sciences in 1771 and during the next two years he presented four papers to the academy—his total mathematical production. It was his fourth paper in which he gave the first connected exposition of determinants. Thomas Muir, who wrote *The Theory of Determinants in the Historical order of their Development* in 1906, stated that Vandermonde was "the only one fit to be viewed as the founder of determinants." Curiously, Vandermonde is best remembered for the determinant which bears his name but which does not seem to occur in any of his works.

To construct the inverse of $\mathbf{V}_3$, we first define the three second-order polynomials

$$P_2^0(p) = (p - p_1)(p - p_2) = \beta_0^0 + \beta_1^0 p + \beta_2^0 p^2$$
$$P_2^1(p) = (p - p_0)(p - p_2) = \beta_0^1 + \beta_1^1 p + \beta_2^1 p^2$$
$$P_2^2(p) = (p - p_0)(p - p_1) = \beta_0^2 + \beta_1^2 p + \beta_2^2 p^2$$

Then, the inverse can be formulated as

$$\mathbf{V}_3^{-1} = \begin{bmatrix} \dfrac{\beta_0^0}{P_2^0(p_0)} & \dfrac{\beta_1^0}{P_2^0(p_0)} & \dfrac{\beta_2^0}{P_2^0(p_0)} \\ \dfrac{\beta_0^1}{P_2^1(p_1)} & \dfrac{\beta_1^1}{P_2^1(p_1)} & \dfrac{\beta_2^1}{P_2^1(p_1)} \\ \dfrac{\beta_0^2}{P_2^2(p_2)} & \dfrac{\beta_1^2}{P_2^2(p_2)} & \dfrac{\beta_2^2}{P_2^2(p_2)} \end{bmatrix}$$

which is easily verified by computing the product

$$\mathbf{V}_3^{-1}\mathbf{V}_3 = \begin{bmatrix} \dfrac{P_2^0(p_0)}{P_2^0(p_0)} & \dfrac{P_2^0(p_1)}{P_2^0(p_0)} & \dfrac{P_2^0(p_2)}{P_2^0(p_0)} \\ \dfrac{P_2^1(p_0)}{P_2^1(p_1)} & \dfrac{P_2^1(p_1)}{P_2^1(p_1)} & \dfrac{P_2^1(p_2)}{P_2^1(p_1)} \\ \dfrac{P_2^2(p_0)}{P_2^2(p_2)} & \dfrac{P_2^2(p_1)}{P_2^2(p_2)} & \dfrac{P_2^2(p_2)}{P_2^2(p_2)} \end{bmatrix} = \begin{bmatrix} 1 & 0 & 0 \\ 0 & 1 & 0 \\ 0 & 0 & 1 \end{bmatrix}$$

Therefore,

$$\mathbf{V}_3^{-1} = \begin{bmatrix} \dfrac{p_1 p_2}{(p_0 - p_1)(p_0 - p_2)} & -\dfrac{p_1 + p_2}{(p_0 - p_1)(p_0 - p_2)} & \dfrac{1}{(p_0 - p_1)(p_0 - p_2)} \\ \dfrac{p_0 p_2}{(p_1 - p_0)(p_1 - p_2)} & -\dfrac{p_0 + p_2}{(p_1 - p_0)(p_1 - p_2)} & \dfrac{1}{(p_1 - p_0)(p_1 - p_2)} \\ \dfrac{p_0 p_1}{(p_2 - p_0)(p_2 - p_1)} & -\dfrac{p_0 + p_1}{(p_2 - p_0)(p_2 - p_1)} & \dfrac{1}{(p_2 - p_0)(p_2 - p_1)} \end{bmatrix}$$

This scheme is also readily generalized for the $\ell^{\text{th}}$-order matrix by defining the following $\ell$ polynomials of order $\ell - 1$

$$P_{\ell-1}^i(p) = \prod_{j=0}^{\ell-1}{}' (p - p_j) = \sum_{j=0}^{\ell-1} \beta_j^i p^j \qquad (i = 0, 1, \ldots, \ell - 1) \quad (12.35)$$

where the prime on the product symbol indicates that the factor for which $j = i$ is to be omitted. Then the inverse of the Vandermonde matrix is given by

$$\mathbf{V}_\ell^{-1} = \left\| \frac{\beta_j^i}{P_{\ell-1}^i(p_i)} \right\| \qquad (12.36)$$

Return now to the main problem—namely, the consistency of the equations in $(\alpha)$ for the coefficients $b_0$, $b_1$, $b_2$. The necessary and sufficient condition for these four equations in three unknowns to be consistent is that the determinant

$$D_4 = \begin{vmatrix} 1 & 1 & 1 & 1 \\ p_0 & p_1 & p_2 & \frac{1}{2} \\ p_0^2 & p_1^2 & p_2^2 & \frac{1}{3} \\ p_0^3 & p_1^3 & p_2^3 & \frac{1}{4} \end{vmatrix}$$

be identically zero. By simple row and column operations of the type used in the previous section for the reduction of the determinant $D_3$, we may express $D_4$ in the form

$$D_4 = (p_1 - p_0)(p_2 - p_0) \begin{vmatrix} 1 & 1 & \frac{1}{2} - p_0 \\ p_1 & p_2 & \frac{1}{3} - \frac{1}{2}p_0 \\ p_1^2 & p_2^2 & \frac{1}{4} - \frac{1}{3}p_0 \end{vmatrix}$$

When this is compared with the expressions (12.19) and (12.21) for $D_3$ and $L_3$, we can easily deduce

$$D_4 = V_3 L_4(p_0, p_1, p_2)$$

where $L_4$, called the *constraint function* for the fourth-order algorithm, is determined as

$$\begin{aligned} L_4(p_0, p_1, p_2) &= (\tfrac{1}{4} - \tfrac{1}{3}p_0) - (\tfrac{1}{3} - \tfrac{1}{2}p_0)(p_1 + p_2) + (\tfrac{1}{2} - p_0)p_1 p_2 \\ &= \tfrac{1}{4} - \tfrac{1}{3}(p_0 + p_1 + p_2) + \tfrac{1}{2}(p_0 p_1 + p_0 p_2 + p_1 p_2) - p_0 p_1 p_2 \end{aligned}$$

The equations will be consistent if and only if $p_0$, $p_1$, and $p_2$ are so chosen that

$$L_4(p_0, p_1, p_2) = 0$$

As in the previous section, we can also show that

$$D_4 = V_3 [\tfrac{1}{4} - (p_0^3 b_0 + p_1^3 b_1 + p_2^3 b_2)]$$

Hence again, $L_4$ can be interpreted as the residue of the fourth equation for $b_0$, $b_1$, $b_2$.

In general, when considering an algorithm for which the number of stages $n$ is one less than the order $m$, i.e., $n = m - 1$, the $(\alpha)$ equations

$$(\alpha) \qquad \sum_{j=0}^{m-2} p_j^i b_j = \frac{1}{i+1} \qquad (i = 0, 1, \ldots, m-1) \qquad (12.37)$$

will be inconsistent unless the $m \times m$ augmented determinant

$$D_m = \left| p_j^i \quad \frac{1}{i+1} \right| \quad (i = 0, 1, \ldots, m-1;\ j = 0, 1, \ldots, m-2) \quad (12.38)$$

vanishes. This determinant can always be expressed in the form

$$D_m = V_{m-1} L_m(p_0, p_1, \ldots, p_{m-2}) \quad (12.39)$$

and the constraint function obtained from

$$L_m(p_0, p_1, \ldots, p_{m-2}) = \sum_{j=0}^{m-1} \frac{\beta_j}{m-j} \quad (12.40)$$

with $\beta_0, \beta_1, \ldots, \beta_{m-1}$ defined by means of

$$\prod_{j=0}^{m-2} (p - p_j) = \sum_{j=0}^{m-1} \beta_j p^{m-j-1} \quad (12.41)$$

It is important to note that the constraint functions are *multilinear* and symmetric in their arguments.

### Problem 12–3
The $\beta$'s, which are needed to calculate the constraint function, can be generated by starting with

$$\beta_0 = 1$$

and calculating $\beta_1, \beta_2, \ldots, \beta_{m-1}$ recursively from

$$\beta_i = -\frac{1}{i} \sum_{j=0}^{i-1} \beta_j s_{i-j} \quad \text{where} \quad s_i = p_0^i + p_1^i + \cdots + p_{m-2}^i$$

NOTE: This same algorithm, with proper attention given to signs, can be used to generate the coefficients of the characteristic equation of a matrix $\mathbf{M}$. [See Sect. 13.3.] In that case $s_i$ would be determined as the trace of the $i^{\text{th}}$-power of the matrix, i.e., $s_i = \text{tr}\,\mathbf{M}^i$.

### Problem 12–4
The constraint functions satisfy the recursive relation

$$L_m(p_0, p_1, \ldots, p_{m-2}) = L_m(p_0, p_1, \ldots, p_{m-3}, 0) - p_{m-2} L_{m-1}(p_0, p_1, \ldots, p_{m-3})$$

NOTE: Because the $L$ functions are symmetrical in their arguments, there are, in fact, $m-1$ such identities which can be generated by a cyclical permutation of the parameters $p_0, p_1, \ldots, p_{m-2}$.

## Solution of the Condition Equations

The solution of the condition equations $(\alpha)$, $(\beta)$, and $(\gamma)$ are now

$$a_0 = (1-p_0)b_0 \qquad b_0 = \frac{\frac{1}{3} - \frac{1}{2}(p_1 + p_2) + p_1 p_2}{(p_0 - p_1)(p_0 - p_2)}$$

$$a_1 = (1-p_1)b_1 \qquad b_1 = \frac{\frac{1}{3} - \frac{1}{2}(p_0 + p_2) + p_0 p_2}{(p_1 - p_0)(p_1 - p_2)} \qquad (12.42)$$

$$a_2 = (1-p_2)b_2 \qquad b_2 = \frac{\frac{1}{3} - \frac{1}{2}(p_0 + p_1) + p_0 p_1}{(p_2 - p_0)(p_2 - p_1)}$$

$$q_1 = \frac{\frac{1}{3}p_2 - \frac{1}{4}}{2(p_2 - p_1)b_1}$$
$$c_{21} = \frac{\frac{1}{4} - p_0}{6(p_1 - p_0)b_2} \qquad (12.43)$$
$$q_2 = \frac{\frac{1}{3}p_1 - \frac{1}{4}}{2(p_1 - p_2)b_2}$$

provided that $p_0$, $p_1$, and $p_2$ are so chosen that

$$\tfrac{1}{4} - \tfrac{1}{3}(p_0 + p_1 + p_2) + \tfrac{1}{2}(p_0 p_1 + p_0 p_2 + p_1 p_2) - p_0 p_1 p_2 = 0 \qquad (12.44)$$

In his fundamental memoir, Nyström chose

$$p_0 = 0 \qquad p_1 = \tfrac{1}{2} \qquad p_2 = 1$$

and gave the fourth-order algorithm

$$\begin{aligned}\mathbf{x} &= \mathbf{x}_0 + h\mathbf{y}_0 + \tfrac{1}{6}h^2(\mathbf{k}_0 + 2\mathbf{k}_1) + O(h^5) \\ \mathbf{y} &= \mathbf{y}_0 + \tfrac{1}{6}h(\mathbf{k}_0 + 4\mathbf{k}_1 + \mathbf{k}_2) + O(h^5)\end{aligned} \qquad (12.45)$$

where

$$\begin{aligned}\mathbf{k}_0 &= \mathbf{f}(t_0, \mathbf{x}_0) \\ \mathbf{k}_1 &= \mathbf{f}(t_0 + \tfrac{1}{2}h, \mathbf{x}_0 + \tfrac{1}{2}h\mathbf{y}_0 + \tfrac{1}{8}h^2\mathbf{k}_0) \\ \mathbf{k}_2 &= \mathbf{f}(t_0 + h, \mathbf{x}_0 + h\mathbf{y}_0 + \tfrac{1}{2}h^2\mathbf{k}_1)\end{aligned} \qquad (12.46)$$

◇ **Problem 12–5**

Construct a third-order algorithm from Eqs. (12.29)–(12.31) by deriving the condition equations

$$(\alpha) \quad \begin{bmatrix} 1 & 1 & 1 \\ p_0 & p_1 & p_2 \end{bmatrix} \begin{bmatrix} a_0 \\ a_1 \\ a_2 \end{bmatrix} = \begin{bmatrix} \tfrac{1}{2} \\ \tfrac{1}{6} \end{bmatrix} \qquad \begin{bmatrix} 1 & 1 & 1 \\ p_0 & p_1 & p_2 \\ p_0^2 & p_1^2 & p_2^2 \end{bmatrix} \begin{bmatrix} b_0 \\ b_1 \\ b_2 \end{bmatrix} = \begin{bmatrix} 1 \\ \tfrac{1}{2} \\ \tfrac{1}{3} \end{bmatrix}$$

$(\beta)$ $\qquad\qquad\qquad\qquad\qquad\qquad\qquad q_1 b_1 + q_2 b_2 = \tfrac{1}{2}(\tfrac{1}{3})$

With an appropriate choice for the free parameters, develop the third-order algorithm

$$\begin{aligned}\mathbf{x} &= \mathbf{x}_0 + h\mathbf{y}_0 + h^2(a_0 \mathbf{k}_0 + a_1 \mathbf{k}_1) + O(h^4) \\ \mathbf{y} &= \mathbf{y}_0 + h(b_0 \mathbf{k}_0 + b_1 \mathbf{k}_1 + b_2 \mathbf{k}_2) + O(h^4)\end{aligned}$$

where
$$k_0 = f(t_0, x_0)$$
$$k_1 = f(t_0 + hp, x_0 + hpy_0 + \tfrac{1}{2}h^2 p^2 k_0)$$
$$k_2 = f[t_0 + h, x_0 + hy_0 + h^2(a_0 k_0 + a_1 k_1)]$$

and the coefficients are obtained as the following functions of $p \equiv p_1$:

$$b_0 = \frac{3p-1}{6p}$$
$$b_1 = \frac{1}{6p(1-p)} \qquad a_0 = b_0$$
$$b_2 = \frac{2-3p}{6(1-p)} \qquad a_1 = \frac{1}{6p}$$

This is the third-order version of the type of algorithm considered in Prob. 12-2. There are three stages for the first step only. After that, the last $k$, i.e., $k_2$, is the same as the $k_0$ of the next step.

## 12.4 Fourth-Order R-K Algorithms

In this section, we develop Runge-Kutta algorithms for the equation†

$$\frac{d\mathbf{y}}{dt} = \mathbf{f}(t, \mathbf{y}) \tag{12.47}$$

which is typical of those encountered in the method of variation of parameters. We can parallel the arguments of the previous section, using the indicial notation to write the differential equation in the form

$$\frac{dy^i}{dt} = f^i(y^i)$$

and treating the explicit dependence of the function $\mathbf{f}$ on the independent variable $t$ as was done in Prob. 12-1.

The vector $f^i$ and its first three derivatives are expressed as

$$f^i = \alpha_0^i \qquad \frac{d^2 f^i}{dt^2} = \alpha_2^i + \beta_2^i$$
$$\frac{df^i}{dt} = \alpha_1^i \qquad \frac{d^3 f^i}{dt^3} = \alpha_3^i + 3\beta_3^i + \gamma_3^i + \delta_3^i \tag{12.48}$$

---

† Although the bulk of his publications were in spectroscopy, Carl David Tolmé Runge (1856–1927) regarded himself as an applied mathematician—indeed, he was the first full Professor of Applied Mathematics in Germany at Göttingen. His interests were in the theory, practice, and instruction of numerical and graphical computations as typified by his fundamental paper "Üeber die numerische Auflösung von Differentialgleichungen" published in 1895 in *Mathematische Annalen*, Vol. 46, pp. 167–178.

Sect. 12.4]     Fourth-Order R-K Algorithms     585

where we have defined

$$
\begin{aligned}
&\alpha_0^i \equiv f^i \\
&\alpha_1^i \equiv f_j^i \alpha_0^j \\
&\alpha_2^i \equiv f_{jk}^i \alpha_0^j \alpha_0^k \qquad \beta_2^i \equiv f_j^i \alpha_1^j \\
&\alpha_3^i \equiv f_{jk\ell}^i \alpha_0^j \alpha_0^k \alpha_0^\ell \qquad \beta_3^i \equiv f_{jk}^i \alpha_1^j \alpha_0^k \qquad \gamma_3^i \equiv f_j^i \alpha_2^j \qquad \delta_3^i \equiv f_j^i \beta_2^j
\end{aligned} \tag{12.49}
$$

Therefore, the fourth-order Taylor series expansions may be written as

$$
\mathbf{y} = \mathbf{y}_0 + h[\boldsymbol{\alpha}_0 + \tfrac{1}{2}h\boldsymbol{\alpha}_1 + \tfrac{1}{6}h^2(\boldsymbol{\alpha}_2 + \boldsymbol{\beta}_2) \\
+ \tfrac{1}{24}h^3(\boldsymbol{\alpha}_3 + 3\boldsymbol{\beta}_3 + \boldsymbol{\gamma}_3 + \boldsymbol{\delta}_3)] + O(h^5) \tag{12.50}
$$

Again we seek solutions of the form

$$
\mathbf{y} = \mathbf{y}_0 + h(b_0 \mathbf{k}_0 + b_1 \mathbf{k}_1 + b_2 \mathbf{k}_2 + b_3 \mathbf{k}_3) + O(h^5) \tag{12.51}
$$

with

$$
\begin{aligned}
\mathbf{k}_0 &= \mathbf{f}(\mathbf{y}_0) \\
\mathbf{k}_1 &= \mathbf{f}(\mathbf{y}_0 + h p_1 \mathbf{k}_0) \\
\mathbf{k}_2 &= \mathbf{f}[\mathbf{y}_0 + h(c_{20} \mathbf{k}_0 + c_{21} \mathbf{k}_1)] \\
\mathbf{k}_3 &= \mathbf{f}[\mathbf{y}_0 + h(c_{30} \mathbf{k}_0 + c_{31} \mathbf{k}_1 + c_{32} \mathbf{k}_2)]
\end{aligned} \tag{12.52}
$$

where, for notational convenience in our later equations, we define

$$
c_{20} = p_2 - c_{21} \qquad \text{and} \qquad c_{30} = p_3 - c_{31} - c_{32}
$$

The Taylor series expansions of Eqs. (12.52) are found to be

$$
\begin{aligned}
\mathbf{k}_0 &= \boldsymbol{\alpha}_0 \\
\mathbf{k}_1 &= \boldsymbol{\alpha}_0 + h p_1 \boldsymbol{\alpha}_1 + \tfrac{1}{2} h^2 p_1^2 \boldsymbol{\alpha}_2 + \tfrac{1}{6} h^3 p_1^3 \boldsymbol{\alpha}_3 + O(h^4) \\
\mathbf{k}_2 &= \boldsymbol{\alpha}_0 + h p_2 \boldsymbol{\alpha}_1 + h^2 (\tfrac{1}{2} p_2^2 \boldsymbol{\alpha}_2 + p_1 c_{21} \boldsymbol{\beta}_2) \\
&\quad + h^3 (\tfrac{1}{6} p_2^3 \boldsymbol{\alpha}_3 + p_1 p_2 c_{21} \boldsymbol{\beta}_3 + \tfrac{1}{2} p_1^2 c_{21} \boldsymbol{\gamma}_3) + O(h^4) \\
\mathbf{k}_3 &= \boldsymbol{\alpha}_0 + h p_3 \boldsymbol{\alpha}_1 + h^2 [\tfrac{1}{2} p_3^2 \boldsymbol{\alpha}_2 + (p_1 c_{31} + p_2 c_{32}) \boldsymbol{\beta}_2] \\
&\quad + h^3 [\tfrac{1}{6} p_3^3 \boldsymbol{\alpha}_3 + p_3 (p_1 c_{31} + p_2 c_{32}) \boldsymbol{\beta}_3 \\
&\quad + \tfrac{1}{2} (p_1^2 c_{31} + p_2^2 c_{32}) \boldsymbol{\gamma}_3 + p_1 c_{21} c_{32} \boldsymbol{\delta}_3] + O(h^4)
\end{aligned}
$$

which, when substituted in Eqs. (12.51) and the results compared with Eqs. (12.50), produce the following condition equations for the fourth-order

Runge-Kutta method:

$$(\alpha) \quad \begin{bmatrix} 1 & 1 & 1 & 1 \\ 0 & p_1 & p_2 & p_3 \\ 0 & p_1^2 & p_2^2 & p_3^2 \\ 0 & p_1^3 & p_2^3 & p_3^3 \end{bmatrix} \begin{bmatrix} b_0 \\ b_1 \\ b_2 \\ b_3 \end{bmatrix} = \begin{bmatrix} 1 \\ \frac{1}{2} \\ \frac{1}{3} \\ \frac{1}{4} \end{bmatrix}$$

$$(\beta) \quad \begin{bmatrix} 1 & 1 \\ p_2 & p_3 \end{bmatrix} \begin{bmatrix} b_2 p_1 c_{21} \\ b_3(p_1 c_{31} + p_2 c_{32}) \end{bmatrix} = \frac{1}{2} \begin{bmatrix} \frac{1}{3} \\ \frac{1}{4} \end{bmatrix}$$

$$(\gamma) \quad b_2 p_1^2 c_{21} + b_3(p_1^2 c_{31} + p_2^2 c_{32}) = \frac{1}{12}$$

$$(\delta) \quad b_3 p_1 c_{21} c_{32} = \frac{1}{24}$$

**Solving the Condition Equations**

The first of Eqs. $(\alpha)$ determines $b_0$ as a linear combination of $b_1$, $b_2$, $b_3$. The remainder, together with Eqs. $(\beta)$, $(\gamma)$, and $(\delta)$, can be regrouped as

$$\begin{bmatrix} 1 & 1 & 1 \\ p_1 & p_2 & p_3 \\ p_1^2 & p_2^2 & p_3^2 \end{bmatrix} \begin{bmatrix} b_1 p_1 \\ b_2 p_2 \\ b_3 p_3 \end{bmatrix} = \begin{bmatrix} \frac{1}{2} \\ \frac{1}{3} \\ \frac{1}{4} \end{bmatrix} \qquad \begin{bmatrix} 1 & 1 & 1 \\ p_2 & p_3 & p_3 \\ p_1 & p_1 & p_2 \end{bmatrix} \begin{bmatrix} 2 c_{21} b_2 p_1 \\ 2 c_{31} b_3 p_1 \\ 2 c_{32} b_3 p_2 \end{bmatrix} = \begin{bmatrix} \frac{1}{3} \\ \frac{1}{4} \\ \frac{1}{6} \end{bmatrix}$$

$$(2 c_{21} b_2 p_1)(2 c_{32} b_3 p_2) = \tfrac{1}{6} b_2 p_2$$

From the two vector-matrix equations, we obtain

$$b_2 p_2 = \frac{\frac{1}{4} - \frac{1}{3}(p_1 + p_3) + \frac{1}{2} p_1 p_3}{(p_2 - p_1)(p_2 - p_3)}$$

$$2 c_{21} b_2 p_1 = \frac{\frac{1}{3} p_3 - \frac{1}{4}}{p_3 - p_2} \qquad 2 c_{32} b_3 p_2 = \frac{\frac{1}{6} - \frac{1}{3} p_1}{p_2 - p_1}$$

Therefore, substituting in the last equation, which is equivalent to Eq. $(\delta)$, we find that

$$p_3 = 1 \tag{12.53}$$

is required if the complete set of condition equations is to be consistent. Thus, there are two free parameters, $p_1$ and $p_2$, and we have

$$b_1 = \frac{\frac{1}{6} p_2 - \frac{1}{12}}{p_1(p_1 - p_2)(p_1 - 1)} \qquad b_3 = \frac{\frac{1}{4} - \frac{1}{3}(p_1 + p_2) + \frac{1}{2} p_1 p_2}{(1 - p_1)(1 - p_2)}$$

$$b_2 = \frac{\frac{1}{6} p_1 - \frac{1}{12}}{p_2(p_2 - p_1)(p_2 - 1)} \qquad b_0 = 1 - b_1 - b_2 - b_3$$

(12.54)

together with

$$c_{21} = \frac{\frac{1}{12}}{2b_2 p_1(1-p_2)} \qquad c_{31} = \frac{\frac{1}{12}p_1 - \frac{1}{6} - \frac{1}{3}p_2(p_2 - \frac{5}{4})}{2b_3 p_1(1-p_2)(p_2 - p_1)} \qquad (12.55)$$

$$c_{32} = \frac{\frac{1}{6} - \frac{1}{3}p_1}{2b_3 p_2(p_2 - p_1)}$$

as the parameters for the fourth-order Runge-Kutta algorithm.

◇ **Problem 12–6**

A symmetric fourth-order Runge-Kutta algorithm is possible, i.e., one which uses equally spaced increments in $t$ for each stage. We have

$$\mathbf{y} = \mathbf{y}_0 + \tfrac{1}{8}h(\mathbf{k}_0 + 3\mathbf{k}_1 + 3\mathbf{k}_2 + \mathbf{k}_3) + O(h^5)$$

where

$$\mathbf{k}_0 = \mathbf{f}(t_0, \mathbf{y}_0)$$
$$\mathbf{k}_1 = \mathbf{f}(t_0 + \tfrac{1}{3}h, \mathbf{y}_0 + \tfrac{1}{3}h\mathbf{k}_0)$$
$$\mathbf{k}_2 = \mathbf{f}[t_0 + \tfrac{2}{3}h, \mathbf{y}_0 - \tfrac{1}{3}h(\mathbf{k}_0 - 3\mathbf{k}_1)]$$
$$\mathbf{k}_3 = \mathbf{f}[t_0 + h, \mathbf{y}_0 + h(\mathbf{k}_0 - \mathbf{k}_1 + \mathbf{k}_2)]$$

The Classical Runge-Kutta Algorithm

The general solution of the condition equations just derived is, obviously, valid only in the case for which $p_1$, $p_2$, and $p_3$ are all different. However, if we permit $p_1 = p_2$ and divide Eq. ($\gamma$) by the first of Eqs. ($\beta$), we obtain

$$p_1 = p_2 = \tfrac{1}{2}$$

An additional benefit is that Eq. ($\gamma$) can now be discarded.

As before, the first of Eqs. ($\alpha$) determines $b_0$ after the other $b$'s have been found. The remaining equations are then

$$\begin{bmatrix} \tfrac{1}{2} & p_3 \\ \tfrac{1}{4} & p_3^2 \\ \tfrac{1}{8} & p_3^3 \end{bmatrix} \begin{bmatrix} b_1 + b_2 \\ b_3 \end{bmatrix} = \begin{bmatrix} \tfrac{1}{2} \\ \tfrac{1}{3} \\ \tfrac{1}{4} \end{bmatrix}$$

which will be consistent if and only if

$$\begin{vmatrix} \tfrac{1}{2} & p_3 & \tfrac{1}{2} \\ \tfrac{1}{4} & p_3^2 & \tfrac{1}{3} \\ \tfrac{1}{8} & p_3^3 & \tfrac{1}{4} \end{vmatrix} = \tfrac{1}{24}p_3(p_3 - 1)(\tfrac{1}{2} - p_3) = 0$$

By selecting $p_3 = 1$ (actually, the only practical choice), we have

$$b_1 + b_2 = \tfrac{1}{2} \qquad \text{and} \qquad b_3 = \tfrac{1}{6}$$

so that Eq. ($\beta$) and ($\delta$) now become

$$b_2 c_{21} + \tfrac{1}{6}(c_{31} + c_{32}) = \tfrac{1}{3}$$
$$b_2 c_{21} + \tfrac{1}{3}(c_{31} + c_{32}) = \tfrac{1}{2}$$
$$c_{21} c_{32} = \tfrac{1}{2}$$

The first two of these provide the values

$$b_2 c_{21} = \tfrac{1}{6} \quad \text{and} \quad c_{31} + c_{32} = 1$$

Kutta† chose $p \equiv c_{32}$ as the free parameter so that the solution of the condition equations could be written as

$$\begin{aligned}
p_1 &= p_2 = \tfrac{1}{2} & b_0 &= b_3 = \tfrac{1}{6} & c_{21} &= \tfrac{1}{2} p^{-1} \\
p_3 &= 1 & b_1 &= \tfrac{2}{3} - \tfrac{1}{3}p & c_{31} &= 1 - p \\
& & b_2 &= \tfrac{1}{3} p & c_{32} &= p
\end{aligned} \quad (12.56)$$

The classical Runge-Kutta algorithm, corresponding to $p = 1$ and until recently almost universally used, is then

$$\mathbf{y} = \mathbf{y}_0 + \tfrac{1}{6} h (\mathbf{k}_0 + 2\mathbf{k}_1 + 2\mathbf{k}_2 + \mathbf{k}_3) + O(h^5) \qquad (12.57)$$

where

$$\begin{aligned}
\mathbf{k}_0 &= \mathbf{f}(t_0, \mathbf{y}_0) \\
\mathbf{k}_1 &= \mathbf{f}(t_0 + \tfrac{1}{2}h, \mathbf{y}_0 + \tfrac{1}{2}h\mathbf{k}_0) \\
\mathbf{k}_2 &= \mathbf{f}(t_0 + \tfrac{1}{2}h, \mathbf{y}_0 + \tfrac{1}{2}h\mathbf{k}_1) \\
\mathbf{k}_3 &= \mathbf{f}(t_0 + h, \mathbf{y}_0 + h\mathbf{k}_2)
\end{aligned} \qquad (12.58)$$

The advantage of this form was that the independent variable assumes only values corresponding to the beginning, the midpoint, and end of each step. Therefore, it was particularly valuable in cases where the equations involve a function of $t$ defined by a table with equal intervals.

◊ **Problem 12-7**

Derive the two second-order Runge-Kutta algorithms

$$\mathbf{y} = \mathbf{y}_0 + \tfrac{1}{2} h (\mathbf{k}_0 + \mathbf{k}_1) + O(h^3) \qquad \mathbf{y} = \mathbf{y}_0 + h \mathbf{k}_1 + O(h^3)$$
$$\mathbf{k}_0 = \mathbf{f}(t_0, \mathbf{y}_0) \qquad\qquad\qquad \mathbf{k}_0 = \mathbf{f}(t_0, \mathbf{y}_0)$$
$$\mathbf{k}_1 = \mathbf{f}(t_0 + h, \mathbf{y}_0 + h\mathbf{k}_0) \qquad \mathbf{k}_1 = \mathbf{f}(t_0 + \tfrac{1}{2}h, \mathbf{y}_0 + \tfrac{1}{2}h\mathbf{k}_0)$$

---

† Wilhelm Martin Kutta (1867–1944) investigated processes of various orders of accuracy in his paper "Beitrag zur näherungsweisen Integration totaler Differentialgleichungen" published in 1901 in *Zeitschrift für Mathematik und Physik*, Vol. 46, pp. 435–453. He suggested five special cases for which the solution of the condition equations can be expressed in particularly simple forms with one free parameter retained. The most widely used is the one derived here.

It is interesting to note that Kutta also shares the limelight with the Russian mathematician and aerodynamicist Nikolai Jegórowitch Joukowski (1847–1921) for both the Kutta-Joukowski theorem concerning aerodynamic lift and the Kutta-Joukowski airfoil which results from a particular conformal mapping in the complex plane.

Sect. 12.4]   Fourth-Order R-K Algorithms   589

◊ **Problem 12–8**
Derive the two third-order Runge-Kutta algorithms

$$\mathbf{y} = \mathbf{y}_0 + \tfrac{1}{6}h(\mathbf{k}_0 + 4\mathbf{k}_1 + \mathbf{k}_2) + O(h^4) \qquad \mathbf{y} = \mathbf{y}_0 + \tfrac{1}{4}h(\mathbf{k}_0 + 3\mathbf{k}_2) + O(h^4)$$
$$\mathbf{k}_0 = \mathbf{f}(t_0, \mathbf{y}_0) \qquad\qquad\qquad\qquad \mathbf{k}_0 = \mathbf{f}(t_0, \mathbf{y}_0)$$
$$\mathbf{k}_1 = \mathbf{f}(t_0 + \tfrac{1}{2}h, \mathbf{y}_0 + \tfrac{1}{2}h\mathbf{k}_0) \qquad\quad \mathbf{k}_1 = \mathbf{f}(t_0 + \tfrac{1}{3}h, \mathbf{y}_0 + \tfrac{1}{3}h\mathbf{k}_0)$$
$$\mathbf{k}_2 = \mathbf{f}[t_0 + h, \mathbf{y}_0 - h(\mathbf{k}_0 - 2\mathbf{k}_1)] \qquad \mathbf{k}_2 = \mathbf{f}(t_0 + \tfrac{2}{3}h, \mathbf{y}_0 + \tfrac{2}{3}h\mathbf{k}_1)$$

**Problem 12–9**
A recursive formulation of a fourth-order Runge-Kutta algorithm[†] is possible which is designed to minimize the number of memory locations required in a digital computer mechanization.

Let $\mathbf{y}$, $\mathbf{k}$, and $\mathbf{q}$ be three vector memory registers. Then, this algorithm is programmed as

Step 1:   $\mathbf{y}_0 \to \mathbf{y}$   Step 3: $[2 - \sqrt{2}]\mathbf{k} + [-2 + 3\sqrt{\tfrac{1}{2}}]\mathbf{q} \to \mathbf{q}$

$h\mathbf{f}(t_0, \mathbf{y}) \to \mathbf{k}$   $\qquad h\mathbf{f}(t_0 + \tfrac{1}{2}h, \mathbf{y}) \to \mathbf{k}$

$\mathbf{y} + \tfrac{1}{2}\mathbf{k} \to \mathbf{y}$   $\qquad \mathbf{y} + [1 + \sqrt{\tfrac{1}{2}}](\mathbf{k} - \mathbf{q}) \to \mathbf{y}$

Step 2:   $\mathbf{k} \to \mathbf{q}$   Step 4: $[2 + \sqrt{2}]\mathbf{k} + [-2 - 3\sqrt{\tfrac{1}{2}}]\mathbf{q} \to \mathbf{q}$

$h\mathbf{f}(t_0 + \tfrac{1}{2}h, \mathbf{y}) \to \mathbf{k}$   $\qquad h\mathbf{f}(t_0 + h, \mathbf{y}) \to \mathbf{k}$

$\mathbf{y} + [1 - \sqrt{\tfrac{1}{2}}](\mathbf{k} - \mathbf{q}) \to \mathbf{y}$   $\qquad \mathbf{y} + \tfrac{1}{6}\mathbf{k} - \tfrac{1}{3}\mathbf{q} \to \mathbf{y}$

S. Gill 1951

HINT: This algorithm corresponds to Kutta's solution of the condition equations (12.56) with $p = 1 - \sqrt{\tfrac{1}{2}}$. Also, $\mathbf{k}$ is defined here as $h$ times $\mathbf{f}$.

NOTE: The derivation of this algorithm appeared in the paper "A Process for the Step-by-Step Integration of Differential Equations in an Automatic Digital Computing Machine." It was published in 1951 in the *Proceedings of the Cambridge Philosophical Society*, Vol. 47, pp. 96–108. Gill's algorithm also included a clever method to minimize roundoff errors which the interested reader can find in the paper.

---

[†] The author first programmed this algorithm in the early fifties on the IBM Card Programmed Calculator (CPC) which had an extremely limited random access memory. (These were the "ice boxes" alluded to in the Introduction of this book.) He remembers it with great affection because it made possible the solution of a set of differential equations which could not otherwise have been achieved with the classical Runge-Kutta method.

## 12.5 Fifth-Order R-K-N Algorithms

The general Runge-Kutta-Nyström algorithm of order $m$ with $n$ stages, i.e., $n$ evaluations of $\mathbf{f}$, has the form

$$\mathbf{x} = \mathbf{x}_0 + h\mathbf{y}_0 + h^2 \sum_{i=0}^{n-1} a_i \mathbf{k}_i + O(h^{m+1})$$
$$\mathbf{y} = \mathbf{y}_0 + h \sum_{i=0}^{n-1} b_i \mathbf{k}_i + O(h^{m+1})$$
(12.59)

where

$$\mathbf{k}_0 = \mathbf{f}(t_0 + hp_0, \mathbf{x}_0 + hp_0\mathbf{y}_0)$$
$$\mathbf{k}_1 = \mathbf{f}(t_0 + hp_1, \mathbf{x}_0 + hp_1\mathbf{y}_0 + h^2 q_1 \mathbf{k}_0)$$
$$\vdots$$
$$\mathbf{k}_i = \mathbf{f}(t_0 + hp_i, \mathbf{x}_0 + hp_i\mathbf{y}_0 + h^2 \sum_{j=0}^{i-1} c_{ij} \mathbf{k}_j)$$
(12.60)

Further, to simplify the structure of the condition equations, we define

$$c_{10} = q_1 \quad \text{and} \quad c_{i0} = q_i - \sum_{j=1}^{i-1} c_{ij} \quad \text{for} \quad i = 2, 3, \ldots, n-1$$

The main problem is, of course, to determine the parameters $p_i$, $q_i$, $c_{ij}$, $a_i$, $b_i$ so that there will be an $m^{\text{th}}$-order agreement with the Taylor series expansion of $\mathbf{x}$ and $\mathbf{y}$ for the smallest possible value of $n$.

For a fifth-order algorithm, we need the fourth derivative of $f^i$ and must, therefore, define six more vectors in addition to those of Eqs. (12.28). In terms of these vectors

$$\alpha_4^i \equiv f^i_{jk\ell m} y^j y^k y^\ell y^m \qquad \delta_4^i \equiv f^i_{jk} a_0^j a_0^k$$
$$\beta_4^i \equiv f^i_{jk\ell} a_0^j y^k y^\ell \qquad \varepsilon_4^i \equiv f^i_{jk} a_2^j$$
$$\gamma_4^i \equiv f^i_{jk} a_1^j y^k \qquad \varsigma_4^i \equiv f^i_j b_2^j$$
(12.61)

that derivative is expressed as

$$\frac{d^4 f^i}{dt^4} = \alpha_4^i + 6\beta_4^i + 4\gamma_4^i + 3\delta_4^i + \varepsilon_4^i + \varsigma_4^i$$
(12.62)

The condition equations are developed as for the lower-order cases. The ($\alpha$) equations, given in general form in Eqs. (12.37), will be consistent provided that $p_0$, $p_1$, $p_2$, $p_3$ are chosen so that

$$L_5(p_0, p_1, p_2, p_3) \equiv \tfrac{1}{5} - \tfrac{1}{4}(p_0 + p_1 + p_2 + p_3)$$
$$+ \tfrac{1}{3}(p_0 p_1 + p_0 p_2 + p_0 p_3 + p_1 p_2 + p_1 p_3 + p_2 p_3)$$
$$- \tfrac{1}{2}(p_0 p_1 p_2 + p_0 p_1 p_3 + p_0 p_2 p_3 + p_1 p_2 p_3) + p_0 p_1 p_2 p_3 = 0 \quad (12.63)$$

Sect. 12.5]   Fifth-Order R-K-N Algorithms   591

The remaining condition equations are†

($\beta$)
$$\begin{bmatrix} 1 & 1 & 1 \\ p_1 & p_2 & p_3 \\ p_1^2 & p_2^2 & p_3^2 \end{bmatrix} \begin{bmatrix} q_1 b_1 \\ q_2 b_2 \\ q_3 b_3 \end{bmatrix} = \tfrac{1}{2} \begin{bmatrix} \tfrac{1}{3} \\ \tfrac{1}{4} \\ \tfrac{1}{5} \end{bmatrix}$$

($\gamma$)
$$\begin{bmatrix} 1 & 1 \\ p_2 & p_3 \end{bmatrix} \begin{bmatrix} (p_1 - p_0) b_2 c_{21} \\ (p_1 - p_0) b_3 c_{31} + (p_2 - p_0) b_3 c_{32} \end{bmatrix} = \tfrac{1}{6} \begin{bmatrix} \tfrac{1}{4} - p_0 \\ \tfrac{1}{5} - \tfrac{3}{4} p_0 \end{bmatrix}$$

($\delta$)
$$q_1^2 b_1 + q_2^2 b_2 + q_3^2 b_3 = \tfrac{1}{4}(\tfrac{1}{5})$$

($\varepsilon$)
$$\begin{bmatrix} 1 & 1 \end{bmatrix} \begin{bmatrix} (p_1^2 - p_0^2) b_2 c_{21} \\ (p_1^2 - p_0^2) b_3 c_{31} + (p_2^2 - p_0^2) b_3 c_{32} \end{bmatrix} = \tfrac{1}{12}(\tfrac{1}{5} - 2 p_0^2)$$

($\zeta$)
$$q_1 b_2 c_{21} + q_1 b_3 c_{31} + q_2 b_3 c_{32} = \tfrac{1}{120}$$

Equations ($\beta$) determine $q_1$, $q_2$, $q_3$. Therefore, Eq. ($\delta$) provides an additional constraint on the choice of the $p_i$'s. Similarly, Eqs. ($\gamma$) and ($\varepsilon$) determine $b_2 c_{21}$, $b_3 c_{31}$, $b_3 c_{32}$ with still another constraint implied by Eq. ($\zeta$).

## A Simple Solution of the Condition Equations

If we put

$$p_0 = 0 \quad \text{and} \quad q_i = \tfrac{1}{2} p_i^2 \quad \text{for} \quad i = 1, 2, 3 \qquad (12.64)$$

then Eqs. ($\varepsilon$), ($\zeta$), and the last of Eqs. ($\beta$) are identical. Furthermore, Eqs. ($\beta$) and the last three of Eqs. ($\alpha$) are also identical. Therefore, we may discard Eqs. ($\beta$), ($\delta$), and ($\zeta$).

The equations in ($\alpha$) for determining $b_1$, $b_2$, $b_3$ are

$$\begin{bmatrix} 1 & 1 & 1 \\ p_1 & p_2 & p_3 \\ p_1^2 & p_2^2 & p_3^2 \\ p_1^3 & p_2^3 & p_3^3 \end{bmatrix} \begin{bmatrix} p_1 b_1 \\ p_2 b_2 \\ p_3 b_3 \end{bmatrix} = \begin{bmatrix} \tfrac{1}{2} \\ \tfrac{1}{3} \\ \tfrac{1}{4} \\ \tfrac{1}{5} \end{bmatrix}$$

with $p_1$, $p_2$, $p_3$ chosen subject to the constraint

$$L_5(0, p_1, p_2, p_3) = \tfrac{1}{5} - \tfrac{1}{4}(p_1 + p_2 + p_3) \\ + \tfrac{1}{3}(p_1 p_2 + p_1 p_3 + p_2 p_3) - \tfrac{1}{2} p_1 p_2 p_3 = 0 \quad (12.65)$$

---

† From now on we will not write out the condition equations for the coefficients $a_i$ since they are always redundant with those for $b_i$ provided that $a_i = (1 - p_i) b_i$.

Thus, we have

$$b_1 = \frac{\frac{1}{4} - \frac{1}{3}(p_2 + p_3) + \frac{1}{2}p_2 p_3}{p_1(p_1 - p_2)(p_1 - p_3)}$$
$$b_2 = \frac{\frac{1}{4} - \frac{1}{3}(p_1 + p_3) + \frac{1}{2}p_1 p_3}{p_2(p_2 - p_1)(p_2 - p_3)} \quad (12.66)$$
$$b_3 = \frac{\frac{1}{4} - \frac{1}{3}(p_1 + p_2) + \frac{1}{2}p_1 p_2}{p_3(p_3 - p_1)(p_3 - p_2)}$$

after which $b_0$ is calculated from

$$b_0 = 1 - b_1 - b_2 - b_3 \quad (12.67)$$

Finally, $c_{21}$, $c_{31}$, $c_{32}$ are obtained from the equations

$$\begin{bmatrix} 1 & 1 & 1 \\ p_2 & p_3 & p_3 \\ p_1 & p_1 & p_2 \end{bmatrix} \begin{bmatrix} p_1 b_2 c_{21} \\ p_1 b_3 c_{31} \\ p_2 b_3 c_{32} \end{bmatrix} = \frac{1}{6} \begin{bmatrix} \frac{1}{4} \\ \frac{1}{5} \\ \frac{1}{10} \end{bmatrix}$$

with the result that

$$c_{21} = \frac{\frac{1}{5} - \frac{1}{4}p_3}{6b_2 p_1(p_2 - p_3)}$$
$$c_{31} = \frac{\frac{1}{5}p_1 - \frac{3}{10}p_2 + \frac{1}{10}p_3 + \frac{1}{4}p_2^2 - \frac{1}{4}p_1 p_3}{6b_3 p_1(p_1 - p_2)(p_3 - p_2)} \quad (12.68)$$
$$c_{32} = \frac{\frac{1}{10} - \frac{1}{4}p_1}{6b_3 p_2(p_2 - p_1)}$$

◊ **Problem 12-10**

Nyström gave fifth-order algorithms for the following values of the $p_i$'s:

$$p_1 = \tfrac{1}{5} \quad p_2 = \tfrac{2}{3} \quad p_3 = 1 \quad \text{and} \quad p_1 = \tfrac{2}{5} \quad p_2 = \tfrac{2}{3} \quad p_3 = \tfrac{4}{5}$$

For the first set, the algorithm is

(1)
$$\mathbf{x} = \mathbf{x}_0 + h\mathbf{y}_0 + \tfrac{1}{336}h^2(14\mathbf{k}_0 + 100\mathbf{k}_1 + 54\mathbf{k}_2) + O(h^6)$$
$$\mathbf{y} = \mathbf{y}_0 + \tfrac{1}{336}h(14\mathbf{k}_0 + 125\mathbf{k}_1 + 162\mathbf{k}_2 + 35\mathbf{k}_3) + O(h^6)$$

where

$$\mathbf{k}_0 = \mathbf{f}(t_0, \mathbf{x}_0)$$
$$\mathbf{k}_1 = \mathbf{f}(t_0 + \tfrac{1}{5}h, \mathbf{x}_0 + \tfrac{1}{5}h\mathbf{y}_0 + \tfrac{1}{50}h^2\mathbf{k}_0)$$
$$\mathbf{k}_2 = \mathbf{f}[t_0 + \tfrac{2}{3}h, \mathbf{x}_0 + \tfrac{2}{3}h\mathbf{y}_0 - \tfrac{1}{27}h^2(\mathbf{k}_0 - 7\mathbf{k}_1)]$$
$$\mathbf{k}_3 = \mathbf{f}[t_0 + h, \mathbf{x}_0 + h\mathbf{y}_0 + \tfrac{1}{70}(21\mathbf{k}_0 - 4\mathbf{k}_1 + 18\mathbf{k}_2)]$$

For the second set,

(2)
$$\mathbf{x} = \mathbf{x}_0 + h\mathbf{y}_0 + \tfrac{1}{192}h^2(23\mathbf{k}_0 + 75\mathbf{k}_1 - 27\mathbf{k}_2 + 25\mathbf{k}_3) + O(h^6)$$
$$\mathbf{y} = \mathbf{y}_0 + \tfrac{1}{192}h(23\mathbf{k}_0 + 125\mathbf{k}_1 - 81\mathbf{k}_2 + 125\mathbf{k}_3) + O(h^6)$$

Sect. 12.5]  Fifth-Order R-K-N Algorithms

where
$$\mathbf{k}_0 = \mathbf{f}(t_0, \mathbf{x}_0)$$
$$\mathbf{k}_1 = \mathbf{f}(t_0 + \tfrac{2}{5}h, \mathbf{x}_0 + \tfrac{2}{5}h\mathbf{y}_0 + \tfrac{2}{25}h^2\mathbf{k}_0)$$
$$\mathbf{k}_2 = \mathbf{f}(t_0 + \tfrac{2}{3}h, \mathbf{x}_0 + \tfrac{2}{3}h\mathbf{y}_0 + \tfrac{2}{9}h^2\mathbf{k}_0)$$
$$\mathbf{k}_3 = \mathbf{f}[t_0 + \tfrac{4}{5}h, \mathbf{x}_0 + \tfrac{4}{5}h\mathbf{y}_0 + \tfrac{4}{25}h^2(\mathbf{k}_0 + \mathbf{k}_1)]$$

### The General Solution of the Condition Equations

We now address the solution of the condition equations without imposing any of the assumptions specified by Eqs. (12.64). Instead, we calculate the residues of Eqs. ($\boldsymbol{\delta}$) and ($\boldsymbol{\zeta}$). For this purpose, we first form the four-dimensional determinant $D_1$ of the coefficients of the $q_i$'s in Eqs. ($\boldsymbol{\beta}$), ($\boldsymbol{\delta}$) augmented by the right-hand sides

$$D_1 = \begin{vmatrix} b_1 & b_2 & b_3 & \tfrac{1}{6} \\ p_1 b_1 & p_2 b_2 & p_3 b_3 & \tfrac{1}{8} \\ p_1^2 b_1 & p_2^2 b_2 & p_3^2 b_3 & \tfrac{1}{10} \\ q_1 b_1 & q_2 b_2 & q_3 b_3 & \tfrac{1}{20} \end{vmatrix}$$

Solving Eqs. ($\boldsymbol{\alpha}$) and ($\boldsymbol{\beta}$) for $b_i$ and $q_i b_i$, gives†

$$(p_1 - p_0)(p_1 - p_2)(p_1 - p_3)b_1 = L_4(p_0, p_2, p_3) \quad (0,1,2,3)$$
$$2(p_1 - p_2)(p_1 - p_3)q_1 b_1 = L_5(p_2, p_3) \quad (1,2,3)$$

(12.69)

Substituting for the elements of the determinant and removing the common factors, produces

$$D_1 = \frac{1}{A} \begin{vmatrix} L_4(p_0,p_2,p_3) & L_4(p_0,p_1,p_3) & L_4(p_0,p_1,p_2) & \tfrac{1}{3} \\ p_1 L_4(p_0,p_2,p_3) & p_2 L_4(p_0,p_1,p_3) & p_3 L_4(p_0,p_1,p_2) & \tfrac{1}{4} \\ p_1^2 L_4(p_0,p_2,p_3) & p_2^2 L_4(p_0,p_1,p_3) & p_3^2 L_4(p_0,p_1,p_2) & \tfrac{1}{5} \\ (p_1-p_0)L_5(p_2,p_3) & (p_2-p_0)L_5(p_1,p_3) & (p_3-p_0)L_5(p_1,p_2) & \tfrac{1}{5} \end{vmatrix}$$

where
$$A = 4V_4(p_0)(p_1 - p_2)(p_1 - p_3)(p_2 - p_3)$$

To evaluate the determinant, we use the constraint condition (12.63) and the recursion relation for the $L$ functions given in Prob. 12–4 to derive

---

† The equations for $b_0$, $b_2$, $b_3$ are obtained by a cyclic permutation of the subscripts $(0,1,2,3)$. Similarly, $q_2 b_2$ and $q_3 b_3$ are calculated by permuting the subscripts $(1,2,3)$. We shall use this notation frequently for compactness in expressing our results. Also, for convenience, if any of the arguments of the constraint functions is zero, it will be suppressed in the notation. Thus, without confusion, we write

$$L_5(p_2, p_3) = L_5(0, 0, p_2, p_3)$$

the following identities for the elements of the first column of $D_1$:

$$p_1 L_4(p_0, p_2, p_3) = L_5(p_0, p_2, p_3)$$
$$p_1^2 L_4(p_0, p_2, p_3) = L_6(p_0, p_2, p_3) - L_6(p_0, p_1, p_2, p_3)$$
$$(p_1 - p_0) L_5(p_2, p_3) = L_6(p_0, p_2, p_3) - L_6(p_1, p_2, p_3) \qquad (12.70)$$
$$p_0 L_5(p_1, p_2, p_3) = p_1^2 L_4(p_0, p_2, p_3) - (p_1 - p_0) L_5(p_2, p_3)$$

Similar relations are established for the other columns of $D_1$ as well.

Next we perform the following sequence of elementary row and column operations:

1. Row 3 − Row 4 → Row 3 and factor $p_0 L_5(p_1, p_2, p_3)$.
2. $L_6(p_1, p_2, p_3)$ × Row 3 + Row 4 → Row 4.
3. Column 1 − Column 2 → Column 1 and factor $p_1 - p_2$.
4. Column 2 − Column 3 → Column 2 and factor $p_2 - p_3$.

The determinant is now three dimensional:

$$D_1 = \frac{p_0 L_5(p_1, p_2, p_3)}{4 V_4(p_0)(p_1 - p_3)} \begin{vmatrix} L_3(p_0, p_3) & L_3(p_0, p_1) & \frac{1}{3} \\ L_4(p_0, p_3) & L_4(p_0, p_1) & \frac{1}{4} \\ L_5(p_0, p_3) & L_5(p_0, p_1) & \frac{1}{5} \end{vmatrix}$$

To continue the reduction, the next sequence of operations is

5. Column 1 − Column 2 → Column 1 and factor $p_1 - p_3$.
6. $p_1$ × Column 1 + Column 2 → Column 2.

and the determinant is now a function only of $p_0$:

$$D_1 = \frac{p_0 L_5(p_1, p_2, p_3)}{4 V_4(p_0)} \begin{vmatrix} \frac{1}{2} - p_0 & \frac{1}{3} - \frac{1}{2} p_0 & \frac{1}{3} \\ \frac{1}{3} - \frac{1}{2} p_0 & \frac{1}{4} - \frac{1}{3} p_0 & \frac{1}{4} \\ \frac{1}{4} - \frac{1}{3} p_0 & \frac{1}{5} - \frac{1}{4} p_0 & \frac{1}{5} \end{vmatrix}$$

Finally,

7. Column 2 − Column 3 → Column 2 and factor $-p_0$.
8. Column 1 − Column 2 → Column 1 and factor $-p_0$.

and we have

$$D_1 = \frac{p_0^3 L_5(p_1, p_2, p_3)}{8640 V_4(p_0)}$$

To determine the residue of Eq. ($\delta$), we use the arguments near the end of Sect. 12.2 and obtain

$$\text{Residue}(\delta) = \frac{p_0^3 L_5(p_1, p_2, p_3)}{8640 V_4(p_0) V_3(p_1) b_1 b_2 b_3} \qquad (12.71)$$

The symbols $V_4(p_0)$ and $V_3(p_1)$ denote the Vandermonde determinants

$$V_4(p_0) = \begin{vmatrix} 1 & 1 & 1 & 1 \\ p_0 & p_1 & p_2 & p_3 \\ p_0^2 & p_1^2 & p_2^2 & p_3^2 \\ p_0^3 & p_1^3 & p_2^3 & p_3^3 \end{vmatrix} \quad \text{and} \quad V_3(p_1) = \begin{vmatrix} 1 & 1 & 1 \\ p_1 & p_2 & p_3 \\ p_1^2 & p_2^2 & p_3^2 \end{vmatrix}$$

Sect. 12.5]  Fifth-Order R-K-N Algorithms

To calculate the residue of Eq. ($\zeta$) we form the four-dimensional determinant $D_2$ of the coefficients of $b_2 c_{21}$, $b_3 c_{31}$, $b_3 c_{32}$ in Eqs. ($\gamma$), ($\varepsilon$), and ($\gamma$) augmented by their right-hand sides. Then

1. Column 1 − Column 2 → Column 1 and factor $(p_3 - p_2)(p_1 - p_0)$.

The determinant is now three dimensional of the form

$$D_2 = -\tfrac{1}{6}(p_3-p_2)(p_1-p_0)\begin{vmatrix} p_1 - p_0 & p_2 - p_0 & \tfrac{1}{4} - p_0 \\ p_1^2 - p_0^2 & p_2^2 - p_0^2 & \tfrac{1}{10} - p_0^2 \\ q_1 & q_2 & \tfrac{1}{20} \end{vmatrix}$$

Substitute for $q_1$ and $q_2$ from the second of Eqs. (12.70) and then

2. Row 2 − $p_0$ × Row 1 → Row 2.
3. $b_1$ × Column 1 → Column 1.
4. $b_2$ × Column 2 → Column 2.

The result is

$$D_2 = B \begin{vmatrix} L_4(p_0,p_2,p_3) & L_4(p_0,p_1,p_3) & \tfrac{1}{4} - p_0 \\ L_5(p_0,p_2,p_3) & L_5(p_0,p_1,p_3) & \tfrac{1}{10} - \tfrac{1}{4}p_0 \\ L_5(p_2,p_3) & L_5(p_1,p_3) & \tfrac{1}{10} \end{vmatrix}$$

where

$$B = -\frac{(p_3-p_2)(p_1-p_0)}{12(p_1-p_2)(p_1-p_3)(p_2-p_1)(p_2-p_3)b_1 b_2}$$

Next,

5. Row 2 − Row 3 → Row 2 and factor $-p_0$.
6. Row 1 − Row 2 → Row 1 and factor $-p_0$.

and the determinant no longer contains $p_0$. Thus,

$$D_2 = \frac{p_0^2(p_1-p_0)}{12(p_1-p_2)(p_1-p_3)(p_2-p_1)b_1 b_2} \begin{vmatrix} L_3(p_2,p_3) & L_3(p_1,p_3) & 1 \\ L_4(p_2,p_3) & L_4(p_1,p_3) & \tfrac{1}{4} \\ L_5(p_2,p_3) & L_5(p_1,p_3) & \tfrac{1}{10} \end{vmatrix}$$

Finally,

7. Column 1 − Column 2 → Column 1 and factor $p_1 - p_2$.
8. $p_1$ × Column 1 + Column 2 → Column 2.

and the determinant now involves only $p_3$. Specifically,

$$D_2 = \frac{(p_1-p_0)p_0^2}{12(p_1-p_3)(p_2-p_1)b_1 b_2} \begin{vmatrix} \tfrac{1}{2} - p_3 & \tfrac{1}{3} - \tfrac{1}{2}p_3 & 1 \\ \tfrac{1}{3} - \tfrac{1}{2}p_3 & \tfrac{1}{4} - \tfrac{1}{3}p_3 & \tfrac{1}{4} \\ \tfrac{1}{4} - \tfrac{1}{3}p_3 & \tfrac{1}{5} - \tfrac{1}{4}p_3 & \tfrac{1}{10} \end{vmatrix}$$

which reduces to

$$D_2 = \frac{p_0^2(p_1-p_0)(p_3-1)^2}{8640(p_1-p_3)(p_2-p_1)b_1 b_2}$$

The residue of Eq. ($\zeta$) is found as before

$$\text{Residue}(\zeta) = -\frac{p_0^2(p_3-1)^2}{8640 V_3(p_0) V_3(p_1) b_1 b_2} \tag{12.72}$$

In summary, the condition equations are consistent if and only if the $p_i$'s are selected subject to the following constraints:

$$\begin{aligned} L_5(p_0, p_1, p_2, p_3) &= 0 \\ p_0^3 L_5(p_1, p_2, p_3) &= 0 \\ p_0^2(p_3 - 1)^2 &= 0 \end{aligned} \tag{12.73}$$

The third of Eqs. (12.73) requires that either $p_0 = 0$ or $p_3 = 1$. In the first case, we may select the $p_i$'s subject to

$$p_0 = 0 \quad \text{and} \quad L_5(p_1, p_2, p_3) = 0 \tag{12.74}$$

In the second case, the value of $p_0$ is arbitrary and we must have

$$\begin{aligned} L_5(p_0, p_1, p_2, 1) &= 0 \\ L_5(p_1, p_2, 1) &= 0 \end{aligned}$$

But

$$L_5(p_0, p_1, p_2, 1) = L_5(p_1, p_2, 1) - p_0 L_4(p_1, p_2, 1)$$

so that $p_1$ and $p_2$ are determined subject to

$$\begin{aligned} L_4(p_1, p_2, 1) &= 0 \\ L_5(p_1, p_2, 1) &= 0 \end{aligned} \quad \text{or, equivalently} \quad \begin{aligned} p_1 p_2 &= \tfrac{1}{10} \\ p_1 + p_2 &= \tfrac{4}{5} \end{aligned}$$

As a consequence,

$$p_0 = \text{free parameter} \quad \begin{aligned} p_1 &= \tfrac{1}{10}(4 - \sqrt{6}) \\ p_2 &= \tfrac{1}{10}(4 + \sqrt{6}) \\ p_3 &= 1 \end{aligned} \tag{12.75}$$

In either of these two cases, it can be shown that

$$q_i = \tfrac{1}{2} p_i^2 \quad \text{for} \quad i = 1, 2, 3 \tag{12.76}$$

For this purpose, write

$$\begin{aligned} q_1 &= \tfrac{1}{2} p_1^2 \frac{(p_1 - p_0) L_5(p_2, p_3)}{p_1^2 L_4(p_0, p_2, p_3)} \\ &= \tfrac{1}{2} p_1^2 \frac{L_6(p_0, p_2, p_3) - L_6(p_1, p_2, p_3)}{L_6(p_0, p_2, p_3) - L_6(p_0, p_1, p_2, p_3)} \end{aligned}$$

and the conclusion follows from the second of Eqs. (12.73) since

$$\begin{aligned} L_6(p_0, p_1, p_2, p_3) &= L_6(p_1, p_2, p_3) - p_0 L_5(p_1, p_2, p_3) \\ &= L_6(p_1, p_2, p_3) \end{aligned}$$

Sect. 12.5]    Fifth-Order R-K-N Algorithms    597

The arguments for $q_2$ and $q_3$ are the same. Therefore, it follows that $q_i = \frac{1}{2} p_i^2$ is both necessary and sufficient for the condition equations to be consistent.

◇ **Problem 12–11**

For the fifth-order Runge-Kutta-Nyström algorithm, where the $p_i$'s are given by Eqs. (12.75), show that

$$b_0 = 0 \quad \text{and} \quad b_2, b_3 \neq 0$$

The $b_i$'s and the $q_i$'s are calculated from

$$b_3 = \frac{L_4(p_0, p_1, p_2)}{(p_3 - p_0)(p_3 - p_1)(p_3 - p_2)} \qquad (0,1,2,3)$$

$$q_i = \tfrac{1}{2} p_i^2 \quad \text{for} \quad i = 1, 2, 3$$

with $c_{21}$, $c_{31}$, and $c_{32}$ determined from

$$c_{21} = \frac{p_0 p_3 - \tfrac{1}{4}(3p_0 + p_3) + \tfrac{1}{5}}{6 b_2 (p_1 - p_0)(p_2 - p_3)}$$

$$c_{31} = \frac{p_0 p_2 - \tfrac{1}{4}(p_0 + p_2) + \tfrac{1}{10}}{6 b_3 (p_1 - p_0)(p_1 - p_2)} - \frac{b_2}{b_3} c_{21}$$

$$c_{32} = \frac{p_0 p_1 - \tfrac{1}{4}(p_0 + p_1) + \tfrac{1}{10}}{6 b_3 (p_2 - p_0)(p_2 - p_1)}$$

◇ **Problem 12–12**

Develop the fourth-order algorithm

$$\mathbf{x} = \mathbf{x}_0 + h \mathbf{y}_0 + h^2 (a_0 \mathbf{k}_0 + a_1 \mathbf{k}_1 + a_2 \mathbf{k}_2) + O(h^5)$$
$$\mathbf{y} = \mathbf{y}_0 + h (b_0 \mathbf{k}_0 + b_1 \mathbf{k}_1 + b_2 \mathbf{k}_2 + b_3 \mathbf{k}_3) + O(h^5)$$
$$\mathbf{k}_0 = \mathbf{f}(t_0, \mathbf{x}_0)$$
$$\mathbf{k}_1 = \mathbf{f}(t_0 + h p_1, \mathbf{x}_0 + h p_1 \mathbf{y}_0 + \tfrac{1}{2} h^2 p_1^2 \mathbf{k}_0)$$
$$\mathbf{k}_2 = \mathbf{f}[t_0 + h p_2, \mathbf{x}_0 + h p_2 \mathbf{y}_0 + \tfrac{1}{2} h^2 p_2^2 \mathbf{k}_0 + h^2 c_{21} (\mathbf{k}_1 - \mathbf{k}_0)]$$
$$\mathbf{k}_3 = \mathbf{f}[t_0 + h, \mathbf{x}_0 + h \mathbf{y}_0 + h^2 (a_0 \mathbf{k}_0 + a_1 \mathbf{k}_1 + a_2 \mathbf{k}_2)]$$

The coefficients are obtained as the following functions of the free parameters $p_1$ and $p_2$:

$$b_3 = \frac{L_4(p_0, p_1, p_2)}{(p_3 - p_0)(p_3 - p_1)(p_3 - p_2)} \qquad (0,1,2,3)$$

where

$$p_0 = 0 \quad \text{and} \quad p_3 = 1$$

and

$$c_{21} = \frac{\tfrac{1}{4} - b_3}{6 b_2 p_1} \qquad c_{3i} = a_i = (1 - p_i) b_i \quad \text{for} \quad i = 0, 1, 2$$

NOTE: There are four stages for the first step only since $\mathbf{k}_3$ is the same as the $\mathbf{k}_0$ for the following step.

## 12.6 Sixth-Order R-K-N Algorithms

The task of developing the condition equations for higher-order methods is quite laborious since the number of equations increases dramatically with increasing order. As we have seen, 13 equations [not counting those for the $a_i$'s since we always have $a_i = (1 - p_i)b_i$] evolve for $m = 5$. For $m = 6$, ten more are added; for $m = 7$, an additional 20; and for $m = 8$, the increase is 36 making 79 in all. For a ninth-order set, the number is 151.

To develop a sixth-order algorithm, we require the fifth derivative of $f^i$ and must, for this purpose, define the following ten new vectors in addition to those of Eqs. (12.28) and (12.61):

$$\alpha_5^i \equiv f_{jk\ell mn}^i y^j y^k y^\ell y^m y^n \qquad \varsigma_5^i \equiv f_{jk}^i \beta_2^j y^k$$
$$\beta_5^i \equiv f_{jk\ell m}^i \alpha_0^j y^k y^\ell y^m \qquad \eta_5^i \equiv f_{jk}^i \alpha_0^j \alpha_1^k$$
$$\gamma_5^i \equiv f_{jk\ell}^i \alpha_1^j y^k y^\ell \qquad \iota_5^i \equiv f_j^i \alpha_3^j \qquad (12.77)$$
$$\delta_5^i \equiv f_{jk\ell}^i \alpha_0^j \alpha_0^k y^\ell \qquad \kappa_5^i \equiv f_j^i \beta_3^j$$
$$\varepsilon_5^i \equiv f_{jk}^i \alpha_2^j y^k \qquad \lambda_5^i \equiv f_j^i \gamma_3^j$$

The fifth derivative may then be written as

$$\frac{d^5 f^i}{dt^5} = \alpha_5^i + 10\beta_5^i + 10\gamma_5^i + 15\delta_5^i + 5\varepsilon_5^i + 5\varsigma_5^i + 10\eta_5^i + \iota_5^i + 3\kappa_5^i + \lambda_5^i \quad (12.78)$$

Although 23 condition equations result for the sixth-order case, they may be reduced immediately to 13 under the assumption that

$$p_0 = 0 \quad \text{and} \quad q_i = \tfrac{1}{2} p_i^2 \quad \text{for} \quad i = 1, 2, 3, 4 \quad (12.79)$$

as suggested by the analysis of the fifth-order algorithm. The six ($\alpha$) equations are handled as in the lower-order cases so that there remain seven equations to determine the six parameters $c_{ij}$. They are

$$(\gamma) \quad \begin{bmatrix} 1 & 1 & 1 \\ p_2 & p_3 & p_4 \\ p_2^2 & p_3^2 & p_4^2 \end{bmatrix} \begin{bmatrix} A_2^1 b_2 \\ A_3^1 b_3 \\ A_4^1 b_4 \end{bmatrix} = \frac{1}{3!} \begin{bmatrix} \tfrac{1}{4} \\ \tfrac{1}{5} \\ \tfrac{1}{6} \end{bmatrix}$$

$$(\varepsilon) \quad \begin{bmatrix} 1 & 1 & 1 \\ p_2 & p_3 & p_4 \end{bmatrix} \begin{bmatrix} A_2^2 b_2 \\ A_3^2 b_3 \\ A_4^2 b_4 \end{bmatrix} = \frac{1}{4!} \begin{bmatrix} \tfrac{1}{5} \\ \tfrac{1}{6} \end{bmatrix}$$

$$(\iota) \quad A_2^3 b_2 + A_3^3 b_3 + A_4^3 b_4 = \frac{1}{6!}$$

$$(\lambda) \quad A_3^{33} b_3 + A_4^{33} b_4 = \frac{1}{6!}$$

where we have defined

$$A_i^k = \frac{1}{k!} \sum_{j=1}^{i-1} c_{ij} p_j^k \quad i = 2,\ldots, n-1; \quad k = 1, 2, \ldots \qquad (12.80)$$

and

$$A_3^{33} = c_{32} A_2^1 \qquad A_4^{33} = c_{42} A_2^1 + c_{43} A_3^1 \qquad (12.81)$$

### Reformulation of the Condition Equations

In the above format, the condition equations do not readily yield to further analysis. However, an alternate form of these equations is possible if we define†

$$H_i^k = p_i \sum_{j=i+1}^{n-1} p_j^k b_j c_{ji} \quad i = 1, 2, \ldots, n-2; \quad k = 0, 1, \ldots \qquad (12.82)$$

Then, the condition equations can be written as

$$(\gamma^0) \qquad H_1^0 + H_2^0 + H_3^0 = \frac{1}{4 \cdot 3!}$$

$$(\gamma^1) \qquad H_1^1 + H_2^1 + H_3^1 = \frac{1}{5 \cdot 3!}$$

$$(\gamma^2) \qquad H_1^2 + H_2^2 + H_3^2 = \frac{1}{6 \cdot 3!}$$

$$(\varepsilon^0) \qquad \frac{1}{2!}(p_1 H_1^0 + p_2 H_2^0 + p_3 H_3^0) = \frac{1}{5 \cdot 4!}$$

$$(\varepsilon^1) \qquad \frac{1}{2!}(p_1 H_1^1 + p_2 H_2^1 + p_3 H_3^1) = \frac{1}{6 \cdot 4!}$$

$$(\iota^0) \qquad \frac{1}{3!}(p_1^2 H_1^0 + p_2^2 H_2^0 + p_3^2 H_3^0) = \frac{1}{6 \cdot 5!}$$

$$(\lambda^0) \qquad \frac{3!}{3!}\left(\frac{1}{p_2} A_2^1 H_2^0 + \frac{1}{p_3} A_3^1 H_3^0\right) = \frac{1}{6 \cdot 5!}$$

It is clear from the second form of the condition equations that $(\gamma^0)$, $(\varepsilon^0)$, $(\iota^0)$, and $(\lambda^0)$ provide four equations which must be satisfied by the three quantities $H_1^0$, $H_2^0$, and $H_3^0$. Specifically, we have

$$\begin{bmatrix} 1 & 1 & 1 \\ p_1 & p_2 & p_3 \\ p_1^2 & p_2^2 & p_3^2 \\ 0 & A_2^1/p_2 & A_3^1/p_3 \end{bmatrix} \begin{bmatrix} H_1^0 \\ H_2^0 \\ H_3^0 \end{bmatrix} = \frac{1}{4!} \begin{bmatrix} 1 \\ \frac{2}{5} \\ \frac{1}{5} \\ \frac{1}{30} \end{bmatrix}$$

---

† Successful resolution of the condition equations depends strongly on an appropriate arrangement of the terms. The grouping of the variables here, in particular the equations expressed in terms of the $H_i^k$'s, is the author's own.

which can be valid if and only if the four-dimensional augmented determinant of coefficients vanishes identically. This determinant

$$D = \frac{1}{4!}\begin{vmatrix} 1 & 1 & 1 & 1 \\ p_1 & p_2 & p_3 & \frac{2}{5} \\ p_1^2 & p_2^2 & p_3^2 & \frac{1}{5} \\ 0 & A_2^1/p_2 & A_3^1/p_3 & \frac{1}{30} \end{vmatrix}$$

is readily reduced to three dimensions by the following two elementary row operations:

1. Row 3 $- p_1 \times$ Row 2 $\rightarrow$ Row 3.
2. Row 2 $- p_1 \times$ Row 1 $\rightarrow$ Row 2.

The result is

$$D = \frac{1}{144 p_2 b_2 p_3 b_3}\begin{vmatrix} p_2(p_2-p_1)b_2 & p_3(p_3-p_1)b_3 & \frac{2}{5}-p_1 \\ p_2^2(p_2-p_1)b_2 & p_3^2(p_3-p_1)b_3 & \frac{1}{5}-\frac{2}{5}p_1 \\ 6A_2^1 b_2 & 6A_3^1 b_3 & \frac{1}{5} \end{vmatrix}$$

after multiplying the first and second columns by the factors $p_2 b_2$ and $p_3 b_3$, respectively.

Next, we obtain $A_i^1$ from Eqs. ($\alpha$) subject to the usual constraint equation $L_6(p_1, p_2, p_3, p_4) = 0$. First, we find

$$p_2(p_2 - p_1)b_2 = \frac{L_5(p_1, p_3, p_4)}{(p_2 - p_3)(p_2 - p_4)}$$
$$p_3(p_3 - p_1)b_3 = \frac{L_5(p_1, p_2, p_4)}{(p_3 - p_2)(p_3 - p_4)}$$
(12.83)

and, from the results of Prob. 12–4,

$$L_6(p_1, p_2, p_3, p_4) = L_6(p_1, p_3, p_4) - p_2 L_5(p_1, p_3, p_4) = 0$$
$$= L_6(p_1, p_2, p_4) - p_3 L_5(p_1, p_2, p_4) = 0$$

we also have

$$p_2^2(p_2 - p_1)b_2 = \frac{L_6(p_1, p_3, p_4)}{(p_2 - p_3)(p_2 - p_4)}$$
$$p_3^2(p_3 - p_1)b_3 = \frac{L_6(p_1, p_2, p_4)}{(p_3 - p_2)(p_3 - p_4)}$$
(12.84)

Then, from Eqs. ($\gamma$) we obtain

$$6A_2^1 b_2 = \frac{L_6(p_3, p_4)}{(p_2 - p_3)(p_2 - p_4)}$$
$$6A_3^1 b_3 = \frac{L_6(p_2, p_4)}{(p_3 - p_2)(p_3 - p_4)}$$
(12.85)

## Sect. 12.6]  Sixth-Order R-K-N Algorithms

These are substituted for the elements of the determinant and all common factors removed to give

$$D = \frac{1}{144(p_2 - p_3)V_3(p_2)p_2 b_2 p_3 b_3} \begin{vmatrix} L_5(p_1,p_3,p_4) & L_5(p_1,p_2,p_4) & \frac{2}{5} - p_1 \\ L_6(p_1,p_3,p_4) & L_6(p_1,p_2,p_4) & \frac{1}{5} - \frac{2}{5}p_1 \\ L_6(p_3,p_4) & L_6(p_2,p_4) & \frac{1}{5} \end{vmatrix}$$

The two column operations
3. Column 1 − Column 2 → Column 1 and factor $(p_2 - p_3)$.
4. Column 2 + $p_2$ × Column 1 → Column 2.

are next performed to obtain

$$D = \frac{1}{144 V_3(p_2)p_2 b_2 p_3 b_3} \begin{vmatrix} L_4(p_1,p_4) & L_5(p_1,p_4) & \frac{2}{5} - p_1 \\ L_5(p_1,p_4) & L_6(p_1,p_4) & \frac{1}{5} - \frac{2}{5}p_1 \\ L_5(p_4) & L_6(p_4) & \frac{1}{5} \end{vmatrix}$$

Finally, the two row operations
5. Row 2 − Row 3 → Row 2 and factor $p_1$.
6. Row 1 − Row 2 → Row 1 and factor $p_1$.

produce a determinant

$$D = \frac{p_1^2}{144 V_3(p_2)p_2 b_2 p_3 b_3} \begin{vmatrix} L_3(p_4) & L_4(p_4) & 1 \\ L_4(p_4) & L_5(p_4) & \frac{2}{5} \\ L_5(p_4) & L_6(p_4) & \frac{1}{5} \end{vmatrix}$$

which depends only on $p_4$. Hence,

$$D = \frac{p_1^2 (p_4 - 1)^2}{720^2 V_3(p_2)p_2 b_2 p_3 b_3}$$

so that the residue of Eq. ($\lambda$) is simply

$$\text{Residue}(\lambda) = \frac{p_1^2 (p_4 - 1)^2}{720^2 V_3(p_1) V_3(p_2) p_2 b_2 p_3 b_3} \quad (12.86)$$

Thus, the condition equations can be consistent only if either $p_1 = 0$ or $p_4 = 1$. The first option would require all of the $p_i$'s to be zero. For suppose that $p_1 = 0$; then, since

$$A_2^1 = \tfrac{1}{6}(p_2 - p_1)p_2^2 \frac{L_6(p_3,p_4)}{L_6(p_1,p_3,p_4)} \quad (12.87)$$

as obtained from the ratio of Eqs. (12.85) and (12.84), we have

$$A_2^1 = \tfrac{1}{6} p_2^3 \quad (12.88)$$

But, from the definition $A_2^1 = c_{21} p_1$, it would then follow that $p_2$ is also zero. A similar argument would lead to the conclusion $p_3 = p_4 = 0$ since it can be shown that

$$A_i^1 = \tfrac{1}{6} p_i^3 \quad \text{for} \quad i = 2,3,4 \quad \text{provided that} \quad p_1 = 0$$

Therefore, a necessary and sufficient condition for the consistency of the condition equations is that

$$p_4 = 1 \quad \text{and} \quad L_6(p_1, p_2, p_3, 1) = 0 \qquad (12.89)$$

### Solving the Condition Equations

The solution of the ($\alpha$) equations may be written compactly as

$$b_4 = \frac{L_5(p_0, p_1, p_2, p_3)}{(p_4 - p_0)(p_4 - p_1)(p_4 - p_2)(p_4 - p_3)} \qquad (0,1,2,3,4) \quad (12.90)$$

remembering, of course, that $p_0 = 0$ and $p_4 = 1$. The $q_i$'s are calculated from Eq. (12.79) and, as usual, we also have $a_i = (1 - p_i)b_i$.

To determine the $c_{ij}$'s, we note that Eq. ($\lambda$) may be discarded since its residue vanishes identically when $p_4 = 1$ as we have already demonstrated. Therefore, Eqs. ($\gamma^0$), ($\varepsilon^0$), and ($\iota^0$) provide unique solutions for $H_1^0$, $H_2^0$, and $H_3^0$. In compact form,

$$H_1^0 = \frac{\frac{1}{2}p_2 p_3 - \frac{1}{5}(p_2 + p_3) + \frac{1}{10}}{12(p_2 - p_1)(p_3 - p_1)} \qquad (1,2,3) \quad (12.91)$$

and also, from Eqs. (12.85), we have

$$A_2^1 b_2 = \frac{\frac{1}{2}p_3 - \frac{1}{3}}{60(p_3 - p_2)(1 - p_2)} \quad \text{and} \quad A_3^1 b_3 = \frac{\frac{1}{2}p_2 - \frac{1}{3}}{60(p_2 - p_3)(1 - p_3)}$$

Now, according to the definitions (12.80) and (12.82),

$$A_2^1 = p_1 c_{21} \qquad H_1^0 = p_1(b_2 c_{21} + b_3 c_{31} + b_4 c_{41})$$
$$A_3^1 = p_1 c_{31} + p_2 c_{32} \qquad H_2^0 = p_2(b_3 c_{32} + b_4 c_{42})$$
$$A_4^1 = p_1 c_{41} + p_2 c_{42} + p_3 c_{43} \qquad H_3^0 = p_3 b_4 c_{43}$$

which can be rewritten as

$$p_1 b_2 c_{21} = A_2^1 b_2 \qquad p_1 b_4 c_{41} = H_1^0 - A_2^1 b_2 - A_3^1 b_3 + p_2 b_3 c_{32}$$
$$p_1 b_3 c_{31} = A_3^1 b_3 - p_2 b_3 c_{32} \qquad p_2 b_4 c_{42} = H_2^0 - p_2 b_3 c_{32}$$
$$p_3 b_4 c_{43} = H_3^0$$

so that all of the $c_{ij}$'s are expressed in terms of $c_{32}$. Finally, Eq. ($\varepsilon^1$), written in the expanded form

$$p_1 p_2 (p_1 b_2 c_{21}) + p_1 p_3 (p_1 b_3 c_{31}) + p_2 p_3 (p_2 b_3 c_{32})$$
$$+ p_1 (p_1 b_4 c_{41}) + p_2 (p_2 b_4 c_{42}) + p_3 (p_3 b_4 c_{43}) = \tfrac{1}{72}$$

is used to determine $c_{32}$.

The equations needed to calculate the $c_{ij}$'s are summarized as

$$p_1 b_2 c_{21} = \frac{3p_3 - 2}{360(1 - p_2)(p_3 - p_2)}$$
$$p_1 b_3 c_{31} = \frac{3p_1 p_3 - 3p_2^2 - 2p_1 + 3p_2 - p_3}{360(1 - p_3)(p_2 - p_1)(p_3 - p_2)} \quad (12.92)$$
$$p_2 b_3 c_{32} = \frac{3p_1 - 1}{360(1 - p_3)(p_1 - p_2)}$$

and

$$p_1 b_4 c_{41} = H_1^0 - p_1 b_2 c_{21} - p_1 b_3 c_{31}$$
$$p_2 b_4 c_{42} = H_2^0 - p_2 b_3 c_{32} \quad (12.93)$$
$$p_3 b_4 c_{43} = H_3^0$$

where the $H_i^0$'s are determined from Eqs. (12.91).

◊ **Problem 12–13**
Derive the symmetric sixth-order algorithm

$$\mathbf{x} = \mathbf{x}_0 + h\mathbf{y}_0 + \tfrac{1}{90}h^2(7\mathbf{k}_0 + 24\mathbf{k}_1 + 6\mathbf{k}_2 + 8\mathbf{k}_3) + O(h^7)$$
$$\mathbf{y} = \mathbf{y}_0 + \tfrac{1}{90}h(7\mathbf{k}_0 + 32\mathbf{k}_1 + 12\mathbf{k}_2 + 32\mathbf{k}_3 + 7\mathbf{k}_4) + O(h^7)$$

where

$$\mathbf{k}_0 = \mathbf{f}(t_0, \mathbf{x}_0)$$
$$\mathbf{k}_1 = \mathbf{f}(t_0 + \tfrac{1}{4}h, \mathbf{x}_0 + \tfrac{1}{4}h\mathbf{y}_0 + \tfrac{1}{32}h^2\mathbf{k}_0)$$
$$\mathbf{k}_2 = \mathbf{f}[t_0 + \tfrac{1}{2}h, \mathbf{x}_0 + \tfrac{1}{2}h\mathbf{y}_0 - \tfrac{1}{24}h^2(\mathbf{k}_0 - 4\mathbf{k}_1)]$$
$$\mathbf{k}_3 = \mathbf{f}[t_0 + \tfrac{3}{4}h, \mathbf{x}_0 + \tfrac{3}{4}h\mathbf{y}_0 + \tfrac{1}{32}h^2(3\mathbf{k}_0 + 4\mathbf{k}_1 + 2\mathbf{k}_2)]$$
$$\mathbf{k}_4 = \mathbf{f}[t_0 + h, \mathbf{x}_0 + h\mathbf{y}_0 + \tfrac{1}{14}h^2(6\mathbf{k}_1 - \mathbf{k}_2 + 2\mathbf{k}_3)]$$

*Julius Albrecht* 1955

NOTE: In his paper, Albrecht did not provide general solutions of the condition equations. He apparently tried $p_0 = 0$, $p_1 = \tfrac{1}{4}$, $p_2 = \tfrac{1}{2}$, $p_3 = \tfrac{3}{4}$, $p_4 = 1$ and found that this choice led to a unique solution of these equations.

## 12.7 Seventh-Order R-K-N Algorithms

Erwin Fehlberg[†] derived the condition equations through order nine. The derivation is straightforward but tedious and, at this point, the reader should be capable of producing them for himself. Therefore, without further elaboration, we assume the condition equations and concentrate solely on their solution.

---

† "Classical Eighth- and Lower-Order Runge-Kutta-Nyström Formulas with Stepsize Control for Special Second-Order Differential Equations," NASA TR R-381, March 1972.

A seventh-order, six-stage Runge-Kutta-Nyström algorithm does not exist. The author demonstrated this in his paper cited in the introduction to this chapter. It turns out that there are 15 equations to be satisfied by the ten $c_{ij}$ parameters. However, if the number of stages is increased to seven, the number of parameters increases to 15 but the number of condition equations remains unchanged. Furthermore, the set of parameters in the $(\alpha)$ equations is also expanded to include $p_6$ and $b_6$ so that no constraint condition is required.

With the same assumptions as for the sixth-order method, i.e.,

$$p_0 = 0 \quad \text{and} \quad q_i = \tfrac{1}{2} p_i^2 \quad \text{for} \quad i = 1, 2, \ldots, n-1 \tag{12.94}$$

the condition equations for this case are

$$b_0 = 1 - b_1 - b_2 - b_3 - b_4 - b_5 - b_6$$

where

$$(\alpha) \quad \begin{bmatrix} 1 & 1 & 1 & 1 & 1 & 1 \\ p_1 & p_2 & p_3 & p_4 & p_5 & p_6 \\ p_1^2 & p_2^2 & p_3^2 & p_4^2 & p_5^2 & p_6^2 \\ p_1^3 & p_2^3 & p_3^3 & p_4^3 & p_5^3 & p_6^3 \\ p_1^4 & p_2^4 & p_3^4 & p_4^4 & p_5^4 & p_6^4 \\ p_1^5 & p_2^5 & p_3^5 & p_4^5 & p_5^5 & p_6^5 \end{bmatrix} \begin{bmatrix} p_1 b_1 \\ p_2 b_2 \\ p_3 b_3 \\ p_4 b_4 \\ p_5 b_5 \\ p_6 b_6 \end{bmatrix} = \begin{bmatrix} \tfrac{1}{2} \\ \tfrac{1}{3} \\ \tfrac{1}{4} \\ \tfrac{1}{5} \\ \tfrac{1}{6} \\ \tfrac{1}{7} \end{bmatrix}$$

$$(\gamma) \quad \begin{bmatrix} 1 & 1 & 1 & 1 & 1 \\ p_2 & p_3 & p_4 & p_5 & p_6 \\ p_2^2 & p_3^2 & p_4^2 & p_5^2 & p_6^2 \\ p_2^3 & p_3^3 & p_4^3 & p_5^3 & p_6^3 \end{bmatrix} \begin{bmatrix} A_2^1 b_2 \\ A_3^1 b_3 \\ A_4^1 b_4 \\ A_5^1 b_5 \\ A_6^1 b_6 \end{bmatrix} = \frac{1}{3!} \begin{bmatrix} \tfrac{1}{4} \\ \tfrac{1}{5} \\ \tfrac{1}{6} \\ \tfrac{1}{7} \end{bmatrix}$$

$$(\varepsilon) \quad \begin{bmatrix} 1 & 1 & 1 & 1 & 1 \\ p_2 & p_3 & p_4 & p_5 & p_6 \\ p_2^2 & p_3^2 & p_4^2 & p_5^2 & p_6^2 \end{bmatrix} \begin{bmatrix} A_2^2 b_2 \\ A_3^2 b_3 \\ A_4^2 b_4 \\ A_5^2 b_5 \\ A_6^2 b_6 \end{bmatrix} = \frac{1}{4!} \begin{bmatrix} \tfrac{1}{5} \\ \tfrac{1}{6} \\ \tfrac{1}{7} \end{bmatrix}$$

$$(\iota) \quad \begin{bmatrix} 1 & 1 & 1 & 1 & 1 \\ p_2 & p_3 & p_4 & p_5 & p_6 \end{bmatrix} \begin{bmatrix} A_2^3 b_2 \\ A_3^3 b_3 \\ A_4^3 b_4 \\ A_5^3 b_5 \\ A_6^3 b_6 \end{bmatrix} = \frac{1}{5!} \begin{bmatrix} \tfrac{1}{6} \\ \tfrac{1}{7} \end{bmatrix}$$

Sect. 12.7]  Seventh-Order R-K-N Algorithms  605

$$(\lambda) \quad \begin{bmatrix} 1 & 1 & 1 & 1 \\ p_3 & p_4 & p_5 & p_6 \end{bmatrix} \begin{bmatrix} A_3^{33} b_3 \\ A_4^{33} b_4 \\ A_5^{33} b_5 \\ A_6^{33} b_6 \end{bmatrix} = \frac{1}{5!} \begin{bmatrix} \frac{1}{6} \\ \frac{1}{7} \end{bmatrix}$$

which are the same as for the sixth order but expanded in an obvious manner. However, with the increase in order, four more equations are required which we write as

$$(\pi) \quad A_2^1 A_2^1 b_2 + A_3^1 A_3^1 b_3 + A_4^1 A_4^1 b_4 + A_5^1 A_5^1 b_5 + A_6^1 A_6^1 b_6 = \frac{1}{3!\,3!\cdot 7}$$

$$(\sigma) \quad A_2^4 b_2 + A_3^4 b_3 + A_4^4 b_4 + A_5^4 b_5 + A_6^4 b_6 = \frac{1}{6!\cdot 7}$$

$$(\phi) \quad A_3^{43} b_3 + A_4^{43} b_4 + A_5^{43} b_5 + A_6^{43} b_6 = \frac{1}{6!\cdot 7}$$

$$(\psi) \quad A_3^{44} b_3 + A_4^{44} b_4 + A_5^{44} b_5 + A_6^{44} b_6 = \frac{1}{6!\cdot 7}$$

where we have defined

$$A_i^{43} = \frac{3!}{4!} \sum_{j=2}^{i-1} c_{ij} p_j A_j^1$$

$$A_i^{44} = \frac{4!}{4!} \sum_{j=2}^{i-1} c_{ij} A_j^2$$

for $i = 3, 4, \ldots, n-1$. [Refer to Eqs. (12.80) and (12.81) for the other definitions.]

As in the sixth-order case, it is convenient to formulate certain of these equations in terms of the functions $H_i^k$ defined in Eq. (12.82). We have

$$(\gamma^i) \quad \frac{1}{1!} \sum_{j=1}^{n-2} H_j^i = \frac{1}{3!(i+4)} \qquad (\sigma^i) \quad \frac{1}{4!} \sum_{j=1}^{n-2} p_j^3 H_j^i = \frac{1}{6!(i+7)}$$

$$(\varepsilon^i) \quad \frac{1}{2!} \sum_{j=1}^{n-2} p_j H_j^i = \frac{1}{4!(i+5)} \qquad (\phi^i) \quad \frac{3!}{4!} \sum_{j=2}^{n-2} A_j^1 H_j^i = \frac{1}{6!(i+7)}$$

$$(\iota^i) \quad \frac{1}{3!} \sum_{j=1}^{n-2} p_j^2 H_j^i = \frac{1}{5!(i+6)} \qquad (\psi^i) \quad \frac{4!}{4!} \sum_{j=2}^{n-2} \frac{1}{p_j} A_j^2 H_j^i = \frac{1}{6!(i+7)}$$

$$(\lambda^i) \quad \frac{3!}{3!} \sum_{j=2}^{n-2} \frac{1}{p_j} A_j^1 H_j^i = \frac{1}{5!(i+6)}$$

**Solving the Condition Equations**

Although we have a sufficient number of equations to determine the $c_{ij}$'s, the task is far from elementary primarily because Eq. ($\pi$) involves products of these parameters. To circumvent this difficulty, we recall from the argument in the previous section leading to Eq. (12.88) that $A_i^1$ will equal $\frac{1}{6}p_i^3$ if $p_1 = 0$. With this form for $A_i^1$, Eq. ($\pi$) would then be identical to Eq. ($\gamma^3$) and could, therefore, be discarded. Even though this is not an appropriate choice for $p_1$, the same effect is possible by requiring that $b_1$ be zero. The ($\alpha$) equations can then be consistent if and only if

$$L_7(p_2, p_3, p_4, p_5, p_6) = 0 \tag{12.95}$$

Now, if we assume that

$$A_i^1 = \sum_{j=1}^{i-1} c_{ij} p_j = \tfrac{1}{6} p_i^3 \quad \text{for} \quad i = 2, 3, \ldots, n-1 \tag{12.96}$$

then Eqs. ($\gamma$) and ($\pi$) are identical to certain of the ($\alpha$) equations and can be discarded. However, by subtracting equations ($\iota^i$) and ($\lambda^i$) we conclude that we must also have

$$H_1^0 = p_1 \sum_{j=2}^{n-1} b_j c_{j1} = 0 \quad \text{and} \quad H_1^1 = p_1 \sum_{j=2}^{n-1} p_j b_j c_{j1} = 0 \tag{12.97}$$

as requirements. Equations ($\iota$) and ($\lambda$) will then be the same as will also Eqs. ($\sigma$) and ($\phi$). The total number of condition equations is now 14 to determine the 15 parameters $c_{ij}$.

It is also convenient, but clearly not necessary, to assume that

$$A_i^2 = \tfrac{1}{2} \sum_{j=1}^{i-1} c_{ij} p_j^2 = \tfrac{1}{24} p_i^4 \quad \text{for} \quad i = 2, 3, \ldots, n-1 \tag{12.98}$$

so that Eq. ($\varepsilon$) will then be identical to the last three of the ($\alpha$) equations and Eq. ($\psi$) will be identical to ($\phi$). However, in so doing, it follows from the definitions of $A_2^1$ and $A_2^2$ that

$$p_1 = \tfrac{1}{2} p_2 \quad \text{and} \quad c_{21} = \tfrac{1}{3} p_2^2 \tag{12.99}$$

There are now 13 condition equations to determine the 14 parameters $c_{ij}$. (Remember that $c_{21}$ has already been established.) They are Eqs. (12.96) and (12.98) for $i = 3, 4, 5, 6$ together with Eqs. (12.97), ($\lambda$), and ($\phi$). These equations can be written as follows:

$$b_0 = 1 - b_2 - b_3 - b_4 - b_5 - b_6 \quad \text{and} \quad b_1 = 0 \tag{12.100}$$

where $b_2, \ldots, b_6$ are determined from

$$b_2 = \frac{L_6(p_3, p_4, p_5, p_6)}{p_2(p_2 - p_3)(p_2 - p_4)(p_2 - p_5)(p_2 - p_6)} \quad (2, 3, 4, 5, 6) \tag{12.101}$$

## Sect. 12.7]  Seventh-Order R-K-N Algorithms

Then the parameters $c_{31}$, $c_{32}$, $c_{42}$, $c_{43}$, $c_{51}$, and $c_{61}$ are obtained from

$$\begin{bmatrix} 1 & 1 \\ \frac{1}{2}p_2 & p_2 \end{bmatrix} \begin{bmatrix} \frac{1}{2}p_2 c_{31} \\ p_2 c_{32} \end{bmatrix} = \begin{bmatrix} \frac{1}{6}p_3^3 \\ \frac{1}{12}p_3^4 \end{bmatrix}$$

$$\begin{bmatrix} 1 & 1 \\ p_2 & p_3 \end{bmatrix} \begin{bmatrix} p_2 c_{42} \\ p_3 c_{43} \end{bmatrix} = \begin{bmatrix} \frac{1}{6}p_4^3 - \frac{1}{2}p_2 c_{41} \\ \frac{1}{12}p_4^4 - \frac{1}{4}p_2^2 c_{41} \end{bmatrix} \qquad (12.102)$$

$$\begin{bmatrix} 1 & 1 \\ p_5 & p_6 \end{bmatrix} \begin{bmatrix} b_5 c_{51} \\ b_6 c_{61} \end{bmatrix} = -\begin{bmatrix} 1 & 1 & 1 \\ p_2 & p_3 & p_4 \end{bmatrix} \begin{bmatrix} \frac{1}{3}b_2 p_2^2 \\ b_3 c_{31} \\ b_4 c_{41} \end{bmatrix}$$

where $c_{41}$ is to be regarded as a free parameter. Next, the intermediate parameters $A_5^{33}$ and $A_6^{33}$ are calculated from

$$\begin{bmatrix} 1 & 1 \\ p_5 & p_6 \end{bmatrix} \begin{bmatrix} b_5 A_5^{33} \\ b_6 A_6^{33} \end{bmatrix} = \frac{1}{120} \begin{bmatrix} \frac{1}{6} \\ \frac{1}{7} \end{bmatrix} - \begin{bmatrix} 1 & 1 \\ p_3 & p_4 \end{bmatrix} \begin{bmatrix} b_3 A_3^{33} \\ b_4 A_4^{33} \end{bmatrix}$$

where

$$A_3^{33} = \tfrac{1}{6} c_{32} p_2^3$$
$$A_4^{33} = \tfrac{1}{6}(c_{42} p_2^3 + c_{43} p_3^3)$$

Then, $c_{52}$, $c_{53}$, and $c_{54}$ follow from

$$\begin{bmatrix} 1 & 1 & 1 \\ p_2 & p_3 & p_4 \\ p_2^2 & p_3^2 & p_4^2 \end{bmatrix} \begin{bmatrix} p_2 c_{52} \\ p_3 c_{53} \\ p_4 c_{54} \end{bmatrix} = \begin{bmatrix} \frac{1}{6}p_5^3 - \frac{1}{2}p_2 c_{51} \\ \frac{1}{12}p_5^4 - \frac{1}{4}p_2^2 c_{51} \\ 6 A_5^{33} \end{bmatrix} \qquad (12.103)$$

Finally, we calculate another set of intermediate parameters

$$A_3^{43} = \tfrac{1}{24} c_{32} p_2^4$$
$$A_4^{43} = \tfrac{1}{24}(c_{42} p_2^4 + c_{43} p_3^4)$$
$$A_5^{43} = \tfrac{1}{24}(c_{52} p_2^4 + c_{53} p_3^4 + c_{54} p_4^4)$$
$$b_6 A_6^{43} = \tfrac{1}{5040} - b_3 A_3^{43} - b_4 A_4^{43} - b_5 A_5^{43}$$

which are used to determine $c_{62}$, $c_{63}$, $c_{64}$, $c_{65}$ according to

$$\begin{bmatrix} 1 & 1 & 1 & 1 \\ p_2 & p_3 & p_4 & p_5 \\ p_2^2 & p_3^2 & p_4^2 & p_5^2 \\ p_2^3 & p_3^3 & p_4^3 & p_5^3 \end{bmatrix} \begin{bmatrix} p_2 c_{62} \\ p_3 c_{63} \\ p_4 c_{64} \\ p_5 c_{65} \end{bmatrix} = \begin{bmatrix} \frac{1}{6}p_6^3 - \frac{1}{2}p_2 c_{61} \\ \frac{1}{12}p_6^4 - \frac{1}{4}p_2^2 c_{61} \\ 6 A_6^{33} \\ 24 A_6^{43} \end{bmatrix} \qquad (12.104)$$

There are five free parameters—any four of the five quantities $p_2$, $p_3$, $p_4$, $p_5$, $p_6$, subject to the constraint (12.95), together with $c_{41}$.

## 12.8 Eighth-Order R-K-N Algorithms

For an eighth-order algorithm with eight stages the condition equations ($\alpha$), ($\gamma$), ($\varepsilon$), ($\iota$), and ($\lambda$) are expanded, from the corresponding ones for the seventh-order, in an obvious manner to yield 22 equations. The additional equations ($\pi$), ($\sigma$), ($\phi$), and ($\psi$), specific to the seventh-order, double in number. Finally, six new ones are required for the eighth-order giving a grand total of 36. Of these, 29 equations must be satisfied by the 21 parameters $c_{ij}$.

To cope with this large overdetermined set of condition equations, we will as before, assume that

$$b_1 = 0 \qquad A_i^1 = \tfrac{1}{6} p_i^3 \qquad A_i^2 = \tfrac{1}{24} p_i^4 \quad \text{for} \quad i = 2, 3, \ldots, 7$$

and, thereby, impose the constraints

$$L_8(p_2, p_3, p_4, p_5, p_6, p_7) = 0 \qquad p_1 = \tfrac{1}{2} p_2 \qquad c_{21} = \tfrac{1}{3} p_2^2$$

Again, Eqs. ($\gamma$) and ($\varepsilon$) are identical to certain of the ($\alpha$) equations which we record here for later use

$$(\alpha) \quad \begin{bmatrix} 1 & 1 & 1 & 1 & 1 & 1 \\ p_2 & p_3 & p_4 & p_5 & p_6 & p_7 \\ p_2^2 & p_3^2 & p_4^2 & p_5^2 & p_6^2 & p_7^2 \\ p_2^3 & p_3^3 & p_4^3 & p_5^3 & p_6^3 & p_7^3 \\ p_2^4 & p_3^4 & p_4^4 & p_5^4 & p_6^4 & p_7^4 \\ p_2^5 & p_3^5 & p_4^5 & p_5^5 & p_6^5 & p_7^5 \\ p_2^6 & p_3^6 & p_4^6 & p_5^6 & p_6^6 & p_7^6 \end{bmatrix} \begin{bmatrix} p_2 b_2 \\ p_3 b_3 \\ p_4 b_4 \\ p_5 b_5 \\ p_6 b_6 \\ p_7 b_7 \end{bmatrix} = \begin{bmatrix} \tfrac{1}{2} \\ \tfrac{1}{3} \\ \tfrac{1}{4} \\ \tfrac{1}{5} \\ \tfrac{1}{6} \\ \tfrac{1}{7} \\ \tfrac{1}{8} \end{bmatrix}$$

Therefore, they can be discarded. Furthermore, by subtracting Eqs. ($\iota^i$) and ($\lambda^i$) for $i = 0, 1, 2$ we find that, in addition to Eqs. (12.97), we have also†

$$H_1^2 = p_1 \sum_{j=2}^{n-1} p_j^2 b_j c_{j1} = 0 \tag{12.105}$$

As a consequence, Eq. ($\iota$) can be dropped from the set.

Next, we examine the equations ($\pi$), ($\sigma$), ($\phi$), and ($\psi$) which, for the eighth-order algorithm with eight stages, are written in matrix form

---

† The condition equations, expressed in terms of the $H_i^k$ functions, were given in the previous section in a form which is valid for all higher-order algorithms. They are dependent on the number of stages $n$.

## Sect. 12.8]  Eighth-Order R-K-N Algorithms  609

as

$$(\pi) \quad \begin{bmatrix} 1 & 1 & 1 & 1 & 1 & 1 \\ p_2 & p_3 & p_4 & p_5 & p_6 & p_7 \end{bmatrix} \begin{bmatrix} A_2^1 A_2^1 b_2 \\ A_3^1 A_3^1 b_3 \\ A_4^1 A_4^1 b_4 \\ A_5^1 A_5^1 b_5 \\ A_6^1 A_6^1 b_6 \\ A_7^1 A_7^1 b_7 \end{bmatrix} = \frac{1}{3!\,3!} \begin{bmatrix} \frac{1}{7} \\ \frac{1}{8} \end{bmatrix}$$

$$(\sigma) \quad \begin{bmatrix} 1 & 1 & 1 & 1 & 1 & 1 \\ p_2 & p_3 & p_4 & p_5 & p_6 & p_7 \end{bmatrix} \begin{bmatrix} A_2^4 b_2 \\ A_3^4 b_3 \\ A_4^4 b_4 \\ A_5^4 b_5 \\ A_6^4 b_6 \\ A_7^4 b_7 \end{bmatrix} = \frac{1}{6!} \begin{bmatrix} \frac{1}{7} \\ \frac{1}{8} \end{bmatrix}$$

$$(\phi) \quad \begin{bmatrix} 1 & 1 & 1 & 1 & 1 \\ p_3 & p_4 & p_5 & p_6 & p_7 \end{bmatrix} \begin{bmatrix} A_3^{43} b_3 \\ A_4^{43} b_4 \\ A_5^{43} b_5 \\ A_6^{43} b_6 \\ A_7^{43} b_7 \end{bmatrix} = \frac{1}{6!} \begin{bmatrix} \frac{1}{7} \\ \frac{1}{8} \end{bmatrix}$$

$$(\psi) \quad \begin{bmatrix} 1 & 1 & 1 & 1 & 1 \\ p_3 & p_4 & p_5 & p_6 & p_7 \end{bmatrix} \begin{bmatrix} A_3^{44} b_3 \\ A_4^{44} b_4 \\ A_5^{44} b_5 \\ A_6^{44} b_6 \\ A_7^{44} b_7 \end{bmatrix} = \frac{1}{6!} \begin{bmatrix} \frac{1}{7} \\ \frac{1}{8} \end{bmatrix}$$

We see that Eqs. ($\pi$) can also be discarded since they are identical to two of the ($\alpha$) equations. Further, Eqs. ($\sigma^i$), ($\phi^i$), and ($\psi^i$), expressed in terms of $H_j^k$ as given in the previous section, are identical so that only ($\phi$) need be retained.

There are six additional equations specific to the eighth-order case:

($e_1$)  $p_2 A_2^1 A_2^2 b_2 + p_3 A_3^1 A_3^2 b_3 + p_4 A_4^1 A_4^2 b_4 + \cdots + p_7 A_7^1 A_7^2 b_7 = \dfrac{1}{3!\,4!\cdot 8}$

($e_2$)  $p_2 A_2^5 b_2 + p_3 A_3^5 b_3 + p_4 A_4^5 b_4 + p_5 A_5^5 b_5 + p_6 A_6^5 b_6 + p_7 A_7^5 b_7 = \dfrac{1}{7!\cdot 8}$

($e_3$)  $p_3 A_3^{53} b_3 + p_4 A_4^{53} b_4 + p_5 A_5^{53} b_5 + p_6 A_6^{53} b_6 + p_7 A_7^{53} b_7 = \dfrac{1}{7!\cdot 8}$

($e_4$)  $p_3 A_3^{54} b_3 + p_4 A_4^{54} b_4 + p_5 A_5^{54} b_5 + p_6 A_6^{54} b_6 + p_7 A_7^{54} b_7 = \dfrac{1}{7!\cdot 8}$

($e_5$)  $p_3 A_3^{55} b_3 + p_4 A_4^{55} b_4 + p_5 A_5^{55} b_5 + p_6 A_6^{55} b_6 + p_7 A_7^{55} b_7 = \dfrac{1}{7!\cdot 8}$

**610**  Numerical Integration of Differential Equations  [Chap. 12]

$$(\mathbf{e_6}) \qquad p_4 A_4^{533} b_4 + p_5 A_5^{533} b_5 + p_6 A_6^{533} b_6 + p_7 A_7^{533} b_7 = \frac{1}{7! \cdot 8}$$

where we have defined, for $i = 3, 4, \ldots, n-1$,

$$A_i^{53} = \frac{3!}{5!} \sum_{j=2}^{i-1} c_{ij} p_j^2 A_j^1 \qquad A_i^{54} = \frac{4!}{5!} \sum_{j=2}^{i-1} c_{ij} p_j A_j^2 \qquad A_i^{55} = \frac{5!}{5!} \sum_{j=2}^{i-1} c_{ij} A_j^3$$

and

$$A_i^{533} = \frac{5!}{5!} \sum_{j=3}^{i-1} c_{ij} A_j^{33} \qquad \text{with} \qquad A_i^{33} = \frac{3!}{3!} \sum_{j=2}^{i-1} c_{ij} A_j^1$$

These can also be written in terms of $H_i^k$ functions as

$$(e_2^i) \qquad \frac{1}{5!} \sum_{j=1}^{n-2} p_j^4 H_j^i = \frac{1}{7!(i+8)} \qquad (e_5^i) \qquad \frac{5!}{5!} \sum_{j=2}^{n-2} \frac{1}{p_j} A_j^3 H_j^i = \frac{1}{7!(i+8)}$$

$$(e_3^i) \qquad \frac{3!}{5!} \sum_{j=2}^{n-2} p_j A_j^1 H_j^i = \frac{1}{7!(i+8)} \qquad (e_6^i) \qquad \frac{5!}{5!} \sum_{j=3}^{n-2} \frac{1}{p_j} A_j^{33} H_j^i = \frac{1}{7!(i+8)}$$

$$(e_4^i) \qquad \frac{4!}{5!} \sum_{j=2}^{n-2} A_j^2 H_j^i = \frac{1}{7!(i+8)}$$

We see that Eq. $(\mathbf{e_1})$ is equivalent to one of the $(\boldsymbol{\alpha})$ equations. Also, the three equations $(e_2^i)$, $(e_3^i)$, and $(e_4^i)$ are identical so that any two may be discarded. There are now 21 condition equations remaining to determine the 20 parameters $c_{ij}$—specifically, three from $(\boldsymbol{\lambda})$, two from $(\boldsymbol{\phi})$, one each from Eqs. $(\mathbf{e_4})$, $(\mathbf{e_5})$, and $(\mathbf{e_6})$, three from $H_1^i = 0$, and five each from the assumptions $A_i^1 = \frac{1}{6} p_i^3$ and $A_i^2 = \frac{1}{24} p_i^4$.

**Eliminating Equations (w) and (z)**

Two of the condition equations, $(\mathbf{e_5})$ and $(\mathbf{e_6})$, still involve products of the $c_{ij}$'s. It happens that they too may be discarded by the simple expediency of setting $p_7 = 1$. To show this, we form the following five equations obtained by adding Eqs. $(\alpha^i)$ and $(\alpha^{i+2})$ and subtracting two times Eq. $(\alpha^{i+1})$ for $i = 1, 2, \ldots, 5$:

$$\begin{bmatrix} 1 & 1 & 1 & 1 & 1 & 1 \\ p_2 & p_3 & p_4 & p_5 & p_6 & p_7 \\ p_2^2 & p_3^2 & p_4^2 & p_5^2 & p_6^2 & p_7^2 \\ p_2^3 & p_3^3 & p_4^3 & p_5^3 & p_6^3 & p_7^3 \\ p_2^4 & p_3^4 & p_4^4 & p_5^4 & p_6^4 & p_7^4 \end{bmatrix} \begin{bmatrix} p_2(1-p_2)^2 \\ p_3(1-p_3)^2 \\ p_4(1-p_4)^2 \\ p_5(1-p_5)^2 \\ p_6(1-p_6)^2 \\ p_7(1-p_7)^2 \end{bmatrix} = \begin{bmatrix} \frac{1}{12} \\ \frac{1}{30} \\ \frac{1}{60} \\ \frac{1}{105} \\ \frac{1}{168} \end{bmatrix}$$

Sect. 12.8]     Eighth-Order R-K-N Algorithms     611

When these are compared to Eqs. $(\gamma^0)$, $(\varepsilon^0)$, $(\iota^i)$, $(\sigma^i)$, and $(e_2^i)$, and we recall that $H_1^0 = 0$, it follows that

$$H_j^0 = \tfrac{1}{2} p_j (1 - p_j)^2 b_j \quad \text{for} \quad j = 2, 3, \ldots, 6$$

provided we set $p_7 = 1$.

Equations $(e_5)$ and $(e_6)$ can now be written as

$(e_5)$  $(1 - p_2)^2 A_2^3 b_2 + (1 - p_3)^2 A_3^3 b_3 \cdots + (1 - p_6)^2 A_6^3 b_6 = \dfrac{2}{8!}$

$(e_6)$  $(1 - p_3)^2 A_3^{33} b_3 + (1 - p_4)^2 A_4^{33} b_4 \cdots + (1 - p_6)^2 A_6^{33} b_6 = \dfrac{2}{8!}$

Equation $(e_5)$ is identical to the equation formed by adding Eqs. $(\iota^0)$ and $(\iota^2)$ and subtracting two times Eq. $(\iota^1)$. Therefore, it may be discarded if the three $(\iota)$ equations are satisfied by the Nyström parameters. The same operations applied to Eqs. $(\lambda^0)$, $(\lambda^1)$, and $(\lambda^2)$ show that Eq. $(e_6)$ may also be discarded.

Solution of the Condition Equations

The number of condition equations for the eighth-order method is now 19 to determine the 20 parameters $c_{ij}$ subject to the constraints of Eqs. (12.94) and

$$\begin{array}{lll} p_1 = \tfrac{1}{2} p_2 & c_{21} = \tfrac{1}{3} p_2^2 & L_8(p_2, p_3, p_4, p_5, p_6, 1) = 0 \\ p_7 = 1 & & \end{array} \qquad (12.106)$$

A recipe for the complete solution, which parallels the seventh-order case, is given in the following. For the $b_i$ coefficients, we have

$$b_0 = 1 - b_2 - b_3 - b_4 - b_5 - b_6 - b_7 \quad \text{and} \quad b_1 = 0 \qquad (12.107)$$

where $b_2, \ldots, b_7$ are determined from

$$b_2 = \frac{L_7(p_3, p_4, p_5, p_6, p_7)}{p_2(p_2 - p_3)(p_2 - p_4) \cdots (p_2 - p_6)(p_2 - p_7)} \quad (2, 3, 4, 5, 6, 7) \qquad (12.108)$$

Then the parameters $c_{31}$, $c_{32}$ and $c_{42}$, $c_{43}$ are obtained from the first two of Eqs. (12.102) while $c_{51}$, $c_{61}$, $c_{71}$ follow from

$$\begin{bmatrix} 1 & 1 & 1 \\ p_5 & p_6 & 1 \\ p_5^2 & p_6^2 & 1 \end{bmatrix} \begin{bmatrix} b_5 c_{51} \\ b_6 c_{61} \\ b_7 c_{71} \end{bmatrix} = - \begin{bmatrix} 1 & 1 & 1 \\ p_2 & p_3 & p_4 \\ p_2^2 & p_3^2 & p_4^2 \end{bmatrix} \begin{bmatrix} b_2 c_{21} \\ b_3 c_{31} \\ b_4 c_{41} \end{bmatrix} \qquad (12.109)$$

with $c_{41}$ again regarded as a free parameter. Next, the intermediate quantities $A_5^{33}$, $A_6^{33}$, $A_7^{33}$ are calculated from†

$$\begin{bmatrix} 1 & 1 & 1 \\ p_5 & p_6 & 1 \\ p_5^2 & p_6^2 & 1 \end{bmatrix} \begin{bmatrix} b_5 A_5^{33} \\ b_6 A_6^{33} \\ b_7 A_7^{33} \end{bmatrix} = \frac{1}{120} \begin{bmatrix} \frac{1}{6} \\ \frac{1}{7} \\ \frac{1}{8} \end{bmatrix} - \begin{bmatrix} 1 & 1 \\ p_3 & p_4 \\ p_3^2 & p_4^2 \end{bmatrix} \begin{bmatrix} b_3 A_3^{33} \\ b_4 A_4^{33} \end{bmatrix}$$

Then, $c_{52}$, $c_{53}$, $c_{54}$ follow from Eq. (12.103).

The intermediate parameters $A_6^{43}$ and $A_7^{43}$ are determined from

$$\begin{bmatrix} 1 & 1 \\ p_6 & 1 \end{bmatrix} \begin{bmatrix} b_6 A_6^{43} \\ b_7 A_7^{43} \end{bmatrix} = \frac{1}{6!} \begin{bmatrix} \frac{1}{7} \\ \frac{1}{8} \end{bmatrix} - \begin{bmatrix} 1 & 1 & 1 \\ p_3 & p_4 & p_5 \end{bmatrix} \begin{bmatrix} b_3 A_3^{43} \\ b_4 A_4^{43} \\ b_5 A_5^{43} \end{bmatrix}$$

The first is used to determine $c_{62}$, $c_{63}$, $c_{64}$, and $c_{65}$ from Eqs. (12.104).

Finally, we calculate the intermediate quantities

$$A_3^{53} = \tfrac{1}{120} c_{32} p_2^5$$
$$A_4^{53} = \tfrac{1}{120}(c_{42} p_2^5 + c_{43} p_3^5)$$
$$A_5^{53} = \tfrac{1}{120}(c_{52} p_2^5 + c_{53} p_3^5 + c_{54} p_4^5)$$
$$A_6^{53} = \tfrac{1}{120}(c_{62} p_2^5 + c_{63} p_3^5 + c_{64} p_4^5 + c_{65} p_5^5)$$

and

$$b_7 A_7^{53} = \tfrac{1}{5760} - b_3 A_3^{53} - b_4 A_4^{53} - b_5 A_5^{53} - b_6 A_6^{53}$$

which, together with $A_7^{43}$, are then used to obtain $c_{72}$ through $c_{76}$ from

$$\begin{bmatrix} 1 & 1 & 1 & 1 & 1 \\ p_2 & p_3 & p_4 & p_5 & p_6 \\ p_2^2 & p_3^2 & p_4^2 & p_5^2 & p_6^2 \\ p_2^3 & p_3^3 & p_4^3 & p_5^3 & p_6^3 \\ p_2^4 & p_3^4 & p_4^4 & p_5^4 & p_6^4 \end{bmatrix} \begin{bmatrix} p_2 c_{72} \\ p_3 c_{73} \\ p_4 c_{74} \\ p_5 c_{75} \\ p_6 c_{76} \end{bmatrix} = \begin{bmatrix} \frac{1}{6} - \frac{1}{2} p_2 c_{71} \\ \frac{1}{12} - \frac{1}{4} p_2^2 c_{71} \\ 6 A_7^{33} \\ 24 A_7^{43} \\ 120 A_7^{53} \end{bmatrix} \quad (12.110)$$

There are again five free parameters—any four of the five parameters $p_2$, $p_3$, $p_4$, $p_5$, $p_6$ subject to the constraint (12.106), together with $c_{41}$.

A note of caution in obtaining the $c_{ij}$'s from these equations (as well as those for the seventh-order case) is appropriate. High-order Vandermonde matrices can be ill-conditioned so that the explicit analytic inverse, given in Eq. (12.36), should be used rather than a general-purpose numerical matrix inversion algorithm.

---

† The quantities $A_3^{33}$ and $A_4^{33}$ as well as $A_3^{43}$, $A_4^{43}$, and $A_5^{43}$ which appear shortly, are calculated as in the seventh-order case of the previous section.

## 12.9 Integration Step-Size Control

In his 1972 report already cited, Erwin Fehlberg derived Runge-Kutta-Nyström formulas with built-in, automatic step-size control based on the leading term in the truncation error of the **x** variable only. His formulas, which were created by expanding the number of stages, in fact, represent a pair of integration formulas for **x** which differ in order by one. Then, by requiring that the difference between these two expressions remain between preset limits, an automatic step-size control for the lower-order formulas is established. Fehlberg presented results for orders four through eight.

Subsequently, Dale Bettis developed an algorithm[†] which utilizes an estimate of the local truncation error of both **x** and **y** for achieving step-size control. In his method, Bettis generated simultaneously two solutions, one of order four and the other of order five with the difference giving the desired control.

In both the Fehlberg and Bettis formulas, step-size control is inherent in the method with the sacrifice of achieving a minimal stage algorithm. It would seem preferable if one could exercise such control when desired, say, periodically, rather than during each step in the cycle, if computational efficiency could be, thereby, enhanced.

In this section the problem of step-size control is addressed[‡] in the spirit of Fehlberg but with the control feature left to the discretion of the user. The algorithms derived, with the single exception of the sixth-order algorithm with seventh-order control, are equally efficient compared with the Fehlberg and Bettis formulas even when step-size control is exercised during each integration interval. Furthermore, the control check may be skipped at any time with the result that all of the advantages of a minimal-stage algorithm are still retained.

Since many of the problems to which Runge-Kutta-Nyström methods are applied are characterized by fairly complicated right-hand-side functions $\mathbf{f}(t,\mathbf{x})$, the overwhelming portion of the total computation time is spent in evaluating the $\mathbf{k}_i$ functions. Therefore, if a higher-order algorithm is utilized to control the integration step size $h$ for a lower-order algorithm, it would be desirable, to enhance efficiency, if as many of the corresponding $\mathbf{k}_i$'s as possible were identical for a pair of such algorithms. In many cases, the free parameters at our disposal from the general solution of the condition equations can be chosen for the purpose of forcing certain of the $\mathbf{k}_i$'s to be the same for algorithms of different order.

---

[†] "A Runge-Kutta Nyström Algorithm," *Celestial Mechanics*, Vol. 8, No. 2, September, 1973, pp. 229–233.

[‡] The material comprising this section is based on the author's paper "Minimal-Stage Step-Size Control of Runge-Kutta-Nyström Integration Algorithms," *Acta Astronautica*, Vol. 13, No. 6/7, June/July, 1986, pp. 277–283. It was presented in Stockholm, Sweden in 1985 at the Thirty-Sixth Congress of the International Astronautical Federation.

**614**   Numerical Integration of Differential Equations   [Chap. 12]

Despite the fact that specific and useful Runge-Kutta-Nyström algorithms are developed here, the author feels that the chief contribution of this section is to illustrate the practical applications which can result from the general solutions of the condition equations which constitute the main body of this chapter.

### Second-Order Algorithm with Third-Order x-Control

As a simple example, consider the second-order, single-stage algorithm with third-order x-control. We have seen that the solution of the condition equations for the second-order case is

$$L_2(p_0) = \tfrac{1}{2} - p_0 = 0 \quad \text{and} \quad b_0 = 1$$

while the corresponding solution for the third-order, two-stage method with one free parameter is

$$L_3(p_0, p_1) = \tfrac{1}{3} - \tfrac{1}{2}(p_0 + p_1) + p_0 p_1 = 0$$
$$(p_0 - p_1)b_0 = L_2(p_1)$$
$$(p_1 - p_0)b_1 = L_2(p_0)$$
$$b_1 q_1 = \tfrac{1}{6}$$

We see at once that the parameter $p_0$ cannot be the same for both $m = 2$ and $m = 3$. Since we must have $L_2(p_0) = 0$, as required for the second-order method, this would make $b_1' = 0$ for the third-order case. Hence, the equation for $q_1$ would be rendered meaningless. Also, we wish to choose $p_1' = 1$ so that the coefficient $a_1' = (1 - p_1')b_1'$ will be zero, thereby avoiding the necessity of computing $\mathbf{k}_1'$. Indeed, this will be our strategy for all higher-order methods—to set to unity the $p$ parameter for the last stage of the x-control formula.

As a result of this analysis, we find there is only one minimal-stage, second-order algorithm with third-order x-control. Using primes for those quantities required solely for control, the algorithm is

$$\mathbf{x} = \mathbf{x}_0 + h\mathbf{y}_0 + \tfrac{1}{2}h^2 \mathbf{k}_0 + O(h^3)$$
$$\mathbf{y} = \mathbf{y}_0 + h\mathbf{k}_0 + O(h^3)$$
$$\mathbf{k}_0 = \mathbf{f}(t_0 + \tfrac{1}{2}h, \mathbf{x}_0 + \tfrac{1}{2}h\mathbf{y}_0)$$
$$\mathbf{x}' = \mathbf{x}_0 + h\mathbf{y}_0 + \tfrac{1}{2}h^2 \mathbf{k}_0' + O(h^4)$$
$$\mathbf{k}_0' = \mathbf{f}(t_0 + \tfrac{1}{3}h, \mathbf{x}_0 + \tfrac{1}{3}h\mathbf{y}_0)$$

Observe that two evaluations of the function $\mathbf{f}$ are required per integration step when x-control is desired and only one otherwise.

## Third-Order Algorithm with Fourth-Order x-Control

When the third- and fourth-order condition equations are compared, we again see a potential conflict. The constraint function $L_3(p_0, p_1)$ must vanish for the third-order method but cannot vanish for the fourth as the equation for $q_2$ would then be violated. Hence, the parameters $p_0$ and $p_1$ cannot be the same for the two cases; however, we can have $p_0$ identical for both.

As mentioned in the previous subsection, we shall choose $p'_2 = 1$ to optimize the control. Then, from the requirement that the two constraint functions $L_3(p_0, p_1)$ and $L_4(p_0, p'_1, 1)$ be zero, we find that

$$p_0 = \frac{1}{2} \frac{p_1 - \frac{2}{3}}{p_1 - \frac{1}{2}} \quad \text{and} \quad p'_1(1 - p_1) = \frac{1}{6}$$

For arbitrary values of $p_1$ ($\neq \frac{1}{2}$, of course) we have a one-parameter family of algorithms with three evaluations of $\mathbf{f}$ per step using fourth-order x-control.

As an example, if we choose $p_0 = 0$, then a third-order algorithm with fourth-order x-control is

$$\mathbf{x} = \mathbf{x}_0 + h\mathbf{y}_0 + \tfrac{1}{4}h^2(\mathbf{k}_0 + \mathbf{k}_1) + O(h^4)$$
$$\mathbf{y} = \mathbf{y}_0 + \tfrac{1}{4}h(\mathbf{k}_0 + 3\mathbf{k}_1) + O(h^4)$$
$$\mathbf{k}_0 = \mathbf{f}(t_0, \mathbf{x}_0)$$
$$\mathbf{k}_1 = \mathbf{f}(t_0 + \tfrac{2}{3}h, \mathbf{x}_0 + \tfrac{2}{3}h\mathbf{y}_0 + \tfrac{2}{9}h^2\mathbf{k}_0)$$

and, for step-size control, we calculate

$$\mathbf{x}' = \mathbf{x}_0 + h\mathbf{y}_0 + \tfrac{1}{6}h^2(\mathbf{k}_0 + 2\mathbf{k}'_1) + O(h^5)$$
$$\mathbf{k}'_1 = \mathbf{f}(t_0 + \tfrac{1}{2}h, \mathbf{x}_0 + \tfrac{1}{2}h\mathbf{y}_0 + \tfrac{1}{8}h^2\mathbf{k}_0)$$

which, incidentally, are the original Nyström algorithms for the third- and fourth-order cases (if, of course, we include the formula for $\mathbf{y}'$).

## Fourth-Order Algorithm with Fifth-Order x-Control

With $p_0 = 0$, the constraint equation for the fourth-order algorithm is $L_4(p_1, p_2) = 0$. Since $b_3$, which is proportional to $L_4(p_1, p_2)$, must not vanish for the fifth-order case, then $p_1$ and $p_2$ cannot be the same for both. However, we may have $p_1 = p'_1$. Then, since we again wish to have $p'_3 = 1$, the parameters $p_2$ and $p'_2$ are obtained from

$$p_1 = \frac{2}{3} \frac{p_2 - \frac{3}{4}}{p_2 - \frac{2}{3}} \quad \text{and} \quad p'_2(1 - p_2) = \tfrac{1}{5}(\tfrac{3}{2} - p_2)$$

The number of evaluations per step is four including the fifth-order control and there is also a one-parameter family of these algorithms.

**616**  Numerical Integration of Differential Equations  [Chap. 12

◊ **Problem 12-14**
Derive the following fourth-order algorithm with fifth-order step-size control:

$$\mathbf{x} = \mathbf{x}_0 + h\mathbf{y}_0 + \tfrac{1}{30}h^2(3\mathbf{k}_0 + 10\mathbf{k}_1 + 2\mathbf{k}_2) + O(h^5)$$
$$\mathbf{y} = \mathbf{y}_0 + \tfrac{1}{10}h(\mathbf{k}_0 + 5\mathbf{k}_1 + 4\mathbf{k}_2) + O(h^5)$$

where

$$\mathbf{k}_0 = \mathbf{f}(t_0, \mathbf{x}_0)$$
$$\mathbf{k}_1 = \mathbf{f}(t_0 + \tfrac{1}{3}h, \mathbf{x}_0 + \tfrac{1}{3}h\mathbf{y}_0 + \tfrac{1}{18}h^2\mathbf{k}_0)$$
$$\mathbf{k}_2 = \mathbf{f}[t_0 + \tfrac{5}{6}h, \mathbf{x}_0 + \tfrac{5}{6}h\mathbf{y}_0 + \tfrac{1}{144}h^2(5\mathbf{k}_0 + 45\mathbf{k}_1)]$$

With step-size control activated, we compute

$$\mathbf{x}' = \mathbf{x}_0 + h\mathbf{y}_0 + \tfrac{1}{336}h^2(35\mathbf{k}_0 + 108\mathbf{k}_1 + 25\mathbf{k}_2') + O(h^6)$$
$$\mathbf{k}_2' = \mathbf{f}[t_0 + \tfrac{4}{5}h, \mathbf{x}_0 + \tfrac{4}{5}h\mathbf{y}_0 - \tfrac{2}{125}h^2(\mathbf{k}_0 - 21\mathbf{k}_1)]$$

◊ **Problem 12-15**
Derive the following fourth-order algorithm with fifth-order x-control:

$$\mathbf{x} = \mathbf{x}_0 + h\mathbf{y}_0 + \tfrac{1}{120}h^2(13\mathbf{k}_0 + 36\mathbf{k}_1 + 9\mathbf{k}_2 + 2\mathbf{k}_3) + O(h^5)$$
$$\mathbf{y} = \mathbf{y}_0 + \tfrac{1}{8}h(\mathbf{k}_0 + 3\mathbf{k}_1 + 3\mathbf{k}_2 + \mathbf{k}_3) + O(h^5)$$
$$\mathbf{x}' = \mathbf{x}_0 + h\mathbf{y}_0 + \tfrac{1}{120}h^2(13\mathbf{k}_0 + 36\mathbf{k}_1 + 9\mathbf{k}_2 + 2\mathbf{k}_4) + O(h^6)$$

where

$$\mathbf{k}_0 = \mathbf{f}(t_0, \mathbf{x}_0)$$
$$\mathbf{k}_1 = \mathbf{f}(t_0 + \tfrac{1}{3}h, \mathbf{x}_0 + \tfrac{1}{3}h\mathbf{y}_0 + \tfrac{1}{18}h^2\mathbf{k}_0)$$
$$\mathbf{k}_2 = \mathbf{f}(t_0 + \tfrac{2}{3}h, \mathbf{x}_0 + \tfrac{2}{3}h\mathbf{y}_0 + \tfrac{2}{9}h^2\mathbf{k}_1)$$
$$\mathbf{k}_3 = \mathbf{f}[t_0 + h, \mathbf{x}_0 + h\mathbf{y}_0 + \tfrac{1}{6}h^2(2\mathbf{k}_0 + \mathbf{k}_2)]$$
$$\mathbf{k}_4 = \mathbf{f}[t_0 + h, \mathbf{x}_0 + h\mathbf{y}_0 + \tfrac{1}{120}h^2(13\mathbf{k}_0 + 36\mathbf{k}_1 + 9\mathbf{k}_2 + 2\mathbf{k}_3)]$$

*Erwin Fehlberg* 1972

NOTE: It should be remarked that there are five evaluations of **f** in this algorithm for the first step only since $\mathbf{k}_4$ is the same as $\mathbf{k}_0$ for every step after the first. Even so, the algorithm of Prob. 12–14 has also four stages with step-size control but only three if the control feature is omitted.

Sect. 12.9]                    Integration Step-Size Control                    617

◇ **Problem 12–16**
Derive the following fourth-order algorithm with fifth-order **x**- and **y**-control:

$$\mathbf{x} = \mathbf{x}_0 + h\mathbf{y}_0 + \tfrac{1}{6}h^2(\mathbf{k}_0 + 2\mathbf{k}_3) + O(h^5)$$
$$\mathbf{y} = \mathbf{y}_0 + \tfrac{1}{3}h(2\mathbf{k}_2 - \mathbf{k}_3 + 2\mathbf{k}_4) + O(h^5)$$
$$\mathbf{x}' = \mathbf{x}_0 + h\mathbf{y}_0 + \tfrac{1}{90}h^2(7\mathbf{k}_0 + 24\mathbf{k}_2 + 6\mathbf{k}_3 + 8\mathbf{k}_4) + O(h^6)$$
$$\mathbf{y}' = \mathbf{y}_0 + \tfrac{1}{90}h(7\mathbf{k}_0 + 32\mathbf{k}_2 + 12\mathbf{k}_3 + 32\mathbf{k}_4 + 7\mathbf{k}_5) + O(h^6)$$

where

$$\mathbf{k}_0 = \mathbf{f}(t_0, \mathbf{x}_0)$$
$$\mathbf{k}_1 = \mathbf{f}(t_0 + \tfrac{1}{8}h, \mathbf{x}_0 + \tfrac{1}{8}h\mathbf{y}_0 + \tfrac{1}{128}h^2\mathbf{k}_0)$$
$$\mathbf{k}_2 = \mathbf{f}[t_0 + \tfrac{1}{4}h, \mathbf{x}_0 + \tfrac{1}{4}h\mathbf{y}_0 + \tfrac{1}{96}h^2(\mathbf{k}_0 + 2\mathbf{k}_1)]$$
$$\mathbf{k}_3 = \mathbf{f}[t_0 + \tfrac{1}{2}h, \mathbf{x}_0 + \tfrac{1}{2}h\mathbf{y}_0 + \tfrac{1}{24}h^2(\mathbf{k}_0 + 2\mathbf{k}_2)]$$
$$\mathbf{k}_4 = \mathbf{f}[t_0 + \tfrac{3}{4}h, \mathbf{x}_0 + \tfrac{3}{4}h\mathbf{y}_0 + \tfrac{1}{128}h^2(9\mathbf{k}_0 + 18\mathbf{k}_2 + 9\mathbf{k}_3)]$$
$$\mathbf{k}_5 = \mathbf{f}[t_0 + h, \mathbf{x}_0 + h\mathbf{y}_0 + \tfrac{1}{90}h^2(7\mathbf{k}_0 + 24\mathbf{k}_2 + 6\mathbf{k}_3 + 8\mathbf{k}_4)]$$

*Dale Bettis* 1973

NOTE: Actually, Bettis suggests that this algorithm be regarded as fifth-order with fourth-order control.

◇ **Problem 12–17**
Show that, by one more evaluation, i.e., $\mathbf{k}'_3$ in Prob. 12–14, the same control as used in Prob. 12–16 can be achieved at the same cost. Specifically, derive the expressions

$$\mathbf{y} = \mathbf{y}_0 + \tfrac{1}{336}h^2(35\mathbf{k}_0 + 162\mathbf{k}_1 + 125\mathbf{k}'_2 + 14\mathbf{k}'_3) + O(h^6)$$
$$\mathbf{k}'_3 = \mathbf{f}[t_0 + h, \mathbf{x}_0 + h\mathbf{y}_0 + \tfrac{1}{28}h^2(21\mathbf{k}_0 - 12\mathbf{k}_1 + 5\mathbf{k}'_2)]$$

Fifth-Order Algorithm with Sixth-Order **x**-Control

If $p_1$ and $p_2$ are the same for a fifth- and sixth-order method, the parameter $c_{21}$ cannot be the same for both. To show this, we first observe that

$$L_4(p_1, p_3)c_{21} = \frac{p_2(p_2 - p_1)}{6p_1}(\tfrac{1}{4}p_3 - \tfrac{1}{5})$$
$$L_5(p_1, p'_3, 1)c'_{21} = \frac{p_2(p_2 - p_1)}{60p_1}(\tfrac{1}{2}p'_3 - \tfrac{1}{3})$$

Now, in order to have $c_{21} = c'_{21}$, we must require that

$$\begin{vmatrix} p_3 L_3(p_1) + L_4(p_1) & \tfrac{1}{4}p_3 - \tfrac{1}{5} \\ p'_3 L_4(p_1, 1) + L_5(1, p_1) & \tfrac{1}{20}p'_3 - \tfrac{1}{30} \end{vmatrix} = 0$$

where we have used the properties

$$L_4(p_1,p_3) = p_3 L_3(p_1) + L_4(p_1)$$
$$L_5(p_1,p'_3,1) = p'_3 L_4(p_1,1) + L_5(p_1,1)$$

The parameters $p_3$ and $p'_3$ are determined from their respective constraint conditions

$$p_3 = -\frac{L_5(p_1,p_2)}{L_4(p_1,p_2)} \quad \text{and} \quad p'_3 = -\frac{L_6(p_1,p_2,1)}{L_5(p_1,p_2,1)}$$

By direct calculation, it is readily shown that this determinant will vanish if and only if

$$p_1 p_2 - \tfrac{1}{2}(p_1+p_2) + \tfrac{3}{10} = 0$$

which is identical to

$$L_5(p_1,p_2,1) = 0$$

But, this condition would make $b'_3 = 0$ and, hence, would result in a violation of the conditions for the existence of a sixth-order method. Therefore, only $\mathbf{k}_0$ and $\mathbf{k}_1$ can be identical for the two algorithms and six evaluations of $\mathbf{f}$ are required for a fifth-order method with sixth-order $\mathbf{x}$-control. Even though this is equivalent to the Fehlberg algorithm of the same order, the number of stages is only four if the $\mathbf{x}$-control is omitted.

◊ **Problem 12–18**
Derive the following fifth-order algorithm with sixth-order $\mathbf{x}$-control:

$$\mathbf{x} = \mathbf{x}_0 + h\mathbf{y}_0 + \tfrac{1}{378}h^2(27\mathbf{k}_0 + 112\mathbf{k}_1 + 50\mathbf{k}_2) + O(h^6)$$
$$\mathbf{y} = \mathbf{y}_0 + \tfrac{1}{1134}h(81\mathbf{k}_0 + 448\mathbf{k}_1 + 500\mathbf{k}_2 + 105\mathbf{k}_3) + O(h^6)$$

with

$$\mathbf{k}_0 = \mathbf{f}(t_0,\mathbf{x}_0)$$
$$\mathbf{k}_1 = \mathbf{f}(t_0 + \tfrac{1}{4}h, \mathbf{x}_0 + \tfrac{1}{4}h\mathbf{y}_0 + \tfrac{1}{32}h^2 \mathbf{k}_0)$$
$$\mathbf{k}_2 = \mathbf{f}[t_0 + \tfrac{7}{10}h, \mathbf{x}_0 + \tfrac{7}{10}h\mathbf{y}_0 - \tfrac{1}{1000}h^2(7\mathbf{k}_0 - 252\mathbf{k}_1)]$$
$$\mathbf{k}_3 = \mathbf{f}[t_0 + h, \mathbf{x}_0 + h\mathbf{y}_0 + \tfrac{1}{14}h^2(4\mathbf{k}_0 + 3\mathbf{k}_2)]$$

Then, for step-size control,

$$\mathbf{x}' = \mathbf{x}_0 + h\mathbf{y}_0 + \tfrac{1}{90}h^2(7\mathbf{k}_0 + 24\mathbf{k}_1 + 6\mathbf{k}'_2 + 8\mathbf{k}'_3) + O(h^7)$$
$$\mathbf{k}'_2 = \mathbf{f}[t_0 + \tfrac{1}{2}h, \mathbf{x}_0 + \tfrac{1}{2}h\mathbf{y}_0 - \tfrac{1}{24}h^2(\mathbf{k}_0 - 4\mathbf{k}_1)]$$
$$\mathbf{k}'_3 = \mathbf{f}[t_0 + \tfrac{3}{4}h, \mathbf{x}_0 + \tfrac{3}{4}h\mathbf{y}_0 + \tfrac{1}{32}h^2(3\mathbf{k}_0 + 4\mathbf{k}_1 + 2\mathbf{k}'_2)]$$

## Sixth-Order Algorithm with Seventh-Order x-Control

If $p_1$ and $p_2$ are identical for a sixth- and seventh-order method, then there exists a unique pair of values for $p_1$, $p_2$ for which $c_{21} = c'_{21}$. To prove this assertion, we first note that

$$L_5(p_1, p_3, 1) c_{21} = \frac{p_2(p_2 - p_1)}{60 p_1} (\tfrac{1}{2} p_3 - \tfrac{1}{3}) \quad \text{and} \quad c'_{21} = \tfrac{1}{3} p_2^2$$

Therefore, if $c_{21} = c'_{21}$, we must have

$$10 p_2 p_3 (p_2 - 1) - 5 p_2^2 + 6 p_2 + 3 p_3 = 2$$

Also, since we require $p_1 = \tfrac{1}{2} p_2$, the constraint equation for the sixth-order method

$$L_6(\tfrac{1}{2} p_2, p_2, p_3, 1) = 0$$

reduces to

$$5 p_2 p_3 (2 p_2 - 3) - 5 p_2^2 + 9 p_2 + 6 p_3 = 4$$

Now, the two relations between $p_2$ and $p_3$ just derived are equivalent to

$$5 p_2 p_3 - 3(p_2 + p_3) + 2 = 0$$

and

$$5 p_2^2 - 2 p_2 + 3 p_3 - 2 = 0$$

Combining the two, we find that $p_2$ and $p_3$ must be the roots of the quadratic equation

$$p^2 - p + \tfrac{1}{5} = 0$$

Therefore, we have

$$\begin{aligned} p_2 &= \tfrac{1}{10}(5 - \sqrt{5}) \\ p_3 &= \tfrac{1}{10}(5 + \sqrt{5}) \end{aligned} \quad \text{so that} \quad p_1 = \tfrac{1}{2} p_2 = \tfrac{1}{20}(5 - \sqrt{5})$$

In passing, we note that these values result in $L_5(p_2, p_3, 1) = 0$ which causes $b_1$ to be identically zero.

As a result of this analysis, we conclude that $\mathbf{k}_0$, $\mathbf{k}_1$, and $\mathbf{k}_2$ can be the same for both a sixth- and a seventh-order method. Hence, the number of evaluations of $\mathbf{f}$ per step will be eight for a sixth-order method with seventh-order x-control. Fehlberg's algorithm requires only seven evaluations but, again, our algorithm has the minimum number of stages if the step-size control is omitted.

◊ **Problem 12-19**

Since all of the $p$'s are determined for $m=6$, $n=5$ in the above analysis, the sixth-order method is unique. Show that

$$\mathbf{x} = \mathbf{x}_0 + h\mathbf{y}_0 + \tfrac{1}{24}h^2[2\mathbf{k}_0 + (5+\sqrt{5})\mathbf{k}_2 + (5-\sqrt{5})\mathbf{k}_3] + O(h^7)$$
$$\mathbf{y} = \mathbf{y}_0 + \tfrac{1}{12}h(\mathbf{k}_0 + 5\mathbf{k}_2 + 5\mathbf{k}_3 + \mathbf{k}_4) + O(h^7)$$

with

$$\mathbf{k}_0 = \mathbf{f}(t_0, \mathbf{x}_0)$$
$$\mathbf{k}_1 = \mathbf{f}[t_0 + \tfrac{1}{20}(5-\sqrt{5})h, \mathbf{x}_0 + \tfrac{1}{20}(5-\sqrt{5})h\mathbf{y}_0 + \tfrac{1}{80}(3-\sqrt{5})h^2\mathbf{k}_0]$$
$$\mathbf{k}_2 = \mathbf{f}\{t_0 + \tfrac{1}{10}(5-\sqrt{5})h, \mathbf{x}_0 + \tfrac{1}{10}(5-\sqrt{5})h\mathbf{y}_0$$
$$\qquad + \tfrac{1}{60}h^2[(3-\sqrt{5})\mathbf{k}_0 + (6-2\sqrt{5})\mathbf{k}_1]\}$$
$$\mathbf{k}_3 = \mathbf{f}\{t_0 + \tfrac{1}{10}(5+\sqrt{5})h, \mathbf{x}_0 + \tfrac{1}{10}(5+\sqrt{5})h\mathbf{y}_0$$
$$\qquad + \tfrac{1}{60}h^2[(6+2\sqrt{5})\mathbf{k}_0 - (8+4\sqrt{5})\mathbf{k}_1 + (11+5\sqrt{5})\mathbf{k}_2]\}$$
$$\mathbf{k}_4 = \mathbf{f}\{t_0 + h, \mathbf{x}_0 + h\mathbf{y}_0 - \tfrac{1}{12}h^2[(3+\sqrt{5})\mathbf{k}_0$$
$$\qquad - (2+6\sqrt{5})\mathbf{k}_1 + (2+2\sqrt{5})\mathbf{k}_2 - (9-3\sqrt{5})\mathbf{k}_3]\}$$

NOTE: Again, to minimize the number of additional $\mathbf{k}$'s for the control feature, we set $p'_6 = 1$ so that two free parameters may be selected from among $p'_3$, $p'_4$, $p'_5$ together with a free choice for $c'_{41}$.

### Seventh-Order Algorithm with Eighth-Order x-Control

Frank Hriadil[†] analyzed the seventh-order algorithm with eighth-order x-control of the step size. Such an algorithm is possible using nine evaluations of the function $\mathbf{f}(t,\mathbf{x})$ which is the same as that devised by Fehlberg. However, if step-size control is not exercised, the seventh-order method requires just seven function evaluations.

For the first five $\mathbf{k}$'s to be the same, we must have $p_0, \ldots, p_4$ and $c_{21}, \ldots, c_{43}$ the same for both algorithms. For the eight-order control, Frank selected values for the free parameters $p'_3, \ldots, p'_6$ to provide approximate equal spacing from zero to one. Specifically,

$$p_0 = p'_0 = 0 \qquad\qquad p_5 = 0.8327915849$$
$$p_1 = p'_1 = 0.0792766611 \qquad p_6 = 1$$
$$p_2 = p'_2 = 0.1585533223 \qquad p'_5 = 0.7143000000$$
$$p_3 = p'_3 = 0.4286000000 \qquad p'_6 = 0.8571000000$$
$$p_4 = p'_4 = 0.5714000000 \qquad p'_7 = 1$$

---

[†] "Solution of a Special Class of Second-Order Differential Equations through the use of Higher Order Runge-Kutta-Nyström Techniques," MIT, M.S. Thesis, June 1975.

He also chose the other free parameters $c_{41}$ and $c'_{41}$ to be zero. Then, the rest of the identical $c_{ij}$'s are

$$c_{21} = c'_{21} = 0.008379719 \qquad c_{41} = c'_{41} = 0$$
$$c_{31} = c'_{31} = -0.116394789 \qquad c_{42} = c'_{42} = 0.103773748$$
$$c_{32} = c'_{32} = 0.140959190 \qquad c_{43} = c'_{43} = 0.034157295$$

The remaining $c_{ij}$ and $c'_{ij}$ parameters are

$$c_{51} = 0.294810543 \qquad c'_{51} = 1.415901519$$
$$c_{52} = 0.001974362 \qquad c'_{52} = -0.778773705$$
$$c_{53} = 0.053757319 \qquad c'_{53} = 0.142985941$$
$$c_{54} = 0.086695149 \qquad c'_{54} = 0.018704349$$

$$c_{61} = -1.070725551 \qquad c'_{61} = -0.838031914$$
$$c_{62} = 0.770395923 \qquad c'_{62} = 0.708867422$$
$$c_{63} = 0.313053911 \qquad c'_{63} = 0.067707125$$
$$c_{64} = -0.114914297 \qquad c'_{64} = 0.017420055$$
$$c_{65} = 0.073113735 \qquad c'_{65} = 0.028014167$$

$$c'_{71} = 1.401311110$$

The parameters $c'_{72}, \ldots, c'_{76}$ are not calculated since $p'_7 = 1$ eliminates their need in the eighth-order algorithm.

Finally, we list the $b_i$ and $b'_i$ coefficients. (The $q_i$'s, and $a_i$'s are not recorded since they are so easily determined.)

$$b_0 = 0.0460182740 \qquad b'_0 = 0.0463900363$$
$$b_1 = 0 \qquad b'_1 = 0$$
$$b_2 = 0.2483385750 \qquad b'_2 = 0.2464481173$$
$$b_3 = 0.2172603598 \qquad b'_3 = 0.2330101312$$
$$b_4 = 0.1815261108 \qquad b'_4 = 0.1337579954$$
$$b_5 = 0.2576030440 \qquad b'_5 = 0.0939411688$$
$$b_6 = 0.4925363631 \qquad b'_6 = 0.2024308894$$
$$\qquad b'_7 = 0.0440217028$$

It would be better, of course, to have these coefficients and parameters expressed as rational fractions. This should be possible using the *MACSYMA* program which has proven to be such a valuable tool. Surprisingly, without any computing aids of that kind, Erwin Fehlberg was able to represent his parameters as fractions; but in fairness to Frank Hriadil, Fehlberg did not have to satisfy the constraint equations $L_7 = L_8 = 0$.

### Higher-Order x-Control Algorithms

In an earlier subsection, we produced a third-order algorithm with fourth-order x-control which required three evaluations of the function **f**. It is interesting to note that a fifth-order x-control is also possible with the same number of functional evaluations.

Here the algorithm is unique if $p_0 = p'_0 = 0$. With $p_1 = p'_1 = \frac{2}{3}$ and $p'_3 = 1$, then $p'_2$ must be $\frac{1}{5}$. The third-order algorithm is the same as before and the fifth-order x-control is given by:

$$\mathbf{x}' = \mathbf{x}_0 + h\mathbf{y}_0 + \tfrac{1}{672}h^2(733\mathbf{k}_0 - 63\mathbf{k}_1 + 200\mathbf{k}'_2) + O(h^6)$$
$$\mathbf{k}'_2 = \mathbf{f}[t_0 + \tfrac{1}{5}h, \mathbf{x}_0 + \tfrac{1}{5}h\mathbf{y}_0 - \tfrac{1}{500}h^2(11\mathbf{k}_0 - 21\mathbf{k}_1)]$$

With this successful analysis, we are tempted to try an extension to the fourth-order case. If the first three stages of a fourth- and sixth-order algorithm were the same, a sixth-order x-control would be possible for a fourth-order method with just one additional function evaluation. Unfortunately, this is not possible as can be seen from the following argument.

The simultaneous satisfaction of the constraint equations for $m = 4$ and $m = 6$, namely,

$$L_4(p_1, p_2) = 0 \quad \text{and} \quad L_6(p_1, p_2, p'_3, 1) = 0$$

would require that $p_1$ and $p_2$ be the roots of

$$p^2 - \frac{6p'_3 - \tfrac{3}{2}}{5p'_3 - 1}p + \frac{3p'_3 - 1}{10p'_3 - 2} = 0$$

The parameters $c_{21}$ for the fourth-order and $c'_{21}$ for the sixth-order methods are obtained from

$$L_3(p_1)c_{21} = \frac{p_2(p_2 - p_1)}{24p_1}$$

and

$$L_5(p_1, p'_3, 1)c'_{21} = \frac{p_2(p_2 - p_1)}{60p_1}(\tfrac{1}{2}p'_3 - \tfrac{1}{3})$$

Now, if we require $c_{21} = c'_{21}$, then $p_1$ and $p'_3$ must be related by

$$p_1 p'_3 - \tfrac{1}{4}(p_1 + p'_3) + \tfrac{1}{12} = 0$$

Solving this for $p_1$ in terms of $p'_3$ and substituting in the quadratic equation just obtained, we conclude that $p'_3$ must be $\frac{1}{3}$. This is not possible, for then $p_1$ would have to be zero and, thus, equal to $p_0$.

Hence, a fourth-order algorithm with sixth-order x-control requires five function evaluations which is two more than the minimal-stage fourth-order algorithm.

Chapter 13

# The Celestial Position Fix

IN MANY RESPECTS THE NAVIGATION POSITION FIX OBTAINED FROM celestial observations made aboard a spacecraft is similar to the problem encountered by the seagoing and airborne navigators. The fundamental differences are (1) the spacecraft problem is truly three dimensional and (2) the forces governing the motion of the spacecraft are far better known than the motions of the terrestrial seas and air masses. Therefore, although the first difference noted tends to complicate the problem, the second makes the task somewhat easier and the resulting computations and extrapolations capable of greater precision.

As the first step in the formulation of a navigation theory, we consider the processes involved in determining the position of a spacecraft by means of a celestial fix. For a completely onboard determination, the operation may comprise a sequence of any or all of the following types of measurements: (1) the angle between the lines of sight to selected pairs of celestial bodies, (2) the observation of star occultations, (3) the measurement of the apparent diameter of a planet, and (4) radar measurements. One further operation is implied in the fix, namely, the recording of time as indicated by the spacecraft clock. The intended result of these observations is the determination of the coordinates of spacecraft position together with, perhaps, a correction to the current clock reading. Here, we shall describe several possible forms of the required calculations and then relate the resultant errors in position and time estimates to the errors in the primary measurements.

In the first part of the analysis the celestial fix is studied primarily from a geometrical point of view. Later for computational advantage, it is assumed that approximations to spacecraft position and to time are already known, so that perturbation techniques may be employed. In many important applications, no real restriction is implied by this assumption, since deviations from a selected reference trajectory must be kept small in order to complete the mission with a reasonable fuel supply. Specifically, we assume the existence of a reference time $t$ for the fix and a reference position vector $\mathbf{r}$ for the spacecraft at time $t$. We further assume exact knowledge of the position and velocity of all relevant astronomical objects at time $t$.

**624**  The Celestial Position Fix  [Chap. 13

Secondary effects arising from the finite speed of light, the finite distance of stars, etc., are ignored in this analysis. Such corrections can be lumped together for a particular reference point as a modification to the reference values for the various angles to be measured at that point.

The celestial fix is first analyzed by assuming that only a sufficient number of measurements are made to establish the position unambiguously. Gauss' method of least squares is then employed to permit incorporation of redundant measurements to compensate for instrumentation inaccuracies. A number of specific examples is included to illustrate the effectiveness of the method.

Finally, the method of processing the measurements for the celestial fix is recast in a recursive form which eliminates the need for the accumulation of large quantities of data and the inversion of correspondingly large order matrices.

## 13.1 Geometry of the Navigation Fix

The measurement of the angle subtended at the spacecraft between the line of sight to a near body, e.g., the sun or a planet, and the line of sight to a star establishes the position of the vehicle on the surface of a cone whose apex is the position of the near body. The axis of the cone has the direction of the line of sight to the star, and the angle of the cone is twice the supplement of the measured angle. The star is assumed to be at such a large distance that its direction is independent of the point of observation.

A second angle measurement, involving the same near body and a different star, establishes a second cone of position with a different axis and apex angle. The two cones intersect in two straight lines one of which is a line of position for the spacecraft. Another star measurement made with respect to the same near body would serve merely to distinguish between the two lines of position already determined but would otherwise provide no new information. Actually, this possible ambiguity can easily be resolved, since the two lines of position are generally widely separated so that an approximate knowledge of the vehicle's position will suffice to determine the proper one.

A third measurement is needed to determine the radial distance of the vehicle from the near body. For example, if a second near body is selected, the subtended angle between the lines of sight to it and the original body provide a surface of position which is generated by rotating the arc of a circle about a line connecting the two near bodies. The terminal points of the arc are the two bodies, the center lies on the perpendicular bisector of the connecting line, and the radius is a function of the magnitude of the measured angle and the distance between the two near bodies. The intersection of this third surface of position with the already obtained line

Sect. 13.1]  Geometry of the Navigation Fix  625

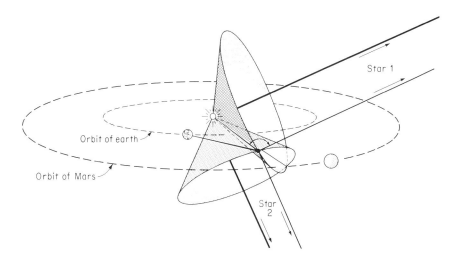

**Fig. 13.1:** Geometry of navigation fix in space.

of position establishes the fix, and the vehicle's position relative to the near body is determined. This particular example is illustrated in Fig. 13.1.

The numerical calculations associated with the geometrical construction just described are easily formulated. Let $\mathbf{r} = r\,\mathbf{i}_r$ be the unknown position vector of the spacecraft, which, for definiteness, we assume to be expressed relative to the sun. Let $\mathbf{r}_p$ be the known position of a planet, and let $\mathbf{i}_1$ and $\mathbf{i}_2$ be two unit vectors in the direction of two selected stars. The three measurements produce three angles $A_1$, $A_2$, $A_3$. Therefore, the position vector of the spacecraft must simultaneously satisfy the following three nonlinear equations:

$$\mathbf{i}_r \cdot \mathbf{i}_1 = -\cos A_1$$
$$\mathbf{i}_r \cdot \mathbf{i}_2 = -\cos A_2$$
$$\mathbf{i}_r \cdot \mathbf{r}_p = r - |\mathbf{r}_p - \mathbf{r}|\cos A_3$$

The solution of these equations for the components of $\mathbf{r}$ results in a fix for the spacecraft.

An alternate method of determining the radial distance from the near body might employ a third cone of position established by measuring the angle between the lines of sight of a second near body and a star. The intersection of this cone with the previously established line of position locates the vehicle.

Still another type of observation would be the measurement of the apparent angular diameter of a nearby planet. In this way a sphere of position is obtained. The observation of a star occultation by a nearby planet establishes a cylinder of position whose axis is in the direction of the star and whose diameter is the diameter of the planet. The angular diameter measurement and the observance of a star occultation are practical only when the vehicle is relatively close to the celestial body involved.

An exact determination of position by methods described in this section has some distinct disadvantages. True, the algebraic equations to be solved are nonlinear, but this is simply a nuisance and not necessarily a significant obstacle to onboard computation. However, the method implies simultaneous measurements which are almost certainly impractical. Finally, and perhaps most important of all, no satisfactory method of incorporating redundant measurements to compensate for instrumentation inaccuracies is known.

All these objections can be circumvented if the determination of spacecraft position is made relative to a known and nearby reference position. When such is the case, the powerful tools of linear perturbation analysis can be brought to bear. Indeed, the rest of this chapter is devoted to an exploitation of linear theory in obtaining a navigation fix.

◇ **Problem 13–1**
The solution of the equations

$$\mathbf{i}_r \cdot \mathbf{i}_1 = -\cos A_1$$
$$\mathbf{i}_r \cdot \mathbf{i}_2 = -\cos A_2$$

can be expressed as

$$\mathbf{i}_r = \alpha\, \mathbf{i}_1 + \beta\, \mathbf{i}_2 + \gamma\, \mathbf{i}_1 \times \mathbf{i}_2$$

where

$$\alpha \sin^2 \varphi = \cos A_2 \cos \varphi - \cos A_1$$
$$\beta \sin^2 \varphi = \cos A_1 \cos \varphi - \cos A_2$$
$$\gamma^2 \sin^2 \varphi = 1 + \alpha \cos A_1 + \beta \cos A_2$$

and $\cos \varphi = \mathbf{i}_1 \cdot \mathbf{i}_2$.

*Carl Grubin* 1976

◇ **Problem 13–2**
A spacecraft and two near celestial bodies are located at $S$, $P_1$, $P_2$, respectively. An $x,y$ coordinate system is set up such that the origin is at $P_1$, the $x$ axis lies along the line connecting $P_1$ and $P_2$, and the $y$ axis lies in the plane established by $S$, $P_1$, and $P_2$. The equation of the locus of points in the $x,y$ plane, for which the angle $A = \angle P_1 S P_2$ is constant, is

$$(x - c)^2 + (y - c \cot A)^2 = (c \csc A)^2$$

where $2c$ is the distance between $P_1$ and $P_2$. The surface generated by rotating this curve about the line connecting $P_1$ and $P_2$ is the surface of position of a spacecraft for which $A$ is the angle between the lines of sight to $P_1$ and $P_2$.

## 13.2 Navigation Measurements

The mathematical processes are considered here in some detail for determining spacecraft position by means of both celestial observation and radar measurements. It is assumed throughout the analysis that approximations to spacecraft position and velocity are already known, so that perturbation techniques can be employed.

As will be shown, each measurement establishes a component of spacecraft position along some direction in space. If $q$ is the quantity to be measured and $\delta q$ is the difference between the true and reference values, then it will be seen that the relation between $\delta q$ and the deviation in spacecraft position $\delta \mathbf{r}$ from the reference position is, to first order,

$$\delta q = \mathbf{h} \cdot \delta \mathbf{r} = \mathbf{h}^T \delta \mathbf{r} \qquad (13.1)$$

regardless of the type of measurement. Thus, the $\mathbf{h}$ vector alone will characterize the measurement.

### Measuring the Angle between a Near Body and a Star

The first type of measurement to be considered is that of the angle between the lines of sight to a near body (e.g., the sun or a planet) and a star. The angle $A$ will be a function $A(\mathbf{r})$ of the position vector $\mathbf{r}$ of the vehicle measured with respect to the near body. Then, $A$ may be expanded in a Taylor series about the reference position $\mathbf{r}_0$ at which point the angle to be measured is $A_0$. We have

$$A(\mathbf{r}) = A(\mathbf{r}_0) + \frac{\partial A}{\partial \mathbf{r}} \delta \mathbf{r} + \cdots = A_0 + \mathbf{h}^T \delta \mathbf{r} + \cdots \qquad (13.2)$$

where the derivative is understood to be evaluated at the reference point.

To calculate the coefficient of $\delta \mathbf{r}$, we differentiate the measurement equation

$$r \cos A = -\mathbf{i}_s^T \mathbf{r}$$

where $\mathbf{i}_s$ is the unit vector in the direction of the star, to obtain

$$\frac{\partial r}{\partial \mathbf{r}} \cos A - r \sin A \frac{\partial A}{\partial \mathbf{r}} = -\mathbf{i}_s^T \frac{\partial \mathbf{r}}{\partial \mathbf{r}}$$

or, equivalently,

$$\mathbf{i}_r^T \cos A - r \sin A \frac{\partial A}{\partial \mathbf{r}} = -\mathbf{i}_s^T$$

Solving for the derivative of $A$ with respect to the components of $\mathbf{r}$ gives

$$\frac{\partial A}{\partial \mathbf{r}} = \frac{1}{r \sin A} (\cos A \, \mathbf{i}_r + \mathbf{i}_s)^T = \frac{1}{r} \mathbf{i}_n^T$$

The vector

$$\mathbf{i}_n = \frac{1}{\sin A} (\cos A \, \mathbf{i}_r + \mathbf{i}_s) \qquad (13.3)$$

**628**                    The Celestial Position Fix                    [Chap. 13

is readily seen to be a unit vector which is in the plane of the measurement, (i.e., the plane determined by the spacecraft, the near body, and the direction to the star), and is normal to the line of sight to the near body.

The so-called *measurement geometry vector* **h** for a near body and star angle measurement is, therefore,

$$\mathbf{h} = \frac{1}{r}\mathbf{i}_n \tag{13.4}$$

Measuring the Apparent Angular Diameter of a Planet

Referring to Fig. 13.2, if $D$ is the actual diameter of a planet, the apparent angular diameter $A$ is found from

$$r \sin \tfrac{1}{2} A = \tfrac{1}{2} D \tag{13.5}$$

Again we assume that the position vector **r** of the spacecraft is measured relative to the planet. We have

$$\frac{\partial r}{\partial \mathbf{r}} \sin \tfrac{1}{2} A + \tfrac{1}{2} r \cos \tfrac{1}{2} A \frac{\partial A}{\partial \mathbf{r}} = \mathbf{0}^{\mathrm{T}}$$

in the same manner as before. Hence,

$$\mathbf{h} = \left[\frac{\partial A}{\partial \mathbf{r}}\right]^{\mathrm{T}} = -\frac{D}{r^2 \cos \tfrac{1}{2} A}\mathbf{i}_r \tag{13.6}$$

Star-Elevation Measurement

Consider next the measurement of the angle between the lines of sight to a star and the edge of a planet disk. From Fig. 13.3 we have

$$r \cos(A + \gamma) = -\mathbf{i}_s^{\mathrm{T}} \mathbf{r} \quad \text{and} \quad r \sin \gamma = \tfrac{1}{2} D$$

where $A$ is the angle to be measured and $\gamma$ is the angle between the lines of sight to the center of the planet and to the planet edge. Therefore,

$$\cos(A + \gamma)\,\mathbf{i}_r^{\mathrm{T}} - r \sin(A + \gamma)\left(\frac{\partial A}{\partial \mathbf{r}} + \frac{\partial \gamma}{\partial \mathbf{r}}\right) = -\mathbf{i}_s^{\mathrm{T}}$$

and

$$\sin \gamma\,\mathbf{i}_r^{\mathrm{T}} + r \cos \gamma \frac{\partial \gamma}{\partial \mathbf{r}} = \mathbf{0}^{\mathrm{T}}$$

so that

$$\frac{\partial A}{\partial \mathbf{r}} = \frac{1}{r \cos \gamma}(\sin \gamma\,\mathbf{i}_r + \cos \gamma\,\mathbf{i}_n)^{\mathrm{T}}$$

where

$$\mathbf{i}_n = \frac{1}{\sin(A + \gamma)}[\cos(A + \gamma)\mathbf{i}_r + \mathbf{i}_s] \tag{13.7}$$

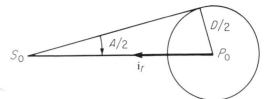

**Fig. 13.2:** Measurement of the apparent diameter of a planet.

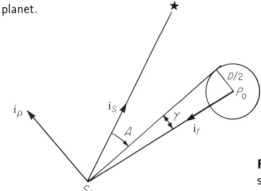

**Fig. 13.3:** Measurement of star elevation angle.

is a vector in the plane of the measurement and perpendicular to the line of sight to the planet.

It is easy to see that $\sin\gamma\,\mathbf{i}_n - \cos\gamma\,\mathbf{i}_r$ is a unit vector in the direction from the spacecraft to the planet edge and that $r\cos\gamma$ is the corresponding distance to that edge. Thus, the measurement geometry vector is simply

$$\mathbf{h} = \frac{1}{r\cos\gamma}\,\mathbf{i}_\rho \quad \text{where} \quad \mathbf{i}_\rho = \cos\gamma\,\mathbf{i}_n + \sin\gamma\,\mathbf{i}_r \tag{13.8}$$

The unit vector $\mathbf{i}_\rho$ lies in the plane of the measurement and perpendicular to the line of sight to the planet edge.

Star-Occultation Measurement

The next type of measurement to be considered is that of noting the time at which a star is occulted by a planet. The analysis depends directly on the star-elevation measurement just considered.

Let $\mathbf{v}_p$ and $\mathbf{v}$ be the respective velocity vectors of the planet and the spacecraft so that

$$\mathbf{v}_r = \mathbf{v} - \mathbf{v}_p$$

is the velocity of the spacecraft relative to the planet. Clearly, the rate of change of the star-elevation angle $A$ is

$$r\cos\gamma\,\frac{dA}{dt} = -\mathbf{v}_r^T\mathbf{i}_\rho \quad \text{or} \quad -\mathbf{v}_r^T\mathbf{i}_\rho\,dt = r\cos\gamma\,dA$$

Hence,
$$-\mathbf{v}_r^T \mathbf{i}_\rho \frac{\partial t}{\partial \mathbf{r}} = r \cos\gamma \frac{\partial A}{\partial \mathbf{r}} = \mathbf{i}_\rho^T$$
so that
$$\mathbf{h} = -\frac{1}{\mathbf{v}_r \cdot \mathbf{i}_\rho} \mathbf{i}_\rho$$
where the unit vector $\mathbf{i}_\rho$ is defined as in the previous subsection.

### Measuring the Angle between Two Near Bodies

For specificity, consider the two near bodies to be the sun and a planet. Let $S_0$ and $P_0$ be, respectively, the reference positions of the spacecraft and the planet at the time of the measurement as shown in Fig. 13.4. Further, let $\mathbf{r}$ be the vector from the sun to $S_0$ and $\mathbf{z}$ the vector from $S_0$ to $P_0$. With $A$ denoting the angle from the sun line to the planet line, we have

$$rz \cos A = -\mathbf{r} \cdot \mathbf{z} = -\mathbf{r}^T \mathbf{z} = -\mathbf{z}^T \mathbf{r} \quad \text{and} \quad \mathbf{r} + \mathbf{z} = \mathbf{r}_p$$

Hence,
$$z \cos A \frac{\partial r}{\partial \mathbf{r}} + r \cos A \frac{\partial z}{\partial \mathbf{r}} - rz \sin A \frac{\partial A}{\partial \mathbf{r}} = -\mathbf{r}^T \frac{\partial \mathbf{z}}{\partial \mathbf{r}} - \mathbf{z}^T \frac{\partial \mathbf{r}}{\partial \mathbf{r}}$$

To dispose of the various partial derivatives in this equation, we first differentiate the scalar relation

$$z^2 = r_p^2 - 2\mathbf{r}_p^T \mathbf{r} + r^2$$

and obtain

$$2z \frac{\partial z}{\partial \mathbf{r}} = -2\mathbf{r}_p^T + 2r\, \mathbf{i}_r^T = -2\mathbf{z}^T \quad \text{so that} \quad \frac{\partial z}{\partial \mathbf{r}} = -\mathbf{i}_z^T$$

Then, from the vector relation between $\mathbf{r}$ and $\mathbf{z}$, we have

$$\frac{\partial \mathbf{r}}{\partial \mathbf{r}} + \frac{\partial \mathbf{z}}{\partial \mathbf{r}} = \mathbf{O} \quad \text{so that} \quad \frac{\partial \mathbf{z}}{\partial \mathbf{r}} = -\mathbf{I}$$

As a consequence, the measurement geometry vector $\mathbf{h} = [\partial A/\partial \mathbf{r}]^T$ for this measurement is

$$\mathbf{h} = \frac{1}{r} \mathbf{i}_n + \frac{1}{z} \mathbf{i}_m \tag{13.9}$$

where

$$\mathbf{i}_n = \frac{1}{\sin A}(\mathbf{i}_z + \cos A\, \mathbf{i}_r) \quad \text{and} \quad \mathbf{i}_m = -\frac{1}{\sin A}(\mathbf{i}_r + \cos A\, \mathbf{i}_z)$$

are unit vectors, each lying in the plane of the measurement, i.e., the plane determined by the spacecraft and the two near bodies. The vector $\mathbf{i}_n$ is

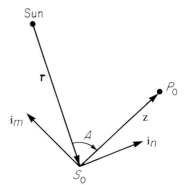

**Fig. 13.4:** Measurement of the angle between the sun and a planet.

normal to the line of sight to the sun whereas $\mathbf{i}_m$ is normal to the line of sight to the planet.

The measurement of the angle between the lines of sight to a near body and to a star may be regarded as a special case of this measurement. If we allow the planet to recede to infinity, the measurement geometry vector becomes the same as that previously considered, i.e., Eq. (13.4).

### Radar-Range, Azimuth, and Elevation Measurements

Assume a radar site on the surface of the earth to be the origin of the coordinate system, and let a cartesian coordinate system be chosen such that the $z$ axis is radially out from the center of the earth through the radar site, the $x$ axis is positive in the direction from which radar azimuths are to be measured, and the $y$ axis completes the coordinate system. Then we may write

$$\mathbf{r} = r \begin{bmatrix} \cos\beta\cos\theta \\ \cos\beta\sin\theta \\ \sin\beta \end{bmatrix} \tag{13.10}$$

where $r$, $\theta$, $\beta$ are, respectively, the range, azimuth, and elevation of the vehicle as observed from the radar site. Hence,

$$\frac{\partial \mathbf{r}}{\partial \mathbf{r}} = \mathbf{I} = \begin{bmatrix} \cos\beta\cos\theta \\ \cos\beta\sin\theta \\ \sin\beta \end{bmatrix} \frac{\partial r}{\partial \mathbf{r}} + \begin{bmatrix} -\sin\beta\cos\theta \\ -\sin\beta\sin\theta \\ \cos\beta \end{bmatrix} r\frac{\partial \beta}{\partial \mathbf{r}} + \begin{bmatrix} -\sin\theta \\ \cos\theta \\ 0 \end{bmatrix} r\cos\beta \frac{\partial \theta}{\partial \mathbf{r}}$$

The vector coefficients in these equations are recognized as orthogonal unit vectors in the directions of increasing $r$, $\beta$, $\theta$, respectively. Thus, we may solve for the partial derivatives by successively multiplying this last equation by the transpose of these unit vectors to obtain

$$\frac{\partial r}{\partial \mathbf{r}} = \begin{bmatrix} \cos\beta\cos\theta & \cos\beta\sin\theta & \sin\beta \end{bmatrix}$$

$$\frac{\partial \beta}{\partial \mathbf{r}} = \frac{1}{r}\begin{bmatrix} -\sin\beta\cos\theta & -\sin\beta\sin\theta & \cos\beta \end{bmatrix} \quad (13.11)$$

$$\frac{\partial \theta}{\partial \mathbf{r}} = \frac{1}{r\cos\beta}\begin{bmatrix} -\sin\theta & \cos\theta & 0 \end{bmatrix}$$

Any three independent measurements can be used to define a navigation fix. Then, since the governing equations for each of the various types of measurements are linear in $\delta q$ and $\delta \mathbf{r}$, we may express the results in the vector-matrix form

$$\delta \mathbf{q} = \mathbf{H}^T \delta \mathbf{r} \quad (13.12)$$

where $\mathbf{H}$ is a 3×3 matrix each of whose columns is composed of the components of the $\mathbf{h}$ vector for an individual measurement. The composite vector $\delta \mathbf{q}$ is the three-dimensional column vector composed of the deviations of observed quantities from their corresponding reference values. If the three measurements represented by the *measurement geometry matrix* $\mathbf{H}$ are, indeed, independent, then the position deviation of the spacecraft at the time of the measurements may be computed from

$$\delta \mathbf{r} = \mathbf{H}^{-T} \delta \mathbf{q} \quad (13.13)$$

## 13.3 Error Analysis of the Navigation Fix

Three independent and precise measurements made at a known instant of time suffice to determine uniquely the position of a spacecraft. Because of the presence of instrumentation errors, there will be an uncertainty associated with the position fix. The best choice of measurements at any time depends on the position of the spacecraft relative to the geometry of the solar system. In order to demonstrate explicitly the effect of different sets of measurements, we shall derive analytic expressions for the errors which result from different combinations of measurements.

For this purpose, it will be necessary to distinguish between the measured and true values of the quantities $\delta \mathbf{q}$ and the estimated and true values of the position deviation $\delta \mathbf{r}$. The notation $\delta \widetilde{\mathbf{q}}$ will represent the measured value of the deviation in $\mathbf{q}$ from its reference value, and $\delta \mathbf{q}$ will be the true value of the deviation. Likewise the notation $\delta \widehat{\mathbf{r}}$ will be used for the estimated or inferred value of the deviation in $\mathbf{r}$ from its reference value and $\delta \mathbf{r}$ will be the actual deviation. We may then write

$$\delta \widetilde{\mathbf{q}} = \delta \mathbf{q} + \boldsymbol{\alpha}$$
$$\delta \widehat{\mathbf{r}} = \delta \mathbf{r} + \boldsymbol{\epsilon}$$

Sect. 13.3]    Error Analysis of the Navigation Fix    633

where $\boldsymbol{\alpha}$ and $\boldsymbol{\epsilon}$ denote explicitly the errors in the measurements and the estimates.

The true deviations $\delta \mathbf{q}$ and $\delta \mathbf{r}$ clearly satisfy Eq. (13.13). With just a sufficient number of measurements being considered, then, for an unbiased estimate, we must have

$$\delta \hat{\mathbf{r}} = \mathbf{H}^{-T} \delta \widetilde{\mathbf{q}} \tag{13.14}$$

Thus, the error vectors $\boldsymbol{\epsilon}$ and $\boldsymbol{\alpha}$ are also related by

$$\boldsymbol{\epsilon} = \mathbf{H}^{-T} \boldsymbol{\alpha} \tag{13.15}$$

The magnitude of the error vector $\boldsymbol{\epsilon}$ is simply the square root of the quantity

$$\epsilon^2 = \boldsymbol{\epsilon}^T \boldsymbol{\epsilon} = \boldsymbol{\alpha}^T \mathbf{H}^{-1} \mathbf{H}^{-T} \boldsymbol{\alpha} = \boldsymbol{\alpha}^T (\mathbf{H}^T \mathbf{H})^{-1} \boldsymbol{\alpha} = \boldsymbol{\alpha}^T \mathbf{M}^{-1} \boldsymbol{\alpha} \tag{13.16}$$

where

$$\mathbf{M} = \mathbf{H}^T \mathbf{H} = \begin{bmatrix} \mathbf{h}_1 \cdot \mathbf{h}_1 & \mathbf{h}_1 \cdot \mathbf{h}_2 & \mathbf{h}_1 \cdot \mathbf{h}_3 \\ \mathbf{h}_2 \cdot \mathbf{h}_1 & \mathbf{h}_2 \cdot \mathbf{h}_2 & \mathbf{h}_2 \cdot \mathbf{h}_3 \\ \mathbf{h}_3 \cdot \mathbf{h}_1 & \mathbf{h}_3 \cdot \mathbf{h}_2 & \mathbf{h}_3 \cdot \mathbf{h}_3 \end{bmatrix} \tag{13.17}$$

Thus, $\epsilon^2$ is a *positive definite quadratic form* in the components of the measurement error vector $\boldsymbol{\alpha}$. Bounds on $\epsilon^2$ can be expressed in terms of the measurement geometry vectors $\mathbf{h}_i$, which constitute the columns of the matrix $\mathbf{H}$, as we shall now demonstrate.

Since the matrix $\mathbf{M}$ is positive definite, it is possible to rotate the system of coordinates in such a way as to diagonalize the matrix $\mathbf{M}$ and to transform the quadratic form to a sum of squares. Denote by $\mathbf{B}$ the modal matrix of $\mathbf{M}$. Then $\mathbf{B}$ is an orthogonal matrix, i.e., $\mathbf{B}^{-1} = \mathbf{B}^T$ and

$$\mathbf{B}^T \mathbf{M} \mathbf{B} \equiv \mathbf{D} = \begin{bmatrix} \lambda_1 & 0 & 0 \\ 0 & \lambda_2 & 0 \\ 0 & 0 & \lambda_3 \end{bmatrix} \quad \text{so that} \quad \mathbf{D}^{-1} = \mathbf{B}^T \mathbf{M}^{-1} \mathbf{B}$$

The characteristic equations of $\mathbf{M}$ and $\mathbf{D}$ are the same:

$$|\mathbf{M} - \lambda \mathbf{I}| = |\mathbf{D} - \lambda \mathbf{I}| = -\lambda^3 + \beta_1 \lambda^2 - \beta_2 \lambda + \beta_3$$
$$= (\lambda_1 - \lambda)(\lambda_2 - \lambda)(\lambda_3 - \lambda)$$

and the coefficients of the characteristic equation are related to the roots as

$$\begin{aligned} \beta_1 &= \lambda_1 + \lambda_2 + \lambda_3 \\ \beta_2 &= \lambda_1 \lambda_2 + \lambda_2 \lambda_3 + \lambda_1 \lambda_3 \\ \beta_3 &= \lambda_1 \lambda_2 \lambda_3 \end{aligned} \tag{13.18}$$

Using Eq. (13.17), these coefficients can also be determined as

$$\begin{aligned} \beta_1 &= \operatorname{tr} \mathbf{M} = |\mathbf{h}_1|^2 + |\mathbf{h}_2|^2 + |\mathbf{h}_3|^2 \\ \beta_2 &= |\mathbf{h}_1 \times \mathbf{h}_2|^2 + |\mathbf{h}_1 \times \mathbf{h}_3|^2 + |\mathbf{h}_2 \times \mathbf{h}_3|^2 \\ \beta_3 &= |\mathbf{M}| = |\mathbf{H}|^2 = (\mathbf{h}_1 \times \mathbf{h}_2 \cdot \mathbf{h}_3)^2 \end{aligned} \tag{13.19}$$

**634**  The Celestial Position Fix  [Chap. 13]

If the coordinate system is rotated using the modal matrix $\mathbf{B}$, then the error vector $\boldsymbol{\alpha}'$ in the rotated coordinate system is related to the original error vector $\boldsymbol{\alpha}$ by the linear transformation

$$\boldsymbol{\alpha} = \mathbf{B}\boldsymbol{\alpha}'$$

As a consequence, from Eq. (13.16) we have

$$\epsilon^2 = \boldsymbol{\epsilon}^T \boldsymbol{\epsilon} = \boldsymbol{\alpha}'^T \mathbf{B}^T \mathbf{M}^{-1} \mathbf{B} \boldsymbol{\alpha}' = \boldsymbol{\alpha}'^T \mathbf{D}^{-1} \boldsymbol{\alpha}' = \frac{\alpha_1'^2}{\lambda_1} + \frac{\alpha_2'^2}{\lambda_2} + \frac{\alpha_3'^2}{\lambda_3}$$

Now, if $\alpha^2$ is the square of the magnitude of the error vector, then

$$\epsilon^2 \leq \left(\frac{1}{\lambda_1} + \frac{1}{\lambda_2} + \frac{1}{\lambda_3}\right)\alpha^2 = \frac{\beta_2}{\beta_3}\alpha^2 \qquad (13.20)$$

according to the definitions (13.18). But, it is easy to see that

$$\beta_2 = \tfrac{1}{2}[(\operatorname{tr}\mathbf{M})^2 - (\lambda_1^2 + \lambda_2^2 + \lambda_3^2)] \leq \tfrac{1}{2}\beta_1^2 \qquad (13.21)$$

Therefore, we have

$$\epsilon^2 \leq \frac{\beta_2}{\beta_3}\alpha^2 \leq \frac{\beta_1^2}{2\beta_3}\alpha^2 \qquad (13.22)$$

and the $\beta$'s involve only the measurement geometry vectors.

We shall now consider several different combinations of measurements and calculate upper bounds for the error in each position estimate.

### Planet-Star, Planet-Star, Planet-Diameter Measurement

For convenience, choose a coordinate system $x, y, z$ centered in the spacecraft with the $z$ axis in the direction of the planet as shown in Fig. 13.5. Let $\mathbf{i}_{n1}$ and $\mathbf{i}_{n2}$ be unit vectors in the respective planes of the planet-star measurements and normal to the direction from the spacecraft to the planet. These vectors will lie in the $x, y$ plane, and we may take $\mathbf{i}_{n1}$ to be along the positive $x$ axis. Then, if $\theta$ is the angle between $\mathbf{i}_{n1}$ and $\mathbf{i}_{n2}$, we have

$$\mathbf{h}_1 = \frac{1}{z}\begin{bmatrix} 1 \\ 0 \\ 0 \end{bmatrix} \qquad \mathbf{h}_2 = \frac{1}{z}\begin{bmatrix} \cos\theta \\ \sin\theta \\ 0 \end{bmatrix} \qquad \mathbf{h}_3 = \frac{D}{z^2 \cos\tfrac{1}{2}A}\begin{bmatrix} 0 \\ 0 \\ 1 \end{bmatrix}$$

Therefore, with Eqs. (13.19) to calculate the $\beta$'s,

$$\beta_1 = \frac{2z^2 \cos^2\tfrac{1}{2}A + D^2}{z^4 \cos^2\tfrac{1}{2}A}$$

$$\beta_2 = \frac{z^2 \sin^2\theta \cos^2\tfrac{1}{2}A + 2D^2}{z^6 \cos^2\tfrac{1}{2}A}$$

$$\beta_3 = \frac{D^2 \sin^2\theta}{z^8 \cos^2\tfrac{1}{2}A}$$

Sect. 13.3]  Error Analysis of the Navigation Fix  635

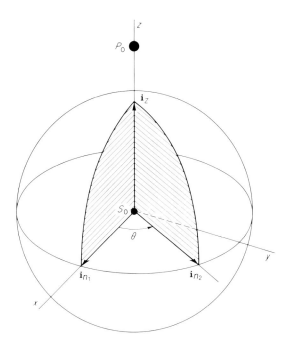

**Fig. 13.5:** Planet-star, planet-star, planet-diameter measurement.

we substitute in (13.22) to obtain

$$\epsilon^2 \leq \frac{z^4 \sin^2\theta \cos^2\frac{1}{2}A + 2D^2 z^2}{D^2 \sin^2\theta} \alpha^2 \leq \frac{(2z^2 \cos^2\frac{1}{2}A + D^2)^2}{2D^2 \cos^2\frac{1}{2}A \sin^2\theta} \alpha^2 \quad (13.23)$$

as upper bounds on the square of the position estimation error. Clearly, the error will be a minimum if two stars can be found such that $\mathbf{i}_{n1}$ and $\mathbf{i}_{n2}$ are orthogonal. We further note that the error is reduced as the distance between the spacecraft and the planet decreases.

### Planet-Star, Planet-Star, Sun-Star Measurement

Choose a coordinate system oriented as described above and illustrated in Fig. 13.6. Let $\mathbf{i}_{n1}$ and $\mathbf{i}_{n2}$ be unit vectors as previously defined, and let $\mathbf{i}_{n3}$ be a unit vector in the plane of the sun-star measurement and normal to the direction from the spacecraft to the sun. Then from the figure we have

$$\mathbf{h}_1 = \frac{1}{z}\begin{bmatrix} 1 \\ 0 \\ 0 \end{bmatrix} \qquad \mathbf{h}_2 = \frac{1}{z}\begin{bmatrix} \cos\theta \\ \sin\theta \\ 0 \end{bmatrix} \qquad \mathbf{h}_3 = \frac{1}{r}\begin{bmatrix} \cos\gamma\cos\phi \\ \cos\gamma\sin\phi \\ \sin\gamma \end{bmatrix}$$

so that

$$\beta_1 = \frac{2}{z^2} + \frac{1}{r^2}$$

$$\beta_2 = \frac{1}{z^4}\sin^2\theta + \frac{1}{r^2 z^2}\{2 - \cos^2\gamma\,[1 + \cos\theta\cos(2\phi - \theta)]\}$$

$$\beta_3 = \frac{1}{r^2 z^4}\sin^2\theta\sin^2\gamma$$

are obtained using Eqs. (13.19).

One of the two possible upper bounds on the squared estimation error is

$$\epsilon^2 \leq \frac{\beta_1^2}{2\beta_3}\alpha^2 = \frac{(2r^2 + z^2)^2}{2r^2\sin^2\theta\sin^2\gamma}\alpha^2 \qquad (13.24)$$

Clearly, the best star to choose for the sun-star measurement is one lying in the plane containing the spacecraft, sun, and planet, for then $\gamma$ will assume its maximum value, which is the angle $A$ between the planet and the sun. Again the two best stars for the planet-star measurements are, apparently, those for which the angle between the planes of measurement is $\theta = \frac{1}{2}\pi$. However, with this analysis, the angle $\phi$ is unspecified. Indeed, $\phi$ appears only in the coefficient $\beta_2$ which is the numerator of the more conservative bound $\beta_2/\beta_3$.

To minimize $\beta_2$, we have two choices, depending on the quadrant of $\theta$. First, if $0 \leq \theta < \frac{1}{2}\pi$, then we want $2\phi - \theta = 0$ or $\phi = \frac{1}{2}\theta$. On the other hand, if $\frac{1}{2}\pi < \theta \leq \pi$, then we want $2\phi - \theta = \pi$ or $\phi = \frac{1}{2}(\pi + \theta)$. In either case,

$$\beta_2 = \frac{1}{z^4}\sin^2\theta + \frac{1}{r^2 z^2}[2 - \cos^2\gamma(1 \pm \cos\theta)]$$

where the choice of upper or lower sign depends on whether $\theta$ is in the first or second quadrant, respectively.

The ratio $\beta_2/\beta_3$ as a function of $\gamma$ has the form

$$\frac{\beta_2}{\beta_3} = \frac{a - b\cos^2\gamma}{\sin^2\gamma}$$

where $a(\theta) > b(\theta)$. Hence,

$$\frac{d}{d\gamma}\left(\frac{\beta_2}{\beta_3}\right) = 2(b - a)\csc^2\gamma\cot\gamma < 0$$

Thus, to minimize $\beta_2/\beta_3$ we should, as before, choose $\gamma$ to be as large as possible, namely, $\gamma = A$. There results

$$\frac{\beta_2}{\beta_3} = r^2\csc^2 A + \frac{2 + \cot^2 A\,(1 \mp \cos\theta)}{\sin^2\theta}z^2$$

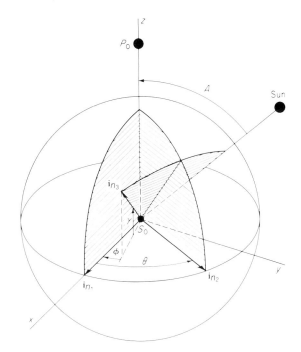

**Fig. 13.6:** Planet-star, planet-star, sun-star measurement.

Finally, to select the optimum value for $\theta$ we set to zero the derivative of $\beta_2/\beta_3$ with respect to $\theta$. When $\theta$ lies in the first quadrant, we obtain

$$(1 - \cos\theta)^2 - 4\tan^2 A \cos\theta = 0 \quad \text{or} \quad \sin^2 A = \left(\frac{1 - \cos\theta}{1 + \cos\theta}\right)^2$$

Therefore,
$$\cos\theta = \frac{1 - \sin A}{1 + \sin A} \quad \text{for} \quad 0 \leq \theta < \tfrac{1}{2}\pi$$

On the other hand, if $\theta$ is in the second quadrant, we have

$$(1 + \cos\theta)^2 + 4\tan^2 A \cos\theta = 0 \quad \text{or} \quad \sin^2 A = \left(\frac{1 + \cos\theta}{1 - \cos\theta}\right)^2$$

so that
$$\cos\theta = -\frac{1 - \sin A}{1 + \sin A} \quad \text{for} \quad \tfrac{1}{2}\pi < \theta \leq \pi$$

In either case, the squared-estimation error is

$$\epsilon^2 \leq \frac{r^2 + \tfrac{1}{2}z^2(1 + \sin A)^2}{\sin^2 A}\alpha^2 \tag{13.25}$$

provided that (1) the star for the sun-star measurement lies in the plane containing the spacecraft, sun, and planet, (2) the two stars for the planet-star measurements are those for which the angle between the planes of

**638**  The Celestial Position Fix  [Chap. 13]

measurement is $\arccos[(1-\sin A)/(1+\sin A)]$, and (3) that angle is bisected by the plane of the sun-star measurement.

Entirely, analogous results are obtained if the three measurements are sun-star, sun-star, and planet-star by simply interchanging $r$ and $z$ in the equations.

### Planet-Star, Planet-Star, Planet-Sun Measurement

A particularly useful set of three observations consists of measuring the angles between the lines of sight to a nearby planet and (1) a star, (2) a different star, and (3) the sun. However, if the planet is much closer to the spacecraft than to the sun, the inverse of the $\mathbf{H}^T$ matrix, as required for the estimation formula (13.14), cannot be obtained by straightforward means.

Before analyzing the effectiveness of this measurement set, we devise a procedure to handle the matrix inversion problem. For this purpose, let unit vectors be defined as follows:

$\mathbf{i}_{n1}$: in the plane of the planet-star$_1$ measurement and normal to the direction from the spacecraft to the planet,

$\mathbf{i}_{n2}$: in the plane of the planet-star$_2$ measurement and normal to the direction from the spacecraft to the planet,

$\mathbf{i}_{n3}$: in the plane of the planet-sun measurement and normal to the direction from the spacecraft to the planet, and

$\mathbf{i}_{n4}$: in the plane of the planet-sun measurement and normal to the direction from the spacecraft to the sun.

Now if $r$ and $z$ are the distances to the sun and to the planet, respectively, then the $\mathbf{H}$ matrix is

$$\mathbf{H} = \begin{bmatrix} \dfrac{\mathbf{i}_{n1}}{z} & \dfrac{\mathbf{i}_{n2}}{z} & \dfrac{\mathbf{i}_{n3}}{z} + \dfrac{\mathbf{i}_{n4}}{r} \end{bmatrix}$$

Since $\mathbf{i}_{n1}$, $\mathbf{i}_{n2}$, and $\mathbf{i}_{n3}$ are each perpendicular to the same direction, they cannot be independent. Therefore, if $z \ll r$, the matrix $\mathbf{H}$ will be ill-conditioned—having a determinant nearly equal to zero.

To counter this difficulty, define two scalar quantities $a$ and $b$ such that

$$\mathbf{i}_{n3} = a\,\mathbf{i}_{n1} + b\,\mathbf{i}_{n2}$$

so that the measurement matrix may be written as

$$\mathbf{H} = \begin{bmatrix} \mathbf{i}_{n1} & \mathbf{i}_{n2} & \mathbf{i}_{n4} \end{bmatrix} \begin{bmatrix} 1/z & 0 & a/z \\ 0 & 1/z & b/z \\ 0 & 0 & 1/r \end{bmatrix}$$

### Sect. 13.3]    Error Analysis of the Navigation Fix

with the coefficients $a$ and $b$ determined from

$$a = \frac{\mathbf{i}_{n1} \cdot \mathbf{i}_{n3} - (\mathbf{i}_{n1} \cdot \mathbf{i}_{n2})(\mathbf{i}_{n2} \cdot \mathbf{i}_{n3})}{1 - (\mathbf{i}_{n1} \cdot \mathbf{i}_{n2})^2}$$

$$b = \frac{\mathbf{i}_{n2} \cdot \mathbf{i}_{n3} - (\mathbf{i}_{n1} \cdot \mathbf{i}_{n2})(\mathbf{i}_{n1} \cdot \mathbf{i}_{n3})}{1 - (\mathbf{i}_{n1} \cdot \mathbf{i}_{n2})^2}$$

Since the vectors $\mathbf{i}_{n1}$, $\mathbf{i}_{n2}$, and $\mathbf{i}_{n4}$ are independent, the inversion of the first matrix factor of $\mathbf{H}$ presents no problem. Since the inversion of the second factor can be obtained explicitly, then

$$\mathbf{H}^{-1} = \begin{bmatrix} z & 0 & -ar \\ 0 & z & -br \\ 0 & 0 & r \end{bmatrix} \begin{bmatrix} \mathbf{i}_{n1} & \mathbf{i}_{n2} & \mathbf{i}_{n4} \end{bmatrix}^{-1} \tag{13.26}$$

which is free of numerical difficulties and may be used without compunction in the estimation formula (13.14).

Refer now to Fig. 13.7 and choose a coordinate system as before. Then the matrix $\mathbf{H}$ may be written as

$$\mathbf{H} = \begin{bmatrix} 1 & \cos\theta & -\cos A \cos\phi \\ 0 & \sin\theta & -\cos A \sin\phi \\ 0 & 0 & \sin A \end{bmatrix} \begin{bmatrix} 1/z & 0 & a/z \\ 0 & 1/z & b/z \\ 0 & 0 & 1/r \end{bmatrix}$$

where

$$a = \frac{\sin(\theta - \phi)}{\sin\theta} \quad \text{and} \quad b = \frac{\sin\phi}{\sin\theta}$$

The inverse matrix is

$$\mathbf{H}^{-1} = \begin{bmatrix} z & -z\cot\theta & az\cot A - ar\csc A \\ 0 & -z\csc\theta & bz\cot A - br\csc A \\ 0 & 0 & r\csc A \end{bmatrix}$$

For the error analysis of this measurement set, it is more convenient to use Eq. (13.16) to write

$$\epsilon^2 \leq \operatorname{tr}(\mathbf{H}^{-1}\mathbf{H}^{-T})\alpha^2$$

since we already have an explicit expression for the inverse of the matrix $\mathbf{H}$. Thus, the factor bounding the squared-estimation error is

$$2z^2\csc^2\theta + r^2\csc^2 A + (z\cot A - r\csc A)^2 \left(\frac{\sin^2(\theta-\phi) + \sin^2\phi}{\sin^2\theta}\right)$$

Again the optimum choice of $\phi$ is one-half $\theta$, so that the minimum value of the upper bound as a function of $\theta$ is

$$2z^2\csc^2\theta + r^2\csc^2 A + \frac{(z\cot A - r\csc A)^2}{1 + \cos\theta}$$

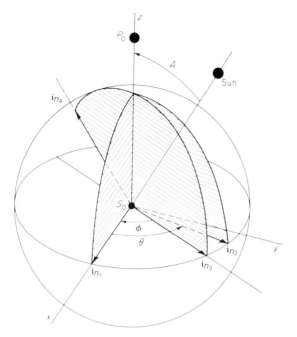

**Fig. 13.7:** Planet-star, planet-star, planet-sun measurement.

The optimum value of $\theta = \theta_o$ is found as the solution of

$$\frac{4z^2 \cos \theta_o}{(1 - \cos \theta_o)^2} = (z \cot A - r \csc A)^2$$

Thus,

$$\cos \theta_o = \frac{\sqrt{1 - 2p \cos A + p^2} - \sin A}{\sqrt{1 - 2p \cos A + p^2} + \sin A}$$

where, for convenience, we have defined

$$p = \frac{r}{z}$$

With this value of $\theta$ the optimum bound on the squared-estimation error is expressed as

$$\epsilon^2 \leq r^2 \csc^2 A + \tfrac{1}{2} z^2 \csc^2 A \left( \sin A + \sqrt{1 - 2p \cos A + p^2} \right)^2 \alpha^2 \quad (13.27)$$

Here again, by interchanging $r$ and $z$, we obtain analogous results for the measurement set consisting of sun-star$_1$, sun-star$_2$, and sun-planet.

◊ **Problem 13-3**

A fix is made using a planet and two stars. One of the measurements is the apparent diameter of the planet. For the other two measurements we can use (1) the angles between the stars and the center of the planet or (2) the elevation angles of the stars above the planet horizon or (3) the angles between the stars and a planet landmark. Compare the effectiveness of each of the three choices.

◊ **Problem 13-4**

Analyze the effectiveness of a fix obtained by measuring the angles between two stars and a planet and observing a star occultation by the same planet. How should the stars be oriented to minimize the squared measurement error in the position fix?

◊ **Problem 13-5**

Determine the locus in the $p$-$A$ plane for which the measurement sets planet-star, planet-star, sun-star and planet-star, planet-star, planet-sun give equal upper limits for the squared-measurement errors. In other words, determine the relationship between $p$ and $A$ for which the bounds on $\epsilon^2$, as computed by Eqs. (13.25) and (13.27), are equal. In which section of the $p$-$A$ plane is one set of measurements to be preferred over the other?

## 13.4 A Method of Correcting Clock Errors

Thus far, the navigation fix has comprised three measurements of the type previously described. However, with the addition of a fourth independent measurement, it is possible not only to determine the coordinates of the spacecraft position vector but also to infer other information such as, for example, a correction to the spacecraft clock.

For this analysis, we assume that when the spacecraft clock indicates the reference time $t_0$ the sighting process commences. Let the clock be in error at this time by the amount $\delta t_c$ so that the sighting actually begins at the time $t_0 - \delta t_c$. We consider only one type of measurement, namely, the measurement of the angle between the lines of sight to the sun and to a planet. The other measurement types are left as exercises.

As before, $S_0$ and $P_0$ are the reference positions of the spacecraft and the planet at time $t_0$. The vectors $\mathbf{r}$ and $\mathbf{z}$ are again the position vectors of the vehicle with respect to the sun and the planet with respect to the vehicle. With $A$ denoting the angle from the sun line to the planet line, as shown in Fig. 13.8, we shall derive an expression for the change in angle $\delta A$ arising from the motion of the planet during the interval $\delta t_c$ between the reference time and the actual time of measurement and the initial displacement $\delta \mathbf{r}$ of the spacecraft position with respect to $S_0$ at the time $t_0 - \delta t_c$ when the sighting process begins. We note that the initial displacement $\delta \mathbf{r}$ may arise in part from the motion of the spacecraft during the time interval $\delta t_c$ and in part from its deviation from the reference

# The Celestial Position Fix [Chap. 13]

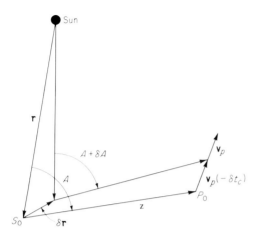

Fig. 13.8: Measurement of the angle between the sun and a planet.

position $S_0$ at time $t_0$. One of the significant tasks in analyzing the fix data is properly to separate these components.

As before, the angle $A$ is given in terms of the scalar product

$$rz \cos A = -\mathbf{r} \cdot \mathbf{z} = -\mathbf{r}^T \mathbf{z} = -\mathbf{z}^T \mathbf{r} \qquad \text{where} \qquad \mathbf{r} + \mathbf{z} = \mathbf{r}_p$$

Differentiate each with respect to time to obtain

$$z \cos A \frac{dr}{dt} + r \cos A \frac{dz}{dt} - rz \sin A \frac{dA}{dt} = -\mathbf{r}^T \frac{d\mathbf{z}}{dt} - \mathbf{z}^T \frac{d\mathbf{r}}{dt}$$

and

$$\frac{d\mathbf{r}}{dt} + \frac{d\mathbf{z}}{dt} = \mathbf{v}_p$$

where $\mathbf{v}_p$ is the velocity vector of the planet at time $t_0$. Then, noting that

$$\frac{dr}{dt} = \mathbf{i}_r^T \frac{d\mathbf{r}}{dt} \qquad \text{and} \qquad \frac{dz}{dt} = \mathbf{i}_z^T \frac{d\mathbf{z}}{dt}$$

the following expression for the time derivative of $A$ results:

$$\frac{dA}{dt} = \mathbf{h}^T \frac{d\mathbf{r}}{dt} - \frac{1}{z} \mathbf{i}_m^T \mathbf{v}_p$$

where $\mathbf{h}$ and $\mathbf{i}_m$ are defined in Eq. (13.9). Thus, we have

$$\delta A = \mathbf{h}^T \delta \mathbf{r} + \frac{\mathbf{i}_m \cdot \mathbf{v}_p}{z} \delta t_c$$

For convenience of notation, it is advantageous to work with a four-dimensional deviation vector $\delta \mathbf{x}$ defined as

$$\delta \mathbf{x} = \begin{bmatrix} \delta \mathbf{r} \\ \delta t_c \end{bmatrix}$$

so that we may express the measured deviation $\delta A$ in the form

$$\delta A = \mathbf{b}^T \delta \mathbf{x}$$

Here, of course, the measurement vector, now denoted by $\mathbf{b}$, is also four dimensional. For the particular type of measurement discussed we have

$$\mathbf{b} = \begin{bmatrix} \mathbf{h} \\ \mathbf{i}_m \cdot \mathbf{v}_p/z \end{bmatrix} = \begin{bmatrix} \mathbf{i}_n/r + \mathbf{i}_m/z \\ \mathbf{i}_m \cdot \mathbf{v}_p/z \end{bmatrix}$$

Then, in matrix form, the expression analogous to Eq. (13.12) is simply

$$\delta \mathbf{q} = \mathbf{H}^T \delta \mathbf{x}$$

where $\mathbf{H}$ is now a $4 \times 4$ matrix each of whose columns consists of the components of the four-dimensional $\mathbf{b}$ vectors for the four individual measurements. As before, the vector $\delta \mathbf{q}$ is composed of the deviations of the observed quantities from their reference values and is also four dimensional. If the matrix $\mathbf{H}$ is nonsingular, i.e., if the measurements provide independent information, then

$$\delta \mathbf{x} = \mathbf{H}^{-T} \delta \mathbf{q}$$

We again distinguish measured, true, and estimated quantities with the same notational conventions of Sect. 13.3, so that

$$\delta \widehat{\mathbf{x}} = \mathbf{H}^{-T} \delta \widetilde{\mathbf{q}}$$

However, with the data available, a better estimate of $\delta \mathbf{r}$ is possible. Once an estimate for $\delta t_c$ has been obtained, a correction may be applied to the position-deviation estimate to account for the fact that the time of starting the fix is in error. This may be accomplished by adding to the position estimate the vector distance traveled by the spacecraft with velocity $\mathbf{v}$ in the time $\widehat{\delta t_c}$. Thus, the best estimate of the four-dimensional deviation vector is obtained from

$$\delta \widehat{\mathbf{x}} = \mathbf{X} \mathbf{H}^{-T} \delta \widetilde{\mathbf{q}} \tag{13.28}$$

where $\mathbf{X}$ is a matrix defined by

$$\mathbf{X} = \begin{bmatrix} \mathbf{I} & \mathbf{v} \\ \mathbf{0}^T & 1 \end{bmatrix}$$

and $\mathbf{I}$ is the three-dimensional identity matrix.

◊ **Problem 13-6**

For an angular diameter measurement in which $\delta t_c$ is the clock error, then

$$\delta A = -\frac{D}{r^2 \cos \frac{1}{2} A} \, \mathbf{i}_r^T \left( \delta \mathbf{r} + \mathbf{v}_p \, \delta t_c \right)$$

where $\mathbf{v}_p$ is the velocity vector of the planet relative to the sun.

## 13.5 Processing Redundant Measurements

The procedure for determining position deviations is called *deterministic* if only a sufficient number of measurements to define the parameters uniquely is assumed. If the instrumentation is perfect, the computed deviations will be exact to within the assumptions inherent in a linear analysis. However, instrumentation inaccuracies do exist, so that it is advantageous to include redundant measurements in the fix to achieve a further reduction in the uncertainties in the quantities to be determined beyond that which can be had using only a minimum number of measurements.

### The Pseudo-Inverse of a Matrix

Assume that a total of $m$ ($m > 3$) measurements has been made for a single position fix. An example of a fix consisting of six angle measurements is illustrated on pages 21 and 22 in the Introduction of this book. The first three measurements involve the nearest planet, and the angles are from the sun and two stars. The fourth measurement is from the sun to a star, and the fifth from the sun to the second nearest planet. Finally, a sixth measurement is the subtended angle of the nearest planet.

The linear equations in vector-matrix form relating the deviations in the measured quantities $\delta \mathbf{q}$ and the deviation in the position vector $\delta \mathbf{r}$ from the reference value is the same as Eq. (13.12), that is,

$$\delta \mathbf{q} = \mathbf{H}^T \delta \mathbf{r} \qquad (13.29)$$

except that now the measurement geometry matrix $\mathbf{H}$ is $3 \times m$ while the vector $\delta \mathbf{q}$ is $m \times 1$. Since $\mathbf{H}$ is not a square matrix, its inverse is not defined in the ordinary sense. However, if the three-dimensional matrix $\mathbf{H}\mathbf{H}^T$ is not singular, then the set of over-determined equations describing the position fix can be "solved" using the so-called *pseudo-inverse* of a matrix.

For this purpose, we multiply Eq. (13.29), first by $\mathbf{H}$ and then by $(\mathbf{H}\mathbf{H}^T)^{-1}$, to obtain

$$\delta \mathbf{r} = (\mathbf{H}\mathbf{H}^T)^{-1} \mathbf{H} \, \delta \mathbf{q} \qquad (13.30)$$

It will soon be apparent what significance, if any, can be attached to this result and if it is ever proper in any sense to speak of Eq. (13.30) as the "solution" of Eq. (13.29).

First, however, let us address a somewhat different but related issue. When we attempt to utilize more than a sufficient number of measurements for determining a position fix, we may also wish to attach different levels of importance to each of the various measurements. Some measurements in the set might, indeed, be more accurate than others in some quantifiable

## Sect. 13.5] Processing Redundant Measurements

way. Consider, for example, a position fix consisting of four measurements.

(1) $\delta q_1 = \mathbf{h}_1^T \delta \mathbf{r}$
(2) $\delta q_2 = \mathbf{h}_2^T \delta \mathbf{r}$
(3) $\delta q_3 = \mathbf{h}_3^T \delta \mathbf{r}$
(4) $\delta q_4 = \mathbf{h}_4^T \delta \mathbf{r}$

which, of course, is equivalent to

$$\delta \mathbf{q} = \mathbf{H}^T \delta \mathbf{r} \quad \text{where} \quad \mathbf{H} = \begin{bmatrix} \mathbf{h}_1 & \mathbf{h}_2 & \mathbf{h}_3 & \mathbf{h}_4 \end{bmatrix}$$

We assume that there is associated with each measurement a weighting factor $1/\sigma_i^2$ such that the smaller the value of $\sigma_i$, the more significant will be the $i^{\text{th}}$ measurement.

Multiply each Eq. (i) in the above set by $\mathbf{h}_i/\sigma_i^2$ and add together the resulting four equations to obtain

$$\frac{\mathbf{h}_1}{\sigma_1^2} \delta q_1 + \frac{\mathbf{h}_2}{\sigma_2^2} \delta q_2 + \frac{\mathbf{h}_3}{\sigma_3^2} \delta q_3 + \frac{\mathbf{h}_4}{\sigma_4^2} \delta q_4 = \left( \frac{\mathbf{h}_1 \mathbf{h}_1^T}{\sigma_1^2} + \frac{\mathbf{h}_2 \mathbf{h}_2^T}{\sigma_2^2} + \frac{\mathbf{h}_3 \mathbf{h}_3^T}{\sigma_3^2} + \frac{\mathbf{h}_4 \mathbf{h}_4^T}{\sigma_4^2} \right) \delta \mathbf{r}$$

This equation may be written in the form

$$\begin{bmatrix} \mathbf{h}_1 & \mathbf{h}_2 & \mathbf{h}_3 & \mathbf{h}_4 \end{bmatrix} \mathbf{A}^{-1} \delta \mathbf{q} = \begin{bmatrix} \mathbf{h}_1 & \mathbf{h}_2 & \mathbf{h}_3 & \mathbf{h}_4 \end{bmatrix} \mathbf{A}^{-1} \begin{bmatrix} \mathbf{h}_1^T \\ \mathbf{h}_2^T \\ \mathbf{h}_3^T \\ \mathbf{h}_4^T \end{bmatrix} \delta \mathbf{r}$$

where $\mathbf{A}$ is the diagonal matrix

$$\mathbf{A} = \begin{bmatrix} \sigma_1^2 & 0 & 0 & 0 \\ 0 & \sigma_2^2 & 0 & 0 \\ 0 & 0 & \sigma_3^2 & 0 \\ 0 & 0 & 0 & \sigma_4^2 \end{bmatrix} \quad (13.31)$$

Thus, we have

$$\mathbf{H A}^{-1} \delta \mathbf{q} = \mathbf{H A}^{-1} \mathbf{H}^T \delta \mathbf{r}$$

Regardless of what these manipulations and the end result really mean, we can certainly regard

$$\delta \widehat{\mathbf{r}} = (\mathbf{H A}^{-1} \mathbf{H}^T)^{-1} \mathbf{H A}^{-1} \delta \widetilde{\mathbf{q}} \quad (13.32)$$

as an estimate of the position deviation vector. The matrix coefficient of $\delta \widetilde{\mathbf{q}}$ is called an *estimator* and, clearly, this estimator is linear.

The estimator is also *unbiased* in the sense that if the measurements are exact then the estimate will be error free. That is, if $\boldsymbol{\alpha} = \mathbf{0}$, then

$$\delta \widetilde{\mathbf{q}} = \delta \mathbf{q} = \mathbf{H}^T \delta \mathbf{r}$$

so that
$$\delta \hat{\mathbf{r}} = (\mathbf{H}\mathbf{A}^{-1}\mathbf{H}^T)^{-1}(\mathbf{H}\mathbf{A}^{-1}\mathbf{H}^T)\,\delta \mathbf{r} = \delta \mathbf{r}$$
The estimator also reduces to the deterministic case
$$\delta \hat{\mathbf{r}} = \mathbf{H}^{-T}\,\delta \widetilde{\mathbf{q}}$$
if there are no redundant measurements. For the proof, assume that $\mathbf{H}$ is a square matrix and nonsingular. Then
$$\delta \hat{\mathbf{r}} = (\mathbf{H}\mathbf{A}^{-1}\mathbf{H}^T)^{-1}\mathbf{H}\mathbf{A}^{-1}\,\delta \widetilde{\mathbf{q}} = (\mathbf{H}^{-T}\mathbf{A}\mathbf{H}^{-1})\mathbf{H}\mathbf{A}^{-1}\,\delta \widetilde{\mathbf{q}} = \mathbf{H}^{-T}\,\delta \widetilde{\mathbf{q}}$$

Finally, as we shall see in the next subsection, this estimator is identical to that obtained using *Gauss' method of weighted least squares*. When this assertion is validated, the somewhat bizarre manipulations which led to this form of the estimator will be better appreciated.

### Gauss' Method of Least Squares

In its simplest form, the method of least squares† is applied to approximate the solution of an overdetermined set of linear algebraic equations of the form
$$\sum_{j=1}^{n} m_{ij} x_j = c_i \quad \text{with} \quad i = 1, 2, \ldots, N > n$$
where $m_{ij}$ and $c_i$ are given quantities. The problem is to determine $x_j$ so that these equations are "as nearly satisfied as possible." Specifically, we define a set of $N$ *residuals* as the differences
$$e_i = \sum_{j=1}^{n} m_{ij} x_j - c_i$$
and choose a set of *weighting factors*, denoted by $w_1, w_2, \ldots, w_N$. Then we determine $x_1, x_2, \ldots, x_n$ so that the weighted sum $R$ of squares of the residuals will be a minimum where
$$R = w_1 e_1^2 + w_2 e_2^2 + \cdots + w_N e_N^2$$

---

† Carl Friedrich Gauss invented and first used the method of least squares in 1795 when he was but 18 years of age. However, 14 years elapsed before publication in his book *Theoria Motus* which we have referred to many times. Meanwhile, Adrien-Marie Legendre independently invented the method and published his results in 1806. Gauss acknowledged Legendre's work in *Theoria Motus* by stating

> "Our principle, which we have made use of since the year 1795, has lately been published by Legendre in the work *Nouvelles méthodes pour la determination des orbites des comètes*, Paris, 1806, where several other properties of this principle have been explained, which, for the sake of brevity, we here omit."

This served only to anger Legendre who wrote to Gauss saying "*You, who are already so rich in discoveries, might have had the decency not to appropriate the method of least squares.*" Despite the evidence which substantiated Gauss' priority, he was indeed magnanimous in bowing apologetically to Legendre.

## Sect. 13.5] Processing Redundant Measurements

In vector-matrix form the basic set of equations is

$$\mathbf{Mx} = \mathbf{c} \qquad (13.33)$$

where $\mathbf{M}$ is an $N \times n$ matrix. The vector of residuals is

$$\mathbf{e} = \mathbf{Mx} - \mathbf{c}$$

and, in terms of the weighting matrix

$$\mathbf{W} = \begin{bmatrix} w_1 & 0 & \cdots & 0 \\ 0 & w_2 & \cdots & 0 \\ \vdots & \vdots & \ddots & \vdots \\ 0 & 0 & \cdots & w_N \end{bmatrix}$$

the weighted sum of squares of the residuals is

$$R = \mathbf{e}^T \mathbf{W} \mathbf{e} = (\mathbf{x}^T \mathbf{M}^T - \mathbf{c}^T) \mathbf{W} (\mathbf{Mx} - \mathbf{c})$$
$$= \mathbf{x}^T \mathbf{M}^T \mathbf{W} \mathbf{Mx} - \mathbf{c}^T \mathbf{W} \mathbf{Mx} - \mathbf{x}^T \mathbf{M}^T \mathbf{W} \mathbf{c} + \mathbf{c}^T \mathbf{W} \mathbf{c}$$

To minimize $R$, we set to zero the derivative

$$\frac{\partial R}{\partial \mathbf{x}} = 2\mathbf{x}^T \mathbf{M}^T \mathbf{W} \mathbf{M} - 2\mathbf{c}^T \mathbf{W} \mathbf{M} = \mathbf{0}^T$$

and develop, thereby, the equation

$$\mathbf{M}^T \mathbf{W} \mathbf{M} \mathbf{x} = \mathbf{M}^T \mathbf{W} \mathbf{c}$$

to be solved for $\mathbf{x}$. We have, then

$$\mathbf{x} = (\mathbf{M}^T \mathbf{W} \mathbf{M})^{-1} \mathbf{M}^T \mathbf{W} \mathbf{c} \qquad (13.34)$$

which is identical in form to the expression (13.32) obtained using the pseudo-inverse matrix. Therefore, *solving a set of overdetermined linear algebraic equations using the pseudo-inverse method is formally equivalent to Gauss' method of least squares.*

Furthermore, we observe that if the coefficient matrix $\mathbf{M}$ is a nonsingular square matrix, then Eq. (13.34) reduces to Cramer's rule

$$\mathbf{x} = \mathbf{M}^{-1} \mathbf{c}$$

as, of course, must be the case.

Finally, before closing this section, we note that when calculating the derivative of $R$, there is a quadratic term of the form

$$y = \mathbf{x}^T \mathbf{B} \mathbf{x} = (\mathbf{x}^T \mathbf{B} \mathbf{x})^T = \mathbf{x}^T \mathbf{B}^T \mathbf{x}$$

whose derivative is

$$\frac{\partial y}{\partial \mathbf{x}} = \mathbf{x}^T \mathbf{B} \mathbf{I} + \mathbf{x}^T \mathbf{B}^T \mathbf{I} = \mathbf{x}^T (\mathbf{B} + \mathbf{B}^T)$$

## 13.6 Recursive Formulations

The formulation of the weighted least-squares estimator as a recursive operation, in which the current estimate is combined with newly acquired information to produce an improved estimate, is the subject of this section. Its importance is fundamental to the general problem of space navigation to be treated in the next chapter and is developed here with pedagogical motives. Significant advantages accrue from a recursive formulation of the navigation problem in that measurement data may be incorporated sequentially as they are obtained. The necessity of batch processing and matrix inversion with its associated numerical pitfalls are, thereby, avoided.

### The Matrix Inversion Lemma

Basic to the development of recursive formulas is a certain matrix identity generally attributed to Georg Ferdinand Frobenius which can be derived most conveniently from the results of Prob. B-20 in Appendix B. There are given two expressions for the inversion of the block partitioned matrix

$$\mathbf{M} = \begin{bmatrix} \mathbf{A}_{nn} & \mathbf{B}_{nm} \\ \mathbf{C}_{mn} & \mathbf{D}_{mm} \end{bmatrix}$$

The first of these is from part (c) of that problem

$$\mathbf{M}^{-1} = \begin{bmatrix} \mathbf{A}_{nn}^{-1} + \mathbf{A}_{nn}^{-1}\mathbf{B}_{nm}\mathbf{E}_{mm}^{-1}\mathbf{C}_{mn}\mathbf{A}_{nn}^{-1} & -\mathbf{A}_{nn}^{-1}\mathbf{B}_{nm}\mathbf{E}_{mm}^{-1} \\ -\mathbf{E}_{mm}^{-1}\mathbf{C}_{mn}\mathbf{A}_{nn}^{-1} & \mathbf{E}_{mm}^{-1} \end{bmatrix}$$

where

$$\mathbf{E}_{mm} = \mathbf{D}_{mm} - \mathbf{C}_{mn}\mathbf{A}_{nn}^{-1}\mathbf{B}_{nm}$$

and the second is from part (f)

$$\mathbf{M}^{-1} = \begin{bmatrix} \mathbf{F}_{nn}^{-1} & -\mathbf{F}_{nn}^{-1}\mathbf{B}_{nm}\mathbf{D}_{mm}^{-1} \\ -\mathbf{D}_{mm}^{-1}\mathbf{C}_{mn}\mathbf{F}_{nn}^{-1} & \mathbf{D}_{mm}^{-1}\mathbf{C}_{mn}\mathbf{F}_{nn}^{-1}\mathbf{B}_{nm}\mathbf{D}_{mm}^{-1} + \mathbf{D}_{mm}^{-1} \end{bmatrix}$$

where

$$\mathbf{F}_{nn} = \mathbf{A}_{nn} - \mathbf{B}_{nm}\mathbf{D}_{mm}^{-1}\mathbf{C}_{mn}$$

Important identities ensue by equating corresponding blocks of these two forms of the inverse. In particular, we have

$$(\mathbf{A}_{nn} - \mathbf{B}_{nm}\mathbf{D}_{mm}^{-1}\mathbf{C}_{mn})^{-1} =$$
$$\mathbf{A}_{nn}^{-1} + \mathbf{A}_{nn}^{-1}\mathbf{B}_{nm}(\mathbf{D}_{mm} - \mathbf{C}_{mn}\mathbf{A}_{nn}^{-1}\mathbf{B}_{nm})^{-1}\mathbf{C}_{mn}\mathbf{A}_{nn}^{-1} \quad (13.35)$$

and

$$(\mathbf{D}_{mm} - \mathbf{C}_{mn}\mathbf{A}_{nn}^{-1}\mathbf{B}_{nm})^{-1} =$$
$$\mathbf{D}_{mm}^{-1} + \mathbf{D}_{mm}^{-1}\mathbf{C}_{mn}(\mathbf{A}_{nn} - \mathbf{B}_{nm}\mathbf{D}_{mm}^{-1}\mathbf{C}_{mn})^{-1}\mathbf{B}_{nm}\mathbf{D}_{mm}^{-1} \quad (13.36)$$

There is also an important determinant identity which we will require that is obtained by equating parts (b) and (e) of that same problem. We obtain

$$|\mathbf{A}_{nn}||\mathbf{D}_{mm} - \mathbf{C}_{mn}\mathbf{A}_{nn}^{-1}\mathbf{B}_{nm}| = |\mathbf{D}_{mm}||\mathbf{A}_{nn} - \mathbf{B}_{nm}\mathbf{D}_{mm}^{-1}\mathbf{C}_{mn}| \quad (13.37)$$

◊ **Problem 13–7**
Another form of the matrix inversion lemma is

$$(\mathbf{I}_{nn} + \mathbf{X}_{nm}\mathbf{Y}_{mn})^{-1} = \mathbf{I}_{nn} - \mathbf{X}_{nm}(\mathbf{I}_{mm} + \mathbf{Y}_{mn}\mathbf{X}_{nm})^{-1}\mathbf{Y}_{mn}$$

NOTE: If $m < n$, there is a tradeoff between the inversion of an $n \times n$ matrix and the easier inversion of an $m \times m$ matrix.

**Problem 13–8**
Use the matrix inversion lemma to obtain

$$\begin{bmatrix} \mathbf{A} & \mathbf{B} \\ \mathbf{C} & \mathbf{D} \end{bmatrix}^{-1} = \begin{bmatrix} (\mathbf{A} - \mathbf{B}\mathbf{D}^{-1}\mathbf{C})^{-1} & (\mathbf{C} - \mathbf{D}\mathbf{B}^{-1}\mathbf{A})^{-1} \\ (\mathbf{B} - \mathbf{A}\mathbf{C}^{-1}\mathbf{D})^{-1} & (\mathbf{D} - \mathbf{C}\mathbf{A}^{-1}\mathbf{B})^{-1} \end{bmatrix}$$

provided that all block partitions are square and have the same dimension.

**Problem 13–9**
The matrix inversion lemma can be used to invert a matrix of the form

$$\mathbf{M} = \mathbf{I} + [\mathbf{a} \quad \mathbf{b}]\begin{bmatrix} \mathbf{c}^T \\ \mathbf{d}^T \end{bmatrix}$$

to obtain

$$\mathbf{M}^{-1} = \mathbf{I} - \frac{1}{\gamma}[\mathbf{a}\mathbf{c}^T + \mathbf{b}\mathbf{d}^T + (\mathbf{a}\mathbf{b}^T - \mathbf{b}\mathbf{a}^T)(\mathbf{d}\mathbf{c}^T - \mathbf{c}\mathbf{d}^T)]$$

where

$$\gamma = 1 + \operatorname{tr}\mathbf{N} + \det \mathbf{N} \quad \text{and} \quad \mathbf{N} = \begin{bmatrix} \mathbf{b}^T\mathbf{d} & \mathbf{b}^T\mathbf{c} \\ -\mathbf{a}^T\mathbf{d} & \mathbf{a}^T\mathbf{c} \end{bmatrix}$$

NOTE: The explicit forms of the fundamental perturbation matrices derived in Sect. 9.7 are precisely of the form of $\mathbf{M}$ in this problem.

## The Information Matrix and its Inverse

The matrix $\Sigma$ defined as

$$\Sigma \equiv \mathbf{H}\mathbf{A}^{-1}\mathbf{H}^T = \sum_{i=1}^{N} \frac{\mathbf{h}_i \mathbf{h}_i^T}{\sigma_i^2} \qquad (13.38)$$

is sometimes called the *information matrix*. It is the sum of terms of the form $\mathbf{h}\mathbf{h}^T/\sigma^2$—one for each measurement. In effect, including another measurement adds information and "increases" the information matrix. The matrix

$$\mathbf{P} \equiv \Sigma^{-1} \qquad (13.39)$$

appears in Eq. (13.32) as a part of the weighted least-squares estimator which can be written as

$$\delta\hat{\mathbf{r}} = \mathbf{F}\,\delta\tilde{\mathbf{q}} \quad \text{where} \quad \mathbf{F} = \mathbf{P}\mathbf{H}\mathbf{A}^{-1} \qquad (13.40)$$

The information matrix can also be expressed recursively in the form

$$\Sigma^* = \Sigma + \mathbf{h}\left(\frac{1}{\sigma^2}\right)\mathbf{h}^T \qquad (13.41)$$

The asterisk denotes the new information matrix obtained by incorporating a new measurement, characterized by the measurement vector $\mathbf{h}$ and the associated weighting factor $1/\sigma^2$, with the old information matrix.

Actually, it is the inverse of the information matrix that is required in the estimator and it would be convenient to have a recursive formula for $\mathbf{P}$ in addition to the one for $\Sigma$. This is precisely the reason for introducing the matrix inversion lemma in the previous subsection. For with $n = 3$, $m = 1$, and

$$\Sigma = \mathbf{P}^{-1} \to \mathbf{A}_{33} \qquad \mathbf{h}^T \to \mathbf{C}_{13}$$
$$\mathbf{h} \to \mathbf{B}_{31} \qquad -\sigma^2 \to \mathbf{D}_{11}$$

we can use Eq. (13.35) to obtain

$$\mathbf{P}^* = \mathbf{P} - \mathbf{P}\mathbf{h}(\sigma^2 + \mathbf{h}^T\mathbf{P}\mathbf{h})^{-1}\mathbf{h}^T\mathbf{P}$$

For convenience, here and in the sequel, we define

$$a = \sigma^2 + \mathbf{h}^T\mathbf{P}\mathbf{h} \quad \text{and} \quad \mathbf{w} = \frac{1}{a}\mathbf{P}\mathbf{h} \qquad (13.42)$$

Then we have

$$\mathbf{P}^* = (\mathbf{I} - \mathbf{w}\mathbf{h}^T)\mathbf{P} \qquad (13.43)$$

as the desired recursive formula for updating the inverse of the information matrix.

### Recursive Form of the Estimator

One form of the Gaussian least-squares estimator is given in Eqs. (13.40). The corresponding formulas, which include one additional measurement, are

$$\delta \hat{\mathbf{r}}^* = \mathbf{F}^* \, \delta \tilde{\mathbf{q}}^* \quad \text{where} \quad \mathbf{F}^* = \mathbf{P}^* \mathbf{H}^* \mathbf{A}^{*-1} \tag{13.44}$$

and the relations between the starred and unstarred quantities are readily seen to be

$$\delta \tilde{\mathbf{q}}^* = \begin{bmatrix} \delta \tilde{\mathbf{q}} \\ \delta q \end{bmatrix} \qquad \mathbf{A}^* = \begin{bmatrix} \mathbf{A} & \mathbf{0} \\ \mathbf{0}^T & \sigma^2 \end{bmatrix}$$

$$\mathbf{P}^* = (\mathbf{I} - \mathbf{w}\mathbf{h}^T)\mathbf{P} \qquad \mathbf{H}^* = \begin{bmatrix} \mathbf{H} & \mathbf{h} \end{bmatrix}$$

A recursive form of the estimator can be obtained using the following sequence of steps with the block partitions of the factors of $\mathbf{F}^*$:

$$\mathbf{F}^* = (\mathbf{I} - \mathbf{w}\mathbf{h}^T)\mathbf{P}\begin{bmatrix} \mathbf{H} & \mathbf{h} \end{bmatrix}\begin{bmatrix} \mathbf{A}^{-1} & \mathbf{0} \\ \mathbf{0}^T & \sigma^{-2} \end{bmatrix} = (\mathbf{I} - \mathbf{w}\mathbf{h}^T)\mathbf{P}\begin{bmatrix} \mathbf{H}\mathbf{A}^{-1} & \dfrac{\mathbf{h}}{\sigma^2} \end{bmatrix}$$

$$= \begin{bmatrix} (\mathbf{I} - \mathbf{w}\mathbf{h}^T)\mathbf{F} & \dfrac{\mathbf{Ph} - \mathbf{w}(\mathbf{h}^T\mathbf{Ph})}{\sigma^2} \end{bmatrix} = \begin{bmatrix} (\mathbf{I} - \mathbf{w}\mathbf{h}^T)\mathbf{F} & \dfrac{a\mathbf{w} - \mathbf{w}(a - \sigma^2)}{\sigma^2} \end{bmatrix}$$

Therefore,

$$\mathbf{F}^* = \begin{bmatrix} (\mathbf{I} - \mathbf{w}\mathbf{h}^T)\mathbf{F} & \mathbf{w} \end{bmatrix} \tag{13.45}$$

so that

$$\delta \hat{\mathbf{r}}^* = \mathbf{F}^* \, \delta \tilde{\mathbf{q}}^* = \begin{bmatrix} (\mathbf{I} - \mathbf{w}\mathbf{h}^T)\mathbf{F} & \mathbf{w} \end{bmatrix} \begin{bmatrix} \delta \tilde{\mathbf{q}} \\ \delta \tilde{q} \end{bmatrix} = (\mathbf{I} - \mathbf{w}\mathbf{h}^T)\delta \hat{\mathbf{r}} + \mathbf{w}\,\delta \tilde{q}$$

$$= \delta \hat{\mathbf{r}} + \mathbf{w}(\delta \tilde{q} - \mathbf{h}^T \, \delta \hat{\mathbf{r}}) \tag{13.46}$$

Since $\delta q = \mathbf{h}^T \, \delta \mathbf{r}$, then $\mathbf{h}^T \, \delta \hat{\mathbf{r}}$ provides an estimate of the new measurement before it is actually made. As a consequence, if we denote this quantity by

$$\delta \hat{q} = \mathbf{h}^T \, \delta \hat{\mathbf{r}} \tag{13.47}$$

then the recursive estimation equation is

$$\delta \hat{\mathbf{r}}^* = \delta \hat{\mathbf{r}} + \mathbf{w}(\delta \tilde{q} - \delta \hat{q}) \tag{13.48}$$

with the vector

$$\mathbf{w} = \dfrac{1}{\sigma^2 + \mathbf{h}^T \mathbf{Ph}}\, \mathbf{Ph} \tag{13.49}$$

playing the role of a weighting factor. To obtain the updated estimate $\delta \hat{\mathbf{r}}^*$ from the old estimate $\delta \hat{\mathbf{r}}$ we simply add the weighted difference between what we actually measure and what we would expect to measure as anticipated from the old estimate. (If that difference is zero, there is no necessity to change the estimate.) Finally, we must update the $\mathbf{P}$ matrix by using Eq. (13.43) in preparation for processing the next measurement.

## ◊ Problem 13-10

A space vehicle is en route to Mars, and one-tenth year after departure from earth the vector positions in astronomical units of the vehicle and the two planets are as follows:

$$\mathbf{r} = 0.104\,\mathbf{i}_x + 0.998\,\mathbf{i}_y + 0.018\,\mathbf{i}_z$$
$$\mathbf{r}_E = 0.155\,\mathbf{i}_x + 0.972\,\mathbf{i}_y$$
$$\mathbf{r}_M = -1.076\,\mathbf{i}_x + 1.251\,\mathbf{i}_y + 0.053\,\mathbf{i}_z$$

in a heliocentric, ecliptic oriented coordinate system.

(a) A position fix is made by measuring the following angles:

1. Angle between Mars and Sirius
2. Angle between Mars and Beta Centauri
3. Angle between earth and Beta Centauri

Determine an upper bound on the squared position estimation error. For simplicity, assume that the measurements are all made simultaneously and at precisely the reference time.

(b) If the additional measurement of the angle between the sun and Mars is made, determine the decrease in the upper bound on the squared estimation error. Assume that all four measurements have equal weight.

NOTE: The lines of sight to the stars Sirius and Beta Centauri are given by the following sets of direction cosines:

Sirius: $(-0.180, 0.749, -0.637)$
Beta Centauri: $(-0.430, -0.575, -0.696)$

## The Characteristic Polynomial of the P Matrix

The quadratic form associated with the information matrix is at least positive semidefinite since

$$\mathbf{x}^T \Sigma \mathbf{x} = \sum_{i=1}^{N} \frac{\mathbf{x}^T \mathbf{h}_i \mathbf{h}_i^T \mathbf{x}}{\sigma_i^2} = \sum_{i=1}^{N} \frac{(\mathbf{x} \cdot \mathbf{h}_i)^2}{\sigma_i^2}$$

If $N = 1$ or $N = 2$, the quadratic form can be zero simply by choosing the $\mathbf{x}$ vector to be normal to $\mathbf{h}_1$ or normal to $\mathbf{h}_1$ and $\mathbf{h}_2$ as the case may be. On the other hand, if $N \geq 3$ and if $\mathbf{h}_1$, $\mathbf{h}_2$, $\mathbf{h}_3$ span the measurement space, then no $\mathbf{x} \neq \mathbf{0}$ can be chosen which is normal to all three $\mathbf{h}$ vectors. Hence, $\Sigma$ (and, of course, also $\mathbf{P} = \Sigma^{-1}$) will be positive definite for $N \geq 3$ if the scalar product of at least one set of three of the measurement vectors is not zero.

The characteristic polynomial of $\mathbf{P}$ is

$$\det(\mathbf{P} - \lambda \mathbf{I}) = -\lambda^3 + \beta_1 \lambda^2 - \beta_2 \lambda + \beta_3 \quad (13.50)$$

with the coefficients $\beta_1$, $\beta_2$, and $\beta_3$ determined from

$$\beta_1 = \operatorname{tr} \mathbf{P}$$
$$\beta_2 = \tfrac{1}{2}\left[(\operatorname{tr} \mathbf{P})^2 - \operatorname{tr} \mathbf{P}^2\right] \quad (13.51)$$
$$\beta_3 = \det \mathbf{P}$$

Recursion formulas for these coefficients can also be obtained which will be particularly useful in the next chapter.

To develop the formula for $\beta_1$, we require the identity

$$\text{tr}(\mathbf{A}_{mn}\mathbf{B}_{nm}) = \text{tr}(\mathbf{B}_{nm}\mathbf{A}_{mn}) \tag{13.52}$$

The proof of (13.52) follows immediately from the fact that

$$\text{tr}(\mathbf{A}_{mn}\mathbf{B}_{nm}) = \sum_{i=1}^{m}\sum_{j=1}^{n} a_{ij}b_{ji} \quad \text{and} \quad \text{tr}(\mathbf{B}_{nm}\mathbf{A}_{mn}) = \sum_{j=1}^{n}\sum_{i=1}^{m} b_{ji}a_{ij}$$

are, obviously, equal.

Then, from the recursion formula (13.43) for $\mathbf{P}$ written in the form

$$\mathbf{P}^* = (\mathbf{I} - \mathbf{wh}^T)\mathbf{P} = \mathbf{P} - \frac{1}{a}\mathbf{Phh}^T\mathbf{P}$$

we have

$$\text{tr}\,\mathbf{P}^* = \text{tr}\,\mathbf{P} - \frac{1}{a}\text{tr}(\mathbf{Phh}^T\mathbf{P}) = \text{tr}\,\mathbf{P} - \frac{1}{a}\text{tr}(\mathbf{h}^T\mathbf{PPh})$$

$$= \text{tr}\,\mathbf{P} - \frac{1}{a}\text{tr}(a^2\mathbf{w}^T\mathbf{w}) = \text{tr}\,\mathbf{P} - aw^2$$

Here, we have used the identity just derived and the fact that the trace of a scalar is simply the scalar itself. In this way, the desired recursion formula

$$\beta_1^* = \beta_1 - aw^2 \tag{13.53}$$

is established.

Next, we develop a corresponding formula for $\beta_3$ by first writing the recursion formula (13.43) as

$$\mathbf{P}^* = (\mathbf{I} - \mathbf{wh}^T)\mathbf{P} = \mathbf{P}\left(\mathbf{I} - \mathbf{h}\frac{\mathbf{P}^{-1}}{a}\mathbf{h}^T\mathbf{P}\right)$$

Then we employ the determinant identity (13.37) together with the first of Eqs. (13.42). The required steps are

$$|a\mathbf{P}^*| = |a\mathbf{P}|\left|\mathbf{I} - \mathbf{h}\frac{\mathbf{P}^{-1}}{a}\mathbf{h}^T\mathbf{P}\right| = |\mathbf{I}||a\mathbf{P} - \mathbf{h}^T\mathbf{PI}^{-1}\mathbf{hP}|$$

$$= (a - \mathbf{h}^T\mathbf{Ph})|\mathbf{P}| = \sigma^2|\mathbf{P}|$$

Hence,†

$$|\mathbf{P}^*| = \frac{\sigma^2}{a}|\mathbf{P}| \tag{13.54}$$

---

† This result was first published by James E. Potter and Donald C. Fraser in a note titled "A Formula for Updating the Determinant of the Covariance Matrix" which appeared in the July, 1967 issue of the *AIAA Journal*. Their derivation is considerably more involved than the one presented here.

or, equivalently,

$$\beta_3^* = \frac{\sigma^2}{\sigma^2 + \mathbf{h}^T \mathbf{P} \mathbf{h}} \beta_3 \qquad (13.55)$$

Finally, for $\beta_2$, we first show that

$$\operatorname{tr} \Sigma = \frac{\beta_2}{\beta_3} \qquad (13.56)$$

For this purpose, the Cayley-Hamilton theorem† can be put to good use. If we multiply the matrix form of the characteristic equation

$$-\mathbf{P}^3 + \beta_1 \mathbf{P}^2 - \beta_2 \mathbf{P} + \beta_3 \mathbf{I} = \mathbf{O}$$

by $\mathbf{P}^{-3}$, we have the characteristic equation for the information matrix

$$-\Sigma^3 + \frac{\beta_2}{\beta_3}\Sigma^2 - \frac{\beta_1}{\beta_3}\Sigma + \frac{1}{\beta_3}\mathbf{I} = \mathbf{O}$$

Then, from a kind of reverse application of the Cayley-Hamilton theorem, we conclude that the characteristic polynomial of $\Sigma$ must be

$$\det(\Sigma - \lambda \mathbf{I}) = -\lambda^3 + \frac{\beta_2}{\beta_3}\lambda^2 - \frac{\beta_1}{\beta_3}\lambda + \frac{1}{\beta_3} \qquad (13.57)$$

Equation (13.56) is, therefore, substantiated.

To obtain the recursion formula for $\beta_2$, recall the recursive form of the information matrix

$$\Sigma^* = \Sigma + \frac{1}{\sigma^2}\mathbf{h}\mathbf{h}^T$$

Then, from Eq. (13.57),

$$\operatorname{tr}\Sigma^* = \frac{\beta_2^*}{\beta_3^*} = \operatorname{tr}\Sigma + \frac{1}{\sigma^2}\operatorname{tr}(\mathbf{h}\mathbf{h}^T) = \frac{\beta_2}{\beta_3} + \frac{1}{\sigma^2}\operatorname{tr}(\mathbf{h}^T\mathbf{h})$$
$$= \frac{\beta_2}{\beta_3} + \frac{h^2}{\sigma^2}$$

Here again, we have used the identity (13.52) for the trace of a matrix product. As a consequence,

$$\beta_2^* = \beta_2 \frac{\beta_3^*}{\beta_3} + \frac{h^2}{\sigma^2}\beta_3^* = \beta_2 \frac{\sigma^2}{a} + \frac{h^2}{\sigma^2} \times \frac{\sigma^2}{a}\beta_3$$

so that

$$\beta_2^* = \frac{\sigma^2 \beta_2 + h^2 \beta_3}{\sigma^2 + \mathbf{h}^T \mathbf{P} \mathbf{h}} \qquad (13.58)$$

results as the desired formula.

---

† The Cayley-Hamilton theorem asserts that every symmetric matrix satisfies its own characteristic equation.

## 13.7 Square-Root Formulation of the Estimator

For any positive semidefinite matrix $\mathbf{M}$ there is a matrix $\mathbf{B}$ whose columns are the orthonormal characteristic unit vectors of $\mathbf{M}$ such that

$$\mathbf{B}^T \mathbf{M} \mathbf{B} = \mathbf{D} \tag{13.59}$$

The elements of the diagonal matrix $\mathbf{D}$ are the characteristic values of $\mathbf{M}$. The square root of $\mathbf{D}$, written $\mathbf{D}^{\frac{1}{2}}$, is a diagonal matrix with the square roots of the characteristic values on the main diagonal. Since there are no negative characteristic values, the matrix $\mathbf{D}^{\frac{1}{2}}$ is guaranteed to be real. The square root matrix $\mathbf{W}$ of the matrix $\mathbf{M}$ is defined as that matrix for which

$$\mathbf{W}\mathbf{W}^T = \mathbf{M} \tag{13.60}$$

It is apparent that one such square root matrix is

$$\mathbf{W} = \mathbf{B}\mathbf{D}^{\frac{1}{2}} \tag{13.61}$$

### Symmetric Square Roots of a Matrix

It is also possible to determine a symmetric square root matrix $\mathbf{W} = \mathbf{W}^T$ by noting that Eq. (13.59) can be written as

$$\mathbf{B}^T \mathbf{M} \mathbf{B} = \mathbf{B}^T \mathbf{W} \mathbf{W} \mathbf{B} = \mathbf{B}^T \mathbf{W} \mathbf{B} \mathbf{B}^T \mathbf{W} \mathbf{B} = \mathbf{D} = \mathbf{D}^{\frac{1}{2}} \mathbf{D}^{\frac{1}{2}}$$

Obviously, then, for a three-dimensional matrix as an example, we have

$$\mathbf{W} = \mathbf{B} \begin{bmatrix} \pm\sqrt{\lambda_1} & 0 & 0 \\ 0 & \pm\sqrt{\lambda_2} & 0 \\ 0 & 0 & \pm\sqrt{\lambda_3} \end{bmatrix} \mathbf{B}^T \tag{13.62}$$

With all the possible combinations of sign and permutations of elements on the main diagonal, the number of different square roots of this kind are large indeed.

Since the calculation of characteristic values and characteristic vectors is not simple for large dimensional matrices, it is worthwhile exploring other possibilities for calculating a square root. Consider, as an example, the problem of determining the general square root of a two-dimensional matrix. Specifically, if

$$\mathbf{W}\mathbf{W}^T = \begin{bmatrix} w_{11} & w_{12} \\ w_{21} & w_{22} \end{bmatrix} \begin{bmatrix} w_{11} & w_{21} \\ w_{12} & w_{22} \end{bmatrix} = \begin{bmatrix} m_{11} & m_{12} \\ m_{12} & m_{22} \end{bmatrix} \tag{13.63}$$

then we must have

$$w_{11}^2 + w_{12}^2 = m_{11}$$
$$w_{11}w_{21} + w_{12}w_{22} = m_{12}$$
$$w_{21}^2 + w_{22}^2 = m_{22}$$

These equations do not have a unique solution. Indeed, we obtain

$$w_{11} = \pm\sqrt{m_{11} - w_{12}^2}$$
$$w_{21} = \frac{m_{12}}{m_{11}}w_{11} \pm \frac{1}{m_{11}}\sqrt{\det \mathbf{M}}\, w_{12} \qquad (13.64)$$
$$w_{22} = \frac{m_{12}}{m_{11}}w_{12} \mp \frac{1}{m_{11}}\sqrt{\det \mathbf{M}}\, w_{11}$$

with $w_{12}$ as a free parameter.

Specializing to a symmetric square root requires $w_{12} = w_{21}$ and, in this case, $w_{21}$ must be determined as the solution of

$$\left[(m_{11}^2 + m_{22}^2 - 2\sqrt{\det \mathbf{M}})w_{21}^2 - (m_{11}+m_{22})m_{12}^2\right]^2$$
$$= 4m_{12}^4\left[m_{11}m_{22} - (m_{11}+m_{22})w_{21}^2 + w_{21}^4\right] \qquad (13.65)$$

with $w_{11}$ and $w_{22}$ calculated as before. Of course, higher-order matrices involve an even greater amount of algebra which tends to compromise the practicality of the method.

◇ **Problem 13-11**

Use the Cayley-Hamilton theorem to find a symmetric square root of a two-dimensional matrix $\mathbf{M}$. Specifically, find scalar constants $c_1$ and $c_2$ such that

$$\mathbf{W} = c_1\mathbf{I} + c_2\mathbf{M} \quad \text{with} \quad \mathbf{W}^2 = \mathbf{M}$$

HINT: Derive the relations

$$\beta_1 = \frac{1 - 2c_1c_2}{c_2^2} = \operatorname{tr}\mathbf{M} \quad \text{and} \quad \beta_2 = \frac{c_1^2}{c_2^2} = \det \mathbf{M}$$

Test for a Positive Definite Matrix

A necessary and sufficient condition for a real symmetric matrix to have a real square root is that it have no negative characteristic values. This is equivalent to requiring that the quadratic form associated with the matrix be either positive definite or at least positive semidefinite. Symbolically, this means

$$Q \equiv \mathbf{x}^T\mathbf{M}\mathbf{x} \geq 0 \quad \text{for all} \quad \mathbf{x} \neq \mathbf{0} \qquad (13.66)$$

Consider a two-dimensional symmetric matrix $\mathbf{M}$ and its associated quadratic form

$$Q = \begin{bmatrix} x_1 & x_2 \end{bmatrix} \begin{bmatrix} m_{11} & m_{12} \\ m_{12} & m_{22} \end{bmatrix} \begin{bmatrix} x_1 \\ x_2 \end{bmatrix} = m_{11}x_1^2 + 2m_{12}x_1x_2 + m_{22}x_2^2 \qquad (13.67)$$

which we write in the form

$$Q = m_{11}\left(x_1 + \frac{m_{12}}{m_{11}}x_2\right)^2 + \left(m_{22} - \frac{m_{12}^2}{m_{11}}\right)x_2^2$$

Therefore, if $Q$ is to be positive for all values of $x_1$ and $x_2$ (except, of course, $x_1 = x_2 = 0$), then we must have

$$m_{11} > 0 \quad \text{and} \quad \begin{vmatrix} m_{11} & m_{12} \\ m_{12} & m_{22} \end{vmatrix} > 0 \qquad (13.68)$$

For the three-dimensional quadratic form

$$Q = \begin{bmatrix} x_1 & x_2 & x_3 \end{bmatrix} \begin{bmatrix} m_{11} & m_{12} & m_{13} \\ m_{12} & m_{22} & m_{23} \\ m_{13} & m_{23} & m_{33} \end{bmatrix} \begin{bmatrix} x_1 \\ x_2 \\ x_3 \end{bmatrix} \qquad (13.69)$$

we have

$$\begin{aligned} Q &= m_{11}x_1^2 + m_{22}x_2^2 + m_{33}x_3^2 + 2(m_{12}x_1x_2 + m_{13}x_1x_3 + m_{23}x_2x_3) \\ &= \frac{1}{m_{11}}(m_{11}x_1 + m_{12}x_2 + m_{13}x_3)^2 + \frac{1}{m_{11}}(m_{11}m_{22} - m_{12}^2)x_2^2 \\ &\quad + \frac{2}{m_{11}}(m_{11}m_{23} - m_{12}m_{13})x_2x_3 + \frac{1}{m_{11}}(m_{11}m_{33} - m_{13}^2)x_3^2 \end{aligned}$$

Provided that $m_{11} > 0$, the first term can never be negative. What remains is a quadratic form in $x_2$ and $x_3$ which will be always positive if

$$\begin{vmatrix} m_{11} & m_{12} \\ m_{12} & m_{22} \end{vmatrix} > 0 \quad \text{and} \quad \begin{vmatrix} m_{11}m_{22} - m_{12}^2 & m_{11}m_{23} - m_{12}m_{13} \\ m_{11}m_{23} - m_{12}m_{13} & m_{11}m_{33} - m_{13}^2 \end{vmatrix} > 0$$

according to conditions (13.68) derived for the two-dimensional quadratic form. But this second condition is equivalent to requiring that the determinant of the matrix $\mathbf{M}$ be positive since it is easy to show that

$$\begin{vmatrix} m_{11} & m_{12} & m_{13} \\ m_{12} & m_{22} & m_{23} \\ m_{13} & m_{23} & m_{33} \end{vmatrix} = \frac{1}{m_{11}} \begin{vmatrix} 1 & m_{12} & m_{13} \\ 0 & m_{11}m_{22} - m_{12}^2 & m_{11}m_{23} - m_{12}m_{13} \\ 0 & m_{11}m_{23} - m_{13}m_{12} & m_{11}m_{33} - m_{13}^2 \end{vmatrix}$$

Thus, for the quadratic form (13.69) to be always positive, we must have

$$m_{11} > 0 \quad \begin{vmatrix} m_{11} & m_{12} \\ m_{12} & m_{22} \end{vmatrix} > 0 \quad \begin{vmatrix} m_{11} & m_{12} & m_{13} \\ m_{12} & m_{22} & m_{23} \\ m_{13} & m_{23} & m_{33} \end{vmatrix} > 0 \qquad (13.70)$$

This can be generalized to provide a test for positive definiteness of a matrix of any size—the *principal minor test*. The $k^{\text{th}}$ *leading principal minor* $\Delta_k$ is defined as the determinant of the array formed by deleting the last $n - k$ rows and columns of an $n$-dimensional matrix. As can be shown, a necessary and sufficient condition† for an $n^{\text{th}}$-order symmetric

---

† John E. Prussing recently emphasized that the analogous statement, to wit, a necessary and sufficient condition that a matrix be positive semidefinite is that all $n$ leading principal minors $\Delta_k$ are nonnegative—is not true. The correct necessary and sufficient condition is that *all possible* principal minors be nonnegative. His paper titled "The Principal Minor Test for Semidefinite Matrices" appeared in the *Journal of Guidance, Control, and Dynamics*, Vol. 9, Jan.-Feb. 1986.

matrix to be positive definite is that all leading principal minors $\Delta_k$ are positive.

### Triangular Square Roots of a Matrix

The simplest algebraic method of calculating a square root of a matrix is predicated on the assumption that the square root is to be a triangular matrix. For example, a three-dimensional triangular matrix would have the form

$$\mathbf{W} = \begin{bmatrix} w_{11} & 0 & 0 \\ w_{21} & w_{22} & 0 \\ w_{31} & w_{32} & w_{33} \end{bmatrix} \quad (13.71)$$

with obvious extensions to higher-order cases.

If $\mathbf{W}$ is a square root of $\mathbf{M}$, i.e., satisfying Eq. (13.60), then

$$\begin{bmatrix} w_{11} & 0 & 0 \\ w_{21} & w_{22} & 0 \\ w_{31} & w_{32} & w_{33} \end{bmatrix} \begin{bmatrix} w_{11} & w_{21} & w_{31} \\ 0 & w_{22} & w_{32} \\ 0 & 0 & w_{33} \end{bmatrix} = \begin{bmatrix} m_{11} & m_{12} & m_{13} \\ m_{12} & m_{22} & m_{23} \\ m_{13} & m_{23} & m_{33} \end{bmatrix}$$

As a consequence, we must have

$$w_1^2 = m_{11} \qquad w_2^2 + w_3^2 = m_{22}$$
$$w_1 w_3 = m_{12} \qquad w_2 w_5 + w_3 w_6 = m_{23}$$
$$w_1 w_6 = m_{13} \qquad w_4^2 + w_5^2 + w_6^2 = m_{33}$$

It is because of the assumption of a triangular form for $\mathbf{W}$ that the solution of these equations is straightforward. Indeed, we readily obtain

$$w_1 = \pm\sqrt{\Delta_1} \qquad w_4 = \pm\sqrt{\frac{\Delta_3}{\Delta_2}}$$

$$w_2 = \pm\sqrt{\frac{\Delta_2}{\Delta_1}} \qquad w_5 = \frac{m_{23} - w_3 w_6}{w_2} \quad (13.72)$$

$$w_3 = \frac{m_{12}}{w_1} \qquad w_6 = \frac{m_{13}}{w_1}$$

where

$$\Delta_1 = m_{11}$$
$$\Delta_2 = m_{11} m_{22} - m_{12}^2$$
$$\Delta_3 = \det \mathbf{M}$$

All of the radicals involve only the leading principal minors of the positive definite matrix $\mathbf{M}$ and are, therefore, guaranteed to be nonnegative.

## Problem 13-12

For an $n \times n$ positive semidefinite matrix $\mathbf{M} = \|m_{ij}\|$, derive the following recursive algorithm† for computing the triangular square root matrix $\mathbf{W} = \|w_{ij}\|$ where

$$w_{ij} = 0 \quad \text{for} \quad i < j$$

For $i = 1, 2, \ldots, n$, the elements of $\mathbf{W}$ are calculated from

$$w_{ii} = \sqrt{m_{ii} - \sum_{j=1}^{i-1} w_{ij}^2}$$

$$w_{ji} = \begin{cases} 0 & \text{for } j < i \\ \dfrac{1}{w_{ii}}\left(m_{ji} - \sum_{k=1}^{i-1} w_{jk} w_{ik}\right) & \text{for } j = i+1, i+2, \ldots, n \end{cases}$$

As an example, for

$$\mathbf{M} = \begin{bmatrix} 1 & 2 & 3 \\ 2 & 8 & 2 \\ 3 & 2 & 14 \end{bmatrix} \quad \text{we have} \quad \mathbf{W} = \begin{bmatrix} 1 & 0 & 0 \\ 2 & 2 & 0 \\ 3 & -2 & 1 \end{bmatrix}$$

*Cholesky* and *Banachiewicz*

### Recursion Formula for the Square Root of the P Matrix

For practical reasons which are discussed in the next chapter, it may be desirable to formulate the recursive estimator in terms of the square root of the $\mathbf{P}$ matrix.‡ Indeed, this is our sole motive for introducing the concept of the matrix square root.

Let $\mathbf{W}$ be a square root of $\mathbf{P}$ and $\mathbf{W}^*$ be the square root of the updated $\mathbf{P}^*$ matrix. Then, since the recursion formula (13.43) for $\mathbf{P}$ may be written in the form

$$\mathbf{P}^* = \mathbf{P} - \frac{1}{a}\mathbf{P}\mathbf{h}\mathbf{h}^T\mathbf{P}$$

---

† For a variety of other algorithms on this same subject, see the paper on "Discrete Square Root Filtering: A Survey of Current Techniques" by Paul G. Kaminski, Arthur E. Bryson, and Stanley F. Schmidt published in the *IEEE Transactions on Automatic Control*, December 1971. It was reprinted in *Kalman Filtering: Theory and Application*, edited by H. W. Sorenson for the IEEE PRESS Selected Reprint Series, 1985.

‡ The idea of the square-root estimator and its derivation originated with James E. Potter in 1962. It appeared in the author's book *Astronautical Guidance* as Prob. 9.11. The book was in galley proof form at the time or it would certainly have been given greater prominence. The square-root estimator was of the utmost importance in the Apollo navigation system. It also spawned a number of technical papers by various authors over the years since then. The previous footnote references one such paper.

we have

$$\mathbf{W}^*\mathbf{W}^{*T} = \mathbf{W}\left(\mathbf{I} - \frac{1}{a}\mathbf{W}^T\mathbf{h}\mathbf{h}^T\mathbf{W}\right)\mathbf{W}^T$$

$$= \mathbf{W}\left(\mathbf{I} - \frac{1}{a}\mathbf{z}\mathbf{z}^T\right)\mathbf{W}^T$$

in terms of the square root matrices $\mathbf{W}$ and $\mathbf{W}^*$. For convenience, we have defined the vector $\mathbf{z}$ as

$$\mathbf{z} = \mathbf{W}^T\mathbf{h}$$

The problem of finding a recursion formula for $\mathbf{W}$ is equivalent to the problem of obtaining a square root of $\mathbf{I} - \mathbf{z}\mathbf{z}^T/a$. For this purpose, we attempt a factorization of the form

$$\mathbf{I} - \frac{1}{a}\mathbf{z}\mathbf{z}^T = (\mathbf{I} - \kappa\mathbf{z}\mathbf{z}^T)(\mathbf{I} - \kappa\mathbf{z}\mathbf{z}^T)$$

suggested by a unique property of the matrix $\mathbf{z}\mathbf{z}^T$. Specifically, powers of this matrix result in the same matrix multiplied by a scalar, i.e.,

$$(\mathbf{z}\mathbf{z}^T)(\mathbf{z}\mathbf{z}^T) = \mathbf{z}(\mathbf{z}^T\mathbf{z})\mathbf{z}^T = z^2\mathbf{z}\mathbf{z}^T$$

Therefore,

$$\mathbf{I} - \frac{1}{a}\mathbf{z}\mathbf{z}^T = \mathbf{I} - (2\kappa - \kappa^2 z^2)\mathbf{z}\mathbf{z}^T$$

Now,

$$z^2 = \mathbf{z}^T\mathbf{z} = \mathbf{h}^T\mathbf{W}\mathbf{W}^T\mathbf{h} = \mathbf{h}^T\mathbf{P}\mathbf{h} = a - \sigma^2$$

according to the first of Eqs. (13.42). Hence, $\kappa$ is a solution of the quadratic equation

$$2\kappa - \kappa^2(a - \sigma^2) = \frac{1}{a}$$

from which

$$\kappa = \frac{a \pm \sqrt{a\sigma^2}}{a^2 - a\sigma^2} = \frac{1}{a \mp \sqrt{a\sigma^2}}$$

The choice of sign is at our disposal. Since $a$ is never negative, we choose the plus sign in the denominator so that

$$\kappa = \frac{1}{a + \sqrt{a\sigma^2}}$$

Thereby, any possibility of dividing by zero is avoided.

Finally, then we have established the factorization

$$\mathbf{W}^*\mathbf{W}^{*T} = \mathbf{W}(\mathbf{I} - \kappa\mathbf{z}\mathbf{z}^T)(\mathbf{I} - \kappa\mathbf{z}\mathbf{z}^T)\mathbf{W}^T = [\mathbf{W}(\mathbf{I} - \kappa\mathbf{z}\mathbf{z}^T)][\mathbf{W}(\mathbf{I} - \kappa\mathbf{z}\mathbf{z}^T)]^T$$

so that

$$\mathbf{W}^* = \mathbf{W}\left(\mathbf{I} - \frac{\mathbf{z}\mathbf{z}^T}{a + \sqrt{a\sigma^2}}\right) \quad \text{where} \quad \mathbf{z} = \mathbf{W}^T\mathbf{h} \quad (13.73)$$

is the recursion formula for the $\mathbf{W}$ matrix.

Chapter 14

# Space Navigation

NOT UNTIL EIGHT YEARS AFTER JAMES BERNOULLI DIED IN 1705 was his main work, the *Ars Conjectandi*, published. This was the first significant book on probability and the most important new result that it contained is still named for Bernoulli. Specifically, if $p$ and $q$ are the respective probabilities that a single event will or will not occur, then the probability that this event will occur at least $m$ times in $n$ trials is the sum of the terms in the expansion of $(p+q)^n$ from the term involving $p^m q^{n-m}$ to $p^n$. Indeed, the $k^{\text{th}}$ term in the binomial expansion

$$(p+q)^n = \sum_{k=0}^{n} \binom{n}{k} p^k q^{n-k}$$

is the probability that the event in question will occur exactly $k$ times in the $n$ trials. The "trials" are called *Bernoulli trials*.

The binomial expansion, for positive integral $n$, was familiar to the Arabs of the thirteenth century. The term "binomial coefficient" was first introduced by the German mathematician Michael Stifel (1486?–1567) and the pattern of integers

```
         1
        1 1
       1 2 1
      1 3 3 1
     1 4 6 4 1
     . . . . . . . . . .
```

in which each number is the sum of the two immediately above it, was used by Blaise Pascal (1623–1662) in 1654 to obtain these coefficients. Although this arrangement was known to many of his predecessors, including both Stifel and Tartaglia (who is famous for his solution of the cubic equation), it is, nevertheless, called "Pascal's triangle" since he derived from it the greatest number of applications which he published in *Traité du triangle arithmétique*. Indeed, the very beginnings of the theory of probability are rooted in the work of both Pascal and the renowned mathematical hobbyist Pierre de Fermat (1601–1665) who made his living as a lawyer in the French city of Toulouse.

It all began in the summer of 1654 when Pascal and Fermat corresponded on two particular problems concerning games of chance. Antoine Gombaud, the Chevalier de Méré (1610–1685), a gambling friend, proposed these to Pascal—the first, to determine the probability that a player will obtain a particular face on a die in a given number of throws. The second was more complex—to determine how to divide the stake for any game involving several players if the game is interrupted.†

Fermat succeeded in solving the problems using only combinatorial analysis while Pascal concentrated on the use of mathematical induction. Fermat relied on direct computations rather than mathematical formulas and his methods were rendered obsolete with the publication in 1657 of the more sophisticated work *De ratiociniis in ludo aleae* by Christiaan Huygens.‡ Nevertheless, the effort of these two men during those few months marks the beginning of the calculus of probability.

Nicholas Bernoulli,§ the nephew of James and John, was responsible for another famous gambling problem called the "St. Petersburg paradox" which first appeared in 1713 in the paper *Essai d'analyse sur les jeux de hazard* by Pierre Rémond de Montmort (1678–1719). In the Petersburg game a player tosses a coin until it falls heads; if this occurs on the $n^{th}$ toss, the player receives $2^n$ dollars. What are his expected winnings—i.e., what is a fair price to charge a player for playing this game?

*The answer is an infinite amount of money!* Since the result of each play is a random variable $X$ assuming values $x_1 = 2^1$, $x_2 = 2^2$, $x_3 = 2^3$, ... with corresponding probabilities $p_1 = 2^{-1}$, $p_2 = 2^{-2}$, $p_3 = 2^{-3}$, ..., then by the definition of "mathematical expectation" the gambler can expect to win $\sum_{i=1}^{\infty} x_i p_i$ dollars. Each term of the series is 1 so that his expected gain is infinite. Although the fallacy lies in assuming that the "bank" has infinite resources, the philosophical controversy this problem generated elevates it to one of the classical paradoxes of history.¶

---

† Such problems had been treated with only minor success by the great Renaissance mathematicians Jerome Cardan, Tartaglia, and Luca di Pacioli (c. 1445–c. 1514) who was a friend and teacher of Leonardo Da Vinci.

‡ Christiaan Huygens (1629–1695) and Robert Hooke (1635–1703) did basic work on the motion of the pendulum. (Hooke, of course, is best remembered for "Hooke's law" in mechanics.) The work of both men had considerable influence on Sir Isaac Newton when he began, in 1665, to apply the law of gravitation to planetary motion.

§ It is interesting to note that Nicholas Bernoulli (1687–1759) obtained the degree of Doctor of Jurisprudence in 1709 with a dissertation on the application of the calculus of probability to questions of law.

¶ Two other famous ones are
(1) Bertrand Russell's paradox: If a barber shaves every one in a town who does not shave himself, then who shaves the barber?
(2) Zeno's paradox: The hare (or Achilles) can never catch the tortoise since the pursuer must first arrive at the point from which the pursued started so that, necessarily, the tortoise is always ahead.

## Sect. 14.1] Review of Probability Theory

The research in probability by the eighteenth-century mathematicians was summarized by Pierre-Simon de Laplace in his great *Théorie analytique des probabilités* published in 1812. His view of probability, which dominated the subject throughout the nineteenth century, was essentially a theory of errors—a method of systematically studying the deviations from a mean that occur with the repeated measurements of a quantity. This point of view was further developed by Legendre and Gauss and it will form the basis of our treatment of the space navigation problem in this chapter.

As an introduction, it seems appropriate to review certain of the fundamentals of probability theory which are required for our treatment of the problem of space navigation. Some familiarity with the concepts will be assumed, so that the following brief treatment of the subject is intended only to orient the reader for the particular applications to be made here. For a more "in depth" review the reader is referred to Appendix H.

### 14.1 Review of Probability Theory

The set of all possible chance outcomes of a random experiment is called a *sample space*. The term *random variable* is used to denote a real valued function $\alpha$ whose value is determined by the outcome of the random experiment. With the random variable $X$ is associated a *probability distribution function* $F(x)$ defined as

$$F(x) = \text{Prob}(X \leq x)$$

so that to each $x$ there is a corresponding probability $F(x)$ that the experiment will result in a value of $X$ which does not exceed $x$. If $F(x)$ is differentiable, the *frequency function* or *probability density function* $f(x)$ is defined by

$$f(x) = \frac{dF(x)}{dx} \qquad \text{so that} \qquad F(x) = \int_{-\infty}^{x} f(u)\,du$$

Furthermore,

$$\text{Prob}(a < X \leq b) = \int_{a}^{b} f(x)\,dx \qquad \text{and} \qquad \int_{-\infty}^{\infty} f(x)\,dx = 1$$

The *average*, *mean*, or *mathematical expectation* of the random variable $X$ is defined by

$$\overline{X} = \int_{-\infty}^{\infty} x f(x)\,dx$$

and the *mean-squared value* of $X$ by

$$\overline{X^2} = \int_{-\infty}^{\infty} x^2 f(x)\,dx$$

The terminology *variance* of $X$ is used to denote the mean-squared value of $X$ about the mean. The notation $\sigma^2$ is used for this quantity, so that

$$\sigma^2 = \overline{(X - \overline{X})^2} = \overline{X^2} - \overline{X}^2$$

The term *standard deviation* is applied to the square root $\sigma$ of the variance.

These concepts can be extended to two or more random variables. For two variables $X_1$ and $X_2$, the *joint distribution function* $F(x_1, x_2)$ is defined by

$$F(x_1, x_2) = \text{Prob}(X_1 \leq x_1 \text{ and } X_2 \leq x_2)$$

and the *joint density function* by

$$f(x_1, x_2) = \frac{\partial^2 F(x_1, x_2)}{\partial x_1 \partial x_2}$$

The two random variables are called *statistically independent* if

$$F(x_1, x_2) = F_1(x_1) F_2(x_2)$$

or, equivalently,

$$f(x_1, x_2) = f_1(x_1) f_2(x_2)$$

If $X_1$ and $X_2$ are independent, it can be shown that

$$\overline{X_1 X_2} = \int_{-\infty}^{\infty} \int_{-\infty}^{\infty} x_1 x_2 f(x_1, x_2) \, dx_1 \, dx_2 = \overline{X}_1 \overline{X}_2$$

However, the relation

$$\overline{X_1 + X_2} = \int_{-\infty}^{\infty} \int_{-\infty}^{\infty} (x_1 + x_2) f(x_1, x_2) \, dx_1 \, dx_2 = \overline{X}_1 + \overline{X}_2$$

is true whether or not $X_1$ and $X_2$ are independent. The extension to more than two random variables is made in the obvious way.

Let $X_1, X_2, \ldots, X_m$ be $m$ random variables with zero mean values. The joint distribution is said to be an *m-dimensional normal distribution* if

$$f(x_1, x_2, \ldots, x_m) = \frac{1}{\sqrt{(2\pi)^m |\mathbf{X}|}} \exp\left(-\tfrac{1}{2} \mathbf{x}^T \mathbf{X}^{-1} \mathbf{x}\right) \tag{14.1}$$

where $\mathbf{x}$ is an $m$-dimensional vector whose components are $x_1, \ldots, x_m$ and $\mathbf{X}$, called the *moment matrix* or *covariance matrix* of the distribution, is

$$\mathbf{X} = \begin{bmatrix} \overline{x_1^2} & \overline{x_1 x_2} & \cdots & \overline{x_1 x_m} \\ \overline{x_2 x_1} & \overline{x_2^2} & \cdots & \overline{x_2 x_m} \\ \vdots & \vdots & \ddots & \vdots \\ \overline{x_m x_1} & \overline{x_m x_2} & \cdots & \overline{x_m^2} \end{bmatrix} \tag{14.2}$$

The notation $|\mathbf{X}|$ is used to indicate the determinant of the matrix $\mathbf{X}$.

## Problem 14–1

The *joint characteristic function* $\phi(\mathbf{t}_m)$ of a vector of random variables $\mathbf{X}_m$ of dimension $m$ is defined by

$$\phi(\mathbf{t}_m) = \overline{\exp(it_m^T \mathbf{X}_m)} = \int_{-\infty}^{\infty} f(\mathbf{x}_m) \exp(it_m^T \mathbf{x}_m) \, d\mathbf{x}_m$$

where $i = \sqrt{-1}$ and the notation $d\mathbf{x}_m$ stands for $dx_1 \, dx_2 \ldots dx_m$, so that the integration is really an $m$-fold integral.

The joint characteristic function of an $m$-dimensional normal distribution with zero mean is

$$\phi(\mathbf{t}_m) = \exp(-\tfrac{1}{2} \mathbf{t}_m^T \mathbf{X} \mathbf{t}_m)$$

where $\mathbf{X}$ is the moment matrix of the random vector $\mathbf{X}_m$.

HINT: Since $\mathbf{X}$ is a positive definite matrix, there exists an orthogonal transformation $\mathbf{B}$ such that

$$\mathbf{B X B}^T = \mathbf{D}$$

where $\mathbf{D}$ is a diagonal matrix

$$\mathbf{D} = \begin{bmatrix} \mu_1 & 0 & \cdots & 0 \\ 0 & \mu_2 & \cdots & 0 \\ \vdots & \vdots & \ddots & \vdots \\ 0 & 0 & \cdots & \mu_m \end{bmatrix}$$

Also

$$|\mathbf{X}| = \mu_1 \mu_2 \ldots \mu_m$$

Therefore, a change of variables

$$\mathbf{x}'_m = \mathbf{B} \mathbf{x}_m$$

will separate the variables of integration.

## 14.2  Maximum-Likelihood Estimate

Let us return now to the problem of redundant measurements which was developed in Sect. 13.5 and assume that a total of $m$ $(m > 3)$ measurements has been made for a single position fix. As before, we have the relation between the true values of the measured quantities and position deviation, which can be written in the form

$$\delta \mathbf{q} = \mathbf{H}^T \delta \mathbf{r}$$

where the measurement geometry matrix $\mathbf{H}$ is rectangular with three rows and $m$ columns. Then, since

$$\delta \widetilde{\mathbf{q}} = \delta \mathbf{q} + \boldsymbol{\alpha}$$

it follows that

$$\boldsymbol{\alpha} = \delta \widetilde{\mathbf{q}} - \mathbf{H}^T \delta \mathbf{r} \tag{14.3}$$

Again consider the problem of estimating a value for $\delta \mathbf{r}$ when a set of measurements $\delta \widetilde{\mathbf{q}}$ is given. As an alternative to Gauss' method of least squares, the estimate will be based on the assumption that the measurement errors, i.e., the components of the vector $\boldsymbol{\alpha}$, are random variables having a joint normal distribution.

In the deterministic case ($m = 3$) developed in Sect. 13.3, the estimated value for $\delta \widehat{\mathbf{r}}$ was equivalent to the assumption that $\boldsymbol{\alpha} = \mathbf{0}$. That is, in the absence of any known bias, we must assume that the measurements are free of errors. As we shall see, such an estimate will maximize the probability density function and is, in a sense, the "most likely value" of $\boldsymbol{\alpha}$ for any typical set of measurements.

With redundant measurements, it is not possible to choose an estimate for $\delta \mathbf{r}$ which would be consistent with estimating all measurement errors as zero. On the other hand, it is possible to estimate $\delta \mathbf{r}$ in such a manner that the corresponding $\boldsymbol{\alpha}$ will also maximize the associated probability density function. The resulting $\delta \widehat{\mathbf{r}}$ is known as the *maximum-likelihood estimate* and is obtained in the following manner.

By substituting Eq. (14.3) into Eq. (14.1) we obtain a function of $\delta \mathbf{r}$ for a given set of measurements $\delta \widetilde{\mathbf{q}}$. The resulting function

$$L(\delta \mathbf{r}) = \frac{1}{\sqrt{(2\pi)^m |\mathbf{A}|}} \exp\left[-\tfrac{1}{2}(\delta \widetilde{\mathbf{q}} - \mathbf{H}^T \delta \mathbf{r})^T \mathbf{A}^{-1} (\delta \widetilde{\mathbf{q}} - \mathbf{H}^T \delta \mathbf{r})\right] \quad (14.4)$$

is called the *likelihood function*. To maximize $L(\delta \mathbf{r})$ we need simply to set to zero the derivative of $L$ with respect to $\delta \mathbf{r}$. Thus, we have

$$(\delta \widetilde{\mathbf{q}} - \mathbf{H}^T \delta \mathbf{r})^T \mathbf{A}^{-1} \mathbf{H}^T = \mathbf{0}^T$$

Hence, the maximum-likelihood estimate is obtained as the solution of

$$\mathbf{H} \mathbf{A}^{-1} \mathbf{H}^T \delta \widehat{\mathbf{r}} = \mathbf{H} \mathbf{A}^{-1} \delta \widetilde{\mathbf{q}}$$

Therefore,

$$\delta \widehat{\mathbf{r}} = \mathbf{F} \delta \widetilde{\mathbf{q}} \quad \text{where} \quad \mathbf{F} = (\mathbf{H} \mathbf{A}^{-1} \mathbf{H}^T)^{-1} \mathbf{H} \mathbf{A}^{-1} \quad (14.5)$$

This is exactly the same result as Eq. (13.32) which was derived using the method of least squares. Formerly, the matrix $\mathbf{A}$ was a diagonal matrix of weighting factors but now it is the covariance matrix of measurement errors which need not be diagonal. In other words, the measurements need not be independent so that cross correlations can be different from zero.

Note that when $m = 3$, the deterministic case, then $\mathbf{F} = \mathbf{H}^{T^{-1}}$. Furthermore, the estimate is unbiased in the following sense. If there exist no errors in the measurements, the position deviation estimate will be exact. To see this, we substitute

$$\delta \widetilde{\mathbf{q}} = \delta \mathbf{q} = \mathbf{H}^T \delta \mathbf{r}$$

Sect. 14.2]  Maximum-Likelihood Estimate

into Eq. (14.5) and find that $\delta\hat{\mathbf{r}} = \delta\mathbf{r}$. In general, we define an unbiased estimator as one for which
$$\mathbf{FH}^T = \mathbf{I} \tag{14.6}$$
and refer to (14.6) as the *condition of unbias*.

Now since
$$\delta\hat{\mathbf{r}} = \delta\mathbf{r} + \boldsymbol{\epsilon} = \mathbf{F}\,\delta\tilde{\mathbf{q}}$$
$$= \mathbf{F}(\delta\mathbf{q} + \boldsymbol{\alpha})$$
$$= \mathbf{FH}^T\,\delta\mathbf{r} + \mathbf{F}\boldsymbol{\alpha} = \delta\mathbf{r} + \mathbf{F}\boldsymbol{\alpha}$$

it follows that
$$\boldsymbol{\epsilon} = \mathbf{F}\boldsymbol{\alpha} \tag{14.7}$$

The covariance matrices of the estimation errors and the measurement errors are thus related by
$$\mathbf{P} = \mathbf{FAF}^T = (\mathbf{HA}^{-1}\mathbf{H}^T)^{-1} \tag{14.8}$$
where $\mathbf{P} = \overline{\boldsymbol{\epsilon}\boldsymbol{\epsilon}^T}$ and $\mathbf{A} = \overline{\boldsymbol{\alpha}\boldsymbol{\alpha}^T}$. In Chapter 13, the matrix $\mathbf{P}$ was defined as the inverse of the information matrix and its relation to $\mathbf{H}$ and $\mathbf{A}$ was the same as Eq. (14.8). Finally, note that the estimator may be written as
$$\mathbf{F} = \mathbf{PHA}^{-1} \tag{14.9}$$
which is the same as Eq. (13.39).

◇ **Problem 14–2**

Two measurements $\tilde{x}_1$ and $\tilde{x}_2$ of a scalar quantity $x$ are made with errors $\alpha_1$ and $\alpha_2$.

(a) If the errors are normally distributed with joint probability density function
$$f(\alpha_1, \alpha_2) = \frac{1}{2\pi\sigma_1\sigma_2(1-\rho^2)} \exp\left[-\frac{1}{2(1-\rho^2)}\left(\frac{\alpha_1^2}{\sigma_1^2} - \frac{2\rho\alpha_1\alpha_2}{\sigma_1\sigma_2} + \frac{\alpha_2^2}{\sigma_2^2}\right)\right]$$
derive the maximum-likelihood estimate directly in the form
$$\hat{x} = \frac{(\sigma_2^2 - \rho\sigma_1\sigma_2)\tilde{x}_1 + (\sigma_1^2 - \rho\sigma_1\sigma_2)\tilde{x}_2}{\sigma_1^2 + \sigma_2^2 - 2\rho\sigma_1\sigma_2}$$
where $\sigma_1$, $\sigma_2$ are the standard deviations of the errors and $\rho$ is the correlation coefficient.

(b) If the errors are independent with a *Cauchy distribution* for which
$$f(\alpha_1, \alpha_2) = \frac{1}{\pi^2(1+\alpha_1^2)(1+\alpha_2^2)}$$
derive the maximum-likelihood estimate. Note that the estimate is *not* a linear function of the measurements in this case.

668                    Space Navigation                    [Chap. 14

### Problem 14-3
Use the results of Prob. 14-1 to prove that the set of random variables $\epsilon$ will be normally distributed with covariance matrix $\mathbf{P}$ given by

$$\mathbf{P} = \mathbf{FAF}^T \quad \text{if} \quad \epsilon = \mathbf{F}\alpha_m$$

where $\alpha_m$ is a set of jointly normal random variables.
HINT: Show that

$$\phi(\mathbf{t}) = \overline{\exp(i\mathbf{t}^T \epsilon)} = \exp(-\tfrac{1}{2}\mathbf{t}^T \mathbf{P}\mathbf{t})$$

by making the following change of variable:

$$\mathbf{t}_m = \mathbf{F}^T \mathbf{t}$$

### Problem 14-4
According to the theorem proved in Prob. 14-3, the probability density function of the estimate errors in a fix is

$$f(\epsilon) = \frac{1}{\sqrt{(2\pi)^3 |\mathbf{P}|}} \exp(-\tfrac{1}{2} \epsilon^T \mathbf{P}^{-1} \epsilon)$$

Surfaces of equal probability are obtained from

$$\epsilon^T \mathbf{P}^{-1} \epsilon = k^2$$

where $k$ is a constant. Such surfaces are called *equiprobability ellipsoids*.
(a) The lengths of the three principal semiaxes of the equiprobability ellipsoid are $k\lambda_1$, $k\lambda_2$, $k\lambda_3$, where the $\lambda$'s are roots of the equation

$$|\mathbf{P} - \lambda^2 \mathbf{I}| = 0$$

(b) Show that for a particular fix, the probability of the error vector $\epsilon$ lying within the equiprobability ellipsoid is

$$\sqrt{\frac{2}{\pi}} \int_0^k r^2 e^{-\tfrac{1}{2}r^2} dr$$

by integrating the probability density $p(\epsilon)$ over the volume of the ellipsoid.
(c) The probability of an estimate falling within an ellipsoid whose principal axes are one, two, and three times the principal rms estimate errors are, respectively, 0.20, 0.74, and 0.97.

### Problem 14-5
The volume of the equiprobability ellipsoid is

$$\tfrac{4}{3}\pi k^3 \sqrt{|\mathbf{P}|}$$

With this result, show that for a fix using three measurements, the volume of the ellipsoid is minimized by maximizing the determinant $|\mathbf{P}|$. The determinant is simply the triple scalar product of the vectors $\mathbf{h}_1$, $\mathbf{h}_2$, and $\mathbf{h}_3$. Thus, if the lengths of the $\mathbf{h}$ vectors are given, the volume of the equiprobability ellipsoid is minimized by choosing their directions as nearly orthogonal as possible.

## Maximum-Likelihood Estimate

### The Gauss-Markov Theorem

If $\mathbf{A}$ and $\mathbf{B}$ are symmetric matrices, $\mathbf{A}$ is said to be greater than $\mathbf{B}$, and written $\mathbf{A} > \mathbf{B}$, if the difference matrix $\mathbf{A} - \mathbf{B}$ is positive definite. Similarly, the notation $\mathbf{A} \geq \mathbf{B}$ will indicate that $\mathbf{A} - \mathbf{B}$ is positive semidefinite. It is important to remember that $\mathbf{A} > \mathbf{B}$ does not imply that every element of $\mathbf{A}$ is greater than the corresponding one of $\mathbf{B}$. For example,

$$\begin{bmatrix} 4 & 0 \\ 0 & 4 \end{bmatrix} > \begin{bmatrix} 2 & 1 \\ 1 & 2 \end{bmatrix}$$

even though the off-diagonal elements of the right-hand matrix are larger than those of the left-hand matrix.

The *Gauss-Markov theorem* asserts that the covariance matrix of the maximum-likelihood estimator never exceeds the covariance matrix of any other of the possible linear unbiased estimators.†

For the proof, let $\mathbf{F}^\diamond$ be an arbitrary unbiased linear estimator for which

$$\delta \hat{\mathbf{r}} = \mathbf{F}^\diamond \delta \tilde{\mathbf{q}} \quad \text{with} \quad \mathbf{F}^\diamond \mathbf{H}^T = \mathbf{I}$$

and, as was the case for $\mathbf{P}$, the covariance matrix $\mathbf{P}^\diamond$ of the associated estimation errors is

$$\mathbf{P}^\diamond = \mathbf{F}^\diamond \mathbf{A} \mathbf{F}^{\diamond T}$$

Now, if we define the matrix $\mathbf{J}$ as the difference between $\mathbf{F}^\diamond$ and the maximum-likelihood estimator $\mathbf{F}$, i.e.,

$$\mathbf{J} = \mathbf{F}^\diamond - \mathbf{F}$$

then we have

$$\begin{aligned} \mathbf{P}^\diamond &= (\mathbf{J} + \mathbf{F}) \mathbf{A} (\mathbf{J} + \mathbf{F})^T \\ &= \mathbf{P} + \mathbf{J} \mathbf{A} \mathbf{F}^T + \mathbf{F} \mathbf{A} \mathbf{J}^T + \mathbf{J} \mathbf{A} \mathbf{J}^T \\ &= \mathbf{P} + \mathbf{J} \mathbf{H}^T \mathbf{P} + \mathbf{P} \mathbf{H} \mathbf{J}^T + \mathbf{J} \mathbf{A} \mathbf{J}^T \\ &= \mathbf{P} + \mathbf{J} \mathbf{A} \mathbf{J}^T \end{aligned}$$

The last step follows from the condition of unbias since

$$\mathbf{J} \mathbf{H}^T \mathbf{P} = (\mathbf{F}^\diamond - \mathbf{F}) \mathbf{H}^T \mathbf{P} = \mathbf{O}$$

and

$$\mathbf{P} \mathbf{H} \mathbf{J}^T = (\mathbf{J} \mathbf{H}^T \mathbf{P})^T = \mathbf{O}$$

---

† The material in the rest of this section is from the paper "Statistical Filtering of Space Navigation Measurements" by James E. Potter and Robert G. Stern. It was presented at the AIAA Guidance and Control Conference held in Cambridge, Mass., August, 1963 and appeared in *Progress in Astronautics and Aeronautics*, Volume 13, 1964 published by Academic Press Inc. of New York.

But $\mathbf{A} > \mathbf{O}$, so that† $\mathbf{JAJ}^T \geq \mathbf{O}$ and, therefore,

$$\mathbf{P}^\diamond \geq \mathbf{P} \tag{14.10}$$

The equality can occur only if $\mathbf{J} = \mathbf{O}$, that is, only if the estimators are identical.

### Properties of the Maximum-Likelihood Estimate

There are two corollaries of the Gauss-Markov theorem which depend on some matrix properties to be derived as required in their proofs.

First, consider the scalar

$$b \equiv \overline{\boldsymbol{\epsilon}^T \mathbf{Q} \boldsymbol{\epsilon}} \tag{14.11}$$

where $\mathbf{Q}$ is an arbitrary positive semidefinite matrix and $\boldsymbol{\epsilon}$ the maximum likelihood estimation error vector. In particular, if $\mathbf{Q}$ is the identity matrix then $b$ is the mean-squared estimation error. We now compare $b$ with the corresponding quantity $b^\diamond$ for an arbitrary unbiased linear estimate.

Since $b$ is a scalar, it may also be considered as the trace of the average value of a one-dimensional matrix—specifically, the matrix obtained as the product of the $1 \times 3$ matrix $\boldsymbol{\epsilon}^T \mathbf{Q}$ and the $3 \times 1$ matrix $\boldsymbol{\epsilon}$. Thus,

$$b = \overline{\boldsymbol{\epsilon}^T \mathbf{Q} \boldsymbol{\epsilon}} = \overline{\text{tr}(\boldsymbol{\epsilon}^T \mathbf{Q} \boldsymbol{\epsilon})} = \overline{\text{tr}(\boldsymbol{\epsilon} \boldsymbol{\epsilon}^T \mathbf{Q})} = \text{tr}(\overline{\boldsymbol{\epsilon} \boldsymbol{\epsilon}^T} \mathbf{Q})$$

using the commutative property of the trace of a matrix given in Eq. (13.52) of the previous chapter. Then, since the covariance matrix of estimation errors $\overline{\boldsymbol{\epsilon} \boldsymbol{\epsilon}^T}$ is denoted by $\mathbf{P}$, we have

$$b = \text{tr}(\mathbf{PQ}) \tag{14.12}$$

The same development holds also for $b^\diamond$; therefore,

$$b^\diamond - b = \text{tr}[(\mathbf{P}^\diamond - \mathbf{P})\mathbf{Q}]$$

According to the Gauss-Markov theorem, the matrix $\mathbf{P}^\diamond - \mathbf{P}$ is positive definite. Therefore, it has a real square root $\mathbf{S}$ and

$$b^\diamond - b = \text{tr}[(\mathbf{P}^\diamond - \mathbf{P})\mathbf{Q}] = \text{tr}(\mathbf{SS}^T \mathbf{Q}) = \text{tr}(\mathbf{S}^T \mathbf{QS})$$

But $\mathbf{Q} \geq \mathbf{O}$. Hence, using the reasoning of the last step in the proof of the Gauss-Markov theorem, we conclude that $\text{tr}(\mathbf{S}^T \mathbf{QS}) \geq 0$. As a consequence, we establish

$$b \leq b^\diamond \tag{14.13}$$

i.e., *the maximum-likelihood estimate minimizes the statistic* $b = \overline{\boldsymbol{\epsilon}^T \mathbf{Q} \boldsymbol{\epsilon}}$ *for any positive semidefinite matrix* $\mathbf{Q}$.

---

† If $\mathbf{x}$ is an arbitrary vector and $\mathbf{y} = \mathbf{J}^T \mathbf{x}$, then

$$q = \mathbf{y}^T \mathbf{A} \mathbf{y} = \mathbf{x}^T \mathbf{JAJ}^T \mathbf{x} \geq 0$$

provided that $\mathbf{A} \geq \mathbf{O}$.

Sect. 14.2]  Maximum-Likelihood Estimate  671

Furthermore, we note that if $\mathbf{Q}$ is positive definite, then $\text{tr}[(\mathbf{P}^\diamond - \mathbf{P})\mathbf{Q}]$ will be zero if and only if $\mathbf{P}^\diamond = \mathbf{P}$. For the proof, let $\mathbf{s}_i$ be the $i^{\text{th}}$ column of the $n \times n$ square root matrix $\mathbf{S}$ of the matrix difference $\mathbf{P}^\diamond - \mathbf{P}$. Then

$$\text{tr}[(\mathbf{P}^\diamond - \mathbf{P})\mathbf{Q}] = \text{tr}(\mathbf{S}^T \mathbf{Q} \mathbf{S}) = \sum_{i=1}^{n} \mathbf{s}_i^T \mathbf{Q} \mathbf{s}_i = 0$$

implies $\mathbf{s}_i = \mathbf{0}$ if $\mathbf{Q} > \mathbf{O}$. Therefore, $\mathbf{S} = \mathbf{O}$ so that $\mathbf{P}^\diamond = \mathbf{P}$.

The second property involves the determinant of the covariance matrix of estimation errors. We will show that

$$\mathbf{P}^\diamond \geq \mathbf{P} > \mathbf{O} \quad \text{implies} \quad \det \mathbf{P}^\diamond \geq \det \mathbf{P} > 0 \qquad (14.14)$$

and also $\det \mathbf{P} = \det \mathbf{P}^\diamond$ only if $\mathbf{P} = \mathbf{P}^\diamond$. Since the volume of the equiprobability ellipsoid is proportional to the square root of the determinant of the covariance matrix, we can conclude that *the maximum-likelihood estimate also minimizes the volume of the error ellipsoid.*

The proof depends on a transformation which will, simultaneously, reduce both $\mathbf{P}^\diamond$ and $\mathbf{P}$ to diagonal form. Consider the two quadratic forms

$$q^\diamond = \mathbf{x}^T \mathbf{P}^\diamond \mathbf{x} \quad \text{and} \quad q = \mathbf{x}^T \mathbf{P} \mathbf{x}$$

and let $\mathbf{B}$ be the modal matrix of $\mathbf{P}$. Then the transformation $\mathbf{x} = \mathbf{B}\mathbf{y}$ will reduce $q$ to the canonical form

$$q = \mathbf{y}^T \mathbf{D} \mathbf{y} = \mu_1 y_1^2 + \mu_2 y_2^2 + \cdots + \mu_n y_n^2$$

where the characteristic values $\mu_i$ are all positive. Next, the transformation $\mathbf{y} = \mathbf{D}^{-\frac{1}{2}} \mathbf{z}$ results in

$$q = \mathbf{z}^T \mathbf{z} = z_1^2 + z_2^2 + \cdots + z_n^2$$

This same pair of transformations applied to the quadratic form $q^\diamond$ gives

$$q^\diamond = \mathbf{z}^T \mathbf{U} \mathbf{z} \quad \text{where} \quad \mathbf{U} = \mathbf{D}^{-\frac{1}{2}} \mathbf{B}^T \mathbf{P}^\diamond \mathbf{B} \mathbf{D}^{-\frac{1}{2}}$$

Finally, the transformation $\mathbf{z} = \mathbf{V}\mathbf{w}$, where $\mathbf{V}$ is the modal matrix of $\mathbf{U}$, reduces $q^\diamond$ to the canonical form

$$q^\diamond = \mathbf{w}^T \mathbf{V}^T \mathbf{U} \mathbf{V} \mathbf{w} = \mathbf{w}^T \mathbf{D}^\diamond \mathbf{w} = \lambda_1 w_1^2 + \lambda_2 w_2^2 + \cdots + \lambda_n w_n^2$$

but does not alter the form of $q$ since

$$q = \mathbf{z}^T \mathbf{z} = \mathbf{w}^T \mathbf{V}^T \mathbf{V} \mathbf{w} = \mathbf{w}^T \mathbf{w} = w_1^2 + w_2^2 + \cdots + w_n^2$$

Therefore, we conclude that $\mathbf{M} = \mathbf{B} \mathbf{D}^{-\frac{1}{2}} \mathbf{V}$ has the desired property

$$\mathbf{M}^T \mathbf{P}^\diamond \mathbf{M} = \mathbf{D}^\diamond \quad \text{and} \quad \mathbf{M}^T \mathbf{P} \mathbf{M} = \mathbf{I}$$

where $\mathbf{D}^\diamond$ is a diagonal matrix.

The inequality for the determinants (14.14) is now readily established. Since $\mathbf{P}^\diamond - \mathbf{P}$ is positive semidefinite, then

$$\mathbf{M}^T(\mathbf{P}^\diamond - \mathbf{P})\mathbf{M} = \mathbf{D}^\diamond - \mathbf{I} \geq \mathbf{O}$$

Hence, the elements of $\mathbf{D}^\diamond$ cannot be less than unity. But

$$|\mathbf{M}|^2|\mathbf{P}^\diamond| = |\mathbf{D}^\diamond| \quad \text{and} \quad |\mathbf{M}|^2|\mathbf{P}| = 1$$

so that

$$|\mathbf{P}^\diamond| = |\mathbf{P}||\mathbf{D}^\diamond| \quad \text{or} \quad |\mathbf{P}^\diamond| \geq |\mathbf{P}|$$

## 14.3 Position and Velocity Estimation

Consider the deterministic problem of obtaining a fix in six dimensions by a sequence of individual celestial observations made at six discrete times. Under the assumptions of a linearized theory, a single observation serves to fix the position of a spacecraft in one coordinate. For example, as shown in Chapter 13, if $A$ is the angle measured at time $t$ and is defined by the lines of sight from the vehicle to a star and to a nearby celestial body, the position of the vehicle is established along a line normal to the direction toward the near body and in the plane of the measurement. In general, for all the various measurements considered in Chapter 13, we may express the results as

$$\delta q = \mathbf{b}^T \delta \mathbf{x} \tag{14.15}$$

where the six-dimensional vectors $\delta \mathbf{x}$ and $\mathbf{b}$ are defined by

$$\delta \mathbf{x}(t) = \begin{bmatrix} \delta \mathbf{r}(t) \\ \delta \mathbf{v}(t) \end{bmatrix} \quad \text{and} \quad \mathbf{b} = \begin{bmatrix} \mathbf{h} \\ \mathbf{0} \end{bmatrix}$$

The vector $\mathbf{h}$, determined as described in Chapter 13, depends on the geometrical configuration of the relevant celestial objects at time $t$ as well as on the particular type of measurement made.

The state-transition matrix

$$\Phi(t_n, t_{n-1}) = \Phi_{n,n-1}$$

introduced in Sect. 9.5 provides the relationship

$$\delta \mathbf{x}_n = \Phi_{n,n-1} \delta \mathbf{x}_{n-1} \tag{14.16}$$

between $\delta \mathbf{x}(t_n) = \delta \mathbf{x}_n$ and $\delta \mathbf{x}(t_{n-1}) = \delta \mathbf{x}_{n-1}$. Therefore, by combining Eqs. (14.15) and (14.16) so that

$$\delta q(t_{n-1}) = \delta q_{n-1} = \mathbf{b}_{n-1}^T \delta \mathbf{x}_{n-1} = \mathbf{b}_{n-1}^T \Phi_{n,n-1}^{-1} \delta \mathbf{x}_n$$

it is clear that the effect at time $t_n$ of an observation made at time $t_{n-1}$ is to determine the component of the six-dimensional deviation vector $\delta \mathbf{x}_n$

Sect. 14.3]   Position and Velocity Estimation   673

in the direction defined by $\Phi_{n,n-1}^{-T}\mathbf{b}_n$. Six observations made at different times $t_1, t_2, \ldots, t_6$ would provide a set of six equations of the form:†

$$\delta q_1 = \mathbf{b}_1^T \, \delta \mathbf{x}_1 = \mathbf{b}_1^T \Phi_{6,1}^{-1} \, \delta \mathbf{x}_6$$
$$\delta q_2 = \mathbf{b}_2^T \, \delta \mathbf{x}_2 = \mathbf{b}_2^T \Phi_{6,2}^{-1} \, \delta \mathbf{x}_6$$
$$\vdots$$
$$\delta q_6 = \mathbf{b}_6^T \, \delta \mathbf{x}_6$$

If the component directions span the state space, then the deviation vector $\delta \mathbf{x}(t_6)$ could be obtained by inverting the six-dimensional coefficient matrix in the vector-matrix form of these equations:

$$\delta \mathbf{q} = \begin{bmatrix} \mathbf{b}_1^T \Phi_{6,1}^{-1} \\ \mathbf{b}_2^T \Phi_{6,2}^{-1} \\ \vdots \\ \mathbf{b}_6^T \end{bmatrix} \delta \mathbf{x}_6$$

### Range-Rate Measurement

If the basic source of navigation information is range rate measured, for example, from a ground-based tracking radar, then the measurement vector $\mathbf{b}$ will be a full six-dimensional vector. The range-rate $\dot{r}$ will be a function $f(\mathbf{x})$ where $\mathbf{x}$ is the six-dimensional state vector. We expand $\dot{r}$ in a Taylor series about the reference state $\mathbf{x}_0$ at which the quantity to be measured is $\dot{r}_0$. Thus,

$$\dot{r}(\mathbf{x}) = \dot{r}(\mathbf{x}_0) + \frac{\partial f}{\partial \mathbf{x}} \delta \mathbf{x} + \cdots = \dot{r}_0 + \mathbf{b}^T \, \delta \mathbf{x} + \cdots \quad (14.17)$$

where it is understood that the derivative is to be evaluated at the reference point.

To calculate the vector $\mathbf{b}$ we differentiate the measurement equation

$$r\dot{r} = \mathbf{r} \cdot \mathbf{v} = \mathbf{r}^T \mathbf{v} = \mathbf{v}^T \mathbf{r}$$

to obtain

$$\frac{\partial r}{\partial \mathbf{x}} \dot{r} + r \frac{\partial \dot{r}}{\partial \mathbf{x}} = \mathbf{r}^T \frac{\partial \mathbf{v}}{\partial \mathbf{r}} + \mathbf{v}^T \frac{\partial \mathbf{r}}{\partial \mathbf{x}}$$

Substituting for the derivatives, we have

$$\begin{bmatrix} \mathbf{i}_r^T & \mathbf{0}^T \end{bmatrix} \dot{r} + r \frac{\partial \dot{r}}{\partial \mathbf{x}} = \mathbf{r}^T \begin{bmatrix} \mathbf{O} & \mathbf{I} \end{bmatrix} + \mathbf{v}^T \begin{bmatrix} \mathbf{I} & \mathbf{O} \end{bmatrix} = \begin{bmatrix} \mathbf{v}^T & \mathbf{r}^T \end{bmatrix}$$

---

† It should be remembered that the state-transition matrix is symplectic so that its inverse is particularly simple to obtain.

Hence,

$$\frac{\partial \dot{r}}{\partial \mathbf{x}} = \frac{1}{r}[\mathbf{v}^T \quad \mathbf{r}^T] - \frac{\dot{r}}{r^2}[\mathbf{r}^T \quad \mathbf{0}] = \left[\frac{1}{r}\mathbf{v}^T - \frac{\dot{r}}{r^2}\mathbf{r}^T \quad \frac{1}{r}\mathbf{r}^T\right]$$

$$= \left[\frac{1}{r}\mathbf{v}^T - \frac{\mathbf{v}^T\mathbf{r}}{r^3}\mathbf{r}^T \quad \frac{1}{r}\mathbf{r}^T\right] = \left[\frac{1}{r^3}[\mathbf{r} \times (\mathbf{v} \times \mathbf{r})]^T \quad \frac{1}{r}\mathbf{r}^T\right]$$

Therefore, the appropriate form for the six-dimensional $\mathbf{b}$ vector is

$$\mathbf{b} = \frac{1}{r^3}\begin{bmatrix} \mathbf{r} \times (\mathbf{v} \times \mathbf{r}) \\ r^2\mathbf{r} \end{bmatrix} \quad (14.18)$$

Recursive Estimation

Assume that an estimate $\delta\hat{\mathbf{x}}$ of the deviation in the state and the corresponding covariance matrix $\mathbf{P}$ of the estimation error vector $\mathbf{e}$ are known and that a single measurement is made of the type discussed in the previous chapter. The observed deviation in the measured quantity $q$ is $\delta\tilde{q}$, and the best estimate of $\delta q$, as obtained from the estimate $\delta\hat{\mathbf{x}}$, is

$$\delta\hat{q} = \mathbf{b}^T \delta\hat{\mathbf{x}}$$

With the development of the recursive form of the Gaussian least squares estimator as a guide, we are led to express the new linear estimate $\delta\hat{\mathbf{x}}^*$ for the deviation vector as a linear combination of the old estimate $\delta\hat{\mathbf{x}}$ and the difference between the observed and estimated deviations in the measured quantity $q$. Thus,

$$\delta\hat{\mathbf{x}}^* = \delta\hat{\mathbf{x}} + \mathbf{w}(\delta\tilde{q} - \delta\hat{q}) \quad (14.19)$$

where the vector $\mathbf{w}$ is a weighting factor to be chosen so as to minimize the variance in the error of the estimate.

As before, we distinguish between the measured value of $q$ and its true value by writing

$$\delta\tilde{q} = \delta q + \alpha$$

where $\alpha$ is the error in the measurement which we shall assume to be a random variable with zero mean and independent of the old estimation error vector $\mathbf{e}$. In symbols, $\overline{\alpha\mathbf{e}} = \overline{\alpha}\,\overline{\mathbf{e}} = \mathbf{0}$.

To address the optimization problem we expand the error vector as

$$\begin{aligned} \mathbf{e}^*(\mathbf{w}) &= \delta\hat{\mathbf{x}}^* - \delta\mathbf{x} \\ &= \delta\hat{\mathbf{x}} + \mathbf{w}(\delta q + \alpha - \delta\hat{q}) - \delta\mathbf{x} \\ &= (\mathbf{I} - \mathbf{w}\mathbf{b}^T)(\delta\hat{\mathbf{x}} - \delta\mathbf{x}) + \mathbf{w}\alpha \\ &= (\mathbf{I} - \mathbf{w}\mathbf{b}^T)\mathbf{e} + \mathbf{w}\alpha \end{aligned} \quad (14.20)$$

## Sect. 14.3]    Position and Velocity Estimation    675

where **I** is the six-dimensional identity matrix. Then the covariance matrix **P*** may be expressed as a function of the weighting vector **w** as

$$\mathbf{P}^*(\mathbf{w}) = (\mathbf{I} - \mathbf{w}\mathbf{b}^T)\mathbf{P}(\mathbf{I} - \mathbf{b}\mathbf{w}^T) + \mathbf{w}\mathbf{w}^T\overline{\alpha^2} \qquad (14.21)$$

### Partitioning and Propagating the Covariance Matrix

The six-dimensional covariance matrix is partitioned into three-dimensional blocks as

$$\mathbf{P} = \overline{\mathbf{e}\mathbf{e}^T} = \overline{\begin{bmatrix} \boldsymbol{\epsilon} \\ \boldsymbol{\delta} \end{bmatrix}\begin{bmatrix} \boldsymbol{\epsilon}^T & \boldsymbol{\delta}^T \end{bmatrix}} = \begin{bmatrix} \overline{\boldsymbol{\epsilon}\boldsymbol{\epsilon}^T} & \overline{\boldsymbol{\epsilon}\boldsymbol{\delta}^T} \\ \overline{\boldsymbol{\delta}\boldsymbol{\epsilon}^T} & \overline{\boldsymbol{\delta}\boldsymbol{\delta}^T} \end{bmatrix} = \begin{bmatrix} \mathbf{P}_1 & \mathbf{P}_2 \\ \mathbf{P}_3 & \mathbf{P}_4 \end{bmatrix} \qquad (14.22)$$

The variances in the estimation errors of position and velocity deviations $\overline{\epsilon^2}$ and $\overline{\delta^2}$ are simply the respective traces of the submatrices $\mathbf{P}_1$ and $\mathbf{P}_4$. If the six-dimensional measurement vector **b** and weighting vector **w** are each partitioned into two three-dimensional vectors

$$\mathbf{b} = \begin{bmatrix} \mathbf{b}_1 \\ \mathbf{b}_2 \end{bmatrix} \quad \text{and} \quad \mathbf{w} = \begin{bmatrix} \mathbf{w}_1 \\ \mathbf{w}_2 \end{bmatrix}$$

then from Eq. (14.21) we have

$$\begin{bmatrix} \mathbf{P}_1^* & \mathbf{P}_2^* \\ \mathbf{P}_3^* & \mathbf{P}_4^* \end{bmatrix} = \overline{\alpha^2}\begin{bmatrix} \mathbf{w}_1\mathbf{w}_1^T & \mathbf{w}_1\mathbf{w}_2^T \\ \mathbf{w}_2\mathbf{w}_1^T & \mathbf{w}_2\mathbf{w}_2^T \end{bmatrix}$$
$$+ \begin{bmatrix} \mathbf{I} - \mathbf{w}_1\mathbf{b}_1^T & -\mathbf{w}_1\mathbf{b}_2^T \\ -\mathbf{w}_2\mathbf{b}_1^T & \mathbf{I} - \mathbf{w}_2\mathbf{b}_2^T \end{bmatrix}\begin{bmatrix} \mathbf{P}_1 & \mathbf{P}_2 \\ \mathbf{P}_3 & \mathbf{P}_4 \end{bmatrix}\begin{bmatrix} \mathbf{I} - \mathbf{b}_1\mathbf{w}_1^T & -\mathbf{b}_1\mathbf{w}_2^T \\ -\mathbf{b}_2\mathbf{w}_1^T & \mathbf{I} - \mathbf{b}_2\mathbf{w}_2^T \end{bmatrix}$$

When expanded, we see that $\mathbf{P}_1^*$ is a function only of $\mathbf{w}_1$ and $\mathbf{P}_4^*$ is a function only of $\mathbf{w}_2$. Therefore, for the purposes of the following discussion, it is legitimate formally to treat the variance in the estimation error vector $\mathbf{e}^*(\mathbf{w})$ as the trace of the six-dimensional covariance matrix $\mathbf{P}^*(\mathbf{w})$. The subvectors of the optimum weighting vector **w** will then be optimum for the respective estimates of position and velocity deviations.

Consider next the problem of propagating the estimation error vector and the covariance matrix. Suppose that an estimate of the deviation vector $\delta\mathbf{x}_{n-1}$ has been made by some method, and let $\mathbf{e}_{n-1}$ be the error in the estimate. Then

$$\delta\widehat{\mathbf{x}}_{n-1} = \delta\mathbf{x}_{n-1} + \mathbf{e}_{n-1}$$

If no new observations are made, the estimate $\delta\widehat{\mathbf{x}}_{n-1}$ extrapolated to a later time $t_n$ is obtained from

$$\delta\widehat{\mathbf{x}}_n = \boldsymbol{\Phi}_{n,n-1}\,\delta\widehat{\mathbf{x}}_{n-1} \qquad (14.23)$$

Since the actual deviation vector as well as the estimate propagates via the transition matrix, then so also does the error vector. The extrapolated error vector is determined from

$$\mathbf{e}_n = \mathbf{\Phi}_{n,n-1}\mathbf{e}_{n-1} \tag{14.24}$$

so that the extrapolated covariance matrix $\mathbf{P}_n$ is related to $\mathbf{P}_{n-1}$ by

$$\mathbf{P}_n = \mathbf{\Phi}_{n,n-1}\mathbf{P}_{n-1}\mathbf{\Phi}_{n,n-1}^T \tag{14.25}$$

### The Minimum-Variance Estimator

To determine the weighting vector which minimizes the estimation error variance, we apply the usual technique of the variational calculus. Let $\mathbf{w}$ take on a variation $\delta\mathbf{w}$, and obtain from Eq. (14.21)

$$\delta\overline{e^{*2}(\mathbf{w})} = 2\operatorname{tr}\left[-\delta\mathbf{w}\,\mathbf{b}^T\mathbf{P}(\mathbf{I} - \mathbf{b}\mathbf{w}^T) + \delta\mathbf{w}\,\mathbf{w}^T\overline{\alpha^2}\right]$$

If $\delta\overline{e^{*2}(\mathbf{w})}$ is to vanish for all variations $\delta\mathbf{w}$, it follows that†

$$a\mathbf{w} = \mathbf{P}\mathbf{b} \quad \text{where} \quad a = \mathbf{b}^T\mathbf{P}\mathbf{b} + \overline{\alpha^2} \tag{14.26}$$

It can be readily shown that this value of $\mathbf{w}$ actually does minimize $\overline{e^{*2}(\mathbf{w})}$. For suppose that the optimum $\mathbf{w}$ is replaced by another weighting factor $\mathbf{w} - \mathbf{y}$; then, from Eq. (14.21), we have

$$\overline{e^{*2}(\mathbf{w}-\mathbf{y})} = \operatorname{tr}[\mathbf{P} - 2(\mathbf{w}-\mathbf{y})\mathbf{b}^T\mathbf{P} + a(\mathbf{w}-\mathbf{y})(\mathbf{w}^T - \mathbf{y}^T)]$$

Using the first of Eqs. (14.26), this becomes

$$\overline{e^{*2}(\mathbf{w}-\mathbf{y})} = \operatorname{tr}[\mathbf{P} - a(\mathbf{w}-\mathbf{y})(\mathbf{w}^T + \mathbf{y}^T)]$$

so that

$$\overline{e^{*2}(\mathbf{w}-\mathbf{y})} = \overline{e^2(\mathbf{w})} + a\operatorname{tr}(\mathbf{y}\mathbf{y}^T)$$

Thus, the variance is not decreased by perturbing $\mathbf{w}$ if Eqs. (14.26) hold.

The optimum weighting vector having been obtained, the expression for the covariance matrix of the estimation errors $\mathbf{P}^*$ given by Eq. (14.21) may be written in a more compact form. Thus, from the definition of $a$ in the second of Eqs. (14.26), we obtain

$$\mathbf{P}^* = \mathbf{P}(\mathbf{I} - \mathbf{b}\mathbf{w}^T) - \mathbf{w}\mathbf{b}^T\mathbf{P} + a\mathbf{w}\mathbf{w}^T$$

Substituting from the first of Eqs. (14.26), the alternate expression is

$$\mathbf{P}^* = (\mathbf{I} - \mathbf{w}\mathbf{b}^T)\mathbf{P} \tag{14.27}$$

From the standpoint of numerical accuracy, the question arises as to which of the two equivalent expressions, (14.21) or (14.27), should be used for updating the covariance matrix.

---

† In this manner, we have established that the minimum-variance estimate, the maximum-likelihood estimate, and Gauss's method of least squares are all equivalent.

To address this question, suppose that a small error $\delta \mathbf{w}$ is made in the calculation of the weighting vector $\mathbf{w}$. The resulting error in $\delta \mathbf{P}^*$ using Eq. (14.21) is found to be

$$\delta \mathbf{P}^* = -\delta \mathbf{w}\, \mathbf{b}^T \mathbf{P}(\mathbf{I} - \mathbf{b}\mathbf{w}^T) - (\mathbf{I} - \mathbf{w}\mathbf{b}^T)\mathbf{P}\mathbf{b}\,\delta \mathbf{w}^T$$
$$+ \overline{\alpha^2}(\delta \mathbf{w}\,\mathbf{w}^T + \mathbf{w}\,\delta \mathbf{w}^T) = \mathbf{O}$$

since

$$\overline{\alpha^2}\mathbf{w}^T - \mathbf{b}^T \mathbf{P}(\mathbf{I} - \mathbf{b}\mathbf{w}^T) = a\mathbf{w}^T - \mathbf{b}^T\mathbf{P} = \mathbf{0}^T$$

On the other hand, if Eq. (14.27) is used, then the error is

$$\delta \mathbf{P}^* = -\delta \mathbf{w}\, \mathbf{b}^T \mathbf{P} = -a\,\delta \mathbf{w}\,\mathbf{w}^T$$

Clearly then, the first relationship, even though more complex, is to be preferred for numerical work.

### A Property of the Optimum Estimator

An important property of the optimum estimate, which is needed for the development of the statistical analysis procedures described in Sect. 14.4, will now be derived. The result may be simply stated as

$$\overline{\mathbf{e}\,\delta \widehat{\mathbf{x}}^T} = \mathbf{O} \qquad (14.28)$$

if $\delta \widehat{\mathbf{x}}$ is the optimum estimate. In words, *the optimum estimate and the associated error in that estimate are uncorrelated.*

For the proof, we note that since

$$\delta \widetilde{q} = \mathbf{b}^T\, \delta \mathbf{x} + \alpha \quad \text{and} \quad \delta \widehat{q} = \mathbf{b}^T\, \delta \widehat{\mathbf{x}}$$

then

$$\delta \widehat{\mathbf{x}}^* = \delta \widehat{\mathbf{x}} + \mathbf{w}(\delta \widetilde{q} - \delta \widehat{q})$$
$$= \delta \widehat{\mathbf{x}} + \mathbf{w}(\alpha - \mathbf{b}^T \mathbf{e}) \qquad (14.29)$$

Subtracting $\delta \mathbf{x}$ from both sides gives

$$\mathbf{e}^* = (\mathbf{I} - \mathbf{w}\mathbf{b}^T)\mathbf{e} + \alpha \mathbf{w} \qquad (14.30)$$

Now form the dyadic product of Eqs. (14.29) and (14.30) and compute the average. There results

$$\overline{\mathbf{e}^*\, \delta \widehat{\mathbf{x}}^{*T}} = (\mathbf{I} - \mathbf{w}\mathbf{b}^T)\, \overline{\mathbf{e}\,\delta \widehat{\mathbf{x}}^T} - (\mathbf{I} - \mathbf{w}\mathbf{b}^T)\mathbf{P}\mathbf{b}\mathbf{w}^T + \overline{\alpha^2}\mathbf{w}\mathbf{w}^T$$

or

$$\overline{\mathbf{e}^*\, \delta \widehat{\mathbf{x}}^{*T}} = (\mathbf{I} - \mathbf{w}\mathbf{b}^T)\, \overline{\mathbf{e}\,\delta \widehat{\mathbf{x}}^T} + (a\mathbf{w} - \mathbf{P}\mathbf{b})\mathbf{w}^T \qquad (14.31)$$

Equation (14.31) may be viewed from two perspectives. First, if (14.28) is required for each consecutive measurement, then we must have

$$\mathbf{w} = \frac{1}{a}\mathbf{P}\mathbf{b} \qquad (14.32)$$

On the other hand, if Eq. (14.32) holds, then

$$\overline{e^* \delta \widehat{x}^{*T}} = (I - wb^T) \overline{e \delta \widehat{x}^T}$$

obtains as a recursion formula for the matrix in question. If $\delta \widehat{x}$ or $e$ is initially zero, it follows that $\overline{e \delta \widehat{x}^T}$ would then always be zero.

### Energy and Angular Momentum Pseudo-Measurements

The application of recursive filtering techniques to space navigation requires that the associated state vector and error covariance matrix be specified both initially and at those times when reinitialization of the filter is required to cope with problems of divergence and instability. A fairly common initialization method, and one that was used in Apollo navigation, is to postulate that the initial errors in the state are uncorrelated and spherically distributed, which implies that the initial covariance matrix of estimation errors is diagonal. The true state vector errors are, of course, correlated and several measurements must be processed before the covariance matrix begins to exhibit proper cross correlations.

The purpose of this subsection is to describe an initialization technique which partially accounts for these correlations, considerably reduces the undesirable transient effects of the first few measurements, and acts to inhibit filter divergence when the interval between measurements is inordinately large. The improvement in navigation performance is quite dramatic.†

The method consists of including in the initial covariance matrix the effect of a number of pseudo-measurements of certain orbital quantities $q$. Although no direct measurement of these quantities is made, this measurement information can be adjoined to the error covariance matrix by assuming a reasonable value for a measurement variance and updating the covariance matrix as though an actual measurement had been performed.

Two orbital quantities that have proved to be particularly useful for this purpose are the total energy

$$E = \frac{1}{2}v^2 - \frac{\mu}{r}$$

and the angular momentum

$$h = |\mathbf{h}| = |\mathbf{r} \times \mathbf{v}|$$

The determination of the $\mathbf{b}$ vectors for these "measurements" is left as an exercise in the following problem.

---

† Examples are given in the paper "Recursive Filter Initialization" for both midcourse navigation in an Apollo-type voyage to the moon and for the Space Shuttle navigating with the aid of somewhat sparsely distributed ground beacons. The article was coauthored by Richard H. Battin, Steven R. Croopnick, and James M. Habbe and published in the *Journal of Spacecraft and Rockets*, Vol. 9, June 1972.

## ◇ Problem 14-6

The measurement vectors for pseudo-measurements of energy and angular velocity are

$$\mathbf{b}_E = \begin{bmatrix} \mu \mathbf{r}/r^3 \\ \mathbf{v} \end{bmatrix} \quad \text{and} \quad \mathbf{b}_h = \frac{1}{h}\begin{bmatrix} \mathbf{v} \times \mathbf{h} \\ \mathbf{h} \times \mathbf{r} \end{bmatrix}$$

## Square-Root Filtering with Plant Noise

Consider a system whose state is described by a linear difference equation

$$\delta \mathbf{x}_n = \mathbf{\Phi}_{n,n-1} \delta \mathbf{x}_{n-1} + \mathbf{n}_{n-1} \tag{14.33}$$

and which is observed through measurement data $\delta \widetilde{q}_n$ obtained at discrete instants of time $t_n$. These data are assumed to be linearly related to the state according to

$$\delta \widetilde{q}_n = \mathbf{b}_n^T \delta \mathbf{x}_n + \alpha_n$$

The noise sequences† $\mathbf{n}_n$ and $\alpha_n$ are assumed to be Gaussian and uncorrelated between sampling times with zero means and covariances

$$\overline{\mathbf{n}_n \mathbf{n}_n^T} = \mathbf{N}_n \qquad \overline{\alpha_n^2} = \sigma_n^2 \qquad \overline{\alpha \mathbf{n}} = \mathbf{0} \tag{14.34}$$

For this system, the unbiased, minimum variance estimate of $\delta \mathbf{x}_n$, given the data $\delta \widetilde{q}_n$, is described by the following system. The estimate of the state $\delta \mathbf{x}_n$ is

$$\delta \widehat{\mathbf{x}}_n^* = \delta \widehat{\mathbf{x}}_n + \mathbf{w}_n (\delta \widetilde{q}_n - \mathbf{b}_n^T \delta \widehat{\mathbf{x}}_n) \tag{14.35}$$

where

$$\delta \widehat{\mathbf{x}}_n = \mathbf{\Phi}_{n,n-1} \delta \widehat{\mathbf{x}}_{n-1} \qquad \mathbf{w}_n = \frac{1}{a_n} \mathbf{P}_n \mathbf{b}_n \qquad a_n = \mathbf{b}_n^T \mathbf{P}_n \mathbf{b}_n + \sigma_n^2$$

and

$$\mathbf{P}_n = \mathbf{\Phi}_{n,n-1} \mathbf{P}_{n-1}^* \mathbf{\Phi}_{n,n-1}^T + \mathbf{N}_{n-1} \tag{14.36}$$

$$\mathbf{P}_n^* = (\mathbf{I} - \mathbf{w}_n \mathbf{b}_n^T) \mathbf{P}_n (\mathbf{I} - \mathbf{w}_n \mathbf{b}_n^T)^T + \mathbf{w}_n \mathbf{w}_n^T \sigma_n^2 \tag{14.37}$$

The covariance matrix is at least positive semidefinite so that there exists a real matrix $\mathbf{W}_n$ for which

$$\mathbf{W}_n \mathbf{W}_n^T = \mathbf{P}_n \tag{14.38}$$

We desire analogous equations for extrapolating and updating the square-root matrix $\mathbf{W}_n$.

To this end, observe that (14.36) and (14.37) are special cases of

$$\mathbf{Y}\mathbf{Y}^T = \mathbf{X}\mathbf{X}^T + \mathbf{x}\mathbf{x}^T \tag{14.39}$$

---

† These are examples of what we generally call *white noise*. It is also customary today to refer to the random variable $\mathbf{n}$ in Eq. (14.33) as "plant noise" just as though the matrix $\mathbf{\Phi}$ were the model of some manufacturing plant.

which can be written as
$$\mathbf{Y}\mathbf{Y}^T = \mathbf{X}(\mathbf{I} + \mathbf{z}\mathbf{z}^T)\mathbf{X}^T \tag{14.40}$$
with the vector $\mathbf{z}$ defined by
$$\mathbf{z} = \mathbf{X}^{-1}\mathbf{x} \tag{14.41}$$
Using the technique developed in the last subsection of Sect. 13.7, we can obtain the factorization
$$\mathbf{I} + \mathbf{z}\mathbf{z}^T = \left(\mathbf{I} + \frac{\mathbf{z}\mathbf{z}^T}{1 + \sqrt{c}}\right)\left(\mathbf{I} + \frac{\mathbf{z}\mathbf{z}^T}{1 + \sqrt{c}}\right)$$
where†
$$c = 1 + \mathbf{z}^T\mathbf{z} = 1 + \operatorname{tr}(\mathbf{z}\mathbf{z}^T)$$
Therefore, we obtain
$$\mathbf{Y} = \mathbf{X}\left(\mathbf{I} + \frac{\mathbf{z}\mathbf{z}^T}{1 + \sqrt{a}}\right) \tag{14.42}$$
The corresponding equation for the inverse of the $\mathbf{Y}$ matrix
$$\mathbf{Y}^{-1} = \left(\mathbf{I} - \frac{\mathbf{z}\mathbf{z}^T}{\sqrt{a}(1 + \sqrt{a})}\right)\mathbf{X}^{-1} \tag{14.43}$$
is found using the result of Prob. 13–7.

We can apply these results to the case at hand. First, to extrapolate the square-root matrix, we define an intermediate matrix
$$\mathbf{C}_{n,n-1} = \mathbf{W}_{n-1}^{*-1}\mathbf{\Phi}_{n,n-1}^{-1}\mathbf{N}_{n-1}\mathbf{\Phi}_{n,n-1}^{-T}\mathbf{W}_{n-1}^{*-T} \tag{14.44}$$
in terms of which we have
$$\mathbf{W}_n = \mathbf{\Phi}_{n,n-1}\mathbf{W}_{n-1}^*\left(\mathbf{I} + \frac{\mathbf{C}_{n,n-1}}{1 + \sqrt{c_n}}\right) \tag{14.45}$$
and‡
$$\mathbf{W}_n^{-1} = \left(\mathbf{I} - \frac{\mathbf{C}_{n,n-1}}{\sqrt{c_n}(1 + \sqrt{c_n})}\right)\mathbf{W}_{n-1}^{*-1}\mathbf{\Phi}_{n,n-1}^{-1} \tag{14.46}$$
where
$$c_n = 1 + \operatorname{tr}\mathbf{C}_{n,n-1}$$
The update equation can be derived in the same manner or it can be obtained directly from Eq. (13.73) as
$$\mathbf{W}_n^* = \mathbf{W}_n\left(\mathbf{I} - \frac{\mathbf{Z}_n^*}{a_n + \sqrt{a_n\sigma_n^2}}\right) \tag{14.47}$$

---

† The alternate form of the constant $c$ is to be preferred when the term $\mathbf{x}\mathbf{x}^T$ is to represent the covariance matrix $\mathbf{N} = \overline{\mathbf{n}\mathbf{n}^T}$.

‡ It is necessary to keep track of both the square-root matrix and its inverse in order to avoid the computation of a matrix square root.

in terms of the intermediate matrix
$$\mathbf{Z}_n^* = \mathbf{W}_n^T \mathbf{b}_n \mathbf{b}_n^T \mathbf{W}_n \quad (14.48)$$
For the inverse,
$$\mathbf{W}_n^{*-1} = \left(\mathbf{I} - \frac{\mathbf{Z}_n^*}{\sqrt{a_n \sigma_n^2}}\right)\mathbf{W}_n^{-1} \quad (14.49)$$
we can again use the result of Prob. 13–7.

## 14.4 Statistical Error Analysis

In Sect. 11.6 the fixed-time-of-arrival velocity correction was derived and expressed in Eq. (11.41) in terms of the vector deviations in position and velocity from the reference path at the time the correction was made. Actually, of course, all that is available is an estimate of these quantities so that the estimated velocity correction is

$$\Delta\hat{\mathbf{v}} = \mathbf{C}^* \, \delta\hat{\mathbf{r}} - \delta\hat{\mathbf{v}} = \mathbf{B} \, \delta\hat{\mathbf{x}} \quad (14.50)$$

where we have defined the $3 \times 6$ dimensional matrix

$$\mathbf{B} = \begin{bmatrix} \mathbf{C}^* & -\mathbf{I} \end{bmatrix}$$

For the purpose of a statistical analysis of the guidance problem, we need a convenient expression for the covariance matrix of the velocity-correction vector. As we shall see, it is possible to express this covariance matrix directly in terms of the covariance matrix of estimation errors $\mathbf{P}$ and the covariance matrix of the actual deviation vector. This latter matrix is defined as

$$\mathbf{X} = \overline{\delta\mathbf{x}\,\delta\mathbf{x}^T} \quad (14.51)$$

From Eq. (14.50) we have

$$\Delta\hat{\mathbf{v}} = \mathbf{B}(\delta\mathbf{x} + \mathbf{e})$$

so that

$$\overline{\Delta\hat{\mathbf{v}}\,\Delta\hat{\mathbf{v}}^T} = \mathbf{B}(\overline{\delta\mathbf{x}\,\delta\mathbf{x}^T} + \overline{\mathbf{e}\,\delta\mathbf{x}^T} + \overline{\delta\mathbf{x}\,\mathbf{e}^T} + \mathbf{P})\mathbf{B}^T$$

On the other hand,

$$\delta\hat{\mathbf{x}} = \delta\mathbf{x} + \mathbf{e}$$

from which

$$\overline{\mathbf{e}\,\delta\hat{\mathbf{x}}^T} = \overline{\mathbf{e}\,\delta\mathbf{x}^T} + \mathbf{P} = \mathbf{O}$$

according to the property (14.28) proved in Sect. 14.3. Hence

$$\overline{\Delta\hat{\mathbf{v}}\,\Delta\hat{\mathbf{v}}^T} = \mathbf{B}(\mathbf{X} - \mathbf{P})\mathbf{B}^T \quad (14.52)$$

The inaccuracy in establishing a commanded velocity correction $\Delta\hat{\mathbf{v}}$ is attributed to errors in both magnitude and orientation. In the following analysis these two sources of error will be assumed to be independent random variables with zero means.

Consider a coordinate system in which the estimated vector-velocity correction is along one of the coordinate axes. Then if $\mathbf{T}$ is the transformation matrix which relates the selected axis system and the original reference system, we write

$$\Delta\hat{\mathbf{v}} = \Delta\hat{v}\,\mathbf{T} \begin{bmatrix} 0 \\ 0 \\ 1 \end{bmatrix}$$

Now define a random variable $\kappa$ such that

$$\Delta v = (1+\kappa)\,\Delta\hat{v}$$

and let $\gamma$ be the random angle between $\Delta\hat{\mathbf{v}}$ and $\Delta\mathbf{v}$. It will be assumed that $\kappa$ and $\gamma$ are small quantities, so that their powers and products are negligible. The actual vector-velocity correction is then

$$\Delta\mathbf{v} = (1+\kappa)\,\Delta\hat{v}\,\mathbf{T} \begin{bmatrix} \gamma\cos\beta \\ \gamma\sin\beta \\ 1 \end{bmatrix}$$

where the polar angle $\beta$ is measured in the plane normal to $\Delta\hat{\mathbf{v}}$ from the coordinate axis to the projection of $\Delta\mathbf{v}$. Hence, the uncertainty vector $\boldsymbol{\eta}$ can be expressed as

$$\boldsymbol{\eta} = \Delta\hat{\mathbf{v}} - \Delta\mathbf{v} = -\Delta\hat{v}\,\mathbf{T} \begin{bmatrix} \gamma\cos\beta \\ \gamma\sin\beta \\ \kappa \end{bmatrix}$$

Assume that $\kappa$, $\gamma$, $\beta$ are statistically independent random variables with zero means. Further, assume that $\beta$ is uniformly distributed over the interval from $-\pi$ to $\pi$. Then we obtain for the covariance matrix of the velocity-correction uncertainty

$$\mathbf{N} = \overline{\boldsymbol{\eta}\boldsymbol{\eta}^T} = \overline{\kappa^2}\,\overline{\Delta\hat{\mathbf{v}}\Delta\hat{\mathbf{v}}^T} + \tfrac{1}{2}\overline{\gamma^2}\,\overline{(\Delta\hat{v})^2}\,\mathbf{T}\begin{bmatrix} 1 & 0 & 0 \\ 0 & 1 & 0 \\ 0 & 0 & 0 \end{bmatrix}\mathbf{T}^T$$

so that

$$\mathbf{N} = \overline{\kappa^2}\,\overline{\Delta\hat{\mathbf{v}}\Delta\hat{\mathbf{v}}^T} + \tfrac{1}{2}\overline{\gamma^2}\left(\overline{\Delta\hat{\mathbf{v}}^T\Delta\hat{\mathbf{v}}}\,\mathbf{I} - \overline{\Delta\hat{\mathbf{v}}\Delta\hat{\mathbf{v}}^T}\right) \quad (14.53)$$

where $\overline{\kappa^2}$ and $\overline{\gamma^2}$ are the variances of $\kappa$ and $\gamma$.

The actual change in the velocity-deviation vector $\delta\mathbf{v}$ as a result of the correction is

$$\Delta\mathbf{v} = \mathbf{B}\,\delta\hat{\mathbf{x}} - \boldsymbol{\eta}$$

Sect. 14.4]  Statistical Error Analysis  683

To serve the requirements of our present notation, we need to introduce a compatibility matrix $\mathbf{M}$ having six rows and three columns and defined by

$$\mathbf{M} = \begin{bmatrix} \mathbf{O} \\ \mathbf{I} \end{bmatrix}$$

Then at the time of a velocity-correction we have

$$\delta \mathbf{x}^* = \delta \mathbf{x} + \mathbf{MB}\,\delta \widehat{\mathbf{x}} - \mathbf{M}\eta$$
$$= (\mathbf{I} + \mathbf{MB})\,\delta \mathbf{x} + \mathbf{MB}\mathbf{e} - \mathbf{M}\eta$$

Hence,

$$\overline{\delta \mathbf{x}^* \delta \mathbf{x}^{*\,T}} = (\mathbf{I} + \mathbf{MB})\,\overline{\delta \mathbf{x}\,\delta \mathbf{x}^T}(\mathbf{I} + \mathbf{MB})^T + \mathbf{MBP}(\mathbf{MB})^T + \mathbf{MNM}^T$$
$$+ (\mathbf{I} + \mathbf{MB})\,\overline{\delta \mathbf{x}\,\mathbf{e}^T}(\mathbf{MB})^T + \mathbf{MB}\,\overline{\mathbf{e}\,\delta \mathbf{x}^T}(\mathbf{I} + \mathbf{MB})^T$$

This expression may be further reduced using Eq. (14.28). There results

$$\mathbf{X}^* = (\mathbf{I} + \mathbf{MB})(\mathbf{X} - \mathbf{P})(\mathbf{I} + \mathbf{MB})^T + \mathbf{P} + \mathbf{MNM}^T \qquad (14.54)$$

The $\mathbf{P}$ matrix changes also when a velocity correction is imperfectly implemented. Thus, the estimation error is altered according to

$$\mathbf{e}^* = \mathbf{e} + \mathbf{M}\eta$$

and the corresponding change in the covariance matrix is

$$\mathbf{P}^* = \mathbf{P} + \mathbf{MNM}^T$$

To summarize, the following collection of recursion formulas is all that is needed for a statistical simulation of a guided flight:

$$\mathbf{P}^* = \begin{cases} \mathbf{P} - a^{-1}\mathbf{Pbb}^T\mathbf{P} & \text{measurement} \\ \mathbf{P} + \mathbf{MNM}^T & \text{correction} \\ \mathbf{P} & \text{no action} \end{cases} \qquad (14.55)$$

and

$$\mathbf{X}^* = \begin{cases} \mathbf{S}(\mathbf{X} - \mathbf{P})\mathbf{S}^T + \mathbf{P} + \mathbf{MNM}^T & \text{correction} \\ \mathbf{X} & \text{no correction} \end{cases} \qquad (14.56)$$

with $\mathbf{X} = \mathbf{P}$ at the time of injection into orbit. For convenience of notation we have defined

$$\mathbf{S} = (\mathbf{I} + \mathbf{MB})$$

In the spacecraft computer the following recursion formula must be processed in addition to Eq. (14.55):

$$\delta \widehat{\mathbf{x}}^* = \begin{cases} \delta \widehat{\mathbf{x}} + a^{-1}\mathbf{Pb}(\delta \widetilde{q} - \delta \widehat{q}) & \text{measurement} \\ \mathbf{S}\,\delta \widehat{\mathbf{x}} & \text{correction} \\ \delta \widehat{\mathbf{x}} & \text{no action} \end{cases} \qquad (14.57)$$

with $\delta \widehat{\mathbf{x}}$ equal to zero at the time of injection. However, it should be emphasized that Eq. (14.57) is not needed for the statistical simulation.

The variance in the estimate of the velocity correction is determined as the trace of the matrix $\overline{\Delta \widehat{\mathbf{v}} \Delta \widehat{\mathbf{v}}^T}$ computed in Eq. (14.52) and is important to the navigation process. Clearly, a velocity correction having a large uncertainty should not be commanded if it is possible to improve the estimate substantially by future observations.

The uncertainty $\mathbf{d}$ in the estimate $\Delta \widehat{\mathbf{v}}$ is simply

$$\mathbf{d} = \Delta \widehat{\mathbf{v}} - \mathbf{B}\,\delta \mathbf{x} = \mathbf{B}\mathbf{e}$$

Hence, the mean-squared uncertainty is determined as the trace of the matrix

$$\overline{\mathbf{d}\mathbf{d}^T} = \mathbf{B}\mathbf{P}\mathbf{B}^T \tag{14.58}$$

Turning finally to the problem of guidance accuracy, the determination of the position-deviation vector at the nominal time of arrival at the target is made by extrapolating the deviation vector from the point of the final velocity correction. Thus, if $t_N$ is the time of the last correction and $\delta \mathbf{x}_A$ is the deviation vector at the time of arrival $t_A$, then

$$\delta \mathbf{x}_A = \mathbf{\Phi}_{A,N}\, \delta \mathbf{x}_N^+$$

But from the results of Prob. 9–22 and the terminal conditions for the perturbation matrices, we have

$$\mathbf{\Phi}_{A,N} = \begin{bmatrix} \mathbf{R}_A \mathbf{\Lambda}_N^{-1} & \mathbf{O} \\ \mathbf{V}_A \mathbf{\Lambda}_N^{-1} & \mathbf{\Lambda}_N^{\star\,-1} \end{bmatrix} \begin{bmatrix} \mathbf{C}_N^{\star} & -\mathbf{I} \\ \mathbf{C}_N & -\mathbf{I} \end{bmatrix}$$

Hence, the position-deviation vector at the target $\delta \mathbf{r}_A$ may be written as

$$\delta \mathbf{r}_A = \mathbf{R}_A \mathbf{\Lambda}_N^{-1} \mathbf{B}_N\, \delta \mathbf{x}_N^+$$

with a similar expression obtainable for the velocity deviation at time $t_A$.

The target position error may be written ultimately in terms of the error vector $\mathbf{e}_N$ according to the following self-evident steps:

$$\begin{aligned}
\delta \mathbf{r}_A &= \mathbf{R}_A \mathbf{\Lambda}_N^{-1} \mathbf{B}_N (\delta \mathbf{x}_N^- + \mathbf{M}\,\Delta \mathbf{v}_N) \\
&= \mathbf{R}_A \mathbf{\Lambda}_N^{-1} (\mathbf{B}_N\, \delta \mathbf{x}_N^- - \Delta \mathbf{v}_N) \\
&= -\mathbf{R}_A \mathbf{\Lambda}_N^{-1} (\mathbf{B}_N \mathbf{e}_N - \boldsymbol{\eta}_N) \\
&= -\mathbf{R}_A \mathbf{\Lambda}_N^{-1} \mathbf{B}_N \mathbf{e}_N^{\star}
\end{aligned} \tag{14.59}$$

The position error variance at the target is then computed as the trace of the matrix $\overline{\delta \mathbf{r}_A\, \delta \mathbf{r}_A^T}$.

## Error Propagation during Planetary Contact

In the analysis of the interplanetary round-trip mission, it was possible to consider the outbound and return portions of the voyage independently with the sole connecting link between the two being the random initial-velocity error for the return trip. This error provides an $\overline{\eta_0^2}$ for the return trip in the same manner as does the injection-velocity error for the outbound trip.

For the purpose of an error analysis† we require an expression for the deviations in the outbound relative-velocity vector as a result of variations in the inbound relative-velocity vector and the vector point of aim during planetary contact. It is clear from Fig. 9.1 that

$$\mathbf{v}_{\infty o} = \cos 2\nu \, \mathbf{v}_{\infty i} - v_\infty \frac{\sin 2\nu}{r_a} \mathbf{r}_a$$

Therefore, by calculating variations, we obtain

$$\delta \mathbf{v}_{\infty o} = \cos 2\nu \, \delta \mathbf{v}_{\infty i} - v_\infty \frac{\sin 2\nu}{r_a} \delta \mathbf{r}_a - \frac{\sin 2\nu}{r_a} \mathbf{r}_a \, \delta v_\infty$$
$$+ v_\infty \frac{\sin 2\nu}{r_a^2} \mathbf{r}_a \, \delta r_a - 2 \left( \sin 2\nu \, \mathbf{v}_{\infty i} + v_\infty \frac{\cos 2\nu}{r_a} \mathbf{r}_a \right) \delta \nu$$

But from Eq. (9.2), we have

$$\delta \nu = -\frac{\sin 2\nu}{2} \left( \frac{\delta r_a}{r_a} + 2 \frac{\delta v_\infty}{v_\infty} \right)$$

where

$$\delta r_a = \frac{\mathbf{r}_a \cdot \delta \mathbf{r}_a}{r_a} \quad \text{and} \quad \delta v_\infty = \frac{\mathbf{v}_{\infty i} \cdot \delta \mathbf{v}_{\infty i}}{v_\infty}$$

Combining these results, we write

$$\delta \mathbf{v}_{\infty o} = \mathbf{K} \, \delta \mathbf{v}_{\infty i} + \mathbf{L} \, \delta \mathbf{r}_a \tag{14.60}$$

where the matrices $\mathbf{K}$ and $\mathbf{L}$ are defined by

$$\mathbf{K} = \cos 2\nu \, \mathbf{I} + 2 \frac{\sin^2 2\nu}{v_\infty^2} \mathbf{v}_{\infty i} \mathbf{v}_{\infty i}^T + \frac{\sin 2\nu}{v_\infty r_a} (2 \cos 2\nu - 1) \mathbf{r}_a \mathbf{v}_{\infty i}^T$$

$$\mathbf{L} = -v_\infty \frac{\sin 2\nu}{r_a} \mathbf{I} + \frac{\sin^2 2\nu}{r_a^2} \mathbf{v}_{\infty i} \mathbf{r}_a^T + v_\infty \frac{\sin 2\nu}{r_a^3} (1 + \cos 2\nu) \mathbf{r}_a \mathbf{r}_a^T$$

An alternate and somewhat simpler expression was obtained in Prob. 9-1, in which only variations in the magnitude of the velocity change due to variations in the magnitude of the point of aim were considered.

---

† The development here and in the next subsection is based on material contained in the first section of Chapter 9.

### Variation in the Point of Impact

In a similar manner we analyze the variation in the point of impact at a target planet. From Fig. 9.2 we see that

$$\mathbf{r}_s = r_s \frac{\sin \beta}{r_a} \mathbf{r}_a - r_s \frac{\cos \beta}{v_\infty} \mathbf{v}_{\infty i}$$

Calculating variations as before gives

$$\delta \mathbf{r}_s = r_s \frac{\sin \beta}{r_a} \delta \mathbf{r}_a - r_s \frac{\cos \beta}{v_\infty} \delta \mathbf{v}_{\infty i} - r_s \frac{\sin \beta}{r_a^2} \mathbf{r}_a \, \delta r_a$$
$$+ r_s \frac{\cos \beta}{v_\infty^2} \mathbf{v}_{\infty i} \, \delta v_\infty + r_s \left( \frac{\sin \beta}{v_\infty} \mathbf{v}_{\infty i} + \frac{\cos \beta}{r_a} \mathbf{r}_a \right) \delta \beta$$

Now, from Eqs. (9.7) and (9.8),

$$\delta \beta = \frac{\tan \psi}{r_a^2} \left[ (2 r_a - r_s \sin \beta) \delta r_a + 2 \frac{\mu r_s}{v_\infty^3} (1 - \cos \beta) \delta v_\infty \right]$$

and combining these results produces

$$\delta \mathbf{r}_s = \mathbf{Q} \, \delta \mathbf{v}_{\infty i} + \mathbf{R} \, \delta \mathbf{r}_a \tag{14.61}$$

where the matrices $\mathbf{Q}$ and $\mathbf{R}$ are defined by

$$\mathbf{Q} = -r_s \frac{\cos \beta}{v_\infty} \mathbf{I} + r_s \frac{\cos \beta}{v_\infty^3} \mathbf{v}_{\infty i} \mathbf{v}_{\infty i}^T$$
$$+ 2 r_s \frac{\tan \psi}{r_a v_\infty^2} (r_a - r_s \sin \beta) \left( \frac{\cos \beta}{r_a} \mathbf{r}_a + \frac{\sin \beta}{v_\infty} \mathbf{v}_{\infty i} \right) \mathbf{v}_{\infty i}^T$$

$$\mathbf{R} = r_s \frac{\sin \beta}{r_a} \mathbf{I} - r_s \frac{\sin \beta}{r_a^3} \mathbf{r}_a \mathbf{r}_a^T$$
$$+ r_s \frac{\tan \psi}{r_a^3} (2 r_a - r_s \sin \beta) \left( \frac{\cos \beta}{r_a} \mathbf{r}_a + \frac{\sin \beta}{v_\infty} \mathbf{v}_{\infty i} \right) \mathbf{r}_a^T$$

In Prob. 9–2 an alternate expression was obtained in which were considered only variations in the impact point due to changes in the magnitude of the point of aim.

## 14.5 Optimum Selection of Measurements

In Sect. 13.3 we treated the problem of selecting the most appropriate stars for navigation measurements to minimize the error in the determination of a position fix. The purpose of this section is to address the associated problem of selecting those measurements which are, in some sense, most effective for space navigation. For example, the requirement might be to select the measurement to be made at a time $t_n$ which results in the maximum reduction in the variance in position or velocity uncertainty at time $t_n$. Of perhaps greater significance would be the requirement of selecting the measurement which minimizes the uncertainty in any linear combination of position and velocity deviations. Specifically, one might select the measurement which minimizes the uncertainty in the required velocity correction.

Consider first the simplest case, i.e., minimizing the variance in position uncertainty at a particular time. From Eq. (14.27) this uncertainty is expressible as

$$\overline{\epsilon^2} = \text{tr}\,\mathbf{P}_1 - \frac{\mathbf{h}^T\mathbf{P}_1\mathbf{P}_1\mathbf{h}}{\mathbf{h}^T\mathbf{P}_1\mathbf{h}+\overline{\alpha^2}} \tag{14.62}$$

In the absence of any measurement error ($\overline{\alpha^2} = 0$), the problem of minimizing the mean-squared error is equivalent to finding a direction for the $\mathbf{h}$ vector which maximizes the ratio of two quadratic forms. For this case, the geometrical interpretation is clear. Since the principal directions of $\mathbf{P}_1$ and $\mathbf{P}_1\mathbf{P}_1$ are the same, the optimum direction for $\mathbf{h}$ coincides with the major principal direction of $\mathbf{P}_1$.

The problem of minimizing the variance in velocity uncertainty by proper choice of the $\mathbf{h}$ vector is not so easily solved or interpreted. Again, from Eq. (14.27) the velocity uncertainty variance can be written as

$$\overline{\delta^2} = \text{tr}\,\mathbf{P}_4 - \frac{\mathbf{h}^T\mathbf{P}_2\mathbf{P}_3\mathbf{h}}{\mathbf{h}^T\mathbf{P}_1\mathbf{h}+\overline{\alpha^2}} \tag{14.63}$$

Denote by $p$ and $q$ the two quadratic forms

$$p = \mathbf{h}^T\mathbf{P}_2\mathbf{P}_3\mathbf{h} \quad \text{and} \quad q = \mathbf{h}^T\mathbf{P}_1\mathbf{h}$$

From the theory of quadratic forms an orthogonal transformation exists which will reduce $q$ to a diagonal form. Thus

$$\mathbf{h} = \mathbf{Q}\mathbf{d} \quad \text{gives} \quad q = \mathbf{d}^T\mathbf{Q}^T\mathbf{P}_1\mathbf{Q}\mathbf{d} = \mu_1 d_1^2 + \mu_2 d_2^2 + \mu_3 d_3^2$$

where $\mu_1$, $\mu_2$, $\mu_3$ are the characteristic roots of the matrix $\mathbf{P}_1$ and the columns of the $\mathbf{Q}$ matrix are the associated characteristic unit vectors. Since $\mathbf{P}_1$ is a positive definite matrix, the characteristic roots are positive and a further transformation

$$\mathbf{f} = \mathbf{D}\mathbf{d} \quad \text{gives} \quad q = \mathbf{f}^T\mathbf{f} = f_1^2 + f_2^2 + f_3^2$$

where $\mathbf{D}$ is a diagonal matrix whose diagonal elements are $\sqrt{\mu_1}$, $\sqrt{\mu_2}$, $\sqrt{\mu_3}$.

The same transformation from $\mathbf{h}$ to $\mathbf{f}$ applied to the quadratic form $p$ produces

$$p = \mathbf{f}^T \mathbf{D}^{-1} \mathbf{Q}^T \mathbf{P}_2 \mathbf{P}_3 \mathbf{Q} \mathbf{D}^{-1} \mathbf{f}$$

One final transformation applied to $\mathbf{f}$ will reduce $p$ to a diagonal form. Thus

$$\mathbf{f} = \mathbf{S}\mathbf{m} \quad \text{results in} \quad p = \lambda_1 m_1^2 + \lambda_2 m_2^2 + \lambda_3 m_3^2$$

where the columns of the $\mathbf{S}$ matrix are the characteristic unit vectors of the matrix $\mathbf{D}^{-1}\mathbf{Q}^T\mathbf{P}_2\mathbf{P}_3\mathbf{Q}\mathbf{D}^{-1}$ and $\lambda_1$, $\lambda_2$, $\lambda_3$, the corresponding characteristic roots. The same transformation from $\mathbf{f}$ to $\mathbf{m}$ applied to the diagonal form for $q$ gives

$$q = \mathbf{m}^T \mathbf{S}^T \mathbf{S}\mathbf{m} = m_1^2 + m_2^2 + m_3^2$$

since $\mathbf{S}$ is an orthogonal matrix.

In summary, then, the transformation

$$\mathbf{h} = \mathbf{Q}\mathbf{D}^{-1}\mathbf{S}\mathbf{m} \tag{14.64}$$

produces for the ratio of the two quadratic forms in Eq. (14.63)

$$\frac{p}{q} = \frac{\lambda_1 m_1^2 + \lambda_2 m_2^2 + \lambda_3 m_3^2}{m_1^2 + m_2^2 + m_3^2} \tag{14.65}$$

Furthermore, if the matrix $\mathbf{P}_2$ is nonsingular, the product $\mathbf{P}_2\mathbf{P}_3 = \mathbf{P}_2\mathbf{P}_2^T$ is positive definite and it would then follow that $\lambda_1$, $\lambda_2$, $\lambda_3$ are all real and positive.

The problem of maximizing the ratio $p/q$ is now readily solved. Since no measurement error is assumed, we cannot hope to determine more than the direction for the optimum $\mathbf{h}$ or, equivalently, the optimum $\mathbf{m}$. Therefore, it may be assumed that $\mathbf{m}$ is a unit vector. Let

$$\lambda_k = \max(\lambda_1, \lambda_2, \lambda_3)$$

Then the optimum value of $\mathbf{m}$ is

$$m_j = \begin{cases} 1 & j = k \\ 0 & j \neq k \end{cases}$$

The same technique can be used to select that direction for $\mathbf{h}$ which minimizes the uncertainty in any linear combination of position and velocity deviations. Specifically, consider the selection of that measurement which minimizes the uncertainty in the velocity correction which would be required immediately following the measurement.

The covariance matrix of the velocity-correction uncertainty is

$$\overline{\mathbf{d}\mathbf{d}^T} = \mathbf{B}\mathbf{P}\mathbf{B}^T$$

Sect. 14.5]    Optimum Selection of Measurements    689

and the variance in the uncertainty can be expressed as

$$\overline{d^2} = \text{tr}(\mathbf{BPB}^T) - \frac{\mathbf{h}^T \mathbf{Wh}}{\mathbf{h}^T \mathbf{P}_1 \mathbf{h} + \overline{\alpha^2}} \tag{14.66}$$

Here $\mathbf{W}$ is a symmetric matrix defined by

$$\mathbf{W} = \begin{bmatrix}\mathbf{P}_1 & \mathbf{P}_2\end{bmatrix} \mathbf{B}^T \mathbf{B} \begin{bmatrix}\mathbf{P}_1 \\ \mathbf{P}_2^T\end{bmatrix}$$

so that, if $\begin{bmatrix}\mathbf{P}_1 & \mathbf{P}_2\end{bmatrix}\mathbf{B}^T$ is nonsingular, the matrix $\mathbf{W}$ will be positive definite. Under any circumstances, if the identification

$$\begin{bmatrix}\mathbf{P}_1 & \mathbf{P}_2\end{bmatrix}\mathbf{B}^T \to \mathbf{P}_2$$

is made, then the exact same procedure may be used to select the optimum direction for the $\mathbf{h}$ vector as was used previously to minimize the mean-squared velocity uncertainty.

In all cases of practical interest the determination of the optimum direction for the $\mathbf{h}$ vector must be made subject to certain constraints. For example, we might wish to select the "best" star to be used in measuring the angle between the line of sight to the center of a planet disk and the line of sight to the star. For such a measurement the $\mathbf{h}$ vector is required to be perpendicular to the line of sight to the planet. If $\mathbf{z}$ is the position vector of the planet from the space vehicle, then we must have

$$\mathbf{h}^T \mathbf{z} = 0$$

Applying the transformation defined in Eq. (14.64) gives

$$\mathbf{m}^T \mathbf{S}^T \mathbf{D}^{-1} \mathbf{Q}^T \mathbf{z} = 0$$

Let $\mathbf{p}$ be a unit vector in the direction of $\mathbf{S}^T \mathbf{D}^{-1} \mathbf{Q}^T \mathbf{z}$. Then the problem of selecting the optimum direction for $\mathbf{h}$ or, equivalently, for $\mathbf{m}$ is to maximize $\lambda_1 m_1^2 + \lambda_2 m_2^2 + \lambda_3 m_3^2$ subject to the conditions of constraint

$$\mathbf{m}^T \mathbf{p} = 0 \quad \text{and} \quad \mathbf{m}^T \mathbf{m} = 1$$

In terms of the Lagrange multipliers $\rho$ and $\sigma$, this is equivalent to the problem of obtaining a free maximum for

$$\sum_{j=1}^{3} \lambda_j m_j^2 - 2\rho \sum_{j=1}^{3} p_j m_j - \sigma \left(\sum_{j=1}^{3} m_j^2 - 1\right)$$

Setting the partial derivatives with respect to each of the $m_j$'s equal to zero, we have

$$m_j = \frac{\rho p_j}{\lambda_j - \sigma} \quad \text{for} \quad j = 1, 2, 3 \tag{14.67}$$

where $\rho$ and $\sigma$ are to be determined from the conditions of constraint.

The condition that **m** be orthogonal to **p** leads to a quadratic equation for $\sigma$:

$$\sigma^2 - [p_1^2(\lambda_2 + \lambda_3) + p_2^2(\lambda_1 + \lambda_3) + p_3^2(\lambda_1 + \lambda_2)]\sigma$$
$$+ p_1^2 \lambda_2 \lambda_3 + p_2^2 \lambda_1 \lambda_3 + p_3^2 \lambda_1 \lambda_2 = 0 \quad (14.68)$$

If the $\lambda$'s are ordered $\lambda_1 < \lambda_2 < \lambda_3$, then the two roots $\sigma_1$ and $\sigma_2$ will be such that $\lambda_1 < \sigma_1 < \lambda_2 < \sigma_2 < \lambda_3$. The other Lagrange multiplier $\rho$ is determined so that **m** will be a unit vector. With the optimum vector **m** selected, the corresponding value for **h** is found from Eq. (14.64).

It is easy to show that $\sigma_2$ provides the desired maximum while $\sigma_1$ gives the minimum. We rewrite Eqs. (14.67) as

$$m_j(\lambda_j - \sigma) = \rho p_j \quad \text{for} \quad j = 1, 2, 3$$

Now multiply each equation by the corresponding $m_j$ and add to obtain

$$\sum_{j=1}^{3} m_j^2(\lambda_j - \sigma) = \rho \sum_{j=1}^{3} m_j p_j$$

Then, using the conditions of constraint, it follows that

$$\sigma = \sum_{j=1}^{3} \lambda_j m_j^2$$

Hence, $\sigma_1$ and $\sigma_2$ are the respective minimum and maximum of the original expression to be maximized.

## 14.6 Optimization of the Measurement Schedule

The uncertainties in position and velocity at the target point clearly depend on the entire measurement schedule. Selecting an optimum schedule is far more complicated than the optimization of a single measurement in which only currently available information is exploited. A measurement schedule, developed according to the principles presented in the last section, may well prove satisfactory for most purposes, but there is no justification to assume that it will be optimum. In this section we present a method which can be used to improve a measurement schedule iteratively in such a manner that terminal uncertainties will be reduced.

To have a tractable problem, we ignore the effect of uncertainties in any velocity corrections which may be applied. If this assumption is reasonable, the navigation and guidance problems are uncoupled and navigation accuracy is independent of guidance implementation errors. Furthermore, for simplicity, we assume that the times at which measurements are to be made are prescribed and that no correlation between measurement errors

Sect. 14.6]     Optimization of the Measurement Schedule

exists. Within the framework of these postulates, the problem involves only the information matrix† $\Sigma$ which is propagated according to the equation

$$\Sigma_n = \Phi_{n,n-1}^{-T} \Sigma_{n-1}^* \Phi_{n,n-1}^{-1} \tag{14.69}$$

and updated after a measurement by the formula

$$\Sigma_n^* = \Sigma_n + \frac{1}{\sigma_n^2} \mathbf{b}_n \mathbf{b}_n^T \tag{14.70}$$

The change in a measurement at time $t_n$ propagates via these equations to affect the information matrix at the target point $\Sigma_A$; however, the computation is tedious and impractical. Fortunately, it is possible to express a change in the elements of $\Sigma_A$ directly in terms of a change in $\mathbf{b}_n$ which avoids continuous use of the recurrence relations (14.69) and (14.70). With such a formulation, we can develop a scheme for systematically improving the selected elements of $\Sigma_A$ by a variation in each of the measurement vectors $\mathbf{b}_1, \mathbf{b}_2, \ldots, \mathbf{b}_N$.

A small change $\delta \mathbf{b}_n$ in the measurement at time $t_n$ will produce a small change $\delta \Sigma_n^*$ at time $t_n$ as well as at all later times. The propagation of these changes may then be studied by calculating the total differential of Eqs. (14.69) and (14.70). We have

$$\delta \Sigma_n = \Phi_{n,n-1}^{-T} \delta \Sigma_{n-1}^* \Phi_{n,n-1}^{-1}$$
$$\delta \Sigma_n^* = \delta \Sigma_n + \frac{1}{\sigma_n^2} (\delta \mathbf{b}_n \mathbf{b}_n^T + \mathbf{b}_n \delta \mathbf{b}_n^T) \tag{14.71}$$

Now define a matrix $\mathbf{L}$, called the *adjoint* of the matrix $\Sigma$, which is propagated by the relation

$$\mathbf{L}_n = \Phi_{n,n-1} \mathbf{L}_{n-1} \Phi_{n,n-1}^T \tag{14.72}$$

Then,

$$\mathbf{L}_n \delta \Sigma_n = \Phi_{n,n-1} \mathbf{L}_{n-1} \Phi_{n,n-1}^T \Phi_{n,n-1}^{-T} \delta \Sigma_{n-1}^* \Phi_{n,n-1}^{-1}$$
$$= \Phi_{n,n-1} \mathbf{L}_{n-1} \delta \Sigma_{n-1}^* \Phi_{n,n-1}^{-1}$$

so that

$$\text{tr}(\mathbf{L}_n \delta \Sigma_n) = \text{tr}(\Phi_{n,n-1} \mathbf{L}_{n-1} \delta \Sigma_{n-1}^* \Phi_{n,n-1}^{-1})$$
$$= \text{tr}(\delta \Sigma_{n-1}^* \Phi_{n,n-1}^{-1} \Phi_{n,n-1} \mathbf{L}_{n-1})$$
$$= \text{tr}(\delta \Sigma_{n-1}^* \mathbf{L}_{n-1})$$
$$= \text{tr}(\mathbf{L}_{n-1} \delta \Sigma_{n-1}^*)$$

where we have freely exploited the commutative multiplication property of the trace. Finally, premultiply the second of Eqs. (14.71) by $\mathbf{L}_n$, calculate

---

† The solution of this same problem was developed in the author's book *Astronautical Guidance* using the covariance matrix rather than the information matrix. The new approach results in a significant reduction in the required computations.

the trace, and use the relation just derived to obtain

$$\text{tr}(\mathbf{L}_n \delta\mathbf{\Sigma}_n^* - \mathbf{L}_{n-1}\delta\mathbf{\Sigma}_{n-1}^*) = \frac{1}{\sigma_n^2}\text{tr}(\mathbf{L}_n\delta\mathbf{b}_n\mathbf{b}_n^T + \mathbf{L}_n\mathbf{b}_n\delta\mathbf{b}_n^T)$$

$$= \frac{1}{\sigma_n^2}\text{tr}[(\mathbf{b}_n^T\mathbf{L}_n + \mathbf{b}_n^T\mathbf{L}_n^T)\delta\mathbf{b}_n]$$

$$= \lambda_n \cdot \delta\mathbf{b}_n \qquad (14.73)$$

where

$$\lambda_n = \frac{1}{\sigma_n^2}(\mathbf{L}_n + \mathbf{L}_n^T)\mathbf{b}_n \qquad (14.74)$$

Hence

$$\text{tr}(\mathbf{L}_A \delta\mathbf{\Sigma}_A) = \sum_{n=1}^{N} \lambda_n \cdot \delta\mathbf{b}_n \qquad (14.75)$$

since $\delta\mathbf{\Sigma}_L = \mathbf{O}$.

The covariance matrix $\mathbf{P}$ is the inverse of the information matrix $\mathbf{\Sigma}$. Therefore, from

$$\mathbf{P}_A\mathbf{\Sigma}_A = \mathbf{I} \quad \text{we have} \quad \mathbf{P}_A\delta\mathbf{\Sigma}_A + \delta\mathbf{P}_A\mathbf{\Sigma}_A = \mathbf{O}$$

so that

$$\delta\mathbf{P}_A = -\mathbf{P}_A\delta\mathbf{\Sigma}_A\mathbf{P}_A$$

Fortunately, the terminal value of the $\mathbf{L}$ matrix is at our disposal. For example, by choosing

$$\mathbf{L}_A = -\mathbf{P}_A\mathbf{M}\mathbf{P}_A \quad \text{where} \quad \mathbf{M} = \begin{bmatrix} \mathbf{I} & \mathbf{O} \\ \mathbf{O} & \mathbf{O} \end{bmatrix} \qquad (14.76)$$

we have

$$\text{tr}(-\mathbf{P}_A\mathbf{M}\mathbf{P}_A\delta\mathbf{\Sigma}_A) = \text{tr}(-\mathbf{M}\mathbf{P}_A\delta\mathbf{\Sigma}_A\mathbf{P}_A)$$

$$= \text{tr}(\mathbf{M}\delta\mathbf{P}_A) = \text{tr}(\delta\overline{\boldsymbol{\epsilon}\boldsymbol{\epsilon}^T}) = \delta\overline{\epsilon^2}$$

In this way, Eq. (14.75) relates changes in terminal position uncertainties to changes in the measurement schedule. A variety of other choices is possible, so that many combinations of terminal conditions can be selected for minimization.

The reference trajectory together with a given measurement schedule is sufficient to determine the vectors $\lambda_1, \lambda_2, \ldots, \lambda_N$. These vectors determine the sensitivity of the terminal uncertainties in position to small changes in the measurement geometry. For example, if the original schedule calls for a star-landmark measurement at time $t_n$, we may examine the effect of using an alternate nearby landmark and/or a different star. The only limitation which governs our freedom of choice is that the changes we make must be kept small enough that the linearization leading to the expression for $\delta\overline{\epsilon^2}$ remains valid. After small changes have been made in the measurements at each of the times $t_1, t_2, \ldots, t_N$, a new schedule is obtained. Then we may recalculate the vectors $\boldsymbol{\lambda}_n$ and repeat the entire process as often as desired.

For the particular choice of the terminal value of $\mathbf{L}$ given in Eq. (14.76) and the form of the propagation equation (14.72), we see that $\mathbf{L}$ is a symmetric matrix. Therefore, the sensitivity vector for this case is

$$\boldsymbol{\lambda} = \frac{2}{\sigma_n^2} \mathbf{L}_n \mathbf{b}_n$$

Further, since $\mathbf{L}$ is always specified at the terminal end, the propagation equation must be used in reverse, i.e.,

$$\mathbf{L}_{n-1} = \boldsymbol{\Phi}_{n,n-1}^{-1} \mathbf{L}_n \boldsymbol{\Phi}_{n,n-1}^{-T}$$

Again, we should remember that the inverse of the state transition matrix is easily obtained since it is symplectic.

## 14.7 Correlated Measurement Errors

When cross correlation of measurement errors is considered, it is convenient to use an augmented deviation vector having seven dimensions and defined as

$$\delta \mathbf{x} = \begin{bmatrix} \delta \mathbf{r} \\ \delta \mathbf{v} \\ \alpha \end{bmatrix} \quad (14.77)$$

Since, in this case, the error in a measurement can be predicted on the basis of previous observations, we may define

$$\widehat{\alpha} = \alpha + \beta \quad (14.78)$$

as the best estimate of the error to be expected in the measurement of $q$. The term $\beta$ is then the error in the estimate of the measurement error. The error vector $\mathbf{e}$ will, of course, also be seven dimensional and expressible as

$$\mathbf{e} = \begin{bmatrix} \epsilon \\ \delta \\ \beta \end{bmatrix} \quad (14.79)$$

Correspondingly, the covariance matrix becomes

$$\mathbf{P} = \overline{\mathbf{e}\mathbf{e}^T} = \begin{bmatrix} \overline{\epsilon \epsilon^T} & \overline{\epsilon \delta^T} & \overline{\epsilon \beta} \\ \overline{\delta \epsilon^T} & \overline{\delta \delta^T} & \overline{\delta \beta} \\ \overline{\beta \epsilon^T} & \overline{\beta \delta^T} & \overline{\beta^2} \end{bmatrix} \quad (14.80)$$

For purposes of illustration consider the following model for correlated measurement errors. Let the error at time $t_n$ be composed of two parts

$$\alpha_n = \alpha_{n-1} \exp[-\lambda(t_n - t_{n-1})] + \zeta_n \quad (14.81)$$

where $\alpha_n$ and $\varsigma_n$ are independent random numbers, $\lambda$ is a positive constant, and $\bar{\varsigma}_n$ is zero. To relate this mathematical model to a physical model, we can suppose some effect such as temperature to cause a random drift in the calibration of the angle-measuring device. The drift produces an error which is superimposed on the random error of the actual observation due to imperfect horizon determination or other sources of imprecise measurement. The drift component at time $t_n$ is the first term on the right side of Eq. (14.81) and the observation component is $\varsigma_n$, so that the total error $\alpha_n$ is the sum.

For the purpose of assigning a significance to the model parameter $\lambda$, consider the special case for which $\overline{\varsigma_n^2} = \overline{\varsigma^2}$ and $t_n - t_{n-1} = \Delta t$, both constant. Then by squaring and averaging Eq. (14.81) we have

$$\overline{\alpha_n^2} = \overline{\alpha_{n-1}^2} e^{-2\lambda \Delta t} + \overline{\varsigma^2}$$

as a difference equation whose solution is readily found to be

$$\overline{\alpha_n^2} = \frac{\overline{\varsigma^2} e^{-2\lambda \Delta t}}{1 - e^{-2\lambda \Delta t}} + \overline{\varsigma^2}$$

Thus, in this example, the drift component is seen to have a constant mean-squared value. If the random measurement error variance and the steady-state drift of the measuring instrument can be determined by experiment, then the value of the parameter $\lambda$ can be calculated for the model.

The transition matrix appropriate for propagation of the seven-dimensional deviation vector is an augmented form of the ordinary transition matrix $\mathbf{\Phi}_{n+1,n}$. Since $\alpha_n$ is one of the state variables, it follows from the model definition (14.81) that

$$\delta \mathbf{x}_n = \mathbf{\Psi}_{n,n-1} \delta \mathbf{x}_{n-1} + \boldsymbol{\zeta}_n \qquad (14.82)$$

where $\boldsymbol{\zeta}_n$ is a seven-dimensional vector all of whose components are zero except the last, which is the random measurement error $\varsigma_n$. The augmented transition matrix is

$$\mathbf{\Psi}_{n,n-1} = \begin{bmatrix} \mathbf{\Phi}_{n,n-1} & 0 \\ \mathbf{0}^T & \exp[-\lambda(t_n - t_{n-1})] \end{bmatrix} \qquad (14.83)$$

Furthermore, since the mean value of $\varsigma_n$ is zero, the extrapolation of the best estimate of the deviation vector is determined from

$$\delta \widehat{\mathbf{x}}_n = \mathbf{\Psi}_{n,n-1} \delta \widehat{\mathbf{x}}_{n-1} \qquad (14.84)$$

By comparing this with Eq. (14.82), it follows that the propagation of the error vector is made according to

$$\mathbf{e}_n = \mathbf{\Psi}_{n,n-1} \mathbf{e}_{n-1} - \boldsymbol{\zeta}_n$$

Hence, the augmented extrapolated covariance matrix is computed from

$$\mathbf{P}_n = \mathbf{\Psi}_{n,n-1} \mathbf{P}_{n-1} \mathbf{\Psi}_{n,n-1}^T + \overline{\boldsymbol{\zeta}_n \boldsymbol{\zeta}_n^T} \qquad (14.85)$$

## Correlated Measurement Errors

When the measurement errors are correlated, the deviation of the optimum linear estimate of the deviation vector is only slightly altered from the one developed for the uncorrelated case. The linear estimate of the seven-dimensional deviation vector $\delta\hat{\mathbf{x}}^*$ at a time $t$ is again expressed as a linear combination of the extrapolated estimate $\delta\hat{\mathbf{x}}$ and the difference between the observed and estimated deviations in the measured quantity $q$. However, the estimated deviation in $q$ must also include the estimate of the error in the observation. Thus

$$\delta\hat{\mathbf{x}}^* = \delta\hat{\mathbf{x}} + \mathbf{w}[\delta\widetilde{q} - (\delta\hat{q} + \hat{\alpha})]$$

where now the weighting vector $\mathbf{w}$ has seven dimensions.

The estimation formula may be put in a somewhat more convenient form by properly augmenting the vector $\mathbf{b}$. A seven-dimensional $\mathbf{b}$ vector is defined whose first six components are those of the ordinary $\mathbf{b}$ vector and the seventh component is unity. Then we have

$$\delta\hat{\mathbf{x}}^* = \delta\hat{\mathbf{x}} + \mathbf{w}(\delta\widetilde{q} - \mathbf{b}^T \delta\hat{\mathbf{x}}) \tag{14.86}$$

Furthermore, the measured quantity $\delta\widetilde{q}$ is

$$\delta\widetilde{q} = \delta q + \alpha = \mathbf{b}^T \begin{bmatrix} \delta\mathbf{r} \\ \delta\mathbf{v} \\ \alpha \end{bmatrix}$$

From the definition of the error model, Eq. (14.81), it is seen that the true value of the seventh component of the state vector undergoes a step change as a result of a measurement. Thus, in the case of cross correlation, one must distinguish between the true value of the state just prior to and just after a measurement. Hence, the post-measurement error in the estimate is

$$\mathbf{e}^* = \delta\hat{\mathbf{x}}^* - \begin{bmatrix} \delta\mathbf{r} \\ \delta\mathbf{v} \\ \alpha + \zeta \end{bmatrix} = (\mathbf{I} - \mathbf{w}\mathbf{b}^T)\mathbf{e} - \boldsymbol{\zeta}$$

and the covariance matrix, expressed as a function of the weighting vector $\mathbf{w}$, is†

$$\mathbf{P}^*(\mathbf{w}) = (\mathbf{I} - \mathbf{w}\mathbf{b}^T)\mathbf{P}(\mathbf{I} - \mathbf{b}\mathbf{w}^T) + \overline{\boldsymbol{\zeta}\boldsymbol{\zeta}^T} \tag{14.87}$$

Again, if we require $\overline{\delta e^2(\mathbf{w})}$ to vanish for all variations $\delta\mathbf{w}$, it is readily shown that

$$a\mathbf{w} = \mathbf{P}\mathbf{b} \quad \text{where} \quad a = \mathbf{b}^T\mathbf{P}\mathbf{b} \tag{14.88}$$

---

† The second term in the recursion formula for the covariance matrix was omitted in the author's book *Astronautical Guidance*. This subject of cross correlation of measurement errors was first discussed by this author in his paper "A Statistical Optimizing Navigation Procedure for Space Flight" published in the *Journal of the American Rocket Society* Vol. 32, November 1962. The treatment in the paper was correct although the notation was quite different. It seems strange that no one has reported the error in the book during the last 25 years.

The derivation of the minimum-variance linear estimate is now complete. If we have an estimate of the state vector (14.77) and the corresponding covariance matrix $\mathbf{P}$ at some time $t_{n-1}$, then Eqs. (14.84) and (14.85) can be used to propagate these quantities to the time $t_n$ at which a measurement is to be made. After the measurement the state vector estimate and the covariance matrix are updated using Eqs. (14.86) and (14.87).

## 14.8 Effect of Parameter Errors

The results obtained from a statistical analysis of navigation accuracy are, of course, heavily dependent on certain quantities whose mean-squared values may not always be accurately known. For example, suppose that the value assigned to the measurement error variance is $\overline{\alpha^{\diamond 2}}$ when, in fact, the real value is $\overline{\alpha^2}$. If the measurements are correlated, a value $\lambda^\diamond$ assumed for the cross-correlation parameter might be grossly different from the true value $\lambda$. In this section we develop a method of analyzing the effects of erroneously assigned error parameter values on the accuracy of the estimate.

### Effect of Incorrect Measurement Variance

Suppose that the deviation $\delta\widetilde{q}$ measured onboard a spacecraft at time $t$ is composed of three parts

$$\delta\widetilde{q} = \delta q + \alpha + \gamma \tag{14.89}$$

where $\alpha$ is a random variable with zero mean, statistically independent of previous measurement errors, and $\gamma$ is a constant but random bias with nonzero mean and statistically independent of $\alpha$. Further suppose that the estimation process performed in the spacecraft computer is made assuming the measurement error variance to be $\sigma^2 = \overline{\alpha^2}$. Then the onboard estimate would be made in accordance with the following equations:

$$\delta\widehat{\mathbf{x}}^* = \delta\widehat{\mathbf{x}} + \mathbf{w}(\delta\widetilde{q} - \mathbf{b}^T \delta\widehat{\mathbf{x}})$$

where

$$\mathbf{w} = a^{-1}\mathbf{Pb} \qquad a = \mathbf{b}^T\mathbf{Pb} + \sigma^2 \qquad \mathbf{P}^* = \mathbf{P} - a^{-1}\mathbf{Pbb}^T\mathbf{P}$$

The matrix $\mathbf{P}$ is not the covariance matrix of the real estimation errors, since an erroneous value of the variance of measurement errors has been used in the calculation. The real estimation errors are, in fact,

$$\mathbf{e}^\diamond = \delta\widehat{\mathbf{x}} - \delta\mathbf{x}$$

and our task is to find a recursion relation for the covariance matrix

$$\mathbf{P}^\diamond = \overline{\mathbf{e}^\diamond \mathbf{e}^{\diamond T}}$$

We expand the real estimation error vector in terms of the more basic quantities as

$$\mathbf{e}^{\diamond *} = \delta\hat{\mathbf{x}} + \mathbf{w}(\mathbf{b}^T\delta\mathbf{x} + \alpha + \gamma - \mathbf{b}^T\delta\hat{\mathbf{x}}) - \delta\mathbf{x}$$
$$= \mathbf{e}^{\diamond} - \mathbf{w}\mathbf{b}^T\mathbf{e}^{\diamond} + \mathbf{w}(\alpha + \gamma)$$

so that

$$\mathbf{P}^{\diamond *} = \mathbf{D}\mathbf{P}^{\diamond}\mathbf{D}^T + \mathbf{D}\overline{\gamma\mathbf{e}^{\diamond}}\mathbf{w}^T + \mathbf{w}\overline{\gamma\mathbf{e}^{\diamond T}}\mathbf{D}^T + \mathbf{w}\mathbf{w}^T(\sigma^2 + \overline{\gamma^2})$$

where the notation

$$\mathbf{D} = (\mathbf{I} - \mathbf{w}\mathbf{b}^T) \tag{14.90}$$

is used for convenience. Now define the vector

$$\boldsymbol{\phi} = \overline{\gamma\mathbf{e}^{\diamond}} \tag{14.91}$$

and obtain a recurrence formula for it by multiplying the recurrence formula for $\mathbf{e}^{\diamond}$ by $\gamma$ and averaging. We have

$$\boldsymbol{\phi}^* = \mathbf{D}\boldsymbol{\phi} + \mathbf{w}\overline{\gamma^2} \tag{14.92}$$

The statistical analysis proceeds as before by a recursive evaluation of the spacecraft estimation equations together with Eqs. (14.92) and

$$\mathbf{P}^{\diamond *} = \mathbf{D}\mathbf{P}^{\diamond}\mathbf{D}^T + \boldsymbol{\phi}^*\mathbf{w}^T + \mathbf{w}\boldsymbol{\phi}^{*T} + \mathbf{w}\mathbf{w}^T(\sigma^2 - \overline{\gamma^2}) \tag{14.93}$$

The initial values for $\mathbf{P}^{\diamond}$ and $\boldsymbol{\phi}$ are

$$\mathbf{P}_0^{\diamond} = \overline{\delta\mathbf{x}_0\delta\mathbf{x}_0^T} \quad \text{and} \quad \boldsymbol{\phi}_0 = \mathbf{0}$$

The true estimation error variances at any time $t$ are then obtained as the elements on the main diagonal of the matrix $\mathbf{P}^{\diamond}$.

### Effect of Incorrect Cross-Correlation Error Model Parameters

A similar method can be applied to analyze the effect of assigning incorrect values to $\lambda$ and $\overline{\varsigma_n^2}$ when cross correlation between measurement errors exists. As a special case, with $\lambda = \infty$, the technique to be developed can be applied directly to determine the resulting estimation errors if one assumes that measurement errors are uncorrelated when, in fact, some degree of correlation actually does exist.

The method of analysis parallels exactly that of the previous subsection with one very important exception. The assumed transition matrix $\boldsymbol{\Psi}_{n,n-1}$ differs from the actual transition matrix $\boldsymbol{\Psi}_{n,n-1}^{\diamond}$ in the right corner element because of the erroneously assumed propagation of measurement errors. Noting this exception, we develop the true error in the estimate as

follows:
$$\begin{aligned}
\mathbf{e}_n^{\diamond*} &= \delta\hat{\mathbf{x}}_n^* - \delta\mathbf{x} \\
&= \mathbf{\Psi}_{n,n-1}\,\delta\hat{\mathbf{x}}_{n-1}^* + \mathbf{w}_n(\mathbf{b}_n^T\,\delta\mathbf{x}_n - \mathbf{b}_n^T\mathbf{\Psi}_{n,n-1}\,\delta\hat{\mathbf{x}}_{n-1}^*) - \delta\mathbf{x}_n \\
&= \mathbf{D}_n[\mathbf{\Psi}_{n,n-1}\,\delta\hat{\mathbf{x}}_{n-1}^* - (\mathbf{\Psi}_{n,n-1}^*\,\delta\mathbf{x}_{n-1} + \boldsymbol{\zeta}_n)] \\
&= \mathbf{D}_n\mathbf{\Psi}_{n,n-1}\mathbf{e}_{n-1}^{\diamond*} + \mathbf{D}_n[(\mathbf{\Psi}_{n,n-1} - \mathbf{\Psi}_{n,n-1}^*)\,\delta\mathbf{x}_{n-1} - \boldsymbol{\zeta}_n] \\
&= \mathbf{A}_{n,n-1}\mathbf{e}_{n-1}^{\diamond*} + \mathbf{B}_{n,n-1}\,\delta\mathbf{x}_{n-1} - \mathbf{D}_n\boldsymbol{\zeta}_n
\end{aligned}$$
where, for compactness of notation, we have defined
$$\mathbf{D}_n = (\mathbf{I} - \mathbf{w}_n\mathbf{b}_n^T)$$
$$\mathbf{A}_{n,n-1} = \mathbf{D}_n\mathbf{\Psi}_{n,n-1}$$
$$\mathbf{B}_{n,n-1} = \mathbf{D}_n(\mathbf{\Psi}_{n,n-1} - \mathbf{\Psi}_{n,n-1}^*)$$

It is convenient to combine $\mathbf{e}_n^{\diamond*}$ with $\delta\mathbf{x}_n$ and define a 14-dimensional vector $\mathbf{y}_n$ as
$$\mathbf{y}_n = \begin{bmatrix} \mathbf{e}_n^{\diamond*} \\ \delta\mathbf{x}_n \end{bmatrix} = \mathbf{S}_{n,n-1}\mathbf{y}_{n-1} + \mathbf{S}_n\boldsymbol{\zeta}_n \quad (14.94)$$
where
$$\mathbf{S}_{n,n-1} = \begin{bmatrix} \mathbf{A}_{n,n-1} & \mathbf{B}_{n,n-1} \\ \mathbf{O} & \mathbf{\Psi}_{n,n-1}^* \end{bmatrix} \quad \text{and} \quad \mathbf{S}_n = \begin{bmatrix} -\mathbf{D}_n \\ \mathbf{I} \end{bmatrix}$$
Therefore, the 14-dimensional covariance matrix
$$\mathbf{Y}_n = \overline{\mathbf{y}_n\mathbf{y}_n^T}$$
of the vector $\mathbf{y}_n$ satisfies the recurrence relation
$$\mathbf{Y}_n = \mathbf{S}_{n,n-1}\mathbf{Y}_{n-1}\mathbf{S}_{n,n-1}^T + \mathbf{S}_n\overline{\boldsymbol{\zeta}_n\boldsymbol{\zeta}_n^T}\mathbf{S}_n^T \quad (14.95)$$

The seven-dimensional partition of $\mathbf{Y}_n$ in the upper left-hand corner is the desired covariance matrix of the true estimate errors. Again the statistical analysis proceeds as before by adjoining the recurrence formula for $\mathbf{Y}_n$ to the standard equations required for the simulation.

Appendix H

# Probability Theory and Applications

We assume that the reader has had some exposure to the basic principles of probability. In the first section of Chapter 14, a brief review of that topic was given to establish the concepts and notation relevant to our treatment of the space navigation problem. In this appendix we provide a more detailed account of probability theory† for those readers whose background in this fundamental subject may be somewhat skimpy.

## H.1 Sampling and Probabilities

Consider an experiment whose outcome depends upon chance—tossing a coin, rolling a pair of dice, drawing a card from a bridge deck, or sampling a population. Basic to any analysis is the set of elements consisting of all possible distinct outcomes of the experiment. We call this set the *sample space* for the experiment, using the term *space* as a synonym for the word *set* in this connection. The individual elements or points of the sample space are often called *sample points*.

Let $S$ represent the sample space (for example, the 52 cards in a deck) and assume that the experiment in question (drawing a single card) is performed a large number of times $N$. Then for any event $A$ (such as obtaining the ace of spades) let $n_A$ be the number of occurrences of $A$ in the $N$ trials and define

$$p_A = \frac{n_A}{N}$$

Clearly, $0 \leq n_A \leq N$, so that $0 \leq p_A \leq 1$. Furthermore, let us assume that $p_A$ tends to a limit as $N$ becomes infinite. This limit, which we shall denote by $\text{Prob}(A)$ (intuitively, we anticipate the limit to be $\frac{1}{52}$), is a nonnegative real number and is defined for all sets of the sample space $S$. It is called a *probability function* of $S$.

---

† The author borrowed heavily from the the second chapter of the book *Random Processes in Automatic Control* by J. H. Laning, Jr. and R. H. Battin published in 1956 by the McGraw-Hill Book Co. He also found the first volume of the second edition of *An Introduction to Probability Theory and Its Applications* by William Feller published by John Wiley & Sons, Inc. in 1957 to be quite helpful.

For a more elaborate illustration, let $N$ be the total number of people in a population, i.e., sample space, $S$ under consideration and let

$N_S$ : Number of people who smoke
$N_M$ : Number of males
$N_{SM}$ : Number of males who smoke

The experiment consists of selecting a person at random and the events of interest to us are defined as

$S$ : The selected person is a smoker
$M$ : The selected person is a male
$SM$ : The selected person is a male smoker

Then the probabilities of occurrence of these events are

$$\text{Prob}(S) = \frac{N_S}{N} \qquad \text{Prob}(M) = \frac{N_M}{N}$$

$$\text{Prob}(S \text{ and } M) \equiv \text{Prob}(S \cap M) \equiv \text{Prob}(SM) = \frac{N_{SM}}{N}$$

The probability that the selected person is either a smoker or a male or both is

$$\text{Prob}(S \text{ or } M \text{ or both}) \equiv \text{Prob}(S \cup M) \equiv \text{Prob}(S + M)$$
$$= \frac{N_S + N_M - N_{SM}}{N}$$
$$= \text{Prob}(S) + \text{Prob}(M) - \text{Prob}(SM)$$

If $SM = O$, i.e., the null set or empty set, there are no male smokers and we have

$$\text{Prob}(S + M) = \text{Prob}(S) + \text{Prob}(M)$$

In this case, the events $S$ and $M$ are said to be *mutually exclusive*.

In general, for any two mutually exclusive events $A$ and $B$

$$\text{Prob}(A \text{ or } B) \equiv \text{Prob}(A + B) = \text{Prob}(A) + \text{Prob}(B) \qquad \text{(H.1)}$$

This is the *additive property* of probabilities and applies only to mutually exclusive events.

Let the notation $S \mid M$ represent the event that the person selected is a smoker when it is already known that the selected person is a male. Then the *conditional probability* of the occurrence of $S \mid M$ is obtained from

$$\text{Prob}(S \mid M) = \frac{N_{SM}}{N_M} = \frac{N_{SM}/N}{N_M/N} = \frac{\text{Prob}(SM)}{\text{Prob}(M)}$$

Thus we have established that, for any two events $A$ and $B$,

$$\text{Prob}(AB) = \text{Prob}(A \mid B)\,\text{Prob}(B) = \text{Prob}(B \mid A)\,\text{Prob}(A) \qquad \text{(H.2)}$$

More generally, suppose that $E_1, E_2, \ldots, E_n$ are mutually exclusive events which exhaust the sample space; that is $S = \sum E_i$. Then the occurrence of any event $E$ is equivalent to the occurrence of

$$E = E \text{ and } E_1 \text{ or } E \text{ and } E_2 \text{ or } \cdots \text{ or } E \text{ and } E_n$$
$$= EE_1 + EE_2 + \cdots + EE_n$$

Since the sets $EE_j$ are disjoint, we obtain *Bayes' rule*†

$$\text{Prob}(E) = \text{Prob}(E \mid E_1)\text{Prob}(E_1) + \text{Prob}(E \mid E_2)\text{Prob}(E_2) + \cdots$$
$$+ \text{Prob}(E \mid E_n)\text{Prob}(E_n) \quad \text{(H.3)}$$

If $\text{Prob}(A \mid B) = \text{Prob}(A)$ or $\text{Prob}(B \mid A) = \text{Prob}(B)$, then the events $A$ and $B$ are said to be *independent*. In this case, we have

$$\text{Prob}(A \text{ and } B) \equiv \text{Prob}(AB) = \text{Prob}(A)\text{Prob}(B) \quad \text{(H.4)}$$

called the *multiplicative property* of probabilities.

In summary,

- Probabilities are *additive* for *mutually exclusive* events.
- Probabilities are *multiplicative* for *independent* events.

◇ **Problem H–1**

Consider a sample space $S$ composed of four mutually exclusive events $E_1$, $E_2$, $E_3$, and $E_4$, each occurring with probability $\frac{1}{4}$. Define three compound events as follows:

$$A = E_1 + E_2 \qquad B = E_1 + E_3 \qquad C = E_1 + E_4$$

Then

$$\text{Prob}(AB) = \text{Prob}(A)\text{Prob}(B)$$
$$\text{Prob}(AC) = \text{Prob}(A)\text{Prob}(C)$$
$$\text{Prob}(BC) = \text{Prob}(B)\text{Prob}(C)$$

but

$$\text{Prob}(ABC) \neq \text{Prob}(A)\text{Prob}(B)\text{Prob}(C)$$

ANSWER:
$$\text{Prob}(A) = \text{Prob}(B) = \text{Prob}(C) = \tfrac{1}{2}$$

and
$$\text{Prob}(AB) = \text{Prob}(AC) = \text{Prob}(BC) = \text{Prob}(E_1) = \tfrac{1}{4}$$

NOTE: To avoid any absurdities as a consequence of the definition of independence for the three events $A$, $B$, $C$, it is necessary to require that all four of the above equations be satisfied.

---

† Thomas Bayes (1702–1761) is remembered for his brief paper "Essay Towards Solving a Problem in the Doctrine of Chances." It was published in the *Philosophical Transactions of the Royal Society of London* in 1763 two years after his death. His only other mathematical publication was a defense of Newton's method of fluxions—now called the calculus.

## ◇ Problem H-2

Among a set of $n$ people, determine the probability that at least two people will have the same birthday—assuming, of course, that all birthdays are equally likely. How small can $n$ be if the probability is $\frac{1}{2}$ that at least two will have the same birthday?

ANSWER: The number of ways in which $n$ people can have distinct birthdays is

$$[365][365-1][365-2]\ldots[365-(n-1)]$$

while $(365)^n$ is the total number of possible birthdays. Hence, the probability that no two people have the same birthday is

$$\left(1-\frac{1}{365}\right)\left(1-\frac{2}{365}\right)\left(1-\frac{3}{365}\right)\ldots\left(1-\frac{n-1}{365}\right)$$

If this number is one half and if $n$ is to be an integer, then it is easier to solve the equation by first writing

$$\sum_{k=1}^{n-1}\log\left(1-\frac{k}{365}\right)\approx \log\tfrac{1}{2}=-0.69\ldots$$

Next, observe that $\log(1-x)\approx -x$ for small $x$ so that

$$\frac{1+2+3+\cdots+(n-1)}{365}=\frac{\tfrac{1}{2}(n-1)n}{365}\approx 0.69$$

Therefore, *if $n=23$, the probability exceeds $\frac{1}{2}$ that at least two people will have a common birthday.*

## H.2 Coin-tossing Experiment

Consider the experiment which consists of tossing a coin two times with probability $p$ of heads and $q$ of tails such that $p+q=1$. There are $2^2=4$ points in the sample space and they are, of course, *not* equally likely to occur. Then define the following sets of events:

$$A_1, A_2, A_3, A_4 = \text{basic events}$$
$$B_1, B_2 = \text{heads, tails on the first toss}$$
$$C_1, C_2 = \text{heads, tails on the second toss}$$
$$D_k = \text{exactly } k \text{ heads in 2 tosses}$$

from which we can construct the table

| Sample Point | Event A | Event B | Event C | Event D | Prob. |
|---|---|---|---|---|---|
| HH | $A_1$ | $B_1$ | $C_1$ | $D_2$ | $pp$ |
| HT | $A_2$ | $B_1$ | $C_2$ | $D_1$ | $pq$ |
| TH | $A_3$ | $B_2$ | $C_1$ | $D_1$ | $qp$ |
| TT | $A_4$ | $B_2$ | $C_2$ | $D_0$ | $qq$ |

## Probability Theory and Applications

For this experiment we see that the following are examples of mutually exclusive events:

$$A_1 \text{ and } A_2 \text{ and } A_3 \text{ and } A_4 = A_1 A_2 A_3 A_4 = O \text{ (Null set)}$$
$$A_1 \text{ and } B_2 = A_1 B_2 = O$$
$$B_2 \text{ and } D_2 = B_2 D_2 = O$$

We also have
$$B_1 = A_1 \text{ or } A_2 = A_1 + A_2$$
so that
$$\text{Prob}(B_1) = \text{Prob}(A_1) + \text{Prob}(A_2) = pp + pq = p(p+q) = p$$

Similarly,
$$D_1 = A_2 \text{ or } A_3 = A_2 + A_3$$
so that
$$\text{Prob}(D_1) = \text{Prob}(A_2) + \text{Prob}(A_3) = pq + qp = 2pq$$

The following are examples of events which *are* equally likely to occur, called *events of equal likelihood*:

$$\text{Prob}(A_2) = \text{Prob}(A_3) = pq$$
$$\text{Prob}(B_1) = \text{Prob}(C_1) = p$$
$$\text{Prob}(B_2) = \text{Prob}(C_2) = q$$

We shall return to this example often as occasions arise.

### ◊ Problem H-3

Three urns are filled with a mixture of black and white balls. The first urn contains six black balls and three white ones. The second contains six black and nine white, while the third contains three black and three white. An urn is picked at random, and a ball is selected at random from this urn.
(a) What is the probability that the chosen ball is black?
(b) If it is known that the chosen ball is black, what is the probability that it came from the first urn?

ANSWER: Let $E_k$ be the event: "the $k^{\text{th}}$ urn is selected" and let $B$ be the event: "the chosen ball is black." Then

(a) $$\text{Prob}(B) = \sum_{k=1}^{3} \text{Prob}(B \mid E_k) \text{Prob}(E_k) = \frac{47}{90}$$

(b) $$\text{Prob}(E_1 \mid B) = \frac{\text{Prob}(E_1 B)}{\text{Prob}(B)} = \frac{\frac{1}{3} \cdot \frac{6}{9}}{\frac{47}{90}} = \frac{20}{47}$$

## ◇ Problem H-4

A machine can fail if any of three independent parts fail. If the probabilities of failure during a year's operation of parts $A$, $B$, and $C$ are $\frac{1}{3}$, $\frac{1}{4}$, and $\frac{1}{5}$, respectively, what is the probability of the machine failing during the year?

ANSWER: Let $E_A$ be the event: "part $A$ fails during the year," etc. Then the complementary event $E_A^*$ is the event: "part $A$ does not fail." Let $E_F$ and $E_F^*$ be the events that the machine does and does not fail during the year. Then

$$\text{Prob}(E_F^*) = \text{Prob}(E_A^* E_B^* E_C^*) = \text{Prob}(E_A^*)\,\text{Prob}(E_B^*)\,\text{Prob}(E_C^*)$$
$$= (1 - \tfrac{1}{3})(1 - \tfrac{1}{4})(1 - \tfrac{1}{5}) = \tfrac{2}{5}$$

Hence,

$$\text{Prob}(E_F) = 1 - \text{Prob}(E_F^*) = 1 - \tfrac{2}{5} = \tfrac{3}{5}$$

NOTE: The probability that the machine will fail is substantially *greater* than the probabilities that any of its independent parts will fail.

## ◇ Problem H-5

Consider the four dice $A$, $B$, $C$, $D$ with faces numbered as shown in the table.† Prove that if pair $A$ and $B$ are rolled together, then $A$ will beat $B$ with probability $\frac{2}{3}$. Similarly, show that if the pair $B$ and $C$ are rolled together or the pair $C$ and $D$ are rolled together, then in each case, $B$ will beat $C$ and $C$ will beat $D$ each with probability $\frac{2}{3}$. BUT *if $A$ and $D$ are rolled together, then D will beat A with probability* $\frac{2}{3}$. In short, Die $A$ beats $B$, $B$ beats $C$, $C$ beats $D$—and $D$ beats $A$!!

| Die $A$ | Die $B$ | Die $C$ | Die $D$ |
|---|---|---|---|
| 2 | 0 | 5 | 4 |
| 3 | 1 | 5 | 4 |
| 3 | 7 | 6 | 4 |
| 9 | 8 | 6 | 4 |
| 10 | 8 | 6 | 12 |
| 11 | 8 | 6 | 12 |

ANSWER:

$$\text{Prob}(A \text{ beats } B) = \text{Prob}(A \text{ beats } B \mid A = 9, 10, 11)\,\text{Prob}(A = 9, 10, 11)$$
$$+ \text{Prob}(A \text{ beats } B \mid A = 2, 3)\,\text{Prob}(A = 2, 3) = 1 \times \tfrac{1}{2} + \tfrac{1}{3} \times \tfrac{1}{2} = \tfrac{2}{3}$$

$$\text{Prob}(B \text{ beats } C) = \text{Prob}(B \text{ beats } C \mid B = 7, 8)\,\text{Prob}(B = 7, 8)$$
$$+ \text{Prob}(B \text{ beats } C \mid B = 0, 1)\,\text{Prob}(B = 0, 1) = 1 \times \tfrac{2}{3} + 0 \times \tfrac{1}{3} = \tfrac{2}{3}$$

---

† These dice were invented by Bradley Efron, a statistician at Stanford University, to dramatize some discoveries concerning a general class of probability paradoxes that violate transitivity. They were the subject of an article by Martin Gardner in his "Mathematical Games" section of the *Scientific American* for December 1970.

Prob($C$ beats $D$) = Prob($C$ beats $D \mid C = 6$) Prob($C = 6$)
+ Prob($C$ beats $D \mid C = 5$) Prob($C = 5$) = $\frac{2}{3} \times \frac{2}{3} + \frac{2}{3} \times \frac{1}{3} = \frac{2}{3}$

BUT

Prob($D$ beats $A$) = Prob($D$ beats $A \mid D = 12$) Prob($D = 12$)
+ Prob($D$ beats $A \mid D = 4$) Prob($D = 4$) = $1 \times \frac{1}{3} + \frac{1}{2} \times \frac{2}{3} = \frac{2}{3}$

◊ **Problem H-6**
A game is played in which two men each toss a coin in turn, the one obtaining "heads" first being the winner. Find the probability that the man who plays first will win.

ANSWER: Let $E_k$ be the event: "the first player wins on his $k^{\text{th}}$ toss." Then

$$\text{Prob(first player wins)} = \sum_{k=1}^{\infty} \text{Prob}(E_k) = \frac{1}{2} + \frac{1}{2^3} + \frac{1}{2^5} + \cdots = \frac{2}{3}$$

NOTE: The purpose of this example is to show that even elementary problems can require consideration of infinite sample spaces.

## H.3 Combinatorial Analysis

The number of subsets of $k$ items in a set of $n$ items, with the order of the items in the subsets disregarded, is given by the *binomial coefficient*

$$\binom{n}{k} = \frac{n!}{k!\,(n-k)!} \tag{H.5}$$

The classical terminology is "the number of combinations of $n$ things taken $k$ at a time."

Euler's *Gamma function* defined by

$$\Gamma(x) = \int_0^{\infty} t^{x-1} e^{-t}\, dt \tag{H.6}$$

is a generalization of the *factorial function* $n!$. More specifically, using integration by parts, we can establish the important property

$$\Gamma(x+1) = x\Gamma(x) \tag{H.7}$$

from which, when $x$ is an integer $n$, it follows that

$$\Gamma(n+1) = n! \tag{H.8}$$

When $n$ is large the task of computing $n!$ is formidable. However, the Gamma function can be used to derive a simple approximation which is useful even for moderate size values of $n$. For this purpose we write Eq. (H.6) as

$$n! = \int_0^{\infty} t^n e^{-t}\, dt = \int_0^{\infty} \exp(-t + n \log t)\, dt$$

Then, since the function
$$y = -t + n \log t$$
has a maximum at $t = n$, we are led to make the change of variable
$$t = n + x$$
so that the new integrand will have its maximum at $x = 0$. Thus,
$$n! = \int_{-n}^{\infty} \exp[-n - x + n \log(n + x)]\, dx$$
$$= e^{-n} n^n \int_{-n}^{\infty} \exp\left[-x + n \log\left(1 + \frac{x}{n}\right)\right] dx$$
But
$$\log\left(1 + \frac{x}{n}\right) = \frac{x}{n} - \frac{1}{2}\frac{x^2}{n^2} + \frac{1}{3}\frac{x^3}{n^3} - \cdots$$
Hence, for large $n$, we have
$$n! \approx e^{-n} n^n \int_{-n}^{\infty} e^{-\frac{1}{2}x^2/n}\, dx = e^{-n} n^n \sqrt{n} \int_{-n}^{\infty} e^{-\frac{1}{2}\xi^2}\, d\xi$$
and, replacing the lower limit by infinity, we obtain *Stirling's formula*†
$$n! \approx \sqrt{2\pi n}\, e^{-n} n^n \qquad (H.9)$$
Indeed, it can be shown that the ratio of the two sides of (H.9) approach unity as $n$ becomes infinite.

Stirling's asymptotic formula for the Gamma function is
$$\Gamma(x) \simeq \sqrt{2\pi}\, e^{-x} x^{x-\frac{1}{2}} \left(1 + \frac{1}{12x} + \frac{1}{288 x^2}\right.$$
$$\left. - \frac{139}{51,840 x^3} - \frac{571}{2,488,320 x^4} + \cdots\right) \quad (H.10)$$
which has the property that for any positive $x$ each partial sum of terms on the right approximates the left member with an error numerically less than the last term retained. Regarded as an infinite series, the right member diverges for all $x$, so that it is, indeed, an asymptotic expansion as defined in Sect. 10.6.

---

† James Stirling (1692–1770) was one of Scotland's best known mathematicians. His major work *Methodus differentialis: sive tractatus de summatione et interpolatione serierum infinitarum* published in 1730 was reprinted twice in his lifetime, in 1753 and 1764. An English translation also appeared in 1749. The important thrust of the first part of this work was the transformation of series to improve convergence. Here he gave the logarithmic form of the asymptotic expansion of the Gamma function to five terms and the recurrence formula for determining the succeeding ones. In the interpolation portion of his book he also calculated $\Gamma(\frac{1}{2})$ to ten decimal places. Its true value is $\sqrt{\pi}$.

It is interesting that the logarithmic form of (H.10) involves the Bernoulli numbers which were defined in Sect. 1.2. Specifically,

$$\log \Gamma(x) = \tfrac{1}{2}\log(2\pi) + (x - \tfrac{1}{2})\log x - x + \sum_{k=1}^{\infty} \frac{B_{2k}}{2k(2k-1)x^{2k-1}} \quad \text{(H.11)}$$

### ◊ Problem H–7

Three dice are rolled. Given that no two faces are the same, what is the probability that one of the faces is a four?

ANSWER: Define the events

    A: One face is a four.
    B: No two faces are the same.

and denote by $n_B$ and $n_{AB}$ the number of different ways in which the events $B$ and $AB$ can be realized. Clearly,

$$n_B = \binom{6}{3} \quad \text{and} \quad n_{AB} = \binom{5}{2}$$

so that

$$\text{Prob}(A \mid B) = \frac{n_{AB}}{n_B} = \frac{1}{2}$$

### ◊ Problem H–8

From three dozen eggs one dozen is selected. If it is known that there are four bad eggs in the three dozen, what is the probability that the one dozen selected will all be good?

ANSWER:

$$\frac{\binom{32}{12}}{\binom{36}{12}} = \frac{46}{255} \approx 0.18$$

### ◊ Problem H–9

Determine the number of possible bridge hands and the number of different situations at the bridge table.

ANSWER: The number of way of selecting 13 cards from 52 is

$$\binom{52}{13} = \frac{52!}{39!\,13!} = 635{,}013{,}559{,}600$$

and the number of possible deals at the bridge table is

$$\binom{52}{13}\binom{39}{13}\binom{26}{13} = \frac{52!}{39!\,13!} \cdot \frac{39!}{26!\,13!} \cdot \frac{26!}{13!\,13!} = \frac{52!}{(13!)^4} = 5.3645\ldots \times 10^{28}$$

NOTE: This is such a large number that if every living person played one game every second, day and night, it would require thousands of billions of years to exhaust all of the possible hands!!

# Appendix H

◊ **Problem H–10**

Find the probability that 13 cards contain exactly $k$ aces and that West is dealt exactly $k$ aces in a game of bridge.

ANSWER: The probabilities are the same:

$$\frac{\binom{4}{k}\binom{48}{13-k}}{\binom{52}{13}} = \frac{\binom{4}{k}\binom{48}{13-k}\binom{39}{13}\binom{26}{13}}{\binom{52}{13}\binom{39}{13}\binom{26}{13}}$$

◊ **Problem H–11**

What is the probability of all of the aces falling in one hand in a game of bridge?

ANSWER:

$$4 \times \frac{\binom{48}{9}}{\binom{52}{13}} = \frac{44}{4165} \approx 0.01$$

◊ **Problem H–12**

Find the probability that each player at the bridge table is dealt one ace.

ANSWER:

$$\frac{\binom{4}{1}\binom{48}{12}\binom{3}{1}\binom{36}{12}\binom{2}{1}\binom{24}{12}}{\binom{52}{13}\binom{39}{13}\binom{26}{13}} = 4! \times \frac{\frac{48!}{(12!)^4}}{\frac{52!}{(13!)^4}} \approx \frac{1}{10}$$

◊ **Problem H–13**

What is the probability that South has no ace when it is known that North has no ace?

ANSWER: The number of hands for which North has no ace is

$$n_N = \binom{48}{13}\binom{39}{13}\binom{26}{13}$$

and the number of hands for which North and South each have no ace is

$$n_{NS} = \binom{48}{13}\binom{35}{13}\binom{26}{13}$$

Hence, the probability in question is

$$\frac{n_{NS}}{n_N} = \frac{\binom{48}{13}\binom{35}{13}\binom{26}{13}}{\binom{48}{13}\binom{39}{13}\binom{26}{13}} \approx 0.18$$

… Probability Theory and Applications

◇ **Problem H–14**

In a game of bridge, North and South have ten trumps between them.
(a) Find the probability that all three remaining trumps are in the same hand.
(b) If the king of trumps is among the three, what is the probability that it is unguarded?

ANSWER: Let $E$ and $W$ be the events that East and West, respectively, have three trumps. Then

(a) $$\text{Prob}(E+W) = \text{Prob}(E) + \text{Prob}(W) = 2 \times \frac{\binom{23}{10}}{\binom{26}{13}} = \frac{11}{50}$$

Let $E$ and $W$ be the events that East and West, respectively, have the king of trumps and no other. Then

(b) $$\text{Prob}(E+W) = \text{Prob}(E) + \text{Prob}(W) = 2 \times \frac{\binom{23}{12}}{\binom{26}{13}} = \frac{13}{50}$$

◇ **Problem H–15**

Find the probability for a poker hand to be
(a) a royal flush (ten, jack, queen, king, ace in a single suit)
(b) a straight flush (five cards of the same suit in a sequence)
(c) four of a kind (four cards of equal face values)
(d) a full house (one pair and one triple of cards with equal face values)
(e) a flush (five cards of one suit)
(f) a straight (five cards in a sequence regardless of suit)
(g) three of a kind (three cards of equal face values plus two extra cards)
(h) two pairs (two pairs of cards of equal face values plus one extra card)
(i) a pair (one pair of cards of equal face values plus three different cards)

ANSWER: Let $N$ be the number of different poker hands, i.e.,

$$N = \binom{52}{5} = 2{,}598{,}960$$

Then

(a) $$\binom{4}{1} \cdot \frac{1}{N} = \frac{1}{649{,}740} \approx 0.0000015$$

(b) $$\binom{9}{1}\binom{4}{1} \cdot \frac{1}{N} = \frac{9}{649{,}740} \approx 0.0000135$$

(c) $$\binom{13}{1}\binom{12}{1}\binom{4}{1} \cdot \frac{1}{N} = \frac{1}{4165} \approx 0.00024$$

(d) $\quad\binom{13}{1}\binom{4}{3}\binom{12}{1}\binom{4}{2}\cdot\dfrac{1}{N}=\dfrac{6}{4165}\approx 0.00144$

(e) $\quad\binom{4}{1}\binom{13}{5}\cdot\dfrac{1}{N}=\dfrac{429}{216,580}\approx 0.002$

(f) $\quad\binom{9}{1}\binom{4}{1}^{5}\cdot\dfrac{1}{N}=\dfrac{192}{54,145}\approx 0.0035$

(g) $\quad\binom{13}{1}\binom{4}{3}\binom{12}{2}\binom{4}{1}^{2}\cdot\dfrac{1}{N}=\dfrac{88}{4165}\approx 0.02$

(h) $\quad\binom{13}{2}\binom{4}{2}^{2}\binom{11}{1}\binom{4}{1}\cdot\dfrac{1}{N}=\dfrac{198}{4165}\approx 0.0475$

(i) $\quad\binom{13}{1}\binom{4}{2}\binom{12}{3}\binom{4}{1}^{3}\cdot\dfrac{1}{N}=\dfrac{1760}{4165}=\dfrac{352}{833}\approx 0.42$

## H.4  Random Variables

Consider all of the possible outcomes of a random experiment and to each of these outcomes or events assign a real number. Then, when the experiment is performed, we may identify the outcome solely by the associated real number rather than by giving a physical description of the event which has occurred. For example, in the coin-tossing experiment the sample space for a single toss consists of the two mutually exclusive events "heads" and "tails." If we assign the number 1 to the event "heads" and 0 to the event "tails," then when we discuss the outcome of a single performance of this experiment we need no longer use the descriptive phrases "toss a coin" and "heads or tails," but may simply say that the result is the number 1 or 0. Pursuing this idea further, we may introduce a variable $X$ and define its values to be one and zero corresponding to each of the two possible results of the experiment. The quantity $X$ is commonly called a *random variable* in probability theory. Thus, *the term random variable is used to denote a real number whose value is determined by the outcome of a random experiment.*

For example, in our coin-tossing experiment we might define three random variables as

$X =$ Number of heads in the first toss
$Y =$ Number of heads in the second toss
$Z =$ Number of heads in both tosses

so that

| Basic Event | Value of $X = x_k$ | $Y = y_k$ | $Z = z_k$ | Prob.$= p_k$ |
|---|---|---|---|---|
| $A_1 = (HH)$ | 1 | 1 | 2 | pp |
| $A_2 = (HT)$ | 1 | 0 | 1 | pq |
| $A_3 = (TH)$ | 0 | 1 | 1 | qp |
| $A_4 = (TT)$ | 0 | 0 | 0 | qq |

We can also extend the notion of random variable to the case of a continuous sample space. Consider the experiment of a random selection of a point on the real line from zero to one. The sample space is the unit interval itself and the events or sample points are the points on the line. The function which assigns a number to each of these points in the usual way is a random variable. Another random variable defined over the same sample space is the function which has the value zero if the point selected corresponds to a rational number and the value one if the number is irrational.

When the sample space is continuous, the problem of assigning a probability distribution to a random variable is more complicated. In our example of randomly selecting a point on a line, the first random variable defined assumes a nondenumerable number of values so that the probability of selecting any particular point at random is certainly zero. Therefore, it makes sense to speak *only* of the probability that the random variable assumes a value lying in some subinterval of the unit interval. Thus, we could say the probability of a randomly selected point belonging to the interval $\frac{1}{4} \leq x \leq \frac{3}{4}$ is $\frac{1}{2}$. For the random variable defined in the second part of this example, the appropriate probability distribution to assign is zero when the value of the function is zero, and one when the value of the function is one. This follows because, although the rational numbers are everywhere dense in the real line, they constitute a set of *zero length*.

## H.5 Probability Distribution and Density Functions

Let $X$ be a random variable which assumes values in a set $S$ of real numbers. As we have seen, if $S$ is not discrete it makes little sense to speak of the probability that $X$ will have any particular value. Thus, we are obliged to consider the probability that $X$ will assume a value lying in a subset of $S$. For this reason it is convenient to define the *distribution function* $F(x)$ as

$$F(x) = \text{Prob}(X \leq x) \qquad (\text{H}.12)$$

If $a$ and $b$ are real numbers such that $a < b$, we have

$$\text{Prob}(X \leq b) = \text{Prob}(X \leq a) + \text{Prob}(a < X \leq b)$$

so that
$$\text{Prob}(a < X \leq b) = F(b) - F(a) \quad \text{(H.13)}$$
From this relation and the fact that the probability of any event is always nonnegative, we have
$$F(x_1) \leq F(x_2) \quad \text{if} \quad x_1 \leq x_2$$
In other words, $F(x)$ is a monotone nondecreasing function. [It is understood that $F(x)$ is constant in any interval that contains no points of $S$.] Also it is always true that
$$F(-\infty) = 0 \quad \text{and} \quad F(+\infty) = 1$$
Thus $F(x)$ is positive and has values between zero and one.

One may interpret the distribution function physically in terms of the distribution of one unit of mass over the real line $-\infty < x < \infty$. If the mass allotted to any set of points represents the probability that $X$ will assume a value lying in that set, then $F(x)$ is the total mass associated with the point $x$ and all points lying to the left of $x$. With this interpretation we may imagine concentrations of mass at certain points on the line in addition to or in place of the *continuous* mass distribution. The latter condition is described as a *discrete* distribution, while the former characterizes a *mixed* distribution, i.e., one that is both continuous and discrete.

As an example of distribution functions of the discrete type, consider the random variable associated with the single toss of a coin
$$X = \begin{cases} 1 & \frac{1}{2} \\ 0 & \frac{1}{2} \end{cases}$$
and illustrated in Fig. H.1. The random selection of a point in the unit interval offers an illustration of a continuous distribution. For the random variable defined first in that example, the distribution function is shown in Fig. H.2. The second random variable described is defined over a continuous sample space but gives rise to a discrete distribution function. In fact, it is the unit step function with its jump at $x = 1$.

If $F(x)$ is differentiable, we define the *frequency function* or *probability density function* $f(x)$ by
$$f(x) = \frac{dF(x)}{dx} \quad \text{(H.14)}$$
Since $F(x)$ is monotone nondecreasing, for all $x$ we have
$$f(x) \geq 0$$
Also from the definitions we have the further properties
$$F(x) = \int_{-\infty}^{x} f(u)\, du \qquad \int_{-\infty}^{\infty} f(x)\, dx = 1 \quad \text{(H.15)}$$

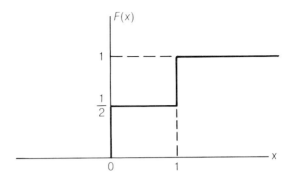

**Fig. H.1:** Distribution function for a true coin.

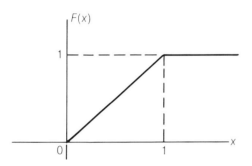

**Fig. H.2:** Distribution function for selecting a point on a line.

and
$$\text{Prob}(a < X \leq b) = \int_a^b f(x)\,dx \qquad (H.16)$$

If the random variable $X$ takes on only discrete values, say $x_1$, $x_2$, $x_3$, ..., then the distribution function $F(x)$ is not differentiable in the ordinary sense. To handle this situation in a manner consistent with the previous discussion, we may assign to each point $x_j$ a probability density given by $p_j\,\delta(x - x_j)$ where $\delta(x)$ is the *Dirac $\delta$-function*† (or, in other applications, called the *unit impulse function*), defined by the properties

$$\delta(x) = 0 \qquad x \neq 0$$
$$\int_a^b \delta(x)\,dx = 1 \qquad a < 0 < b$$

The delta function is not a mathematical function in the strict sense. In all legitimate applications, it is visualized as a result of a limiting process

---

† Named for the English physicist Paul Adrien Maurice Dirac, born in Bristol in 1902. He, as did Sir Isaac Newton and Sir Edmond Halley, held the Lucasian chair of mathematics at Cambridge University.

involving a function $\delta(x,\epsilon)$ defined, for example, by

$$\delta(x,\epsilon) = \begin{cases} \dfrac{1}{2\epsilon} & -\epsilon \leq x \leq +\epsilon \\ 0 & |x| > |\epsilon| \end{cases}$$

Then, we have

$$\delta(x) = \lim_{\epsilon \to 0} \delta(x,\epsilon)$$

Thus, in terms of the delta function we define the probability density for the discrete case by

$$f(x) = \sum_j p_j\, \delta(x - x_j) \tag{H.17}$$

The choice of the name "probability density function" is a consequence of our analogy between probability distribution functions and mass distributions. In the regions in which the mass distribution $F(x)$ is continuous, the mass density $f(x)$ is the mass per unit length. When discrete masses are attached to certain points, the mass density at these points is infinite. In any case the function $F(x)$ specifies the total mass lying to the left of and including the point $x$.

As another example of the discrete distribution, consider again the random variable $Z$ whose value is the number of heads in two tosses of a coin. Then,

$$Z = \begin{cases} 2 & p^2 \\ 1 & 2pq \\ 0 & q^2 \end{cases} \quad \text{where} \quad p^2 + 2pq + q^2 = (p+q)^2 = 1$$

so that

$$f(z) = p^2\, \delta(z-2) + 2pq\, \delta(z-1) + q^2\, \delta(z)$$

Therefore, the density function consists of three mass particles on the $z$ axis at the points 2, 1, 0 with masses $p^2$, $2pq$, $q^2$, respectively.

◊ **Problem H-16**

The random variable having a *Cauchy distribution* is defined as

$$X = \tan \Theta$$

where $\Theta$ is a random variable which is uniformly distributed over the interval $-\tfrac{1}{2}\pi < \theta < \tfrac{1}{2}\pi$. Compute the frequency function.

NOTE: See Sect. H.6 for the definition of a uniform distribution.
ANSWER:

$$\text{Prob}(X \leq x) = \text{Prob}(\Theta \leq \theta) = \frac{\theta + \tfrac{1}{2}\pi}{\pi} = \frac{\arctan x + \tfrac{1}{2}\pi}{\pi} = F(x)$$

so that

$$f(x) = \frac{d}{dx} F(x) = \frac{1}{\pi(1+x^2)}$$

## H.6 Expectation, Mean, and Variance

Consider a random variable $X$, associated with the outcome of a certain experiment, which can assume any one of the discrete values $x_1$, $x_2$, ... with respective probabilities $p_1$, $p_2$, ... Suppose that the experiment is repeated a large number $N$ times and that the event $X = x_i$ occurs $m_i$ times. Then the numerical average of $X$ over the $N$ trials is

$$\frac{1}{N}(m_1 x_1 + m_2 x_2 + \cdots) = \frac{m_1}{N} x_1 + \frac{m_2}{N} x_2 + \cdots$$

According to the empirical notion of probability, as $N \to \infty$ the ratio $m_i/N$ approaches the probability $p_i$. Therefore, we are led to define the *average*, *mean*, or *mathematical expectation* $E(X)$ of the random variable $X$ by the limiting expression

$$E(X) \equiv \overline{X} = p_1 x_1 + p_2 x_2 + \cdots \tag{H.18}$$

The expectation of a random variable is a probability-weighted average of the values which the variable can take on, and represents the anticipated numerical average of the observed values of this variable in a very large number of trials. To calculate the mathematical expectation of a random variable $X$ in the discrete case, we multiply each of the values which $X$ can assume by the corresponding probabilities with which they are assumed and sum over all possible cases.

We may extend this notion of mathematical expectation in a natural way to include continuous distributions. If $f(x)$ is the frequency function of $X$, then the probability that $X$ will assume a value in the interval $(x, x+dx)$ is approximately given by

$$\text{Prob}(x < X \leq x + dx) \simeq f(x)\,dx$$

Thus, if we regard the continuous distribution function as the limit of a suitable discrete distribution, we are tempted to write $x$ as the value expected multiplied by the probability $f(x)\,dx$ with which it is to be expected and to integrate (rather than sum) over all possible values. Accordingly, we define

$$E(X) \equiv \overline{X} = \int_{-\infty}^{\infty} x f(x)\,dx \tag{H.19}$$

Finally, we note that for any real-valued function $g$, such that $g(X)$ is a random variable, it can be shown that

$$E[g(X)] \equiv \overline{g(X)} = \int_{-\infty}^{\infty} g(x) f(x)\,dx \tag{H.20}$$

For a given function $g$, $E[g(X)]$ depends on $f(x)$ and hence helps to characterize $f(x)$. For this reason one sometimes refers to an expectation generated by a certain function of $g$ as a *statistical parameter*. We now define the most common of these parameters. (Similar definitions apply also to the discrete case.)

- The *mean* or *average* of $X$ (the *center of gravity* of the distributed mass) is defined as

$$E(X) \equiv \overline{X} = \int_{-\infty}^{\infty} x f(x)\, dx \qquad \text{(H.21)}$$

- The *mean-squared value* of $X$ is

$$E(X^2) \equiv \overline{X^2} = \int_{-\infty}^{\infty} x^2 f(x)\, dx \qquad \text{(H.22)}$$

- The *variance* of $X$ (the *moment of inertia* of the distributed mass about its center of gravity) is defined as

$$\text{Var}(X) \equiv E\big[(X - \overline{X})^2\big] = \int_{-\infty}^{\infty} (x - \overline{X})^2 f(x)\, dx \qquad \text{(H.23)}$$

By expanding the integrand we obtain the convenient form

$$\text{Var}(X) = \overline{X^2} - \overline{X}^2 \qquad \text{(H.24)}$$

- We also define the *standard deviation* $\sigma$ as the square root of the variance so that

$$\text{Var}(X) = \sigma^2 \qquad \text{(H.25)}$$

As an illustration, consider again the random variable $X$ whose value is determined by the random selection of a point on the real line from zero to one. The probability density function $f(x)$ is

$$f(x) = \begin{cases} 1 & \text{for } 0 < x \leq 1 \\ 0 & \text{otherwise} \end{cases}$$

so that, for example,

$$\text{Prob}(\tfrac{1}{4} < X \leq \tfrac{1}{2}) = \int_{\frac{1}{4}}^{\frac{1}{2}} f(x)\, dx = \tfrac{1}{4}$$

The probability distribution is said to be *uniform* and the random variable $X$ is, therefore, called *uniformly distributed*.

For the statistical parameters, we have

$$\overline{X} = E(X) = \int_{-\infty}^{\infty} x f(x)\, dx = \int_{0}^{1} x\, dx = \tfrac{1}{2}$$

$$\sigma^2 = \text{Var}(X) = \int_{-\infty}^{\infty} (x - \tfrac{1}{2})^2 f(x)\, dx = \int_{0}^{1} (x - \tfrac{1}{2})^2\, dx = \tfrac{1}{12}$$

As an example for the discrete case, consider again the random variable $Z$ whose value is the number of heads in two tosses of the coin. In this case, the statistical parameters are

$$E(Z) = \overline{Z} = \sum_{k=1}^{3} z_k p_k = 2 \cdot p^2 + 1 \cdot 2pq + 0 \cdot q^2$$
$$= 2p(p+q) = 2p$$
$$E(Z^2) = \overline{Z^2} = \sum_{k=1}^{3} z_k^2 p_k = 2^2 \cdot p^2 + 1^2 \cdot 2pq + 0^2 \cdot q^2$$
$$= (2p)^2 + 2pq = 2p(1+p)$$
$$\text{Var}(Z) = E\big[(Z - \overline{Z})^2\big] = \sum_{k=1}^{3}(z_k - 2p)^2 p_k$$
$$= (2 - 2p)^2 \cdot p^2 + (1 - 2p)^2 \cdot 2pq + (0 - 2p)^2 \cdot q^2$$
$$= 2pq$$

## H.7  Independence and Covariance of Random Variables

Again we reference the experiment of tossing a coin twice. Recall that the random variables $X$ and $Y$ represent the number of heads on the first and second tosses, respectively so that

$$X = \begin{cases} 1 & p \\ 0 & q \end{cases} \quad \text{and} \quad Y = \begin{cases} 1 & p \\ 0 & q \end{cases}$$

The sum of $X$ and $Y$ is the total number of heads which we called $Z$ and is determined as

$$Z = X + Y = \begin{cases} 2 & p^2 \\ 1 & 2pq \\ 0 & q^2 \end{cases}$$

Then, since

$$E(X) = p \qquad E(Y) = p \qquad E(Z) = 2p$$

we see that

$$E(X + Y) = E(X) + E(Y)$$

The random variable $Y + Z$ provides another example. To obtain its values and corresponding probabilities

$$Y + Z = \begin{cases} 3 & p^2 \\ 1 & pq \\ 2 & qp \\ 0 & q^2 \end{cases}$$

we can use the table developed in Sect. H.4. Hence,

$$E(Y + Z) = 3 \cdot p^2 + 1 \cdot pq + 2 \cdot qp + 0 \cdot q^2 = 3p = E(Y) + E(Z)$$

In general, it can be proved that the expectation of the sum of *any* two random variables $X_1$ and $X_2$ is the sum of their separate expectations. Thus,

$$E(X_1 + X_2) = E(X_1) + E(X_2) \qquad (H.26)$$

The analogous proposition for products of random variables requires that they be independent. We have already seen in Eq. (H.4) that the probability of the simultaneous occurrence of independent events is the same as the product of their individual probabilities. To define independence of random variables, let $X$ take on the values $x_i$ with probabilities $p_i$ and $Y$, the values $y_j$ with probabilities $q_j$. Then $X$ and $Y$ are said to be *independent* if

$$\text{Prob}(X = x_i \text{ and } Y = y_j) = p_i q_j \quad \text{for all } i \text{ and } j \qquad (H.27)$$

For the coin-tossing experiment, the two random variables $X$ and $Y$, representing "heads on the first toss" and "heads on the second toss," are certainly independent. Also,

$$XY = \begin{cases} 1 & p^2 \\ 0 & 2pq + q^2 \end{cases}$$

so that

$$E(XY) = p^2 = E(X)E(Y) = p \cdot p$$

On the other hand, the random variables $Y$ and $Z$ are not independent since $Y$ is "heads on the second toss" while $Z$ is the "total number of heads." In this case,

$$YZ = \begin{cases} 2 & p^2 \\ 0 & pq + q^2 \\ 1 & pq \end{cases}$$

but

$$E(YZ) = 2p^2 + pq \neq E(Y)E(Z) = p \cdot 2p$$

In general, it can be shown that if two random variables $X_1$ and $X_2$ are *independent*, then

$$E(X_1 X_2) = E(X_1) E(X_2) \qquad (H.28)$$

In Eq. (H.23) we defined the term "variance" of a random variable. For two random variables, the notion of correlation is of particular importance. To explore this concept, we define the *covariance* of two random variables $X$ and $Y$ as

$$\text{Cov}(X, Y) \equiv E[(X - \overline{X})(Y - \overline{Y})] \qquad (H.29)$$

If variance may be considered to be the moment of inertia of the distributed mass about its center of gravity, then covariance plays the role of the *product of inertia.* Also, from the expansion

$$\begin{aligned}\text{Cov}(X,Y) &= E(XY) - \overline{X}E(Y) - \overline{Y}E(X) + \overline{X}\,\overline{Y}\\&= E(XY) - E(X)E(Y)\end{aligned} \quad \text{(H.30)}$$

we see that the covariance will be zero if $X$ and $Y$ are independent. The random variables $Y$ and $Z$ for the coin-tossing experiment are *not independent* so that

$$\text{Cov}(Y,Z) = E(YZ) - E(Y)E(Z) = 2p^2 + pq - p(2p) = pq \neq 0$$

Furthermore, by calculating the variance of $X + Y$, we obtain

$$\begin{aligned}\text{Var}(X+Y) &= E\{[(X-\overline{X}) + (Y-\overline{Y})]^2\}\\&= \text{Var}(X) + \text{Var}(Y) + 2\,\text{Cov}(X,Y)\end{aligned} \quad \text{(H.31)}$$

Therefore, if $X$ and $Y$ are independent, then

$$\text{Var}(X+Y) = \text{Var}(X) + \text{Var}(Y) \quad \text{(H.32)}$$

The normalized covariance, denoted by $\rho$, is called the *correlation coefficient* of $X$ and $Y$ and is defined by

$$\rho \equiv \frac{\text{cov}(X,Y)}{\sqrt{\text{var}(X)\,\text{var}(Y)}} = \frac{\text{cov}(X,Y)}{\sigma_x \sigma_y} \quad \text{so that} \quad |\rho| \leq 1 \quad \text{(H.33)}$$

It may be interpreted as a measure of the interdependence of $X$ and $Y$.

For example, suppose that we seek to determine that linear function $aX + b$ of the random variable $X$ which gives the best fit to the variable $Y$ in the least-squares sense. To be more specific, we seek those values of $a$ and $b$ which minimize the expression $E[(Y - aX - b)^2]$. We assume, for convenience, that $\overline{X} = \overline{Y} = 0$.

By a simple calculation we have

$$E[(Y - aX - b)^2] = \overline{Y^2} - 2a\overline{XY} + a^2\overline{X^2} + b^2$$

For any value of $a$ the best selection of $b$ is clearly zero. To determine $a$, we set to zero the derivative with respect to $a$ to obtain

$$a = \frac{\overline{XY}}{\overline{X^2}}$$

Therefore, the

Minimum value of $E\{[Y - (aX + b)]^2\} = \sigma_y^2(1 - \rho^2)$

so that the correlation coefficient is directly related to the minimum mean-squared error in fitting $Y$ by a linear function of $X$.

Finally, note that if two random variables are uncorrelated ($\rho = 0$), it does *not* necessarily follow that they are independent. For example, consider two random variables defined as

$$X = \begin{cases} 1 & \frac{1}{2} \\ -1 & \frac{1}{2} \end{cases} \quad \text{and} \quad Y = |X| = \begin{cases} 1 & \frac{1}{2} \\ 1 & \frac{1}{2} \end{cases}$$

so that $\overline{X} = 0$, $\overline{Y} = 1$, and

$$XY = \begin{cases} 1 & \frac{1}{2} \\ -1 & \frac{1}{2} \end{cases}$$

Clearly, $X$ and $Y$ are not independent; but,

$$\text{Cov}(X, Y) = E(XY) - E(X)E(Y) = (1 \cdot \tfrac{1}{2} - 1 \cdot \tfrac{1}{2}) - 0 \cdot 1 = 0$$

Here, then, we have a case in which the covariance and hence the correlation coefficient are zero; yet $Y$ is functionally related to $X$. Thus, a zero value for $\rho$ does not of itself imply independence.

In summary,

- Expectations are *additive* for *any* random variables.
- Expectations are *multiplicative* for *independent* random variables.
- Variances are *additive* for *independent* random variables.
- Covariances are zero for *independent* random variables.

## H.8 Applications to Coin-tossing and Card-matching

To illustrate the basic concepts of the last two sections, we will develop two applications and, incidentally, show how statistical parameters can often be obtained without full knowledge of the probability distributions.

First, for the experiment in which a coin is tossed with the events "heads" and "tails" occurring with probabilities $p$ and $q$ $(p + q = 1)$, we wish to calculate the average number of heads and the variance if the coin is tossed $n$ times. This problem may be easily solved by introducing a random variable $X_k$ to represent the outcome of the $k^{\text{th}}$ toss. To be precise, we define

$$X_k = \begin{cases} 1 & \text{if "heads" occurs on } k^{\text{th}} \text{ toss} \\ 0 & \text{if "tails" occurs on } k^{\text{th}} \text{ toss} \end{cases}$$

Then the variable $X$ defined by

$$X = X_1 + X_2 + \cdots + X_n$$

has the desirable property that its value for each experiment is the total number of heads occurring in the $n$ tosses. The average of $X_k$ is

$$E(X_k) = 1 \cdot \text{Prob}(X_k = 1) + 0 \cdot \text{Prob}(X_k = 0) = p \quad k = 1, 2, \ldots, n$$

Therefore,

$$\overline{X} = E(X) = E(X_1) + E(X_2) + \cdots + E(X_n) = np$$

Probability Theory and Applications

To compute $\text{Var}(X)$, note that

$$X^2 = \sum_{k=1}^{n} X_k^2 + 2\sum_{j=1}^{n-1}\sum_{k=j+1}^{n} X_j X_k$$

Now,

$$E(X_k^2) = 1^2 \cdot \text{Prob}(X_k = 1) + 0^2 \cdot \text{Prob}(X_k = 0) = p$$

and $X_j$ and $X_k$ are independent for $j \neq k$ so that

$$E(X_j X_k) = E(X_j)E(X_k) = p^2$$

Hence,

$$E(X^2) = nE(X_k^2) + n(n-1)E(X_j X_k)$$
$$= np + n(n-1)p^2 = np(1-p) + n^2 p^2$$
$$= npq + n^2 p^2$$

and the variance in question is, therefore,

$$\text{Var}(X) = \sigma^2 = E[(X - \overline{X})^2] = \overline{X^2} - \overline{X}^2$$
$$= n^2 p^2 + npq - (np)^2 = npq$$

The next example goes back to Montmort† who posed it in 1708. The problem has many forms and was later generalized by Laplace and many others.

Two identical decks of $n$ different cards are put in random order and matched against each other. We say that a match has occurred if a card occupies the same place in both decks. Of course, matches can occur at any of the $n$ places and at several places simultaneously. Our problem is to determine the average number of matches and the variance.

For this purpose, let $X_k$ be a random variable whose value is one if card number $k$ is in the correct location and zero otherwise. Thus, $X_k = 1$ with probability $1/n$ if there are $n$ cards in the experiment.

---

† The book on probability by Pierre Rémond de Montmort (1678–1719), *Essay d'analyse sur les jeux de hazard*, was published in 1708 after James Bernoulli's death in 1705 but before the appearance of his *Ars Conjectandi*. The theory of games of chance had not been treated mathematically since Christiaan Huygens's monograph of 1657— some 50 years earlier. Montmort's work aroused the interest of Nicholas Bernoulli and provided an incentive for Nicholas to publish his uncle's book.

In his book, Montmort analyses a game he called "Treize" in which the thirteen cards of one suit are shuffled and then drawn one after the other. The player who is drawing cards wins the round if the $n^{\text{th}}$ card drawn is itself the card $n$. The chance of winning is shown to be

$$\sum_{k=1}^{13} \frac{(-1)^{k-1}}{k!}$$

Leibnitz provided Montmort with a rough idea of the limit to which this tends as the number of cards increases, but Euler was first to give this limit as $1 - e^{-1}$.

Furthermore, if $j \neq k$, then

$$\text{Prob}(X_j = 1 \text{ and } X_k = 1)$$
$$= \text{Prob}(X_j = 1 \mid X_k = 1)\text{Prob}(X_k = 1) = \frac{1}{n-1} \cdot \frac{1}{n}$$

so that the product $X_j X_k$ has unit value with probability $1/n(n-1)$ if cards numbered $j$ and $k$ are in their correct places and is zero otherwise. The random variable

$$X = X_1 + X_2 + \cdots + X_n$$

is the total number of cards in their correct places. Then, since

$$E(X) = E(X_1) + E(X_2) + \cdots + E(X_n) = \overline{X} = n \cdot \frac{1}{n} = 1$$

we have the surprising result that *the average number of matches in a single performance of the experiment is precisely one—no matter how large $n$ may be!*

The mean-squared value and the variance are computed as for the coin-tossing example. We have

$$E(X^2) = n \cdot \frac{1}{n} + n(n-1) \cdot \frac{1}{n(n-1)} = 2$$

and

$$\text{Var}(X) = \overline{X^2} - \overline{X}^2 = 2 - 1 = 1$$

which are again independent of $n$. Therefore, in one application of card matching, the average number of matches is one with a standard deviation of one.

The probability of obtaining at least one match turns out to be

$$1 - \frac{1}{2!} + \frac{1}{3!} - \frac{1}{4!} + \cdots \pm \frac{1}{n!}$$

Therefore, with good approximation,

$$\text{Prob}(X \geq 1) \approx 1 - e^{-1} = 0.63212\ldots$$

as Euler first demonstrated.

◊ **Problem H–17**

A die is rolled $n$ times. Find the average number of occurrences of the event "a six is followed by a number no smaller than a three."

ANSWER: Let $X_k$ be a random variable whose value is one if the stated event occurs on the $k^{\text{th}}$ and following roll and zero otherwise. Then

$$X_k = \begin{cases} 1 & \frac{1}{6} \cdot \frac{4}{6} = \frac{1}{9} \\ 0 & — \end{cases} \quad \text{so that} \quad \overline{X} = \overline{X_1 + X_2 + \cdots + X_{n-1}} = \frac{n-1}{9}$$

◇ **Problem H–18**

A coin is tossed $n$ times with equal probabilities for heads and tails. Find the average and the variance of the number of times a head is followed by a tail.

ANSWER: As before, define

$$X_k = \begin{cases} 1 & \frac{1}{2} \cdot \frac{1}{2} = \frac{1}{4} \\ 0 & — \end{cases} \quad \text{and} \quad X = X_1 + X_2 + \cdots + X_{n-1}$$

There are $n - 1$ terms of the form $X_j^2$ and $n^2 - n - 2(n - 1)$ terms of the form $X_j X_k$ or $X_k X_j$ where $j < k + 1$. Hence

$$\overline{X} = \frac{n-1}{4} \quad \text{and} \quad \text{Var}(X) = \frac{n-1}{4} + \frac{(n-3)(n-2)}{16} - \frac{(n-1)^2}{16} = \frac{3(n-1)}{16}$$

◇ **Problem H–19**

Two men decide to meet between 12 o'clock and 1 o'clock, each not waiting more than 10 minutes for the other. If all times of arrival within the hour are equally likely for each person, and if their times of arrival are independent, find the probability that they will meet.

ANSWER: The straight lines $y = x \pm \frac{1}{6}$ define a band within the unit square. The probability that the men will meet is the ratio of the area of the band to the total area which is $\frac{11}{36}$.

◇ **Problem H–20**

In the interval $(0, 1)$ $n$ points are distributed uniformly and independently. Find:
(a) The probability of the event $A$: "the point lying farthest to the right is to the right of the number $x$."
(b) The probability of the event $B$: "the point lying farthest to the left is to the left of the number $y$."
(c) The probability of the event $C$: "the point lying next farthest to the right is to the left of the number $z$."

ANSWER: The event $A$ is the complement of the event that all lie to the left of $x$. Since this probability is $x^n$, then

(a) $$\text{Prob}(A) = 1 - x^n$$

An alternate approach is to observe that the event $A$ is the same as the event $D$: "at least one point is to the right of $x$." Then if $E_i$ is the event: "the $i^{\text{th}}$ point is to the right of $x$," we have

$$\text{Prob}(D) = \text{Prob}\left(\sum E_i\right) = \binom{n}{1}(1 - x) - \binom{n}{2}(1 - x)^2 + \cdots$$
$$= 1 - [1 - (1 - x)]^n = 1 - x^n$$

In the same manner,

(b) $$\text{Prob}(B) = 1 - (1 - y)^n$$

The event $C$ is equivalent to the event $E$: "all points are to the left of $z$" or the event $F$: "one point is to the right of $z$." Hence,

(c)  $\text{Prob}(C) = \text{Prob}(E + F) = \text{Prob}(E) + \text{Prob}(F) = z^n + n(1-z)z^{n-1}$

## H.9  Characteristic Function of a Random Variable

The *characteristic function* $\phi(t)$ of a random variable $X$ with probability density function $f(x)$ is defined by

$$\phi(t) = E(e^{itX}) = \int_{-\infty}^{\infty} e^{itx} f(x)\, dx \qquad (\text{H}.34)$$

where $i = \sqrt{-1}$. Equation (H.34) also defines $\phi(t)$ as the inverse Fourier transform of $f(x)$, but it is customary in probability theory to refer to it as the characteristic function. The characteristic function always exists; that is, the integral defining $\phi(t)$ converges for every density function $f(x)$. Furthermore, it is known that $f(x)$ is uniquely determined by its characteristic function.

The inversion formula corresponding to Eq. (H.34) follows at once from the theory of Fourier transforms, but it is interesting to derive this relation directly from a probabilistic point of view.

For this purpose, we use the classical integral

$$\int_{-\infty}^{\infty} \frac{\sin t}{t}\, dt = \pi$$

to show that

$$\frac{1}{\pi}\int_{-\infty}^{\infty} \frac{\sin at}{t}\, dt = \begin{cases} 1 & \text{for } a > 0 \\ 0 & \text{for } a = 0 \\ -1 & \text{for } a < 0 \end{cases}$$

The function $(1 - \cos at)/t$ is an odd function of $t$ so that its integral over a symmetric range vanishes. Using these facts, it is easy to establish the equation

$$\frac{1}{2} - \frac{1}{2\pi}\int_{-\infty}^{\infty} \frac{\sin at + i(1 - \cos at)}{t}\, dt = \begin{cases} 0 & \text{for } a > 0 \\ \frac{1}{2} & \text{for } a = 0 \\ 1 & \text{for } a < 0 \end{cases}$$

Next, replace $a$ by $X - x$ and introduce complex exponentials for the sine and cosine functions to obtain

$$\frac{1}{2} - \frac{i}{2\pi}\int_{-\infty}^{\infty} \frac{1 - e^{i(X-x)t}}{t}\, dt = \begin{cases} 1 & X < x \\ \frac{1}{2} & X = x \\ 0 & X > x \end{cases} \qquad (\text{H}.35)$$

For a fixed value of $x$, the left side of Eq. (H.35) is a function of $X$ which we call $Y(X)$. If $X$ is regarded as a random variable, then $Y(X)$

is also a random variable defined over the same sample space as the one associated with $X$. Hence,

$$E(Y) = \frac{1}{2} - \frac{i}{2\pi} \int_{-\infty}^{\infty} \frac{1 - E(e^{i(X-x)t})}{t} \, dt = \frac{1}{2} - \frac{i}{2\pi} \int_{-\infty}^{\infty} \frac{1 - e^{-ixt}\phi(t)}{t} \, dt$$

using Eq. (H.34) to obtain the second form.

Now, the random variable $Y$ takes on the discrete values 0, $\frac{1}{2}$, and 1 so that

$$E(Y) = 1 \cdot \text{Prob}(X < x) + \tfrac{1}{2} \cdot \text{Prob}(X = x) + 0 \cdot \text{Prob}(X > x)$$
$$= \text{Prob}(X < x)$$

(The probability is zero that $X$ will exactly equal $x$ if $x$ is a point of continuity of the distribution function of $X$.) But the distribution function is by definition $F(x) \equiv \text{Prob}(X < x)$ which we have shown to be the same as $E(Y)$. Therefore, by differentiating $F(x) \equiv E(Y)$ with respect to $x$, we obtain the desired inversion formula

$$f(x) = \frac{1}{2\pi} \int_{-\infty}^{\infty} e^{-ixt} \phi(t) \, dt \tag{H.36}$$

For discrete distributions we apply these results in a formal manner using the delta-function concept introduced in Sect. H.5. Thus, the characteristic function of $\delta(x, \epsilon)$

$$\phi_\epsilon(t) = \int_{-\epsilon}^{\epsilon} \frac{1}{2\epsilon} e^{itx} \, dx = \frac{\sin \epsilon t}{\epsilon t}$$

is found directly using Eq. (H.34). Hence,

$$\lim_{\epsilon \to 0} \phi_\epsilon(t) = 1$$

so that the characteristic function of $\delta(x)$ is one. Similarly, we find the characteristic function of $\delta(x - x_1)$ to be $e^{itx_1}$. Therefore, if we wish to use the inversion formula (H.36), at least in a formal way, we must have

$$\delta(x - x_1) = \frac{1}{2\pi} \int_{-\infty}^{\infty} e^{-i(x-x_1)t} \, dt \tag{H.37}$$

Consider, for example, the single toss of a coin with $X$ defined as

$$X = \begin{cases} 1 & p \\ 0 & q \end{cases}$$

The characteristic function is

$$\phi(t) = E(e^{itX}) = \exp(it \cdot 1) \cdot p + \exp(it \cdot 0) \cdot q = e^{it}p + q$$

and, from the inversion formula (H.36), the density function is

$$f(x) = p\,\delta(x - 1) + q\,\delta(x)$$

One of the most important uses of characteristic functions follows from the inversion formula (H.36). In many problems when it is required to find the density function of a certain random variable, it is easier to compute the characteristic function first and from this find the density function.

In particular, the importance of the characteristic function is evident when we consider the sum of two *independent* random variables. Suppose that $Z = X + Y$; then the characteristic function of $Z$ is

$$\phi_z(t) = E(e^{itZ}) = E(e^{it(X+Y)}) = E(e^{itX}e^{itY})$$
$$= E(e^{itX})E(e^{itY}) \quad \text{(since } X \text{ and } Y \text{ are independent)}$$
$$= \phi_x(t)\phi_y(t)$$

In general,

- The characteristic function of the sum of independent random variables is equal to the product of the characteristic functions of the individual variables.

Continuing the coin tossing example, let $X$ and $Y$ be the random variables representing the outcomes of two successive tosses of a coin. Each has the same characteristic function so that the characteristic function of the sum is the product of the characteristic functions:

$$\phi_z(t) = (e^{it}p + q)^2 = e^{i2t}p^2 + e^{it}2pq + q^2$$

from which, using Eq. (H.36),

$$f(z) = p^2\,\delta(x-2) + 2pq\,\delta(x-1) + q^2\,\delta(x)$$

## H.10 The Binomial Distribution

We are now in a position to generalize the coin-tossing experiment to include an arbitrary number $n$ of tosses so that there are $2^n$ points in the sample space. Let the random variable $X$ be the total number of heads (also called "successes") in one performance of the experiment. It is the distribution function for $X$, called the *binomial distribution function*, which we shall derive.

One aspect of this problem has already been considered—finding the average number of successes in $n$ trials rather than the actual distribution of these successes. The latter problem in most practical instances is a rather formidable one and usually we must be satisfied with a computation of a few of the various statistical parameters such as the mean or standard deviation. However, for this simple problem the distribution function is relatively easy to determine. We carry through the computation in both a direct and indirect manner, the latter illustrating the use of characteristic functions.

Probability Theory and Applications

The random variable $X$ can assume the values 0, 1, 2, ..., $n$. The probability that $X = k$ is precisely the probability of $k$ successes and $n - k$ failures in $n$ trials. Now, since the trials are independent and the probability of success in each is the same, the probability of $k$ successes and $n - k$ failures occurring in a particularly prescribed order is $p^k q^{n-k}$. The number of distinct ways of obtaining precisely $k$ successes is $\binom{n}{k}$, and therefore the probability of $k$ successes and $n - k$ failures is $\binom{n}{k} p^k q^{n-k}$. Then, from Eq. (H.17),

$$f(x) = \sum_{k=0}^{n} \binom{n}{k} p^k q^{n-k} \delta(x - k) \qquad (\text{H.38})$$

so that the distribution function is

$$F(x) = \sum_{k=0}^{[x]} \binom{n}{k} p^k q^{n-k} \qquad (\text{H.39})$$

where the symbol $[x]$ denotes the greatest integer less than or equal to $x$.

The same result can be obtained in a more routine fashion using characteristic functions. As we saw in the last section, the characteristic function of the random variable $X_k$, denoting success or failure on the $k^{\text{th}}$ toss, is

$$\phi_k(t) = pe^{it} + q$$

Since the $X_k$'s are mutually independent random variables, the characteristic function of their sum is

$$\phi(t) = E[e^{itX}] = \prod_{k=1}^{n} \phi_k(t) = (pe^{it} + q)^n \qquad (\text{H.40})$$

which may be expanded using the binomial theorem to give

$$\phi(t) = \sum_{k=0}^{n} \binom{n}{k} p^k q^{n-k} e^{itk} \qquad (\text{H.41})$$

Now apply the inversion formula (H.36) and the result is again Eq. (H.38). We can also write the random variable $X$ in the form

$$X = \begin{cases} 0 & q^n \\ 1 & npq^{n-1} \\ 2 & \frac{1}{2}n(n-1)p^2 q^{n-2} \\ \vdots & \vdots \\ k & \binom{n}{k} p^k q^{n-k} \\ \vdots & \vdots \\ n & p^n \end{cases}$$

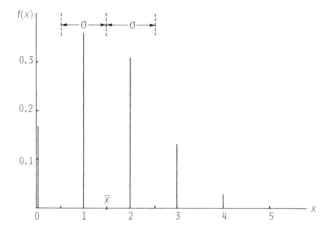

**Fig. H.3:** The binomial probability density function.

and derive

$$E(X) = np \qquad E(X^2) = n^2p^2 + npq \qquad \text{Var}(X) = npq$$

For a specific example of the *binomial distribution*, we assign the values $p = \frac{3}{10}$, $q = \frac{7}{10}$, $n = 5$ and plot the density function of $X$

$$X = \begin{cases} 0 & 0.16807 \\ 1 & 0.36015 \\ 2 & 0.30870 \\ 3 & 0.13230 \\ 4 & 0.02835 \\ 5 & 0.00243 \end{cases}$$

in Fig. H.3. Here, the mean and variance are

$$E(X) = \overline{X} = np = 1.5 \quad \text{and} \quad \text{Var}(X) = \sigma^2 = npq = 1.05$$

A convenient approximation to the binomial distribution may be had when we consider a certain limiting case. Specifically, suppose that we let $p \to 0$ and $n \to \infty$ but maintain the average number of successes $\lambda = np$ as constant. Then for $k = 0$ we have

$$\lim_{\substack{p \to 0 \\ n \to \infty}} \text{Prob}(X = 0) = \lim_{n \to \infty} \left(1 - \frac{\lambda}{n}\right)^n = e^{-\lambda}$$

and for $k = 1$

$$\lim_{\substack{p \to 0 \\ n \to \infty}} \text{Prob}(X = 1) = \lim_{n \to \infty} \frac{n\left(\frac{\lambda}{n}\right)}{1 - \frac{\lambda}{n}} \left(1 - \frac{\lambda}{n}\right)^n = \lambda e^{-\lambda}$$

In general, we can show that

$$\lim_{\substack{p \to 0 \\ n \to \infty}} \text{Prob}(X = k) = \frac{\lambda^k}{k!} e^{-\lambda}$$

In summary, the binomial distribution can be approximated by

$$X = \begin{cases} 0 & e^{-\lambda} \\ 1 & \lambda e^{-\lambda} \\ \vdots & \vdots \\ k & \dfrac{\lambda^k}{k!} e^{-\lambda} \end{cases}$$

for large values of $n$ and small values of $p$. This limiting form is called the *Poisson distribution*.†

## H.11 The Poisson Distribution

The probability that an event will occur exactly $k$ times when it is known to occur $\lambda$ times on the average is governed by the Poisson distribution. If the random variable $X$ has this distribution, its density function is

$$f(x) = \sum_{k=0}^{\infty} \frac{\lambda^k}{k!} e^{-\lambda} \delta(x - k) \tag{H.42}$$

In the preceding section we used the Poisson expression merely as a convenient approximation to the binomial distribution in the case of large $n$ and small $p$. However, it should be remarked that the Poisson distribution, as well as the binomial distribution and the normal distribution (to be discussed shortly), occur in a surprisingly large variety of problems.

Since

$$\sum_{k=0}^{\infty} \text{Prob}(x = k) = e^{-\lambda} \sum_{k=0}^{\infty} \frac{\lambda^k}{k!} = e^{-\lambda} e^{\lambda} = 1$$

it should be possible to conceive of an experiment for which $\lambda^k e^{-\lambda}/k!$ would be the probability of exactly $k$ successes. With $X$ denoting the random variable, then

$$E(X) = \sum_{k=0}^{\infty} k \frac{\lambda^k}{k!} e^{-\lambda} = \lambda \sum_{k=1}^{\infty} \frac{\lambda^{k-1}}{(k-1)!} e^{-\lambda} = \lambda e^{\lambda} e^{-\lambda} = \lambda$$

as, of course, it should be. The mean-squared value is

$$E(X^2) = \sum_{k=0}^{\infty} k^2 \frac{\lambda^k}{k!} e^{-\lambda} = \lambda^2 + \lambda$$

---

† Siméon-Denis Poisson's book *Recherches sur la probabilité des jugements en matière criminelle et en matière civile, précédées des règles générales du calcul des probabilités* was published in 1837.

so that
$$\text{Var}(X) = E(X^2) - E(X)^2 = \lambda$$
(Curiously, the mean and the variance are the same.) Finally, the characteristic function of the Poisson distribution is
$$\phi(t) = E(e^{itX}) = \sum_{k=0}^{\infty} e^{itk} \frac{\lambda^k}{k!} e^{-\lambda} = \exp(\lambda e^{it} - \lambda)$$

For an example using the Poisson distribution, consider the problem of a hardware store owner who sells boxes each containing 100 screws. Extra screws are inserted in the boxes to account for the possibility that some may be defective. Experience has shown that in the manufacturing process the probability that a screw will be defective is $p = 0.015$.

Let the random variable $X$ be the number of defective screws in a box. If it is desired to keep this smaller than some number $w$, then what is the probability that the number of defective screws will not exceed $w$? For customer satisfaction we will put $100 + w$ screws in the box and, hopefully, $w$ need not be very large.

In essence, we must calculate
$$\text{Prob}(X \leq w) = \text{Prob}(X = 0 \text{ or } X = 1 \text{ or } \cdots \text{ or } X = w)$$
$$= \text{Prob}(X = 0) + \text{Prob}(X = 1) + \cdots + \text{Prob}(X = w)$$
If $X$ has a Poisson distribution with frequency
$$\lambda = (100 + w)p \approx 1.5$$
then
$$\text{Prob}(X \leq w) = e^{-1.5} \left(1 + \frac{1.5}{1!} + \frac{(1.5)^2}{2!} + \cdots + \frac{(1.5)^w}{w!}\right)$$
$$= \begin{cases} 0.8088 & \text{for } w = 2 \\ 0.9344 & \text{for } w = 3 \\ 0.9814 & \text{for } w = 4 \end{cases}$$
Therefore, if the store owner wishes the customer to have a box of at least 100 good screws better than 98% of the time, he must include four extra screws in each box.

## H.12 Example of the Central Limit Theorem

Let $X_1, X_2, \ldots, X_n$ be $n$ independent random variables each having a Poisson distribution with a mean and variance of $\lambda$. The characteristic function of the random variable
$$X = X_1 + X_2 + \cdots + X_n$$
is just the product of the individual characteristic functions. Thus,
$$\phi_x(t) = [\exp(\lambda e^{it} - \lambda)]^n = \exp(n\lambda e^{it} - n\lambda)$$

Probability Theory and Applications 731

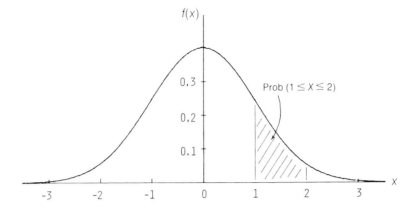

**Fig. H.4:** The normal probability density function.

which demonstrates that $X$ has also a Poisson distribution but with a mean and variance of $n\lambda$.

Define the random variable

$$Y = \frac{X - n\lambda}{\sqrt{n\lambda}}$$

which will have a Poisson distribution with zero mean and unit variance. Then the characteristic function of $Y$ is

$$\phi_y(t) = E(e^{itY}) = \sum_{k=0}^{\infty} e^{it(k-n\lambda)/\sqrt{n\lambda}} \frac{(n\lambda)^k}{k!} e^{-n\lambda}$$

$$= \exp(-it\sqrt{n\lambda} - n\lambda + n\lambda e^{it/\sqrt{n\lambda}})$$

$$= \exp(-\tfrac{1}{2}t^2 - it^3\sqrt{n\lambda} + \cdots)$$

so that

$$\lim_{n \to \infty} \phi_y(t) = e^{-\frac{1}{2}t^2}$$

To find the probability density function from this limiting form of the characteristic function, we can use the inversion formula (H.36) to obtain

$$f(x) = \frac{1}{2\pi} \int_{-\infty}^{\infty} e^{-ixt} \phi(t)\, dt = \frac{1}{2\pi} e^{-\frac{1}{2}x^2} \int_{-\infty}^{\infty} e^{-\frac{1}{2}(t-ix)^2}\, dt$$

The value of this last integral is $\sqrt{2\pi}$. (See the note in Prob. H–24.) Therefore,

$$f(x) = \frac{1}{\sqrt{2\pi}} e^{-\frac{1}{2}x^2} \qquad (\text{H.43})$$

which is the probability density function of a random variable having a *normal distribution* with zero mean and unit variance shown in Fig. H.4.

A similar result is obtained in many practical applications when we are dealing with a large number of steps each of which contributes only a small amount to the outcome of an experiment. This statement is well substantiated by experience. Indeed, it is a fact that many distributions which are encountered in the physical world are either normal or approximately normal. This remarkable state of affairs has some basis mathematically in the so called *central-limit theorem*.

◇ **Problem H–21**

Let $X_1$, $X_2$, ..., $X_n$ be mutually independent random variables, each possessing a *Cauchy distribution* whose frequency function is

$$f(x) = \frac{1}{\pi(1+x^2)}$$

By means of characteristic functions, compute the frequency function for the random variable $S_n = X_1 + X_2 + \cdots + X_n$, and thus show that the random variable

$$Y = \frac{1}{n}(X_1 + X_2 + \cdots + X_n)$$

is independent of $n$.

NOTE: The application of the central-limit theorem is not valid here since the moments of $X_k$ do not exist.

## H.13  The Gaussian Probability Density Function

The normal distribution is also called the *Gaussian distribution*. The more general form of the normal or Gaussian density function is

$$f(x) = \frac{1}{\sqrt{2\pi}\,\sigma} e^{-\frac{1}{2}(x-m)^2/\sigma^2} \qquad (H.44)$$

and, again referencing the note in Prob. H–23, we have

$$\int_{-\infty}^{\infty} f(x)\,dx = 1$$

Furthermore, if a random variable $X$ is normally distributed, then

$$E(X) = \int_{-\infty}^{\infty} x f(x)\,dx = m$$

$$E\left[X^2 - E^2(X)\right] = \int_{-\infty}^{\infty} (x-m)^2 f(x)\,dx = \sigma^2$$

so that in Eq. (H.44), $m$ is the mean and $\sigma^2$ is the variance. Graphs of the density function for several values of the standard deviation are shown in Fig. H.5.

Just as the Poisson distribution was approximated using the normal distribution, so also can the binomial distribution be so approximated. In the Poisson approximation, $\lambda = np$ is constant so that as $n$ grows,

$p$ tends to zero. However, for the normal approximation to the binomial distribution, as $n$ grows so also does $np$.

Specifically, if $X$ is a random variable having the binomial distribution

$$f(x) = \sum_{k=0}^{n} \binom{n}{k} p^k (1-p)^k \delta(x-k)$$

then the random variable

$$Y = \frac{X - np}{\sqrt{np(1-p)}}$$

will also have a binomial distribution with zero mean and unit variance. If $g(y)$ is the probability density function of $Y$, it can be shown that†

$$\lim_{n \to \infty} g(y) = \frac{1}{\sqrt{2\pi}} e^{-\frac{1}{2} y^2}$$

just as for the Poisson distributed variables. Also, for large $n$

$$\text{Prob}(X = k) = \binom{n}{k} p^k (1-p)^{n-k} \approx \frac{1}{\sqrt{2\pi}\,\sigma} e^{-\frac{1}{2}(k-m)^2/\sigma^2} \qquad \text{(H.45)}$$

with

$$\text{Mean: } m = np$$

and

$$\text{Variance: } \sigma^2 = np(1-p)$$

---

† The most memorable discovery by Abraham De Moivre (1667–1754) is his approximation to the binomial probability distribution by the normal distribution. To this end, he first developed the approximation

$$n! \approx c n^n e^{-n} \sqrt{n}$$

now called *Stirling's formula* after James Stirling who discovered that $c = \sqrt{2\pi}$ and used it to sum the terms of the distribution. (Stirling was referenced earlier in Sect. H.3.) Here, indeed, was the first occurrence of the normal probability integral—the *Gaussian distribution*. It was later that Pierre-Simon de Laplace and Carl Friedrich Gauss gave the formula in its modern form.

De Moivre was born and educated in France but emigrated to England in 1686 where he took up a lifelong but unprofitable occupation as a tutor in mathematics. Edmond Halley became his mentor and Isaac Newton, his friend. Indeed, he dedicated his masterpiece *The Doctrine of Chances* to Newton—a Latin version of which appeared in *Philosophical Transactions of the Royal Society* in 1711. The only earlier *published* treatises were the ones by Huygens and Montmort. James Bernoulli's *Ars Conjectandi* had been written but not published.

Considering his many fundamental contributions to probability, it is somewhat ironic that he is best remembered for *De Moivre's theorem*:

$$(\cos \phi + i \sin \phi)^n = \cos n\phi + i \sin n\phi$$

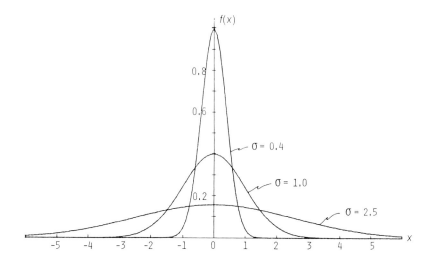

**Fig. H.5:** Examples of the Gaussian probability density function.

As an example, consider the binomial probability density function with $p = \frac{3}{10}$ and $n = 5$ which is plotted in Fig. H.3. In Fig. H.6 we have, for comparison, overlaid this density function and the normal density function. Thus, the approximation is quite good even for moderate values of $p$ and $n$.

Many applications are facilitated through the formula

$$\text{Prob}(a \leq X \leq b) = \sum_{k=a}^{b} \binom{n}{k} p^k (1-p)^k$$

$$\approx \frac{1}{\sqrt{2\pi}} \int_{(a-\frac{1}{2}-m)/\sigma}^{(b+\frac{1}{2}-m)/\sigma} e^{-\frac{1}{2}x^2} \, dx \qquad \text{(H.46)}$$

For illustration, consider the following problem:

In 200 tosses of a true coin ($p = \frac{1}{2}$), what is the probability that the number of heads deviates from 100 by at most 5?

In this case, $n = 200$, $m = 100$, $\sigma = \sqrt{50}$, $a = 95$, and $b = 105$ so that

$$\frac{1}{\sqrt{2\pi}} \int_{-5.5/\sqrt{50}}^{5.5/\sqrt{50}} e^{-\frac{1}{2}x^2} \, dx = 0.56331$$

Therefore, most of the time ($\approx 56\%$) we can expect the number of heads to fall within this range.

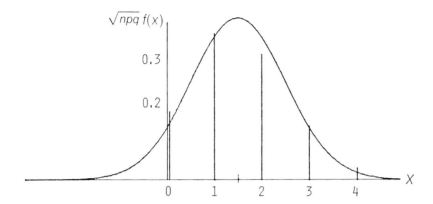

**Fig. H.6:** Normal approximation to the binomial distribution.

## H.14 The Law of Large Numbers

Our intuitive notion of probability rests on a basic assumption: If in $n$ identical repetitions of an experiment, e.g., tossing a coin, the event $E$ occurs $n_E$ times, then $n_E/n$ should differ very little from the probability $p$ associated with the event $E$. Fortunately, we can translate this vague remark into a more precise statement.

Let $X$ be a random variable having a binomial distribution with mean $m = np$ and variance $\sigma^2 = npq$. Then

$$\text{Prob}(m - \epsilon n \leq X \leq m + \epsilon n) = \text{Prob}(-\epsilon n \leq X - np \leq \epsilon n)$$
$$= \text{Prob}\left(-\epsilon \leq \frac{X}{n} - p \leq \epsilon\right)$$
$$= \text{Prob}\left(\left|\frac{X}{n} - p\right| \leq \epsilon\right)$$

Using the normal approximation, we obtain

$$\text{Prob}(m - \epsilon n \leq X \leq m + \epsilon n) \approx \frac{1}{\sqrt{2\pi}\,\sigma} \int_{m-\epsilon n}^{m+\epsilon n} e^{-\frac{1}{2}(x-m)^2/\sigma^2}\, dx$$
$$\approx \frac{1}{\sqrt{2\pi}} \int_{-\epsilon n/\sigma}^{\epsilon n/\sigma} e^{-\frac{1}{2}x^2}\, dx$$

Therefore,

$$\text{Prob}\left(\left|\frac{X}{n} - p\right| \leq \epsilon\right) \approx \frac{1}{\sqrt{2\pi}} \int_{-\epsilon\sqrt{n/pq}}^{\epsilon\sqrt{n/pq}} e^{-\frac{1}{2}x^2}\, dx \qquad \text{(H.47)}$$

and as $n$ increases the right side of (H.47) approaches one. Hence,

$$\lim_{n\to\infty} \text{Prob}\left(\left|\frac{X}{n} - p\right| \le \epsilon\right) = 1 \quad \text{(H.48)}$$

This is a form of the *law of large numbers*—as $n$ increases, the probability that the average number of successes deviates from $p$ by more than any preassigned $\epsilon$ tends to zero.

An interesting application of these ideas can be had in the context of the following problem:

> An unknown fraction $p$ of a particular population are smokers. Let $X$ be a random variable whose value is the number of smokers observed in a sample size of $n$. The problem is to find the value that $n$ must have to be assured that
>
> $$\text{Prob}\left(\left|\frac{X}{n} - p\right| \le 0.005\right) \ge 0.95$$

The number 0.95 is referred to as the *confidence level*.

First we consult a table of values of the normal distribution function

$$F(x) = \frac{1}{\sqrt{2\pi}} \int_{-\infty}^{x} e^{-\frac{1}{2}u^2} \, du \quad \text{(H.49)}$$

to determine that

$$\frac{1}{\sqrt{2\pi}} \int_{-1.96}^{1.96} e^{-\frac{1}{2}x^2} \, dx \approx 0.9750021$$

By comparison with Eq. (H.47), we must have

$$0.005 \sqrt{\frac{n}{pq}} \ge 1.96$$

or

$$n \ge (392)^2 pq$$

But, of course, $pq \le \frac{1}{4}$ so that

$$n \ge \frac{1}{4} \cdot 153664 = 38416$$

or

$$n \approx 40{,}000$$

Thus, using a sample size of 40,000 people, we can say, with a confidence level of 0.95, that the observed fraction of smokers will differ from the true fraction $p$ by no more than five parts out of a thousand.

## Problem H-22

The normal distribution function $F(x)$, which is defined in Eq. (H.49), can be expressed in terms of confluent hypergeometric functions as

$$F(x) = \frac{1}{2} + \frac{1}{\sqrt{2\pi}} x M(\tfrac{1}{2}, \tfrac{3}{2}, -\tfrac{1}{2}x^2) = \frac{1}{2} + \frac{x}{\sqrt{2\pi}} e^{-\frac{1}{2}x^2} M(1, \tfrac{3}{2}, \tfrac{1}{2}x^2)$$

the second form of which leads to the following continued fraction expansions valid for positive values of $x$

$$F(x) = 1 - \frac{1}{\sqrt{2\pi}} e^{-\frac{1}{2}x^2} \cfrac{1}{x + \cfrac{1}{x + \cfrac{2}{x + \cfrac{3}{x + \cfrac{4}{x + \cdots}}}}}$$

and

$$F(x) = \frac{1}{2} + \frac{1}{\sqrt{2\pi}} e^{-\frac{1}{2}x^2} \cfrac{x}{1 - \cfrac{x^2}{3 + \cfrac{2x^2}{5 - \cfrac{3x^2}{7 + \cfrac{4x^2}{9 - \cdots}}}}}$$

## H.15 The Chi-square Distribution

Let $X$ be a random variable having a normal distribution with zero mean and unit standard deviation. Then

$$\text{Prob}(X \le x) = F(x) \quad \text{and} \quad \frac{d}{dx} F(x) = f(x) = \frac{1}{\sqrt{2\pi}} e^{-\frac{1}{2}x^2}$$

Define another random variable $Y = X^2$ so that, for $y > 0$,

$$G(y) = \text{Prob}(Y \le y) = \text{Prob}(-\sqrt{y} \le X \le \sqrt{y})$$
$$= F(\sqrt{y}) - F(-\sqrt{y})$$

Hence,

$$\frac{d}{dy} G(y) = g(y) = \frac{1}{2\sqrt{y}} [F'(\sqrt{y}) + F'(-\sqrt{y})]$$
$$= \frac{1}{\sqrt{2\pi y}} e^{-\frac{1}{2}y}$$

The random variable whose density function is

$$f_1(x) = \begin{cases} \dfrac{1}{\sqrt{2\pi x}} e^{-\frac{1}{2}x} & x > 0 \\ 0 & x \le 0 \end{cases} \tag{H.50}$$

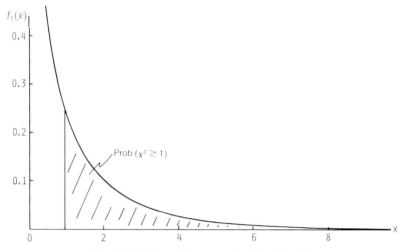

**Fig. H.7:** The chi-square density function.

is called *chi-square* and written $\chi^2$. A graph of this density function is shown in Fig. H.7 and the characteristic function of $\chi^2$ is

$$\phi_1(t) = \int_0^\infty e^{itx} f_1(x)\,dx = (1 - 2it)^{-\frac{1}{2}} \tag{H.51}$$

The *chi-square test* is used by statisticians to check the validity of the results of an experiment. For example, suppose that a coin is tossed $n = 4040$ times with the result $a = 2048$ heads and $b = 1992$ tails. Are these data consistent with the hypothesis that $p = q = \frac{1}{2}$? In other words: Is this a true coin?

For this set of data

$$X = \frac{a - np}{\sqrt{npq}} = \frac{2048 - 2020}{\sqrt{1010}}$$

so that the random variable $\chi^2$ has the value

$$\chi^2 = X^2 = \frac{(a-np)^2}{npq} = \frac{28^2}{1010} = 0.776$$

and

$$\text{Prob}(\chi^2 \geq 0.776) = \int_{0.776}^\infty f_1(x)\,dx = 0.38$$

This means that there is a probability of 38% of obtaining a deviation from the expected result at least as great as that actually observed. The test is not, of course, conclusive. We can never really know if the coin is true.

We can extend the definition of the chi-square distribution to include the sum of squares of normally distributed random variables. Let $X_1$, $X_2$,

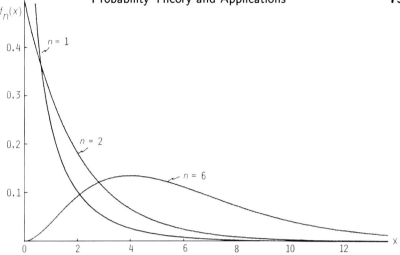

**Fig. H.8:** Chi-square density function with various degrees of freedom.

..., $X_n$ be $n$ independent and normal random variables with zero mean and unit standard deviation. Then, the $\chi^2$ random variable

$$\chi^2 = X_1^2 + X_2^2 + \cdots + X_n^2$$

is said to have $n$ degrees of freedom and its characteristic function, from Eq. (H.51), is

$$\phi_n(t) = (1 - 2it)^{-\frac{1}{2}n}$$

Therefore, the probability density function is

$$f_n(x) = \frac{1}{2\pi} \int_{-\infty}^{\infty} e^{-ixt}(1 - 2it)^{-\frac{1}{2}n}\, dt$$

Carrying out the integration results in

$$f_n(x) = \begin{cases} \dfrac{1}{2^{\frac{1}{2}n}\Gamma(\frac{1}{2}n)} x^{\frac{1}{2}n-1} e^{-\frac{1}{2}x} & x > 0 \\ 0 & x \leq 0 \end{cases} \quad \text{(H.52)}$$

as the density function of the *chi-square distribution with n degrees of freedom*. The function $\Gamma(\frac{1}{2}n)$ is Euler's Gamma function. In this connection it is useful to know that $\Gamma(\frac{1}{2}) = \sqrt{\pi}$ as developed in the next problem.

Finally, we can show that the chi-square distribution with $n$ degrees of freedom has a mean of $n$ and a variance of $2n$. Furthermore, the associated density function $f_n(x)$ tends to the normal density function as $n \to \infty$. This is another example of the central-limit theorem. For illustration we have plotted the density functions for the chi-square distributions of one, two, and six degrees of freedom in Fig. H.8.

## ◊ Problem H–23

For positive values of $m$ and $n$ the integral

$$B(m,n) = \int_0^1 x^{m-1}(1-x)^{n-1}\,dx = 2\int_0^{\frac{1}{2}\pi} \sin^{2m-1}\theta \cos^{2n-1}\theta\,d\theta$$

defines the *Beta function*. It was first investigated by John Wallis; however, because of the extensive work by Leonhard Euler, Legendre called it the *first Eulerian integral*. The *second Eulerian integral*†

$$\Gamma(n+1) = \int_0^\infty x^n e^{-x}\,dx = 2\int_0^\infty x^{2n-1} e^{-x^2}\,dx$$

was later named the Gamma function by Legendre.

(a) Derive the relation between the Beta and Gamma functions

$$B(m,n) = \frac{\Gamma(m)\Gamma(n)}{\Gamma(m+n)}$$

which was discovered by Euler in 1771.

HINT: Use the second form of the Gamma function above and change to polar coordinates. Then

$$\Gamma(n)\Gamma(m) = 2\int_0^\infty r^{2(n+m)-1} e^{-r^2}\,dr \times 2\int_0^\infty \sin^{2m-1}\theta \cos^{2n-1}\theta\,d\theta$$

(b) Use Euler's relation to obtain

$$\Gamma(\tfrac{1}{2}) = \sqrt{\pi}$$

(c) The integrals

$$W_n = \int_0^{\frac{1}{2}\pi} \cos^n\theta\,d\theta = \int_0^{\frac{1}{2}\pi} \sin^n\theta\,d\theta$$

are called *Wallis' integrals*. Show that

$$W_n = \tfrac{1}{2} B[\tfrac{1}{2},\tfrac{1}{2}(n+1)] = \frac{\sqrt{\pi}\,\Gamma[\tfrac{1}{2}(n+1)]}{2\Gamma(\tfrac{1}{2}n+1)}$$

---

† The "interpolation problem" posed to Euler by Christian Goldbach (1690–1764) was to give meaning to $n!$ for nonintegral values of $n$. Euler announced his solution in a letter to Goldbach on October 13, 1729.

Euler gave the solution in several forms

$$n! = \lim_{m\to\infty} \frac{m!\,m^{n+1}}{(n+1)(n+2)\ldots(n+m+1)} = \int_0^1 (-\log x)^n\,dx = \int_0^\infty x^n e^{-x}\,dx$$

Another form he obtained is the infinite product for the Gamma function

$$\frac{1}{\Gamma(x)} = x e^{\gamma x} \prod_{k=1}^\infty \left(1+\frac{x}{k}\right) e^{-x/k} \quad \text{where} \quad \gamma = \lim_{m\to\infty}(1+\tfrac{1}{2}+\tfrac{1}{3}+\tfrac{1}{4}+\cdots+\tfrac{1}{m}-\log m)$$

The quantity $\gamma = 0.5772156649\ldots$ is known as *Euler's constant*.

Hence,
$$W_{2n+1} = \frac{2 \cdot 4 \cdots (2n-2) \cdot 2n}{3 \cdot 5 \cdots (2n-1) \cdot (2n+1)}$$
$$W_{2n} = \frac{1 \cdot 3 \cdots (2n-3) \cdot (2n-1)}{2 \cdot 4 \cdots (2n-2) \cdot 2n} \times \frac{\pi}{2}$$

(d) Derive Wallis' infinite product representation of $\pi$ as given in a footnote in Sect. 1.1.

HINT: First establish
$$\frac{\pi}{2} > \frac{\pi}{2} \times \frac{W_{2n+1}}{W_{2n}} > \frac{\pi}{2} \times \frac{W_{2n+2}}{W_{2n}}$$

Then show that
$$\lim_{n \to \infty} \frac{W_{2n+2}}{W_{2n}} = \lim_{n \to \infty} \frac{2n+1}{2n+2} = 1$$

Wallis' product follows as the limit of
$$\frac{\pi}{2} \times \frac{W_{2n+1}}{W_{2n}} = \frac{2 \cdot 2 \cdot 4 \cdot 4 \cdots (2n-2) \cdot (2n-2) \cdot 2n \cdot 2n}{1 \cdot 3 \cdot 3 \cdot 5 \cdots (2n-3) \cdot (2n-1) \cdot (2n-1) \cdot (2n+1)}$$

The Rayleigh distribution† is derived from the chi-square distribution with two degrees of freedom. Let $X$ and $Y$ be independent and normally distributed random variables with zero mean and a variance of $\sigma^2$. Then $X/\sigma$ and $Y/\sigma$ are normal with unit variance and the random variable
$$\chi^2 = \frac{X^2}{\sigma^2} + \frac{Y^2}{\sigma^2}$$
has a chi-square distribution with two degrees of freedom.

Now define a new random variable
$$R = \sigma\sqrt{\chi^2} = \sqrt{X^2 + Y^2}$$
whose distribution function will be
$$F(r) = \text{Prob}(R \leq r) = \text{Prob}\left(\chi^2 \leq \frac{r^2}{\sigma^2}\right)$$

The density function is derived as follows:
$$f(r) = \frac{d}{dr}\text{Prob}(R \leq r) = \frac{d}{dr}\text{Prob}\left(\chi^2 \leq \frac{r^2}{\sigma^2}\right) = \frac{2r}{\sigma^2} f_2\left(\frac{r^2}{\sigma^2}\right)$$

The random variable $R$ has a *Rayleigh distribution* with
$$f(r) = \begin{cases} \dfrac{r}{\sigma^2} e^{-\frac{1}{2}r^2/\sigma^2} & r > 0 \\ 0 & r \leq 0 \end{cases} \tag{H.53}$$

---

† Named for John William Strutt (1842–1919), the third Baron Rayleigh. He is best known as Lord Rayleigh and was England's foremost mathematician and physicist during the last half of the nineteenth century.

742                                    Appendix H

so that

$$\text{Prob}(0 \le R \le r_0) = \int_0^{r_0} \frac{r}{\sigma^2} e^{-\frac{1}{2}r^2/\sigma^2}\, dr = 1 - e^{-\frac{1}{2}r_0^2/\sigma^2} \tag{H.54}$$

Examples, of the Rayleigh density function for various values of the standard deviation are plotted in Fig. H.9.

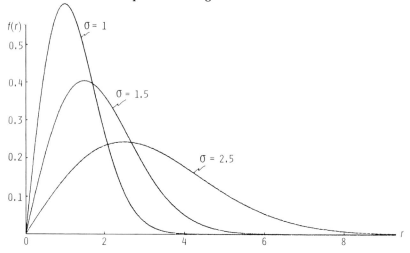

**Fig. H.9:** Rayleigh density functions.

◇ **Problem H–24**

A gun is fired at a target. Taking the origin of coordinates as the point of aim, it is known that due to dispersion effects the $x$ and $y$ coordinates of the hit are independent and each may be specified in a probabilistic sense by the same frequency function $f$ where

$$f(x) = \frac{1}{\sqrt{2\pi}\,\sigma} e^{-\frac{1}{2}x^2/\sigma^2}$$

Show that the probability of a point of hit lying within a circle of radius $R$ centered at the origin is

$$1 - e^{-\frac{1}{2}R^2/\sigma^2}$$

HINT: The probability in question is

$$\iint\limits_{\text{Circle}} f(x)f(y)\,dx\,dy = \frac{1}{2\pi\sigma^2} \int_0^{2\pi} \int_0^R e^{-\frac{1}{2}r^2/\sigma^2} r\, dr\, d\theta$$

NOTE: This is an easy way of establishing the result

$$\frac{1}{\sqrt{2\pi}\,\sigma} \int_{-\infty}^{\infty} e^{-\frac{1}{2}x^2/\sigma^2}\, dx = 1$$

## H.16 The Markov Chain

A sequence of random variables $X_1, X_2, \ldots, X_n, \ldots$ is called a *Markov chain*† if, given the value of the present variable $X_n$, the future $X_{n+1}$ is independent of the past $X_1, X_2, \ldots, X_{n-1}$.

As an example, let the value of the random variable $X_n$ be the result of the $n^{\text{th}}$ random selection of a number on the real line from zero to one. One could imagine a game in which a wheel is spun to select the number and we could define the random variable $Y_n$ to be the cumulative score after $n$ spins. Then

$$Y_n = Y_{n-1} + X_n \tag{H.55}$$

is a Markov chain—also called a *Markov sequence* or a *Markov process*. We have already seen that the probability density function of $X_n$ is

$$f(x) = \begin{cases} 1 & \text{for } 0 \leq x \leq 1 \\ 0 & \text{otherwise} \end{cases}$$

The mean and variance of the cumulative score $Y_n$ are readily obtained:

$$E(Y_n) = \sum_{k=1}^{n} E(X_k) = \tfrac{1}{2} n \qquad \text{Var}(Y_n) = \sum_{k=1}^{n} \text{var}(X_k) = \tfrac{1}{12} n$$

To find the probability density function of $Y_n$ we note that

$$\text{Prob}(a < Y_n \leq b) = \int_a^b f_n(y)\, dy = \iint_\Gamma f(x) f_{n-1}(y)\, dx\, dy$$

where $\Gamma$ is the diagonal strip: $a < X_n + Y_{n-1} \leq b$. Then

$$\int_a^b f_n(y)\, dy = \int_{-\infty}^{\infty} dx \int_{a-x}^{b-x} f(x) f_{n-1}(y)\, dy$$

$$= \int_{-\infty}^{\infty} dx \int_a^b f(x) f_{n-1}(y - x)\, dy$$

$$= \int_a^b dy \int_{-\infty}^{\infty} f(x) f_{n-1}(y - x)\, dx$$

Since this holds for any $a$ and $b$, the density function for $Y_n$ is

$$f_n(y) = \int_{-\infty}^{\infty} f(x) f_{n-1}(y - x)\, dx \tag{H.56}$$

This is a *convolution integral*. The technique has general applicability to find the density function for the sum of two random variables.

---

† In his efforts to establish general laws of probability Andrei Andreevich Markov (1856–1922), the great Russian mathematician, introduced such sequences for the first time in his 1906 paper "The Extension of the Law of Large Numbers on Mutually Dependent Variables." Markov was a student of Pafnuti L. Tschebycheff, who headed the mathematics department of the St. Petersburg University. He spent his entire career as a professor at that University in the city of Petrograd.

For the particular case at hand, Eq. (H.56) can be reduced to

$$f_n(y) = \int_{-\infty}^{\infty} f(x) f_{n-1}(y-x)\,dx = \int_{-\infty}^{\infty} f(y-x) f_{n-1}(x)\,dx$$

$$= \int_{y-1}^{y} f_{n-1}(x)\,dx \qquad (H.57)$$

Further, as another example of the central limit theorem, it can be shown that $f_n(y)$ tends to the normal density function as $n \to \infty$.

A *Gaussian random sequence* or a *Gaussian random process* is a sequence of vectors whose components are random variables having a jointly normal distribution. In the absence of contradictory evidence it is generally desirable to assume a normal or Gaussian distribution. The advantages for this are summarized below:

- The distribution function depends only on the mean and covariance.
- A linear transformation of a Gaussian process is itself Gaussian.
- A Gaussian process passed through a linear filter remains Gaussian.
- The optimum estimate of a Gaussian random process is linear.

In the space navigation problem, we model the dynamics as a Markov process. If the random variables have a Gaussian distribution, the process is termed a *Gauss-Markov process*. Thus if $\mathbf{x}_0, \mathbf{x}_1, \ldots$ is a sequence of Gaussian random vectors and if $\boldsymbol{\beta}_0, \boldsymbol{\beta}_1, \ldots$ is a sequence of *purely random* Gaussian vectors, i.e.,

$$E(\boldsymbol{\beta}_n) = \overline{\boldsymbol{\beta}_n} = \mathbf{0} \qquad E(\boldsymbol{\beta}_n \boldsymbol{\beta}_n^T) = \overline{\boldsymbol{\beta}_n \boldsymbol{\beta}_n^T} \qquad E(\boldsymbol{\beta}_i \boldsymbol{\beta}_j^T) = \mathbf{O} \quad (i \neq j)$$

then the linear system dynamics

$$\mathbf{x}_n = \boldsymbol{\Phi}_{n,n-1} \mathbf{x}_{n-1} + \boldsymbol{\beta}_n$$

comprise a Markov process. The terminology is that $\mathbf{x}_n$ is the *state vector*, $\boldsymbol{\Phi}_{n,n-1}$ is the *state transition matrix*, and $\boldsymbol{\beta}_n$ is the *plant noise* or *process noise*.

The system is *observed* by a sequence of measurements $q_1, q_2, \ldots$ with

$$q_n = \mathbf{h}_n^T \mathbf{x}_n + \alpha_n$$

where $\mathbf{h}_1, \mathbf{h}_2, \ldots$ is a sequence of *measurement vectors* and $\alpha_1, \alpha_2, \ldots$ is a sequence of purely random Gaussian variables.

With these assumptions we are guaranteed that the optimum estimate of the state vector will be linear.

◊ **Problem H-25**

Plot the density functions $f_1(x)$, $f_2(x)$, and $f_3(x)$ of the cumulative score random variable $Y_n$ considered above. Verify, for $n = 3$, the density function already has the familiar "bell-shape" of the normal distribution.

Appendix I

# Miscellaneous Problems

This appendix consists of a collection of problems on guidance and navigation which, to the author, do not seem to fit naturally in the previous chapters. Some are examination questions and some are from the author's earlier book *Astronautical Guidance*.

◊ **Problem I–1**

A spacecraft is in the plane of the ecliptic with the sun at the origin of coordinates. The earth is located on the $x$ axis at a distance of one astronomical unit from the sun and the reference coordinates of the spacecraft are

$$x = y = 1 \text{ a.u.}$$

To obtain a position fix, the distances from the sun and the earth are somehow measured. The measured distance from the sun is $\sqrt{2} \cdot 10^{-6}$ a.u. greater than expected while the measured distance from the earth is $10^{-6}$ a.u. greater than expected.

(a) Find the position deviation vector of the spacecraft from its reference position.

(b) If the measurements are assumed to be statistically independent with a standard deviation for each measurement of $\sigma = 10^{-6}$ a.u., calculate the two-dimensional covariance matrix of the position estimation errors.

(c) Calculate the rms error, i.e., the square root of the mean-squared error, in the estimate of the position deviation from the reference point.

A new estimate of position is made by adding a redundant third measurement of the angle between the earth and the sun as observed from the spacecraft.

(d) If the standard deviation of this angle measurement is $\sigma = 10^{-6}/\sqrt{2}$ radians, calculate the new covariance matrix of the position estimation errors as well as the rms error in the estimate.

(e) If the measured angle is found to be $\frac{1}{2} \cdot 10^{-6}$ radians larger than the reference angle, calculate the new estimate of the position deviation vector of the spacecraft from its reference position.

## ◇ Problem I-2

A vehicle is launched in a parabolic orbit in a constant gravitational force field. Therefore, if $\mathbf{r}(t)$ and $\mathbf{v}(t)$ are, respectively, the position and velocity vectors, we have

$$\frac{d\mathbf{r}}{dt} = \mathbf{v} \qquad \frac{d\mathbf{v}}{dt} = \mathbf{g}$$

where $\mathbf{g}$ is the constant acceleration vector.

(a) Assume that two position fixes are made at times $t_n$ and $t_{n-1}$ with the result that $\widehat{\delta \mathbf{r}}_n$ and $\widehat{\delta \mathbf{r}}_{n-1}$ are the indicated deviations in position from the reference trajectory. Show that the velocity correction $\Delta \widehat{\mathbf{v}}_n$ to be applied at time $t_n$ to carry the vehicle to the reference target point (fixed in space and time) is given by

$$\Delta \widehat{\mathbf{v}}_n = -\frac{t_f - t_{n-1}}{(t_f - t_n)(t_n - t_{n-1})} \widehat{\delta \mathbf{r}}_n + \frac{1}{t_n - t_{n-1}} \widehat{\delta \mathbf{r}}_{n-1}$$

where $t_f$ is the reference time of flight.

(b) Assume that the vehicle is injected into orbit with an initial velocity error and that only one position fix and associated velocity correction is made. If the mean-squared error in determining position is independent of the time of the fix, show that the optimum time (measured from launch) to make the fix, in order to minimize the magnitude of the resulting velocity correction, is less than $\frac{1}{2} t_f$.

## ◇ Problem I-3

A vehicle is launched in orbit in a constant gravitational force field to intercept a target after a flight time of $t_f = 10$. Position fixes are made at $t = 1$ and $t = 3$, and the following deviations from the reference trajectory are determined:

$$\delta \mathbf{r}_1 = 3\,\mathbf{i}_x + 2\,\mathbf{i}_y - 8\,\mathbf{i}_z$$
$$\delta \mathbf{r}_3 = 7\,\mathbf{i}_x - 14\,\mathbf{i}_z$$

(a) What is the velocity deviation at $t = 3$?
(b) If no correction is made, what are the position and velocity deviations at the target?
(c) What velocity correction is required at $t = 3$ to reduce to zero the position error at $t = t_f$?
(d) What is the resulting velocity deviation at $t = t_f$?
(e) What velocity correction would be required at $t = 3$ if it were desired to reduce the velocity deviation to zero at $t = t_f$ and what would be the resulting position deviation?

◊ **Problem I-4**

A vehicle is launched in a parabolic orbit in a constant gravitational force field. The position and velocity vectors $\mathbf{r}(t)$ and $\mathbf{v}(t)$ satisfy the following vector differential equations

$$\frac{d\mathbf{r}}{dt} = \mathbf{v} \qquad \frac{d\mathbf{v}}{dt} = -g\,\mathbf{i}_y$$

subject to the initial conditions

$$\mathbf{r}(0) = \mathbf{0} \qquad \mathbf{v}(0) = \tfrac{1}{2}gt_f(\mathbf{i}_x + \mathbf{i}_y)$$

Here, $\mathbf{i}_x$ and $\mathbf{i}_y$ are the orthonormal reference coordinate vectors and $t_f$ is the reference time of flight. Assume the target to be moving with a constant velocity given by

$$\mathbf{v}_p = \tfrac{1}{2}gt_f\,\mathbf{i}_x$$

(a) Assume the position and velocity correction uncertainties to be isotropic and statistically independent random variables. Show that the rms required velocity correction at any time for variable-time-of-arrival is just $\sqrt{\tfrac{2}{3}}$ of that for fixed-time-of-arrival guidance.

(b) Assuming only one position fix at time $t = t_1$ and variable-time-of-arrival guidance, show that the mean-squared change in the arrival time is given by

$$\frac{4}{3g^2}\left(\frac{\sigma_\epsilon^2}{t_1^2} + \sigma_\eta^2\right)$$

where $\sigma_\epsilon$ is the rms uncertainty in position at the time of the fix and $\sigma_\eta$ is the rms error in the launch velocity.

◊ **Problem I-5**

Consider the problem of a vehicle moving along the $x$ axis in a force free field. At time $t$ a measurement of the quantity

$$q = x + cvt$$

is made where $x$ is the distance from the origin, $v$ is the velocity, and $c$ is a positive constant at our disposal. Assume that the measurement error is a random variable with zero mean and variance $\sigma^2$. At time $t = 0$, the covariance matrix of the estimation errors of position and velocity is the identity matrix.

(a) At what time $t$ should the measurement be made to minimize the mean-squared error in the estimate of position?

(b) At what time should the measurement be made to minimize the mean-squared error in the estimate of velocity?

(c) At what time should the measurement be made to minimize the correlation between position and velocity estimation errors?

(d) Show that the mean-squared error in position for a measurement taken at any time is minimized by choosing $c = \sigma^2$. With this value of $c$ what are the answers to the first three questions?

(e) After the measurement using the optimum $c$, how do the mean-squared estimates of position and velocity and their correlations change with time?

(f) Determine the sensitivity of the mean-squared error in the velocity estimate (at any time following the measurement) to small changes in $c$ in the vicinity of the optimum value.

(g) Suppose that the measurement at time $t$ is processed (using the optimum value of $c$) under the assumption that no error in the measurement had been made. However, unknown to us, a random error with zero mean and variance $\beta^2$ did actually exist in the measurement. What are the statistics of the actual errors, including correlations, in the estimates of position and velocity?

◇ **Problem I-6**

A satellite is in orbit about a spherical earth. Measurement information consisting of sets of simultaneous range and range-rate data from the center of the earth are available from which to estimate certain orbital elements. The times of the measurements are unknown. The range measurements are error-free but the range-rate measurements have errors which are assumed to be independent normally distributed random variables each having zero mean and a standard deviation of 0.01 miles/sec.

The measurement data are as follows:

| Set Number | Range (miles) | Range-Rate (miles/sec) |
|---|---|---|
| 1 | 5000 | 0.4373328 |
| 2 | 4800 | 0.4175231 |
| 3 | 5300 | 0.3200625 |

and the gravitational constant for the earth is $\mu = 95630 \text{ miles}^3/\text{sec}^2$

(a) From the first two data sets determine an estimate for the orbital parameter $p$ and the semimajor axis $a$.

(b) Calculate the covariance matrix of estimation errors.

(c) Use the third data set to determine an improved estimate of $p$ and $a$ as well as the new covariance matrix of estimation errors.

(d) Sketch the equiprobability error ellipse for the case of two measurements and the case of three measurements.

◇ **Problem I-7**

A spacecraft is in orbit about a spherical planet whose radius is 1,000 km. Altitude above the surface of the planet can be measured with a radar altimeter. The errors in the altitude measurements are assumed to be independent normally distributed random variables each having a zero mean and a standard deviation of 100 meters. The direction of the radar beam is referenced without error to an inertially-stabilized platform which is known to be free of drift but whose attitude relative to inertial space is completely unknown. Thus, only the angular difference between two successive line of sight directions to the planet can be accurately measured.

Three altitude measurements are made

$$r_1 = 3,186,983 \text{ meters}$$
$$r_2 = 3,432,777 \text{ meters}$$
$$r_3 = 3,800,000 \text{ meters}$$

Miscellaneous Problems    749

at three points in the orbit where the angles between successive line of sight directions are exactly 15 degrees.

(a) Calculate an estimate for the parameter, the eccentricity, and the location of pericenter.

(b) From an appropriate linearized relationship calculate the covariance matrix of the estimation errors.

(c) A fourth altitude measurement is made

$$r_4 = 4,312,700 \text{ meters}$$

at a point whose line of sight is exactly 15 degrees displaced from the corresponding direction of the third measurement. Using this new measurement, calculate new estimates of the three orbital quantities and the new covariance matrix of estimation errors.

## Problem I-8

According to mission plans, a spacecraft is to make a soft landing at a specified time and location on the surface of a planet whose gravity field is so small as to negligible.

(a) Show that the optimum commanded acceleration vector, chosen to minimize the integral of the square of the magnitude of the acceleration vector, is of the form

$$\mathbf{a} = f_1(t_{go})\mathbf{r}_g + f_2(t_{go})\mathbf{v}_g$$

where $t_{go}$ is the time remaining before touchdown. The vector $\mathbf{r}_g$ is the target location minus the current position. The vector $\mathbf{v}_g$, called the velocity-to-be-gained, is the desired velocity at the target (assumed here to be zero) minus the current velocity.

Suppose the mission plan is altered so that only a hard landing at the target is required.

(b) Design a control law, using the same performance index as before, with the added requirement that the velocity-to-be-gained is always identically zero. That is, the desired velocity at impact is specified to be the same as the current velocity.

NOTE: This is *not* the same as dropping the second term in the control law of part (b).

(c) If the vehicle is flown using this control law, then by integrating the state equations, obtain an expression for the vector $\mathbf{r}_g$ of the form

$$\mathbf{r}_g = g_1(t_{go})\mathbf{c}_1 + g_2(t_{go})\mathbf{c}_2$$

where $\mathbf{c}_1$ and $\mathbf{c}_2$ are the integration constants. Further, show that the commanded acceleration vector has a fixed direction and a magnitude proportional to $t_{go}^4$.

(d) Is this control law the same as the optimum control law with no requirement on the terminal velocity? If not, how would you design an optimum control law for a hard landing?

# Problem I-9

Consider the problem of a vehicle moving in a straight line along the $x$ axis in a force-free field. At time $t = 0$ a measurement of $x$, the distance from the origin, is made. Assume that the measurement error is a random variable with zero mean and variance equal to $\sigma^2$. At time $t = 0$, prior to the measurement, the covariance matrix of the estimation errors of position and velocity is the identity matrix.

(a) Determine the error covariance matrix immediately following the measurement.

(b) A second measurement of $x$ is made at time $t$ with the measurement error statistically independent of the first measurement error and having zero mean and variance $\sigma^2$. Determine the error covariance matrix immediately following the second measurement.

(c) Assume at time $t$, when the second measurement is made, that the measurement error is composed of a random variable $\alpha$ with zero mean, variance $\sigma^2$, statistically independent of the previous measurement error and a random bias with variance $\beta^2$, statistically independent of $\alpha$. If the measurement had been processed under the assumption that no bias existed, what are the statistics of the actual errors in the estimates of position and velocity?

(d) Let $\alpha_0$ be the measurement error at $t = 0$ and $\alpha(t)$ be the measurement error at time $t$. Assume they are correlated using the following model:

$$\alpha(t) = \alpha_0 e^{-\lambda t} + \sigma(1 - e^{-2\lambda t})^{\frac{1}{2}} n(t)$$

where $n(t)$ is a white noise with a variance of one and $\sigma^2$ is the variance of $\alpha_0$. A measurement of $x$ is made at $t = 0$ and a second measurement at time $t$. Determine the mean-squared position error immediately following the second measurement.

NOTE: This question is independent of part (c), i.e., there is no bias error.

# Epilogue

*Originally appeared as Chapter 14 of the book* Theory and Application of Kalman Filtering — *a publication of the Advisory Group for Aerospace Research and Development of the North Atlantic Treaty Organization published in February 1970 as AGARDograph 139 and edited by Cornelius T. Leondes. It was prepared at the request of the Guidance and Control Panel of AGARD-NATO. The chapter was entitled* "Application of Kalman Filtering Techniques to the Apollo Program" *and was co-authored by Richard H. Battin and Gerald M. Levine.*

The first manned trip to the vicinity of the moon of Apollo 8 during December 1968 gave an excellent test of the Apollo system's onboard navigation capability. Although ground tracking navigation was the primary system, the onboard navigation system had the task of confirming a safe trajectory and providing a back-up for return to earth in the remote chance that ground assistance became unavailable for onboard use.

Apollo 8 was to use sun illuminated visual horizons rather than landmarks for operational simplicity, even though, as confirmed from earth orbit by Astronaut Donn Eisele in Apollo 7, the earth's horizon does not provide a distinct target for visual use. Moreover, the filter in the sextant beamsplitter, designed originally to enhance the contrast between water and land when looking down at the earth, filters out the blue in such a way as to make the horizon even more indistinct. Originally a blue sensitive photometer had been designed for horizon detection in the prototype sextant models, but was removed from the production systems, since a decision had been made that ground tracking would be the primary source of midcourse navigation. Without the photometer, interest in the earth's horizon as a visual target resulted in demonstrations on simulators that, in some subjective way, the human with a little experience can choose an altitude sufficiently repeatable, at least as good as ±3 kilometers. Accordingly, a few weeks before the Apollo 8 launch, the navigator command module pilot, Jim Lovell, spent a few hours on the sextant earth horizon simulator at the Massachusetts Institute of Technology in Cambridge for training and to calibrate the horizon altitude he seemed to prefer. He was remarkably consistent in choosing a location 32.8 kilometers above the sea level horizon. This value was recorded in the command module computer as part of the prelaunch mission load.

The plan for the mission was to examine the sextant angle measurements made early while still near the earth and, based on the spacecraft state vector determined by ground tracking, infer in real-time the horizon altitude Lovell was using. After the first eleven sightings on the earth at distances of about 30,000 nautical miles from earth, it was estimated that he was using an 18.2 kilometers altitude and the onboard computer was reloaded with this new value. (Later during the mission it was agreed that a truer estimate was nearer 23 kilometers, but the value was not changed since the difference then was too small to be of concern.) Following the horizon calibration, the first midcourse correction of almost 25 ft/sec was performed. The large size of this correction was due to trajectory perturbations, resulting from the maneuvers performed in getting the spacecraft safely away from the third stage of the launch vehicle.

After the midcourse correction, the command module computer's version of the state vector was made to agree with the value obtained from ground tracking. The important parameter, predicted perilune altitude, was 69.7 nautical miles—very close to the true value estimated later to be 68.8. The next 31 navigation measurements were made using the earth's horizon modeled at 18.2 kilometers altitude. Being sufficiently far from both earth and moon during this time, it is not surprising that the initially good state vector was degraded. At the end of this period, the indicated perilune was 32 nautical miles below the moon's surface. With the next nine sightings, still using the earth's horizon, the predicted perilune increased to 92.9 nautical miles—about 22 nautical miles too high. The exact altitude of the earth's horizon was unimportant for these sightings since the distance from earth was now approximately 150,000 nautical miles, so that the 10 arc-second accuracy of the sextant was the predominant source of error.

The next group of 16 sightings was made using the moon horizon at a distance of about 50,000 nautical miles. As would be expected, the first few of these resulted in fairly large changes in the estimated state vector, while the remaining had a very small effect. At the end of this group of measurements, the indicated perilune was 67.1 nautical miles. The final set of 15 translunar sightings was made about 35,000 nautical miles from the moon with little additional effect on the perilune estimation. The final estimate was 67.5 nautical miles or about 1.3 nautical miles lower than the value later reconstructed from ground tracking data. At this time the onboard and ground tracking data were practically identical and consideration was given to using the onboard state vector for lunar orbit insertion. Although the state vector update hardly changed the onboard value, it was performed since there was no overriding argument to deviate from the flight plan.

Epilogue                                                             753

The transearth flight of Apollo 8 after 10 lunar orbits also provided a good measure of the onboard navigation capability. The transearth injection maneuver of the service propulsion system was targeted by ground data and executed in back of the moon by the onboard digital autopilot and guidance systems. This 3522.5 ft/sec maneuver was followed 14.7 hours later by a single midcourse correction of 4.8 ft/sec, resulting in entry conditions at 400,000 feet altitude above the earth which were 0.8 ft/sec faster and 0.1° shallower than planned.

Although the primary navigation during this period was again the ground tracking network, 138 onboard navigation measurements were performed by Lovell as a monitor and back-up. In order to determine what would have happened without ground assistance, the actual onboard measurements were incorporated in a simulation with the computer initialized to the actual onboard state vector as it existed when the spacecraft emerged from behind the moon. The single transearth midcourse correction was added appropriately to this simulation in accordance with that actually measured by the inertial guidance system. (In the actual flight a new ground determined state vector was loaded into the computer at the time of this maneuver.)

The last of the 138 measurements was completed 16 hours before entry. The incorporation of these measurements in the simulation left a hypothetical onboard estimate of entry flight path angle at 400,000 feet of $-6.38°$ as compared with the ground tracking estimate of $-6.48°$. This $0.1°$ difference was well within the safe tolerance of $\pm 0.5°$. Another parameter of concern at the entry interface is the error in knowledge of altitude rate. The simulated onboard estimate of this quantity differed from that estimated by ground tracking by 236 ft/sec. However, the conservative allowable tolerance is $\pm 200$ ft/sec.

It should be emphasized that, in the event ground data were not available, the plan was to continue the onboard measurements to optimize the final midcourse correction and state vector for safe earth atmospheric entry. In the absence of actual flight data, a continuation of the simulation using the planned sighting program was made with standard deviation errors in the sextant of 10 arc-seconds and in the horizon of 3 kilometers. In addition, bias errors of 5 arc-seconds in the sextant and 4 kilometers in the horizon were included. The resulting estimation error in the entry angle at the entry interface had a standard deviation of $0.03°$ and a bias of $0.007°$. The corresponding altitude rate uncertainty had a standard deviation of 41.1 ft/sec and a bias of 26.5 ft/sec. The capability of the onboard navigation systems to bring the spacecraft safely back from the moon seems clearly to have been demonstrated.

## Coasting Flight Navigation

Navigating the Apollo spacecraft in coasting flight involves two processes. First, frequent navigation measurements are made to improve the estimate of the spacecraft's position and velocity. Second, a prediction of the orbit is made periodically so that small corrections to the speed and direction of motion can be applied using the propulsion system if the spacecraft is not following the intended course.

Predicting the course of Apollo during prolonged periods of coasting flight is the same as the astronomer's problem of predicting the positions of the moon and planets. There are several considerations which influence the ability to make long-range predictions. First of all are the mathematical techniques used for solving the equations of motion. Unless rather elaborate computational techniques are employed, numerical errors will propagate and rapidly degrade the solution. Second, the accuracy of predicting position and velocity also is subject to our knowledge of the physical properties of the solar system. Finally, and most important of all, is the accuracy of initial conditions—the values of position and velocity at the time from which the prediction is made.

In order to insure accurate initial conditions, it is necessary periodically to correct the estimate of spacecraft position and velocity using data gathered from optical or radar measurements. Use of a space sextant allows the astronaut, for example, to measure the apparent elevation of a star above the earth's horizon or to measure the angle subtended by the directions to a star and to a landmark on the moon. At the time a measurement is made, the best estimate of the spacecraft's position and velocity is contained in the onboard digital computer. Then, since the directions of the stars and the locations of landmarks are known, it is possible to calculate the expected value of the angle to be measured.

When the expected value of this measurement is compared with the value actually measured, the difference can be used to correct the estimate of the spacecraft's position and velocity. A sequence of such measurements separated in time, together with an accurate mathematical description of the solar system, will eventually produce estimates with sufficient precision to permit corrective maneuvers to be made with confidence.

### Navigation Instruments

To accomplish the lunar landing objectives of the Apollo mission, two vehicles are used—the command module (CM) and the lunar module (LM). Each vehicle is equipped with sensors for acquiring navigation measurement data, together with a digital computer for information processing and orbit prediction. By appropriate utilization of these instruments, the Apollo astronauts can solve the three major coasting flight navigation problems of the lunar mission: (1) cislunar-midcourse navigation of the CM to and

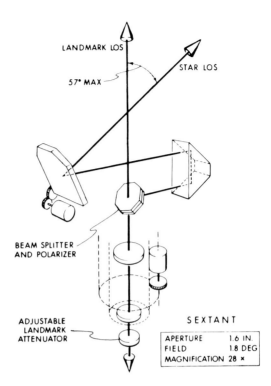

**Fig. 1:** Sextant schematic diagram.

from the moon; (2) navigation of the CM in orbit about the moon; and (3) navigating both the CM and the LM during the rendezvous phase in lunar orbit.

Reference orientation is maintained in both vehicles by means of a device called an inertial measuring unit (IMU). This instrument is basically a small platform supported and pivoted so that the spacecraft is free to rotate about it. On this platform are mounted three gyroscopes that sense and prevent any rotation of the platform occurring.

In the command module a rigid structure mounted to the spacecraft, called the navigation base, provides a common mounting structure for a telescope, a sextant, and the base of the IMU gimbal system. Precision angle transducers on each of the axes of the optical instruments and on each of the axes of the IMU gimbals permit the indicated angles to be processed in the command module computer (CMC) to generate the components of the optical targets in inertial system stable-member coordinates.

The sextant (SXT), shown schematically in Fig. 1, is a 28-power, narrow-field-of-view instrument having two lines of sight. One of these lines of sight is fixed to the spacecraft and is aimed at a landmark or the horizon by turning the vehicle in space using orientation commands to the attitude control system. The second line of sight can be pointed to a star

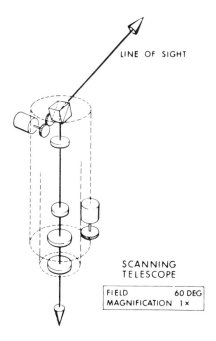

**Fig. 2:** Scanning telescope schematic diagram.

through the use of a two-axis hand controller. One axis of this motion, the shaft axis, is parallel to the landmark line of sight. The shaft drive changes the plane in which the navigation angle is measured by rotating the head of the instrument as a whole. The trunnion drive sets the navigation angle by tilting the trunnion axis mirror. By superimposing the two images in the field of view and signaling this event to the computer by depressing a mark button, the navigator can measure the angle between the lines of sight to a star and either a landmark or the horizon.

The scanning telescope (SCT), shown in Fig. 2, is a unity-power, wide-field-of-view instrument having a single line of sight. Both the shaft and trunnion control of this line of sight is possible. The shaft angle is made always to follow the sextant shaft angle, while the trunnion can be selected so that the astronaut may look either along the star or landmark line of sight of the sextant. The wide field of view aids considerably in star and landmark recognition.

A very high frequency (VHF) link between the two vehicles exists which is normally used for intervehicle voice communication. However, a signal path through this VHF link from the CM to the LM and back to the CM makes possible a measurement of the distance separating the vehicles. Data from this automatic VHF range-link is used in the CMC during rendezvous navigation to complement the manually acquired optical measurements.

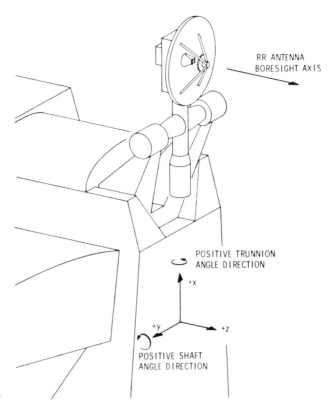

**Fig. 3:** Rendezvous radar.

Optical navigation sightings are not required in the lunar module. However, to aid in a successful rendezvous with the command module, a rendezvous radar (RR), shown in Fig. 3, is provided and mounted near the LM inertial measuring unit so that direction data can be related between the two. With this instrument the range and range-rate of the CM with respect to the LM as well as the direction, in terms of shaft and trunnion angles of the radar antenna, are made available to the lunar module computer (LMC) for state vector updating.

The command and lunar module computers are designed to handle a relatively large and diverse set of onboard data processing and control functions. Some special requirements for this computer include (a) real-time solution of several problems simultaneously on a priority basis, (b) efficient two-way communication with the navigator, (c) capability of ground control radio links, and (d) multiple signal interfaces of both a discrete and continuously variable type. The memory section has a cycle time of 12-$\mu$sec and consists of a fixed (read-only) portion of 36,864 words together with an erasable portion of 2048 words. Each word in memory is

16 bits long (15 data bits and an odd parity bit). Data words are stored as signed 14-bit words.

Most of the computer programs relevant to guidance and navigation are written in a pseudocode notation for economy of storage. This notation is encoded and stored as a list of data words. An "interpreter" program translates this list into a sequence of subroutine linkages. Thus, the small basic instruction set is effectively expanded into a comprehensive mathematical language, which includes matrix and vector operations, using numbers of 28 bits and sign.

The display and keyboard (DSKY), illustrated in Fig. 4, serves as the communication medium between the computer and the navigator. The principal part of the display is a set of three registers, each containing five decimal digits, so that a word of 15 bits can be displayed in one register by five octal digits. Three registers are used because of the frequent need to display the three components of a vector. Data are entered in the computer by the astronaut through the keyboard. When the computer requires a response from the astronaut, certain lights are caused to flash on and off in order to attract his attention.

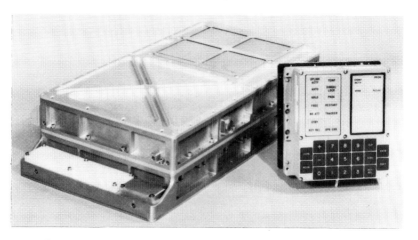

**Fig. 4:** Apollo guidance computer and display and keyboard.

Navigation Data Processing

The recursive formulation of the optimum linear estimator is ideally suited to the space navigation problem—especially when a vehicle-borne computer is utilized to process data obtained from onboard instrumentation. Measurement data may be incorporated sequentially, as they are obtained, without recourse to the batch processing techniques required by other methods. Furthermore, within the framework of a single computational algorithm, estimates of quantities in addition to position and velocity, such as radar biases, may be included by the simple expedient of increasing the dimension of the state vector. Finally, matrix inversion, with all of its numerical pitfalls, may also be avoided by regarding all measurement data as single-dimensional or scalar information.

Each computer in the two vehicles maintains an estimate of the position and velocity vectors of both its own and the other spacecraft. These two state vectors are normally six-dimensional but at times are augmented by certain parameters which must also be estimated as part of the navigation process.

In cislunar-midcourse navigation the command module computer only is involved and the CM state vector is of six dimensions. However, when the command module is navigating in lunar orbit, it is necessary also to estimate the position vector of the particular landmark which is being tracked. This is conveniently accomplished by utilizing a nine-dimensional state, the first six elements of which are the components of the CM state vector in moon-centered, nonrotating rectangular coordinates, while the last three are the components of the lunar landmark position vector.

Two separate rendezvous navigation programs, one in each computer, are used simultaneously during the rendezvous phase of the mission with each solving the navigation problem independently of the other. In the CMC the six-dimensional state vector of either the CM or the LM can be updated from the measurement data obtained with the CM sensors. Normally, it is the LM state vector which is altered, but the mode is at the option of the astronauts. The selection of the updating mode is based primarily on which vehicle's state is more accurately known initially and which vehicle is active in controlling the rendezvous maneuvers.

Since the rendezvous radar of the LM is not structurally mounted with the IMU, significant unknown biases in the knowledge of the direction of the radar antenna are possible. In order to achieve the required accuracy during rendezvous, it is necessary to include the RR angle biases as components of the state vector to be estimated in the LMC. Although the augmented state vector then has eight components, it is treated as nine dimensional for computational convenience with the ninth element zero.

### State Vector and Transition Matrix Extrapolation

The estimates of position and velocity are maintained in the spacecraft computers in nonrotating rectangular coordinates and are referenced to either the earth or the moon. An earth-centered equatorial coordinate system is used when the vehicle is outside of the lunar sphere of influence. Inside of this sphere the center of coordinates coincides with the center of the moon. The extrapolation of position and velocity is made by a direct numerical integration of the equations of motion.

The basic equation may be written in the form as

$$\frac{d^2}{dt^2}\mathbf{r}_{PV} + \frac{\mu_P}{r_{PV}^3}\mathbf{r}_{PV} = \mathbf{a}_d \tag{1}$$

where $\mathbf{r}_{PV}$ is the vector position of the vehicle with respect to the primary body $P$, which is either the earth or moon, and $\mu_P$ is the gravitational constant of $P$. The vector $\mathbf{a}_d$ is the vector acceleration which prevents the motion of the vehicle from being precisely a conic with $P$ at the focus.

If $\mathbf{a}_d$ is small compared with the central force field, direct use of Eq. (1) is inefficient. As an alternative, the integration may be accomplished by employing the technique of differential accelerations suggested by the astronomer Encke.

### Encke's Method

At time $t_0$, the position and velocity vectors $\mathbf{r}_{PV}(t_0)$ and $\mathbf{v}_{PV}(t_0)$ define an osculating conic orbit. The vector difference $\boldsymbol{\delta}(t)$ between the actual and conic orbits satisfies the following differential equation:

$$\frac{d^2\boldsymbol{\delta}}{dt^2} = \frac{\mu_P}{r_{PV(C)}^3}\left[\left(1 - \frac{r_{PV(C)}^3}{r_{PV}^3}\right)\mathbf{r}_{PV} - \boldsymbol{\delta}\right] + \mathbf{a}_d \tag{2}$$

subject to the initial conditions

$$\boldsymbol{\delta}(t_0) = \mathbf{0} \quad \text{and} \quad \left.\frac{d\boldsymbol{\delta}}{dt}\right|_{t=t_0} = \boldsymbol{\nu}(t_0) = \mathbf{0}$$

where $\mathbf{r}_{PV(C)}$ is the osculating conic position vector. The numerical difficulties which would arise from the evaluation of the coefficient of $\mathbf{r}_{PV}$ in Eq. (2) may be avoided. Since

$$\mathbf{r}_{PV}(t) = \mathbf{r}_{PV(C)}(t) + \boldsymbol{\delta}(t) \tag{3}$$

it follows that

$$1 - \frac{r_{PV(C)}^3}{r_{PV}^3} = -f(q_C) = 1 - (1+q_C)^{\frac{3}{2}}$$

where

$$q_C = \frac{(\boldsymbol{\delta} - 2\mathbf{r}_{PV})\cdot\boldsymbol{\delta}}{r_{PV}^2} \tag{4}$$

The function $f(q)$ may be conveniently evaluated from

$$f(q) = q\frac{3 + 3q + q^2}{1 + (1+q)^{\frac{3}{2}}} \tag{5}$$

Encke's method may now be summarized as follows:
1. Position in the osculating orbit is calculated from

$$\mathbf{r}_{PV(C)}(t) = \left[1 - \frac{x^2}{r_{PV}(t_0)} C(\alpha_0 x^2)\right] \mathbf{r}_{PV}(t_0)$$
$$+ \left[(t - t_0) - \frac{x^3}{\sqrt{\mu_P}} S(\alpha_0 x^2)\right] \mathbf{v}_{PV}(t_0) \quad (6)$$

where

$$\alpha_0 = \frac{2}{r_{PV}(t_0)} - \frac{v_{PV}^2(t_0)}{\mu_P} \quad (7)$$

and $x$ is determined as the root of Kepler's equation in the form

$$\sqrt{\mu_P}(t - t_0) = \frac{\mathbf{r}_{PV}(t_0) \cdot \mathbf{v}_{PV}(t_0)}{\sqrt{\mu_P}} x^2 C(\alpha_0 x^2)$$
$$+ [1 - r_{PV}(t_0)\alpha_0] x^3 S(\alpha_0 x^2) + r_{PV}(t_0) x \quad (8)$$

The special transcendental functions $S$ and $C$ are defined by

$$S(x) = \frac{1}{3!} - \frac{x}{5!} + \frac{x^2}{7!} - \cdots \qquad C(x) = \frac{1}{2!} - \frac{x}{4!} + \frac{x^2}{6!} - \cdots \quad (9)$$

2. Deviations from the osculating orbit are obtained by a numerical integration of

$$\frac{d^2}{dt^2} \boldsymbol{\delta}(t) = -\frac{\mu_P}{r_{PV(C)}^3(t)} \left[f(q_C)\mathbf{r}_{PV}(t) + \boldsymbol{\delta}(t)\right] + \mathbf{a}_d(t) \quad (10)$$

The first term on the right-hand side of the last equation must remain small, i.e., of the same order as $\mathbf{a}_d(t)$, if the method is to be efficient. As the deviation vector $\boldsymbol{\delta}$ grows in magnitude, this term will eventually increase in size. Therefore, in order to maintain the efficiency, a new osculating orbit should be defined by the true position and velocity. The process of selecting a new conic orbit from which to calculate deviations is called rectification. When rectification occurs, the initial conditions of the differential equation for $\boldsymbol{\delta}$ are again zero and the right-hand side is simply the perturbation acceleration $\mathbf{a}_d$ at the time of rectification.

3. The position vector $\mathbf{r}_{PV}(t)$ is computed from Eq. (3) using Eq. (6). The velocity vector $\mathbf{v}_{PV}(t)$ is then computed as

$$\mathbf{v}_{PV}(t) = \mathbf{v}_{PV(C)}(t) + \boldsymbol{\nu}(t) \quad (11)$$

where

$$\mathbf{v}_{PV(C)}(t) = \frac{\sqrt{\mu_P}}{r_{PV}(t_0) r_{PV(C)}(t)} [\alpha_0 x^3 S(\alpha_0 x^2) - x] \mathbf{r}_{PV}(t_0)$$
$$+ \left[1 - \frac{x^2}{r_{PV(C)}(t)} C(\alpha_0 x^2)\right] \mathbf{v}_{PV}(t_0) \quad (12)$$

## Disturbing Acceleration

The form of the disturbing acceleration $\mathbf{a}_d$ to be used depends on the phase of the mission. In earth or lunar orbit only the gravitational perturbations arising from the nonspherical shape of the primary body $P$ need be considered. During the translunar and transearth flight, the gravitational attraction of the sun and the secondary body $Q$ (either earth or moon) are relevant forces. A summary of the various cases appears below.

1. Earth orbit

$$\mathbf{a}_d = \frac{\mu_E}{r_{EV}^2} \sum_{k=2}^{4} J_{k,E} \left(\frac{r_E}{r_{EV}}\right)^k \left[P'_{k+1}(\cos\phi)\mathbf{i}_{EV} - P'_k(\cos\phi)\mathbf{i}_z\right] \quad (13)$$

where

$P'_2(\cos\phi) = 3\cos\phi$  $\qquad P'_4(\cos\phi) = \frac{1}{3}(7\cos\phi\, P'_3 - 4P'_2)$
$P'_3(\cos\phi) = \frac{1}{2}(5\cos\phi\, P'_2 - 3)$  $\qquad P'_5(\cos\phi) = \frac{1}{4}(9\cos\phi\, P'_4 - 5P'_3)$

are the derivatives of the Legendre polynomials;

$$\cos\phi = \mathbf{i}_{EV} \cdot \mathbf{i}_z$$

is the cosine of the angle $\phi$ between the unit vector $\mathbf{i}_{EV}$ in the direction of $\mathbf{r}_{EV}$ and the unit vector $\mathbf{i}_z$ in the direction of the north pole; $r_E$ is the equatorial radius of the earth; and $J_{2,E}$, $J_{3,E}$, $J_{4,E}$ are the coefficients of the second, third, and fourth harmonics of the earth's potential function. The subscript $E$ denotes the center of the earth as the origin of coordinates.

2. Translunar and transearth flight†

$$\mathbf{a}_d = -\frac{\mu_Q}{r_{QV}^3}[f(q_Q)\mathbf{r}_{PQ} + \mathbf{r}_{PV}] - \frac{\mu_S}{r_{SV}^3}[f(q_S)\mathbf{r}_{PS} + \mathbf{r}_{PV}] \quad (14)$$

where the subscripts $Q$ and $S$ denote the secondary body and the sun, respectively. Thus, for example, $\mathbf{r}_{PS}$ is the position vector of the sun with respect to the primary body. The arguments $q_{()}$ are calculated from

$$q_{()} = \frac{(\mathbf{r}_{PV} - 2\mathbf{r}_{P()}) \cdot \mathbf{r}_{PV}}{r_{P()}^2} \quad (15)$$

and the function $f$ from Eq. (5).

In the vicinity of the lunar sphere of influence a change in origin of coordinates is made. Thus

$$\begin{aligned}\mathbf{r}_{PV}(t) - \mathbf{r}_{PQ}(t) = \mathbf{r}_{QV}(t) &\to \mathbf{r}_{PV}(t) \\ \mathbf{v}_{PV}(t) - \mathbf{v}_{PQ}(t) = \mathbf{v}_{QV}(t) &\to \mathbf{v}_{PV}(t)\end{aligned} \quad (16)$$

---

† During the periods of transearth and translunar flight when the vehicle is near the primary body, the disturbing acceleration which is used is the sum of Eq. (14) and either (13) or (17), whichever is appropriate.

3. Lunar orbit

$$\mathbf{a}_d = \frac{\mu_M}{r_{MV}^2} \left\{ \sum_{k=2}^{4} J_{kM} \left(\frac{r_M}{r_{MV}}\right)^k \left[ P'_{k+1}(\cos\phi) \mathbf{i}_{MV} - P'_k(\cos\phi) \mathbf{i}_z \right] \right.$$
$$+ 3 J_{22,M} \left(\frac{r_M}{r_{MV}}\right)^2 \left[ 4 \frac{x_{MV} y_{MV}}{x_{MV}^2 + y_{MV}^2} \mathbf{i}_{MV} \times \mathbf{i}_z \right.$$
$$\left. \left. + \frac{x_{MV}^2 - y_{MV}^2}{x_{MV}^2 + y_{MV}^2} \left[ (5\cos^2\phi - 3) \mathbf{i}_{MV} - 2\cos\phi \, \mathbf{i}_z \right] \right] \right\} \quad (17)$$

where $r_M$ is the mean lunar radius; $x_{MV}$ and $y_{MV}$ are the lunar equatorial plane components of $\mathbf{r}_{MV}$; and $J_{22,M}$ is that coefficient of the moon's potential function associated with the asymmetry of the moon about its polar axis.

### Error Transition Matrix

The position and velocity vectors, as maintained in the computer, are only estimates of the true values. As part of the recursive estimation technique, it is also necessary to record statistical data for the processing of navigation measurements.

If $\epsilon(t)$ and $\eta(t)$ are the errors in the estimates of the position and velocity vectors, respectively, then the six-dimensional covariance matrix $\mathbf{P}_{66}(t)$ is defined by

$$\mathbf{P}_{66}(t) = \begin{bmatrix} \overline{\epsilon(t)\epsilon(t)^\mathrm{T}} & \overline{\epsilon(t)\eta(t)^\mathrm{T}} \\ \overline{\eta(t)\epsilon(t)^\mathrm{T}} & \overline{\eta(t)\eta(t)^\mathrm{T}} \end{bmatrix} \quad (18)$$

As previously noted, for certain applications it is necessary to expand the state vector and the covariance matrix to more than six dimensions, in order to include the estimation of the landmark locations in the CMC during orbital navigation and the rendezvous radar tracking biases in the LMC for rendezvous navigation. For this purpose a nine-dimensional covariance matrix is defined as follows:

$$\mathbf{P}(t) = \begin{bmatrix} \mathbf{P}_{66}(t) & \overline{\epsilon(t)\zeta^\mathrm{T}} \\ & \overline{\eta(t)\zeta^\mathrm{T}} \\ \overline{\zeta\epsilon(t)^\mathrm{T}} & \overline{\zeta\eta(t)^\mathrm{T}} & \overline{\zeta\zeta^\mathrm{T}} \end{bmatrix} \quad (19)$$

where the components of the three-dimensional vector $\zeta$ are the estimation errors of the three additional variables to be estimated.

To take full advantage of the operations provided by the interpreter in the spacecraft computers, the covariance matrix is restricted to either six or nine dimensions. In the LMC rendezvous navigation procedure, only two additional quantities are estimated. However, a dummy variable is added to the state vector to make it nine dimensional.

The appropriate relation to be used in extrapolating the matrix $\mathbf{P}_{66}(t)$ is obtained as follows. The basic equation of motion of the vehicle with respect to the primary body may be written in the form:

$$\frac{d^2}{dt^2}\mathbf{r}_{PV} + \frac{\mu_P}{r_{PV}^3}\mathbf{r}_{PV} + \frac{\mu_Q}{r_{QV}^3}\mathbf{r}_{QV} + \frac{\mu_Q}{r_{PQ}^3}\mathbf{r}_{PQ} = \mathbf{g}$$

where the vector $\mathbf{g}$ encompasses the gravitational acceleration of the sun as well as the other disturbances arising from the asymmetrical shapes of the earth and moon. This equation can be linearized about the best estimate of the vehicle's state and, to a first order of approximation, the estimate error vectors will satisfy the resulting linear differential equation. We have

$$\frac{d}{dt}\begin{bmatrix}\epsilon(t)\\\eta(t)\end{bmatrix} = \begin{bmatrix}\mathbf{O} & \mathbf{I}\\\mathbf{G}(t) & \mathbf{O}\end{bmatrix}\begin{bmatrix}\epsilon(t)\\\eta(t)\end{bmatrix}$$

where $\mathbf{I}$ and $\mathbf{O}$ are the three-dimensional identity and zero matrices, respectively.

The matrix $\mathbf{G}(t)$ is the three-dimensional gradient of the gravitational field with respect to the components of the position vector $\mathbf{r}_{PV}$. If we neglect the gradient of the vector $\mathbf{g}$, then it is easy to show that

$$\mathbf{G}(t) = \frac{\mu_P}{r_{PV}^5(t)}\left[3\mathbf{r}_{PV}(t)\mathbf{r}_{PV}(t)^\mathrm{T} - r_{PV}^2(t)\mathbf{I}\right]$$
$$+ \frac{\mu_Q}{r_{QV}^5(t)}\left[3\mathbf{r}_{QV}(t)\mathbf{r}_{QV}(t)^\mathrm{T} - r_{QV}^2(t)\mathbf{I}\right] \quad (20)$$

which is a sufficiently good approximation for translunar and transearth flight. For orbital navigation about the primary body only the first term in the expression for $\mathbf{G}(t)$ need be included.

Since

$$\mathbf{P}_{66}(t) = \overline{\begin{bmatrix}\epsilon(t)\\\eta(t)\end{bmatrix}\begin{bmatrix}\epsilon(t)^\mathrm{T} & \eta(t)^\mathrm{T}\end{bmatrix}}$$

it is easy to show that the six-dimensional covariance matrix satisfies the following differential equation:

$$\frac{d}{dt}\mathbf{P}_{66} = \begin{bmatrix}\mathbf{O} & \mathbf{I}\\\mathbf{G} & \mathbf{O}\end{bmatrix}\mathbf{P}_{66} + \mathbf{P}_{66}\begin{bmatrix}\mathbf{O} & \mathbf{G}\\\mathbf{I} & \mathbf{O}\end{bmatrix}$$

Because of accumulated numerical inaccuracies, it is possible that the covariance matrix may fail to remain positive definite after a large number of computations, as it theoretically must. An innovation to avoid this problem, which has also the advantage of significantly reducing certain computational requirements, is to replace the covariance matrix by a matrix $\mathbf{W}(t)$, called the error transition matrix. The $\mathbf{W}(t)$ matrix has the property:

$$\mathbf{P}(t) = \mathbf{W}(t)\mathbf{W}(t)^\mathrm{T} \quad (21)$$

and thus, in a sense, is the square root of the covariance matrix. If needed, the covariance matrix may be determined as the product of the matrix $\mathbf{W}(t)$ and its transpose, thereby guaranteeing it to be at least positive semi-definite.

The chief computational advantage of the $\mathbf{W}$ matrix is the simplicity of the differential equation which it satisfies. From the differential equation for $\mathbf{P}_{66}$, the fact that the components of the vector $\zeta$ do not change with time, and Eq. (21), it is obvious that

$$\frac{d}{dt}\mathbf{W} = \begin{bmatrix} \mathbf{O} & \mathbf{I} & \mathbf{O} \\ \mathbf{G}(t) & \mathbf{O} & \mathbf{O} \\ \mathbf{O} & \mathbf{O} & \mathbf{O} \end{bmatrix} \mathbf{W} \qquad (22)$$

Now let the nine-dimensional matrix $\mathbf{W}(t)$ be partitioned as

$$\mathbf{W} = \begin{bmatrix} \mathbf{w}_0 & \mathbf{w}_1 & \cdots & \mathbf{w}_8 \\ \mathbf{w}_9 & \mathbf{w}_{10} & \cdots & \mathbf{w}_{17} \\ \mathbf{w}_{18} & \mathbf{w}_{19} & \cdots & \mathbf{w}_{26} \end{bmatrix} \qquad (23)$$

Then, we have

$$\begin{aligned} \frac{d}{dt}\mathbf{w}_i(t) &= \mathbf{w}_{i+9}(t) \\ \frac{d}{dt}\mathbf{w}_{i+9}(t) &= \mathbf{G}(t)\mathbf{w}_i(t) \qquad \text{for} \qquad i = 0, 1, \ldots, 8 \qquad (24) \\ \frac{d}{dt}\mathbf{w}_{i+18}(t) &= \mathbf{0} \end{aligned}$$

Thus, the extrapolation of the $\mathbf{W}$ matrix may be accomplished by successively integrating the vector differential equations

$$\frac{d^2}{dt^2}\mathbf{w}_i(t) = \mathbf{G}(t)\mathbf{w}_i(t) \qquad \text{for} \qquad i = 0, 1, \ldots, 8 \qquad (25)$$

Finally, then, if $D$ is the dimension of the matrix $\mathbf{W}(t)$, the differential equations for the $\mathbf{w}_i(t)$ vectors are simply

$$\begin{aligned} \frac{d^2}{dt^2}\mathbf{w}_i(t) &= \frac{\mu_P}{r_{PV}^3(t)} \left\{ 3\left[\mathbf{i}_{PV}(t) \cdot \mathbf{w}_i(t)\right] \mathbf{i}_{PV}(t) - \mathbf{w}_i(t) \right\} \\ &+ \frac{\mu_Q}{r_{QV}^3(t)} \left\{ 3\left[\mathbf{i}_{QV}(t) \cdot \mathbf{w}_i(t)\right] \mathbf{i}_{QV}(t) - \mathbf{w}_i(t) \right\} \qquad (26) \end{aligned}$$

for $i = 0, 1, \ldots, D-1$ with the second term omitted for orbital navigation about the primary body $P$.

## Numerical Integration

The extrapolation of the state vector and the error transition matrix requires the solution of $D+1$ second-order vector differential equations, specifically Eqs. (10) and (26). These are all special cases of the form

$$\frac{d^2\mathbf{y}}{dt^2} = \mathbf{f}(\mathbf{y}, t) \qquad (27)$$

in which the right-hand side is a function of the independent variable and time only. Nyström's method is particularly well-suited to this form and gives an integration method of fourth-order accuracy, while requiring only three computations of the derivatives per time step. (The usual fourth-order Runge-Kutta integration methods require four derivative computations per time step.)

The second-order system is written as

$$\frac{d\mathbf{y}}{dt} = \mathbf{z} \qquad \frac{d\mathbf{z}}{dt} = \mathbf{f}(\mathbf{y}, t) \qquad (28)$$

and the formulas are summarized as follows:

$$\begin{aligned} \mathbf{y}_{n+1} &= \mathbf{y}_n + \varphi(\mathbf{y}_n)\,\Delta t \\ \mathbf{z}_{n+1} &= \mathbf{z}_n + \psi(\mathbf{y}_n)\,\Delta t \end{aligned} \qquad (29)$$

where

$$\begin{aligned} \varphi(\mathbf{y}_n) &= \mathbf{z}_n + \tfrac{1}{6}(\mathbf{k}_1 + 2\mathbf{k}_2)\,\Delta t \\ \psi(\mathbf{y}_n) &= \tfrac{1}{6}(\mathbf{k}_1 + 4\mathbf{k}_2 + \mathbf{k}_3) \end{aligned}$$

and

$$\begin{aligned} \mathbf{k}_1 &= \mathbf{f}(\mathbf{y}_n, t_n) \\ \mathbf{k}_2 &= \mathbf{f}\left[\mathbf{y}_n + \tfrac{1}{2}\mathbf{z}_n\,\Delta t + \tfrac{1}{8}\mathbf{k}_1(\Delta t)^2,\, t_n + \tfrac{1}{2}\Delta t\right] \\ \mathbf{k}_3 &= \mathbf{f}\left[\mathbf{y}_n + \mathbf{z}_n\,\Delta t + \tfrac{1}{2}\mathbf{k}_2(\Delta t)^2,\, t_n + \Delta t\right] \end{aligned}$$

For efficient use of computer storage as well as computing time, the computations should be performed in the following order:

1. Equation (10) is solved using the Nyström formulas (29). It is necessary to preserve the values of the vectors $\mathbf{r}_{PV}$ at times $t_n$ and $t_n + \tfrac{1}{2}\Delta t$ for use in the solution of Eqs. (26).
2. Equations (26) are solved one at a time using formulas (29), together with the values of $\mathbf{r}_{PV}$ which resulted from the first step.

It has been found experimentally that the maximum value that the integration time step $\Delta t$ can have is either $0.3\sqrt{r_{PV}^3/\mu_P}$ or 4000 seconds, whichever is the smaller.

## Incorporation of Measurement Data

An important feature of the Apollo navigation method is that measurement data from a wide variety of sources may be incorporated within the same framework of computation. Associated with each measurement is a $D$-dimensional vector $\mathbf{b}$ representing, to a first order of approximation, the variation in the quantity $Q$ which would result from variations in the components of the state vector. Thus, each measurement establishes a component of the spacecraft state vector along the direction of the $\mathbf{b}$ vector in state space.

By algebraically combining the $\mathbf{W}$ matrix, the $\mathbf{b}$ vector and a mean-squared *a priori* estimation error $\overline{\alpha^2}$ in the measurement, there are produced a weighting vector $\boldsymbol{\omega}$ and the step change to be made in the error transition matrix to reflect the changes in the uncertainties in the estimated quantities as a result of the measurement. The weighting vector $\boldsymbol{\omega}$ has $D$ components and is determined so that the observation data is utilized in a statistically optimum manner.

According to the recursive estimation algorithm for a one-dimensional scalar measurement, the weighting vector is determined from

$$a\boldsymbol{\omega} = \mathbf{Pb} \tag{30}$$

where $\mathbf{P}$ is the value of the covariance matrix extrapolated to the time of the measurement and

$$a = \mathbf{b}^T \mathbf{Pb} + \overline{\alpha^2} \tag{31}$$

For convenience, define the vector $\mathbf{z}$ as

$$\mathbf{z} = \mathbf{W}^T \mathbf{b} \tag{32}$$

so that Eqs. (30) and (31) may be written as

$$a\boldsymbol{\omega} = \mathbf{W}\mathbf{z} \tag{33}$$

$$a = \mathbf{z}^T \mathbf{z} + \overline{\alpha^2} \tag{34}$$

Then, if $\delta Q$ represents the difference between the quantity actually measured and its expected value based on the extrapolated value of the state vector, the change in the state vector is simply $\boldsymbol{\omega}\,\delta Q$.

As a result of the measurement, the statistics embodied in the covariance matrix $\mathbf{P}$ or its square root $\mathbf{W}$ must be altered. Again, the algorithm dictates that the new value for $\mathbf{P}$, denoted by $\mathbf{P}^*$, is obtained from

$$\mathbf{P}^* = \mathbf{P}\left(\mathbf{I} - \frac{1}{a}\mathbf{b}\mathbf{b}^T \mathbf{P}\right)$$

or, in terms of the $\mathbf{W}$ matrix and the $\mathbf{z}$ vector,

$$\mathbf{P}^* = \mathbf{W}\left(\mathbf{I} - \frac{1}{a}\mathbf{z}\mathbf{z}^T\right)\mathbf{W}^T \tag{35}$$

where $\mathbf{I}$ is the $D$-dimensional identity matrix.

What is desired, of course, is a formula for updating the **W** matrix rather than the **P** matrix. The objective will be achieved if a square root can be found for the parenthesized factor in Eq. (35). Indeed, the desired result is obtained by determining the value of the parameter $\gamma$ such that

$$\left(\mathbf{I} - \frac{1}{a}\mathbf{z}\mathbf{z}^T\right) = \left(\mathbf{I} - \frac{\gamma}{a}\mathbf{z}\mathbf{z}^T\right)\left(\mathbf{I} - \frac{\gamma}{a}\mathbf{z}\mathbf{z}^T\right)^T$$

By straightforward computation it is seen that $\gamma$ must be

$$\gamma = \frac{1}{1 + \sqrt{\overline{\alpha^2}/a}} \tag{36}$$

so that the new value $\mathbf{W}^*$ is computed as

$$\mathbf{W}^* = \mathbf{W} - \gamma \boldsymbol{\omega} \mathbf{z}^T \tag{37}$$

In order to take full advantage of the three-dimensional vector and matrix operations provided by the interpreter in the computer, the nine-dimensional **W** matrix is stored sequentially as follows:

$$\mathbf{w}_0^T \quad \mathbf{w}_1^T \quad \ldots \quad \mathbf{w}_{26}^T$$

Then, by defining the three-dimensional submatrices

$$\mathbf{W}_0 = \begin{bmatrix} \mathbf{w}_0^T \\ \mathbf{w}_1^T \\ \mathbf{w}_2^T \end{bmatrix} \quad \mathbf{W}_1 = \begin{bmatrix} \mathbf{w}_3^T \\ \mathbf{w}_4^T \\ \mathbf{w}_5^T \end{bmatrix} \quad \ldots \quad \mathbf{W}_8 = \begin{bmatrix} \mathbf{w}_{24}^T \\ \mathbf{w}_{25}^T \\ \mathbf{w}_{26}^T \end{bmatrix}$$

so that

$$\mathbf{W} = \begin{bmatrix} \mathbf{W}_0^T & \mathbf{W}_1^T & \mathbf{W}_2^T \\ \mathbf{W}_3^T & \mathbf{W}_4^T & \mathbf{W}_5^T \\ \mathbf{W}_6^T & \mathbf{W}_7^T & \mathbf{W}_8^T \end{bmatrix} \tag{38}$$

and partitioning the possibly nine-dimensional vectors **b**, $\boldsymbol{\omega}$, and **z** as

$$\mathbf{b} = \begin{bmatrix} \mathbf{b}_0 \\ \mathbf{b}_1 \\ \mathbf{b}_2 \end{bmatrix} \quad \boldsymbol{\omega} = \begin{bmatrix} \boldsymbol{\omega}_0 \\ \boldsymbol{\omega}_1 \\ \boldsymbol{\omega}_2 \end{bmatrix} \quad \mathbf{z} = \begin{bmatrix} z_0 \\ \vdots \\ z_8 \end{bmatrix} = \begin{bmatrix} \mathbf{z}_0 \\ \mathbf{z}_1 \\ \mathbf{z}_2 \end{bmatrix} \tag{39}$$

the computations developed in this section are conveniently performed as

$$\mathbf{z}_j = \sum_k \mathbf{W}_{j+3k}\mathbf{b}_k \qquad a = \sum_k \mathbf{z}_k \cdot \mathbf{z}_k + \overline{\alpha^2} \qquad \boldsymbol{\omega}_j^T = \frac{1}{a}\sum_k \mathbf{z}_k^T \mathbf{W}_{3j+k}$$

from which we obtain

$$\mathbf{w}_{i+9j}^* = \mathbf{w}_{i+9j} - \gamma z_i \boldsymbol{\omega}_j \tag{40}$$

where the subscript $i$ ranges from 0 to $D-1$ while the subscripts $j$ and $k$ range from 0 to $\frac{1}{3}D - 1$.

It is worth emphasizing that the particular formulation of the recursive estimator, developed in this and the previous subsection, has achieved what is felt to be a maximum in practical computational efficiency as well as a most sparing use of erasable memory locations in the small vehicle-borne Apollo computers. The latter is possible primarily because of the introduction of the error transition matrix instead of the more conventional covariance matrix.

It should be remarked, however, that straightforward application of the **W** matrix techniques would not have been possible had the dynamics of the state vector model included what is commonly called "process noise". Since we have elected to neglect process noise in favor of computational compactness, the gradual and inevitable decay of the elements of the **W** matrix must be countered in practice by a periodic reinitialization of those elements. It should be clear that, without either including process noise or reinitializing the **W** matrix, eventually all measurement data would be ignored simply because they would be given a zero weighting factor.

### Command Module Cislunar Midcourse Navigation

During the translunar and transearth phases of the Apollo mission, navigation data can be obtained with the sextant by measuring the angle between the lines of sight to a star and an earth or moon horizon or landmark. When the navigator depresses the mark button, indicating to the computer that the two target optical images are properly superimposed in the SXT field of view, the time of the measurement and the measured angle, i.e. the SXT trunnion angle $A$, are automatically recorded in the CMC. The navigator must then inform the computer, through the DSKY, of the identity of the star and the particular feature of the earth or moon involved in the sighting. These data are used by the computer to determine: (1) the CM state vector estimate and its associated error transition matrix extrapolated to the measurement time; (2) the estimated CM position vector $\mathbf{r}_V$ relative to that body used in the measurement (since this may be either the primary body $P$ or the secondary body $Q$, the single subscript will serve without ambiguity); (3) the unit vector $\mathbf{i}_s$ in the direction of the particular star used (there are data for 37 stars in the CMC fixed memory which the navigator can identify by code number); and (4) the position vector $\mathbf{r}_L$ of the landmark, assuming a star-landmark measurement.

### Star-Landmark Measurements

Because of the extremely high accuracy required for midcourse navigation measurements, observation data must be corrected for aberration prior to processing. Aberration is the term used to describe the change in the apparent direction of an object due to the velocity of the observer normal to the line of sight to the object. It is only when this perpendicular velocity has a magnitude of tens of thousands of feet per second that the aberration correction is necessary. It should be remarked that the correction is not required for rendezvous or orbit navigation.

The apparent direction of the star is computed from

$$\mathbf{i}_s^* = \text{Unit}\left(\mathbf{i}_s + \frac{\mathbf{v}_{PV} + \mathbf{v}_{SE}}{c}\right) \tag{41}$$

where the notation Unit $(\mathbf{q})$ is understood to mean a unit vector in the direction of the vector $\mathbf{q}$. The vector $\mathbf{v}_{SE}$ is the velocity of the earth relative to the sun and $c$ is the speed of light. The velocity of the sun relative to the star is already taken into account for the basic star directional data $\mathbf{i}_s$ and the relative velocity between the earth and moon is negligible for the case in which the moon is the primary body. The expected direction of the landmark is obtained from

$$\mathbf{i}_{VL}^* = \text{Unit}\left(\mathbf{i}_{VL} + \frac{\mathbf{v}_{PV}}{c}\right) \tag{42}$$

where

$$\mathbf{i}_{VL} = \text{Unit}(\mathbf{r}_L - \mathbf{r}_V)$$

is the unit vector defining the estimated direction of the landmark relative to the vehicle.

Using the corrected unit vectors corresponding to the directions of the two SXT lines of sight, the CMC computes the measured deviation $\delta Q$ as

$$\delta Q = A - \arccos(\mathbf{i}_s^* \cdot \mathbf{i}_{VL}^*) \tag{43}$$

Although for cislunar-midcourse navigation the CM state vector is six dimensional, only the first three components of the measurement geometry vector $\mathbf{b}$ will be nonzero since the measured quantity is independent of velocity except through the small aberration correction. To determine this nonzero partition $\mathbf{b}_0$ of the $\mathbf{b}$ vector, we compute the first-order differential of the expression

$$r_{VL}^* \cos A = \mathbf{i}_s^* \cdot \mathbf{r}_{VL}^*$$

with the result that

$$\cos A \, \delta r_{VL}^* - r_{VL}^* \sin A \, \delta A = \mathbf{i}_s^* \cdot \delta \mathbf{r}_{VL}^*$$

Then, since

$$\delta r_{VL}^* = \frac{\mathbf{r}_{VL}^* \cdot \delta \mathbf{r}_{VL}^*}{r_{VL}^*} \quad \text{and} \quad \delta \mathbf{r}_{VL}^* = -\delta \mathbf{r}_{PV}$$

we have
$$r^*_{VL} \sin A \, \delta A = (\mathbf{i}^*_s - \cos A \, \mathbf{i}^*_{VL}) \cdot \delta \mathbf{r}_{PV}$$

Hence
$$\mathbf{b}_0 = \frac{1}{r^*_{VL} \sin A} (\mathbf{i}^*_s - \cos A \, \mathbf{i}^*_{VL})$$

or
$$\mathbf{b}_0 = \frac{1}{r^*_{VL}} \operatorname{Unit}[\mathbf{i}^*_s - (\mathbf{i}^*_s \cdot \mathbf{i}^*_{VL}) \mathbf{i}^*_{VL}] \tag{44}$$

Finally, the measurement error variance is computed from
$$\overline{\alpha^2} = \sigma_A^2 + \frac{\sigma_L^2}{r^{*2}_{VL}} \tag{45}$$

where $\sigma_A^2$ and $\sigma_L^2$ are the assumed error variances in the SXT trunnion angle and the landmark, respectively.

### Star-Horizon Measurements

The processing of star-horizon measurement data is the same as that required for star-landmark data, provided the landmark location vector $\mathbf{r}_L$ is replaced by a vector from the spacecraft to the horizon. The determination of this horizon vector is made according to the following geometrical arguments.

Consider first an earth-horizon measurement. The star direction $\mathbf{i}_s$ and the estimated CM position vector $\mathbf{r}_V$ determine a plane. Assuming that the horizon is at a constant altitude above the earth's surface, the intersection of this measurement plane with the horizon is approximately an ellipse. The orientation of this horizon ellipse is defined in terms of three mutually orthogonal unit vectors:

$$\mathbf{i}_2 = \operatorname{Unit}(\mathbf{i}_s \times \mathbf{r}_V) \qquad \mathbf{i}_0 = \operatorname{Unit}(\mathbf{i}_z \times \mathbf{i}_2) \qquad \mathbf{i}_1 = \mathbf{i}_2 \times \mathbf{i}_0 \tag{46}$$

where $\mathbf{i}_z$ is a unit vector in the direction of the earth's polar axis. Referring to Fig. 5, it is clear that $\mathbf{i}_0$ and $\mathbf{i}_1$ coincide with the semimajor and semiminor axes of the horizon ellipse, respectively. The plane containing the horizon ellipse is inclined with respect to the earth's equatorial plane by an angle $I$, where

$$\sin I = \mathbf{i}_1 \cdot \mathbf{i}_z \tag{47}$$

The shape of the horizon ellipse is determined by the lengths of its major and minor axes. Assuming the contour of the earth to be well-approximated by the so-called *Fischer ellipsoid*, the semimajor axis of the horizon ellipse $a_H$ is simply the sum of the semimajor axis of this ellipsoid and the constant horizon altitude. Likewise, the semiminor axis $b_H$ is found by adding the horizon altitude to that value of the radius of the Fischer ellipsoid which corresponds to a latitude equal to the inclination angle $I$.

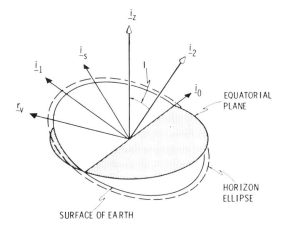

**Fig. 5:** Horizon coordinate system.

The problem of determining the vector $r_L$ is readily solved in the horizon coordinate system for which the $x$ and $y$ axes coincide with the directions $i_0$ and $i_1$, respectively, as illustrated in Fig. 6. The matrix

$$\mathbf{M}^T = \begin{bmatrix} \mathbf{i}_0 & \mathbf{i}_1 & \mathbf{i}_2 \end{bmatrix} \qquad (48)$$

will serve to transform vectors from the original coordinate system to the horizon system. Let $\mathbf{r}_H$ and $\mathbf{i}_{s_H}$ represent the components of $\mathbf{r}_V$ and $\mathbf{i}_s$ in horizon coordinates, so that

$$\mathbf{r}_H = \begin{bmatrix} x_H \\ y_H \\ 0 \end{bmatrix} = \mathbf{M}\mathbf{r}_V \quad \text{and} \quad \mathbf{i}_{s_H} = \mathbf{M}\mathbf{i}_s \qquad (49)$$

Further, define $\mathbf{t}_0$ and $\mathbf{t}_1$ as vectors from the point $(x_H, y_H)$ to the two points of tangency with the horizon ellipse.

The vectors $\mathbf{t}_0$ and $\mathbf{t}_1$ are obtained by solving simultaneously the equation of the horizon ellipse:

$$\frac{x^2}{a_H^2} + \frac{y^2}{b_H^2} = 1$$

and the equation of the line tangent to the ellipse and passing through the point $(x_H, y_H)$:

$$\frac{x x_H}{a_H^2} + \frac{y y_H}{b_H^2} = 1$$

We have, for $i = 0, 1$,

$$\mathbf{t}_i = \frac{1}{d} \begin{bmatrix} x_H \pm (a_H/b_H) y_H \sqrt{d-1} \\ y_H \mp (b_H/a_H) x_H \sqrt{d-1} \\ 0 \end{bmatrix} \qquad (50)$$

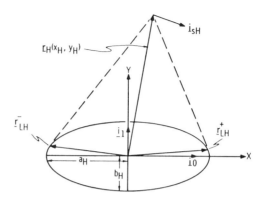

**Fig. 6:** Geometry of star-horizon measurement.

where
$$d = \frac{x_H^2}{a_H^2} + \frac{y_H^2}{b_H^2}$$

The upper and lower signs in Eq. (50) correspond, respectively, with $i = 0$ and $i = 1$.

The two points of tangency, $\mathbf{t}_0$ and $\mathbf{t}_1$, correspond to the two horizon points, $\mathbf{r}_{LH}^+$ (near horizon) and $\mathbf{r}_{LH}^-$ (far horizon), but not necessarily respectively. By definition, the near horizon is associated with that vector $\mathbf{t}_i$ which makes the smaller angle with the star vector. The horizon vector is then

$$\mathbf{r}_L = \mathbf{M}^T \mathbf{t}_i \qquad (51)$$

using for $\mathbf{t}_i$ either $\mathbf{t}_0$ or $\mathbf{t}_1$, whichever is appropriate.

In the CMC, moon-horizon measurements are processed under the assumption that the moon has a spherical shape. The determination of the horizon vector $\mathbf{r}_L$ is made as for the earth-horizon case but with the mean radius of the moon used for both $a_H$ and $b_H$.

### Command Module Navigation during Rendezvous

To accomplish rendezvous with the CM and LM in the Apollo mission, two types of navigation data are obtained with the CM-based sensors. These data are acquired both manually with the optics and automatically through the VHF range-link. They serve in complementary roles for rendezvous navigation. The optical data yields information in the directions normal to the CM-to-LM line of sight, whereas the VHF range-link gives information along the line of sight. Taken together, they provide an excellent rendezvous navigation capability.

It is recognized that the optimum method of applying the recursive estimation technique to rendezvous navigation would be to solve the twelve-dimensional problem for the simultaneous estimation of both the CM and LM states. However, the limitations of a relatively small erasable memory preclude the storage of the required twelve-dimensional error transition matrix. In the CMC rendezvous navigation procedure (and in the LMC as well), the estimated state vector of one vehicle only is altered by the navigation data. Consequently, the error transition matrix is of manageable six-dimensional size and is extrapolated with whichever state vector is being estimated.

Because of this simplification, together with some other related factors, the error transition matrix does not maintain an accurate representation of the covariances of the relative state vectors between the two vehicles. This situation necessitates a reinitialization of the matrix at various times during the rendezvous, essentially to restart the solution in both the CMC and the LMC.

Optical Measurements

Approximately once per minute during the navigation portions of the rendezvous phase, the Apollo navigator sights the LM using the reticle pattern in the SXT star line of sight. When the mark button is depressed, the time of the measurement, the SXT shaft and trunnion angles, and the three IMU gimbal angles, which describe the orientation of the navigation base with respect to the inertially stabilized platform, are all automatically recorded in the CMC.

From these five angles and the known orientation of the inertial platform, the measured direction $\mathbf{i}_m$, of the LM as observed from the CM is obtained. It is convenient to consider the unit vector $\mathbf{i}_m$ as having been found by the simultaneous measurement of the angles between the lines of sight to the LM and two stars. The data are processed by selecting two convenient unit vectors (fictitious star directions), converting the vector $\mathbf{i}_m$ to an equivalent set of two artificial star-LM measurements, and using the measurement incorporation procedure twice—once for each artificial measurement. These two unit vectors are chosen such that they and the estimated line of sight vector form an orthogonal triad.

The processing of each of the two artificial measurements is similar to the cislunar-midcourse navigation procedure. Let $\mathbf{r}_{PC}$ and $\mathbf{r}_{PL}$ be the CM and LM position vectors, respectively, extrapolated to the time of the measurement.

For convenience, the first fictitious star direction is chosen to be

$$\mathbf{i}_{s_1} = \text{Unit}[(\mathbf{i}_{CL} \times \mathbf{i}_m) \times \mathbf{i}_{CL}] \tag{52}$$

where

$$\mathbf{i}_{CL} = \text{Unit}(\mathbf{r}_{PL} - \mathbf{r}_{PC})$$

is the estimated CM-to-LM line of sight. The measured deviation is then given by

$$\delta Q = \arccos(\mathbf{i}_{s_1} \cdot \mathbf{i}_m) - \tfrac{1}{2}\pi \tag{53}$$

since $\mathbf{i}_{s_1}$ and $\mathbf{i}_{CL}$ are perpendicular. This orthogonality also permits a simplification in the calculation of the geometry vector $\mathbf{b}_0$. Indeed, from Eq. (44), we have

$$\mathbf{b}_0 = \pm \frac{1}{r_{CL}} \mathbf{i}_{s_1} \tag{54}$$

The plus or minus sign is selected, respectively, according as the CM or LM state vector is to be updated.

Following the alteration of the appropriate state vector, the second artificial measurement is incorporated by recomputing the vector $\mathbf{i}_{CL}$ and selecting the fictitious star direction as

$$\mathbf{i}_{s_2} = \text{Unit}(\mathbf{i}_{s_1} \times \mathbf{i}_{CL}) \tag{55}$$

The measurement error variance $\overline{\alpha^2}$ for each incorporation is a constant which is the sum of the assumed error variances of the SXT and knowledge of the IMU orientation.

### VHF Range-link Measurements

Asynchronous with the manually acquired optical data, and at approximately the same frequency, VHF range measurements are automatically taken. Again, the time of the measurement and the measured CM-to-LM range $R$ are recorded in the CMC.

The measured deviation for this range measurement is simply

$$\delta Q = R - r_{CL} \tag{56}$$

To determine the geometry vector $\mathbf{b}_0$, we compute the first-order differential of

$$R^2 = (\mathbf{r}_{PL} - \mathbf{r}_{PC}) \cdot (\mathbf{r}_{PL} - \mathbf{r}_{PC})$$

under the assumption that it is the CM state vector which is to be estimated. There results

$$R \, \delta R = -(\mathbf{r}_{PL} - \mathbf{r}_{PC}) \cdot \delta \mathbf{r}_{PC}$$

or

$$\delta R = -\mathbf{i}_{CL} \cdot \delta \mathbf{r}_{PC}$$

On the other hand, if the LM state vector is to be updated, the relationship would be

$$\delta R = \mathbf{i}_{CL} \cdot \delta \mathbf{r}_{PL}$$

Hence

$$\mathbf{b}_0 = \mp \mathbf{i}_{CL} \tag{57}$$

where the upper or lower sign depends, respectively, on whether the CM or LM state vector is to be changed.

## Lunar Module Navigation during Rendezvous

While the CMC is actively acquiring and processing data during the rendezvous phase of Apollo, the same navigation problem is being simultaneously and independently solved in the LMC, using tracking information gathered from the rendezvous radar mounted on the LM vehicle. After RR tracking acquisition of the CM is established, the LMC records, at approximately one minute intervals, the measured range $R$ and range-rate $\dot{R}$ of the CM with respect to the LM, together with the shaft angle $\beta$ and the trunnion angle $\vartheta$ of the gimballed radar dish. In addition to these four measured quantities, the time of the measurement and the three IMU gimbal angles are also noted.

As described earlier, eight variables are estimated as part of the navigation procedure. The usual three components of the position and velocity vectors, $\mathbf{r}_{PV}$ and $\mathbf{v}_{PV}$, of the selected spacecraft, LM or CM, with respect to the primary body $P$, constitute the first six components of the state vector. The estimates of the biases $\delta\beta$ and $\delta\vartheta$, in the RR shaft and trunnion angles are the seventh and eighth elements. A dummy variable is used for the ninth component to facilitate three-dimensional vector operations in the computers.

The measured quantities produce four sequential alterations of the nine-dimensional state vector through four separate applications of the measurement incorporation method. The incorporation is, of course, performed recursively, i.e., at each stage the new components of the state vector, resulting from the previous update, are used in computing the next update in the sequence. The measurement error variances $\overline{\alpha^2}$ are based on *a priori* knowledge of the radar performance.

### Range and Range-Rate Measurements

The measured deviation in range and the associated measurement geometry vector $\mathbf{b}_0$ are the same as for the VHF measurement of range in the CM and need not be further discussed. For the range-rate data, the measured deviation is easily seen to be

$$\delta Q = \dot{R} - (\mathbf{v}_{PC} - \mathbf{v}_{PL}) \cdot \mathbf{i}_{LC} \tag{58}$$

The geometry vector $\mathbf{b}$ for the range-rate measurement provides the first case encountered thus far for which both the $\mathbf{b}_0$ and $\mathbf{b}_1$ partitions are different from zero. To determine the $\mathbf{b}$ vector, we compute the first-order

## Epilogue

differential of the relation

$$\dot{R} = \frac{(\mathbf{r}_{PC} - \mathbf{r}_{PL}) \cdot (\mathbf{v}_{PC} - \mathbf{v}_{PL})}{[(\mathbf{r}_{PC} - \mathbf{r}_{PL}) \cdot (\mathbf{r}_{PC} - \mathbf{r}_{PL})]^{\frac{1}{2}}}$$

with the assumption that it is the CM state vector which is to be estimated. We have

$$\delta\dot{R} = \frac{1}{r_{LC}}(\mathbf{v}_{LC} \cdot \delta\mathbf{r}_{PC} + \mathbf{r}_{LC} \cdot \delta\mathbf{v}_{PC}) - \frac{1}{r_{LC}^3}(\mathbf{r}_{LC} \cdot \mathbf{v}_{LC})\mathbf{r}_{LC} \cdot \delta\mathbf{r}_{PC}$$

or, alternatively,

$$\delta\dot{R} = \frac{1}{r_{LC}^3}\mathbf{r}_{LC} \times (\mathbf{v}_{LC} \times \mathbf{r}_{LC}) \cdot \delta\mathbf{r}_{PC} + \frac{1}{r_{LC}}\mathbf{r}_{LC} \cdot \delta\mathbf{v}_{PC}$$

To obtain the proper relationship if the LM state vector is to be estimated, we need only note that $\delta\mathbf{r}_{PL} = -\delta\mathbf{r}_{PC}$. Thus, the nonzero partitions of the **b** vector are

$$\mathbf{b}_0 = \pm\frac{1}{r_{LC}}\mathbf{i}_{LC} \times (\mathbf{v}_{LC} \times \mathbf{i}_{LC}) \tag{59}$$
$$\mathbf{b}_1 = \pm\mathbf{i}_{LC}$$

with the choice of sign dependent on the particular state vector to be updated (plus for CM and minus for LM).

### Radar Antenna Angle Measurements

The rendezvous radar antenna dish is at the origin of the RR cartesian coordinate system as seen in Fig. 7. Shaft motion takes place about the positive $y$ axis and the shaft angle $\beta$ is measured from the positive $z$ axis. Trunnion motion occurs in a plane normal to the $xz$ plane and containing the shaft axis $y$. The configuration is such that the trunnion axis would coincide with the $x$ axis for zero shaft angle with the trunnion angle $\vartheta$, under those circumstances, measured from the $z$ axis.

Let $\mathbf{i}_x$, $\mathbf{i}_y$, $\mathbf{i}_z$ be unit vectors along the RR coordinate axes. By means of the recorded IMU gimbal angles, together with the knowledge of the inertial orientation of the IMU, the components of $\mathbf{i}_x$, $\mathbf{i}_y$, $\mathbf{i}_z$ are readily obtained in basic reference coordinates. It is easy to verify with reference to the figure that the measured deviation for the shaft angle is

$$\delta Q = \beta - \left[\arctan\left(\frac{\mathbf{i}_x \cdot \mathbf{i}_{LC}}{\mathbf{i}_z \cdot \mathbf{i}_{LC}}\right) + \delta\beta\right] \tag{60}$$

while that for the trunnion angle is

$$\delta Q = \vartheta - [\arcsin(-\mathbf{i}_y \cdot \mathbf{i}_{LC}) + \delta\vartheta] \tag{61}$$

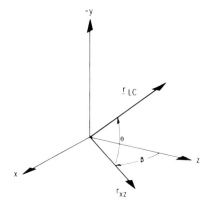

**Fig. 7:** Rendezvous radar coordinate system.

The position vector of the CM with respect to the LM is expressible in terms of components along the RR coordinate axes as

$$\mathbf{r}_{PC} - \mathbf{r}_{PL} = r_{LC} \begin{bmatrix} \sin\beta\cos\vartheta \\ -\sin\vartheta \\ \cos\beta\cos\vartheta \end{bmatrix}$$

The measurement geometry vector corresponding to the shaft angle is computed by again obtaining the differential

$$\delta\mathbf{r}_{PC} = r_{LC} \begin{bmatrix} \cos\beta\cos\vartheta \\ 0 \\ -\sin\beta\cos\vartheta \end{bmatrix} \delta\beta = r_{LC}\cos\vartheta\,\mathrm{Unit}(\mathbf{i}_y \times \mathbf{i}_{LC})\,\delta\beta$$

Hence

$$\delta\beta = \frac{1}{r_{LC}\cos\vartheta}\,\mathrm{Unit}(\mathbf{i}_y \times \mathbf{i}_{LC}) \cdot \delta\mathbf{r}_{PC}$$

Similarly, for the trunnion angle, we have

$$\delta\mathbf{r}_{PC} = r_{LC} \begin{bmatrix} -\sin\beta\sin\vartheta \\ -\cos\vartheta \\ -\cos\beta\sin\vartheta \end{bmatrix} \delta\vartheta = \frac{r_{LC}}{\cos\vartheta}(\mathbf{i}_y \times \mathbf{i}_{LC}) \times \mathbf{i}_{LC}\,\delta\vartheta$$

so that

$$\delta\vartheta = \frac{1}{r_{LC}\cos\vartheta}(\mathbf{i}_y \times \mathbf{i}_{LC}) \times \mathbf{i}_{LC} \cdot \delta\mathbf{r}_{PC}$$

The quantity $r_{LC}\cos\vartheta$, which appears in the expressions for $\delta\beta$ and $\delta\vartheta$, is simply the length of the projection of the $\mathbf{r}_{LC}$ vector in the $xz$ plane. Denoting this length by $r_{xz}$, we may calculate its value from

$$r_{xz} = r_{LC}\sqrt{1 - (\mathbf{i}_y \cdot \mathbf{i}_{LC})^2} \tag{62}$$

Thus, the measurement geometry vector $\mathbf{b}_0$ for the shaft angle is

$$\mathbf{b}_0 = \pm \frac{1}{r_{xz}} \text{Unit}(\mathbf{i}_y \times \mathbf{i}_{LC}) \tag{63}$$

and for the trunnion angle

$$\mathbf{b}_0 = \pm \frac{1}{r_{xz}} (\mathbf{i}_y \times \mathbf{i}_{LC}) \times \mathbf{i}_{LC} \tag{64}$$

where the plus or minus sign, as before, indicates either the CM or LM state vector is to be updated.

Since the antenna angles are independent of velocity, we conclude that $\mathbf{b}_1 = \mathbf{0}$ for both. However, because we are also estimating the angle biases, the $\mathbf{b}_2$ partitions of the $\mathbf{b}$ vectors are not zero. Indeed,

$$\mathbf{b}_2 = \begin{bmatrix} 1 \\ 0 \\ 0 \end{bmatrix} \quad \text{shaft angle} \quad \text{and} \quad \mathbf{b}_2 = \begin{bmatrix} 0 \\ 1 \\ 0 \end{bmatrix} \quad \text{trunnion angle}$$

The estimation of the radar biases is included regardless of whether it is the CM or LM state vector which is being estimated. Since the radar biases do not depend on which vehicle's state vector comprises the first six components of the estimated state, there is no sign selection associated with these $\mathbf{b}_2$ partitions.

## Command Module Orbit Navigation

When the Apollo spacecraft is in either earth or lunar orbit, navigation data can be obtained by optical measurements of the lines of sight to landmarks. These planetary surface features can be either of the "known" or "unknown" variety. A known landmark is an identifiable feature whose coordinates are known and tabulated. In contrast, an unknown landmark is any surface feature which the astronaut may select and optically track in the brief period during which it is visible. The mechanics of the measuring process is quite similar to the CM-to-LM line of sight measurement procedure already described.

### Known Landmark Measurements

As previously noted, orbit navigation involves a nine-dimensional state vector, the last three components of which are the coordinates of the landmark. Since the tracking period for one landmark is very short (less than one minute in the case of the earth and only two or three minutes during lunar orbit), all navigation data for any particular landmark are acquired before the processing begins. At the conclusion of the tracking, the landmark partition of the state vector is initialized from the identification data entered by the navigator into the computer.

Sufficient CMC erasable storage is allocated for five measurements on a single landmark. The data from each of these sightings consists of the time of the measurement and the set of angles described above. From these angles the measured unit vector $\mathbf{i}_m$ along the CM-to-landmark line of sight is computed.

Each of the measured unit vectors is converted to an equivalent set of two artificial star-landmark measurements in exactly the same manner as in the rendezvous navigation procedure. The only difference is that the geometry vector is nine dimensional and is given by

$$\mathbf{b} = \frac{1}{r_{VL}} \begin{bmatrix} \mathbf{i}_s \\ 0 \\ -\mathbf{i}_s \end{bmatrix} \quad (65)$$

where $r_{VL}$ is the current estimated distance between the vehicle and the landmark and $\mathbf{i}_s$ denotes the direction to the artificial stars.

### Unknown Landmark Measurements

The nine-dimensional orbit navigation technique provides a means of mapping on the surface of a planet a point which is designated only by a number of sets of optical tracking data. This process may be used either to locate a desired LM landing site which may have unknown coordinates or to map features on the surface of the moon.

Assume that a landmark has been tracked and $N$ sets of optical navigation data have been acquired. If the navigator cannot (or chooses not to) identify the landmark, it is then treated as an unknown landmark. In this process the data from the first navigation measurement are used to compute an initial estimate of the landmark location. The nine-dimensional state vector is then formed and the data from the remaining $N - 1$ sightings are incorporated exactly as if the optically designated point had been an identified landmark.

The determination of the initial estimate of the landmark position vector $\mathbf{r}_L$ is accomplished as follows. Let $\mathbf{i}_m$ be the measured unit vector from the vehicle to the landmark calculated from the data of the first measurement. Then

$$\mathbf{r}_L = \mathbf{r}_{PV} + r_{VL} \mathbf{i}_m \quad (66)$$

with $r_{VL}$ obtained by applying the law of cosines to the triangle defined by Eq. (66). We have

$$r_{VL}^2 - (2r_{PV} \cos A_L) r_{VL} + (r_{PV}^2 - r_P^2) = 0 \quad (67)$$

where $r_P$ is the radius of the primary body, and $A_L$ is the angle between the directions from the vehicle to the center of the primary body and the landmark and is computed from

$$\cos A_L = -\mathbf{i}_m \cdot \text{Unit}(\mathbf{r}_{PV})$$

Solving Eq. (67) and selecting the appropriate sign yields

$$r_{VL} = r_{PV} \left[ \cos A_L - \left( \frac{r_P^2}{r_{PV}^2} - \sin^2 A_L \right)^{\frac{1}{2}} \right] \quad (68)$$

Thus, the initial estimated location of the unknown landmark is established. The remaining $N-1$ sets of data are then used as for a known landmark.

### Initialization of the Error Transition Matrix

Several different landmarks are tracked by the Apollo navigator to navigate in orbit about the earth or moon. As a consequence, special methods are required to reinitialize the error transition matrix each time a new landmark is acquired. This is necessary to reflect the fact that the initial landmark location errors are not correlated with the errors in the estimated CM position and velocity vectors. Therefore, before processing the measurement data associated with a new landmark, it is necessary to convert the nine-dimensional error transition matrix $\mathbf{W}$ to a six-dimensional matrix having the same CM position and velocity error variances and covariances. A new nine-dimensional matrix is then formed by augmenting appropriate landmark uncertainty information.

As the first stage in the initialization process it is necessary to determine a square root of the six-dimensional partition $\mathbf{P}_{66}$ of the nine-dimensional covariance matrix $\mathbf{P}$ defined in Eq. (19). From Eqs. (21) and (38), it is clear that

$$\mathbf{P}_{66} = \begin{bmatrix} \mathbf{W}_0^T & \mathbf{W}_1^T & \mathbf{W}_2^T \\ \mathbf{W}_3^T & \mathbf{W}_4^T & \mathbf{W}_5^T \end{bmatrix} \begin{bmatrix} \mathbf{W}_0 & \mathbf{W}_3 \\ \mathbf{W}_1 & \mathbf{W}_4 \\ \mathbf{W}_2 & \mathbf{W}_5 \end{bmatrix} \quad (69)$$

The square root of $\mathbf{P}_{66}$, i.e., a six-dimensional matrix $\mathbf{W}_{66}$ such that

$$\mathbf{W}_{66} \mathbf{W}_{66}^T = \mathbf{P}_{66} \quad (70)$$

is not unique. However, a convenient method, which is used in the CMC, is to select a triangular form for $\mathbf{W}_{66}$ so that the resulting set of algebraic equations for the elements of $\mathbf{W}_{66}$ is most easily solved.

The second part of the $\mathbf{W}$ matrix initialization depends on whether the landmark being tracked is known or unknown. For a known landmark, the new nine-dimensional $\mathbf{W}$ matrix is formed as:

$$\mathbf{W} = \begin{bmatrix} \mathbf{W}_0^T & \mathbf{W}_1^T & \mathbf{O} \\ \mathbf{W}_3^T & \mathbf{W}_4^T & \mathbf{O} \\ \mathbf{O} & \mathbf{O} & \mathbf{W}_8^T \end{bmatrix} \quad (71)$$

with the upper left-hand six by six partition being the square root matrix $\mathbf{W}_{66}$ found in part one. The three-dimensional submatrix $\mathbf{W}_8$ is given a value consistent with the expected errors in the knowledge of the landmark location. In particular, $\mathbf{W}_8$ is chosen so that

$$\mathbf{W}_8^T \mathbf{W}_8 = \overline{\boldsymbol{\zeta} \boldsymbol{\zeta}^T} \tag{72}$$

where $\boldsymbol{\zeta}$ is the vector error in the landmark position.

If, on the other hand, the measurements are made using an unknown landmark, the error in the initially computed estimate of the location of the landmark will be a function of the uncertainties in the CM position estimate, the tracking accuracy, and the altitude of the landmark above the gravitational center.† In this case, part two for the initialization of the $\mathbf{W}$ matrix is more complicated.

The basic relationship among the quantities of interest is Eq. (66). Errors in $\mathbf{r}_{PV}$ and $\mathbf{i}_m$, denoted respectively by $\boldsymbol{\epsilon}$ and $\delta\mathbf{i}_m$, will produce an error in the landmark location $\mathbf{r}_L$. However, this error is clearly in the plane of the landmark, i.e., perpendicular to $\mathbf{r}_L$. Also, an error in $r_{VL}$ will result in a landmark position error in the direction of $\mathbf{i}_m$ and of a magnitude $(\sqrt{\mathbf{r}_L^T \mathbf{r}_L}/\mathbf{i}_m^T \mathbf{r}_L) \delta r_L$, where $\delta r_L$ is the error in the landmark altitude. Thus, the total error $\boldsymbol{\zeta}$ in the landmark location is given by

$$\boldsymbol{\zeta} = \mathbf{U}(\boldsymbol{\epsilon} + r_{VL} \delta\mathbf{i}_m) + \delta r_L \frac{\sqrt{\mathbf{r}_L^T \mathbf{r}_L}}{\mathbf{i}_m^T \mathbf{r}_L} \mathbf{i}_m \tag{73}$$

where

$$\mathbf{U} = \mathbf{I} - \frac{\mathbf{i}_m \mathbf{r}_L^T}{\mathbf{i}_m^T \mathbf{r}_L} \tag{74}$$

is the projection operator which assures that only the components of $\boldsymbol{\epsilon}$ and $\delta\mathbf{i}_m$ in the plane normal to $\mathbf{r}_L$ are related to the landmark location error.

Now, consider a coordinate system in which the direction $\mathbf{i}_m$ is along one of the coordinate axes. Then, if $\mathbf{T}$ is the transformation matrix which relates the selected axis system and the original reference system, we have

$$\mathbf{i}_m = \mathbf{T} \begin{bmatrix} 0 \\ 0 \\ 1 \end{bmatrix}$$

The error in $\mathbf{i}_m$ may be expressed as:

$$\delta\mathbf{i}_m = \mathbf{T} \begin{bmatrix} \alpha \cos \chi \\ \alpha \sin \chi \\ 0 \end{bmatrix}$$

where $\alpha$ is the small random angle between the true and measured directions to the landmark. The polar angle $\chi$ is defined in the plane normal

---

† Bellantoni, J. F., "Unidentified Landmark Navigation," published in the *AIAA Journal*, Vol. 5, August 1967, pp. 1478–1483.

to $\mathbf{i}_m$ from the coordinate axis to the projection of the measured direction of $\mathbf{i}_m$.

Assume that $\alpha$ and $\chi$ are statistically independent random variables with zero means. Further, assume that $\chi$ is uniformly distributed over the interval $-\pi$ to $\pi$. Then, for the covariance matrix of the uncertainty in $\mathbf{i}_m$, we obtain

$$\overline{\delta\mathbf{i}_m \delta\mathbf{i}_m^T} = \tfrac{1}{2}\sigma_\alpha^2 \mathbf{T} \begin{bmatrix} 1 & 0 & 0 \\ 0 & 1 & 0 \\ 0 & 0 & 0 \end{bmatrix} \mathbf{T}^T = \tfrac{1}{2}\sigma_\alpha^2(\mathbf{I} - \mathbf{i}_m\mathbf{i}_m^T)$$

where $\sigma_\alpha^2$ is the variance of $\alpha$.

Finally, since $\epsilon$ and the velocity error $\eta$ are statistically independent of $\delta\mathbf{i}_m$ and $\delta\mathbf{r}_L$, we may calculate the following covariance matrices from Eq. (73):

$$\begin{aligned} \overline{\zeta\epsilon^T} &= \mathbf{U}\overline{\epsilon\epsilon^T} \\ \overline{\zeta\eta^T} &= \mathbf{U}\overline{\epsilon\eta^T} \\ \overline{\zeta\zeta^T} &= \mathbf{U}\overline{\epsilon\epsilon^T}\mathbf{U}^T + \tfrac{1}{2}\sigma_\alpha^2 r_{VL}^2 \mathbf{U}(\mathbf{I} - \mathbf{i}_m\mathbf{i}_m^T)\mathbf{U}^T + \sigma_L^2 \frac{\mathbf{r}_L^T \mathbf{r}_L}{(\mathbf{i}_m^T \mathbf{r}_L)^2} \mathbf{i}_m \mathbf{i}_m^T \end{aligned} \quad (75)$$

As a result, the $\mathbf{W}$ matrix is initialized as

$$\mathbf{W} = \begin{bmatrix} \mathbf{W}_0^T & \mathbf{W}_1^T & \mathbf{O} \\ \mathbf{W}_3^T & \mathbf{W}_4^T & \mathbf{O} \\ \mathbf{U}\mathbf{W}_0^T & \mathbf{U}\mathbf{W}_1^T & \mathbf{W}_8^T \end{bmatrix} \quad (76)$$

where $\mathbf{W}_8$ is now determined as the three-dimensional triangular square root of

$$\mathbf{W}_8^T \mathbf{W}_8 = \tfrac{1}{2}\sigma_\alpha^2 r_{VL}^2 \mathbf{U}(\mathbf{I} - \mathbf{i}_m\mathbf{i}_m^T)\mathbf{U}^T + \sigma_L^2 \frac{\mathbf{r}_L^T \mathbf{r}_L}{(\mathbf{i}_m^T \mathbf{r}_L)^2} \mathbf{i}_m \mathbf{i}_m^T \quad (77)$$

After incorporating the data obtained from the tracking of a landmark, all 81 elements of the $\mathbf{W}$ matrix will again, in general, be nonzero, indicating correlations among all components of vehicle position and velocity and landmark position. This matrix is extrapolated until a new landmark is acquired necessitating a new initialization.

---

Consider the mood of America as it approached the end of 1968, by any accounting one of the unhappiest years of the twentieth century. It was a year of riots, burning cities, sickening assassinations, universities forced to close their doors. In Southeast Asia the twelve-month toll of American dead rose 50 percent, to 15,000, and the cost of the war topped $25 billion. By mid-December the country's despair was reflected in the Associated Press's nationwide poll of editors, who chose as the two top stories of 1968 the slayings of Robert Kennedy and Martin Luther King; Time magazine picked a generic symbol, "The Dissenter," as its Man of the Year. The poll and the Man were scheduled for year-end publication. This condition was changed dramatically during the waning days of the year, ...

In the pink dawn of December 21 a quarter million persons lined the approaches to Cape Kennedy, many of them having camped overnight. At 7:51, ... the first manned Saturn V, an alabaster column as big as a naval destroyer, lifted slowly, ever so slowly, from the sea of flame that engulfed Pad 39–A. The upward pace quickened as the first stage's 531,000 gallons of kerosene and liquid oxygen were thirstily consumed, and in 2 minutes 34 seconds the big drink was finished, whereupon the second stage's five J–2 engines lit up. S–II's 359,000 gallons of liquid hydrogen and liquid oxygen boosted the S–IVB and CSM for 6 minutes 10 seconds to an altitude of 108 miles. After the depleted S–II fell away, the S–IVB, this time the third stage, fired for 2 minutes 40 seconds to achieve Earth orbit. Except for slight pogo during the second-stage burn, Commander Frank Borman reported all was smoothness.

During the second orbit, at 2 hours 27 minutes, CapCom Mike Collins sang out "You are go for TLI" (translunar injection), and 23 minutes after that Jim Lovell calmly said "Ignition." The S–IVB had restarted with a long burn over Hawaii that lasted 5 minutes 19 seconds and boosted speed to the 24,200 mph necessary to escape the bonds of Earth. "You are on your way," said Chris Kraft, from the last row of consoles in Mission Control, "you are really on your way."

In Mission Control early in the morning of December 24 ... CapCom Gerry Carr spoke to the three astronauts more than 200,000 miles away, "Ten seconds to go. You are GO all the way." Lovell replied, "We'll see you on the other side," and Apollo 8 disappeared behind the Moon ... For 34 minutes there would be no way of knowing what happened. During that time the 247-second LOI (lunar orbit insertion) burn would take place that would slow down the spacecraft from 5758 to 3643 mph to enable it to latch on to the Moon's field of gravity and go into orbit. If the SPS engine failed, Apollo 8 would whip around the Moon and head back for Earth on a free-return trajectory ... "Longest four minutes I ever spent," said Lovell during the burn ... At 69 hours 15 minutes Apollo 8 went into lunar orbit, ...

At 8:40 p.m. the astronauts were on television ... "For all the people on Earth," said Bill Anders, "the crew of Apollo 8 has a message we would like to send you." He paused a moment and then began reading:

In the beginning God created the Heaven and the Earth ...
... and God saw that it was good.

The commander added: "And from the crew of Apollo 8, we close with good night, good luck, a Merry Christmas, and God bless all of you—all of you on the good Earth."

"At some point in the history of the world," editorialized The Washington Post, "someone may have read the first ten verses of the Book of Genesis under conditions that gave them greater meaning than they had on Christmas Eve. But it seems unlikely ... This Christmas will always be remembered as the lunar one."

As Apollo 8 began its tenth and last orbit, CapCom Ken Mattingly told the astronauts: "We have reviewed all your systems. You have a GO for TEI" (transearth injection). This time the crew was really in thrall to the SPS engine. It had to ignite ... and ignite it did, in a 303-second burn that would effect touchdown in just under 58 hours. Apollo 8 reentered at 25,000 mph and splashed down south of Hawaii two days after Christmas.

During the last week of 1968 the Associated Press repolled its 1278 newspaper editors, who overwhelmingly voted Apollo 8 the story of the year. Time discarded "The Dissenter" in favor of Borman, Lovell, and Anders; and a friend telegraphed Frank Borman, "You have bailed out 1968."

**Samuel C. Phillips**†

---

† From Chapter 9 "The Shakedown Cruises" in the book *Apollo Expeditions to the Moon* edited by Edgar M. Cortright and published by NASA as SP–20 in 1975. Air Force Major General Sam Phillips, who had served as program manager for Minuteman, became Apollo Program Director at NASA Headquarters in 1963.

# Index

Abel, Niels Henrik, 70
Aberration, 769
Aberration correction, 770
**Abramowitz, Milton,** xxxi, 375
Accelerometers, xii
**Adams, John Couch,** 471, 472
Addition theorem
 for the exponential, 405
 for Legendre polynomials, 405
 for the sine, 405
Additive property, 700
Adjoint of a matrix, 691
Adjoint system, 460
Admissible changes, 525
Admissible functions, 559
Air Force Western Development Div., 1
**Airy, Sir George Biddell,** 473
**Albrecht, Julius,** 568, 603
**Aldrin, Edwin E., Jr., "Buzz",** xxviii
**Alonzo, Ramon,** 21, 31
Alpha Centauri, 20
Ames Research Center, 23
Amplitude of elliptic integrals, 69
**Anders, William A.,** 784
Angle of inclination, 124
Angular momentum, scalar, 115
 variation of, 498
Angular momentum vector, 115
 variation of, 498
Angular velocity matrix, 102
 characteristic equation for, 104
Angular velocity vector, 102
Anomaly, generalized, 179
Aphelion, 125
Apoapse, 118
Apocenter, 118, 125
Apogee, 125
Apollo 8, vii–xiii, xxix, 751–753, 783–784
Apollo guidance and navigation system, 25
Apollo guidance computer (AGC), 25, 755
**Apollonius,** 142, 144
Apse, 117
Apsidal line, 124
**Archimedes,** 79, 161
Arcturus, 20
Areal velocity, 103
Argument of latitude, 125
 variation of, 501

Argument of perihelion, 124
 variation of, 501
Arithmetic mean, 72, 343
Arithmetic progression, 343
Arithmetic series, 343
Ascending node, 124
Associated Legendre functions of the first
 kind, 405
 as hypergeometric functions, 408
Astrolabe, vii
Astronomical unit, 114, 118
Asymptotes of the hyperbola, 166
Asymptotic series, 507, 706
Atlas intercontinental ballistic missile, 1
Atmospheric-entry problem, 248
a.u., *See* astronomical unit
Auxiliary circle, 158
Avco Corporation, 23
Average, 663, 715, 716

**Backus, John,** 3
**Baker, Robert M. L., Jr.,** xxxi
BALITAC programming language, 4
Ballistic coefficient, 505
Ballistic Missile Division, USAF, 15
**Banachiewicz,** 659
Barker's equation, 150
 generalized form of, 156
 solution of, 156
 solution of,
  by continued fraction, 153
  by Descartes' method, 154
  by graphical method, 152
  by Lagrange expansion formula, 200
  by Stumpff's method, 151
  by successive substitutions, 198
  by trigonometric formulas, 151
**Barrow, Isaac,** 107
Barycenter, 400
**Battin, Richard H.,** xxxii, 31, 32, 227,
 236, 268, 274, 287, 299, 325, 431, 437,
 457, 465, 513, 568, 613, 659, 678, 691,
 695, 699, 751
Bayes' rule, 701
**Bayes, Thomas,** 701
**Beal, Byron,** 376
**Beckner, F. L.,** 517

Bellantoni, J. F., 784
Benney, David J., 408
Bernoulli, Daniel, 97, 107, 192, 206, 473
Bernoulli, James, 51, 68, 105, 192, 473, 662, 661, 721, 733
Bernoulli, John, 79, 68, 105, 110, 145, 473, 662
Bernoulli, Nicholas, 473, 721, 662
Bernoulli numbers, 51, 707
Bernoulli trials, 661
Bessel, Friedrich Wilhelm, 191, 204, 206, 472
Bessel functions, 191
  continued fraction expansion of, 209
  derivative formula for, 208, 209
  differential equation for, 209
  first kind of order $n$, 207
  first kind of order zero, 192
  generating function for, 208
  integral formula for, 209
  recurrence formula for, 208
  with imaginary arguments, 507
Beta Centauri, 652
Beta function, 740
Bettis, Dale, 613, 617
Bezout, Étienne, 353
Bielliptical transfer, 530
Bilinear transformation, 40, 411
Binomial coefficient, 661, 705
Binomial distribution, 728
Binomial distribution function, 726
  normal approximation to, 734
Binomial theorem, 201
Bipolar coordinates, 376
Birthday problem, 702
Blanchard, Robert. C., 313
Bode, Johann Elbert, 472
Bode's law, 472
Bombelli, Raphael, 46
Bond, George Phillips, 448
Bond, Victor, 121
Bond, William Cranch, 448
Borman, Frank, 784
Born, George H., 510
Bossart, Charles, 6
Bourbaki, Nicolas, xxix
Bouvard, Alexis, 472
Brahe, Tycho, 141
Braking into a circular lunar orbit, 27
Brand, Timothy J., 29, 32
Breakwell, John, 16
Bridge hands, 707-709
Broucke, Roger A., 159, 492
Brouncker, Lord William, 55
Brouncker's formula, 66
Bryson, Arthur E., 659
Bürmann, Heinrich, 213
Bürmann's series, 213

Burn time, 555
Butcher, J. C., 568

Calculus of variations,
  fundamental lemma of, 560
Canonically conjugate, 484
Card matching experiment, 721
Cardan's formula, 150
Cardano, Gerolamo, 150, 662
Carr, Gerry, 784
Catenary, 562
Catherine the Great, 79
Cauchy, Augustine-Louis, 196, 204, 207
Cauchy convergence criterion, 196
Cauchy distribution, 667, 714, 732
Cauchy integral theorem, 204
Cauchy residue theorem, 211
Cauchy's theorem, 196
Cayley, Sir Arthur, 91, 168
Cayley-Hamilton theorem, 654
Cefola, Paul J., 159
Centaur missile, 22
Center of conic, 117
Center of gravity, 716
Center of mass, 96, 110
Central-limit theorem, 732
Centripetal acceleration, 102
Ceres, 296–297, 472
Chain rule, 478
Challis, James, 473
Characteristic,
  of elliptic integral of the third kind, 69
Characteristic equation, 382
Characteristic function, 724
Characteristic values, 382
Characteristic vector,
  of the rotation matrix, 83
Characteristic velocity, 515
Charles Stark Draper Laboratory, Inc., 1
Chevalier de Méré, 662
Chi-square distribution, 738
  with $n$ degrees of freedom, 739
Chi-square test, 738
Cholesky, 659
Chord, 238
Christensen, E. J., 510
Circle, 117
Circular functions, 169
Clairaut, Alexis-Claude, 206
Clarke, Arthur, 4
Colinear libration points, 366
  stability of, 386
Collins, Michael, 784
Command and lunar module computers, 757
Command module (CM), 754

Command module computer (CMC), 755
Common difference, 343
Common ratio, 343
Complementary event, 704
Complete elliptic integrals, 69
  evaluation of,
    first kind, 75
    second kind, 75
Condition equations, 574
Condition of unbias, 666
Conditional probability, 700
Confidence level, 736
Confluent hypergeometric functions, 43, 737
  contiguous, 43
  differential equation for, 43
    singular point of,
      irregular, 43
      regular, 43
Conic sections, 355
  axes of symmetry of, 355
  definition of, 147
  center of, 355
  general equation of,
    in rectangular coordinates, 117, 355
    in polar coordinates, 117
  rotational invariants of, 355
Conjugate axis of hyperbola, 169
Conjugate orbits, 244, 272
  flight-direction angles for, 246
  parameter for, 246
Conservation of
  angular momentum, 97
  linear momentum, 96
  total energy, 98
Constraint functions, 575
  recursive formula for, 582
Continued fractions for
  Bessel functions, 209
  Brouncker's formula for pi, 55
  complete elliptic integrals,
    ratio of second and first kinds, 76
  confluent hypergeometric function, 50
  exponential function, 50
    Euler's fraction for, 66
  Golden Section, 45
  hyperbolic tangent, 52
  hypergeometric functions, 48
    $F(\alpha, 1; \gamma; x)$, 48
    $F(3, 1; \frac{5}{2}; x)$, 310
    $F(\frac{1}{2}, 1; \frac{5}{2}; z)$, 311
    ratio of two contiguous, 48
    ratio of two contiguous confluent, 48
  inverse hyperbolic sine, 52
  inverse hyperbolic tangent, 49
  inverse sine, 52
  inverse tangent, 49
  logarithm, 49
  modified Bessel functions, 507
  normal probability function, 737
  root of quadratic equation, 47
  sine, Euler's fraction for, 66
  sine of a trisected angle, 52
  solution of the cubic equation, 53
  surds, 46
  tangent, 52
  $\tan k\phi$, 49
  $U_1/U_0$, 187
  $U_3$, 189
  $U_5$, 190
Continued fraction algorithm, 46
Continuous distribution, 712
Convair San Diego, 1, 5, 30
Convergence of continued fractions,
  sufficiency test for class I, 58
  sufficiency test for class II, 61
  proof of, for $F(3, 1; \frac{5}{2}; x)$, 62
Convolution integral, 743
**Coolidge, Julian Lowell,** xxxi
Coordinate translation, 81
**Copernicus, Nicholas,** 141
**Copps, Edward M.,** 553
Core rope, 21
Coriolis acceleration, 102
**Coriolis, Gaspard Gustave de,** 102
Correlated flight path, 6
Correlated velocity, 6, 26
Correlation coefficient, 719
**Cortright, Edgar M.,** 784
Co-state, 565
  differential equation for, 565
Cotangential transfer orbits, 527
Covariance, 718
Covariance matrix, 664, 763
Covariance matrix update, 22
**Cowell, Philip Herbert,** 447
Cowell's method, 447, 567
**Cramer, Gabriel,** 132
**Cramér, Harald,** xxxi
Cramer's rule, 132
Critical inclination angle, 504
**Crocco, General Gaetano Arturo,**
  xxvii, 17, 31
**Croopnick, Steven R.,** 465, 678
Cross-product steering, 11, 552
Cubic convergence, 218
Curtate cycloid, 193
Curvature, 104

Dahlgren Naval Weapons Laboratory, 26
**D'Alembert, Jean Le Rond,** 459
**Danby, J. M. A.,** xxxi
**Dandelin, Germinal Pierre,** 148
**Da Vinci, Leonardo,** 662
Declination, 124

Delta guidance method, 5
**De Moivre, Abraham,** 733
De Moivre's theorem, 733
Density scale height of atmosphere, 505
**Deprit, André,** xxxi, 182
**Descartes, René,** 107, 154
Descartes' method
   for Barker's equation, 154
Descartes' rule of signs, 53, 519
Deterministic position fix, 644
Difference equations,
   solution of, 44
Differential corrections,
   method of, 216
**di Pacioli, Luca,** 662
Dirac $\delta$-function, 713
**Dirac, Paul Adrien Maurice,** 713
Direction cosines, 81
Direction of insensitivity, 548
Directrix, 144
Discrete distribution, 712
Display and keyboard (DSKY), 758
Distance-to-be-gained, 6
Distribution function, 711
Disturbing acceleration, 762
Disturbing function, 388
   series expansion for, 391
**Dryden, Hugh,** 18
**Dubyago, Alexander Dmitriyevich,** xxxi
Dyadic products, 130

Eccentric anomaly, 158
   relation to true anomaly, 159
   variation of, 502
Eccentric longitude, 125, 491
Eccentricity, 116
   in terms of $\sin\phi$, 159
   in terms of $\sec\psi$, 166
   variation of, 499
Eccentricity vectors, 116
   locus of, 256
   variation of, 499
Ecliptic system, 123
Economization of power series, 362
**Efron, Bradley,** 704
Eigenvalues, 382
**Eisele, Donn F.,** 751
Elementary functions,
   definition of, 68
Elementary rotation matrices, 85
Elements of an orbit, 123
Eliminating the secular term, 487
Ellipse, 117
   construction of, 357
   definition of,
      as locus of circle centers, 146
      equation of, origin at center, 117
   perimeter of, 73
Elliptic functions, 71
   period of, 71
Elliptic integrals,
   evaluation of, 72–78
   modulus of, 69
   of the first kind, 69
      identity for, 72
   of the second kind, 69
      identity for, 74
   of the third kind, 69
      characteristic of, 69
      special case of, $(n=m)$, 74
   parameter of, 69
**El'yasberg, P. E.,** xxxi
**Encke, Johann Franz,** 217, 320, 420, 448
Encke's comet, 448
Encke's method, 448, 567, 760
Energy integral, 116
   in Jacobi coordinates, 400
Ephemeris time, 437
Epoch, 161
Epoch state vector, variation of, 509
Equation of orbit,
   Euler's universal form of, 143
   in cartesian coordinates, 117
   in polar coordinates, 117
      origin at center, 143
      origin at focus, 143
      origin at pericenter, 143
   in terms of eccentric longitude, 491
   in terms of true longitude, 491
Equations of motion for $n$ bodies, 96
   Jacobi's form of, 400
   Hamilton's canonical form of, 101
   Lagrange's form of, 99
Equation of the center, 212
Equatorial system, 123
Equilateral hyperbola, 119
   as analog of the auxiliary circle, 169
Equilateral libration points,
   stability test for, 384
Equilateral triangle solution, 367
Equinoctial coordinate axes, 494
Equinoctial variables, 492
Equiprobability ellipsoids, 668
   volume of, 668
Equivalent continued fractions, 49, 61
Error transition matrix, 764
   initialization of, 781
Escape velocity, 409
Estimator, 645
**Euclid,** 144
Euler acceleration, 102
Euler angles, 84
   as orbital elements, 123

Euler axis, 83, 89
Eulerian integrals, 740
Euler identity for trigonometric functions, 200
   generalization of, 184
Euler-Lagrange equation, 561
**Euler, Leonhard,** 35, 39, 51, 64, 68, 79, 107, 172, 192, 206, 359, 361, 471, 473, 561, 562, 721, 740
Euler parameters, 88
   in terms of
      Euler angles, 90
      spherical coordinate angles, 90
Euler's constant, 740
Euler's continued fraction for $\tan k\phi$, 49
Euler's series for powers of
   the cosine function, 202
   the sine function, 201
Euler's summation formula, 51
Euler's theorem on rigid body rotation, 81
Euler's time equation for the parabola, 277
Euler's transformation,
   series to continued fraction, 66
Evaluation of continued fractions, 63
   bottom-up method, 64
   top-down method, 68
   Wallis' method, 63
Events of equal likelihood, 703
Excess hyperbolic velocity, 533
Exponential function of a matrix, 87
Extended Kalman filter, 25
External $\Delta v$ mode, 29
Extremals, 561

$F$ and $G$ functions, 113
   *See also* Lagrange $F$ and $G$ functions
**Fagnano, Giulio Carlo de',** 68
**Farquhar, Robert W.,** xxviii
**Fehlberg, Erwin,** 603, 613, 616, 621
**Feller, William,** 699
**Ferdinand, Carl Wilhelm,** 297
**Fermat, Pierre de,** 661–662
Fibonacci series, 44
**Fill, Thomas J.,** 32, 227, 232, 236, 268
Fischer ellipsoid, 771
Fixed-time-of-arrival guidance, 543
**Flamsteed, John,** 472
**Fletcher, James C.,** 1
Flexowriter, 3
Flight-direction angle, 128
Flight-path angle, 128
Focal-radii, 144
   property of, for conic sections, 145
Focus, 117
Focus-directrix,
   property of, for conic sections, 144
Force function, 98

Forcing function, 452
Fortran language, 3
Fourier coefficients, 206
**Fourier, Joseph,** 206
Fourier series, 206
Fourier sine series, 191
Fourier transforms, 724
**Francesco, Jacopo,**
   Count Ricatti of Venice, 459
**Franklin, Philip,** 410
**Fraser, Donald C.,** 465, 513, 653
**Frederick the Great,** 79
Frequency function, 712
**Frey, Elmer J.,** 31
**Frobenius, Georg Ferdinand,** 39, 648
   method of, 39, 43
   orthogonal matrix, equation for, 347
Fundamental ellipse, 258
   conjugate of, 260
   flight-direction angle of, 261
   vacant focus of , 260
Fundamental invariants, 111
Fundamental matrix, 452
Fundamental perturbation matrices, 463
Fundamental plane, 123
Fundamental theorem of algebra, 297

**Galle, Johann Gottfried,** 473
Gamma function, 705, 739, 740
   Stirling's asymptotic formula for, 706
**Gardner, Martin,** 45, 704
**Gauss, Carl Friedrich,** xxxi, 34, 36, 70, 79, 91, 159, 166, 167, 196, 204, 222, 237, 295, 296, 300, 411, 447, 448, 485, 646, 663, 733
Gauss' continued fraction expansion theorem, 48, 66
Gauss' differential equation, 39
   singular points of, 39
Gauss's equation, 140
Gauss-Markov process, 744
Gauss-Markov theorem, 669
Gauss' method of orbit determination, 313
Gauss' method of weighted least squares, 646
Gauss' ratio of the areas of the sector and the triangle, 316
Gauss' variational equations,
   in polar coordinates, 485, 488
   in tangential, normal components, 489
Gaussian distribution, 732, 733
Gaussian gravitational constant, 227
Gaussian random process, 744
Gaussian random sequence, 744
Gaussian sequences, 74
**Gautschi, W.,** 68
**Gedeon, Geza S.,** 528

General perturbations, 419
Generalized anomaly $\chi$, 174
   series expansion for, 218
Generalized coordinates, 99
Generalized momenta, 100
Generating functions,
   for Bernoulli numbers, 51
   for Bessel functions, 208
   for Legendre polynomials, 393
   for modified Bessel functions, 507
   for Tschebycheff polynomials, 360
General perturbations, 419
Geocentric system, 123
Geometric mean, 72, 344
Geometric progression, 343
Geometric series, 343
"George" compiler, 2
Giacobini-Zinner comet, xxviii
**Gibbs, Josiah Willard,** 132, 136
**Gill, S.,** 589
**Gillespie, Rollin,** 16
**Godal, Thore,** 243, 246
**Goldbach, Christian,** 740
Golden Section, 45
**Gombaud, Antoine,** 662
**Goodyear, William H.,** 182
Gravitational potential, 97
   of a distributed mass, 401
Gravity gradient matrix, 451, 764
**Grubin, Carl,** 626
**Gudermann, Christof,** 168
Gudermannian, 168, 280
Guidance matrix, 461
Gyroscope, viii

**Habbe, James M.,** 678
HAL language, 4
Halley's comet, 227
**Halley, Edmond,** 80, 108, 365, 713, 733
**Hamilton, William Rowan,** 91, 93
Hamiltonian function, 101
   for restricted three-body problem, 381
Hamilton's equations of motion,
   canonical form of, 101, 479
   for restricted three-body problem, 381
**Hankins, Philip,** 3
**Hansen, Peter Andreas,** 320
Hansen's continued fraction, 320
Harmonic mean, 247, 344
Harmonic progression, 344
Harmonic series, 344
**Harrison, John,** viii
Heliocentric system, 123
**Henrici, Peter,** xxxi
**Herget, Paul,** xxxi
**Herrick, Samuel,** xxvii, 135

**Herschel, Sir William,** 471
Heuman's Lambda function, 376
**Hoag, David G.,** 25, 32
Hodograph, 126
Hodograph plane, 126, 531
**Hoelker, Rudolf F.,** 531
Hohmann orbits, 530
Hohmann transfer, 427
**Hohmann, Walter,** 427, 527, 530
**Hooke, Robert,** 662
**Hriadil, Francis Michael,** 620–621
**Huygens, Christiaan,** 662, 721, 733
Hyperbola, 117
   construction of tangent to, 171
   construction of,
     using asymptotic coordinates, 173
   definition of,
     as locus of circle centers, 146
   equation for,
     in asymptotic coordinates, 171
     in terms of the asymptotes, 166
Hyperbolic analog of eccentric anomaly,
   relation to true anomaly, 167
Hyperbolic functions, 167
   analogy with circular functions, 169
Hyperbolic locus of velocity vectors,
   equation for, 243, 244, 245, 247, 248
   nonsingular equation for, 245
Hypergeometric functions, 34
   associated Legendre functions, 408
   bilinear transformation formulas for, 41
   contiguous functions, 36
     identities for, 36
   convergence of series for, 36
   complete elliptic integrals, 76
   $\cos nx$, 40
   differential equation for, 39
   Euler's identity for, 41
   geometric series, 34
   inverse hyperbolic sine, 35
   inverse hyperbolic tangent, 35
   inverse sine, 35
   inverse tangent, 35
   Legendre polynomials, 394
   logarithm, 35
   quadratic transformations, 42
   series for, 34
   $\sin nx$, 40
   Tschebycheff polynomials, 360

IBM Card Programmed Calculator, 2, 589
IBM Type 418 Accounting Machine, 13
IBM Type 650 Magnetic Drum Data
   Processing Machine, 3
Inclination angle, 124
   variation of, 500

Incomplete elliptic integrals, 69
  evaluation of,
    first kind, 73
    second kind, 78
Independent events, 701
Independent random variables, 718
Indicial equation, 39
Indicial notation, 571
Inertial measuring unit (IMU), viii, 755
Information matrix, 650
Initial conditions, variation of, 509
Institute of the Aeronautical Sciences, 18
Integrals of the two-body problem, 114
Intermetrics, Inc., 4
International Cometary Explorer, xxviii
International Space Hall of Fame, 25
International Sun-Earth Explorer, xxviii
Intransitive dice, 704
Invariable plane, 97
Inverse linear interpolation, 193
  *See also* Regula falsi
Isothermal atmosphere, 505
"Iterative guidance mode", 566

$J_k$ coefficients for the earth, 407
**Jacobi, Carl Gustav Jacob,** 70, 372, 398
Jacobian elliptic functions, 70
Jacobian matrix, 495
Jacobians, 372, 483
Jacobi coordinates, 398
Jacobi's expansion, 392
Jacobi's form of elliptic integrals, 69
Jacobi's integral, 373
Jacobi's Zeta function, 77
  evaluation of, 78
  identity for, 77
Jet Propulsion Laboratory, 17
Joint characteristic function, 665
Joint density function, 664
Joint distribution function, 664
**Jordan, Camille,** 195
Jordan's inequality, 195
**Joukowski, Nikolai Jegórowitch,** 588
Julian century, 437
Julian date, 437
Julian day number, 437
Julian year, 437

Kalman filter, 22
**Kalman, Rudolf,** 23, 32
**Kaminski, Paul G.,** 659
**Kennedy, John F.,** 24
**Kennedy, Robert F.,** 783
**Kepler, Johanness,** 141, 107, 191, 365, 472

Keplerian orbits, 365
Kepler's equation, 142, 160
  approximate root of, 194
  for an arbitrary epoch, 164
  in terms of eccentric longitude, 491
  generalized form of,
    using universal functions, 178
    solution by Newton's method, 219, 236
  geometrical derivation of, 161
  hyperbolic form of, 166, 167, 168, 170
    approximate root of, 194
    geometric derivation of, 169
    proof of unique solution of, 193
    solution of,
      by Lagrange's expansion formula, 200
      by successive substitutions, 198
      using hyperbolic functions, 168, 170
  in terms of $U_1/U_0$, 220
  proof of unique solution, 192
  solution of,
    by inverse linear interpolation, 193
    by graphical means, 193
      Newton's scheme, 193
    by Fourier sine series, 208
    by regula falsi method, 193
    by successive substitutions, 196
    for near-parabolic orbits, 225
      by Gauss' method, 224
    by successive approximations, 221
    to first order in the eccentricity, 195
    to second order in the eccentricity, 195
    to third order in the eccentricity, 196
    using the extended method of Gauss, 234
  Stumpff's universal form of, 180
Kepler's first law, 142
Kepler's second law, 115, 142
Kepler's third law, 119, 142
**Khrushchev, Nikita,** 18
Kinetic energy, 99, 102, 116
Kinetic potential, 99
**King, Martin Luther, Jr.,** 783
**Kline, Morris,** xxxii
**Klumpp, Allan,** 331
**Knuth, Donald E.,** xxviii, 30
**Kraft, Christopher, C.,** 784
**Kromydas, William M.,** 415
Kutta-Joukowski aerofoil, 588
Kutta-Joukowski theorem, 588
**Kutta, Wilhelm Martin,** 588

Lagrange $F$ and $G$ functions, 112
  for extension of Gauss' method, 233
  for the parabola, 156
  in terms of
    eccentric anomaly difference, 162
    hyperbolic anomaly difference, 170
    true anomaly difference, 130
    universal functions, 179
  series coefficients for, 113, 114
**Lagrange, Joseph-Louis,** xxvii, 16, 80, 191, 237, 365, 471, 474, 476, 561
Lagrange brackets, 478
  properties of, 478
  values of, 482
Lagrange expression, 564
Lagrange interpolation equations, 140
Lagrange matrix, 478, 495
Lagrange multipliers, 564, 689
Lagrange's element set for secular variations, 492
Lagrange's expansion formula, 200
Lagrange's form of the equations of motion, 99
  for restricted three-body problem, 381
Lagrange's fundamental invariants, 111
Lagrange's generalized expansion theorem, 202
Lagrange's planetary equations, 483
Lagrange's quintic equation, 368
Lagrange series for $E$, 200
  convergence criterion of, 205
Lagrange's time equation, 279, 287, 298
Lagrangian coefficients,
  *See also* Lagrange $F$ and $G$ functions
Lagrangian function, 99
  for the two-body problem,
    with respect to center of mass, 110
  for restricted three-body problem, 381
Lagrangian libration points, xxvii, 379
**La Hire, Philippe de,** 145, 358
**Lambert, Johann Heinrich,** 15, 29, 52, 238, 305
Lambert guidance, 29
Lambert's theorem, 276
**Lancaster, E. R.,** 313
**Landen, John,** 71
Landen's transformation, 71, 414
**Laning, J. Halcombe, Jr.,** 2, 30, 31, 457, 699
**Laplace, Pierre-Simon de,** 138, 191, 205, 366, 395, 423, 662, 721, 733
Laplace vector, 116
Laplace's method, 138
Latitude, 124
Latus rectum, 119
Laurent expansion, 207
**Laurent, Pierre-Alphonse,** 207
Law of areas, 103, 141

Law of cosines, 364
  for spherical trigonometry, 347
Law of large numbers, 736
Law of sines, 363
Law of tangents, 363
**Lawden, Derek F.,** 565
Leading principal minor, 657
**Legendre, Adrien-Marie,** 68, 407, 646, 663, 740
Legendre polynomials, 390, 762
  addition theorem for, 405
  as hypergeometric functions, 394
  orthogonality property of, 394
  recurrence formulas for, 393
Legendre's differential equation, 395
Legendre's form of elliptic integrals, 69
**Leibnitz, Gottfried Wilhelm,** 68, 192, 213, 473, 721
Leibnitz's rule
  for differentiating products, 213
**Lenox, Joan (Edwards),** 465, 511
**Leonardo of Pisa,** 44
**Leondes, Cornelius T.,** 32, 751
**Le Verrier, Urbain-Jean-Joseph,** 471–473
**Levine, Gerald M.,** 32, 265, 751
**L'Hospital, Guillaume Antoine François,** 145
Libration points, 379
Likelihood function, 666
Line of apsides, 124
Line of nodes, 124
Linear algebraic systems, 353
Linear fractional transformation, 411
Linear independence, 177
Linear-tangent law, 565
Linearized perturbations, 420, 451
**Liouville, Joseph,** xxviii, 68, 366
**Listing, Johann Benedict,** 411
Lockheed Missiles and and Space Division, 16
Longitude, 124
Longitude of perihelion, 124
Longitude of the ascending node, 124
  variation of, 500
**Lord Rayleigh,** 741
**Lovell, James,** 751–753, 784
Lunar module (LM), 754
Lunar module computer (LMC), 757
**Lundberg, John,** 376

MAC programming language, 3
**MacCullagh, James,** 403
MacCullagh's approximation, 403
**Maclaurin, Colin,** 132
**MacMillan, William Duncan,** xxxii
**MacRobert, Thomas M.,** xxxii

MACSYMA, 215, 225, 621
**Makemson, Maud W.**, xxxi
**Markov, Andrei Andreevich,** 743
Markov chain, 743
Markov process, 743
Markov sequence, 743
**Marscher, William F.**, 461
**Martin, Frederick H.**, 11, 14, 27, 31, 32, 558
Mathematical expectation, 663, 715
Mathematical induction, 56, 202
Mathematical progressions, 343
Matrix algebra, 347
Matrix inversion lemma, 648
Matrix Ricatti equation, 14
Matrizant, 495
**Mattingly, Thomas K.**, 784
Maximum-likelihood estimate, 666
Maximum-likelihood method, 24
**Mayer, Christian Gustav Adolph,** 563
Mayer form, 563
**McCarthy, Senator Joseph,** 366
Mean, 663, 715, 716
Mean anomaly, 160
  approximation of, 162
  variation of, 502
Mean distance, 118, 164
Mean longitude, 161
Mean longitude at the epoch, 161
Mean motion, 119, 160
Mean point of an orbit, 265
  eccentric anomaly of, 268
  flight-direction angle at, 265
  locus of, 265
Mean value theorem of differential calculus, 197, 264
Mean-point radius of the parabola, 269
Mean-squared value, 663, 716
Measurement geometry matrix, 632
Measurement geometry vector, 628
Measurement vectors, 744
**Menaechmus,** 355
Method of adjoints, 420
**Miller, James S.**, 4, 23, 30, 32, 398, 420, 437
Minimum eccentricity ellipse, 258
Minimum-energy orbit, 240
  eccentricity vector for, 263
  parameter for, 246
MIT Instrumentation Laboratory, 1
**Mitchell, Edgar D.**, xxviii
MITILAC programming language, 4
Mixed distribution, 712
**Möbius, August Ferdinand,** 411
Möbius strip, 411
Möbius transformation, 411
Modified Bessel functions, of the first kind, 506
  asymptotic series for, 507
  continued fraction for, 507
  differential equation for, 508
  generating function for, 507
  integral form of, 508
  recurrence formula for, 507
Modulus of elliptic integrals, 69
Moment matrix, 664
Moments of inertia, 403, 716
**Montmort, Pierre Rémond de,** 662, 721
Moon as a triaxial ellipsoid, 404
**Moulton, Forest Ray,** xxxii, 376
**Muir, Thomas,** 177, 579
Multiple-revolution transfer orbits, 305
Multiple rotations of a vector, 91
Multiplicative property, 701
Mutually exclusive, 700

National Aeronautics and Space Administration, 18
Navigation matrix, 461
**Nerem, R. Steven,** 376
**Newman, Charles M.**, 179
**Newton, Sir Isaac,** x, 79, 80, 107, 191, 193, 216, 231, 237, 351, 365, 473, 662, 713, 733
Newton-Raphson method, 216
  *See* Newton's method
Newton's law of gravitation, 95
Newton's method, 205, 216
Newton's root-finding algorithm, 216
Newton's second law of motion, 96
Nominal orbit, 420, 450
Normal distribution, 731
  approximation to the binomial distribution, 734
  as a confluent hypergeometric function, 737
  multidimensional, 664
  two-dimensional, 667
**Nyström, E. J.**, 567, 568
Nyström's method, 766

Obliquity of the ecliptic, 123
Observed acceleration, 102
**O'Keefe, Robert,** 4, 31
Optimality condition, 565
Orbital elements, 123
Orbital tangents,
  construction of, 145
  for a parabola, 156
  equation of, 147
  property of, 145
Orthogonal matrix, 82
  *See also* Rotation matrix

Osculating orbit, 448
Osculating orbital elements, 420

Palermo, 296
**Pappus of Alexandria,** 355
Parabola, 117
  axis of, 355
  construction of, 156
    St. Vincent's method, 157
  vertex of, 355
Parabolic coordinates, 182
Parameter, 116
  for conjugate orbits, 246
  in terms of
    eccentric-anomaly difference, 255
    eccentricity, 263
    flight-direction angle, 248
    mean-point radius, 270
    $\nu$, 275
    semimajor axis, 280
    velocity-components ratio, 246
  of minimum-energy orbit, 246
Parameter of elliptic integrals, 69
**Pardo, L. T.,** 30
Partial convergents, 55
**Pascal, Blaise,** 661, 662
Pascal's triangle, 661
Patched-conic approximation, 419
Peenemuende, 6
Pendulous integrating gyro (PIG), 11
Pendulum, 70, 73
  period of, 70
Periapse, 117
Pericenter, 117, 125
Perigee, 124
Perihelion, 124
Perimeter of
  ellipse, 73
  sine arch, 77
Period, 119
Perturbation matrices, 420
**Petrick, Mary B.,** 4, 31
**Pfaff, Johann Friedrich,** 34, 297, 411
**Phillips, Samuel C.,** xxix, 784
**Piazzi, Giuseppe,** 296, 472
Planetocentric system, 123
Plant noise, 679, 744
**Plummer, Henry Crozier,** xxxii
**Poincaré, Henri,** 398
Point of aim, 422, 429
Point of injection, 534
Poisson bracket, 496
Poisson distribution, 729
Poisson matrix, 495
**Poisson, Siméon-Denis,** 495, 729
Poker hands, 709
Polar coordinates,

  motion referred to rotating, 103
Polaris fleet ballistic missile, 12
Positive definite quadratic form, 633
Potential energy, 116
Potential functions, 97, 98
  expansion of, 405
**Potter, James E.,** 14, 546, 653, 659, 669
Power series, economization of, 362
Primer vector, 565
Principal minor test, 657
Probability density function, 712
Probability distribution function, 663
Probability function, 699
Process noise, 744, 769
Product of inertia, 719
Project Galileo, -27
Project Whirlwind, 2
Projection operator, 546
**Prussing, John E.,** 283, 657
Pseudo-inverse of a matrix, 644
Pseudo-measurements
  of energy and angular velocity, 679
Purely random Gaussian vectors, 744

$Q$ function,
  as a hypergeometric function, 307
$Q$-system, 7
Quadratic convergence, 217
Quarter period, 71
Quaternions, 93
  conjugate of, 94
  elementary, 94
  inverse of, 94
  multiplication of, 94
  scalar part of, 93
  use of, in kinematics, 105
  vector part of, 93

Radius of curvature, 104
Radius vector, mean value of, 164
Ramo-Wooldridge Corporation, 1
Random variable, 663, 710
**Raphson, Joseph,** 216
Rayleigh distribution, 741
Real Time Control Center (RTCC), 29
Rectangular hyperbola, 119
Rectification, 449
Rectilinear motion, 115
Reeves Instrument Company, 2
Reference orbit, 420, 450
Regression of the node, 504
Regula falsi, 193
  See also Inverse linear interpolation
Regularization transformation, 182
Rendezvous radar (RR), 757
Required impulse velocity, 26

Residuals, 646
Residue of an equation, 576
Restricted problem of three bodies, 371
Reverse of a series, 213
Rheonomic system, 101
Ricatti equation, 459
Right ascension, 124
**Robertson, William M.,** 558
Rodrigues' formula, 392
  for Tschebycheff polynomials, 392
**Rodrigues, Olinde,** 91, 392
**Ross, Stanley,** 16
Rotation matrix, 82
  characteristic equation of, 89
  kinematic form of, 88
  in factored form, 92
  in terms of Euler parameters, 89
Rotation of a vector, 87
  multiple rotations, 91
  using quaternions, 94
  using rotation matrices, 91
  using vector operations, 91
**Rouché, Eugène,** 204
Rouché's theorem, 204
**Rousseau, Jean-Jacques,** 238
**Runge, Carl David Tolmé,** 584
Runge-Kutta methods, 567
**Russell, Bertrand,** 662


**Saleh, Adel A. M.,** 152
Sample points, 699
Sample space, 663, 699
**San Vincento, Gregorius a,** 157
Scalar mixing parameter, 27
Scanning telescope (SCT), 756
**Schmidt, Stanley F.,** 23, 32, 659
**Schweidetzky, Walter,** 6
Scleronomic system, 101
**Scott, David R.,** xxviii
Selenocentric system, 123
Self-adjoint system, 460, 461
Self-orthogonal curves, 358
Semilatus rectum, 119
Semimajor axis, 118
  variation of, 497
Semiminor axis, 119
Sensitivity coefficients, 420
Series reversion, 213, 352
Series reversion algorithm, 215
**Seversike, L. K.,** 510
Sextant, viii
Sextant (SXT), 755
**Shepperd, Stanley W.,** 32, 90, 189, 219, 268
**Silber, Paul S.,** 531
Sirius, 20, 652
Skew-symmetric matrix, 87

**Smart, William Marshall,** xxxii, 471
**Smith, Gary R.,** 195
**Smith, Gerald,** 24
**Sorenson, Harold, W.,** 32, 659
Space Task Group, 23
Special perturbations, 419
Sphere of influence, 395
  for the planets, 397
Spherical coordinate system, 83
  motion referred to a rotating, 102
**Spofford, John R.,** 215
Sputnik, 15
Square root of a matrix, 655
St. Petersburg paradox, 662
Stability, definition of, 382
Stages, 568
Standard deviation, 664, 716
Star aberration correction, 770
State transition matrix
  *See* Transition matrix
State vector, 20, 451, 744
State vector update, 23
Statistical parameter, 715
Statistically independent, 664
**Stegun, Irene A.,** xxxi, 375
**Stern, Robert G.,** 162, 669
**Stifel, Michael,** 661
**Stirling, James,** 706, 733
Stirling's formula, 706, 733
St. Petersburg paradox, 662
Straight line solutions, 367
**Strutt, John William,** 741
**Stumpff, Karl,** xxxii, 112, 151, 180
Summation convention, 572
**Sundman, Karl Frithiof,** 174
Sundman transformation, 174, 182
  alternate form of, 182
Surface of zero relative velocity, 376
Symplectic matrix, 14, 129, 453
Synodical period, 431, 433
**Szebehely, Victor,** 376


Tangent-bisector property, 250
Tangent ellipse, 263
Tangential and normal coordinates, 104
**Tartaglia (Niccolò Fontana),** 150, 661, 662
**Taylor, Brook,** 110
Taylor series expansion of a vector, 110
  of a vector function of a vector, 573
Taylor series with remainder, 217
Telescoping series, 59
**Tempelman, Wayne,** 521
TEX , xxviii, xxix
Thor IRBM, 12
Tijuana Mexico, 6
Time of pericenter passage, 120, 150, 160

**Tisserand, François Félix**, 423
Tisserand's criterion for the identification of comets, 424
**Titius, Johann Daniel**, 472
Titius' rule, 472
Total energy constant, 116
Trace of a matrix, 89
  identity for, 653
**Trageser, Milton B.**, 31
Transfer angle, 238
  bisector of, 250
Transition matrix,
  for extended method of Gauss, 235
  for state vector, 452, 744
  for the two-body problem, 129
Transverse axis of hyperbola, 169
"Treize", 721
Trochoid, 193
Trojan asteroids, xxviii, 384
True anomaly, 117
  variation of, 502
True longitude, 125, 160
**Tschebycheff, Pafnuti L.**, 360, 743
Tschebycheff polynomials, 360
  generating function for, 360
  orthogonality property for, 361
  recurrence formula for, 360
Tschebycheff's differential equation, 360
**Tsien, Hsue-shen**, 366, 408, 414
Turn angle, 429
Two-body problem,
  integrals of, 114
  differential equations for,
    in parabolic coordinates, 110, 182
    in polar coordinates, 371
    vector form of, 108
    with respect to center of mass, 110
  transition matrix for, 129

Unbiased estimator, 645, 666
Uniformly distributed, 716
Unit impulse function, 713
Univeral gravitation constant, 95
Universal functions, 464
  alternate set of, 181
  basic identity for, 176
  $F$ and $G$ in terms of, 464
  linear independence of, 177
  relation to elementary functions, 180
  series definition of, 176

Vacant focus, 144
  locus of, 273, 274
**Vandermonde, Alexandre-Théophile**, 579
Vandermonde determinant, 575

Vandermonde matrix, 579
Variable-time-of-arrival guidance, 20, 545
Variance, 663, 716
Variation of constants, 471
Variation of orbital elements, 471
**Vaughan, Robin M.**, 32, 325
Vector algebra, 345
Velocity components along skewed axes, 242
Velocity-to-be-gained, 6
Velocity-to-be-gained vector, 550
  differential equation for, 551
Velocity vector construction, 162
Velocity vector for the Lambert problem, 306
Velocity vector in Gauss' parameters, 319
Vernal equinox, 123
Very high frequency (VHF) link, 756
**Viète, François**, 53
Viète's infinite product for $\pi$, 53
Vis-viva integral, 116
Voltaire, 107, 238
**von Braun, Wernher**, 6
**von Kármán, Theodore**, 366

**Wall, Henry S.**, xxxii
Wallis' infinite product, 55, 741
Wallis' integrals, 740
**Wallis, John**, 34, 55, 740
Wallis' method
  for evaluating continued fractions, 63
Wallis' rule for continued fractions, 56
Wallis' theorem, 34, 741
**Watson, George Neville**, xxxii
Weighted least-squares, 20
Weighting factors, 646
Weighting matrix, 23
**Wen, W. Li-Shu**, 528
**Werner, Charles**, 3
White noise, 679
**Whittaker, Sir Edmund Taylor**, xxxii, 15, 31
Work, 102
**Wroński, Józef Maria Hoené-**, 177
Wronskian determinant, 177, 475
Wronskian matrix, 475
**Wu, Y. T.**, 414

**Yarymovych, Michael**, 23
**Yeomans, D. K.**, 227

**Zach, Baron Franz Xaver von**, 296
Zeno's paradox, 662
**Zierler, Neal**, 2, 30